The Materials Selector

Volume 1

The Materials Selector

Second edition

Volume 1

Edited by

Norman A. Waterman, BSc, PhD
Chief Executive, Quo-Tec Ltd,
Amersham, UK

and

Michael F. Ashby, FRS
Professor of Engineering,
University of Cambridge, UK

CHAPMAN & HALL

London · Weinheim · New York · Tokyo · Melbourne · Madras

Published by Chapman & Hall, 2–6 Boundary Row, London SE1 8HN, UK

Chapman & Hall, 2–6 Boundary Row, London SE1 8HN, UK

Chapman & Hall GmbH, Pappelallee 3, 69469 Weinheim, Germany

Chapman & Hall USA, 115 Fifth Avenue, New York, NY 10003, USA

Chapman & Hall Japan, ITP-Japan, Kyowa Building, 3F, 2-2-1 Hirakawacho, Chiyoda-ku, Tokyo 102, Japan

Chapman & Hall Australia, 102 Dodds Street, South Melbourne, Victoria 3205, Australia

Chapman & Hall India, R. Seshadri, 32 Second Main Road, CIT East, Madras 600 035, India

First edition 1991 published by Elsevier Science Publishers Ltd, UK

Second edition 1997

Typeset in Belgium by Fotek nv

Printed in Great Britain by T.J. International Ltd, Padstow, Cornwall

ISBN 0 412 61550 9 (3 volumes)

A catalogue record for this book is available from the British Library

Library of Congress Catalog Card Number: 96-85468

∞ Printed on acid-free text paper, manufactured in accordance with ANSI/NISO Z39.48–1992 (permanence of paper).

Contents

Volume 1

Foreword

Selection of materials appropriate to the task at hand has been pervasive throughout the history of mankind. Indeed, Cyril Smith (1) has asserted that man appeared on earth and distinguished himself from other ape-like creatures as the culmination of a series of mutations whose survival was favored by an increasing ability to select and manipulate materials. Thus, prehistoric stone-age man learned to select flint for cutting and piercing tools, soapstone for dishes, incised seals and sculptured ornaments, and quartzite and granite for grinding pestles and mortars. Similarly, the unique properties of various woods for particular applications were exploited throughout the centuries – walnut burl for bowls, yew for bows, locust for fence posts, hickory and ash for tool handles, etc. (2) (3).

Today, although not confined to natural materials, man's intelligent use of materials still depends critically on an appreciation that each possesses a whole suite of properties. Thus some particular material will have a *combination* of properties that is optimal for any given application, and it is foolishly simplistic to choose only the strongest, the lightest, or the most degradation resistant. Just as in the past, having chosen a candidate material, the user must still also select a processing technique appropriate to the design of the component part and one to which the candidate material is amenable; failing in this endeavor, his initial choice must be discarded and another made.

Despite the enormously increased understanding we have today of the relationships between properties and composition of materials, and of the influence of processing on structure, proper selection of materials for a given application is still a daunting and non-trivial task because of the sheer volume of data, the inadequacies of existing data, and the advent of new technology. There are literally tens of thousands of materials from which to choose; their number and diversity alone confuse and overwhelm the would-be user. Available handbooks and databases frequently lack data for certain materials and parameters of interest. Furthermore, property values for different classes of materials may not be expressed in comparable terms. An injection-moldable plastic may, indeed, be competitive with a die-castable zinc–base alloy, but how is their relative performance to be predicted? With more varied and demanding market needs prior engineering experience provides little guidance in materials selection. Consider for example the new property requirements which have arisen in recent decades for automobile bodies which do not rust, for larger but lighter aircraft which carry higher payloads and consume less fuel, and for bearings which never need to be lubricated. Drastic changes in the materials selection process may also be imposed by a new technology such as rapid prototyping which permits trial components to be made in days rather than months.

Selection and specification of materials based on their ability to meet the ensemble of properties demanded by the application is a necessary but insufficient screen. We also need information on their manufacturability and cost. Even these subjects are intertwined. Costs can rarely be compared on the basis of dollars per kilogram. A seemingly expensive material, that is readily processed with little scrap generated, will frequently prove to be the more economic choice.

The foregoing discussion has largely been couched in terms of selection of a material for a new product or component. Analogous reasoning may also be required when it is necessary to substitute a new material for one which is already performing satisfactorily. Such circumstances may arise due to regulatory restrictions (asbestos-based insulation), cost reduction programs imposed by competitive pressures, design changes that require new manufacturing routes (cast vs. wrought shapes), or simple unavailability of the traditional material in current markets. To these may be added societal requirements for materials substitution based on recyclability, reduction of environmental impact, improved long-term reliability, or reduced maintenance (plastic fencing for painted wood).

This book aims to address all the problems outlined in the paragraphs above by providing both an analytical system and the necessary information for the selection and specification of engineering materials and their related component manufacturing processes. It is a detailed guide for designers, manufacturing engineers and materials engineers, but will also be an invaluable source of information for metallurgists, ceramists and polymer chemists and technologists. Thus no detailed knowledge of the underlying fundamental materials science is assumed or necessary. Extensive use is made of tabular and graphical presentations so that the reader may readily visualize comparisons of both individual materials and materials classes.

The work is intended to be comprehensive as regards materials: metals, ceramics, carbon and graphite, glasses, and polymeric materials (including thermoplastics, thermosets, elastomers, foamed, laminated and composite materials) are all covered in depth. The design process is recognized as multi-stage, embracing conceptual design, layout design, detail design, and design for manufacture. The iterative and interactive nature of product design and materials selection is emphasized throughout. Attention is also given to the concept that the numeric properties to be found in tables and graphs are those for initial guidance that may be modified by surface treatments, joining processes or the presence of conjunctive materials. Ad hoc testing of candidate materials may also be required for full assessment of their viability in the intended application. Emphasis is also placed on the fact that participants in the materials selection process are not limited to the materials engineer, designer, and manufacturing engineer, but may well also include the materials supplier and even the customer. Finally, the entire selection process is illustrated in detail by presentation of several case studies.

The editors are well equipped to treat the subject. Waterman has extensive industrial experience, both directly in industry and in many years of industrial consultancy practice. Ashby, through both his research and educational experience, has a breadth of view and depth of understanding of engineering materials invaluable to the present work. Both editors, throughout their careers, have devoted much time and thought to a rational and novel approach to the selection of engineering materials, the results of wich are clearly evident.

(1) Smith, Cyril S. (1965) Materials and the development of civilization and science, *Science*, **148**, 908–917.
(2) Delmonte, J. (1985) *Origins of Materials and Processes*, Technomic Publ. Co., Lancaster, PA.
(3) Sloan, E. (1965) *A Reverence for Wood*, W. Funk, New York

J.H. Westbrook
Brookline Technologies
Ballston Spa, NY, USA

Preface

The main aim of these three volumes is to provide a system and the necessary information for the selection and specification of engineering materials and related component manufacturing processes.

These volumes are intended for use by designers, materials engineers and production engineers who are seeking to identify the most suitable materials and manufacturing methods for a specific application.

Throughout the volumes, extensive use has been made of tabular and graphical information to facilitate the comparison of candidate materials for a specific application. Every material and process is described in terms of what it will do for the component designer and product maker and no detailed knowledge of metallurgy, polymer chemistry or materials science is assumed or necessary.

The information is arranged in order of increasing detail with the aim that no section need be read unless it is likely to be of direct relevance to the reader and his quest, whether this is the selection of material for a new product or the search for a substitute material.

Disclaimer

Whilst every effort has been made to check the accuracy of the information contained in these volumes, no material should ever be selected and specified for a component or product on a paper exercise alone. The purpose of these volumes is to provide enough information for a short list of candidate materials for testing and to reduce the number of fruitless tests. No liability can be accepted for loss or damage resulting from the use of information contained herein.

How to use *The Materials Selector*

There will be two main reasons for using *The Materials Selector.*

(a) **To select and specify the materials and manufacturing routes for the components of a new product.**
(b) **To evaluate alternative materials or manufacturing routes for the components of an existing product.**

The method of using *The Materials Selector* for each of these purposes is as follows:

(a) **Selection and specification of materials and manufacturing route for the components of a new product.**

(i) Define the **market specification** for the new product, generate a number of **conceptual designs**, correlate with the **market specification** and select the preferred **conceptual design** through a **design review** (see Sections 1.1.1 and 1.1.2).

(ii) Convert the preferred **conceptual design** into a number of **layout designs**. Analyse the **functions** of each component of the **layout designs**, correlate with the **market specification** and select the preferred **conceptual design** through a **design review**. For the components of the preferred **layout design**, convert the required **functions** (of each component) into the desired combinations of materials properties – **materials desirability indices** (see Sections 1.1.3 to 1.1.7 – Case studies).

(iii) Employ the **Ashby maps/Material selection charts** to identify those **generic classes of materials** which have the most desirable combination of properties (see Sections 1.1.6, 1.1.7 and 1.1.8).

(iv) Convert the preferred **layout design** to a **detailed design** and select the most appropriate grade of materials from generic classes of materials. For this see Section 1.1.14 (costs), Section 1-2 (mechanical properties), 1-3 (physical properties), 1.1.9 and 1-4 (corrosion resistance and corrosion) 1.1.10 and 1-5 (wear resistance) and Volume 2 (metallic materials and ceramics) and Volume 3 (thermoplastics, thermosetting plastics, elastomers, composites, foams and films).

(v) **Manufacture and test prototype designs and candidate materials** (see Sections 1-10 and 1.1.3).

(vi) Convert preferred **detailed design** into **designs for manufacture** (see Section 1.1.5, Sections 1.1.11 and 1-6, 1-7).

(vii) Repeat steps (v) and (vi) to optimize performance and cost of chosen materials and manufacturing route.

(b) **Selection and specification of materials and manufacturing routes for the components of an existing product**

(i) Analyse the **functions** of the component for which new materials or new manufacturing routes are sought and correlate with the reasons for seeking new materials and/or manufacturing routes in terms of existing and future needs of the customers (see Sections 1.1.1 to 1.1.3).

(ii) Convert the **functions** of the components under review to the required combination of materials properties (see Section 1.1.3).

(iii) Examine the properties of the currently-employed materials (Volume 2 for metals or ceramics and Volume 3 for thermoplastics, thermosetting plastics, elastomers or composites).

(iv) Search for generic materials with equivalent or superior properties (see Sections 1.1.6, 1.1.7 and 1.1.8) or lower costs (see Section 1.1.14).

(v) Examine candidate materials in more depth [see Section 1.1.4 (corrosion resistance), 1.1.5 (wear resistance), 1-2 (mechanical properties), 1-3 (physical properties) and Volume 2 (metals or ceramics) and Volume 3 (thermoplastics, thermosetting plastics, elastomers and composites)] and select preferred candidate materials.

(vi) Select component manufacturing routes for preferred materials (see Section 1.1.11, 1-6 and 1-7) and compare with existing manufacturing route.

(vii) **Manufacturing prototype components in new materials (see Section 1.1.12) or by new manufacturing route and test (see Section 1-10, 1.1.13).**

(viii)Repeat stage (vii) for each candidate material and manufacturing route and select preferred option.

Acknowledgements and history of the publication

The origins of *The Materials Selector* may be traced back to the early 1970s. At that time I was employed as a materials engineer in industry and perceived the need for a selection system which would compare the performance and cost of materials in terms which could be understood by designers and production engineers without the benefit of degrees in metallurgy, polymer chemistry or materials science. I am grateful to my employers at that time, Danfoss A/S, for providing a stimulating working environment in which the idea was born.

The Fulmer Research Institute, my next employer, provided the financial and technical resources to convert the idea into reality; the *Fulmer Materials Optimizer*, which was published in 1974.

I am particularly grateful to my colleagues at Fulmer at that time, without whose efforts this current publication would not exist. In 1979 I left Fulmer and since then have earned my living as a consultant helping others to benefit from new and improved materials technology.

In 1988, Elsevier Applied Science acquired the rights to publish the *Fulmer Materials Optimizer* and invited me to convert the information in the 1987 edition of the *Optimizer* into a new three-volume materials selection system to be known as *The CRC-Elsevier Materials Selector*. This was published in 1991 with the help of my co-editor, Professor Michael Ashby FRS.

In 1993, Chapman & Hall purchased the book publishing business of Elsevier Applied Science and in 1994 invited me to begin a substantial revision and updating of *The Materials Selector*.

This current version is the result of the input and insight of my contributors. They are acknowledged on the following page. I would, in addition, like to give a special thank you to the Quo-Tec team who have borne the brunt of the work load, in particular Dr Justin Furness, to Professor Michael Ashby whose materials selection maps and case histories have added very considerable value to my original concept, to Dr Philip Hastings of Chapman & Hall for his advice and encouragement and last, but not least, to Margaret Waterman without whose support I could not have started, let alone finished.

Norman Waterman
Quo-Tec Ltd
Amersham, UK

Contributors to *The Materials Selector*

Contributor	**Section**
Professor M. F. Ashby FRS	1.1.6–1.1.8, 1.1.11 and 3-6
Dr T. J. Baker	2-1 and 2-2
Mr G. B. Brook FEng	2-5–2-14 and 1.7.9
Mr L. Brown	1.4.2.4–1.4.2.9
Dr R. Cayless	3-7
Mr R. J. Davies	2-4
Dr M. Deighton	1-2
Mr R. Ford	3.8.6
Dr G. B. Gibbs	1.4.5
Professor C. Hepburn	3.4.14
Professor D. A. Hills & Associates	1-6
International Tin Research Institute	1.7.3
Dr W. A. Lees	1.7.7
Dr W. Lucas	1.7.5
Dr G. Mahboubian-Jones	1-9 and 3.8.7
Mr J. A. Mead	1-8
Dr R. Morrell	1.4.3 and 2.15
Neale Consulting Engineers	1-5
Paint Research Association	1.4.2.1 and 1.4.2.2
Mr A. Pye	1-9 and 3.8.8
RAPRA Technology Ltd	3-1–3-5
Dr G. Sanderson	1.4.1
Mr A. W. Thompson	3.8.1–3.8.5, 3.8.9
Dr N. A. Waterman	1.1.1–1.1.5, 1.1.9 and 1.1.10, 1.1.12 and 1.1.13 and 2-3

Other sections were edited and updated by Dr N.A. Waterman, Dr J.A.G. Furness and Mr R.J. Davies.

Product design and materials selection

Contents

List of tables

List of figures

List of charts

List of boxes

1.1.1 Relationship of product design to materials and manufacturing process route selection

1.1.1.1 Introduction

A material is selected for a product because:

- It is readily available.
- It can be formed into the desired shape with the required dimensional tolerances.
- When so formed, it will perform the required functions of the product.
- It will!continue to perform these functions satisfactorily for the lifetime of the product.
- It can be disposed of, or recycled, in a way which is environmentally acceptable.

All of the above have to be achieved at a cost which permits the product to be offered at a price which attracts customers **and** gives a profitable return to the manufacturer.

For most products, satisfying all of the product requirements makes conflicting demands on the materials. This is especially true when trying to achieve maximum performance at minimum cost. This section sets out a systematic approach to the selection of materials which should optimize performance and cost and also indicates where to find the relevant information in the three volumes of *The Materials Selector*

1.1.1.2 Market specification, product design and development, materials and manufacturing route selection

The market specification of a product is the starting point for all systematic approaches to the selection of materials. This applies equally to new products, and also to existing products for which alternative materials and manufacturing routes are sought.

The process of systematic materials selection will be illustrated by reference to new product development. If new materials for an existing product are sought, the process may be conducted more quickly as the constraints on choice will be greater.

A systematic process for new product development is summarized in Chart 1. Four stages of design are identified:

- conceptual design,
- layout design (sometimes called embodiment design[1]),
- detail design,
- design for manufacture.

The nature and content of each stage are defined and illustrated with a case history. In practice, several conceptual designs will be proposed, analysed by **design reviews**[2] and eventually narrowed to the preferred **design for manufacture**.

[1]See Ashby, M.F. (1992) *Materials Selection in Mechanical Design*, Pergamon Press, Oxford, UK.
[2]Activities and techniques, designated, in bold type, are defined in:
Table 1.1.1 *Glossary for new product design procedures,* and described in more detail in the boxes to this section.

The evolving market specification, in Chart 1, which accompanies the progression through the design stages, reflects the probability that the ideal product, in terms of performance and costs, as defined by market research, may not be feasible. However, as the design process advances, designs with different combinations of performance and cost will emerge which can satisfy sectors of the market which are sufficient in size and value to justify proceeding to full scale production.

The input of previous experience of similar products, eg in the form of **ethological plots** (see Box 1) to the conceptual stage of new product design reflects the reality that very few products are totally new.

Examples of the **market specification** and the **previous experience of similar products** are given for a small pump intended for multiple applications in beverage vending machines, photographic processing, and small process plant and equipment in:

Table 1.1.2 *Market specification and previous experience of small pumps with multiple applications*

Progress through the four stages of design requires the completion of several activities and the application of a number of techniques. These are listed in Chart 2.

Chart 3 illustrates in more detail the creation and analysis of **conceptual designs,** including the preferred safe design and alternative **higher risk/higher performance/ lower cost designs**.

Chart 4 illustrates the route from **concept design** to **layout design.**

Chart 5 shows the route from **layout design** to **detailed design** and Chart 6 the route from **detailed design** to **design for manufacture**, including feedback loops which underline the iterative nature of the design process.

A glossary of the terms used in Charts 1–6 is given in Table 1.1.1.

For almost all manufacturing companies, to benefit from being first to market with a new product, and to stay ahead of the competition with higher performance and/or lower cost designs, the **preferred safe designs** and **higher risk (higher performance or lower cost) designs** must be pursued in parallel.

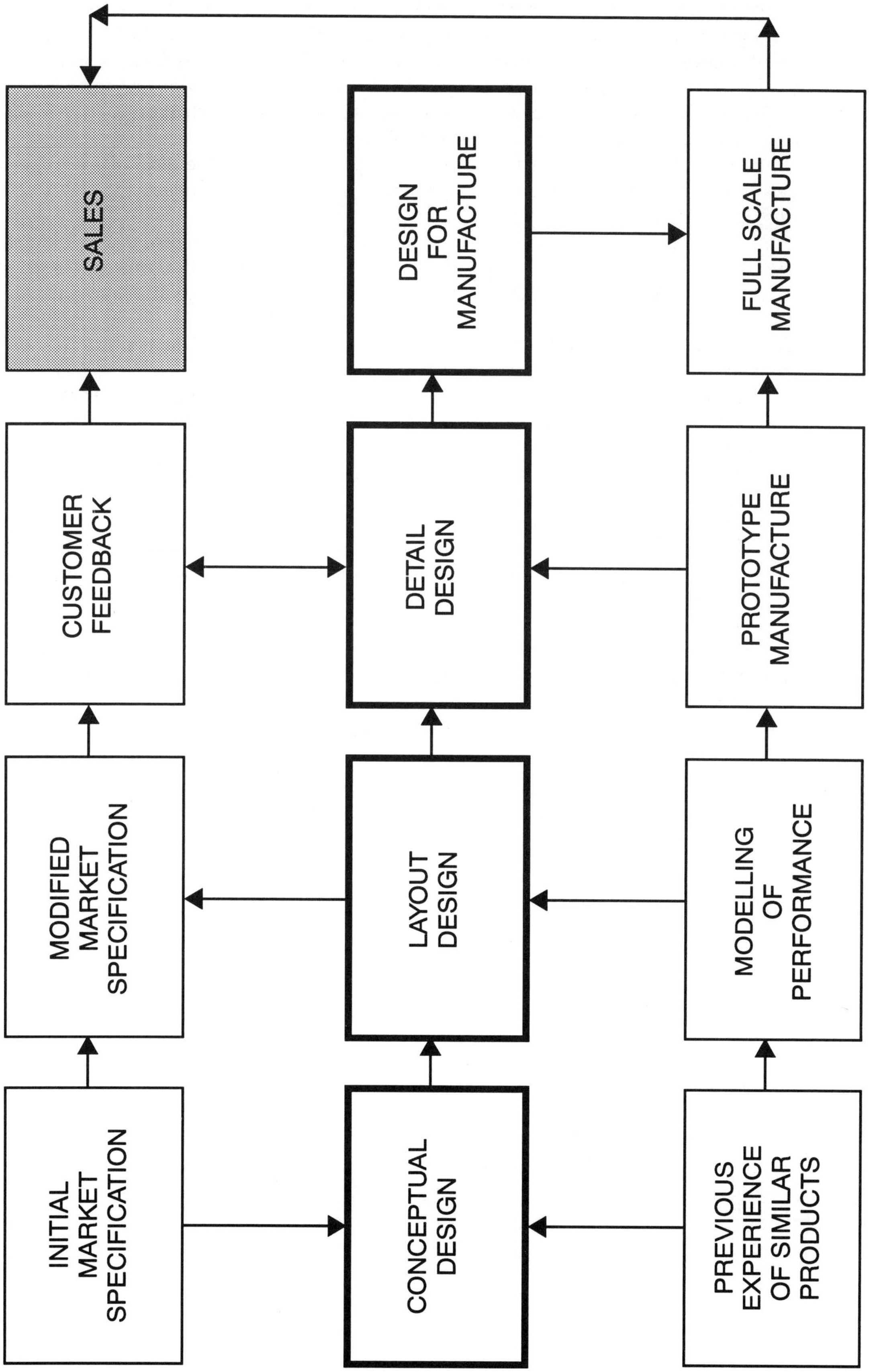

Chart 1 Summary of new product development

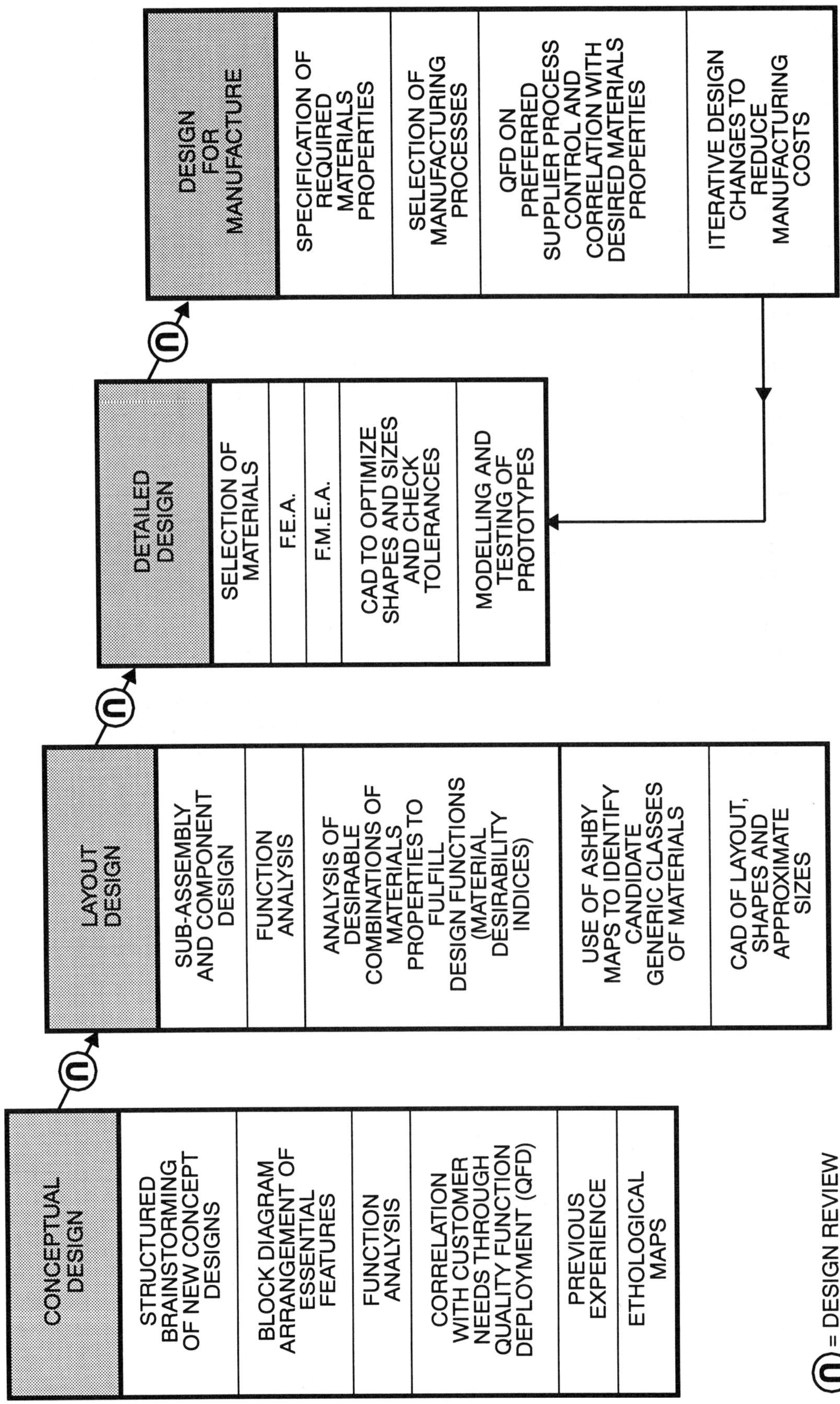

Chart 2 Four design stages of new product development and supporting activities and techniques

Chart 2

Box 1 Ethological maps

Ethological maps plot, in 2-D or 3-D space, the comparative performance parameters of existing products or systems.

Ethological maps can address:

- the comparison of different products or systems which satisfy the same market needs (e.g. batteries, flywheel energy storage systems) to provide uninterrupted power supplies,
- the comparison of similar products from different manufacturers (e.g. centrifugal pumps from a range of suppliers),
- the comparative performance of similar engineering mechanisms in different end use applications (e.g. the relative wear rate of internal combustion engine cylinder liners in products as diverse as chain saw engines and large marine diesel engines).

The reasons for constructing ethological maps are as follows:

- To compare the target performance parameters of the proposed new product against established products.
- If the product is within the performance of **existing** products, why is the new product being developed? What unique performance/cost benefits will it provide?
- If the performance parameters of the new product lie outside the performance envelope of **existing** products, do the unique performance parameters match well defined customer needs? Are the unique performance parameters achievable and, if so, how?

An example of an ethological map for small pumps is given below:

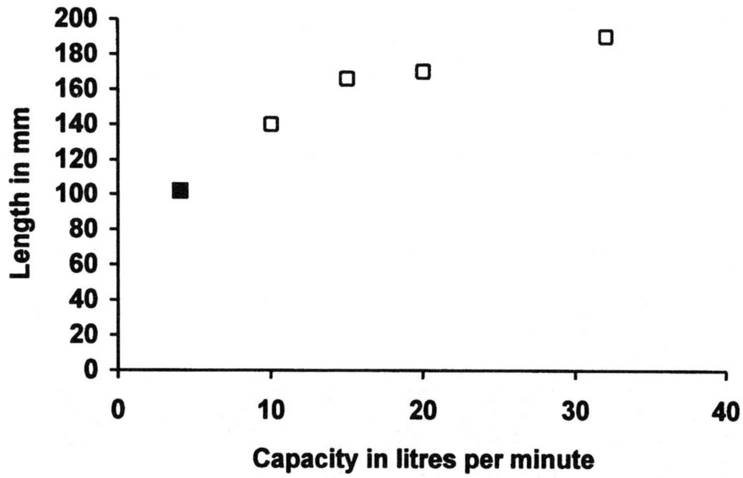

Table 1.1.1 Glossary for new product design procedures

Ashby maps/materials selection charts	Illustrations of the comparative ratios of two properties (e.g. elastic modulus *E* and density ρ) of generic classes of materials, e.g. steels, nylons, alumina ceramics. (Created by Michael F. Ashby – see Sections 1.1.6 and 1.1.7).
CAD	Computer aided design. Modern CAD software packages are capable of creating a 3-D image of a product on a screen and then performing a number of manipulations (e.g. rotating the image, checking that two components are not occupying the same space, calculation of build-up of tolerances).
CAM	Computer aided manufacture. The data defining shape, dimensions and tolerances of components created by the CAD can be employed to control the manufacture of components by computer numerically controlled machines, e.g. milling a shape from a solid form, or by building-up the shape by **rapid prototyping**.
Conceptual design	Description of possible designs in terms of block diagrams which represent the main components or sub-assemblies and illustrate the linkages and juxtaposition of the main components (see Section 1.1.2 and **Charts 2 and 3**).
Design for manufacture	Systematic analysis and modification of the preferred **detailed design** to accommodate well established, low cost manufacturing and assembly techniques with no loss or an acceptable downgrading of product performance (see Section 1.1.5 and **Charts 2 and 6**).
Design review	Formal procedure for evaluating alternative product designs. Employed at each of the four stages of design (**conceptual, layout, detailed and design for manufacture)**. The process normally involves two teams, one of which is responsible for the design and making presentations, the other, the reviewing team which should contain appropriately experienced, independent experts (see Box 3).
Detailed design	Fully dimensioned version of preferred layout design including dimensional tolerances, surface finishes and detailed materials specifications (see Section 1.1.4 and **Charts 2 and 5**).

Table 1.1.1 Glossary for new product design procedures – *continued*

Ethological maps	Two or three dimensional graphs showing the relationship of important parameters of existing products, e.g. pressure versus volumetric capacity versus shaft speed for pumps of different designs (see Box 1).
Finite element analysis (FEA)	Computer based modelling technique which divides a component into small, three dimensional elements for the purposes of mechanical (e.g. stress) or thermal (e.g. heat flow) analyses.
Failure mode effect analysis (FMEA)	(i) Systematic analysis of the effects on product performance of errors or failures in process control.
	(ii) Systematic analysis of the effects on a product or system of failure of a single component or element of the system. This technique is employed to assess the vulnerability of designs and associated manufacturing processes and to introduce additional testing and process control to reduce the risk of the occurrence of the most serious failures/failure modes.
Function analysis	Analysis and definition of the primary and secondary functions of each component in a product design in terms of a verb and a noun. An example would be for a bearing, the primary function is to **allow rotation**, secondary functions are to **minimize wear** and to **reduce friction** (see Tables 1.1.3 and 1.1.4 and Figs 1.1.4 and 1.1.5).
Generic class of material	Material class identified by generic type, e.g. metals, plastics, ceramics or next level of detail, e.g. stainless steels, polycarbonates, aluminas.
Layout design (also called embodiment design[1])	General arrangement version of conceptual designs with approximate dimensions. Permits an early view of proposed designs for marketing, manufacturing and others involved in new product development. Would normally include overall dimensions and shape, possible materials by generic class (see Section 1.1.3 and Charts 2 and 4).
Materials desirability index	Combination of material properties which define the most desirable materials for performing the required functions of a component (Section 1.1.6).
Material grade	Material identified unambiguously by specification in terms of composition and processes.

[1]See Ashby, M.F. (1992) *Materials Selection in Mechanical Design*, Pergamon Press, Oxford, UK.

Table 1.1.1 – *continued*

Table 1.1.1 Glossary for new product design procedures – ***continued***

Preferred safe design	Design chosen at conceptual stage because it involves lowest risk in terms of unfamiliar design features and thereby gives best possibility of shortest time-to-market. Unlikely to be lowest cost or highest performance design.
Quality function deployment (QFD)	(i) Systematic approach to the correlation of customer needs (**voice of the customer**) to the **functions** of the components of the product. The technique employs a matrix format to identify which customer needs correlate strongly, weakly or negatively with the **functions** of the component (see Tables 1.1.3 and 1.1.4).
	(ii) Systematic approach to the correlation of required component functions to the manufacturing processes employed to make the component and, in particular, the quality controls of the process (see Box 5).
Rapid prototyping	Techniques which convert CAD model of component to be made into a solid model, or a tool cavity of the component in a matter of hours (see Section 1.1.12).
Specification of materials	Definition of material in terms of required properties and composition and processing route (see Box 3 and Box 4).
Value analysis	Systematic process of analysing alternative designs to identify areas of poor value. Similar to QFD but focused on function/cost analyses.
Voice of the customer	Customer requirements and preferences for new product characteristics – top row of QFD matrix.

Table 1.1.1 – ***continued***

Table 1.1.2 Market specification and previous experience of small pump with multiple applications

Market specification	***Previous experience***
Reliable operation	Leaking seals
Low maintenance	Failed bearings
No leakage	Noise and vibration
Failsafe	Corrosion
Quiet operation	Wear
Low first cost	
Low cost of ownership	
Easy to clean and sterilize	
Compact size	
No contamination of fluids pumped	
Low energy consumption	

Table 1.1.2

1.1.2 Conceptual design

The process of **conceptual design** is illustrated in Chart 3. Two conceptual designs, in block diagram form, which meet the initial market specification given in Table 1.1.2 for a small pump are shown in:

Fig. 1.1.1 *Block diagram designs for a small pump*

The essential difference between these two designs is that:

Design (i) employs a shaft which penetrates the pump casing and relies on seals to prevent leakage.

Design (ii) employs a magnetic coupling which means that the drive system for the impeller does not provide a possible leakage path for the fluid.

Formal analysis of these alternative design features by the **quality function deployment** technique is given in:

Table 1.1.3 *Quality function deployment analysis of conceptual design*

In this example, only four customer requirements are taken to illustrate the process. In practice, all the requirements should be included, and classified in some order of importance to the customer, e.g. must have, desirable, optional.

In this simple example it is clear that the magnetically coupled design will be preferred if the need to avoid leaks is more important to customers than the higher first cost of magnetic coupling.

In more complex cases it may be necessary to rate all of the customer requirements in order of importance and to investigate seal failure statistics from similar products. This information would be analysed and fed into a formal **design review** as shown in Chart 3. More complex examples of **QFD matrices** and **design reviews** are illustrated below, but for the purposes of this simple example, it will be assumed that the avoidance of leaks dominates the choice of **conceptual design** and the magnetically coupled design is preferred.

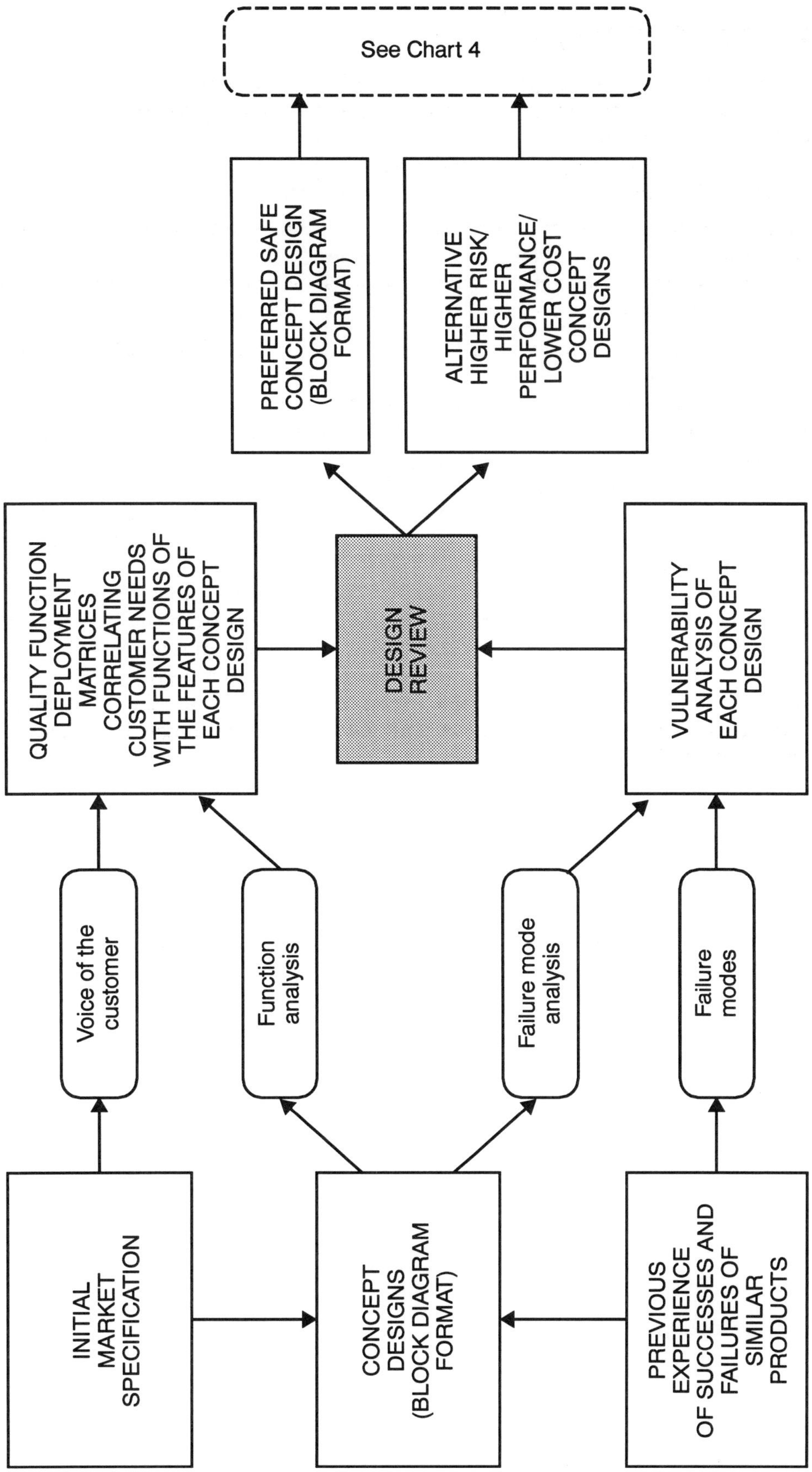

Chart 3 Creation and analysis of concept designs

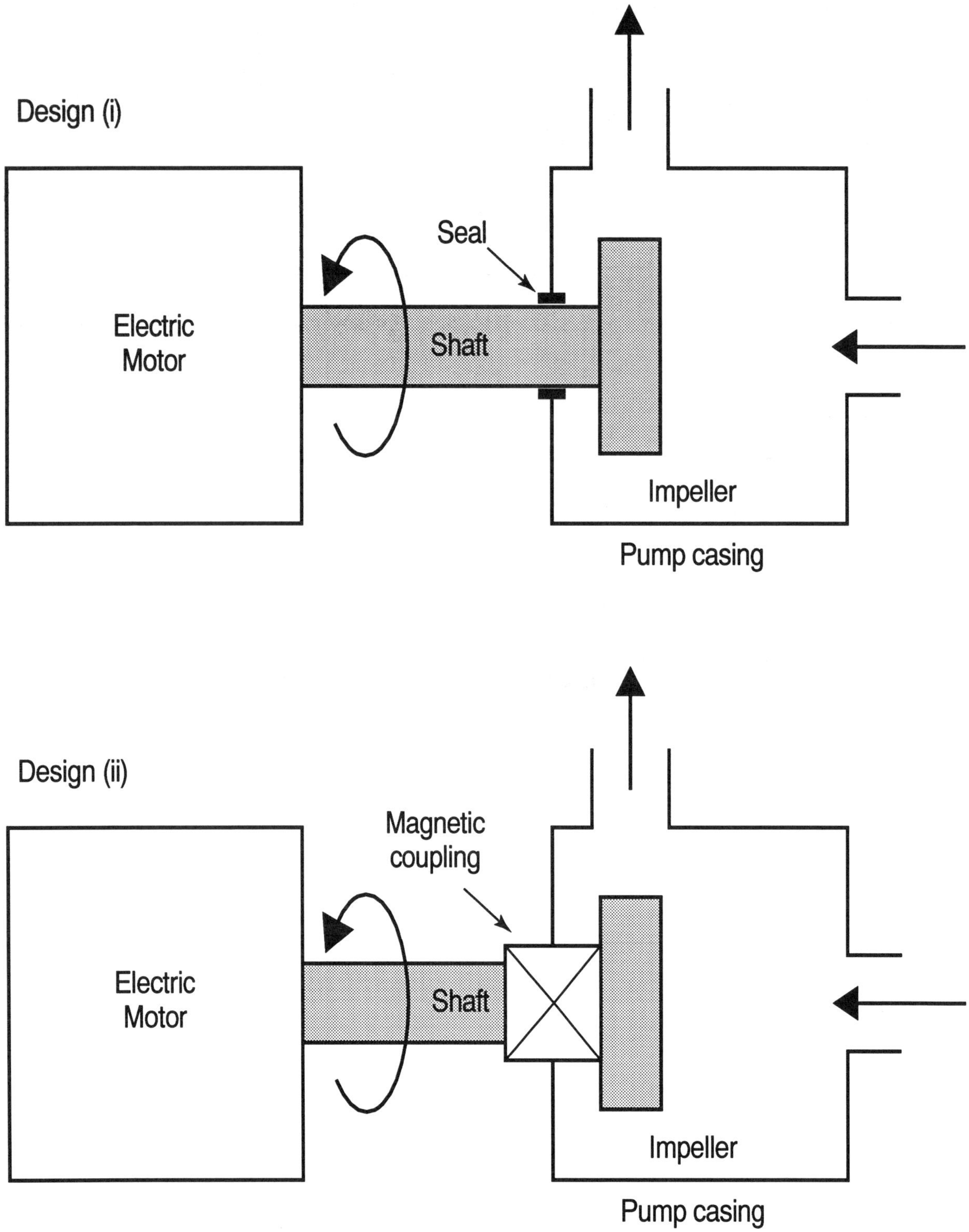

Fig. 1.1.1 Block diagram designs for a small pump

Fig. 1.1.1

Table 1.1.3 Quality function deployment analysis of conceptual design

Functional specification →		Avoid leaks	Fail safe	Minimize cost	Ease of cleaning
Rating →		Must have	Must have	Desirable	Desirable
		A	B	C	D
Alternative designs ↓					
Seals	1	↓	↓	↑	↓
Magnetic coupling	2	↑	↑	↓	↑

Key ↑ Positive correlation
↓ Negative correlation

Examples of correlation

A1	Seals prone to leakage	**A2**	Magnetic coupling cannot leak
B1	Blockage of impeller could cause motor burn-out	**B2**	Blockage of impeller will cause slippage of magnetic coupling
C1	Shaft and seals offer lower cost	**C2**	Magnetic coupling is more expensive
D1	Shaft needs to be decoupled from impeller and seals may need to be replaced	**D2**	Magnetic coupling easily separated to isolate pump from motor

1.1.3 Layout design

The first step in the progression from preferred conceptual design to layout designs (Chart 4) is to convert the block diagram design to an outline design of the components and sub-assemblies showing their general arrangement. The magnetically coupled pump example is shown in:

Fig. 1.1.2 *Diagrammatic view of pump (magnet-coupled drive design),* and

Fig. 1.1.3 *A magnetically coupled centrifugal pump*

Each component or sub-assembly can then be subjected to **function analysis.** This is illustrated in:

Table 1.1.4 *Quality function deployment analysis of component functions against customer requirements*

Table 1.1.4 is a **QFD matrix** correlating the market requirements (the 'voice of the customer') to some of the functions of two of the main components (the pump impeller and casing) of the small pump.

In practice, all market requirements would be correlated with all components of each **layout design.** Then the QFD matrix would be examined for:

(i) component functions which have a strong positive correlation with important (**must-have**) customer requirements. These functions should dominate the search for the materials with the most desirable combination of properties,

(ii) component functions which have a strong negative correlation with important (**must-have**) customer requirements. These designs should be examined for ways of reducing or accommodating the negative effects,

(iii) components which appear to have no influence on important customer requirements. This would raise the question, 'Is this component really necessary?'

The procedure for converting the functions of components into the desired combinations of materials properties (**materials desirability indices**) is illustrated in:

Fig. 1.1.4 *Function analysis of pump impeller,* and

Fig. 1.1.5 *Function analysis of pump impeller (bearing function)*

For some functions, especially those dependent on the strength, stiffness, toughness, fatigue and creep resistance (i.e. mechanical properties of materials), the desired combinations of materials properties are determinable by mathematical modelling. This is illustrated for the pump impeller in Case Study 1 in Section 1.1.7.

For other materials requirements, e.g. corrosion (Section 1.4) and wear resistance (Section 1.5), the preferred materials choice will be based more on empirical or experienced-based data.

When the materials desirability indices have been defined, the **Ashby maps/Materials selection charts** (Section 1.1.6) can be used to identify those materials which meet the minimum requirements and which of them offer the maximum performance or safety margins.

The **Ashby maps** chart the combination of properties of **generic classes** of materials. This level of detail is appropriate for **layout designs,** and identifies those generic classes of materials which will be investigated in greater detail at the **detailed design stage.** Ashby maps and case studies illustrating their application are given in Sections 1.1.6 and 1.1.7.

For the small pump example, the relevant Ashby map which charts the relationship of failure stress σ_f to density ρ, which must be greater than the square of the peripheral tip speed (allowing for a safety factor of 2), is shown in Case Study 1 (Section 1.1.7).

It can be seen that most engineering materials can meet the minimum requirements. Hence in this example, other materials requirements, in particular corrosion resistance (see Section 1.4) and wear resistance (see Section 1.5), are more decisive determining requirements.

An important consideration is the influence of component shape on materials performance. For example, for the same amount of material, an I-beam will be more efficient at resisting bending than a beam of rectangular cross-section. A tube is more efficient at resisting torsion than a solid rod employing the same amount of material. The influence of shape on the selection of the generic class of materials is described in detail in Section 1.1.8.

In the example of the pump impeller the most efficient shape will be dictated by the theory of centrifugal pump design. For this small pump the simple layout design shown in Fig. 1.1.3 is a good compromise between optimum shape and the likely cost of achieving that shape.

A most important decision for each component at this stage is whether one or more than one material is best suited to deliver the combination of required materials properties.

By way of illustration, reference to Fig. 1.1.4 raises the question whether one material can efficiently:

(a) maintain blade integrity,

(b) allow stable rotation (provide bearing surface), and

(c) locate the magnet,

or whether the functions to be performed by the component should be shared among several materials.

The possible options are as follows:

(1) Single material

A single material (e.g. an engineering plastic) could be chosen which:
- maintains blade integrity,
- provides the spindle bearing surface,
- locates the magnet (e.g. it is moulded-on).

(2) Multiple materials
- a material is chosen to maximize blade integrity,
- a different material is chosen as a bearing insert for the spindle bearing,
- the magnet is located by adhesive bonding.

The final decision to choose a single material or split the functions and choose two or more materials, is best made during the **design reviews** at the **detailed design stage**. At this stage all possible candidate materials should be retained for both options.

The required corrosion resistance to the range of fluids to be pumped proves to be the most effective sorting mechanism at this stage. The remaining candidate materials for the impeller after examination of the information contained in Section 1.4 are:

Plastics	Polypropylene	(most fluids)
	PVDF	(all fluids)
Metals	Stainless steel	(most fluids)
	Titanium alloys	(all fluids)
	Nickel alloys	(all fluids)
Ceramics	All engineering ceramics are resistant to all fluids of interest	

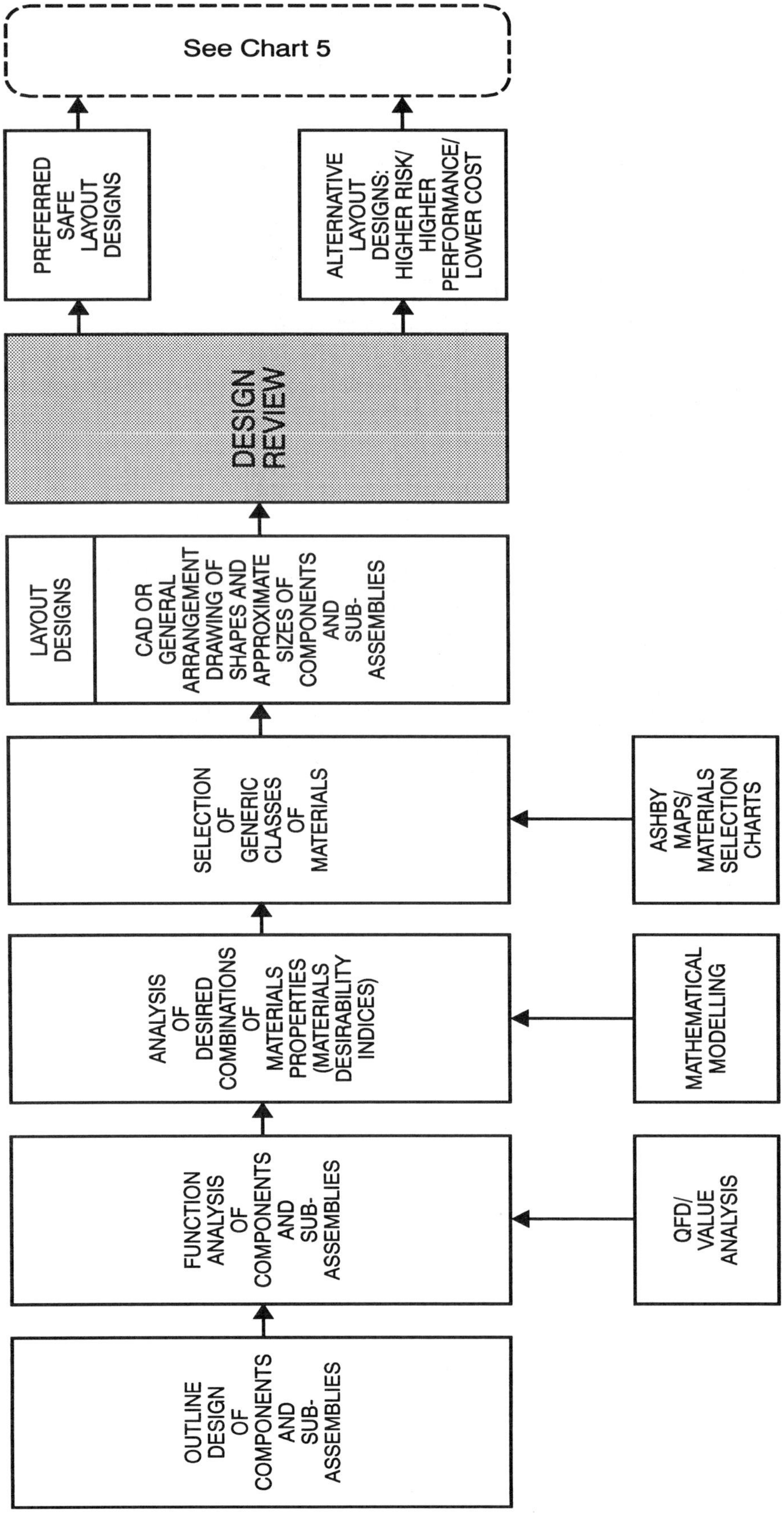

Chart 4 Concept design to layout design

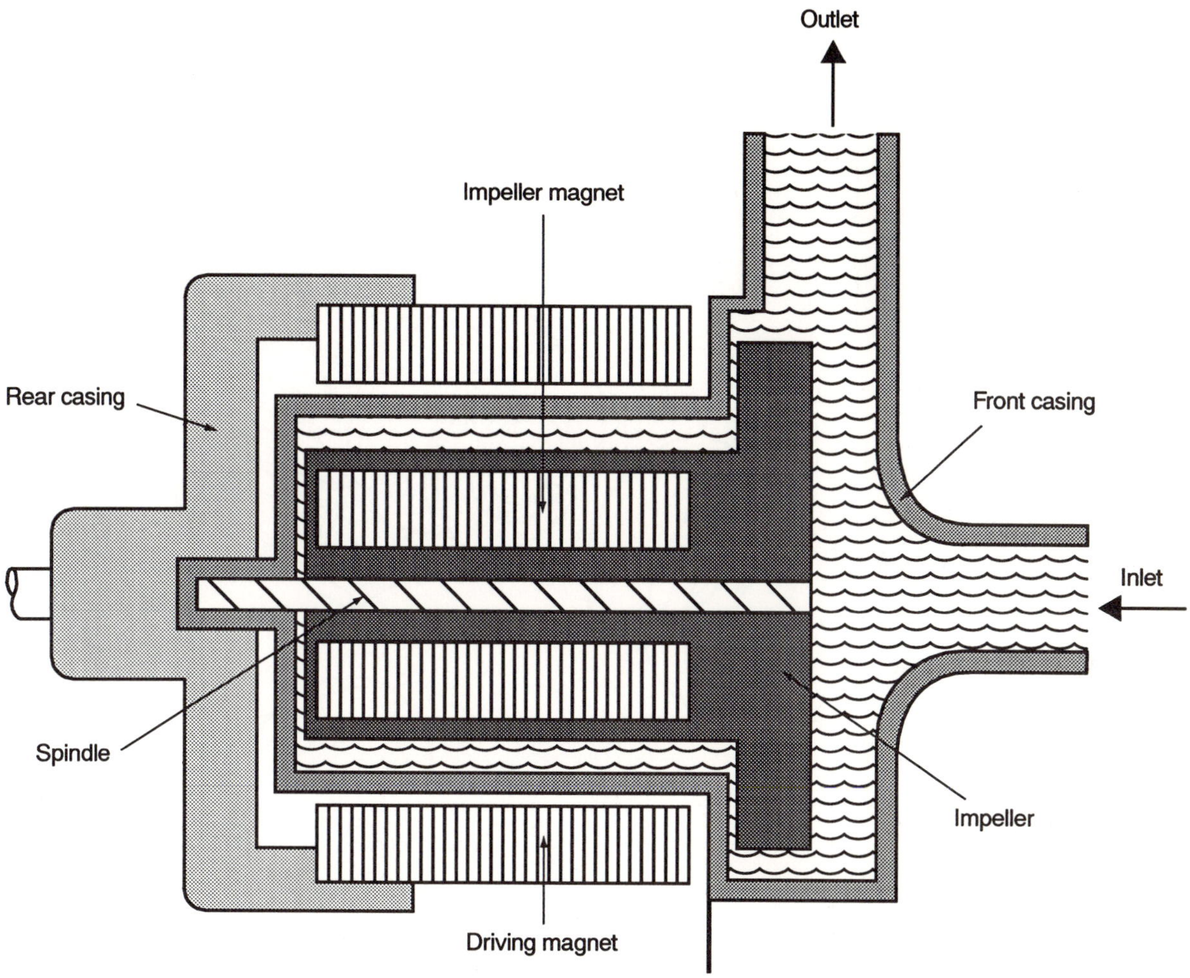

Fig. 1.1.2 Diagrammatic view of pump (magnet-coupled drive design)

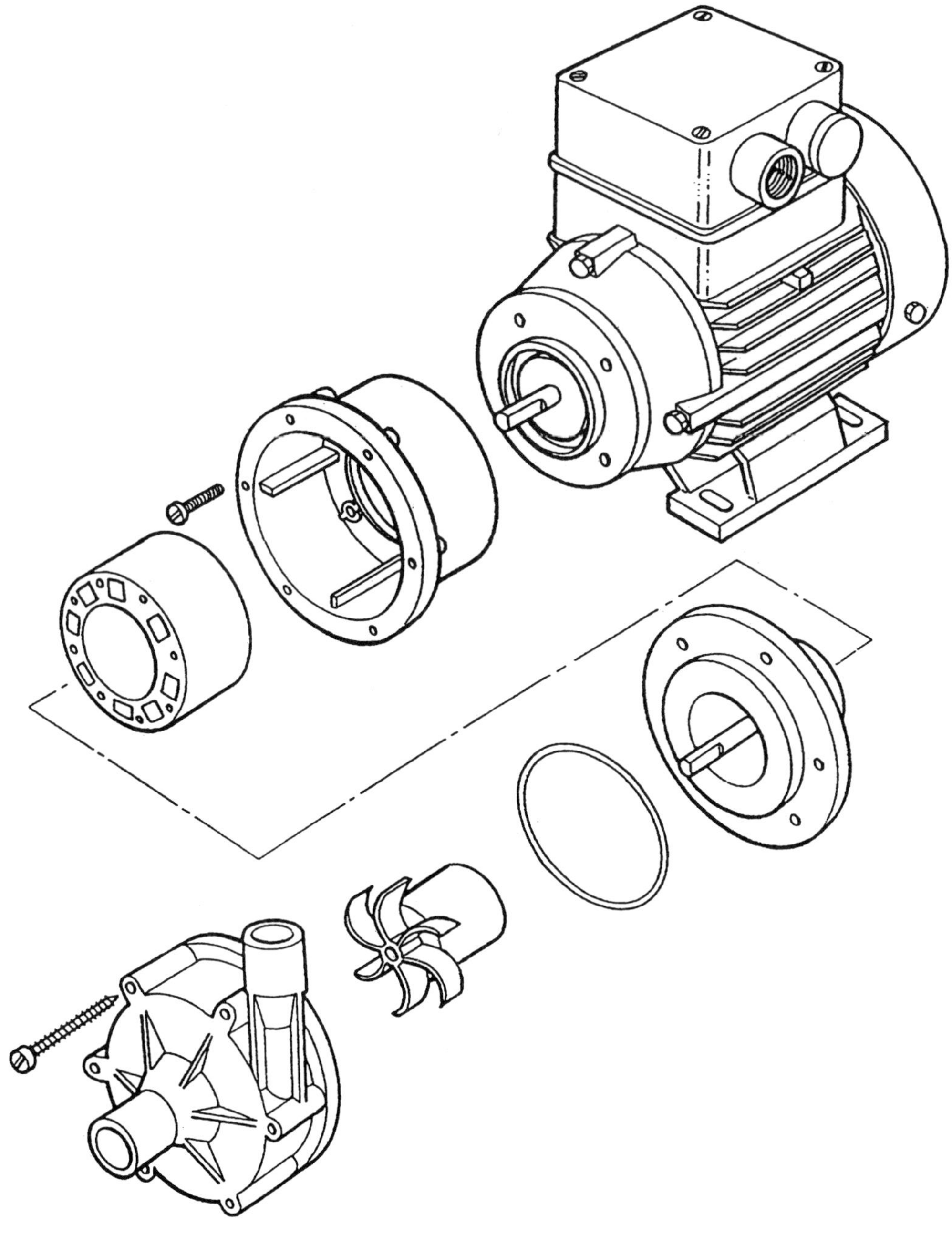

Fig. 1.1.3 A magnetically coupled centrifugal pump. (Courtesy of Totton Pumps.)

Fig. 1.1.3

Table 1.1.4 Quality function deployment analysis of component functions against customer requirements

Market requirement →		Avoid leaks	Failsafe	Maintain pressure	Maintain flow	Minimize size
Rating →		Must have	Must have	Must have	Must have	Desirable
		A	B	C	D	E
Components and functions ↓ IMPELLER						
Transport fluid	1	–	–	↑↑↑	↑↑↑	–
Maintain blade integrity	2	↑↑↑	↑↑↑	↑↑↑	↑↑↑	↓↓↓
Locate magnet	3	↑	↑	–	–	–
Allow stable rotation	4	–	↑↑↑	↑↑↑	↑↑↑	↓↓
CASING						
Contain fluid	5	↑↑↑	↑↑↑	↑↑	↑↑	↓↓
Permit flow	6	↑	↑↑	↑↑	↑↑↑	↓↓

Key
↑↑↑ Strong positive correlation
↓↓↓ Strong negative correlation
↑ Some positive correlation
– No correlation

Examples of correlation

A2 Blade disintegration could cause perforation of the casing. Hence, maintaining blade integrity has strong positive correlation with avoidance of leaks.

E2 The smaller the size of the pump, the higher the required speed of the impeller. The higher the speed, the greater the likelihood of blade disintegration. Hence, a strong negative correlation between minimizing size and maintaining blade integrity.

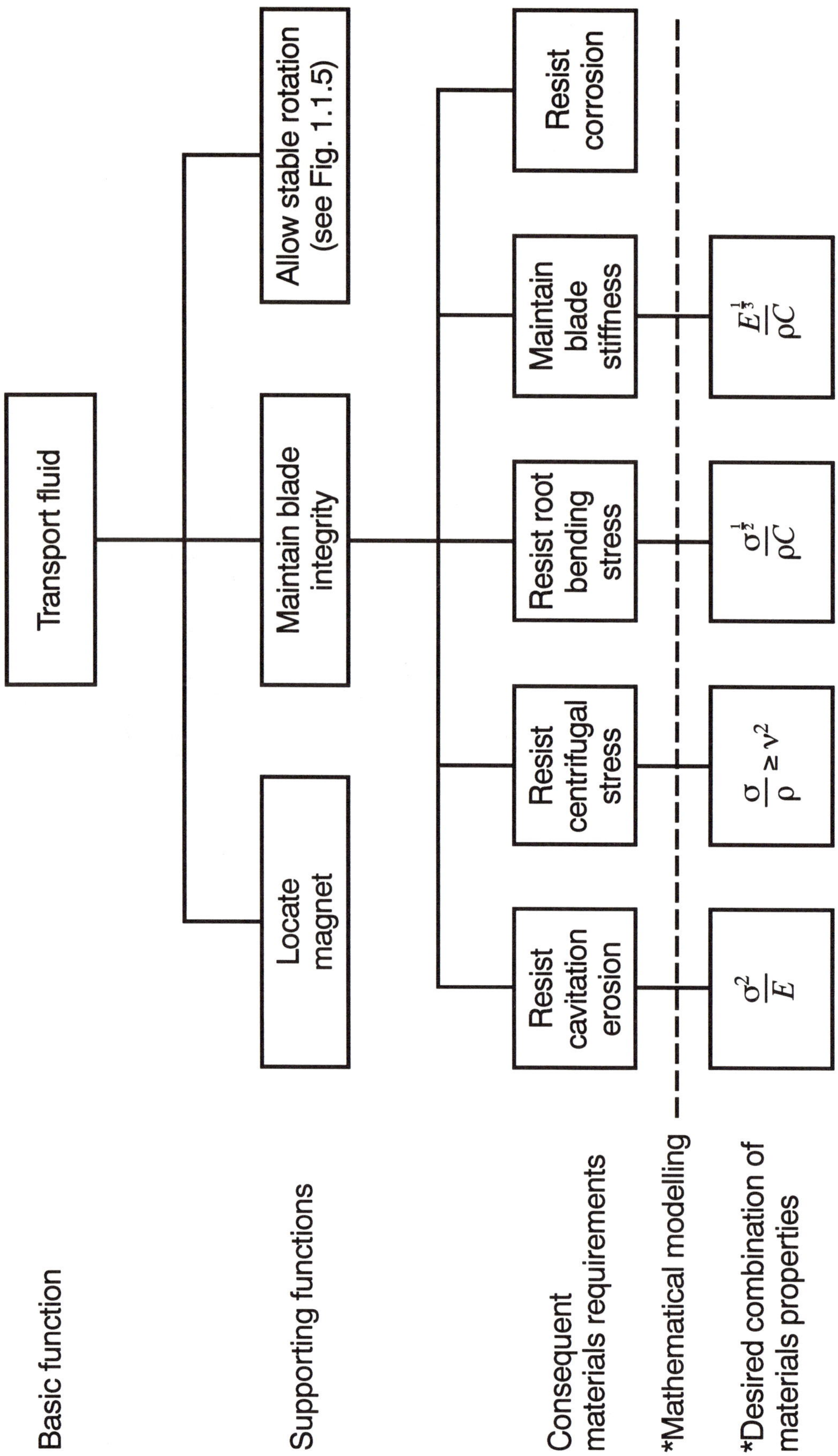

*See case Study 1 Section 1.1.7

Fig. 1.1.4 Function analysis of pump impeller

Fig. 1.1.4

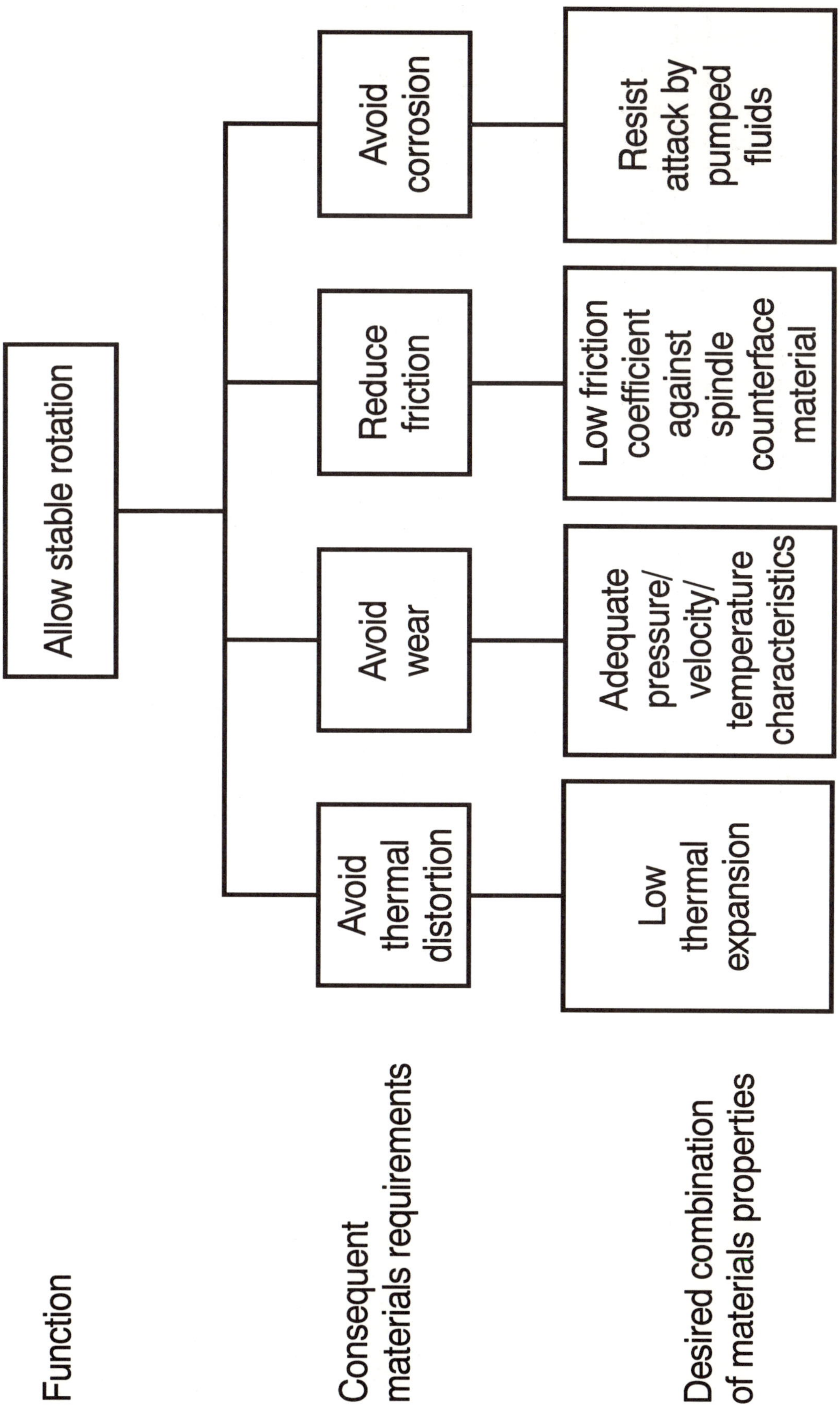

Fig. 1.1.5 Function analysis of pump impeller (bearing function)

Fig. 1.1.5

1.1.4 Detailed design

Progression from **layout design** to **detailed design** depends on the completion of the following activities:

- selection of particular grades of materials which possess the required combination of properties,
- analysis of the stresses and deflections on individual components,
- optimization of shapes, section thickness and dimensional tolerances,
- modelling of likely performance,
- construction and testing of prototypes,
- analysis of likely failure modes,
- input of all of the above information to a **design review** (see Box 2).

Selection of the preferred grades of materials will require the input of information on properties and costs from:

- comparative databases,
- materials suppliers,
- in-house data from testing.

The process is illustrated in Chart 5.

This procedure for the small pump example led to the following designs and materials selections. The materials were subjected to intensive testing.

Component	*Preferred safe design*	*Lower cost design*	*High performance design*
Pump Impeller	Glass fibre-reinforced polypropylene with alumina bearing insert	Acetal or polypropylene without bearing insert	PVDF with alumina bearing insert
Spindle	Alumina	Stainless steel	Alumina

Box 2 Design reviews

WHY?

Design reviews are an objective means of assessing:

- the merits of competing designs,
- the actions necessary to progress the preferred design to market as quickly as possible,
- the actions necessary to progress alternative higher risk, higher performance or low cost designs.

WHEN?

Design reviews should be held at all four stages of design:

- concept design,
- layout design,
- detailed design,
- design for manufacture.

WHAT?

Design reviews require the availability and prior circulation of information on:

- market requirements obtained by quality function deployment (QFD) techniques,
- theoretical calculations and modelling on which the designs are based,
- failure mode effect analyses of each design,
- test programme results,
- cost estimates.

HOW?

Design reviews are best conducted in dedicated, intensive sessions involving two teams.

Team 1. The design team
The design team will make presentations of the information detailed above. (A valuable technique is to display the relevant information (drawings, models, computer simulations) in a secure location where it can be reviewed on a structured basis by the review team prior to the actual design review sessions.)

Team 2. The review team
The review team should be made up of appropriately experienced and expertly qualified individuals who will receive the presentations and interrogate the design team. Members of the review team may include designers of previous similar products, but should not be involved in any of the possible designs being presented. The review team should include external independent experts as appropriate.

Design reviews
These should ideally be video recorded to capture all those ideas which, although not appropriate for the **preferred safe design** which will reduce the time-to-market, may be incorporated in future versions of the product which offer improved performance or lower costs.

Box 2

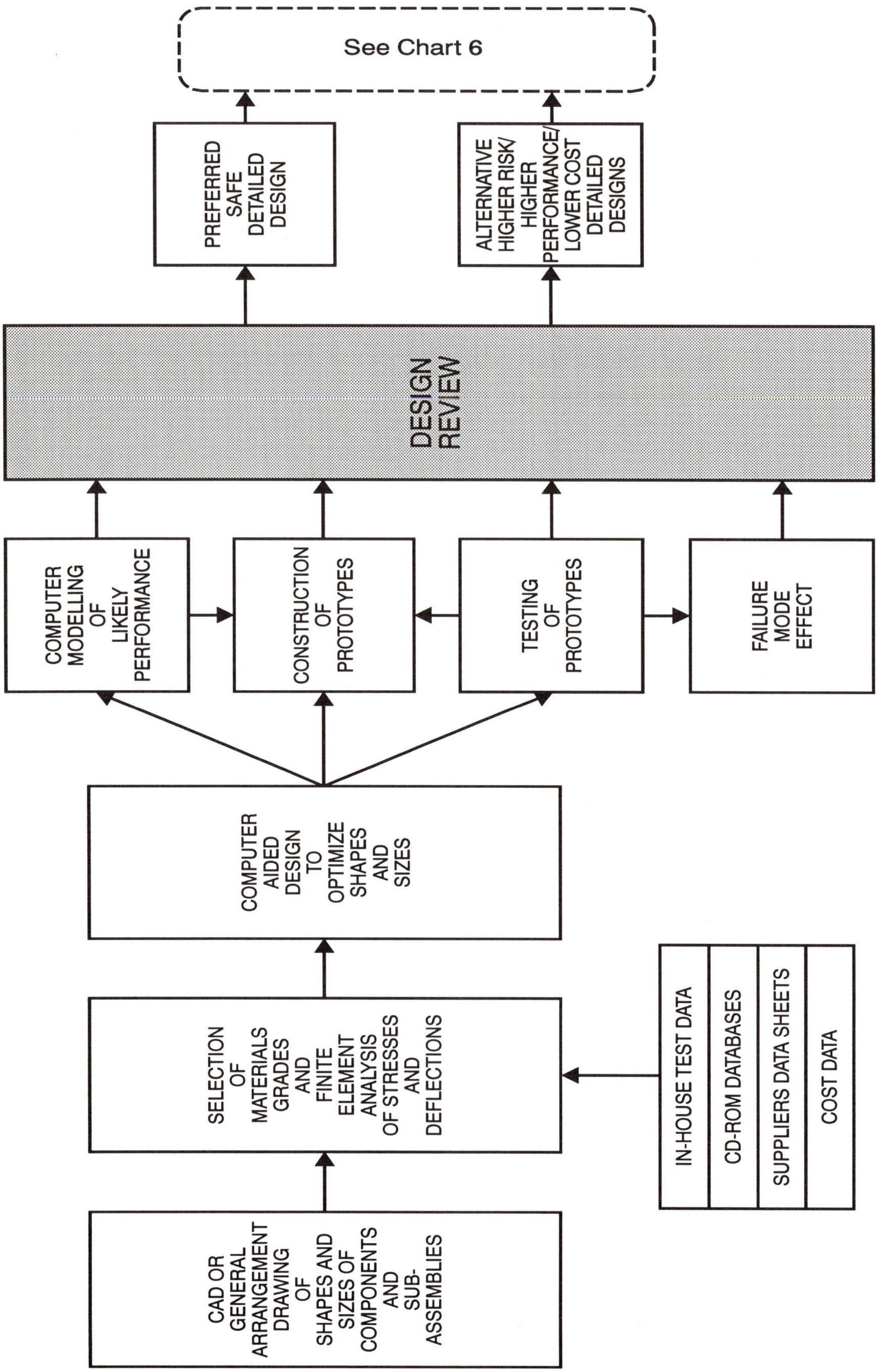

Chart 5 Layout design to detailed design

1.1.5 Design for manufacture

Chart 6 illustrates the procedure to progress from **detailed design** to **design-for-manufacture**.

The first products released to the market will almost certainly not be optimized in terms of maximum performance and minimum cost.

Reduction in time-to-market of a new product with a preferred safe design, the securing of the first sales and subsequent feedback from the market to improve the new product are perceived as likely to be much more profitable than perfecting a product and delaying its launch.

This does not mean that the first product to reach the market is not adequately tested under simulated or actual service conditions. Clearly a rash of early failures could damage the reputation of a new product fatally and incur expensive warranty claims.

It is, however, highly likely that the first product to market after:

- feedback from the market,
- increased testing experience of alternative designs, and
- further cost reduction exercises,

will be replaced by designs which give increased market share and higher profit margins.

Hence the feedback loop in Chart 6 is very important.

A summary of the iterations leading to the current preferred design for manufacture or the small pump is given in:

Table 1.1.5 *Evolution of design for small pump impeller*

A vitally important contribution to economic manufacture is the materials specifications which is placed on suppliers. The specification of material, semi-finished component or component should be detailed and explain why each detail is necessary. Deviation of the required specification and the necessary process controls to ensure that specification is delivered should be conducted in collaboration with the preferred supplier (see Boxes 3, 4 and 5 for examples).

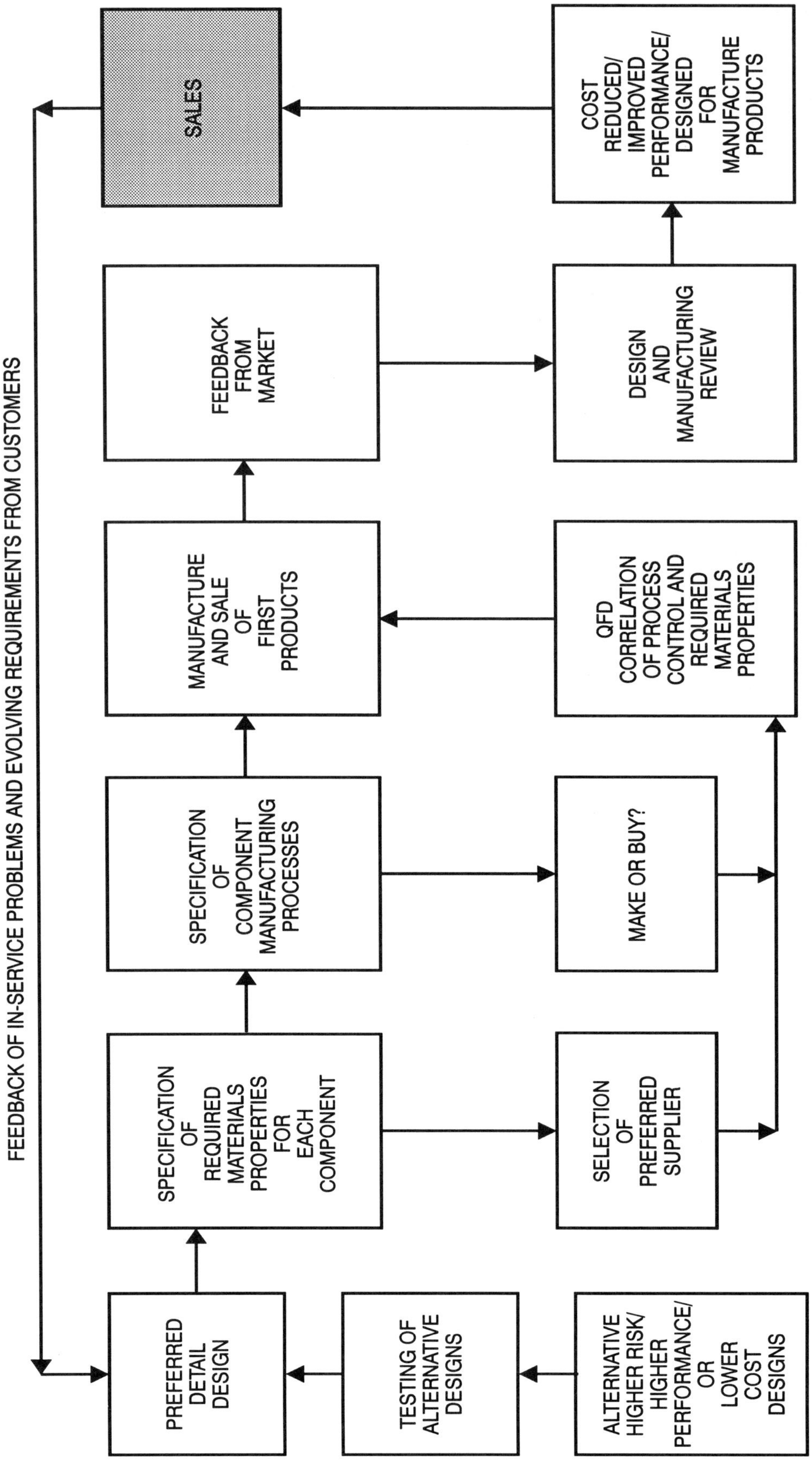

Chart 6 Detailed design to design-for-manufacture

Chart 6

Table 1.1.5 Evolution of design for small pump impeller

	Impeller design and materials	***Spindle material***
First product to market	Monolithic component. Injection moulded acetal copolymer meets requirements of blades and bearing surface	18/8 stainless steel
Feedback from market	Debris embeds in bearing surface and abrades spindle. Spindle attacked by chlorides in cleaning fluids.	
First update	Alumina ceramic bearing insert. Glass-filled polypropylene (cheaper than acetal) provides blade function.	Alumina ceramic
Feedback from market	No failures but bearing insert increases costs	
Second update	Alumina ceramic bearing. Unfilled polypropylene meets requirements of blades and bearing surface.	Alumina ceramic
Feedback from market	No failures. Lower cost. Market also needs a higher performance pump for wider range of aggressive fluids.	
Third update	PVDF plastic impeller with alumina bearing insert.	Alumina ceramic

Table 1.1.5

Box 3 Materials specification

The purpose of a materials specification is to ensure that the properties of the selected materials in the component, as manufactured and in-service, fulfil the required functions. The selected material and manufacturing route will have been chosen on the basis of the closeness of the match between the desired materials properties and the properties given on materials suppliers' data sheets and validated by short-term testing of prototypes. From this it might be inferred that all that needs to be done to specify a material is to quote the supplier's trade name or code or, in the case of most metals, the national specification designation. This procedure may work but it is fraught with dangers and constraints. For example:

- The dependence of the performance of the component (as determined by testing) on variations in the composition and processing parameters will not be clear.
- Small changes in the composition and processing route variables may result in a material which fails in service.
- If it is desired to change suppliers for reasons of increased cost or cessation of supply, the new supplier's equivalent material may not be obvious, and expensive time consuming re-testing may be required.

The two techniques described below lessen dependence on a particular supplier's grade of material and ensure that variations in composition and processing which would adversely influence properties are controlled and eliminated.

(i) Detailed specification with reasons

This is particularly appropriate for metals where the broad designation is the national standard. National standards for most metals guarantee only a minimum level of properties and quality control and a much more exact specification is needed to ensure fitness for purpose. An example for a cast iron hydraulic valve is given in Box 4.

(ii) Quality function deployment matrices correlating component functions (and consequent desired materials properties and microstructure) with composition and process controls

This matrix illustrated in Box 5 is constructed by discussing a detailed specification such as that given in Box 4 with the preferred supplier. The supplier is then invited to define the dependence of the required materials properties and microstructure on variations in composition and process parameters. The resulting QFD matrix is given in Box 5 is based on the correlations given in Section 2.3.8 'Optimization of grey cast iron.'

Box 3

Box 4 Example of a specification for a grey cast iron hydraulic valve which must have a minimum tensile strength of 200 MPa (~14 tons/in^2) and also requires a combination of wear resistance and machinability

Specification	***Reasons and comments***
(a) Graphite cast iron (BS Grade 200) (ASTM Class 30B) (GG 22)	See Section 2.3 for national specifications. Specification covers freedom from porosity, burn-ton sand, and a clause on phosphorus content being agreed between supplier and user (see item (g) below). Specification states that a separately cast test bar of the same iron as that used for the coating will have the minimum tensile strength. An additional requirement might be specified that a test bar machined from the centre of the thickest part of the casting should have a minimum tensile strength of 200 MPa (~14 tons/in^2).
(b) Graphite type A, size 4–5 (ASTM rating) in region of bearing surface. (c) Free ferrite less than 5%.	Recommended for optimum bearing properties, and avoidance of scuffing.
(d) 0.8% Copper	Added to stabilize pearlite in thick sections to ensure minimum strength level is achieved.
(e) Hardness 190–230 HB to be measured 1 mm below as-cast surface in regions to be machined. Apply homogenizing heat treatment if necessary.	To ensure homogeneous machinability (see Section 2.3).
(f) Critical sections (e.g. lugs) must not contain graphite of type C or greater than size 3.	To avoid fatigue failure.
(g) Phosphorus less than 0.2%	Necessary for pressurized application where additional embrittlement caused by phosphorus must be avoided
(h) Dimensional tolerances	Specify on drawing.
(i) Casting to be phosphated as final operation before assembly.	Specify phosphating process. Necessary for running-in assistance (Section 1.5).

Box 5

Box 5 QFD matrix correlating application property requirements and process variables (incomplete, indicative only)

Application property requirements →	*Resist fatigue*			*Resist wear*			*Allow rapid machining*	
Composition variables ↓	Control graphite size	Control pearlite structure	Maintain tensile strength	Reduce free ferrite	Control graphite distribution	Control pearlite structure	Avoid free carbides	Avoid phosphide network
Carbon	↑↑↑		↑↑		↑↑			
Silicon	↑↑							
Phosphorus			↑↑					↑↑↑
Manganese	↑↑		↑↑					
Copper		↑↑↑	↑↑	↑↑↑		↑↑↑		
Process variables ↓								
Pouring temperature							↑	
Innoculation	↑↑↑							
Heat treatment, time and temperature		↑↑				↑↑	↑↑	

Key: ↑↑↑ Very strong correlation ↑↑ Strong correlation ↑ Some correlation

1.1.6 Preliminary materials selection - materials selection charts for mechanical and physical properties

There are over 100 000 different materials available to the design engineer. Hence the main task of **preliminary** materials selection is to reduce this very large number to a much smaller total by eliminating unsuitable materials and focusing on those materials which have the best combination of properties and cost for a particular application. The materials selection charts which follow have been designed for this purpose. They are to be used for conceptual design to identify the candidate materials for detailed design. **They are not appropriate for detailed design calculations.**

The materials of mechanical and structural engineering fall into nine broad classes:

Engineering alloys	– metals and their alloys.
Engineering polymers	– thermoplastics and thermosets.
Engineering ceramics	– 'fine' ceramics.
Engineering composites	– glass, carbon or aramid fibre-reinforced plastics.
Porous ceramics	– brick, cement, concrete, stone.
Glasses	– silicate glasses.
Woods	– structural timbers.
Elastomers	– natural and artificial rubbers.
Foams	– foamed polymers.

Within each class, data are plotted for a representative set of materials, chosen to span the full range of behaviours for that class, and to include the most widely used members of it. In this way the envelope for a class (heavy lines) encloses data not only for the materials listed in Table 1.1.6 but for virtually all other members of the class as well. See:

Table 1.1.6 *Material classes, generic class members and abbreviated names*

Specific materials are not listed on the charts. The aluminium alloy 7075 in the T6 condition (for example) is contained in the property envelopes for Al alloys, nylon 6.6 in those for nylons, etc. For detailed information on the properties of individual material grades, other sections of the Selector should be consulted.

The charts display the properties listed below, for the nine classes of materials.

Basic subset of material properties	*Symbol*	*Units*
Relative cost	C_R	–
Density	ρ	Mg/m^3
Young's modulus	E	GPa
Strength	σ_f	MPa
Fracture toughness	K_{1c}	MPa $m^{1/2}$
Toughness	G_{1c}	J/m^2
Damping coefficient	η	–
Thermal conductivity	λ	W/mK
Thermal diffusivity	a	m^2/s
Volume specific heat	$C_p\rho$	J/m^3K
Thermal expansion coefficient	α	1/K
Thermal shock resistance	ΔT	K
Strength at temperature	σ (T)	MPa

The charts allow selection of the subset of materials with a property within a specified range: materials with modulus E between 100 and 200 GPa, for instance; or materials with thermal conductivity above 100 W/mK.

More usually, performance is maximized by selecting the subset of materials with the greatest value of a grouping of material properties. A light, stiff beam is best made of a material with a high value of $E^{1/2}/\rho$; safe pressure vessels are best made of a material with a high value of K_{1c}^2/σ_f , and so on. Some of these performance-maximizing combinations or 'merit indices' are listed in:

Table 1.1.7 *Performance-maximizing property groups (materials desirability indices)*

The charts are designed to display these, and to allow selection of the subset of materials which
maximize them.

Multiple criteria can be used. The subset of materials with both high $E^{1/2}/\rho$ and high E can be selected from Fig. 1.1.7; that with high σ_f^3/E^2 and high E (good materials for pivots) from Fig. 1.1.9.

Throughout, the goal is to identify from the charts a subset of materials, not a single material. Finding the best material for a given application involves many considerations, many of them (like availability, appearance and feel) not easily quantifiable. The charts do not give the final choice – that requires the use of judgement and experience. Their power is that they lead quickly and efficiently to a subset of materials worth considering, and make sure that a promising candidate is not overlooked.

The data plotted on the charts have been assembled over several years from a wide variety of sources, including the other sections and chapters of the Selector. As far as possible, the data have been validated (cross-checked by comparing values from more than one source) and they have been examined for consistency with physical rules. The charts show a range of values for each property of each material. Sometimes the range is narrow; the modulus of a metal, for instance, varies by only a few percent about its mean value. Sometimes it is wide; the strength of a given ceramic can vary by a factor of 100 or more. The reasons for the range of values vary: heat treatment and mechanical working have a profound effect on yield strength, damping and toughness of metals. Crystallinity and degree of cross-linking greatly influence the modulus of polymers. Grain size and porosity change considerably the fracture strength of ceramics. And so on. These 'structure-sensitive' properties appear as elongated balloons for a class.

A list of the materials selection charts for mechanical and physical properties is given below:

Fig. 1.1.6 *Young's modulus E against density ρ (showing specific stiffness, etc)*

Fig. 1.1.7 *Strength σ_f against density ρ (showing specific strength, etc)*

Fig. 1.1.8 *Fracture toughness K_{1c} against density ρ (showing specific toughness, etc)*

Fig. 1.1.9 *Young's modulus E against strength σ_f (showing max energy storage, etc)*

Fig. 1.1.10 *Specific modulus E/ρ against specific strength $\sigma_f\rho$*

Fig. 1.1.11 *Fracture toughness K_{1c} against Young's modulus E (showing toughness)*

Fig. 1.1.12 *Fracture toughness K_{1c} against strength σ_f (showing yield-before-break, etc)*

Fig. 1.1.13 *Loss coefficient η against Young's modulus E*

Fig. 1.1.14 *Thermal expansion coefficient α against thermal conductivity λ*

Fig. 1.1.15 *Thermal conductivity λ against thermal diffusivity α (showing volumetric specific heat)*

Fig. 1.1.16 *Linear thermal expansion α against Young's modulus E (showing thermal stress)*

Fig. 1.1.17 *Normalized tensile strength σ_t/E against thermal expansion α (showing thermal shock resistance)*

Fig. 1.1.18 *Strength $\sigma_f(T)$ against temperature T*

Table 1.1.6 Material classes, generic class members and abbreviated names

Class	*Members*	*Short name*
Metallic alloys	Aluminium alloys Copper alloys Lead alloys Magnesium alloys Nickel alloys Steels Tin alloys Titanium alloys Zinc alloys	Al alloys Cu alloys Lead alloys Mg alloys Ni alloys Steels Tin alloys Ti alloys Zn alloys
Engineering polymers (Thermoplastics and thermosets)	Epoxies Melamines Polycarbonate Polyesters Polyethylene, high density Polyethylene, low density Polyformaldehyde Polymethylmethacrylate Polypropylene Polytetrafluorethylene Polyvinylchloride	EP MEL PC PEST HDPE LDPE PF PMMA PP PTFE PVC
Engineering ceramics (Fine ceramics capable of load-bearing application)	Alumina Diamond Sialons Silicon carbide Silicon nitride Zirconia	Al_2O_3 C Sialons SiC Si_3N_4 ZrO_2
Engineering composites (A distinction is drawn between the properties of a ply – 'UNIPLY' – and of a laminate 'LAMINATES')	Carbon fibre-reinforced polymer Glass fibre-reinforced polymer Kevlar fibre-reinforced polymer	CFRP GFRP KFRP
Porous ceramics (Traditional ceramics, cements, rocks and minerals)	Brick Cement Common rocks Concrete Porcelain Pottery	Brick Cement Rocks Concrete Pcln Pot
Glasses (Ordinary silicate glass)	Borosilicate glass Soda glass Silica	B–glass Na–glass SiO_2
Woods (Separate envelopes describe properties parallel to the grain and normal to it, and wood products)	Ash Balsa Fir Oak Pine Wood products (ply, etc)	Ash Balsa Fir Oak Pine Wood products
Elastomers (Natural and synthetic rubbers)	Natural rubber Hard butyl rubber Polyurethanes Silicone rubber Soft butyl rubber	Rubber Hard butyl PU Silicone Soft butyl
Polymer foams	These include: Cork Polyester Polystyrene Polyurethane	 Cork PEST PS PU

Table 1.1.6

Table 1.1.7 Performance-maximizing property groups (materials desirability indices)

Mode of loading		***Minimize weight for given***			
		Stiffness	***Ductile strength***	***Brittle strength***	
<u>*TIE*</u> *F*, *l* specified *r* free	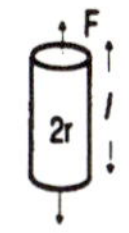	$\frac{E}{\rho}$	$\frac{\sigma_f}{\rho}$	$\frac{K_{1c}}{\rho}$	Figs. 1.1.6, 1.1.7 and 1.1.8
<u>*TORSION BAR*</u> *T*, *l* specified *r* free		$\frac{G^{1/2}}{\rho}$	$\frac{\sigma_f^{2/3}}{\rho}$	$\frac{K_{1c}^{2/3}}{\rho}$	
<u>*TORSION TUBE*</u> *T*, *l*, *r* specified *t* free	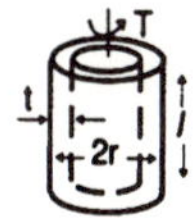	$\frac{G^{1/2}}{\rho}$	$\frac{\sigma_f^{2/3}}{\rho}$	$\frac{K_{1c}^{2/3}}{\rho}$	
<u>*BENDING OF RODS AND TUBES*</u> *F*, *l* specified *r* or *t* free	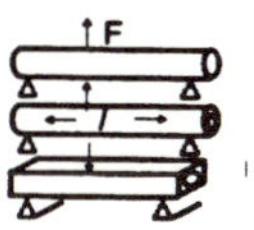	$\frac{E^{1/2}}{\rho}$	$\frac{\sigma_f^{2/3}}{\rho}$	$\frac{K_{1c}^{2/3}}{\rho}$	
<u>*BUCKLING OF SLENDER COLUMN OR TUBE*</u> *F*, *l* specified *r* or *t* free	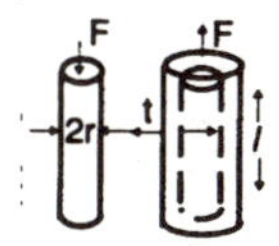	$\frac{E^{1/2}}{\rho}$	–	–	
<u>*BENDING OF PLATE*</u> *F*, *l*, *w* specified *t* free	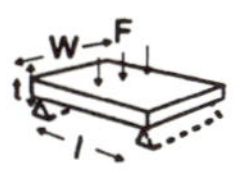	$\frac{E^{1/2}}{\rho}$	$\frac{\sigma_f^{1/2}}{\rho}$	$\frac{K_{1c}^{1/2}}{\rho}$	
<u>*BUCKLING OF PLATE*</u> *F*, *l*, *w* specified *t* free	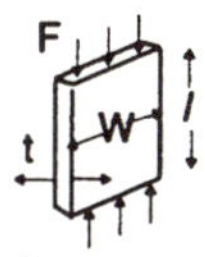	$\frac{E^{1/2}}{\rho}$	–	–	
<u>*CYLINDER WITH INTERNAL PRESSURE*</u> *P*, *r* specified *t* free	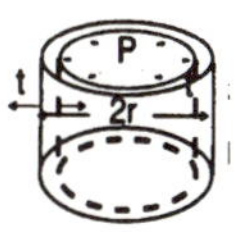	$\frac{E}{\rho}$	$\frac{\sigma_f}{\rho}$	$\frac{K_{1c}}{\rho}$	
<u>*ROTATING CYLINDER*</u> *w*, *r* specified *t* free	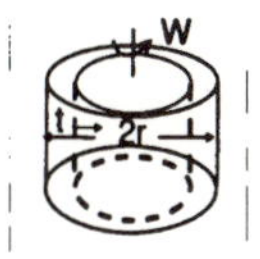	$\frac{E}{\rho}$	$\frac{\sigma_f}{\rho}$	$\frac{K_{1c}}{\rho}$	
<u>*SPHERE WITH INTERNAL PRESSURE*</u> *p*, *r* specified *t* free	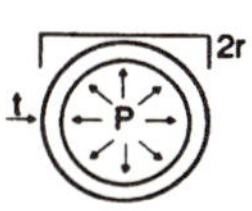	$\frac{E}{(1-\nu)\rho}$	$\frac{\sigma_f}{\rho}$	$\frac{K_{1c}}{\rho}$	

Table 1.1.7 Performance-maximizing property groups – *continued*

ELASTIC DESIGN

			Fig.
SPRINGS 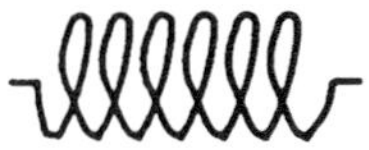	Spring of min. volume Spring of min. weight	Max. σ_f^2/E Max. $\sigma_f^2/\rho E$	1.1.9 1.1.6, 1.1.9
ELASTIC HINGES 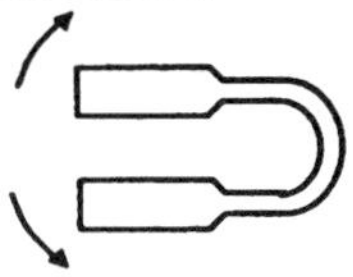	Hinge with no axial load Hinge with axial load	Max. σ_f^2/E Max. σ_f^2/E	1.1.9 1.1.9
KNIFE EDGES, PIVOTS 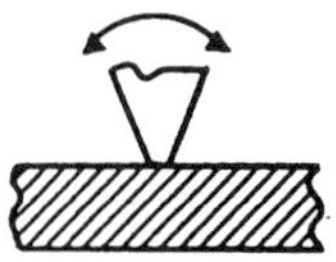	'Point' or 'line' contact with min. friction loss	Max. σ_f^3/E^2 and E	1.1.9

PLASTIC AND FRACTURE–SAFE DESIGN

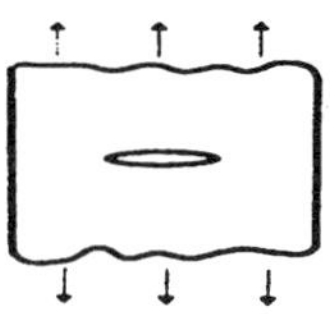	Load-controlled design Displacement-controlled design Yield before break Leak before break	Max. K_{1c} and σ_f Max. K_{1c}/E and σ_f/E Max. K_{1c}/σ_f Max. $K^2{}_{1c}/\sigma_f$	1.1.12 1.1.9, 1.1.10 1.1.12 1.1.12

THERMAL DESIGN

THERMAL FLUX 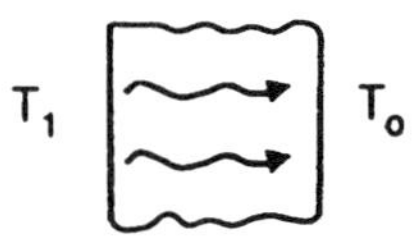	Min. heat flux at steady state Min. temp. rise after time t	Min. λ Min. $\lambda/C_p\rho = a$	1.1.15 1.1.15
THERMAL STRESS, SHOCK 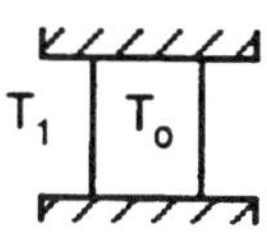	Min. thermal stress Max. thermal shock	Min. $E\alpha$ Max $\sigma_f/E\alpha$	1.1.16 1.1.17

Table 1.1.7 – *continued*

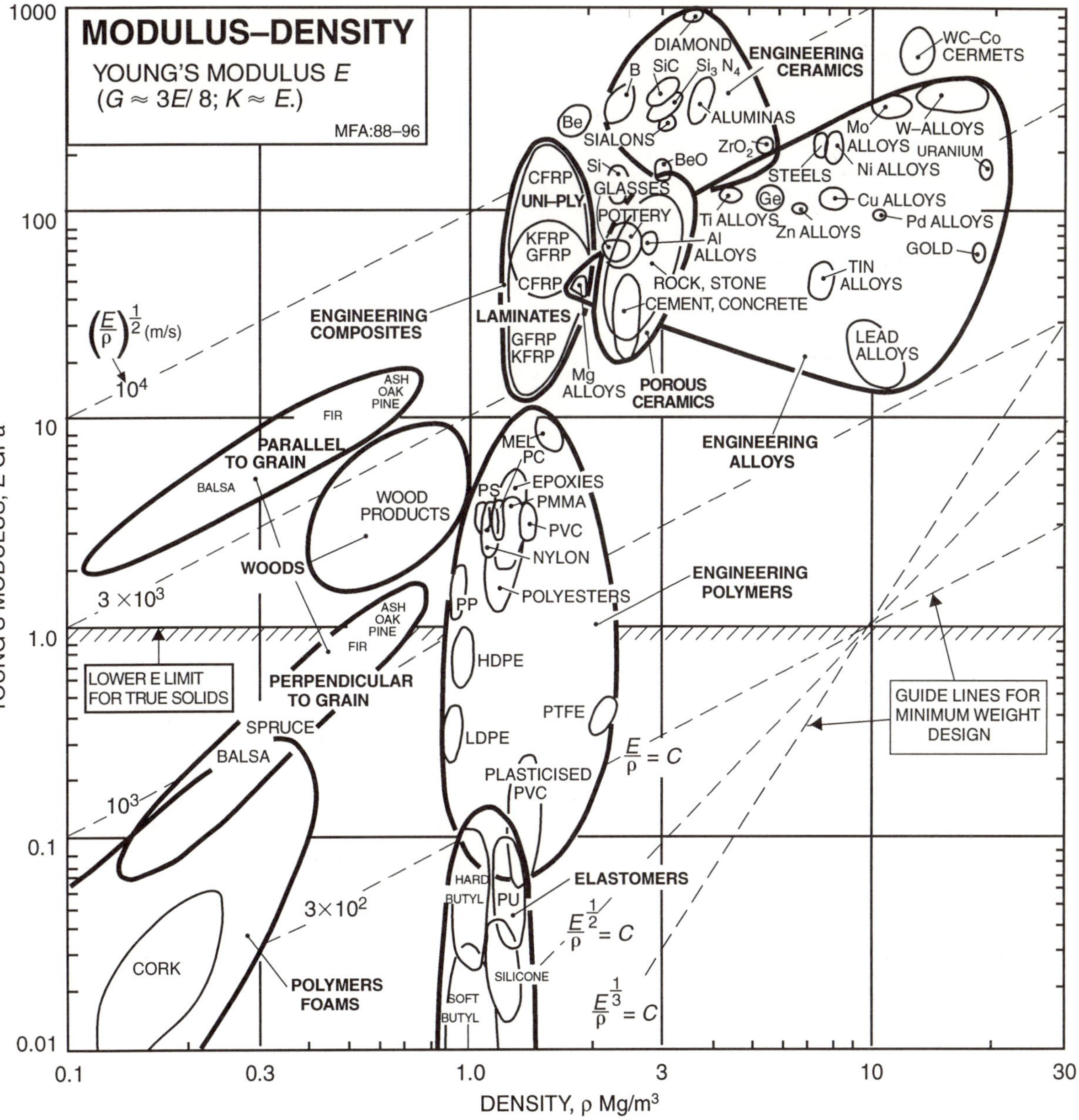

Fig. 1.1.6 Materials selection chart: Young's modulus, *E*, against density, ρ

The chart guides selection of materials for light, stiff, components. The lines show the loci of points for which:

(a) $E/\rho = C$ (criterion for axial tension of ties)
(b) $E^{1/2}/\rho = C$ (criterion for bending, torsion, or buckling of beams, shafts and columns)
(c) $E^{1/3}/\rho = C$ (criterion for bending of plates)

The value of the constant C increases as the lines are displaced upwards and to the left. Materials offering the greatest stiffness-to-weight ratio lie towards the upper left corner.

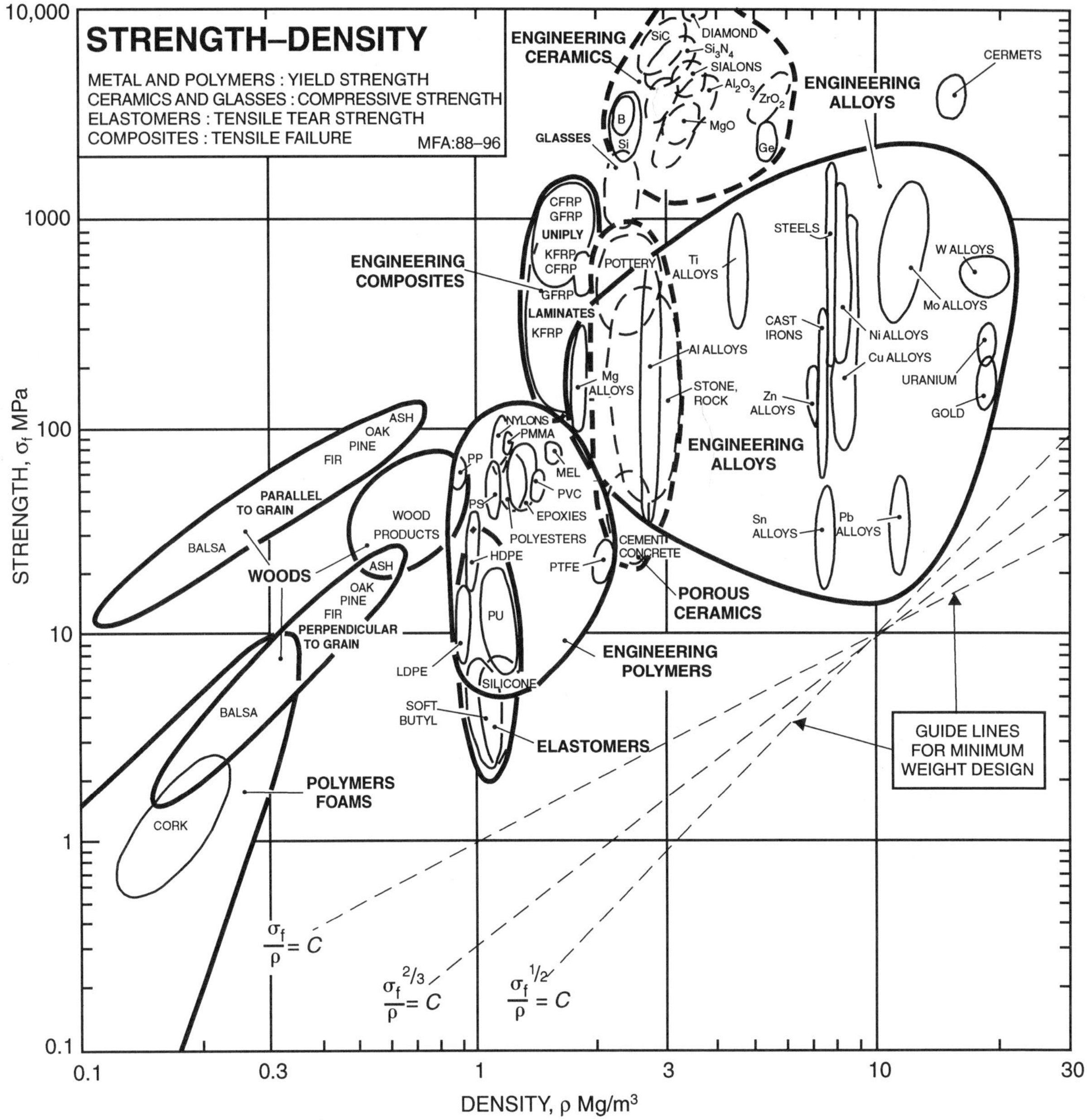

Fig. 1.1.7 Materials selection chart: strength, σ_f against density, ρ

The 'strength' for *metals* is the 0.2% offset yield strength. For *polymers,* it is the stress at which the stress–strain curve becomes markedly non-linear – typically, a strain of about 1%. For ceramics, it is the compressive crushing strength; remember that this is roughly 15 times larger than the tensile (fracture) strength. The chart guides selection of materials for light, strong, components. The lines show the loci of points for which:

(a) $\sigma_f/\rho = C$ (criterion for plastic failure of ties)
(b) $\sigma_f^{2/3}/\rho = C$ (criterion for plastic bending, torsion of beams and shafts)
(c) $\sigma_f^{1/2}/\rho = C$ (criterion for plastic bending of plates)

The value of the constant *C* increases as the lines are displaced upwards and to the left. Materials offering the greatest strength-to-weight ratio lie towards the upper left corner.

Fig. 1.1.7

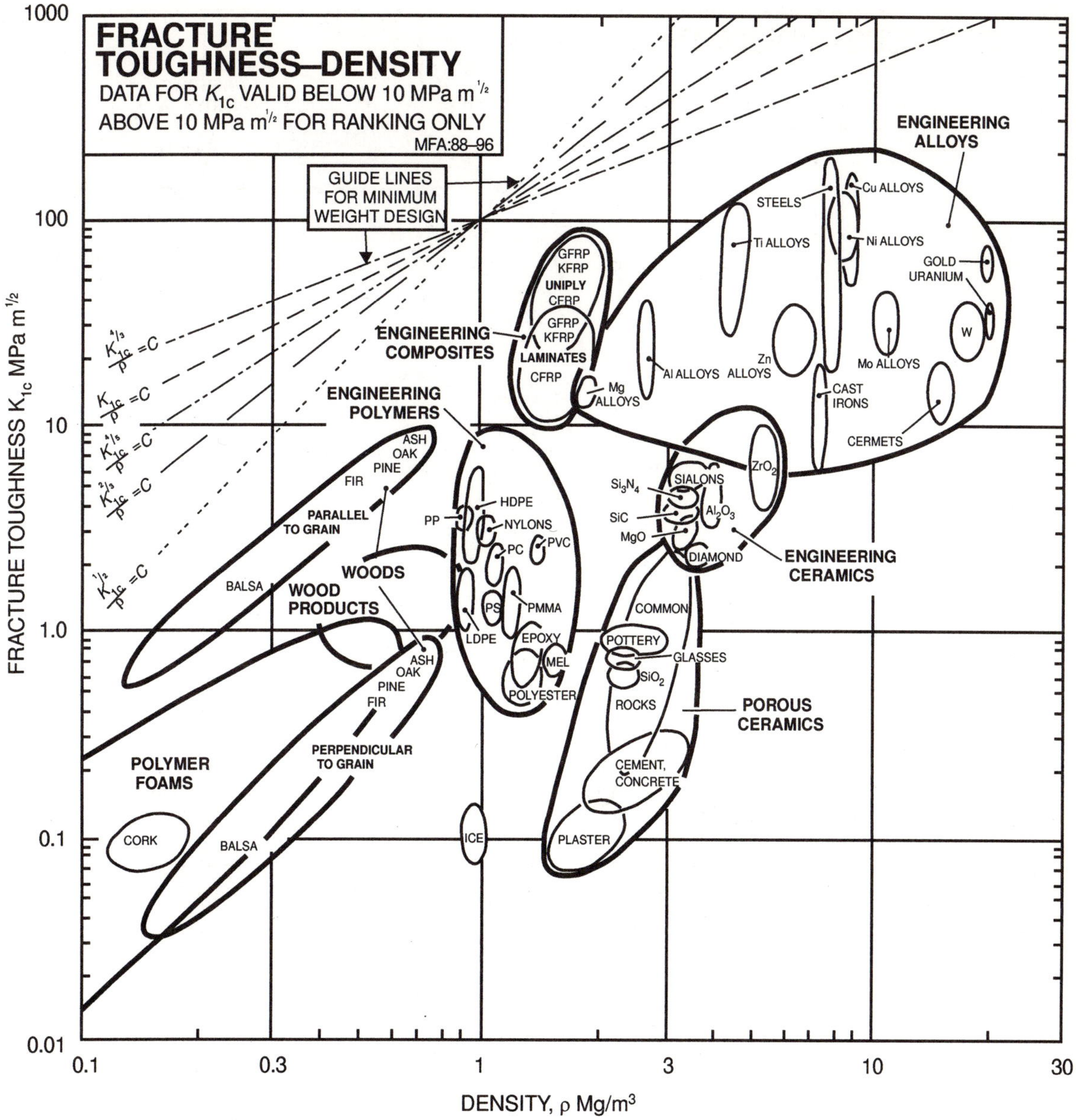

Fig. 1.1.8 Materials selection chart: fracture toughness, K_{1c}, against density, ρ

Linear-elastic fracture mechanics breaks down when the fracture toughness is large and the section is small; then J- integral methods should be used. The data shown here are adequate only for the rough calculations of conceptual design. The chart guides selection of materials for light, fracture-resistant components. The lines show the loci of points for which:

(a) $K_{1c}/\rho = C$ (criterion for brittle failure of ties)

(b) $K_{1c}^{2/3}/\rho = C$ (criterion for plastic bending, torsion of beams and shafts)

(c) $K_{1c}^{1/2}/\rho = C$ (criterion for plastic bending of plates)

The value of the constant C increases as the lines are displaced upwards and to the left. Materials offering the greatest thoughness-to-weight ratio lie towards the upper left corner.

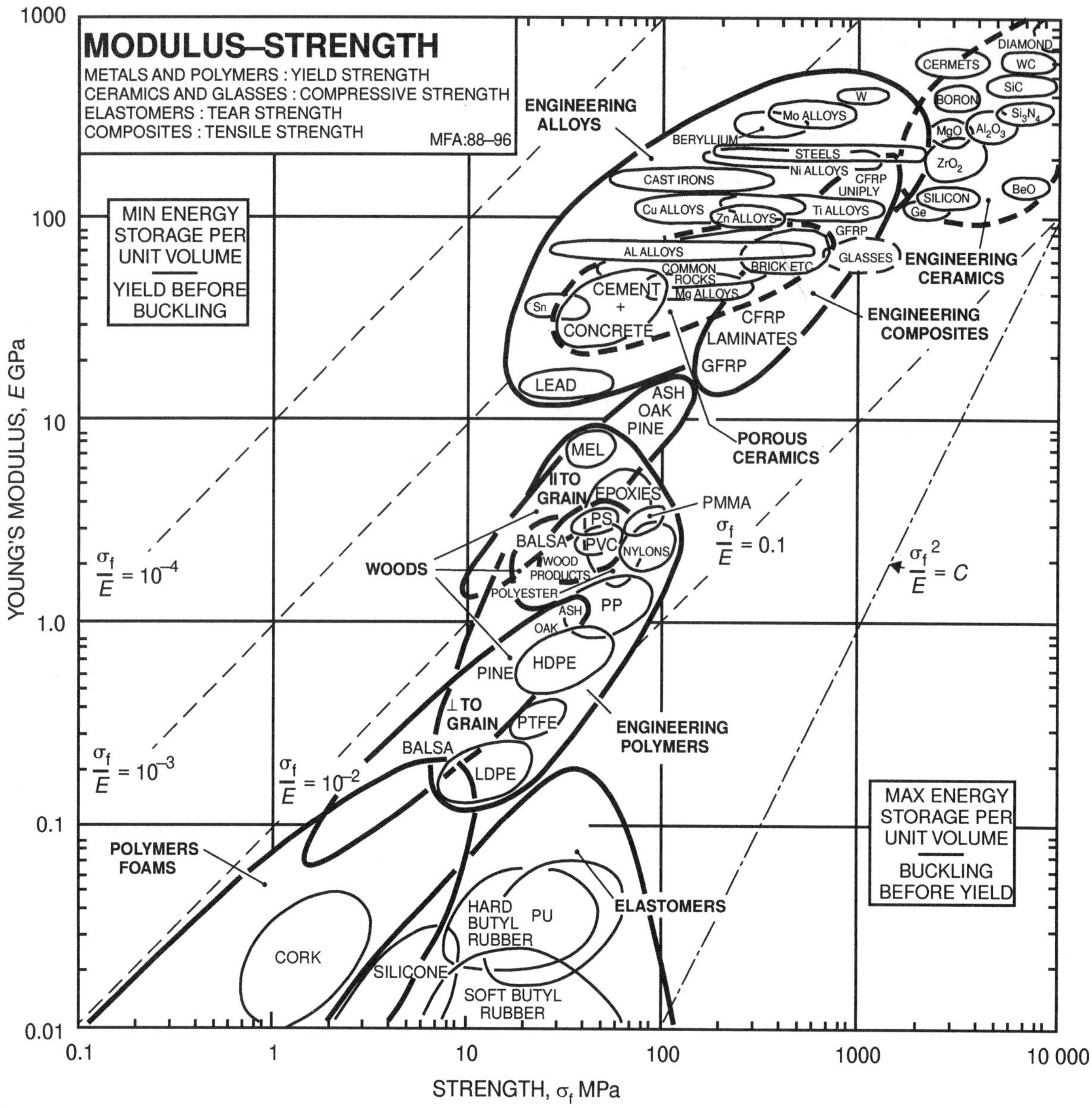

Fig. 1.1.9 Materials selection chart: Young's modulus, E, against strength σ_f

The 'strength' for *metals* is the 0.2% offset yield strength. For *polymers,* it is the 1% yield strength. For *ceramics,* it is the compressive crushing strength; remember that this is roughly 15 times larger than the tensile (fracture) strength. The chart has numerous applications, among them: the selection of materials for springs, elastic hinges, pivots and elastic bearings, and for yield-before-buckling design. The lines show two of these; they are the loci of points for which:

(a) $\sigma_f^2/E = C$ (criterion for spring design)
(b) $\sigma_f/E = C$ (elastic hinge design)

The value of the constant C increases as the lines are displaced downward and to the right.

Fig. 1.1.9

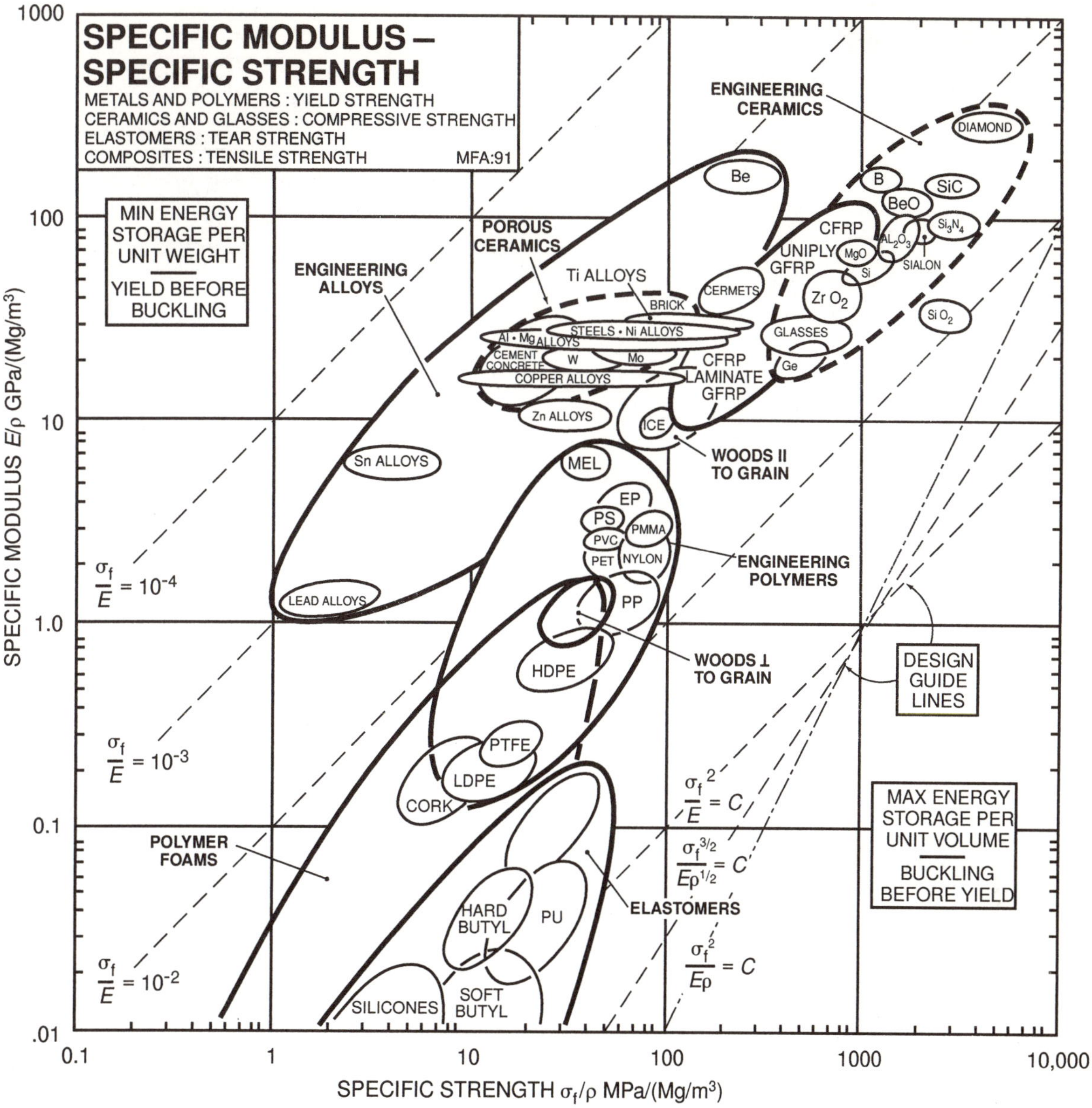

Fig. 1.1.10 Materials selection chart: specific modulus, E/ρ, against specific strength, σ_f/ρ

The chart for specific stiffness and strength. The contours show the yield strain, σ_f/E. The chart finds application in minimum weight design of ties, and springs design of rotating components to maximize rotational speed or energy storage, etc. The guide lines show the loci of points for which:

(a) $\sigma_f^2/E\rho$ = C (ties, springs of minimum weight; maximum rotational velocity of disks)
(b) $\sigma_f^{3/2}/E\rho^{1/2}$ = C
(c) σ_f/E = C (elastic hinge design)

The value of the constant C increases as the lines are displaced downwards and to the right.

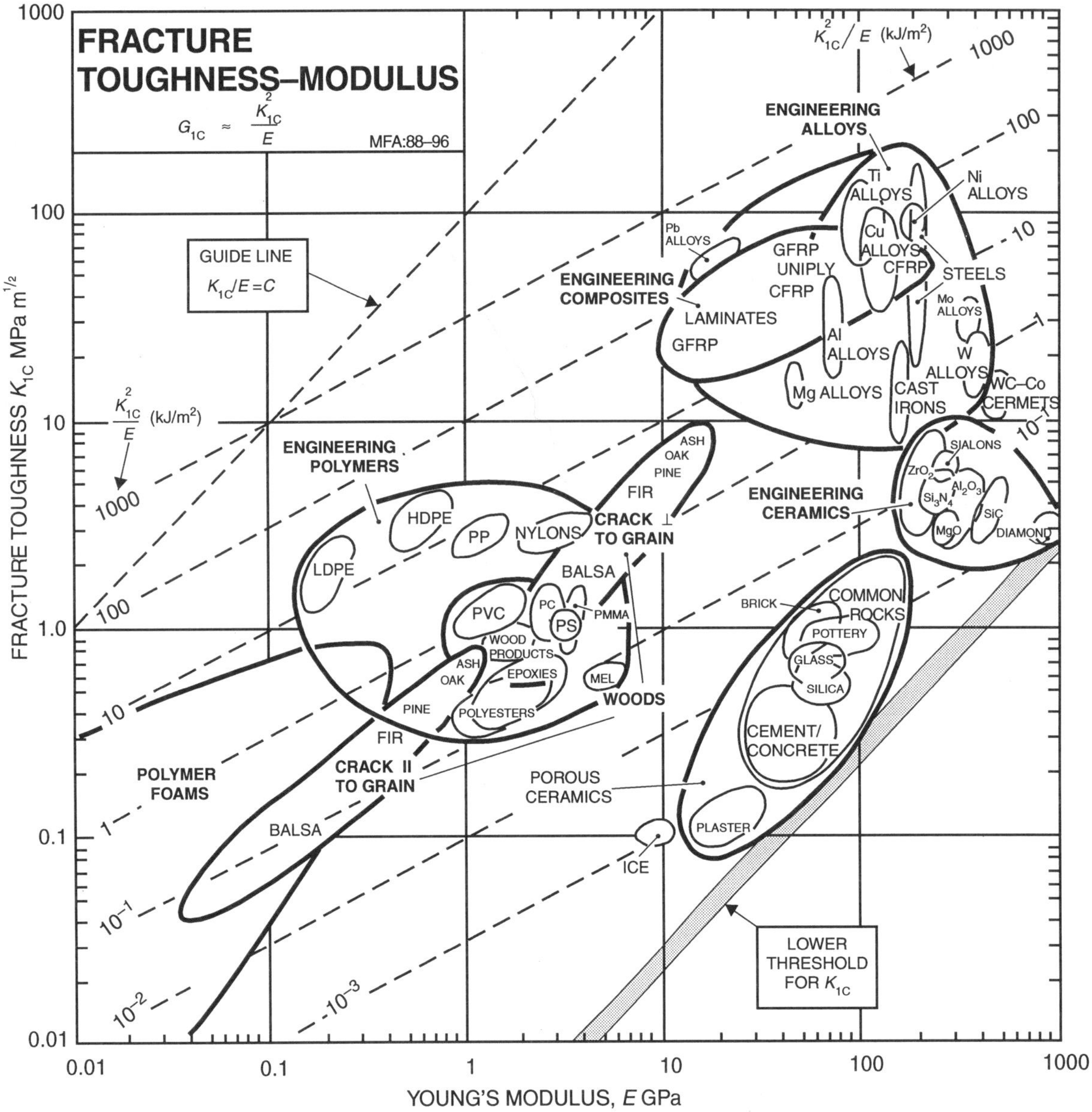

Fig. 1.1.11 Materials selection chart: fracture toughness, K_{1c} against Young's modulus, E

The chart displays both the fracture toughness, K_{1c}, and the toughness, $G_{1c} \approx K_{1c}^2/E$ and it allows criteria for stress and displacement-limited failure criteria (K_{1c} and K_{1c}/E) to be compared. The lines show the loci of points for which:

(a) $K_{1c}^2/E = C$ (lines of constant toughness, G)
(b) $K_{1c}/E = C$ (guideline for displacement-limited failure)

The values of the constant C increase as the lines are displaced upwards and to the left. Tough materials lie towards the upper left corner.

Fig. 1.1.11

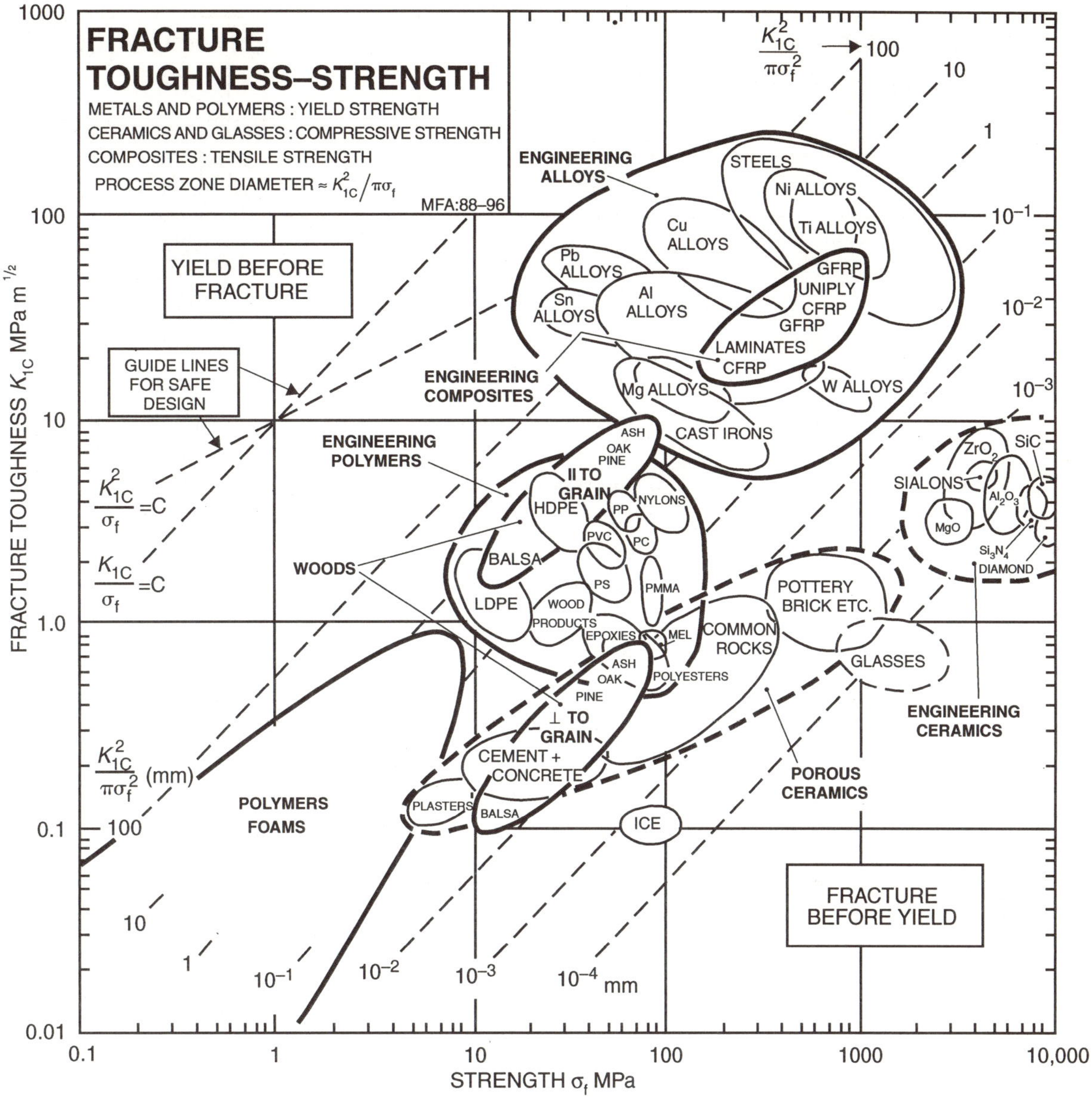

Fig. 1.1.12 Materials selection chart: fracture toughness, K_{1c} against strength σ_f

The 'strength' for metals is the 0.2% offset yield strength. For polymers, it is the 1% yield strength. For ceramics, it is the compressive crushing strength; remember that this is roughly 15 times larger than the tensile fracture strength. The chart guides selection of materials to meet yield-before-break and leak-before-break design criteria, in assessing plastic or process-zone sizes, and in designing samples for valid fracture toughness testing. The lines show the loci of points for which:

(a) $K_{1c}/\sigma_f = C$ (yield-before-break)
(b) $K_{1c}^2/\sigma_f = C$ (leak-before-break)

The value of the constant C increases as the lines are displaced upwards and to the left.

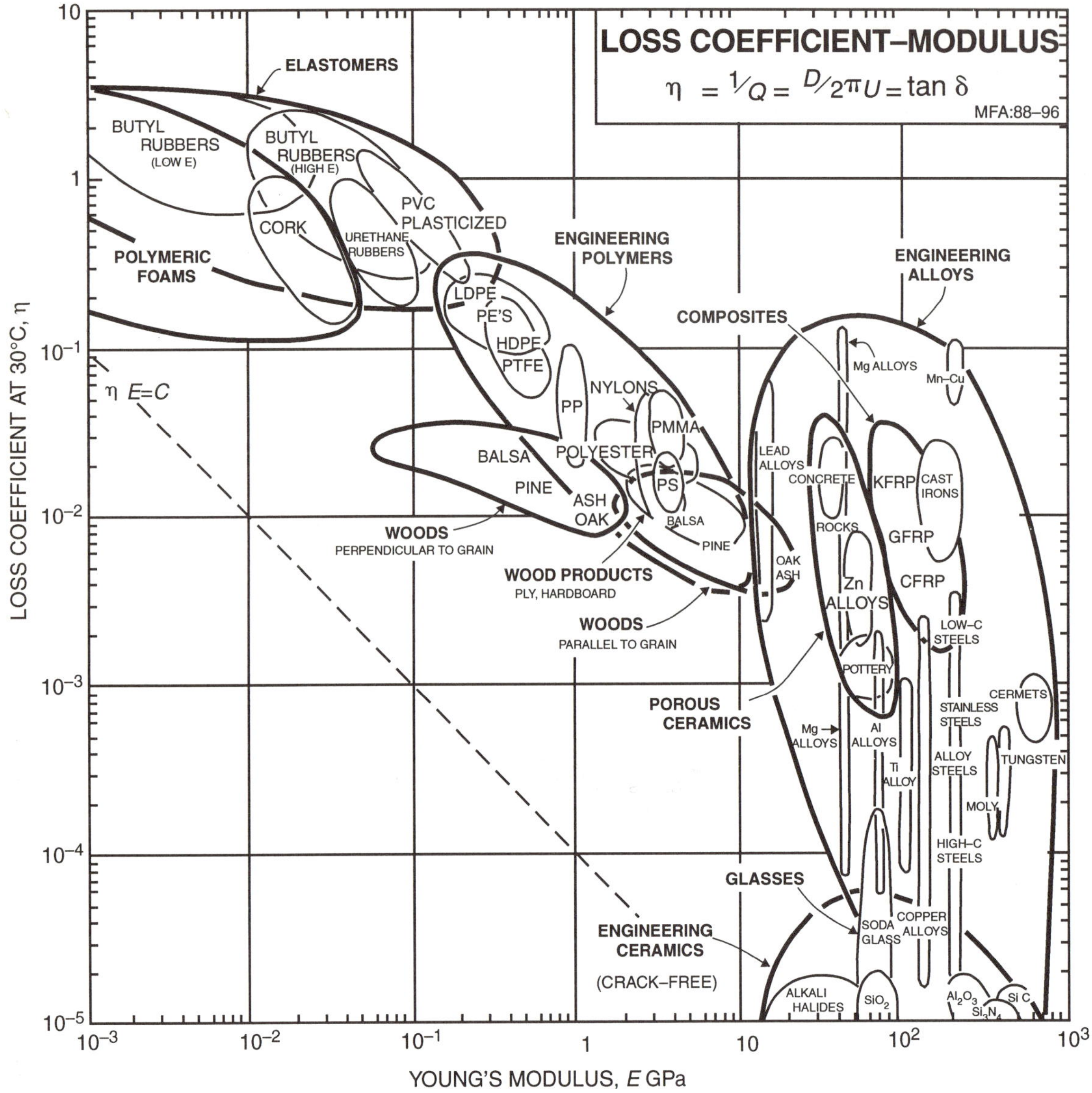

Fig. 1.1.13 Materials selection chart: loss coefficient, η against Young's modulus, *E*

The chart gives guidance in selecting material for low damping (springs, vibrating reeds, etc) and for high damping (vibration-mitigating systems). The line shows the loci of points for which:

(a) $\eta E = C$ (rule-of thumb for polymers)

The value of the constant C increases as the line is displayed upwards and to the right.

Fig. 1.1.13

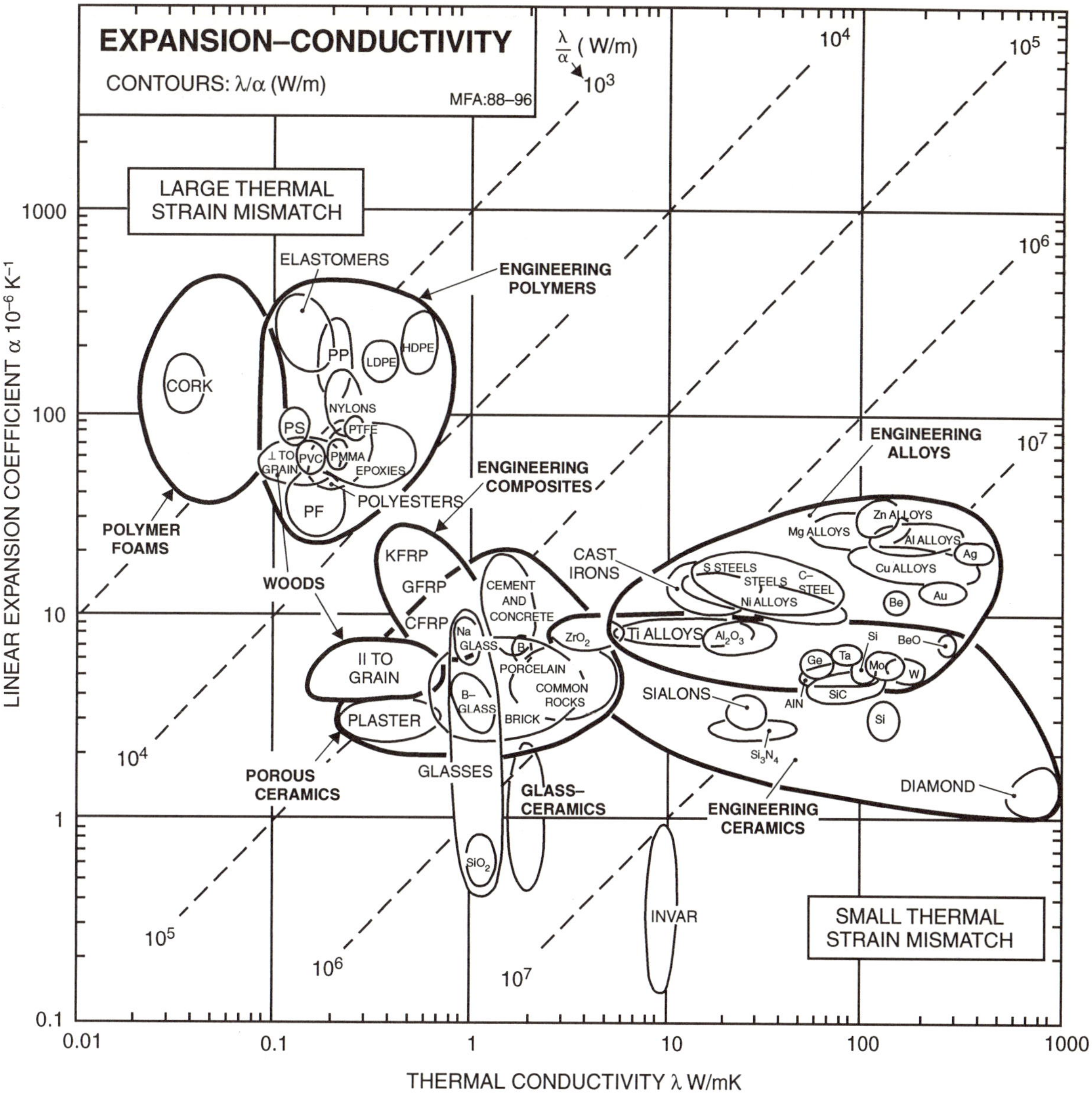

Fig. 1.1.14 Materials selection chart: thermal expansion coefficient, α, against thermal conductivity, λ

The chart guides in selecting materials for low thermal distortion. The lines show the loci of points for which:

(a) $\lambda/\alpha = C \text{W/m}^2$

The value of the constant C increases towards the bottom right.

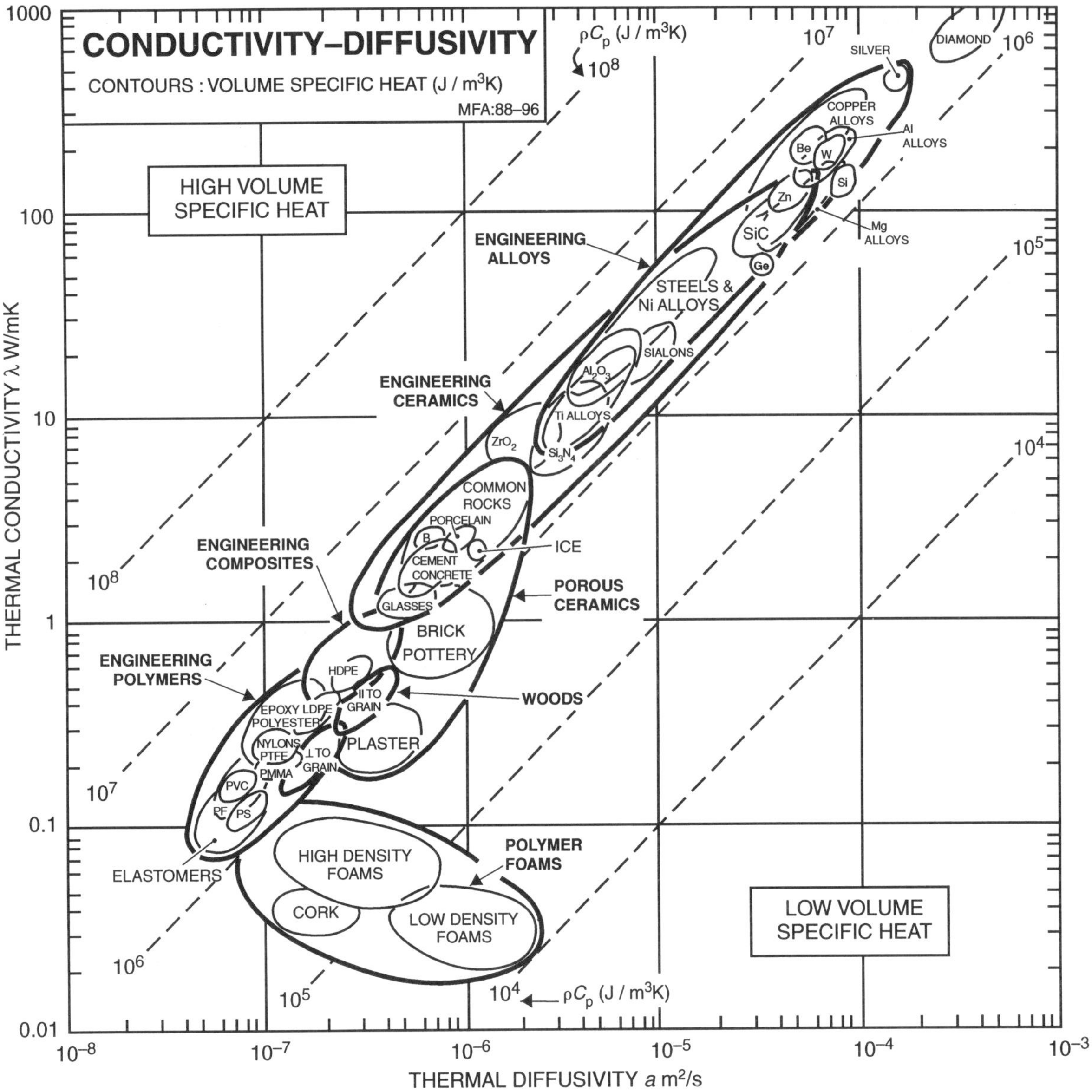

Fig. 1.1.15 Materials selection chart: thermal conductivity, λ against thermal diffusivity, *a*

The chart guides in selecting materials or thermal insulation, for use as heat sinks and so forth, both when heat flow is steady (λ) and when it is transient ($a = \lambda / \rho\, C_p$ where ρ is the density and C_p the specific heat). The lines show the loci of points for which:

(a) $\rho C_p = \lambda / a = C$ J/m³K (constant volumetric specific heat)

The value of the constant *C* increases towards the upper left.

Fig. 1.1.15

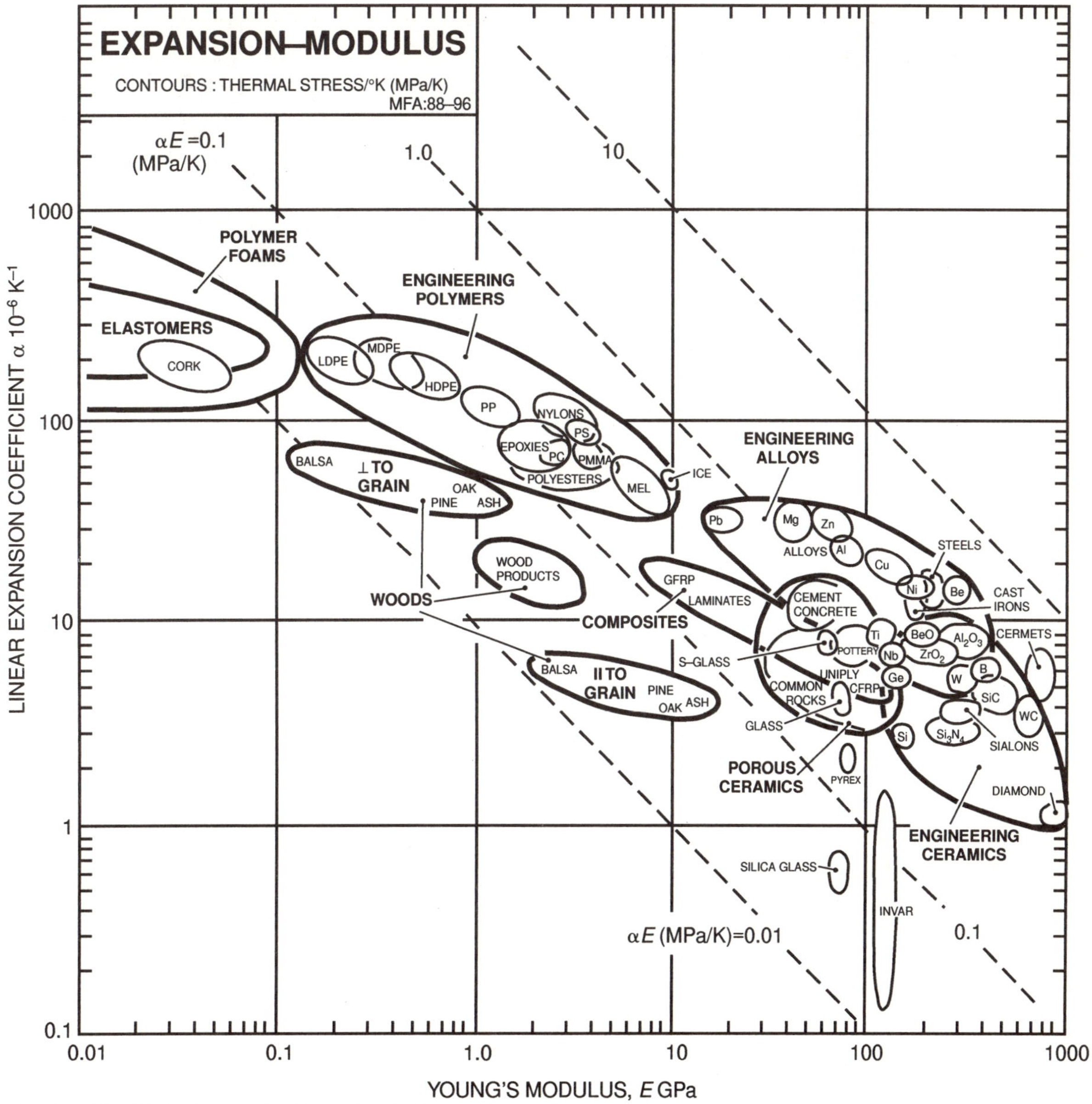

Fig. 1.1.16 Materials selection chart: linear thermal expansion, α, against Young's modulus, *E*

The chart guides in selecting materials when thermal stress is important. The lines show the loci of points for which:

(a) $\alpha E = C$ MPa/K (constant thermal stress per K)

The value of the constant C increases towards the upper right.

Fig. 1.1.16

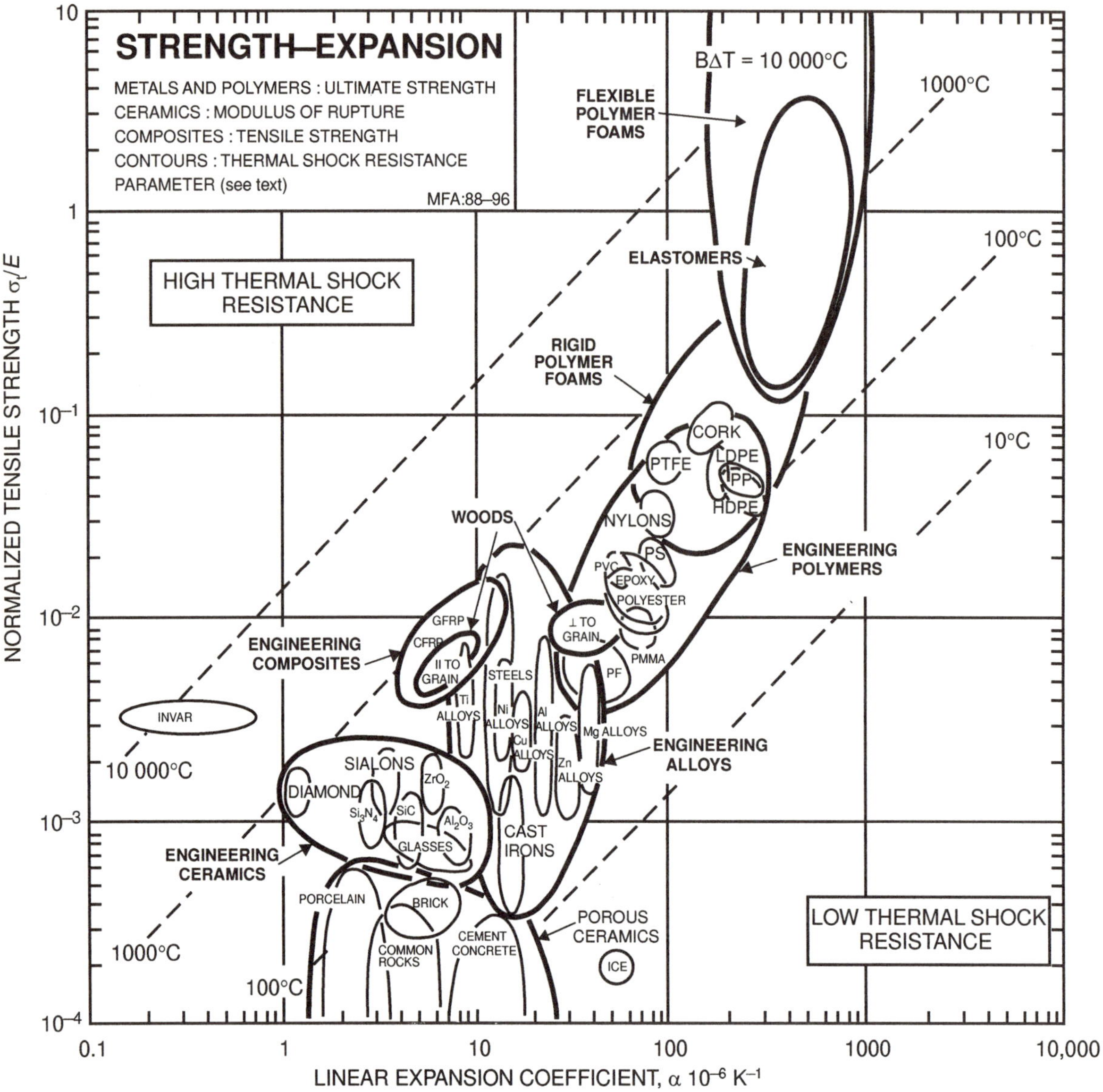

Fig. 1.1.17 Materials selection chart: normalized tensile strength, σ_t/E, against linear expansion coefficient, α

The chart guides in selecting materials to resist sudden changes of temperature ΔT. The lines are the loci of points for which: $B\Delta T = \sigma_t \alpha E = C$. Here σ_t is the tensile failure strength, E is Young's modulus and B is a factor which allows for constraint and for heat-transfer considerations:

$$B = 1/\mathbf{A} \quad \text{(axial constraint)}$$
$$= (1-\nu)/\mathbf{A} \quad \text{(biaxial constraint)}$$
$$= (1-2\nu)/\mathbf{A} \quad \text{(triaxial constraint)}$$

and $\mathbf{A} = \frac{th/\lambda}{1+th/\lambda}$

and ν is Poisson's ratio, t a typical sample dimension, h is the heat transfer coefficient at the sample surface and λ is its thermal conductivity. The value of the constant C increases towards the top left.

Fig. 1.1.17

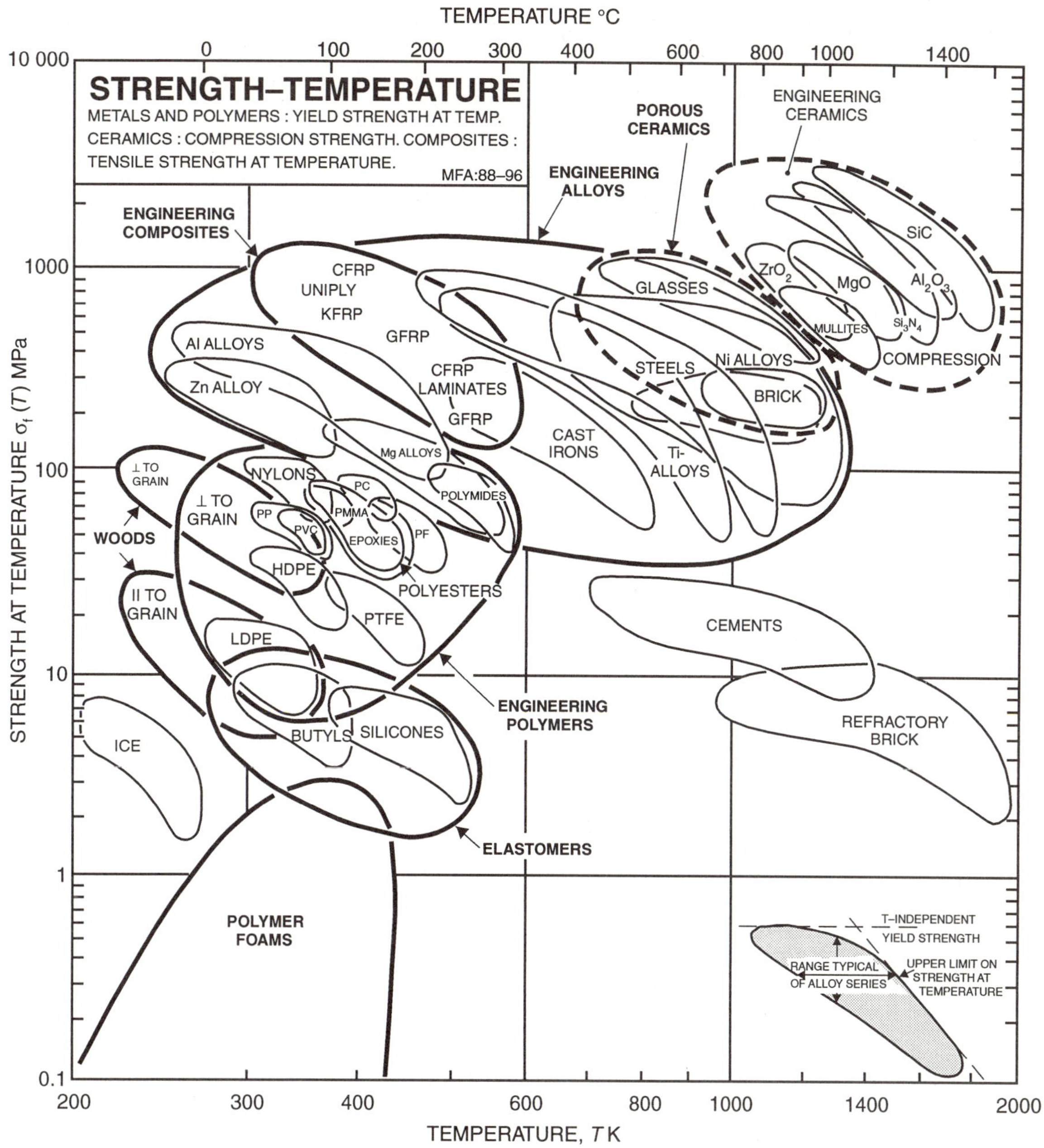

Fig. 1.1.18 Materials selection chart: strength, σ_f (T), against temperature, *T*

Materials tend to show a strength which is almost independent of temperature up to a given temperature (the 'onset of creep' temperature); above this temperature the strength fails, often steeply. The lozenges show this behaviour (see inset at the bottom right). The 'strength' here is a short-term yield strength, corresponding to (roughly) 1 h of loading. For long loading times, (10 000 h, for instance) the strengths are lower.

1.1.7 Case studies

1.1.7.1 General introduction

The case studies which follow illustrate the application of the principles of product design and component function analysis given in Sections 1.1.1 to 1.1.5, the application of the Ashby maps/Materials selection charts given in Section 1.1.6 and 1.1.14 on the comparative performance and costs of generic classes of materials, and the application of the more detailed information to be found in the rest of the Materials Selector.

There are five main steps in the process:

(i) **Function analysis** of the components of the **layout designs**.

(ii) **Mathematical modelling** of the likely performance of the **layout designs** to determine the required combination of materials properties (**materials desirability indices**), including approximate costs which will optimize performance and cost to meet the **market specification**.

(iii) Searching of the **Ashby maps/Materials selection charts** for the generic classes of materials which give the necessary combination of mechanical and thermal properties (Section 1.1.6) and the most desirable combination of properties (i.e. required combination of properties at lowest cost – Section 1.1.14).

(iv) Analysis of the required materials properties which are not readily quantified (e.g. corrosion resistance – Section 1.4 and wear resistance – Section 1.5) and selection of candidate **generic classes** of materials and (where appropriate) surface treatments.

(v) For each of the candidate generic class of material selected by steps (i) to (iv) above, searching of Sections 1.6 and 1.7 for possible manufacturing routes, and Volume 2 (for metals and ceramics) and Volume 3 (for plastics, composites and elastomers) for more detailed information on the most suitable grades of materials within the generic classes.

1.1.7.2 Case study 1 – Materials for a small pump impeller

MARKET REQUIREMENT

The small pump (used as an example in Sections 1.1.1 to 1.1.5) must be capable of pumping a range of fluids including some aggressive chemicals. The main requirements (see Tables 1.1.2, 1.1.3 and 1.1.4) are that it should not leak, it should be reliable and the materials of construction, in addition to resisting corrosion by the fluids and wear, should be suitable for high volume manufacturing methods and as low in cost as possible.

This case study focuses on the pump impeller.

COMPONENT – PUMP IMPELLER

The functions of the pump impeller are:

Primary function	To transport fluid.
Supporting functions and constraints (see Fig. 1.1.4)	(i) To maintain blade integrity. (ii) To allow stable rotation. (iii) To locate magnet.

Mathematical modelling and analysis of desired combinations of materials properties to maintain blade integrity

The centrifugal force at the root of the impeller blade imposes an obligatory tensile strength constraint.

$$\text{Centrifugal force} = \frac{mV^2}{r}$$

For a small element dm at distance r from root

$$\text{Force } F = \mathrm{d}m\ \frac{V^2}{r} = \frac{\rho A \mathrm{d}r\,(2\pi r)^2 n^2}{r}$$

where V = velocity, ρ = density, A = blade area, n = revs/second.

$$\text{Total force on root} \quad = 4\rho A\ \pi^2 n^2 \int_0^R r\,\mathrm{d}r$$

$$= 2\rho A n^2\ \pi^2 R^2 = \frac{\rho A}{2}\ (\text{peripheral tip speed})^2$$

Force = σA

Hence $\sigma_f/\rho \geq {}^1/_2$ (peripheral tip speed)^2

With safety factor of 2, $\sigma_f/\rho \geq$ (peripheral tip speed)^2

MATERIALS SELECTION

Hence, for a given blade tip speed (which depends on the speed of rotation and the blade radius), materials which possess the necessary combination of σ_f/ρ to resist the centrifugal stresses can be identified from the **Ashby map/Materials selection chart of strength against density** (see Fig. CS 1.1). In this example, a maximum speed of rotation of 2900 rpm and maximum blade radius of approximately 7 cm means σ_f/ρ must be greater than a constant C, where

$$C = \frac{\left[2\pi \times 7 \times \frac{2900}{60}\right]^2}{10^7} = 0.45\ \mathrm{J/g}$$

By drawing this line on the Ashby chart, it can be seen that all engineering materials would survive. Hence the basis of selection depends on other criteria.

The next possible failure mode is failure of a blade at its root through bending stresses. For a given application weaker materials will require thicker roots and thus a greater volume of material. The relevant **materials desirability index** is the same as that for beams of a fixed width and length in bending, that is:

$\sigma_f^{1/2}/\rho C_R$ where C_R is the relative cost per unit volume.

Materials with the highest combination of these properties will be preferred.

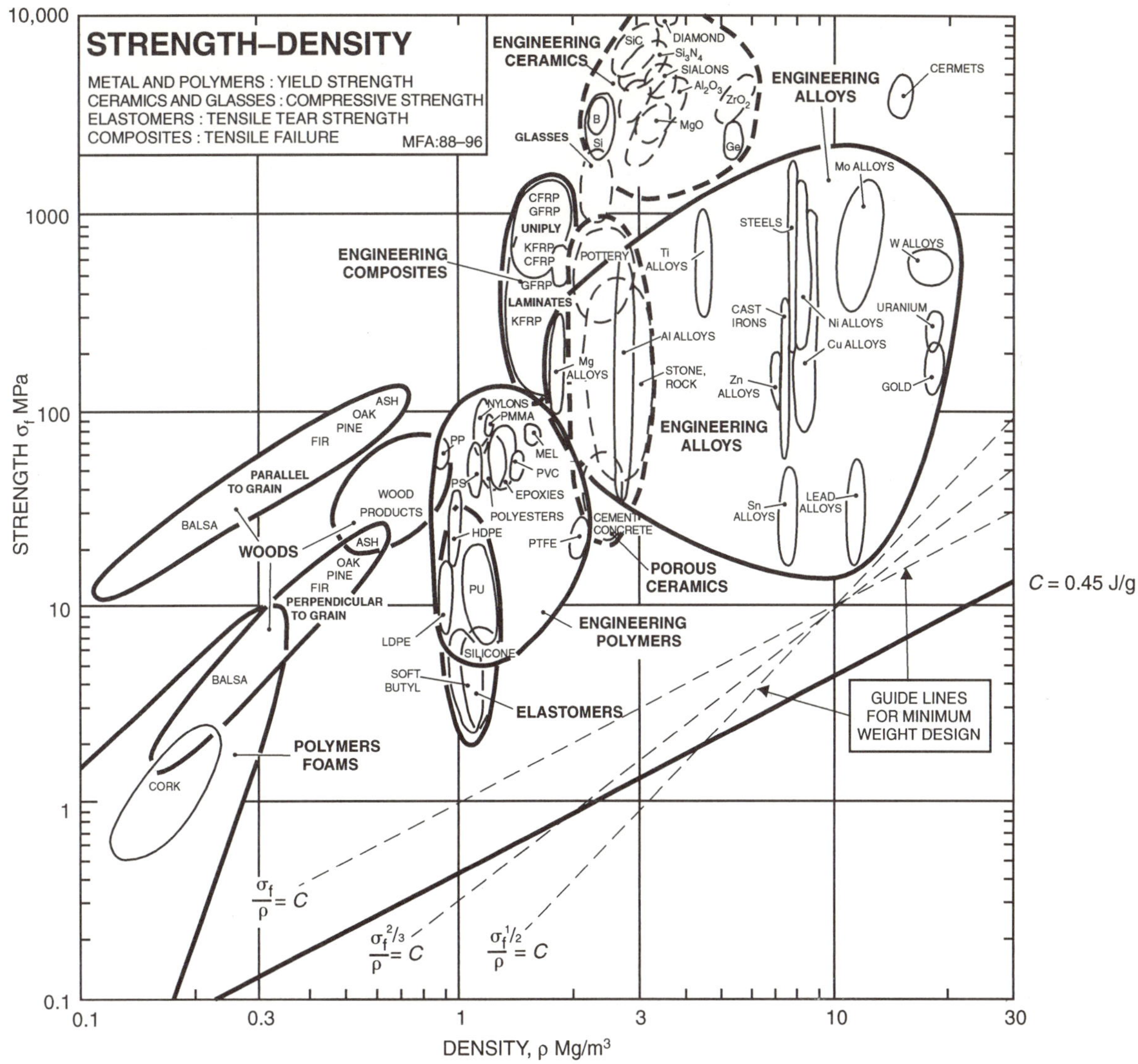

Fig. CS 1.1 Materials selection chart: strength, σ_f, against density, ρ

The ‘strength’ for metals is the 0.2% offset yield strength. For polymers, it is the stress at which the stress–strain curve becomes markedly non-linear – typically, a strain of about 1%. For ceramics, it is the compressive crushing strength; remember that this is roughly 15 times larger than the tensile (fracture) strength. The chart guides selection of materials for light, strong, components. The lines show the loci of points for which:

(a) $\sigma_f/\rho \;= C$ (criterion for plastic failure of ties)

(b) $\sigma_f^{2/3}/\rho = C$ (criterion for plastic bending, torsion of beams and shafts)

(c) $\sigma_f^{1/2}/\rho \;= C$ (criterion for plastic bending of plates)

The value of the constant C increases as the lines are displaced upwards and to the left. Materials offering the greatest strength-to-weight ratio lie towards the upper left corner.

These materials can be identified by reference to the **Ashby map/Materials selection chart** (see Fig. CS 1.2) which relates strength to relative cost per unit volume. This map is reproduced on the next page. Moving the dotted line for $\sigma_f^{1/2}/C_R\rho$ towards the upper left hand corner of the chart to maximize this **materials desirability index** progressively eliminates the more expensive materials. This bold line has been drawn at the point where PP (polypropylene) is the last remaining plastic (i.e. the cheapest in relative cost per unit volume for a given failure strength). This in fact was the material chosen for the pump impeller (see Table 1.1.5).

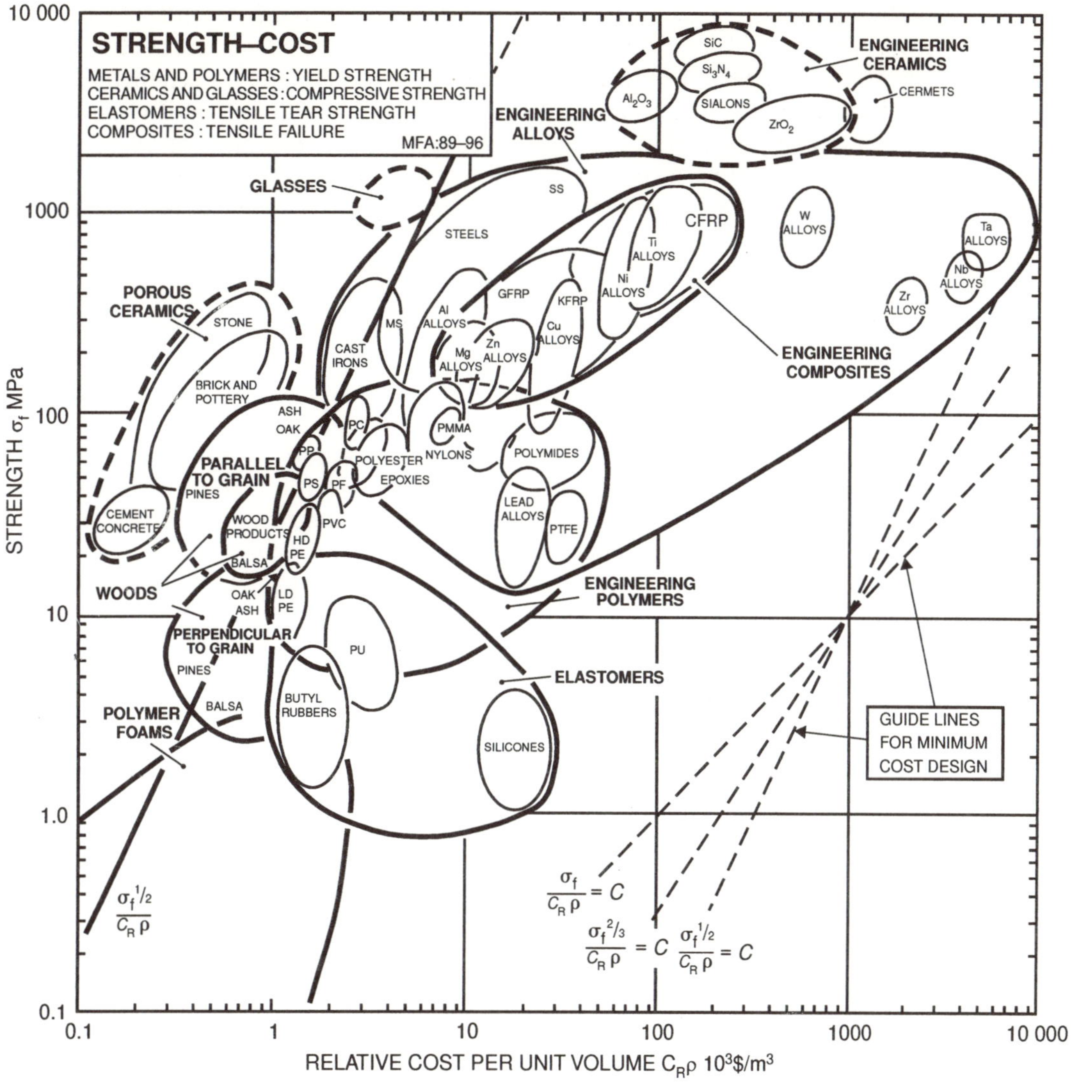

Fig. CS 1.2 Materials selection chart: strength, σ_f, against approximate cost, $C_R\rho$.

The 'strength' for metals is the 0.2% offset yield strength. For polymers, it is the 1% offset yield strength. For ceramics, it is the compressive crushing strength; remember that this is roughly 15 times larger than the tensile (fracture) strength.

The chart guides in selecting materials for cheap, strong components (production, finishing assembly and other costs must be considered separately). The lines show the loci of points for which:

(a) $\sigma_f / C_R\rho$ = C (criterion for axial tension of ties)

(b) $\sigma_f^{2/3} / C_R\rho$ = C (criterion for bending, torsion or buckling of beams, shafts and columns)

(c) $\sigma_f^{1/2} / C_R\rho$ = C (criterion for bending of plates)

The value of the constants C increases as the lines are displaced upwards. Materials offering the greatest strength per unit cost lie towards the upper left corner.

OTHER SUPPORTING FUNCTIONS AND CONSTRAINTS

The other functions of the pump impeller are:

allow-rotation.
accommodate-magnet.

These are not susceptible to such precise mathematical modelling as was conducted above for the integrity of the blades of the impeller.

More detailed analysis of the **allow-rotation** function is given in Fig. 1.1.5.

One of the consequent materials requirements is to **avoid-wear** which, as can be seen in Fig. 1.1.5, means that if the body of the pump impeller is also to provide the bearing surface, then the **pressure-velocity** (PV) characteristics of the material must be adequate for the expected loads and speeds.

Section 1.5 gives detailed guidance on the selection of materials for wear resistance. Polypropylene would not normally be recommended for its bearing properties. Hence the splitting of the functions of **maintain-blade-integrity** and **allow-stable-rotation** and the incorporation of a ceramic bearing insert (see Table 1.1.5).

Satisfaction of the **accommodate-magnet** function requires the examination of information in Sections 1.6 and 1.7 on component manufacturing processes and joining. The information given here would indicate that adhesive bonding of polypropylene is relatively difficult and moulding the magnet to the impeller during the injection moulding process (chosen for its low cost for high volume manufacture) is the preferred option. The bearing insert is also moulded-in during this process.

1.1.7.3 Case study 2 – Materials for short-term insulation

Each member of the crew of a military aircraft carries, for emergencies, a radio beacon. If forced to eject, the crew member could find himself in testing circumstances – in water at 4°C, for example (much of the earth's surface is ocean with a mean temperature of 4°C). The beacon guides friendly rescue services to the crew member, minimizing exposure time.

But microelectronic (like human) metabolisms are upset by low temperatures. In the case of a beacon, the transmission frequency starts to drift. The design specification for the egg-shaped housing containing the electronics (Fig. CS 2.1) requires that, when the temperature of the outer surface is changed by 20°C, the temperature of the inner surface should not change significantly for one hour. To keep the device small, the wall thickness is limited to a thickness *W* of 20 mm. What is the best material for the housing?

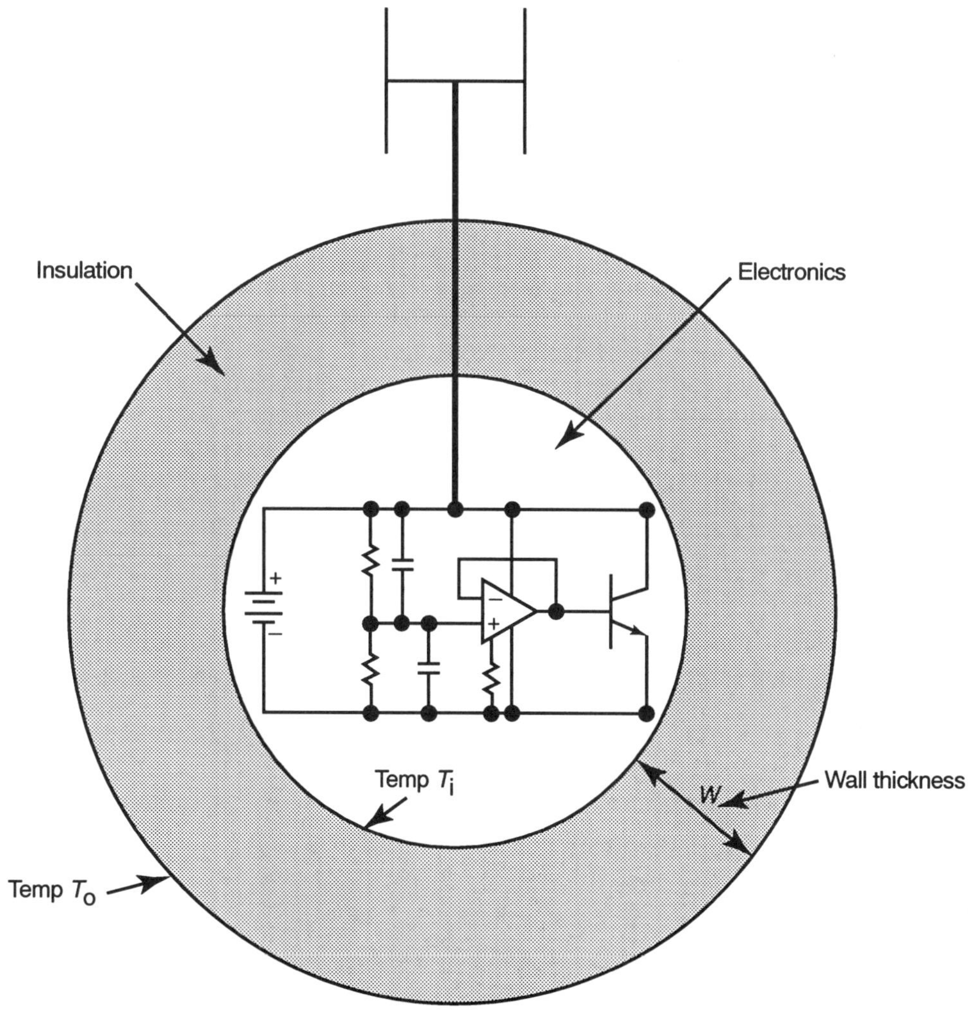

Fig. CS 2.1 An electronic beacon. The insulation should prevent the temperature of the electronics from falling for at least one hour.

A foam of some sort, you might think. But here is a case in which intuition leads you astray. So let us formulate the design requirements and do the job from first principles. The market requirements, and consequent functions to be performed by the device housing, are as follows:

COMPONENT – DEVICE HOUSING

Primary function	Thermal insulation.
Specific performance requirement	Maximize time t before temperature falls 1°C.
Qualifying functions constraints	Wall thickness $\leq W$

MATHEMATICAL MODELLING AND ANALYSIS OF DESIRED COMBINATIONS OF MATERIALS PROPERTIES TO PROVIDE REQUIRED INSULATION

We model the container as a wall of thickness W, and thermal conductivity λ. Once a steady-state has been established, the heat flux J through the wall is:

$$J = \lambda \frac{(T_i - T_o)}{W} \qquad (1)$$

where T_o is the temperature of the outer surface and T_i that of the inner one, Fig. CS 2.1. This heat flux can be minimized by choosing a wall material with the lowest possible value of λ (i.e. lowest thermal conductivity). But we have answered the wrong question. The design brief was not to minimize the heat flux, but maximize the *time* before the temperature of the inner wall changed appreciably (1°C). When the surface temperature of a body is suddenly changed, a temperature wave, so to speak, propagates inwards. The distance X it penetrates in time t is approximately $X = \sqrt{2at}$. Here, a is the thermal diffusivity, defined by:

$$a = \frac{\lambda}{\rho C_p}, \qquad (2)$$

where ρ is the density and C_p is the specific heat. Equating distance X to the wall thickness gives

$$t = \frac{CW^2}{2a} \qquad (3)$$

where C is a constant with a value close to 1. The time is maximized by choosing the smallest value of the thermal diffusivity:

$$a = \frac{\lambda}{\rho C_p} \qquad (4)$$

MATERIALS SELECTION

Reference to the **Ashby map** (Fig. 1.1.15) which plots **thermal conductivity** λ versus **thermal diffusivity** a for all classes of generic materials identifies those materials with the lowest values of λ and a. They are in the bottom right hand corner, i.e. solid elastomers.

Figure 1.1.15 shows that polymer foams (and cork) have low thermal conductivities and are thus good for insulation at steady-state (as in wall insulation). But their thermal diffusivities are not particularly low because they have so little mass and thus low heat capacity C_p. The materials with low **diffusivity** (thermal conductivity divided by density ρ and specific heat C_p) for heat are solid polymers and particularly solid elastomers.

A housing made of a solid elastomer would give the beacon up to a ten times longer transmission life than one made of (say) low density polystyrene foam. The possible materials are summarized in the table below.

Materials for thermal insulation

Material	*Comment*
Elastomers: butyl rubber (BR), polychloroprene (CR), and chlorosulfonated polyethylene (CSM) are examples.	Best choice for short-term insulation.
Commodity polymers: polyethylenes and polypropylenes.	Cheaper than elastomers, but somewhat less good for short-term insulation.
Polymer foams.	Much less good than elastomers for short-term insulation; best choice for long-term insulation at steady state.

DESIGN ITERATION

One can do better than this. The trick is to exploit other ways of absorbing heat. If a liquid – a low-melting wax, for instance – can be found that solidifies at a temperature equal to the minimum desired operating temperature of the transmitter T_i, it can be used as a 'latent heat sink'. Channels in the package are filled with the liquid. The inner temperature can only fall below the desired operating temperature when all the liquid has solidified. The latent heat of solidification must be supplied to do this, giving the package a large (apparent) specific heat, and thus an exceptionally low diffusivity for heat at the temperature T_i. The same idea is used in reverse in 'freezer packs' which solidify when placed in the freezer compartment of a refrigerator and remain cold (by melting, at 4°C) when packed around warm beer cans in a portable cooler.

1.1.7.4 Case Study 3 – Materials for flywheel rotors

MARKET REQUIREMENT

Flywheels store energy. The energy can be recovered, either to smooth an intermittent power supply (as in reciprocating engines) or to drive something (a toy, a dynamo). Small flywheels like those in children's toys are made of lead. Steam, gasoline and diesel engines have cast iron flywheels. Innovative flywheels for power storage and regenerative braking systems for vehicles (Fig. CS 3.1) are made of composites like GFRP. There is a great diversity of materials here. What **is** the best material for a flywheel?

An efficient flywheel stores as much *energy per unit weight* as possible, without *failing*. Failure (were it to occur) is caused by centrifugal loading. When the centrifugal stress reaches the tensile strength (or fatigue strength) the flywheel flies apart. One constraint is that this should not occur. The flywheel of a child's toy is not efficient in this sense. Its velocity is limited by the pulling-power of the child, and never remotely approaches the burst velocity. In this case, and for the flywheel of an automobile engine, we wish to maximize the *energy stored per unit volume* at a constant (specified) or limited maximum *angular velocity*. There is also a constraint on the outer radius, *R*, of the flywheel so that it will fit into a confined space.

The answer therefore depends on the application. The strategy for optimizing flywheel rotors for efficient energy-storing systems differs from that for children's toys. The two alternative sets of design requirements are listed below.

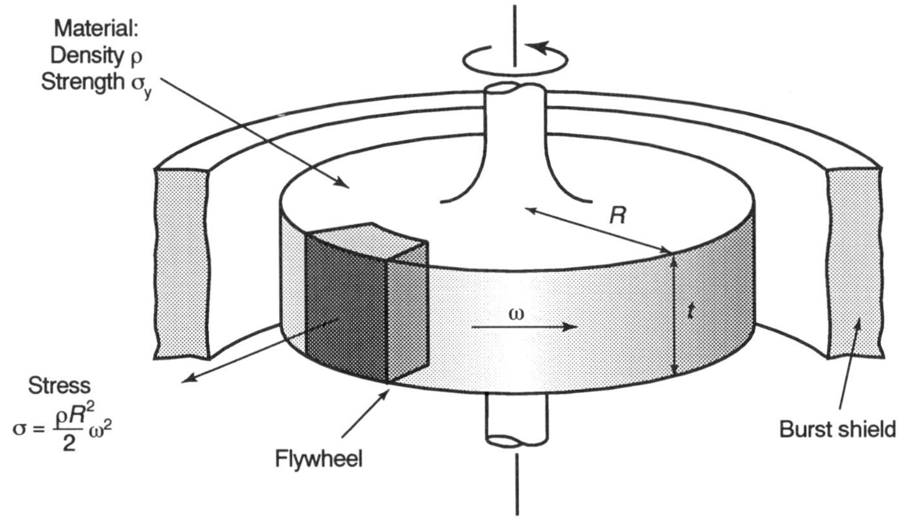

Fig. CS 3.1 A flywheel. Its angular velocity (and thus the kinetic energy it can store) is limited by its strength.

COMPONENT – FLYWHEEL ROTOR

Primary function	**Energy storage:** flywheel to store maximum kinetic energy.	**Energy smoothing:** flywheel to run at constant (low) velocity.
Specific performance requirement	Maximize kinetic energy per unit mass.	Maximize kinetic energy per unit volume.
Constraints/ qualifying functions	Must not burst. Adequate toughness (>20 MPa $m^{1/2}$)	Outer radius, R, fixed. Adequate toughness (> 20 MPa $m^{1/2}$).

MATHEMATICAL MODELLING AND ANALYSIS OF DERIVED COMBINATIONS OF MATERIALS PROPERTIES

The efficient energy-storing flywheel stores as much energy per unit weight as possible, without failing. Think of it as a solid disk of radius R and thickness t, rotating with angular velocity ω (Fig 1). The energy U it stores is:

$$U = \frac{1}{2} J\omega^2 \tag{1}$$

Here $J = \pi \rho R^4 t/2$ is the polar moment of inertia of the disk and ρ the density of the material of which it is made. Therefore:

$$U = \frac{\pi}{4} \rho R^4 t \omega^2 \tag{2}$$

The mass of the disk is:

$$m = \pi R^2 t \rho \tag{3}$$

The quantity to be maximized is the kinetic energy per unit mass, which is the ratio of the last two equations:

$$\frac{U}{m} = \frac{1}{4} R^2 \omega^2 \tag{4}$$

As the flywheel is spun up, the kinetic energy it stores increases, but so does the centrifugal stress. The maximum principal stress in a spinning disk of uniform thickness is:

$$\sigma_{max} = \left(\frac{3+\nu}{8}\right) \rho R^2 \omega^2 \tag{5}$$

where ν is Poisson's ratio. This stress must not exceed the failure stress of the disk. Since the disk is likely to be spun up and down repeatedly, we use the 'endurance limit', σ_e, as a measure of strength; it is the stress below which the fatigue life is infinite. This sets an upper limit to the angular velocity, ω, and disk radius, R (the free variables). Eliminating $R\omega$ between the last two equations (with $\sigma_{max} = \sigma_e$) gives:

$$\frac{U}{m} = \left(\frac{2}{3+\nu}\right)\left(\frac{\sigma_e}{\rho}\right) \tag{6}$$

Poisson's ratio, ν, is roughly 1/3 for most solids, so we can treat it as a constant. The best materials for efficient flywheels are those with high values of the performance index:

$$\frac{\sigma_e}{\rho} \tag{7}$$

It has units of kJ/kg. To avoid failure due to fast fracture, the disk must also have adequate toughness. We achieve this by requiring that K_{1c} is as large as possible within other constraints and at least:

$$> 20\ \text{MPa}\,\text{m}^{1/2} \tag{8}$$

But what of the other sort of flywheel – the child's toy or the energy-smoothing function? Here we seek the material which stores the most energy per unit volume V at constant velocity. The energy per unit volume at a given ω is (from equation 2):

$$\frac{U}{V} = \frac{1}{4}\rho R^2\omega^2 \tag{9}$$

Both R and ω are fixed by the design, so the best material is now that with the greatest value of:

$$\rho \tag{10}$$

MATERIALS SELECTION – HIGH ENERGY STORAGE ROTORS

To a first approximation the failure strength σ_f in the **Ashby maps** can be taken as a measure of the endurance limit σ_e. Hence the **Ashby map** relating σ_f to density ρ (Fig. 1.1.7) can be employed to select the best materials. If the line $\sigma_f/\rho = C$ is moved progressively to the top left hand corner of this map, the materials with the best values of σ_f/ρ can be identified.

The best materials can be seen to be in order of merit:

(i) Engineering ceramics (which will drop out when the toughness criterion is applied).
(ii) Carbon fibre- and glass fibre-reinforced plastics.
(iii) High strength titanium alloys.

Recent designs of flywheels are filament wound glass fibre reinforced rotors and can store around 150 kJ/kg. A 20 kg rotor can therefore store 3 MJ or 800 KW/hour.

The energy density in such a flywheel is considerable; its sudden release in a failure could be catastrophic. The disk must be surrounded by a burst-shield and precise quality control in manufacture is essential to avoid out-of-balance forces. This has been achieved in a number of glass fibre energy storage flywheels intended for use in trucks and buses, and as an energy reservoir for smoothing power flow in wind power generation.

Fracture toughness is therefore a very important material requirement. Ideally the rotor should yield before fracture (and, say, trigger a proximity sensor which would shut down the flywheel).

The **Ashby map** (Fig. 1.1.8) plots fracture toughness against density and can be used to identify those materials with the K_{1c} >20 MPam$^{1/2}$. At this point, the engineering ceramics are eliminated.

For the best combination of good fracture and high fracture strength, the **Ashby map** shown in Fig. 1.1.12 should be examined. Here it can be seen that the materials with the best combination of fracture toughness and strength are metallic materials and advanced composites. The results are summarized in the table below.

Materials for energy-storing flywheels

Material	σ_e/ρ *(kJ/kg)*	*Comment*
Composites: CFRP	200–350	The best performance – a good choice.
Composites: GFRP	150–200	Almost as good as CFRP and considerably cheaper – excellent choice.
Ti alloys	100–160	The best choice among metals – marginally better than high strength steels because of lower ρ.
High strength steels High strength Al alloys High strength Mg alloys	100–150 80–100 80–100	All about equal in performance. Steel and Al alloys are cheaper than Mg and Ti alloys.

MATERIALS SELECTION – CONSTANT VELOCITY ROTORS

The desired combination of materials properties for constant velocity rotors is simply the highest value of density ρ at the lowest cost. Lead is the best if space is limited. Cast iron is less good at storing energy, but cheaper and the first choice if space is not a major constraint.

1.1.7.5 Case Study 4 – Materials for springs

MARKET REQUIREMENT

Springs come in many shapes and have many purposes. One thinks of axial springs (a rubber band for example), leaf springs, helical springs, spiral springs, torsion bars and so on (Fig. CS 4.1). The best material for a spring (regardless of their shape or use) is that which can store the greatest elastic potential energy per unit mass (or volume), without failing. The performance indices derived below can be used to identify materials which satisfy the design specification summarized below.

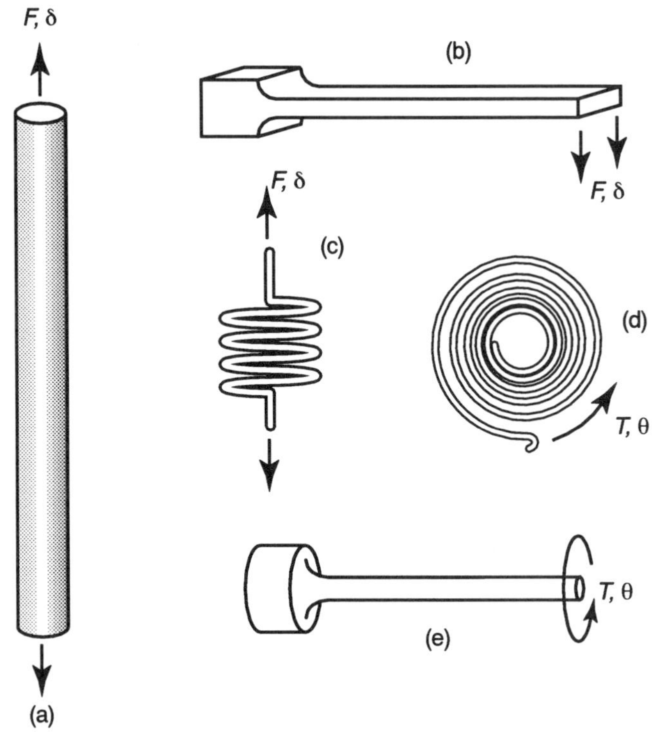

Fig. CS 4.1 Springs have many shapes, but all perform the same function; that of storing elastic energy.

COMPONENT – ELASTIC SPRING

Primary functions	(a) Maximum stored elastic energy/unit volume. (b) Maximum stored elastic energy/unit mass.
Constraints/ qualifying functions	No failure by yield, fatigue or fracture (whichever is more restrictive). Lowest cost per unit weight.

MATHEMATICAL MODELLING AND ANALYSIS OF DESIRED COMBINATION OF MATERIALS PROPERTIES

The primary function of a spring is that of storing elastic energy and releasing it again when required. The elastic energy stored per unit volume in a block of material stressed unformly to a stress σ is:

$$W_v = \frac{1}{2}\frac{\sigma^2}{E} \qquad (1)$$

where E is the Young's modulus. It is this that we wish to maximize. The spring will be damaged if the stress σ exceeds the yield stress or failure stress σ_f. So the constraint $\sigma \leq \sigma_f$. The maximum energy density is therefore:

$$W_v = \frac{1}{2}\frac{\sigma_f^2}{E} \qquad (2)$$

Torsion bars and leaf springs are less efficient than axial springs because some of the material is not fully loaded. The material at the neutral axis, for instance, is not loaded at all. For solid torsion bars:[a]

$$W_v = \frac{1}{3}\frac{\sigma_f^2}{E} \qquad (3)$$

and for leaf springs loaded in pure bending the maximum energy storage is:

$$W_v = \frac{1}{4}\frac{\sigma_f^2}{E} \qquad (4)$$

But, as these results show, this has no influence on the choice of material. The best material for springs, regardless of the way in which they are loaded, is that with the biggest value of:

$$\frac{\sigma_f^2}{E} \qquad (5)$$

If mass matters rather than volume, we must divide this by the density ρ (giving energy stored per unit mass), and seek materials with high values of:

$$\frac{\sigma^2}{\rho E} \qquad (6)$$

[a]The value of the constant (1/3, 1/4) in equations (3) and (4) can be increased by prestressing; the maximum working stress of torsional and leaf springs is increased by straining the material beyond its elastic limit and then releasing the load.

MATERIALS SELECTION (MAXIMUM ENERGY STORAGE – LOW VOLUME)

The Ashby map which plots strength σ_f against modulus E (Fig. 1.1.9) permits the selection of materials with the highest values of σ_f^2/E. If the line $\sigma_f^2/E = C$ (shown on this map) is moved towards the upper left hand corner, then the first material which this line touches will be the best for elastic springs.

In fact, there are materials in most of the main generic classes, i.e, metals, plastics, composites, elastomers and even glasses which give, within a first approximation, equally good values of σ_f^2/E (see table below).

All of these generic classes are used as springs; metals most commonly, glass fibre composites in vehicle leaf springs, plastic springs in children's toys and glass fibres in galvanometers.

Materials for efficient springs of low volume

Material	$\frac{\sigma_f^2}{E}$ *(MJ/m³)*	*Comment*
Spring steel	15–25	The traditional choice; easily formed and heat treated.
Ti alloys	15–20	Expensive, corrosion-resistant.
CFRP	15–20	Comparable in performance with steel; expensive;
GFRP	10–12	Almost as good as CFRP and much cheaper.
Glass fibres	30–60	Brittle in tension, but excellent if protected against damage; very low loss factor
Nylon	1.5–2.5	The least good; cheap and easily shaped, but high loss factor.
Rubber	20–60	Better than spring steel but high loss factor.

MATERIALS SELECTION – LIGHTWEIGHT SPRINGS

The Ashby map which plots specific modulus E/ρ against specific strength σ_f/ρ (Fig. 1.1.10) allows identification of the materials with the maximum value of $\sigma_f^2/E\rho$ which will make the most efficient lightweight spring. Again, by moving the line $\sigma_f^2/E\rho = C$ towards the upper left hand corner of the map, the first materials to touch this line will be the best.

The preferred materials are summarized in the following table

Materials for efficient light springs

Material	$\frac{\sigma_f^2}{E\rho}$ *(kJ/kg)*	*Comment*
Spring steel	2–3	Poor, because of high density.
Ti alloys	2–3	Better than steel; corrosion-resistant; expensive.
GFRP	4–8	Better than steel; expensive.
CFRP	3–5	Better than steel; less expensive than CFRP.
Glass fibres	10–30	Brittle in torsion, but excellent if protected.
Woods	1–2	On a weight basis, wood makes good springs.
Nylon	1.5–2	As good as steel, but with a high loss factor.
Rubber	20–50	Outstanding; 10 times better than steel, but with high loss factor.

Composites, because of their lower densities, are better than metals. Elastomers are better still (you can store almost 8 times more elastic energy per unit weight in a rubber band than in the best spring steel). Elastomeric springs are now widely used in aerospace because of their low weight and high reliability. Wood (the traditional material for archery bows) now appears in the list.

OTHER REQUIREMENTS

Many other considerations enter the choice of a material for a spring. Springs for vehicle suspensions must resist fatigue (the selection should then be made with the endurance limit – see Section 1.2). Valve springs for internal combustion engines must cope with elevated temperatures (see Section 2.1); here a strength-at-temperature is needed. The mechanical loss coefficient is important in springs which are loaded dynamically; polymers have high loss factors and therefore dissipate energy when they vibrate, metals, if strongly hardened, do not. Polymers, because they creep, are unsuitable for springs which carry a steady load, though they are good for catches and locating springs which spend most of their time unstressed. Springs made from unprotected carbon steel fail rapidly in a chemically corrosive environment. The least expensive solutions to this problem is to plate or polymer coat them to provide a corrosion barrier, but if the coating is damaged, failure can follow. The more expensive solution is to make the spring from an intrinsically corrosion-resistant material (stainless steel, copper alloys, nickel and cobalt alloys, titanium alloys, reinforced polymers, GFRP or CFRP). If high thermal or electrical conduction is required, copper–beryllium alloys are the best choice. But these are secondary selection criteria.

1.1.8 The influence of component shape on materials selection – shape factors

Achieving efficiency in load-bearing structures requires the selection of both material and section shape. By efficiency, we mean the ability to carry the design loads safely, whilst minimizing weight, or cost, or optimally meeting some other objective. The choice of material and shape are coupled, because the shape to which a material can be formed depends on the material itself; steels are available as thin-walled tubes, for instance, woods are not.

1.1.8.1 The idea of a shape factor

An example will illustrate the method [1, 2, 3] and define some of the important quantities. Consider selection of a material for a light, stiff beam of length *l*. If the beam is to be efficient we must give it an efficient shape. A tube, I-section or box-section are much lighter, for the same stiffness, than a solid section.

We write an equation for the mass per unit length, *m*, *l*, of the beam:

$$\frac{m}{l} = A\rho \tag{1}$$

where *A* is the area of its cross-section and ρ is the density of the material of which it is made. Equation (1) is called the objective function; it is the quantity we wish to minimize. The beam can be made lighter by reducing its section, but this also makes it less stiff. The specification of stiffness imposes a constraint on *A*. The stiffness, *S*, of a beam in bending is [4]:

$$S = \frac{CEI}{l^3} \tag{2}$$

where *E* is Young's modulus, *I* the second moment of area of the section, *l* the length and C a constant which depends only on the way the loads on the beam are distributed. The quantity *I* depends on both the *shape* and *scale* (size) of the cross-section:

$$I_{xx} = \int_{section} y^2 \, dA \tag{3}$$

where *y* is measured normal to the bending axis and d*A* is the differential element of area at *y*, as in:

Fig. 1.1.19 *A tie, a beam, a shaft and a column*

It is helpful to separate shape from scale. To do this, we define the *shape-factor* [a] for elastic bending:

$$\phi_B^e = \frac{4\pi I}{A^2} \tag{4}$$

The second moment, I, has dimensions of (length)^4; so does the quantity A^2. The ratio of the two is dimensionless; it describes shape, and is independent of scale. Two I-beams which have the same values of ϕ_B^e but which differ in scale have cross-sections which look like photographic enlargements or reductions of each other, as shown in: **Fig. 1.1.20** *The shape factor* ϕ_B^e

The factor 4π is inserted to give a solid, circular section (like that of Fig. 1.1.19a, a shape-factor with a value of 1. Very efficient sections have values of $\phi^e{}_B$ as high as 100.

Substituting equation (4) into (2) gives

$$S = \frac{CE\ \phi_B^e\ A^2}{4\pi l^3} \tag{5}$$

We now eliminate the free variable A in the objective function (1), giving

$$\frac{m}{l} = \left(\frac{4\ \pi S}{C}\right)^{1/2} \left(l^3\right)^{1/2} \frac{\rho}{(\phi_B^e E)^{1/2}} \tag{6}$$

In this equation, the constant C, the stiffness S and the length l are all specified by the design. For beams of a given, fixed, shape, the objective function, equation (1), describing m/l is minimized by maximizing the 'material index' M_1, defined by

$$M_1 = \frac{E^{1/2}}{\rho} \tag{7}$$

but for beams which differ in shape (as aluminium and wood beams generally would), it is minimized by maximizing a modified material index which now includes a term S which relates to shape:

$$M_{1}S = \frac{E^{1/2}}{\rho} (\phi_B^e)^{1/2} \tag{8}$$

[a]This shape factor is related to the radius to the radius of gyration, R_g, by $\phi_B^e = 4\ \pi R_g^2/A$. It is related to the 'shape parameter', k_1, of Shanley [5] by $\phi_B^e = 4\pi k_1$. Finally, it is related to the 'aspect ratio' α and 'sparsity ratio' i of Parkhouse [6, 7] by $\phi_B^e = i\ \alpha$. None of these authors extends the concept to describe torsion, or to include strength, as here, though their approaches would allow it.

The statement can be generalized. The performance P of a load-bearing component, when analysed in the way outlined above, is described by an equation of the form

$$P = f \quad (F, G, M, S) \tag{9}$$

where 'f' means 'a function of..', and F describes the load-bearing capacity of the component, (in the example above, its stiffness), G describes the geometry or scale (here, the length l), M contains the material properties and S characterizes the section-shape. Almost always, the variables F, G, M and S are separable, as they are in the example. Then the equation can be rewritten as:

$$P = f_1 \quad (F) \cdot f_2 \quad (G) \cdot f_3 \quad (M) \cdot f_4 \quad (S) \tag{10}$$

This has the attractive consequence that the material and shape, characterized by M and S, which is the best choice for one set of loads (F) and scale (G) remains the best choice for all F and G of that type; and if shape is fixed, the best choice of material is just that with the largest value of M and if M is fixed, the best choice is that with the largest value of S.

1.1.8.2 Other shape factors

As we have seen, a shape-factor is a dimensionless number which characterizes the efficiency of a section shape, regardless of scale, in a given mode of loading. Thus, as well as the shape-factor, ϕ^e_B, for elastic bending of beams (defined above), there is another, ϕ^e_T, for elastic twisting of shafts (the superscript 'e' means 'elastic'). These are the appropriate shape-factors when design is based on stiffness; when, instead, it is based on strength (that is, on the first onset of plastic yielding or on fracture) two more shape-factors are needed: ϕ^f_B and ϕ^f_T (the superscript 'f' meaning 'failure'). Finally, there is a shape-factor for elastic buckling under axial compression ϕ^e_E. All five shape-factors are defined so that they equal to 1 for a solid bar with a circular cross-section. We briefly examine each in turn.

In the elastic twisting of shafts (Fig. 1.1.19b), shape enters through the torsional moment of area, K. For circular sections it is identical with the polar moment of area, J:

$$K = J = \int_{\text{section}} r^2 \mathrm{d}A \tag{11}$$

where $\mathrm{d}A$ is the differential element of area at the radial distance r, measured from the centre of the section. For non-circular sections, K is less than J; it is defined such that the angle of twist is related to the torque T by

$$\theta = \frac{Tl}{KG}$$

where l is length of the shaft and G the shear modulus of the material of which it is made. Approximate expressions are available to calculate K for common section shapes (2).

The shape-factor for elastic twisting is defined by:

$$\phi_T^e = \frac{2\pi K}{A^2} \tag{12}$$

It, too, has the value 1 for solid circular shafts, and is near 1 for any solid, equiaxed section; but for thin-walled shapes it, too, can be large. As before, sets of sections with the same value of ϕ_T^e differ in size but not shape.

Plasticity starts when the stress, somewhere, first reaches the yield strength, σ_y; fracture occurs when this stress first exceeds the fracture strength, σ_f. Either one of these constitutes failure. We use the symbol σ_f for the failure stress, meaning 'the local stress which will first cause yielding or fracture'. One shape-factor covers both. In bending, the stress σ is largest at the point y_m in the surface of the beam which lies furthest from the neutral axis; it is [4]:

$$\sigma = \frac{M_B\, y_m}{I} = \frac{M_B}{Z} \tag{13}$$

(where M_B is the bending moment and Z is the section modulus $Z = I/y_m$). The shape-factor for failure in bending, ϕ^f_B, is defined by:

$$\phi^f_B = \frac{16\pi Z^2}{A^3} = \frac{16\pi I^2}{y^2_m A^3} \tag{14}$$

Defined in this way, ϕ^f_B has the value 1 for a solid cylinder. It can be as high as 200 for very efficient shapes. Failure occurs if the load exceeds the failure load

$$F_f = C_2 \frac{I}{y_m} \frac{\sigma_f}{l}$$

where C_2 is a constant which depends on the details of the loading. Replacing I/y_m by the appropriate shape-factor ϕ^f_B via equation (14) gives

$$F_f = \frac{C_2}{4\pi^{1/2} l} \left[\phi^f_B A^3\right]^{1/2} \sigma_f \tag{15}$$

Substituting this into equation (1) for the mass of the beam gives

$$\frac{m}{l} = \left[\frac{4\pi^{1/2} F_f l}{C_2}\right]^{2/3} \left[\frac{\rho^{3/2}}{\left[\phi^f_B\right]^{1/2} \sigma_f}\right]^{2/3} \tag{16}$$

The best material-and-shape combination is that with the greatest value of the index

$$M_2 = \frac{\left[\phi^f_B\right]^{1/3} \sigma_f^{\ 2/3}}{\rho} \tag{17}$$

At constant shape the index reduces to the familiar $\sigma_f^{2/3}/\rho$; but when shape as well as material can be chosen, the full index of equation (17) must be used.

In torsion, the problem is more complicated. For circular tubes or cylinders subjected to a torque T (as in Fig. 1.1.19c) the shear stress τ is a maximum at the outer surface, at the radial distance r_m from the axis of bending:

$$\tau = \frac{T r_m}{J} \qquad (18)$$

The quantity J/r_m in twisting has the same character as Iy_m in bending. For non-circular sections with ends that are free to warp, the maximum surface stress is given instead by

$$\tau = \frac{T}{Q} \qquad (19)$$

where Q, with units of m^3, now plays the role in torsion of Z in bending [4]. This allows the definition of a shape-factor, ϕ^f_T for failure in torsion:

$$\phi^f_T = \frac{4\pi Q^2}{A^3} \qquad (20)$$

Like the other shape-factors, ϕ^f_B and ϕ^f_T are dimensionless, and therefore independent of scale and both have the value 1 for a solid circular section.

Fully plastic bending or twisting (such that the yield strength is exceeded throughout the section) involve a further pair of shape-factors. But, generally speaking, shapes which resist the onset of plasticity will resist full plasticity well too. New shape-factors for these are not, at this stage, necessary. The shape-factor for buckling in compression (Fig. 1.1.19d) depends only on I again. The appropriate shape-factor (ϕ^e_E) has the same form as that for elastic bending (equation (4)) except that the I which enters is the lesser of the principal moments I_{xx} and I_{yy} (which we call I_{min}) whereas in bending the loads are usually applied

Table 1.1.8 Definitions of shape factors

Mode of loading	***Constraint***	***Section moment***	***Shape factor***
Bending	Stiffness	I_{max} = greater of I_{xx} and I_{yy}	$\phi^e_B = \frac{4\pi I_{max}}{A^2}$
Bending	Strength	$Z = I/y_{max}$	$\phi^f_B = \frac{16\pi Z^2}{A^3}$
Torsion	Stiffness	J (circular sections) or K	$\phi^e_T = \frac{2\pi K}{A^2}$
Torsion	Strength	J/r (circular sections) or Q	$\phi^f_T = \frac{4\pi Q^2}{A^3}$
Compression	Buckling	I_{min} = lesser of I_{xx} and I_{yy}	$\phi^e_E = \frac{4\pi I_{min}}{A^2}$

in such a way that the larger principal moment resist the deformation (I_{max}). The shape-factors are summarized in:

Table 1.1.8 *Definitions of shape factors*

Expressions for all ϕs for common section shapes are listed by Ashby [1, 2].

1.1.8.3 Shaped materials on selection charts

Shaped materials can be displayed and selected with the Material selection charts. The reasoning, for the case of elastic bending, goes like this. The performance index for elastic bending (equation (8)) can be rewritten as

$$M_1 = \frac{(\phi_B^e E)^{1/2}}{\rho} = \frac{(E/\phi_B^e)^{1/2}}{\rho/\phi_B^e} \tag{21}$$

The equation says, 'a material with modulus E and density ρ, when structured, behaves like a material with effective modulus $E^* = E/\phi_B^e$ and effective density $\rho^* = \rho/\phi_B^e$

The E–ρ chart is shown schematically in:

Fig. 1.1.21 *A material with a shape factor ϕ_B^e behaves like a solid with a modulus E/ϕ_B^e and a density ρ/ϕ_B^e*

The structured material properties E^* and ρ^* can be plotted onto it. Introducing shape ($\phi_B^e = 10$, for example) moves the material M to the lower left along a line of slope 1, from E, ρ to $E/10$, $\rho/10$, as shown in the figure. The selection criteria are plotted onto the figure as before; a constant value of the index of $E^{1/2}/\rho$ for instance plots as a straight line of slope 2; it is shown, for one value of $E^{1/2}/\rho$, as a broken line. The introduction of shape has moved the material from a position below this line to one above; it has improved the performance. Elastic twisting of shafts is treated in the same way.

Materials selection based on strength (rather than stiffness) at a minimum weight uses the chart of strength σ_f against density ρ, shown schematically in:

Fig. 1.1.22 *A material with a shape factor ϕ_B^f behaves like a solid with strength $\sigma_f\ /\phi_B^f$ and a density ρ/ϕ_B^f*

Shape is introduced in a similar way. The performance index for failure in bending (equation 17), can be rewritten as:

$$M_2 = \frac{(\phi_B^f)^{1/3}\ \sigma_f^{2/3}}{\rho} = \frac{(\sigma_f/\phi_B^f)^{2/3}}{\rho/\phi_B^f} \tag{22}$$

The material with strength σ_f and density ρ, when shaped, behaves in bending like a material of strength:

$$\sigma_f^* = \sigma_f/\phi_B^f$$

and density

$$\rho^* = \rho/\phi_B^f$$

The rest (Fig. 1.1.22) will be obvious. Introducing shape ($\phi_B^f = 10$, say) moves a material M along a line of slope 1, taking it, in the schematic, from a position below the merit index line (the broken line) to one above. The performance has improved. Torsional failure is analysed by using ϕ_T^f in place of ϕ_B^f.

1.1.8.4 Summary and conclusions

The designer has two groups of variables with which to optimize the performance of a load-bearing component: the *material properties* and the *shape of the section*. They are not independent. The best choice of material, in a given application, depends on the shapes in which it is available, or to which it could potentially be formed. A procedure is given for simultaneously optimizing the choice of both material and shape.

The contribution of shape is isolated by defining four *shape factors*. The first, ϕ_B^e, is for the elastic bending and buckling of beams; the second, ϕ_T^e, is for the elastic twisting of shafts; the third, ϕ_B^f is for the plastic failure of beams loading in bending; and the last, ϕ_T^f, is for the plastic failure of twisted shafts (Table 1.1.8). The shape factors are dimensionless numbers which characterize the efficiency of use of the material in each mode of loading. They are defined such that all four have the value 1 for solid circular sections. With this definition, all equiaxed solid sections have shape factors of about 1, but efficient shapes which disperse the material far from the axis of bending or twisting (I-beams, hollow tubes, sandwich structures, etc.) have large values of the shape factors.

The case studies in Section 1.1.7 demonstrate the procedure for selecting the best material when shape is not a variable. For each mode of loading, a performance-limiting combination of material properties, or 'material index', was identified. Performance is maximized by making this combination as large as possible. For a light, stiff beam the combination is $E^{1/2}/\rho$; it becomes $\left(E\phi_B^e\right)^{1/2}/\rho$ when shape, described by ϕ_B^e, is a variable. A similar combination (involving ϕ_T^e) gives the lightest stiff shaft. The material index for a light beam with a prescribed strength at fixed shape is $M_2 = \sigma_f^{2/3}/\rho$; it becomes $M_2 = \sigma_f^{2/3}\left(\phi_B^f\right)^{1/3}/\rho$ when shape is a variable. A similar combination involving ϕ_T^f gives the lightest strong shaft. Here, the criterion of 'performance' was that of meeting a design specification at minimum weight. Other such material–shape combinations maximize other performance criteria; minimizing cost rather than weight, for example, or maximizing energy storage.

The selection is most easily carried out by creating materials property charts (Section 1.1.6) onto which indices can be plotted, allowing those materials which have particularly large values of an index to be identified in the way illustrated here.

1.1.8.5 References

1. Ashby, M.F. (1991) Materials and shape, *Acta Metallurgica et Materialia,* **39** (6), June, 1025–1039.
2. Ashby, M.F. (1992) *Materials Selection in Mechanical Design,* Pergamon Press, Oxford, UK.
3. Weaver, P.M. and Ashby, M.F. (1996) The optimal selection of material and shape, *Journal of Engineering Design,* **7** (2) June.
4. Young, W.C. (1989) *Roark's Formulas for Stress and Strain,* McGraw-Hill, New York, NY, USA.
5. Shanley, F.R. (1960) *Weight-Strength Analysis of Aircraft Structures* Second edition, Dover Publications, New York, U.S.A.
6. Parkhouse, J.G. (1984) *Structuring: A Process of Material Dilution,* in *Third Int. Conf. on Space Structures,* (ed. by H. Nooshin), Elsevier, London, UK, 367.
7. Parkhouse, J.G. (1987) Damage accumulation in structures, *Reliability Engineering,* **17**, 97–109.

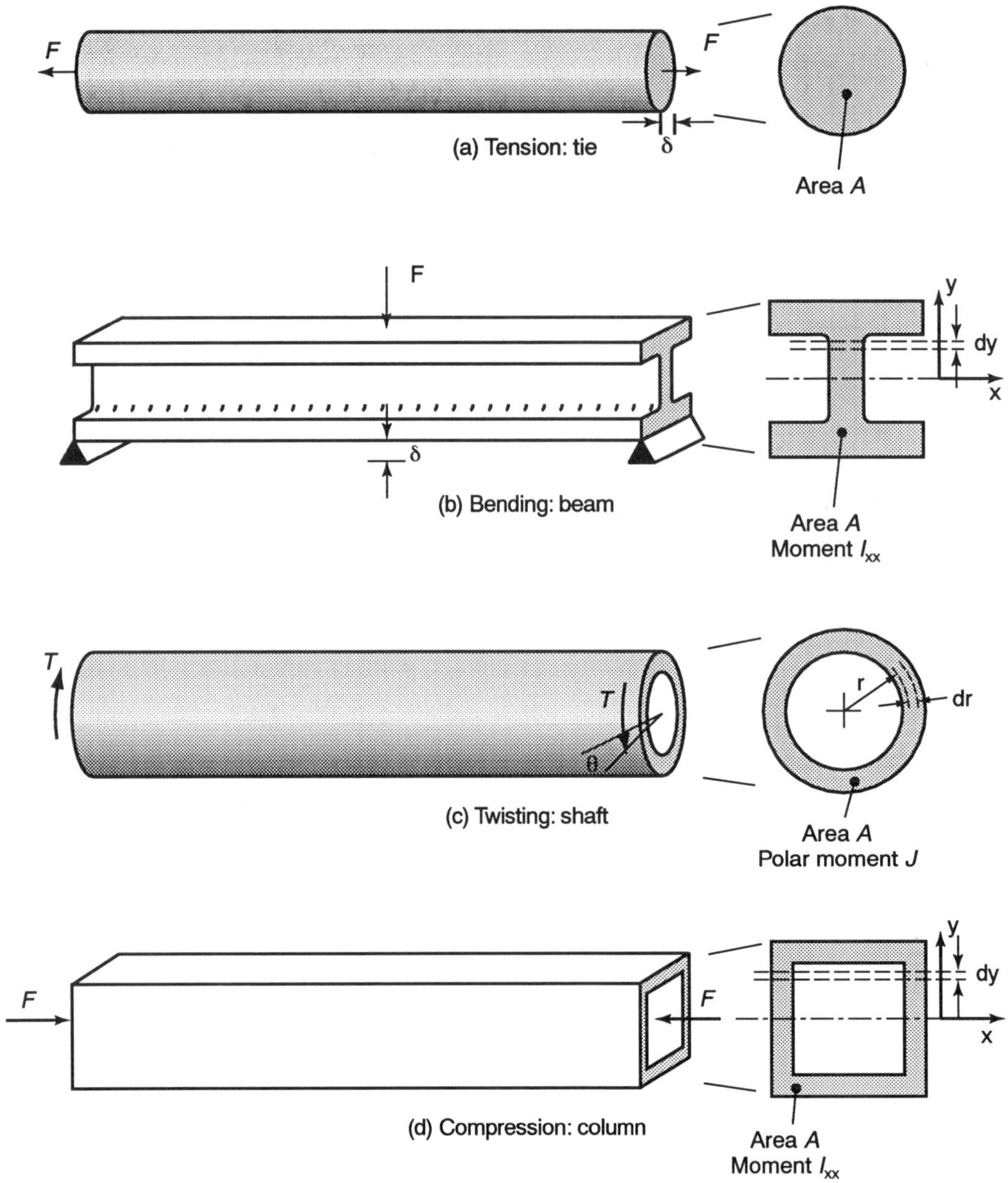

Fig. 1.1.19 A tie, a beam, a shaft and a column. (Efficient shapes are shown for each.)

Fig. 1.1.19

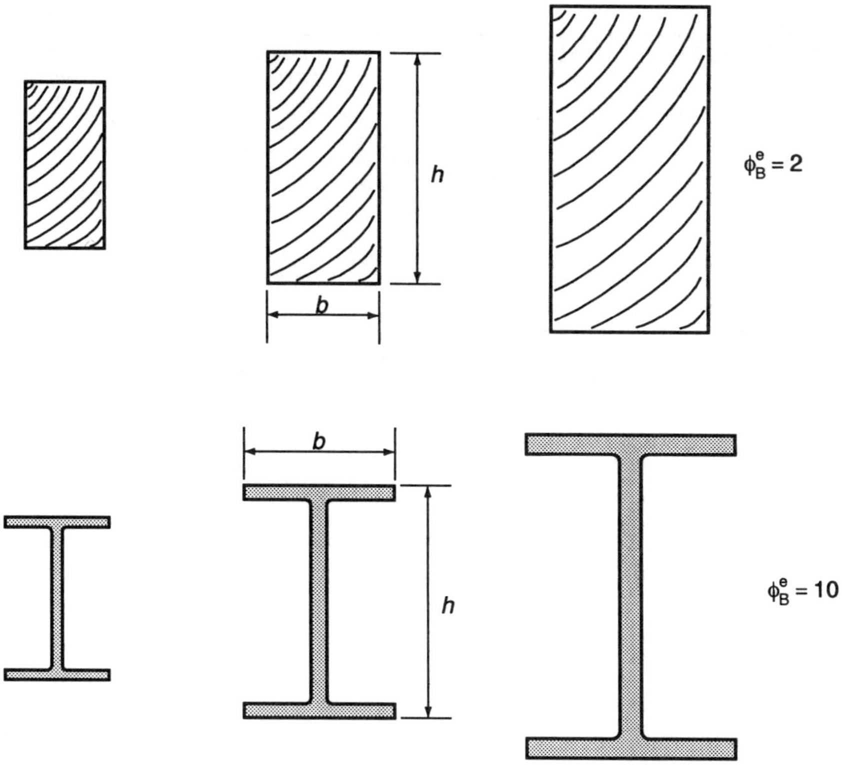

Fig. 1.1.20 The shape factor ϕ_B^e characterizes shape and is independent of scale. (The shapes shown here have values of ϕ_B^e of 2 and 10. Shape factors as high as 100 are possible.)

Fig. 1.1.20

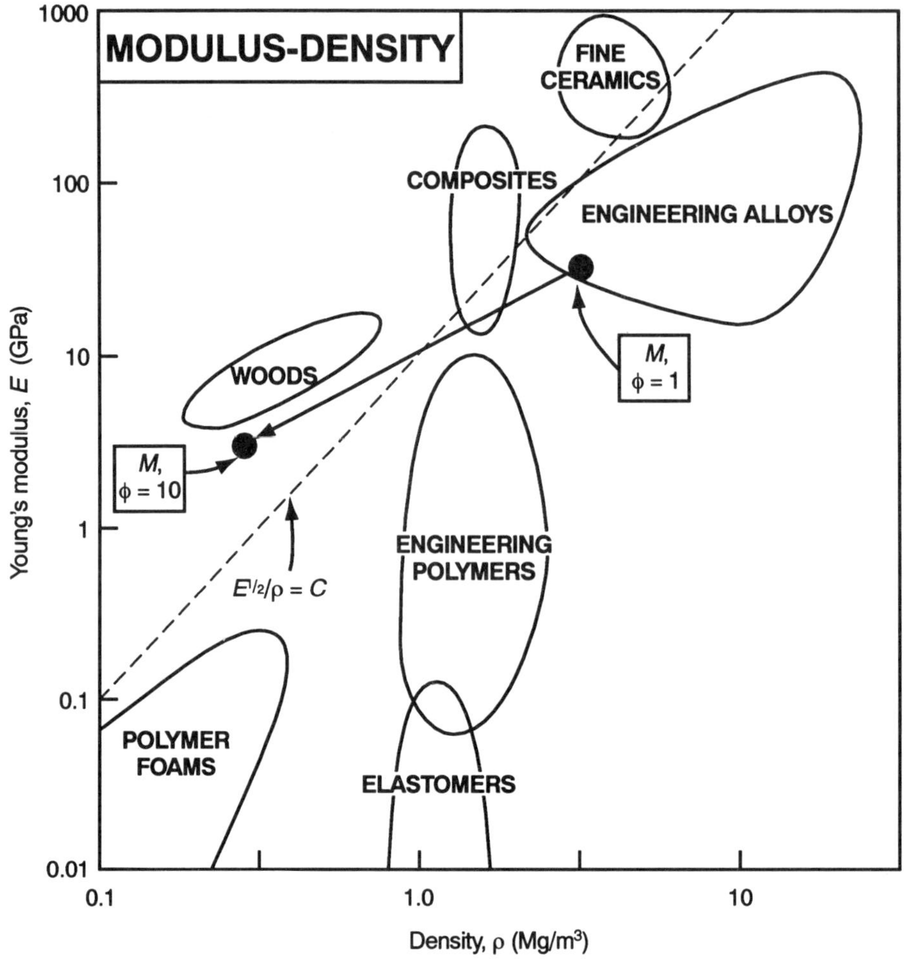

Fig. 1.1.21 A material with a shape-factor ϕ_B^e behaves like a solid with a modulus E/ϕ_B^e and a density ρ/ϕ_B^e. (Although both the effective modulus and effective density are reduced by shaping, the efficiency, measured by $(E\phi)^{1/2}/\rho$ is increased – see Fig. 1.1.6)

Fig. 1.1.21

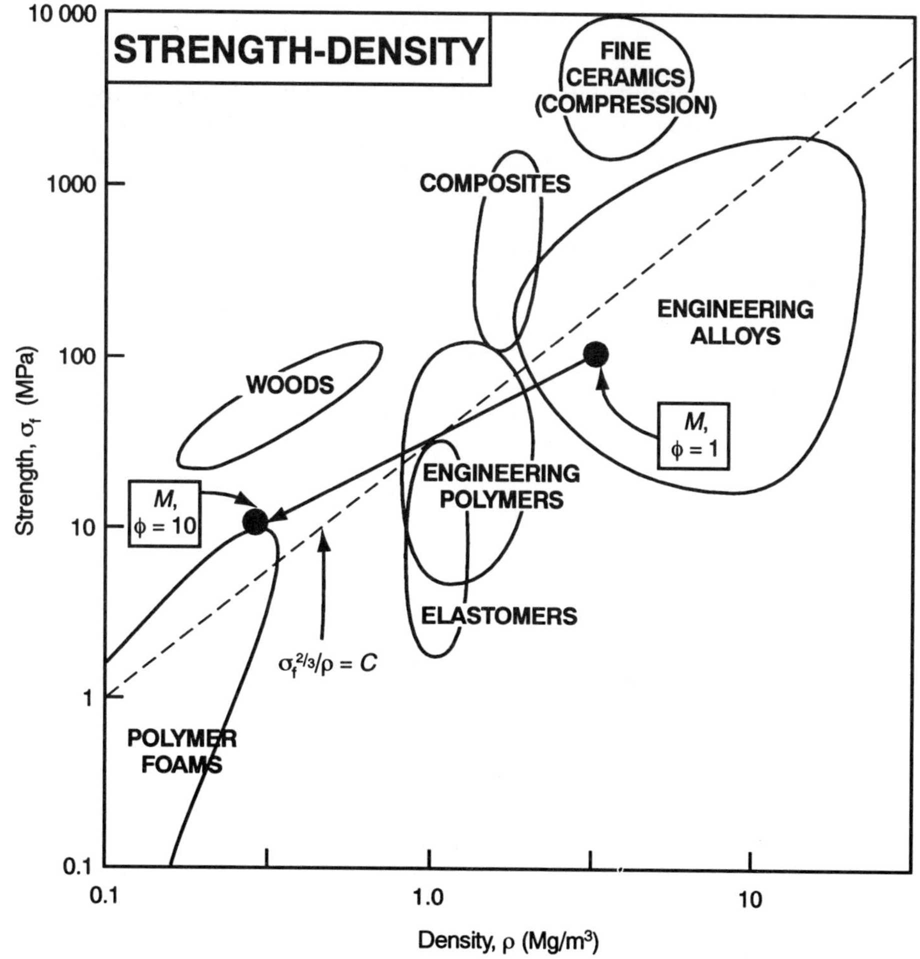

Fig. 1.1.22 A material with a shape-factor ϕ_B^f behaves like a solid with strength σ_f/ϕ_B^f and density ρ/ϕ_B^f. (Although both the effective strength and effective density are reduced by shaping, the efficiency, measured by $M_3 = \sigma_f^{2/3}\ (\phi_B^f)^{1/3}/\rho$ is increased – see Fig. 1.1.7)

Fig. 1.1.22

1.1.9 Selection for resistance to corrosion and degradation

Different components will vary widely in their requirement for resistance to degradation. For example, small pits which may be unimportant in metallic components of large section thickness would be catastrophic in pressurized piping, and thin protective corrosion deposits that may be of no consequence on purely functional parts would be unacceptable where an unblemished appearance is a principal requirement. Corrosion rates are published by many reference sources, including materials suppliers. Actual service conditions, however, are often different from those under which the data were collected. Prediction of corrosion rates for a material in an unfamiliar environment is extremely difficult and testing is a necessity.

It is therefore important that, before any consideration is given to selection of materials for resistance to degradation, a thorough analysis is made of the likely service environment of the product. The analysis should cover the lifecycle of the product from manufacture, through storage and transportation, to in-service conditions. It is, of course, impossible to anticipate the exact service conditions for the majority of products. However, it is rare that some environmental experience, with perhaps different materials, is not available. Alternative materials may be evaluated initially by comparison with these data.

The corrosion and degradation susceptibilities of metals, plastics and ceramics are outlined below and described in detail in Section 1.4.

METALS

Some types of metallic corrosion, such as crevice corrosion, stress-corrosion cracking, and the corrosion caused by contact of dissimilar metals may be avoidable by correct design. Others can be prevented by surface coatings (such as paint systems, plastics, electro-deposited, hot-dip), chemical immersion/conversion, sprayed coatings, or by cathodic protection. The use of a coating is often a cost effective solution to metallic corrosion. The advantages and limitations of commonly employed anti-corrosion coatings are given in Section 1.4.2.

The following factors should be considered when selecting the type of coating most suitable for a given application:

- the necessary pre-treatment of substrate,
- the consequence of damage to a small area of the coating,
- the technical feasibility and comparative economics of protection for life as compared with that of planned maintenance,
- the need to fulfil two functions, such as resistance to both corrosion and wear, with one coating and,
- the adverse effects of the coating on other functions of the component (for example dimensional changes, hydrogen embrittlement from plating baths, and changes in surface conductivity and appearance).

A comparison of organic coating methods is given in Section 1.4.2.2. This can be used to assist in the identification of the most suitable processing route for a product once it has been decided that the performance of an organic coating is acceptable. The processing route selected will depend on other factors, such as existing in-house plant and expertise, but cost will ultimately be the deciding factor.

Cathodic protection (see 1.4.2.10) is sometimes used as an alternative to a coating or to improve its performance.

PLASTICS

Plastics are subject to various types of degradation. As with metals, it is extremely difficult to predict the behaviour of plastics in unfamiliar environments and most of the information provided by suppliers on the effects of different environments on their materials is obtained from laboratory tests. No guarantees are given for particular service conditions. Testing under simulated service conditions should always be carried out as a precaution against premature failure. It should also be noted that some corrective action to prevent or minimize degradation will be taken by the manufacturer. Thus anti-oxidants and stabilizers, for example, are standard additives, which are used as required depending on the polymer and the end use envisaged. In no case can the defect be completely eliminated, but it can be sufficiently minimized for most practical and industrial purposes. (For more details, see Section 1.4.4 and Section 3.1.)

CERAMICS

A ceramic material is often selected for resistance to corrosion, particularly that caused by oxidation at high temperature. However, in some environments certain forms of degradation can occur (see Section 1.4.3).

1.1.10 Selection for resistance to wear

Wear occurs at the surface of components which are subject to relative movement under load. This movement may occur when one component is in contact with another, with bulk fluids or particulate solids.

Section 1.5 provides guidance on conditions causing wear and the selection of materials to avoid the various types of wear.

As with corrosion, wear resistance requirements vary according to the component. For certain parts, such as gears, piston rings and liners, etc, a small amount of running-in wear is desirable to produce an even distribution of stresses over the mating surfaces. However, the wear debris should be removed, or absorbed safely so that local stress concentrations are not developed. The total wear should be such that the designed dimensional tolerances are not exceeded. In other components, longer-term wear, such as cavitation erosion (see Section 1.6) in impellers or propellers, progressive pitting fatigue in rolling-element bearings or gears, and erosion and abrasive wear in process plant or earth moving equipment must be delayed as long as possible, ideally to just beyond the intended service life or at least until replacement or maintenance is convenient and economic.

Although the maintenance of a lubricant film between the mating surfaces of two components will ensure the lowest wear or frictional losses, this is not always possible to achieve (for example during start-up and shut-down) nor is it the most economic solution. In some cases adequate wear resistance may be obtained by a suitable coating on a relatively cheap substrate or by selection of a lubricant with anti-wear additives.

A general guide to the selection of treatments to reduce friction and wear on bearing surfaces is given in Section 1.5.7, which also provides guidance on weld-deposited coatings for dimension restoration and abrasion resistance. Further information on the selection of coatings for resistance to corrosion and wear is given in Section 1.6.

1.1.11 Selection of component manufacturing processes

Selection of the manufacturing process route which will deliver components of the required quality and performance at the lowest cost will depend on the following factors:

Component related
- overall size,
- maximum and minimum section thickness,
- complexity,
- numbers to be made (batch size and production rate),
- surface finish,
- dimensional tolerances.

Materials related
- melting point,
- hardness,
- possibilities to produce shapes from:
  gas (e.g. vapour deposition),
  liquid (e.g. casting, electroforming),
  solid (e.g. deformation processes, etching).

Local conditions
- Investment in production plant and equipment,
- In-house production skills,
- Established relationships with preferred suppliers.

In the systematic approach described in Sections 1.1.1 to 1.1.6, the possible manufacturing process routes are constrained implicitly by:

1. The outline shapes considered at the **layout design stage**. Certain shapes can only be made economically by certain processes (and in certain materials).
2. The properties of the generic classes of materials given in the Ashby maps/ Materials selection charts given in Section 1.1.6. The values shown within the ringed regions cover the range of properties of the materials in component form (even if only in the shape of test bars) made by the available process routes for those materials.

These constraints are, however, unlikely to lead directly to the process which will deliver the required components at the minimum costs. Most components can be made by a range of processes and more detailed examination of the characteristics of component manufacturing processes for each of the candidate materials (or combination of materials and surface coating is required).

Preliminary selection can be made by reference to the Ashby process selection maps.

COMPLEXITY

Complexity is defined here as the presence of features such as holes, threads, undercuts, bosses, re-entrant shapes, etc, which cause manufacturing difficulties or require additional operations to produce. Complexity in relative terms (1 = least complex, 5 = most complex) is plotted in:

Figure 1.1.23 *Complexity level vs size map for component processes* and in:

Table 1.1.9 *Complexity level table for component manufacturing processes*

TOLERANCE – ROUGHNESS

No part can be made exactly to a specified dimension. Some deviation from a desired dimension is permitted, though undesirable, and this is referred to as the tolerance, T. Manufacturing processes vary in the levels of tolerances they can achieve economically. Closely related to tolerance is surface roughness, R, which is defined as the presence of irregularities on the surface. Again surface roughness is unavoidable and its desired level depends on the particular application. Exceptionally smooth surfaces (for example, a lapped surface) are neither necessary nor economical in many applications. Tolerance and surface roughness are obviously related. The best achievable tolerance cannot be less than $2R$ (R is the root-mean square roughness) – it is more like $5R$. The relationship between T and R is shown in:

Fig. 1.1.24 *Tolerance vs surface roughness*

This shows typical tolerances and surface finishes attainable by each manufacturing process. Unnecessarily tight tolerances and surface finish specifications are not economical. Manufacturing costs increase almost exponentially with tighter tolerances and surface finish. The map shows contours of relative cost; an increase in precision corresponding to the separation of two neighbouring contours gives an increase in cost, for a given process, by a factor of 2.

Moulded polymers inherit the finish of the moulds and thus can be very smooth, but tolerances better than 0.1 mm are rarely achievable because of the internal stresses left by moulding and because of the creep caused by service stresses. In this respect, polymers vary from metals and ceramics as the map shows.

It is important to point out that tolerances depend greatly on the dimension (the tolerances given apply to a 25 mm dimension) and do not scale linearly. In addition, tolerances also depend on the material. Typical ranges of tolerances and surface finish are sufficient (with no need for further complications), and as the map shows, discriminate clearly between various processes. Machining processes which are capable of achieving the most accurate dimensions and the smoothest surface finish are commonly used after casting or deformation processes to bring the tolerance or surface finish to the level dictated by the design.

SIZE – SHAPE

There are many ways of describing shapes. The first, which is particularly useful in process selection, is the aspect ratio or, what we call 'slenderness'. Manufacturing processes vary widely in their capability of making thin, slender sections. *Casting processes*, for instance are limited in the slenderness of the shapes they can produce because surface tension and heat flow problems cause defects when molten metal flows through a thin section with a large surface area. Applying pressure helps overcome the problem, but it is well known that casting processes are used more successfully with compact shapes. Powder methods are also limited in slenderness because of the nature of the process – but they can be used to shape materials such as ceramics and very hard metals, which cannot be shaped in other ways.

Processes in which the material flows in the solid state, *deformation processes*, can make more slender shapes since the problems associated with solidification defects are irrelevant. However, limits on the forging-die pressures set a lower limit on thickness and slenderness. *Machining processes* can make very slender shapes since the shapes are generated by removing unwanted material. Processes which build the shape atom by atom, such as *electroforming* are also capable of making slender shapes. Finally, whenever a thin, slender shape needs to be made and no process can do it, *fabrication methods* make it possible by joining smaller pieces of the material to make the desired shape.

Here, slenderness, S, is measured by the ratio t/l, where t is the minimum section thickness and l is the large dimension of the shape: for flat shapes, it is about $\sqrt{A}$; for cylindrical ones, it is $A/\pi t$ where A is the surface area; thus S lies in the range

$$\frac{t}{\sqrt{A}} > S > \frac{\pi t^2}{A}$$

Size is defined by the minimum and maximum volumes of which the process is capable. The volume, V, for uniform sections, is within a factor of 2,

$$V = At$$

Volume can be converted approximately to weight by using an 'average' material density of 5000 kg/m^3 – most engineering materials have densities within a factor of 2 of this value (polymers are the exception and their densities are all around 1000 kg/m^3).

The relationship between size measured in volumetric terms and the slenderness capabilities of manufacturing processes is illustrated in:

Fig. 1.1.25 *Size versus slenderness*

The horizontal axis is the slenderness, S, the vertical axis is the volume, V. Contours of A and t are shown as families of diagonal lines. An aspect ratio or slenderness of l is the maximum limit since it is not possible to create shapes with smaller surface-to-volume ratios than those of spheres.

HARDNESS AND MELTING POINT

There is a limit on the maximum melting point of the materials which can be processed by conventional casting techniques. Low melting point metals can be cast by any one of many casting techniques. As the melting point rises above 2000 K, conventional casting methods are no longer viable, and special casting techniques, such as electron beam casting and vacuum casting, have to be used. Similarly, the yield strength (or hardness) of the material imposes limitations on the range of application of deformation and machining methods. Forging and rolling pressures are proportional to the flow strength; and the heat generated during machining, which limits tool life, also scales with the ultimate strength or hardness. For this reason, deformation processing is limited to materials with hardness values below 3 GPa. Other manufacturing methods exist which are not limited by either the melting point or the hardness of the material. Examples of such methods are powder methods, CVD and evaporation methods, and electroforming.

The above information is presented in a graphical form in:

Fig. 1.1.26 *Hardness vs melting point*

Real materials occupy only a part of the space covered by the axes; it is the region between the two heavy lines. The hardness and melting point are not independent properties; low melting point materials tend to be soft (polymers and lead, for instance); high melting point materials are hard (diamond is the extreme example). This information is captured by the equation as described in Frost and Ashby (1982) *Deformation-Mechanism Maps* (Pergamon Press, Oxford, UK).

$$0.03 < \frac{H\Omega}{kT_m} < 20$$

where Ω is the atomic or molecular volume and k is Boltzmann's constant (1.38×10^{-23} J/K). This defines the two heavy lines.

More detailed information on the characteristics of component manufacturing processes for metals, plastics and ceramics are given in Section 1.6. The information given in these sections should enable the lowest cost component manufacturing route to be selected.

Before the final decisions can be made on the precise grade of material and the preferred manufacturing route, it is necessary to test prototype components under conditions which simulate the extremes of in-service conditions.

The methods of manufacturing prototypes are given in the next section, 1.1.12, and guidance on testing is given in Section 1.1.13.

Table 1.1.9 Complexity level table for component manufacturing processes, (see also Fig. 1.1.23)

Process class	*Process subclass*	*Complexity level*	*Reasons*
Casting	Gravity Low pressure High pressure Investment	4 3 3-4 4	Expendable pattern allows very complex shapes.
Moulding	Injection Compress Blow Foam	4 2 2 3-4	Relatively simple because of short flow length hollow articles with uniform thin sections.
Deformation	Cold Warm Hot Sheet	3 3 3 2	Shapes with essentially constant section thickness.
Machining	Turn Mill Grind Polish	4-5 4-5 4-5 4-5	
Powder routes	Sinter/HIP Slip cast Spray forming Hydration	3 3 2 2	Parts with parallel straight sides, no re-entrant angles. Solid shapes with re-entrant angles.
Composite forming	Lay-up Mould Squeeze-cast Filament wind	3 3 3 3	Complex internal and re-entrant features possible.
Molecular methods	PVD CVD Sputtering Electroforming	2 3 2 4	Uses 'line of sight' spray, so limiting geometry re-entrant angles possible.
			Bosses, undercuts and re-entrant shapes possible.
Special methods	Electrochemical Ultrasonic Chemical Thermal beam	5 5 5 5	All shapes possible.
Fabrication	Weld/braze Adhesive Fasten Microfabrication	5 5 5 5	

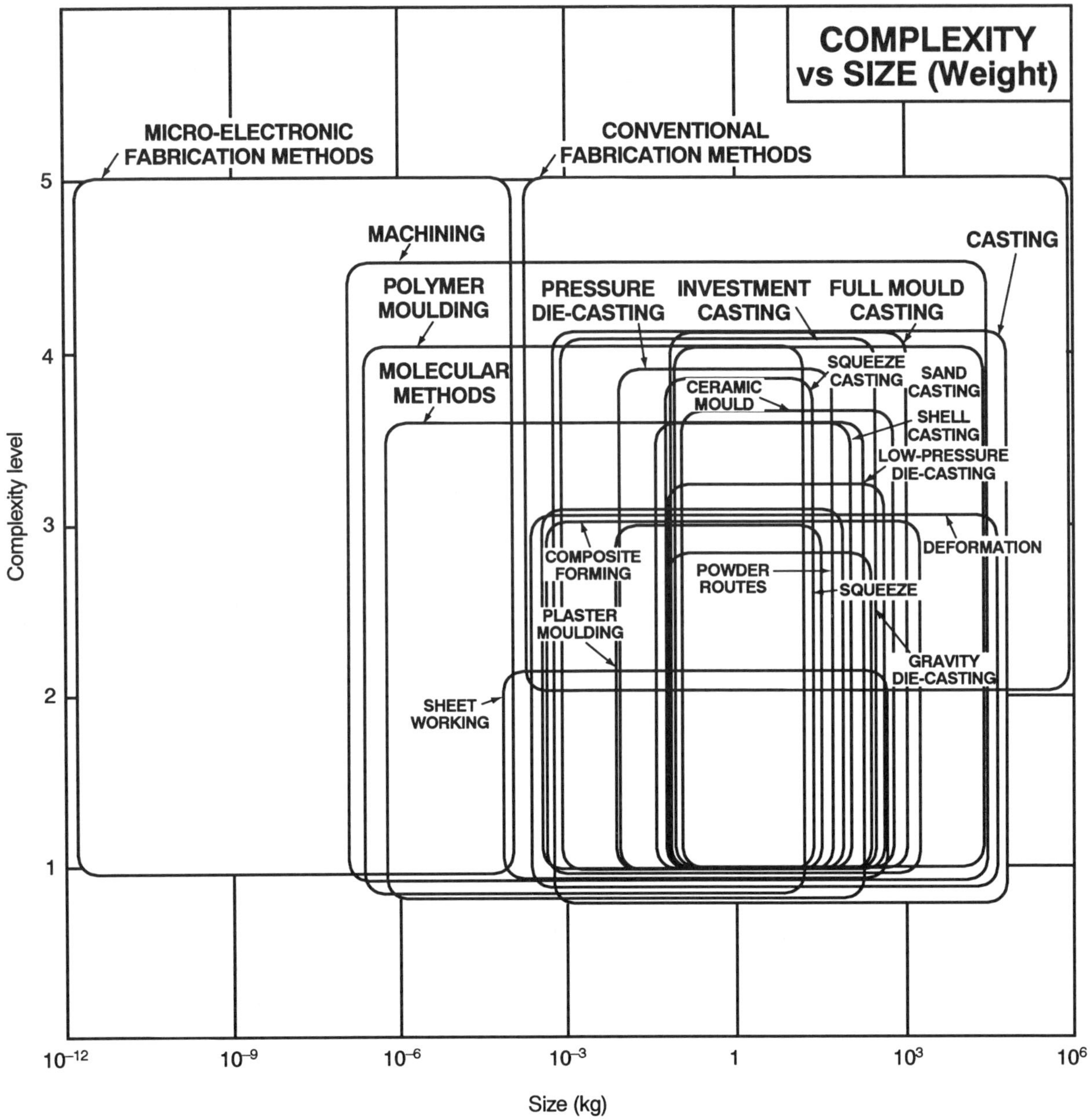

Fig. 1.1.23 Complexity level vs component size for component processes (See also Table 1.1.9)

Fig. 1.1.23

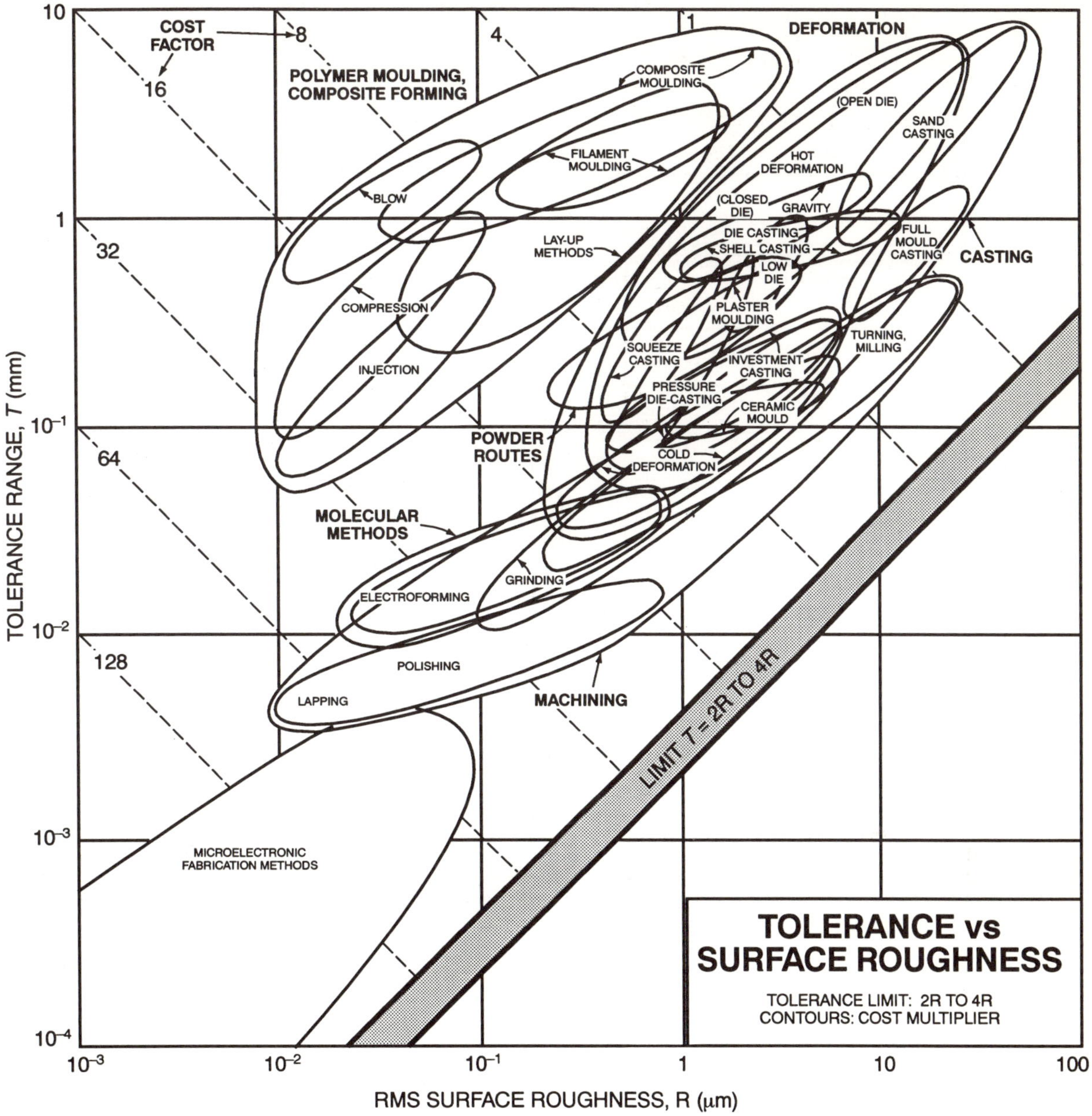

Fig. 1.1.24 Tolerance range vs surface roughness

Fig. 1.1.24

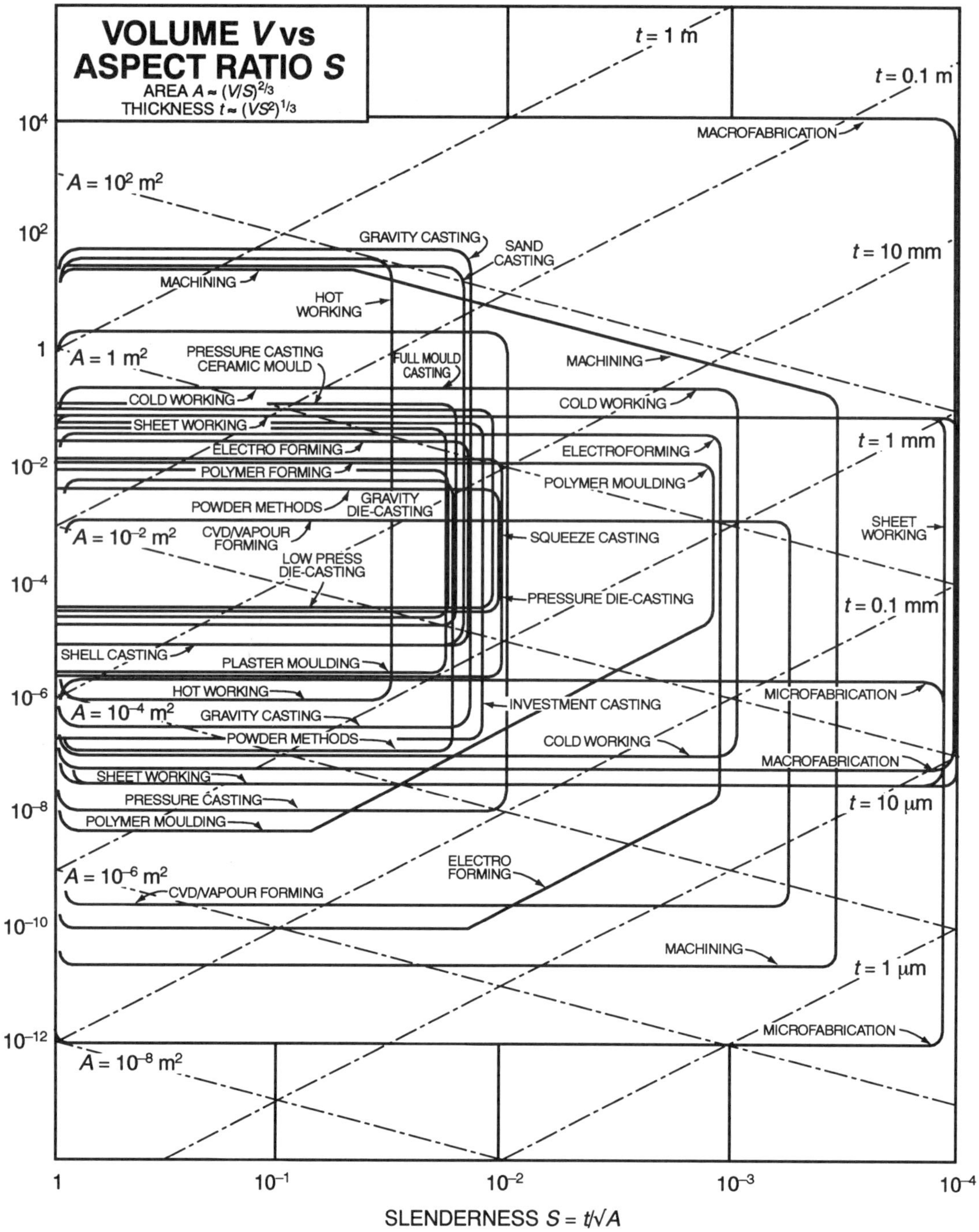

Fig. 1.1.25 Size vs slenderness

Fig. 1.1.25

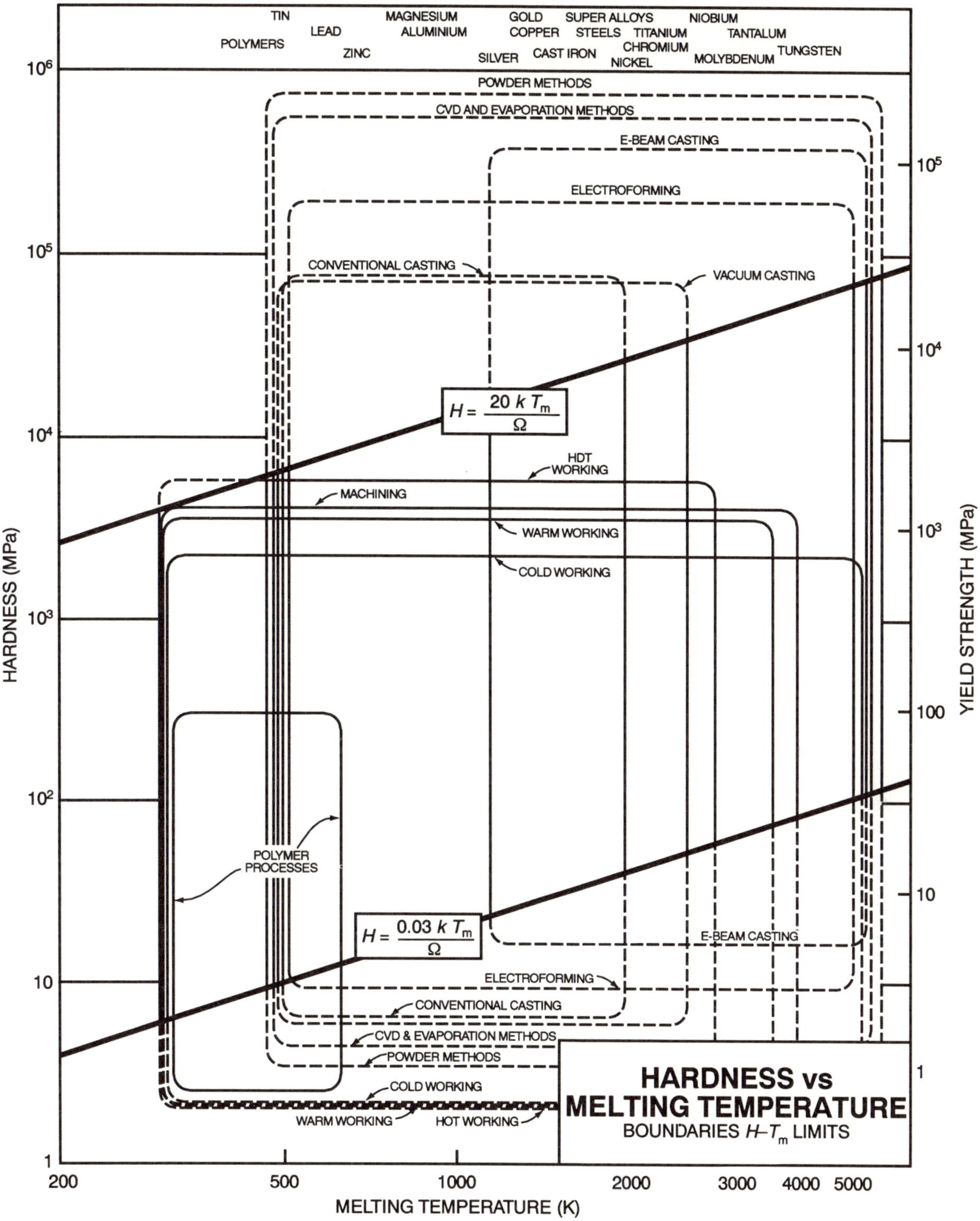

Fig. 1.1.26 Hardness vs melting point

1.1.12 Rapid prototyping techniques

At the **layout design** and **detailed design** stages of new product development, it is valuable to have access to three-dimensional physical models of the components of the product. These models can fulfil a number of functions:

- **Manifestation of conceptual designs**; 'touchy- feely models' which can give senior management a 'feel' for the components to be made. These models can also alert suppliers to the type of components which will be required.
- **Form and fit**; particularly useful for complex components with complicated cavities or piping. Also useful to ensure that design errors do not have two components occupying the same space. Form and fit models can be employed at the detailed design stage to ensure that complex assemblies fit together. Experience has shown when 3-D form and fit solid models accompany engineering drawings, the errors made by suppliers and the number of engineering design changes are reduced by between 30% and 60%.
- **Functional performance**; physical models can be employed to check the function performance of products where fluid flow characteristics are important. Specific examples are car heaters, inlet manifolds, hydraulic and pneumatic systems, air-conditioning units and wind tunnel models.
- **Patterns of the manufacture of castings and mouldings**; traditionally, these models were made either by hand-crafting techniques or by multi-axis computer numerically controlled (CNC) milling. Recently, a new generation of rapid prototyping techniques have evolved. These techniques offer the possibility to translate a 3-D CAD model of a component to a physical model in hours rather than the days, weeks, months required by conventional prototyping techniques.

All rapid prototyping techniques start with a 3-D CAD solid model (preferably) or a surface model of the component to be made. The computer then slices the part into thin layers and feeds information on the shape and dimensions of each layer to the rapid prototyping system. The systems differ in the way the component is built up, layer by layer. Details of rapid prototyping techniques are given in:

Table 1.1.10 *Comparative advantages and limitations of main rapid prototyping systems*, and

Table 1.1.11 *Comparative quantitative characteristics of main rapid prototyping systems.*

Table 1.1.10 Comparative advantages and limitations of main rapid prototyping systems

System	*Advantages*	*Limitations*
Stereolithography	Well proven system. Quickcast^{TM} system permits successful and economic sacrificial patterns. Quickcast^{TM} system gives higher accuracy and better structural integrity of models. Very wide range of application experience; excellent user groups. Best surface finish of currently-available techniques. Most parts can now be 'right first time'.	Expensive raw materials initially, but model material costs comparable with other systems. Limited range of materials (photosensitive resins). Support structures needed but these are automatically supplied from purpose-written software. Care needed with environmentally hazardous solvents used to clean up models.
Selective laser sintering	Range of model materials. Wax models can be invested directly. Polycarbonate models after surface treating with wax can be used as sacrificial patterns. Lower cost of raw materials than stereolithography or solider systems.	Wax models need 12 hour cooling cycle on the machine. Wax models are very fragile. Recycled unsintered powders need careful sieving to avoid 'ball-ups' and 'furrowing' when new layer of powder is laid down. Support structures can be necessary, especially for complex, overhanging features of wax models. Learning curve necessary with most new parts.
Laminated object manufacture (LOM)	Relatively lower capital cost system. Low cost of raw materials. Finished models resemble wood – popular with model makers.	Inherent fire risk. Most models require subsequent hand working to improve surface finish. Inherent difficulties with undercuts and re-entrant features. Waste material can be difficult to remove.
Fused deposition modelling	Lower initial cost of system. Low cost materials. Can be operated in office environment. Highly reliable machines. Short build time for thin wall parts.	Poor surface finish but can be improved by hand working. Support systems necessary for complex overhanging features.

Table 1.1.10 Comparative advantages and limitations of main rapid prototyping systems – *continued*

System	*Advantages*	*Limitations*
Solid ground curing (solider)	High capacity of model building chamber. When chamber is used to full capacity can be most economic rapid prototyping technique. Complex models possible, including ready assembled multi-component products.	Large, heavy, noisy, expensive system. Prone to breakdowns. Requires constant manning. High consumption of expensive raw materials. Model cannot be used as sacrificial patterns for conventional investment casting.
3-D printing systems	Direct route to ceramic shell production for investment casting. Wide range of materials possible soon (in principle, any material available in powder form). Potential for low cost machinery and low cost model materials. Potential for purpose designed computer-controlled microstructures.	Shells very fragile, especially in thin sections. Removal of powder from narrow cavities can be difficult. Ceramic powder needs to be stabilized with mist of water droplets.

Table 1.1.10 – *continued*

Table 1.1.11 Comparative quantitative characteristics of main rapid prototyping systems

System	*Supplier*	*Initial cost*	*No. of systems sold*	*Maintenance costs per year*	*Materials*	*Raw material costs*	*Capacity: Maximum model size in cm*	*Layer thickness (μm)*	*Accuracy (μm)*
Stereolithography	3-D systems	SLA 250 $250,000 SLA 500 $450,000	400	High, approximately 15% of purchase price	Acrylate and epoxy photosensitive resins	High cost to fill vat. Model costs low	SLA 250 25x25x25 SLA 500 50x50x50	62.5 to 250	±100 typical with epoxy ±150 typical with acrylate (±50 best)
Selective laser sintering	DTM	$300,000	25	High, approximately 20% of purchase price	Powdered wax, nylon, polycarbonate, metals (soon)	Relatively high	Sinterstation 2000 30 diameter x 38	125	±125 to ±250
Laminated object manufacture (LOM)	Helisys	LOM-1015 $95,000 LOM-2030 $180,000	38 (all types)	Relatively high, approximately 10% of purchase price	Kraft paper	Low	LOM-1015 25x36x33 LOM-2030 56x81x50	140 min	±250
Fused deposition modelling	Stratasys	FDM 1500 $78,000 3-D Modeller $180,000	30 (all types)	Low, around 5% of purchase price	Investment casting and machinable. Wax, polyolefins and polyamides	Low	FDM 1500 30x30x30 3D Modeller 32x30x22	125	±200
Solid ground curing (solider)	Cubital	$550,000	20	High, around 10–15% of purchase price	Special photosensitive resin	High	Solider 5600 35x50x50	280	±100 best
3-D printing systems	Soligen	$250,000	30 sites in operation	Not known	Ceramic powder	Relatively low	Soligen DSP system 40x40x40	175	Not available

1.1.13 Materials component and product testing

1.1.13.1 Materials testing

Materials testing is, for the most part, conducted on simple test bars machined from metals and ceramics, or moulded from plastics. Most materials tests, for example ultimate tensile strength (the strength at which a round bar separates under uniaxial tensile stress), or impact strength (measured in terms of the energy absorbed when a notched or unnotched sample separates under impact loading) are useful only as a quality control measure. If these tests yield the same result (within agreed error limits laid down by National Standards) for different production batches of materials, then it is reasonable to assume that the composition and the manufacturing process, inasmuch as these variables affect the ultimate tensile strength or impact strength, have remained unaltered. Much lower results (or even much higher results) would lead to an investigation of the process controls and a possible rejection of that batch of material.

The significance of the test results for the design of the components to be made from the material batch from which the test bar has been obtained is, however, much less apparent. Very few components are subject to simple uniaxial stress. Most components fail as a result of repeated stressing which causes a defect to grow until a critical crack is formed and this then propagates rapidly causing the component to disintegrate. Fracture toughness testing, rather than impact loading, measures the resistance of a material to crack propagation. The significance of materials tests for design is discussed in Section 1.8.

- Most of these tests are useful for incoming materials quality control.
- Some of the tests are useful for establishing a league table of materials performance, provided the mechanism of failure is not significantly different from that which occurs in service.
- Most of these tests are not suitable for predicting the long-term behaviour of materials in the in-service product environment.

Hence simulated service testing of materials components, sub-assemblies of components and of complete products is necessary before the selection of materials and manufacturing route is finalised and new products are launched on the market.

1.1.13.2 Component testing

Component tests under conditions which simulate in-service conditions offer the following advantages over tests of simple test bar shaped materials.

- The properties of the materials in the component are representative of the as-manufactured component. (This is, of course, true only if the component has been made by the intended process route, e.g. components machined from the solid may perform better or worse than moulded cast or forged components.)
- Built-in stresses or defects which are a consequence of the chosen manufacturing process should become manifest during the testing.
- Variations in dimensional tolerances and surface finish and their effect on performance should become evident.

1.1.13.3 Product testing

New product testing under service conditions is necessary because the testing of individual components is unlikely to simulate accurately the in-service conditions likely to be experienced by the complete product. This is particularly true of vibration, wear and corrosion conditions. In addition, the effects on performance of the build-up of tolerance variation of adjacent components can be very difficult to predict. Product testing, while necessary, is however expensive and should be undertaken only after extensive component trials.

If the costs are to be kept to a minimum and the information gained from product testing is to be maximized, a carefully planned testing programme is essential. The following guidelines are offered to ensure that product testing delivers the required information for the minimum cost.

- The vulnerabilities of the product design, identified in the **design reviews**, should be the focus of product testing.
- If products are to be tested in-house under extreme conditions, to manifest a potential or expect problems in a reasonably short timescale, it is important not to create conditions of stress, wear or corrosion in the materials and components which are of a different nature to those likely to be experienced by the product. For example, in order to simulate the effects of wear, increasing the frequency of relative movements of two surfaces under load may be acceptable but significant increases in the loads or speed of movement may induce a different wear mechanism.
- For tests conducted on customers' sites, it is necessary to monitor the conditions under which the product is tested, preferably with automatic data loggers. Simply providing customers with a new product and inviting them to comment on the performance is not likely to provide the best feedback to product development.
- For all tests of complete products or systems, it is important, ahead of the tests, to be aware of the actual information on the vulnerabilities of the design which are being tested. The running of tests 'to see what happens' is likely to be uninformative and expensive.

1.1.14 Costs

The in-position cost of a component will be the sum of several interdependent factors, each of which will have many cost sources, as shown in:
Fig. 1.1.27 *Materials costs and cost sources*

The material of a component is the common feature in design, manufacture, and performance, and the monetary losses resulting from processing difficulties with a poorly specified inexpensive as-bought material may well exceed the initial cost savings. Many items, such as yield, machine times and costs, quality control costs, etc, that contribute to the total cost will be very difficult to estimate for a component that is still at the layout or detailed design stage. Once the material and processing route have been decided, it will be easier to estimate, since either the component in question will be made in-house by well established manufacturing routes and a fund of experience on costing will have been developed, or, if the component is to be bought out, then many of the costs that are difficult to estimate will be already included in the as-bought figure.

As a first sorting mechanism, relative cost per unit property, where the relative cost of raw materials is compared, can be used. The Materials selection charts given in Section 1.1.6 can be replotted taking the cost of mild steel reinforcing rods to be unity.

The following Materials selection charts plot respectively:
Fig. 1.1.28 *Young's modulus against relative cost,* and
Fig. 1.1.29 *Strength against relative cost*

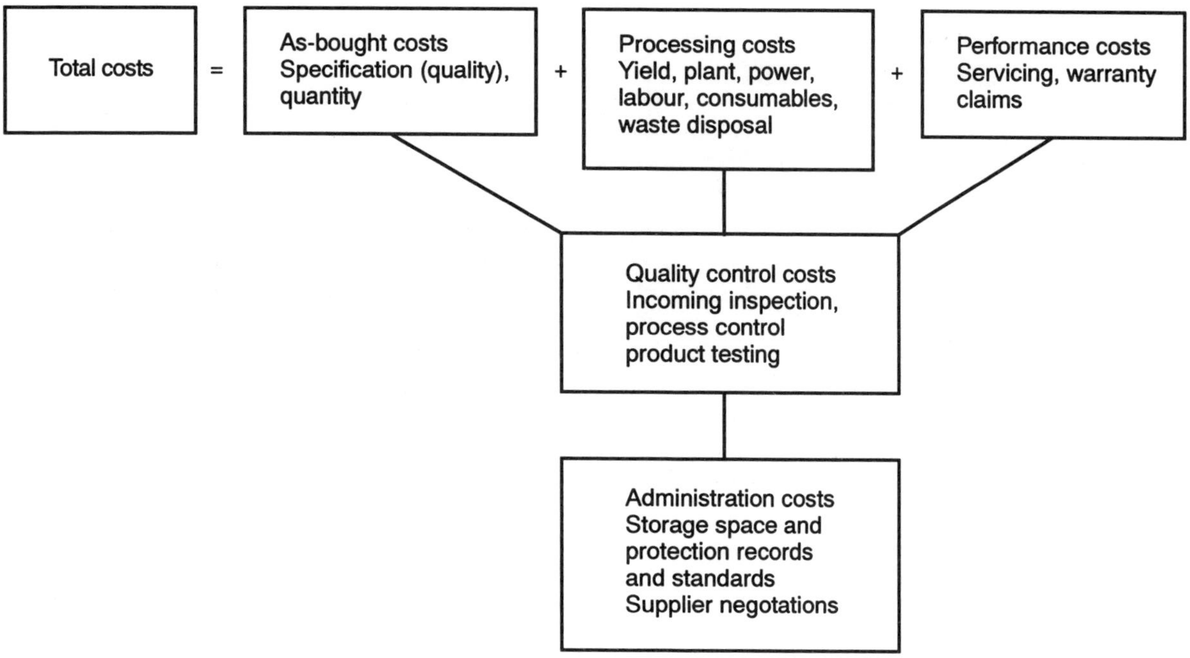

Fig. 1.1.27 Materials costs and cost sources

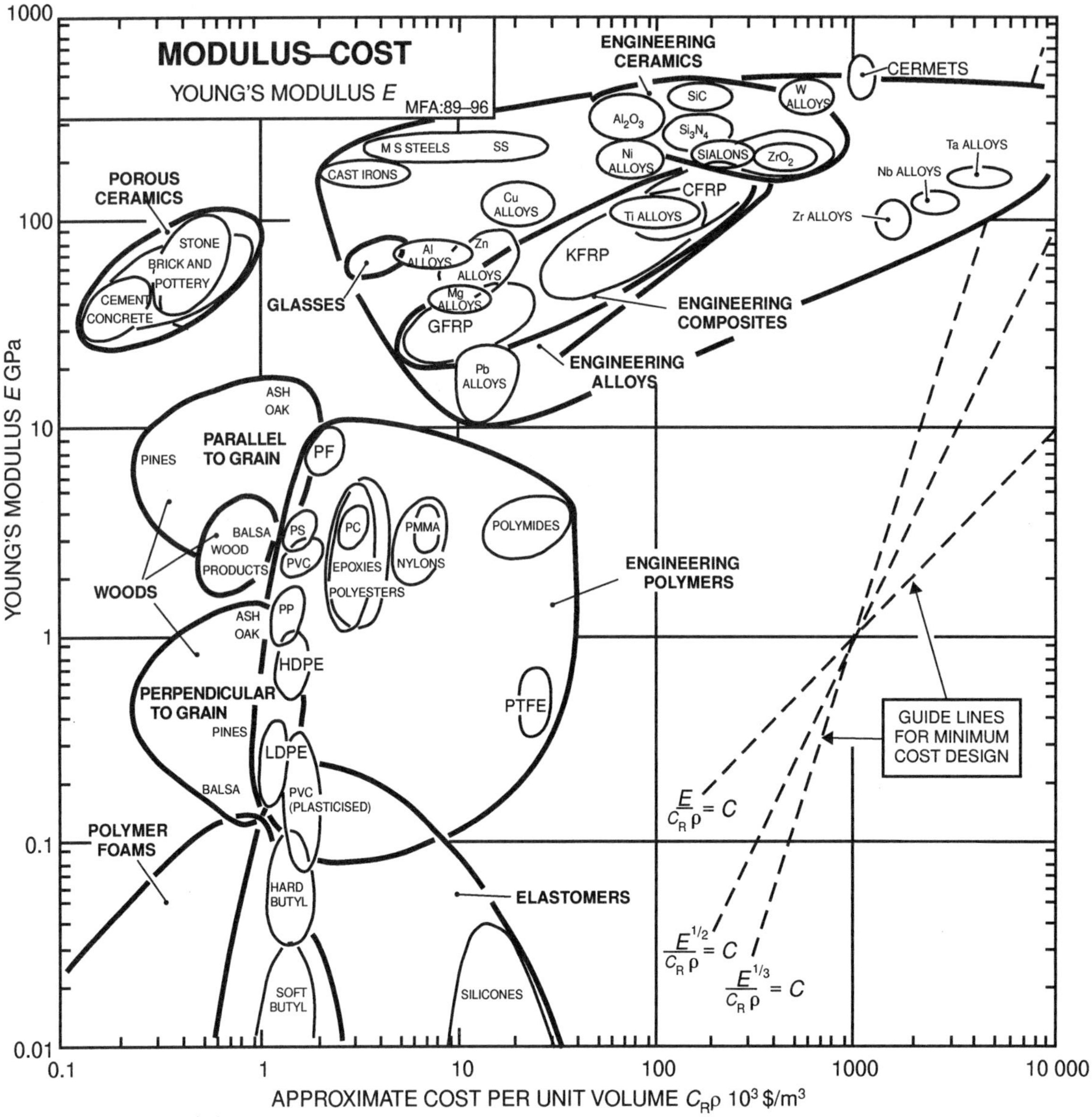

Fig.1.1.28 Materials selection chart: Young's modulus, *E*, against approximate cost, $C_R\rho$

The chart guides in selecting materials for cheap, stiff components (production, finishing, assembling and other costs must be considered separately).

(a) $E^{1/2}/C_R\rho = C$ (criterion for axial tension of ties)
(b) $E^{1/2}/C_R\rho = C$ (criterion for bending, torsion or buckling of beams, shafts and columns)
(c) $E^{1/2}/C_R\rho = C$ (criterion for bending of plates)

The value of the constant *C* increases as the lines are displaced upwards. Materials offering the greatest stiffness per unit cost lie towards the upper left corner.

Fig. 1.1.28

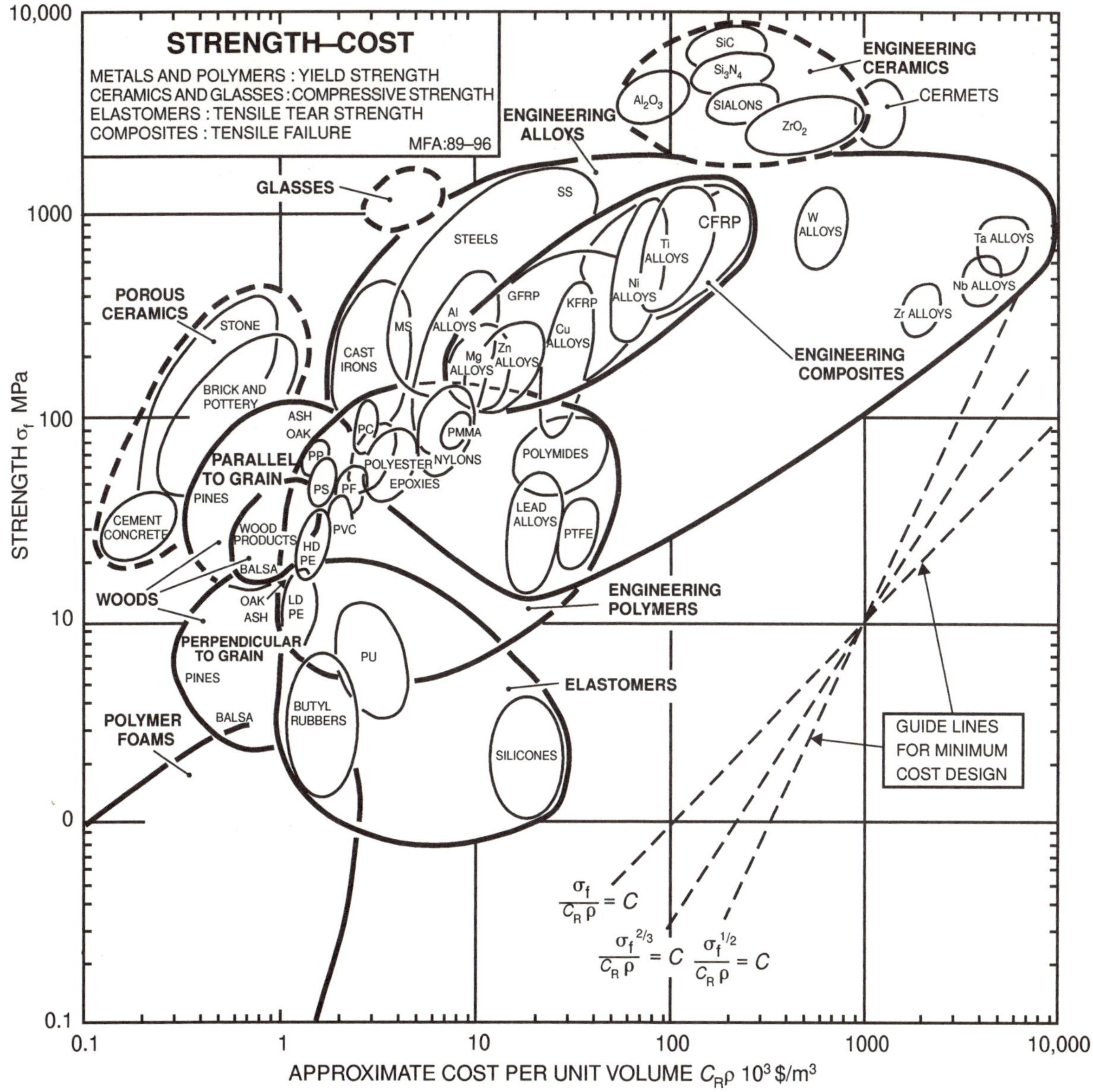

Fig.1.1.29 Materials selection chart: strength, σ_f against approximate cost, $C_R\rho$.
The 'strength' for metals is the 0.2% offset yield strength. For polymers, it is the 1% offset yield strength. For ceramics, it is the compressive crushing strength; remember that this is roughly 15 times larger than the tensile (fracture) strength.

The chart guides in selecting materials for cheap, strong components (production, finishing assemby and other costs must be considered separately). The lines show the loci of points for which:

(a) $\sigma_f / C_R\rho = C$ (criterion for axial tension of ties)
(b) $\sigma_f^{2/3} / C_R\rho = C$ (criterion for bending, torsion or buckling of beams, shafts and columns)
(c) $\sigma_f^{1/2} / C_R\rho = C$ (criterion for bending of plates)

The value of the constant C increases as the lines are displaced upwards. Materials offering the greatest strength per unit cost lie towards the upper left corner.

1–2

Mechanical properties

Contents

List of tables

List of figures

1.2.1 Mechanical properties at temperatures below the creep range

1.2.1.1 Introduction

Specified mechanical properties are not a unique function of a material but of a test piece taken from it and stressed in a specific way. Tensile strength is easily measured and is the most commonly employed criterion, particularly of metallic materials, and being apparently easily comprehended is the basis of many design codes and procedures.

A tensile test-piece used for a ductile material is essentially a right cylinder long enough for the ends which are chamfered to an increased cross-section not to affect the mode of deformation or fracture of the cylindrical gauge length. Normally the gauge length is 5.25 times the square root of the cross sectional area, and this ratio must be specified because variation influences elongation (or reduction in area).

Using such a test for a brittle material such as a grey cast iron or a ceramic would introduce excessive scatter and produce values for ultimate tensile stress which are unacceptably low. These materials are therefore tested by bending a beam of rectangular cross-section and calculating the stress at the surface at failure.

The tests described are used to obtain characteristic values for materials of modulus (compressive and tensile), Poisson's ratio, limit of proportionality (defined as yield or proof stress) and ductility defined as elongation over a standard gauge length or reduction in area at fracture. Shear properties may also be measured.

1.2.1.2 Modulus

The modulus of elasticity, E, of a material is measured by calculating the deflection per unit of stress in the tensile, compression or bend test over the range where strain is effectively proportional to stress (or by calculation from measurements of the specific gravity and velocity of sound in a material). It is important because it controls failure modes such as deflection under load in complex sections and therefore fatigue and resistance to buckling. It also controls the natural frequency of vibration of a component which may cause it to resonate with an applied alternating load and therefore fail (alternatively this may be put to practical use).

Tensile modulus has some correlation with melting point and therefore varies very considerably between materials. Typical values for ceramics, metals and plastics are shown in:

Fig. 1.2.1 *Elastic moduli of materials*

Poisson's ratio ν, which is also important in design is the ratio radial strain/longitudinal strain at a given stress. Its value is 0.3 for most metals in the elastic range.

The shear, or rigidity modulus, G, is significant, particularly for materials stressed in torsion. It may be calculated from the elastic modulus and Poisson's ratio from the formula

$$G = \frac{E}{2(1 + \nu)} \tag{1}$$

1.2.1.3 Tensile strength

LIMIT OF PROPORTIONALITY, YIELD OR PROOF STRESS

As tensile stress is increased the strain/stress relationship of a ductile material departs from a straight line, and this departure is, in most cases, parallelled by a permanent set remaining after the stress is removed.

In the case of carbon steel, and a few similar materials, the departure from linearity and the permanent set takes place at a sharply defined value of stress, the 'yield point'. For most ductile materials the departure from linearity is progressive, and a value, the 'proof stress', is defined at which the permanent set reaches a defined value usually 0.2%.

Yield or proof stress is the basis for most engineering design with ductile materials and, below the creep range, most design codes specify a maximum stress, a defined fraction of the proof stress, usually 62.5% but ranging according to the standardizing authority. This design stress may be exceeded only at those stress concentrations where it can be shown that overstressing will result in reduction of stress concentration, work-hardening of material and/or the generation of an opposing compressive stress.

The tensile yield strengths of a number of engineering metals are illustrated diagrammatically in:

Fig. 1.2.2 *Yield strength of materials*

This illustrates a major advantage of steel as a material of construction.

Yield or proof stress in compression can safely be taken to be at least equivalent to that in tension. For geometrical reasons, test results will usually be found to be higher, as will the behaviour in pure compression of materials in components.

It must, however, be emphasized that the major design problem in compressive loading is the avoidance of buckling.

For many applications (aerospace, transport and where inertia stresses are critical) specific strengths (yield strength/specific gravity ratio) may be of more importance than absolute yields, and are shown diagrammatically in:

Fig. 1.2.3 *Yield strength:specific gravity ratio at 23°C*

The maximum allowable shear stress prior to yield is half the tensile yield stress. Behaviour in torsion is complex because overstressing of the outer fibres of a component generates a system stress which opposes the applied stress.

TENSILE STRENGTH (UTS)

When a ductile material is stressed beyond its yield point at a temperature low compared with its melting point it work-hardens and the stress that it will withstand increases. Simultaneously, however, the cross-sectional area of the component or test-piece decreases, uniformly along the section or at a 'neck'. These two opposing effects cause an initial increase in applied load which reaches a maximum after which the load may decrease. The specimen then fails. The maximum stress is described as the 'ultimate tensile stress' (UTS), which is shown for a variety of materials in:

Fig. 1.2.4 *Tensile strength of materials*

Certain older design codes allow UTS to be used, rather than yield stress, using of course a higher design factor, ranging between 2.4 and 4. This can be advantageous with certain very ductile materials such as, for example, mild steel.

A high ratio of UTS to yield strength is advantageous because high cycle fatigue is related to UTS rather than yield. It may also be considered to provide some degree of security against failure by accidental overload.

As with yield, for some applications the UTS:specific gravity ratio may be more important than the absolute value of UTS. Typical values for metals are shown in:

Fig. 1.2.5 *Ultimate tensile strength:specific gravity ratio at 23°C*

COMPRESSIVE AND TORSIONAL STRENGTHS

A ductile specimen whose geometry is adequate to withstand buckling will, if stressed sufficiently in compression, fail eventually by bursting at the shear planes. A similar specimen stressed in torsion will fail by twisting when the yield stress in shear is exceeded throughout the section. Both these failure modes require significantly higher stresses than those required to produce yield and should not occur with reasonable design. In both cases the most dangerous criterion is buckling.

1.2.1.4 Ductility

The ductility of a material is measured by its elongation (relative to a standard gauge length) or to its reduction in area at fracture. The two parameters would be identical if elongation were uniform along the gauge length, the ratio of reduction in area to elongation being a measure of the degree of necking.

Elongation is a measure of the ability of a material to accommodate stress concentration, and it is usually related directly to resistance to high strain fatigue. The values of elongation for a number of metals are shown in:

Fig. 1.2.6 *Elongation values of materials*

Normally, unless a component is to be subjected to high strain fatigue or exposed to subzero temperatures or multiaxial torsion, an elongation of 10% should be adequate for most design purposes. However, an untypically low value is usually an indication that the material is in some way substandard and justifies rejection or investigation. Ductility is an important criterion in assessing forming capability.

EFFECT OF ELEVATED TEMPERATURES

The design principles enunciated hold for temperatures between ambient and the bottom end of the creep range. Design must, however, allow for the decrease in modulus and yield and ultimate stress with temperature. Ductility usually increases with temperature but some materials exhibit ductility troughs over certain ranges of temperature, and it may be essential to avoid these.

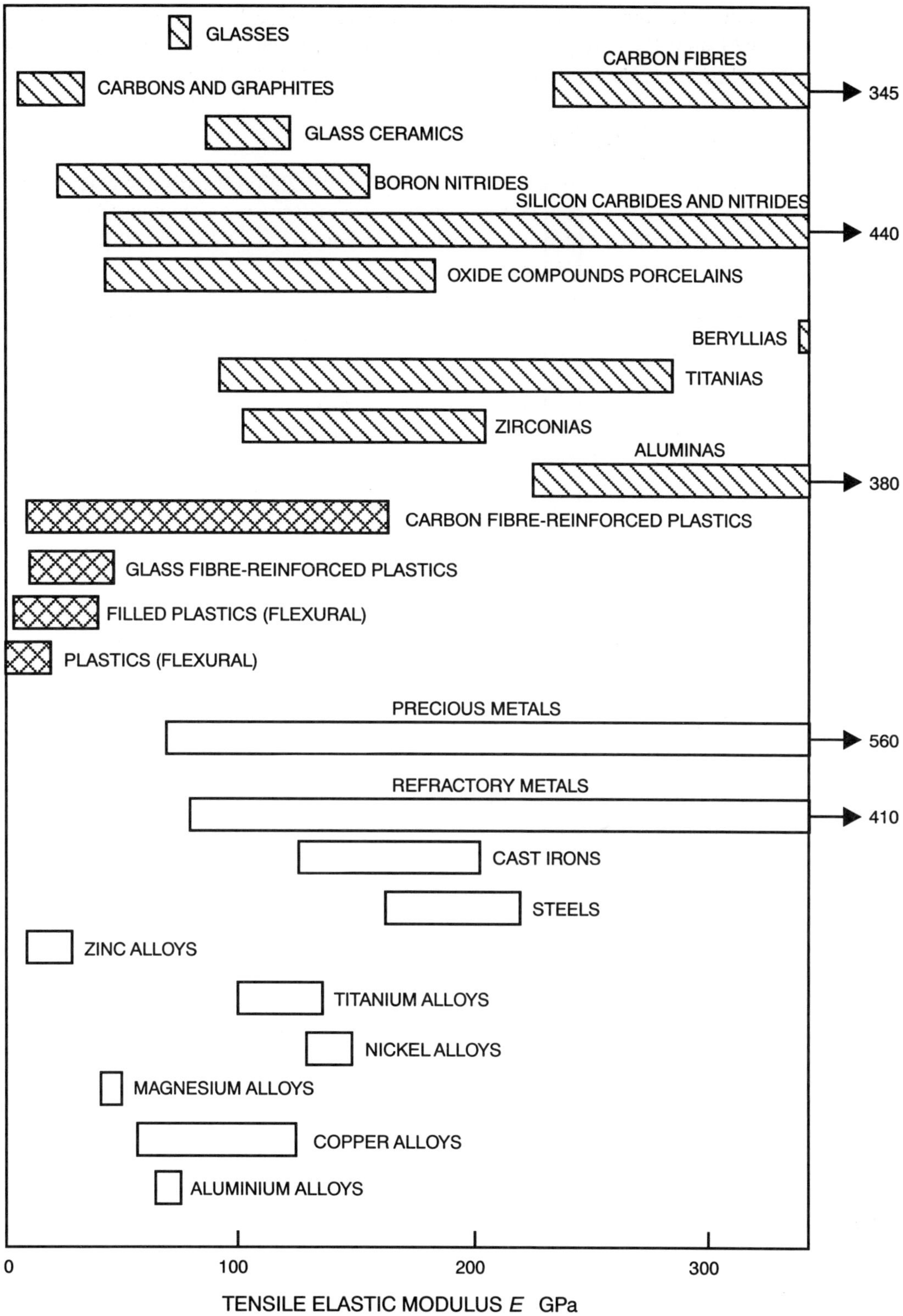

Fig. 1.2.1 Elastic moduli of materials

Fig. 1.2.1

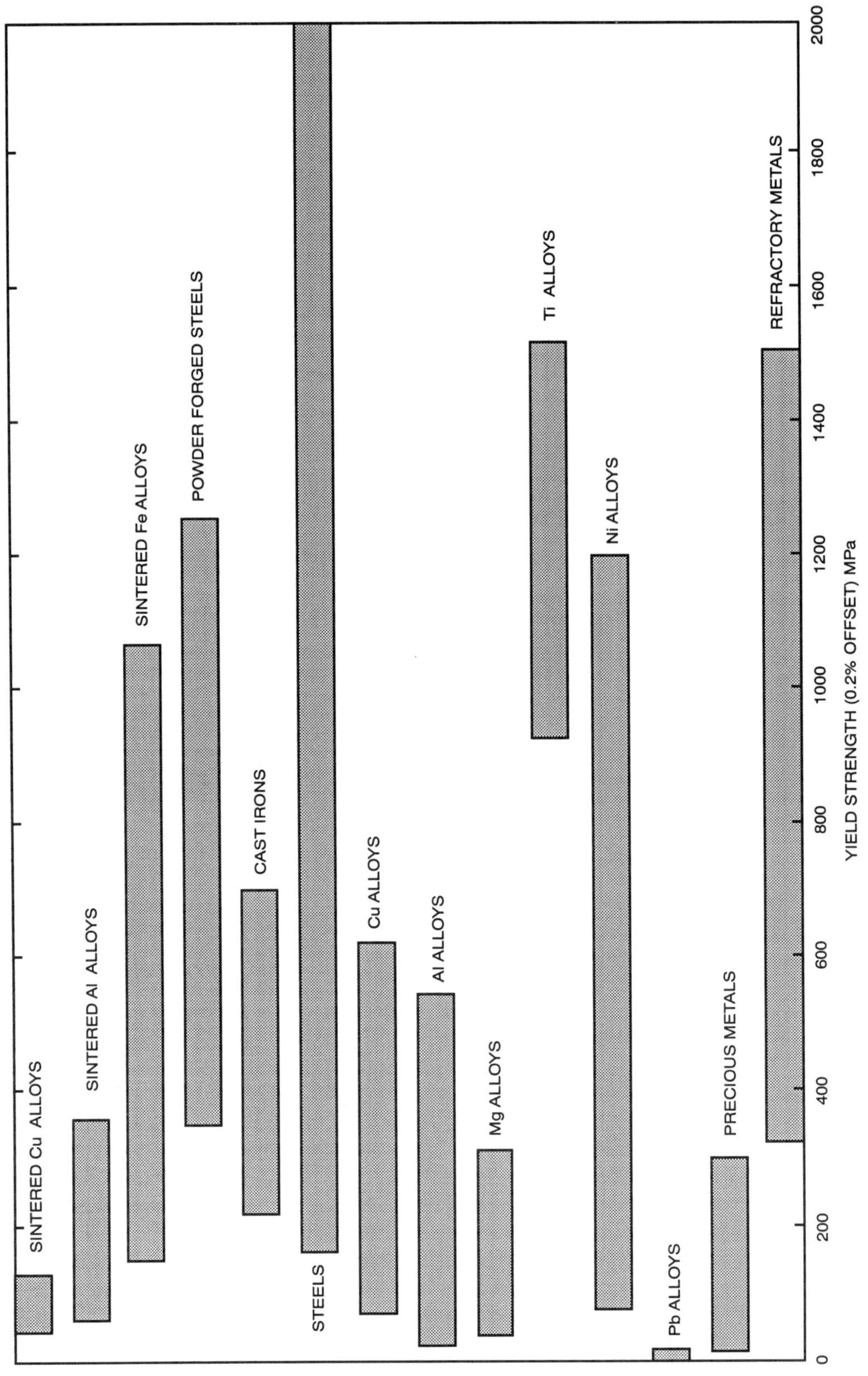

Fig. 1.2.2 Yield strength of materials

Fig. 1.2.2

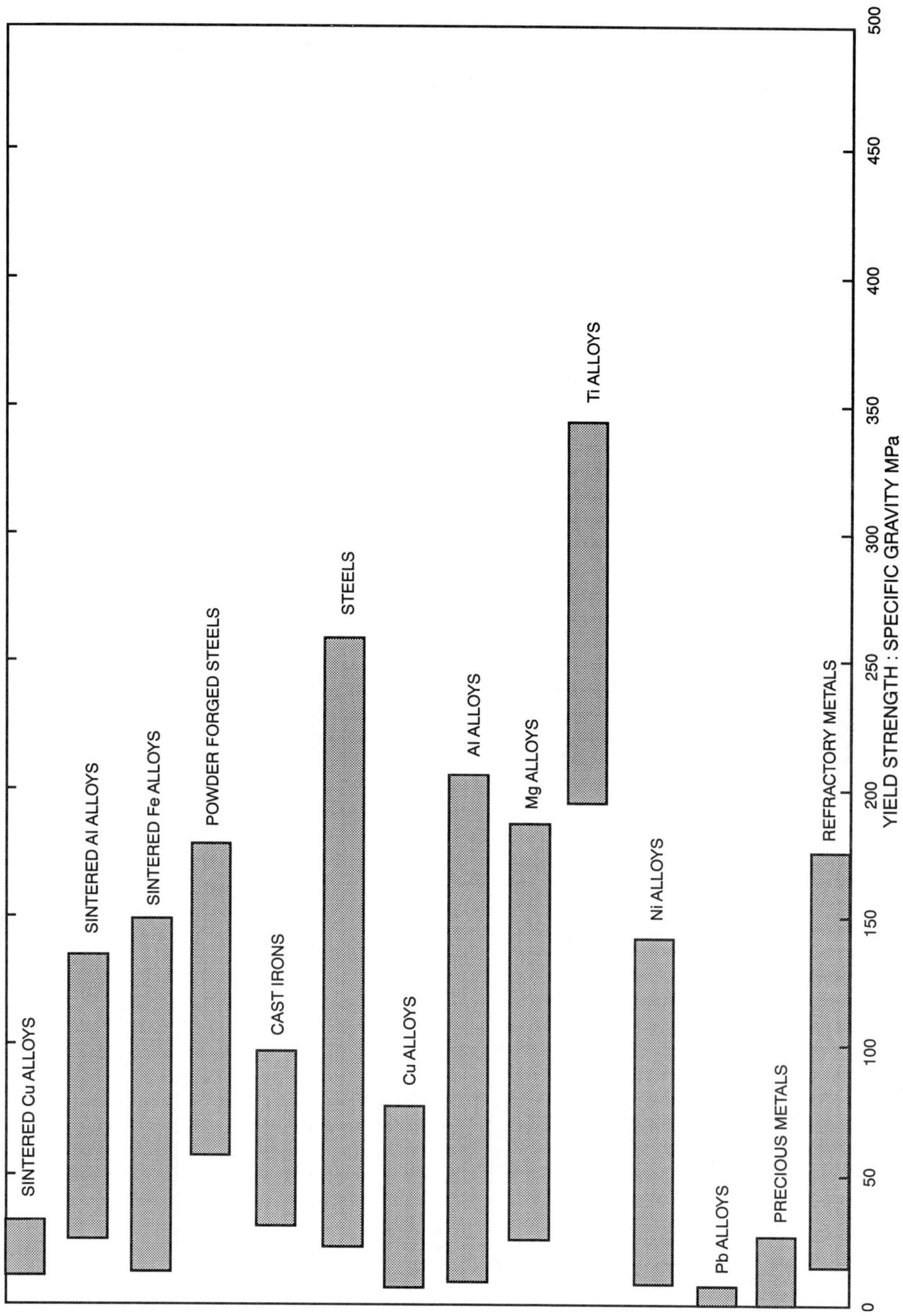

Fig. 1.2.3 Yield strength:specific gravity ratio at 23°C

Fig. 1.2.3

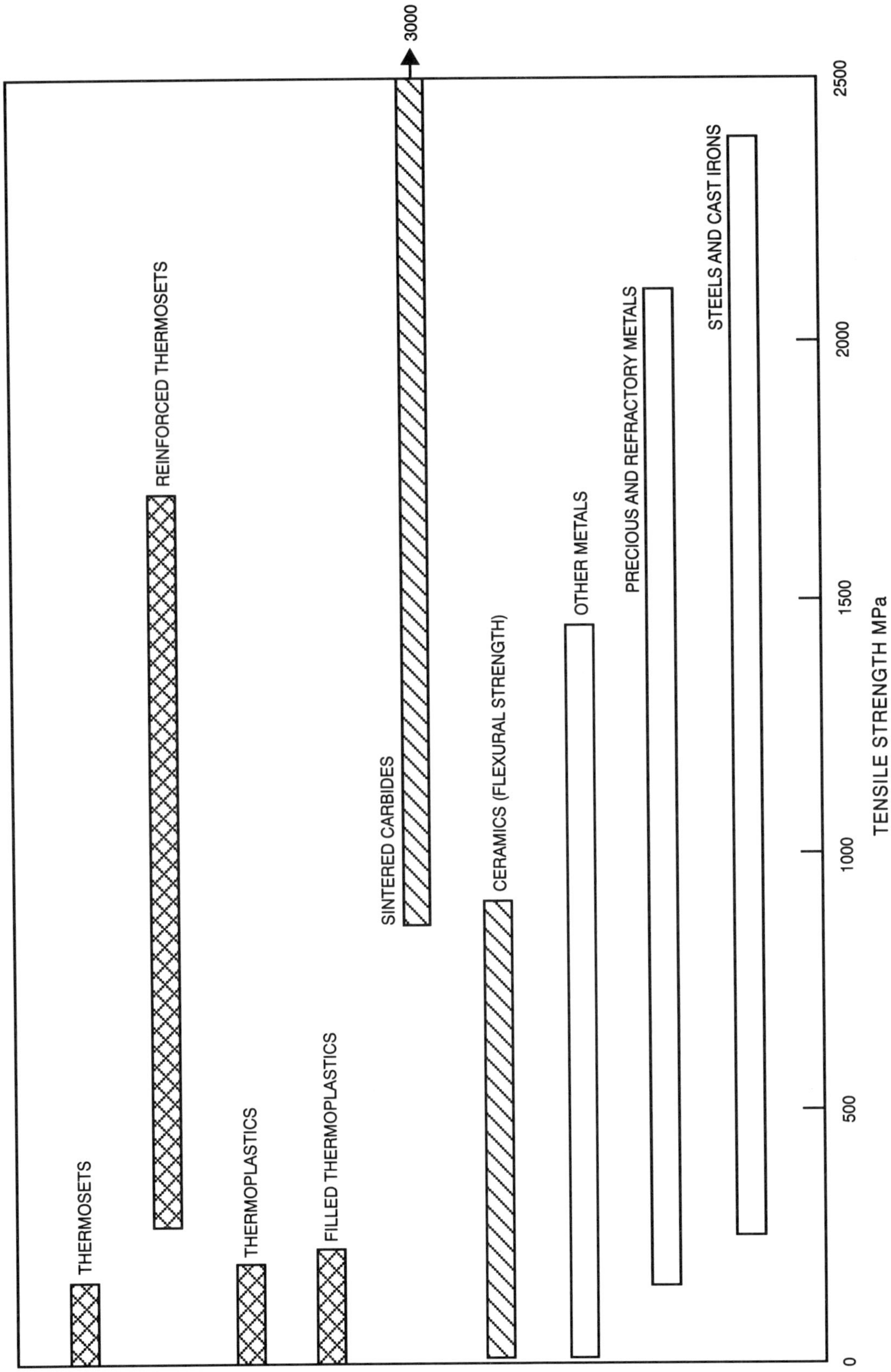

Fig. 1.2.4 Tensile strength of materials

Fig. 1.2.4

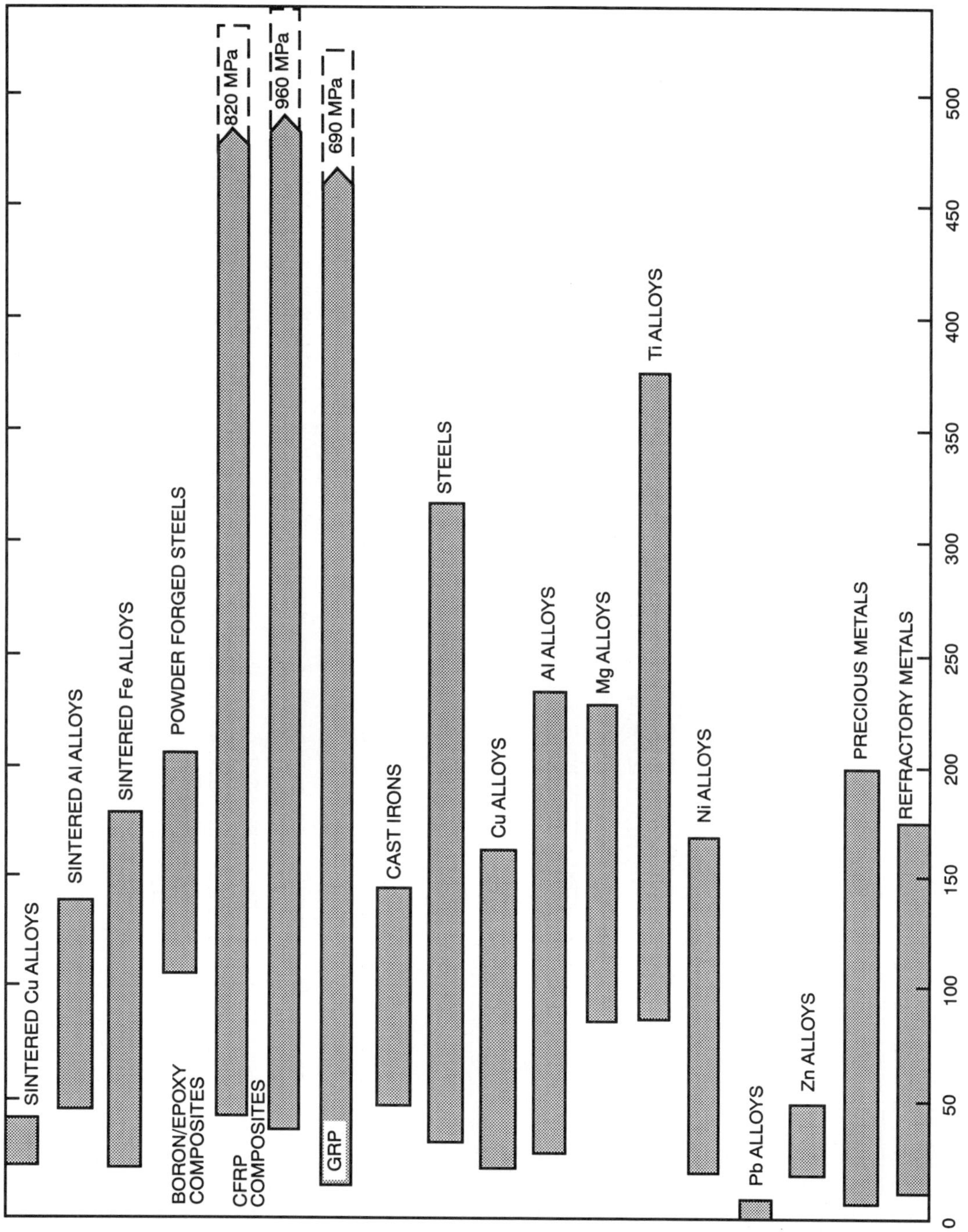

Fig. 1.2.5 Ultimate tensile strength:specific gravity ratio at 23°C

Fig. 1.2.5

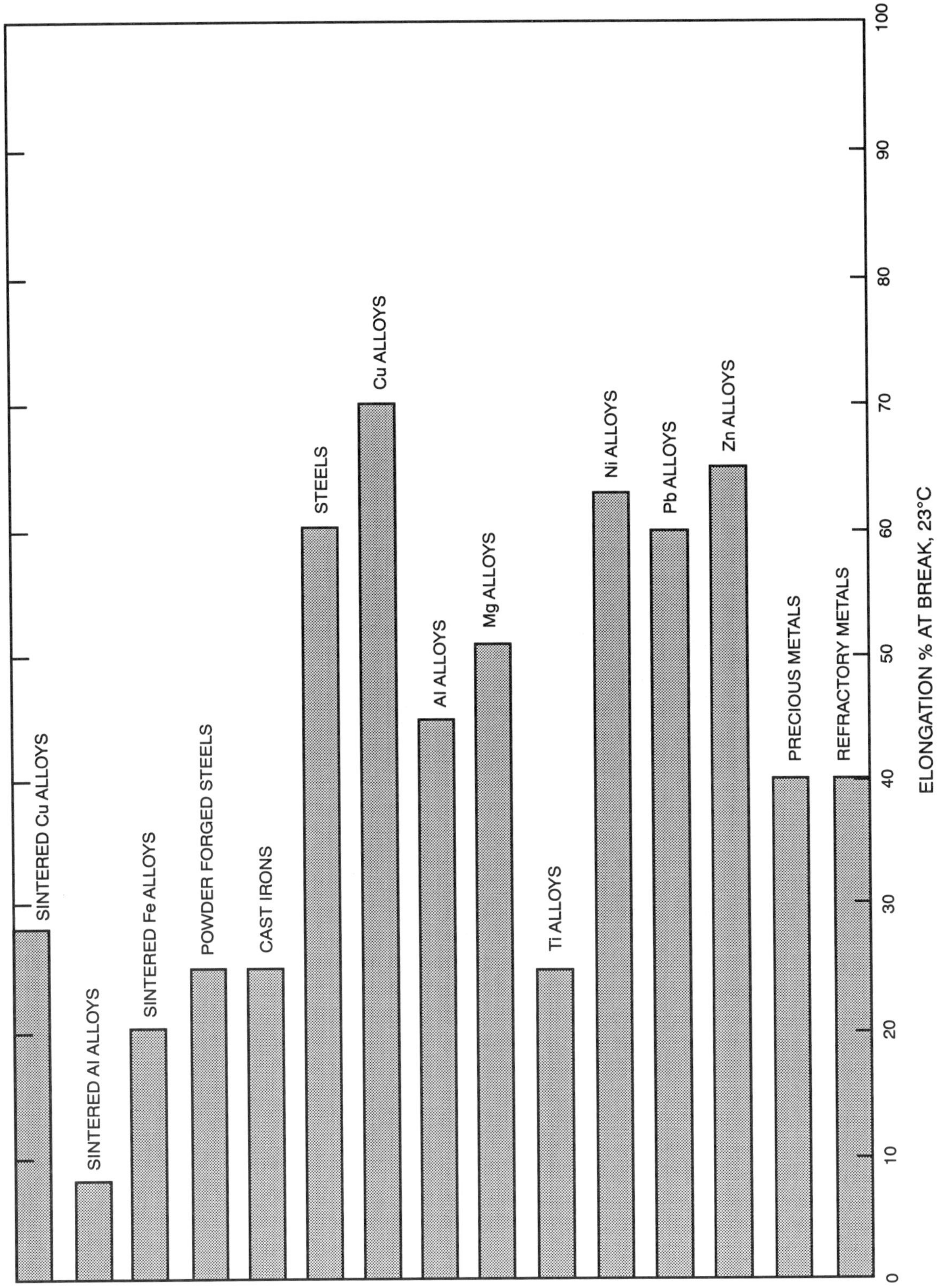

Fig. 1.2.6 Elongation values of materials (a) metals

Fig. 1.2.6

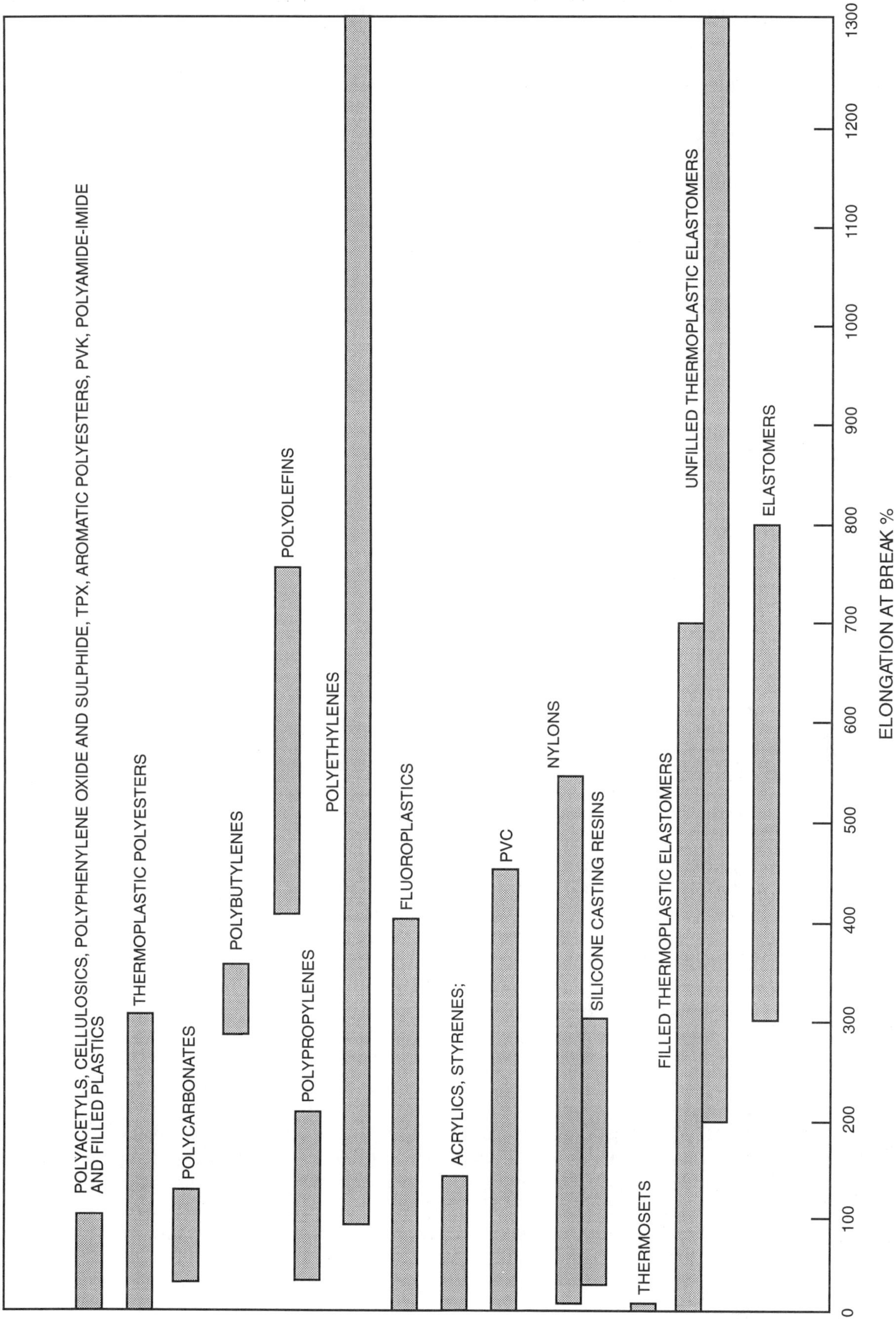

Fig. 1.2.6 Elongation values of materials – *continued* (b) plastics

Fig. 1.2.6 – *continued*

1.2.2 Fracture toughness and fatigue

This section in the *Materials Selector* provides information on the use of theoretical fracture mechanics in the consideration of fracture and fatigue. Fracture mechanics is a topic which has been the subject of continuous development and refinement for many years and so it is recommended that, once the reader has gained an understanding of the theory involved, currently applicable specifications and codes of practice should be consulted.

1.2.2.1 Fracture toughness

The resistance of a material to the rapid propagation of a crack can be defined in terms of a critical stress intensity factor which is a material property known as the **fracture toughness** of the material. In general, the stress intensity factor, K, is related the dimensions of the crack and the gross stress by an expression of the form.

$$K = \sigma\sqrt{Ma} \qquad (1)$$

where K is the stress intensity factor, σ is the stress, M is a flaw-shape and geometry parameter and a is the depth of the crack. The condition for failure, i.e. rapid propagation of the crack is that K reaches a critical value K_{1c}, the plane strain fracture toughness. (The subscript 1 refers to the mode of opening of a crack under tensile loading normal to the plane of the crack.)

1.2.2.2 Fracture safe design

From eqn (1), with a knowledge of K_{1c}, and σ, the critical crack size at failure can be calculated from

$$a_C = \frac{1}{M}\left[\frac{K_{1c}}{\sigma_F}\right]^2 \qquad (2)$$

where σ_F is the failure stress. Given the value of K_{1c}, which is a material property, a designer can then select pairs of values of flaw size and working stress. Figure 1.2.7 shows a schematic plot of equation (2):

Fig. 1.2.7 *Applied stress–crack depth relationship*

Note that by reducing the working stress from the yield stress σ_y (0.2% proof stress) to half of that value the critical crack length for failure becomes 4 times as great. Suppose the working stress were fixed at 0.5 σ_y then the critical crack size would be a_c. Flaw sizes less than a_c would be safe at this working stress. If a suitable non-destructive test method is available for detecting flaws of the size a_c and greater, fracture safe components can be produced.

An alternative form of pre-service inspection is the proof test in which the component, pressure vessels are common examples, is pre-loaded to a higher stress than the normal working stress. In this case, if the component survives the proof test it is apparent that no flaw greater than a_i can be present and there is a margin of safety on the crack size of a_c–a_i. Unfortunately this margin of safety may be reduced by subcritical crack growth, particularly under fatigue loading or aggressive environmental conditions; there is also a possibility of the fracture toughness changing if service conditions are such that they lead to metallurgical changes in the material.

In designing a fracture safe structure it is sometimes possible to make use of the 'leak before break' concept, i.e. the combination of toughness and working stress is arranged so that the critical flaw depth is greater than the thickness of the material so that a through thickness crack will not propagate and the structure would leak before breaking.

1.2.2.3 Flaw size and shape parameter

The relationship $K = \sigma\sqrt{Ma}$ includes M the flaw shape and geometry parameter, which is given by:

$$M = \frac{1.21\pi}{Q} \quad \text{for a surface flaw} \tag{3}$$

and

$$M = \pi/Q \quad \text{for an embedded flaw} \tag{4}$$

Q is the flaw shape parameter given by:

$$Q = (\phi^2 - 0.212\,(\sigma/\sigma_y)^2) \tag{5}$$

where ϕ equals the complete elliptical integral of the second kind, σ is the working stress and σ_y the 0.2% proof stress. Figure 1.2.8 is a graphical representation of eqn (5) for various elliptical shaped flaws having semi-minor axes a, crack depth, and semi-major axes c, crack length:

Fig. 1.2.8 *Flaw shape parameter for surface and internal flaws*

1.2.2.4 Subcritical flaw growth

The possibility of subcritical flaw growth due to fatigue loading or aggressive environmental conditions has already been mentioned.

STRESS CORROSION CRACKING AND K_{1scc}

It is found that if a crack loaded to a stress intensity less than the critical value in air, K_{1c}, is exposed in a liquid environment, such as seawater, the crack grows and failure ensues when the stress intensity factor reaches K_{1c} by virtue of the increased crack size. The time to failure depends on the initial stress intensity factor K_i as shown schematically in:

Fig. 1.2.9 *Effect of stress intensity factor on time to failure in an aggressive environment*

As the initial stress intensity K_i decreases the time to failure increases and becomes infinite at an asymptotic lower limiting value of K_i, known as K_{1scc}. For practical purposes K_{1scc} is a limiting value of K_i below which stress corrosion cracking does not occur, i.e. K_{1scc} may be taken as the wet fracture toughness and is a material property for a particular environment.

Effect of stress corrosion cracking on critical flaw size

Figure 1.2.10 represents the relationship between K_{1scc}, K_{1c}, applied stress and flaw size for a marageing steel (18Ni–5Mo–7Co) with σ_y (0.2% P.S.) of 1920 MPa, K_{1c} of 100 MN/m$^{3/2}$ and K_{1scc} of 40 MN/m$^{3/2}$:

Fig. 1.2.10 *Relationship between K_{1c}, K_{1scc}, stress and flaw size*

The two curves correspond to $K = \sigma\sqrt{2a}$ for K_{1c} and K_{1scc} and separate the stress and flaw size pairs of values into safe and unsafe regions: values to the right of the K_{1c} curve will fail rapidly and values to the left of this curve will be safe in air. However, in the presence of an aggressive environment values between the K_{1c} and K_{1scc} curves will be susceptible to stress corrosion cracking and will eventually fail when the crack grows to a critical size, whereas values to the left of the K_{1scc} curve will be safe. Considering a stress of $\sigma_y/4$, flaws smaller than a_i would be safe under all conditions, flaws greater than a_i, whilst safe in air, would grow to the size of a_c in an aggressive environment and flaws greater than a_c would be unsafe in all conditions. Note that the ratio

$$\frac{a_i}{a_c} = \left[\frac{K_{1scc}}{K_{1c}}\right]^2$$

and in the example cited here the ratio is 0.16. The critical flaw size is 16% of the flaw size in air when the wet toughness, K_{1scc}, is 40% of the dry toughness, K_{1c}.

Susceptibility of alloy types to stress corrosion cracking

The effect of stress corrosion cracking on the critical defect size for the same applied stress level is to reduce the defect size by the factor

$$\left[\frac{K_{1scc}}{K_{1c}}\right]^2$$

If the defect size is kept at the same value then the working stress must be reduced by the factor K_{1scc}/K_{1c}. A convenient way of quantifying the effect of stress corrosion cracking is thus by plotting K_{1scc} against K_{1c} to obtain ranges of values for the ratio K_{1scc}/K_{1c}. Figure 1.2.11 is a plot of K_{1scc} against K_{1c} for aluminium base alloys. Most of the data refer to values of K_{1scc} obtained in a 3% NaCl environment, and in this environment with the exception of some Al–Cu–Mg–Zn alloys (7079 and 7175) the K_{1scc} value is at least 66.6% of the corresponding K_{1c} value:

Fig. 1.2.11 *Susceptibility of aluminium alloys to stress corrosion cracking in 3% NaCl*

Figure 1.2.12 shows the K_{1scc} values of titanium alloys exposed to a 3% NaCl environment; here the K_{1scc} values do not generally exceed 66.6% of the corresponding K_{1c} value and the ratio K_{1scc}/K_{1c} can be as low as 25%. The 6Al–4V titanium alloys are the most resistant, with an average ratio around 66.6%, although extreme values of 52 and 92% occur:

Fig. 1.2.12 *Susceptibility of titanium alloys to stress corrosion cracking in NaCl*

Figures 1.2.13, 1.2.14 and 1.2.15 are plots of K_{1scc} in a 3% NaCl environment against K_{1c} for ferrous alloys. The marageing steels shown in Fig. 1.2.13 have K_{1scc} values between 30 and 60% of the K_{1c} values at strength levels up to 1800 MPa at strength levels greater than 1800 MPa the K_{1scc} value is not proportional to the K_{1c} value and lies between 5 and 20 $MN/m^{3/2}$. Similar behaviour is shown by the K_{1scc} values for the low alloy steels shown in Fig. 1.2.14. The K_{1scc} values for the stainless steels shown in Fig. 1.2.15 vary from 25 to 100% of the corresponding K_{1c} values:

Fig. 1.2.13 *Susceptibility of marageing steel (18Ni–5Mo–7Co) to stress corrosion cracking in 3% NaCl*

Fig. 1.2.14 *Susceptibility of low alloy steels to stress corrosion cracking in 3% NaCl*

Fig. 1.2.15 *Susceptibility of martensitic and precipitation hardening stainless steels to stress corrosion cracking in 3% NaCl*

FATIGUE

The failure of structural components under conditions of cyclic loading by fatigue crack growth is well known to be the cause of many service failures. Most of the available data on the effect of fatigue loading are concerned with nominal stress to cause failure in a given number of loading cycles. These data are usually obtained from laboratory tests on smooth specimens, and, whilst they are of some use as a qualitative guide to materials selection, they do not take account of the effect of discontinuities in the material or stress concentrations on the overall resistance to fracture. The endurance limits obtained from conventional SN curves only indicate the maximum nominal stress to avoid the initiation of a crack in a given number of cycles, and consequently the life of a component or structure will be severely overestimated if no allowance is made for the presence of discontinuities or stress raisers.

The fatigue life of a structure can be divided into three periods:

1. initiation of a crack
2. propagation of the crack
3. fast fracture.

The presence of a pre-existing flaw which acts as a stress raiser will reduce or eliminate the initiation period and the longest part of the fatigue life is then associated with the propagation of the crack. As the crack grows under constant cyclic loading conditions, the stress intensity at the crack tip increases since both the crack length and the applied stress are increasing. Eventually the crack may grow to such an extent that the stress intensity equals the critical value, K_{1c}, when fast fracture ensues.

Crack growth rates

The dependence of the fatigue crack growth rate on stress intensity can be conveniently represented by the expression

$$\frac{\mathrm{d}a}{\mathrm{d}N} = C(\Delta K)^m \qquad (6)$$

where $\mathrm{d}a/\mathrm{d}N$ is the crack growth rate change in crack length for one loading cycle, C is a constant for a particular material and set of loading conditions, ΔK is the change in stress intensity during the loading cycle and m is an empirically determined constant. Figure 1.2.16 represents a log–log plot of $\mathrm{d}a/\mathrm{d}N$ vs ΔK for data obtained on several aluminium, titanium and ferrous alloys, compiled from *Damage Tolerant Design Handbook*. The data conform to a general slope of 4 and the expression can be modified to:

$$\frac{\mathrm{d}a}{\mathrm{d}N} = C(\Delta K)^4 \qquad (7)$$

Fig. 1.2.16 *The effect of stress intensity factor, ΔK , on the rate of fatigue crack propagation, da/dN, for Al, Ti and Fe base alloys*

It should be noted that individual determinations of m have yielded values from 2 to 6. The value of C is seen to increase from the steels to the titanium to the aluminium alloys, although the steel and aluminium fields overlap. The value of C can also be influenced by such factors as:

mean stress level
cycle frequency
environment
alloy composition
orientation of fatigue crack
temperature

As a basis for comparison of the three types of alloy the median lines have been taken for each type:

for steels: $$\frac{\mathrm{d}a}{\mathrm{d}N} = 10^{-12}\,\Delta K^4 \text{ m/cycle}$$

for titanium: $$\frac{\mathrm{d}a}{\mathrm{d}N} = 2\times 10^{-12}\,\Delta K^4 \text{ m/cycle}$$

for aluminium: $$\frac{\mathrm{d}a}{\mathrm{d}N} = 4\times 10^{-11}\,\Delta K^4 \text{ m/cycle}$$

If the three materials were to be used at the **same stress intensity**, the crack growth rate of a titanium alloy would be 0.05 times the growth rate for an aluminium alloy and the crack growth rate for steel would be 0.025 times that for the aluminium alloy or half that of the titanium alloy. However, for the same crack size and geometry, the stress intensity for the three materials would not be the same if the working stress on the three materials was a constant fraction of the yield stress of the material. As an example, we take three high strength alloys of each type and a lower strength steel with high toughness as shown in:

Table 1.2.1 *Strength and toughness data for selected Al, Ti and ferrous alloys*

At a constant fraction of the yield stress ($da/dN\ C\sigma_y{}^4$, eqn (10)) the crack growth rate for the titanium alloy (*T*) would be 1.03 times that of the aluminium alloy (*A*) and the growth rate for steel S2 would be 4.8 times that for the aluminium alloy, whereas the growth rate for steel S1 would be 0.56 times that for the aluminium alloy; a change in stress by a factor of 1.9 for the steels changes the growth rate by a factor of 8.5, i.e. as the fourth power of the ratio of the stresses.

If the same load is applied to equal masses of the three alloys, the growth rate ($\alpha C\rho^4$) (ρ = density) of the titanium alloy would be 0.38 that of aluminium and the growth rate of the steels would be 1.8 times that of the aluminium and 4.75 times that of titanium. Considering the relative masses of the three alloys to give the same load bearing capacity and growth rate ($\alpha C^{1/4}\rho$) the mass of Ti is 0.79 times that of Al and the mass of steel is approximately 1.14 times that of Al or 1.45 times the mass of Ti. These comparisons are shown graphically in Fig. 1.2.17.

Fig. 1.2.17 *Relative growth rates for different design criteria*

Fatigue crack growth rates at extreme values of ΔK

At high and low values of ΔK the fatigue crack growth rate diverges from the straight line behaviour of Fig. 1.2.16.

As K_{MAX} approaches K_{1c} the growth rate becomes fast and if $K_{MAX} = K_{1c}$ failure occurs in 1/4 of a load cycle. The Forman equation makes allowance for this increase in growth rate and also takes into account the mean stress level via the load ratio, $R = K_{MIN}/K_{MAX}$

$$\frac{dA}{dN} = \frac{C\,\Delta K^m}{(1-R\ )K_{1c} - \Delta K}$$

At the other end of the fatigue crack growth rate spectrum, the simple power law breaks down for low ΔK, where the fatigue crack growth rate rapidly decreases to a vanishingly small level. A limiting stress intensity factor range, the threshold level ΔK_{TH}, can be defined, below which fatigue damage is unlikely. ΔK_{TH} is somewhat analogous to K_{1scc}, but is also dependent on environment. The value of ΔK_{TH} is also found to be dependent on the stress ratio, *R*. A range of values for ΔK_{TH} is given for some ferrous and non- ferrous materials in Tables 1.2.2 and 1.2.3 respectively:

Table 1.2.2 *Threshold stress intensity factor for fatigue crack growth, ΔK_{TH}, of some ferrous materials*

Table 1.2.3 *Threshold stress intensity factor for fatigue crack growth, ΔK_{TH}, of some non-ferrous materials*

Life of a flawed component

When considering the life of a flawed component it is not solely a question of the growth rate but also the amount of crack growth that can be tolerated. An initial flaw of size a_i propagates by fatigue crack growth until it reaches a critical size a_c. This critical flaw size can be determined from the expression:

$$K_{1c} = \sigma\sqrt{Ma_c} \qquad (8)$$

if the fracture toughness K_{1c}, the applied stress σ and the flaw shape and component geometry parameter M are known. If the flaw size is small compared with the cross-section of the component the alternating applied stress $\Delta\sigma$ will be independent of the flaw size and the crack growth rate will be given by

$$\frac{\mathrm{d}a}{\mathrm{d}N} = C(\Delta K)^4 \tag{9}$$

substituting for K, $\dfrac{\mathrm{d}a}{\mathrm{d}N} = C(\Delta\sigma)^4 M^2 a^2$ (10)

and integrating between the limits $a_i - a_c$ the number of cycles for the crack to grow to a critical size is given by:

$$N = \frac{1}{CM^2(\Delta\sigma)^4}\left[\frac{1}{a_i} - \frac{1}{a_c}\right] \tag{11}$$

This expression holds when the exponent n in the growth rate law is 4. In the general case for other values of n the life is given by

$$N = \frac{2}{(n-2)CM^{n/2}(\Delta\sigma)^n}\left[\frac{1}{a_i^{(n-2)/2}} - \frac{1}{a_c^{(n-2)/2}}\right]$$

unless $n = 2$, when $N = \dfrac{1}{CM(\Delta\sigma)^2}\left[\ln \dfrac{a_c}{a_i}\right]$

Equation (11) can be used to calculate the life for a known initial flaw size and geometry for a given loading condition. Figure 1.2.18 shows a log–log plot of life against initial flaw size for the materials A, T, S1 and S2 where the value of $M = 3.8$ corresponding to a surface flaw of a 10:1 aspect ratio has been taken:

Fig. 1.2.18 *A log–log plot of cycles to failure, N, for various initial crack lengths, a_i, calculated from* $N = \dfrac{1}{CM^2(\Delta\sigma)^4}[1/a_i - 1/a_c]$ *for* $\Delta\sigma = \sigma_y/2$, $\sigma_y/4$ *and 155 MPa*

The working stress has been taken as $\sigma_y/2$, $\sigma_y/4$ and 155 MPa which is $\sigma_y/4$ for the aluminium alloy A. For example, consider an initial flaw 0.2 mm deep by 2 mm long, then, at a working stress of half the yield stress, the life of a structure containing such a crack would be 100 cycles for steel S2, 630 cycles for the aluminium alloy A, 790 for the titanium alloy T and 1550 cycles for the steel S1. The critical defect sizes plotted as the initial crack size for failure in 1 cycle, are less than the minimum thickness requirement for plane strain conditions so that in materials of sufficient thickness for the application of the linear elastic fracture mechanics approach the critical flaw size for fracture is less than the size of a through thickness flaw, i.e. a break before leak situation.

If the working stress is reduced to a quarter of the yield stress the lives increase to 2500, 15 500, and 25 000 cycles for S2, A&T and S1 respectively. The critical flaw depths are now greater than the minimum thickness so that provided the thickness of the component is less than the critical flaw depth a leak before break situation will be achieved. Note that for the particular aluminium and titanium alloys chosen for this example the lives are the same for crack lengths less than 0.8 mm. As shown previously the fatigue crack growth rates are nearly identical for these alloys at the same fraction of yield, and on the linear portion of the *a–N* curves the life depends only on the fatigue crack growth rate – this is also shown by the coincidence of the S1 and S2 lines at the same stress levels. In these cases the higher values of fracture toughness do not confer a longer life on the component for small initial defect sizes. The higher toughness does, however, provide for a larger critical crack size and affords the opportunity for a leak before break situation or a greater likelihood of detecting a larger crack.

Considering the four alloys at the same stress level, the order of merit is changed; the steel S2 is now always better than the aluminium and better than the titanium for cracks less than 13 mm deep.

Apart from a general comparison of lives of the three types of alloy, graphs of the type shown in Fig. 1.2.18 can also be used to specify inspection procedures. For example, considering the aluminium alloy at a quarter of its yield strength and supposing the smallest flaw which could be detected were 0.3 mm deep by 3 mm long, then, if no flaw were found in a preservice inspection, a life of at least 10 000 cycles could be expected. If the component were inspected again after say 9000 cycles an undetected flaw of just less than 0.3 mm depth could have grown to 1.8 mm depth by 18 mm length and the residual life would be 1000 cycles.

If no flaw greater than 0.3 mm were detected at this stage, a further 9000 cycles could be run safely.

1.2.2.5 Failure analysis diagram

Figure 1.2.19 is a convenient method of presenting strength and toughness data for various materials. It has the merit of permitting a failure analysis of safe defect sizes to be performed. The toughness of the material K_{1c} is plotted as ordinate against the yield strength (0.2% proof stress), σ_y.

CRITICAL FLAW SIZE

The basic relationship $K_{1c} = \sigma\sqrt{Ma}$ can readily be plotted on the diagram as a straight line of slope $\sqrt{Ma}$ passing through the origin. Taking the stress as σ_y and a surface flaw of 10 : 1 aspect ratio $M = (1.21\pi)/Q$, where $Q = 0.875$ from Fig. 1.2.8 so that $M = 4.36$ and $K_{1c} = \sigma_y\sqrt{4.36\,a}$.

Considering a flaw 1 mm deep, then by joining the origin to the scale mark of 1 mm on the σ_y scale it can be seen that steels of greater strength than 1744 MPa have K_{1c} values below the line, i.e. these steels would not bear the load without fast crack propagation. Materials to the left of the line would, however, be fracture safe when stressed to the yield stress. Safe and unsafe materials stressed to yield for other crack sizes can readily be obtained by constructing the appropriate line through the origin.

The basic equation can be written as

$$K_{1c} = f\,\sigma_y\sqrt{Ma} \tag{12}$$

where $f\sigma_y$, the working stress, is a fraction, f, of the yield stress. Crack length scales, for the same flaw geometry, are given in Fig. 1.2.19 for $f = 0.5$ and 0.25. Note that halving the working stress quadruples the critical flaw size. Scales for other flaw geometries and working stress can readily be added to the diagram for the examination of particular cases:

Fig. 1.2.19 *Failure analysis diagram showing a summary of yield strength and fracture toughness data for several alloy systems and selected non-metallics, tested at 20°C*

CRITICAL MATERIAL THICKNESS

In experimental determinations of fracture toughness it is found that the value obtained can depend upon the thickness of test specimen. As the thickness of the specimen increases the experimental value of the fracture toughness decreases until at a minimum thickness, B, the experimental value of toughness becomes constant. This constant value is known as K_{1c}, the plane strain fracture toughness, and it is found that the minimum thickness

$$B = 2.5 \left[\frac{K_{1c}}{\sigma_y}\right]^2 \tag{13}$$

All the data for K_{1c} and K_{1scc} given in the *Materials Selector* satisfy this criterion, i.e. the minimum thickness for plane strain.

Equation (13) can be rewritten in the form $K_{1c} = \sigma_y\sqrt{B/2.5}$ and can then be readily represented in Fig. 1.2.19 for various values of B. Considering a 1 mm crack at σ_y, materials lying above or to the left of this line will not give plane strain conditions at failure unless the thickness is greater than 10 mm. In the event of a design involving one of these materials with a thickness 10 mm or less the linear elastic fracture mechanics approach (LEFM) developed in this chapter is not strictly applicable. The apparent toughness of thinner material will be higher than the K_{1c} value quoted for material thicker than the minimum of 10 mm, so that use of the data should provide a conservative design. The toughness parameter to be used in non-plane strain conditions to produce an efficient design is considered in the section on elastic plastic fracture mechanics, Section 1.2.2.8.

The fracture mechanics principles (LEFM) outlined here for plane strain conditions have become more widely applicable with the advent of higher strength materials and the use of thicker sections. The minimum thickness B decreases with increased strength and, at the same time, K_{1c} is also decreasing so that the high strength materials are particularly amenable to the treatment. The scope is also increased by simply increasing B with more conventional materials. Few fracture toughness data are available for medium and low strength materials since very large test pieces are required. Most data given in Fig. 1.2.19, refer to critical thicknesses less than 25 mm and none refers to a thickness greater than 250 mm, i.e. a K_{1c}/σ_y ratio greater than 0.316.

OPTIMIZATION OF FRACTURE RESISTANCE

The failure analysis diagram, Fig. 1.2.19 contains a summary of the K_{1c} and K_{1scc} data for aluminium, titanium and ferrous alloys† and can be used for the selection of fracture safe materials. More detailed information is given in Section 1.2.2.7. The condition for fast crack propagation is

$$K_{1c} = f\sigma_y\sqrt{Ma} \tag{14}$$

where K_{1c} is the fracture toughness, $f\sigma_y$ the working stress is some fraction, f, of the yield stress, M is the flaw shape parameter and a is the crack depth. The usual way of designing for minimum weight is to maximize the value of $f\sigma_y$ and select a higher strength material. However, in order to avoid catastrophic failure by fast crack propagation, there is an additional condition that the material has adequate toughness for the expected flaw size and stress level. As can be seen from Fig. 1.2.19 the higher toughness and high strength requirements are incompatible and an optimum selection must achieve the appropriate balance between the two properties.

Critical defect size at a common fraction of yield stress

Materials are frequently used at a specific fraction of the yield stress and the selection of a material could be based on the critical crack size. From equation (14) the critical crack size, a_c, is given by

† Specific information on K_{1c}, K_{1scc} and fatigue crack growth rates is available from:
Damage Tolerant Design Handbook: MCIC-HB-01 Metals and Ceramics Information Center, Columbus, Ohio, Battelle Columbus Laboratories, 1972.
Hudson, C.M. and Seward, S.K. (1978). A compendium of sources of fracture toughness and fatigue-crack growth data for metallic alloys. *Int. J. Fract.*, **14**, RISI–R184.

$$a_c = 1/M \left[\frac{K_{1c}}{f\sigma_y}\right]^2 \tag{15}$$

If the same fraction of the yield stress, f, is used and the crack geometry remains constant (i.e. the value of M is constant) the critical crack size will vary as $\left[\frac{K_{1c}}{\sigma_y}\right]^2$. Note that if the critical crack size is to be the same for two materials then the ratio $\frac{K_{1c}}{\sigma_y}$ should also be the same for the two materials, as shown in the example below (load bearing capacity). Taking values from Fig. 1.2.20 for the strongest and toughest beryllium, aluminium, titanium and iron base alloys, the ratios of $\left[\frac{K_{1c}}{\sigma_y}\right]^2$ and $\left[\frac{K_{1scc}}{\sigma_y}\right]^2$ are those given in Table 1.2.4:

Table 1.2.4 *Comparison of* $\left[\frac{K_{1c}}{\sigma_y}\right]^2$ *and* $\left[\frac{K_{1scc}}{\sigma_y}\right]^2$ *ratios for high strength and high toughness, beryllium, aluminium, titanium and iron base alloys*

Working stress at the same defect size

In order to compare two materials, A and B, each with the same crack size and geometry, it can be seen from equation (15) that

$$\frac{K_{1cA}}{f_A\,\sigma_{yA}} = \frac{K_{1cB}}{f_B\,\sigma_{yB}}$$

Taking the highest toughness steel from Fig. 1.2.19 as material A and the highest strength steel for material B we have

$$\frac{160}{f_A \times 1250} = \frac{35}{f_B \times 2350}$$

and

$$\frac{f_A}{f_B} = \frac{160}{1250} \times \frac{2350}{35} = 8.6$$

and the ratio of the working stresses

$$\frac{f_A\sigma_{yA}}{f_B\,\sigma_{yB}} = \frac{1250 \times 8.6}{2340} = 4.6$$

i.e. the higher toughness material, A, can be used at a higher working stress than the higher strength material, B.

Load-bearing capacity

In this section the requirement of minimum thickness and mass of material to support a given load, P, in the presence of a flaw of size, a, is considered. The thickness B_{min} is related to the stress, $f\sigma_y$ and load, P, on unit width by the equation.

$$P = f\,\sigma_y\,B_{min} \tag{16}$$

Note that B_{min} is least for f = 1. From equation (14).

$$f\sigma_y = \frac{K_{1c}}{\sqrt{Ma_c}}$$

and, substituting from equation (16)

$$\frac{P}{B_{min}} = \frac{K_{1c}}{\sqrt{Ma_c}}$$

or

$$B_{min} = \frac{P\sqrt{Ma_c}}{K_{1c}} \tag{17}$$

Since the mass of material, W, = $d\,B_{min}\,XL$ where d is the density, then for unit width, X, and unit length, L,

$$W = \frac{dP\ \sqrt{Ma}}{K_{1c}} \tag{18}$$

where W is the mass of unit area of material with thickness B_{min}, thus from equation (18) the minimum weight varies as the ratio d/K_{1c}.

As an example of the weights of various materials required in air and a sodium chloride solution to support a load of 1MN, consider a 1-mm-deep flaw with a 10 : 1 aspect ratio where the working stress is equal to the yield stress. From Fig. 1.2.19 materials lying above the line joining the origin to the 1 mm mark on the σ_y scale will have adequate toughness. However, to achieve the least thickness the highest strength of the materials must be chosen, i.e. the highest toughness material lying on the line. Such materials selected from Fig. 1.2.19 are shown in Table 1.2.5 and Table 1.2.6 for dry and wet applications, respectively:

Table 1.2.5 *Optimum materials to be fracture safe with a 1 mm flaw stressed to* σ_y *in air*

Table 1.2.6 *Optimum materials to be fracture safe with a 1 mm flaw stressed to* σ_y *in sodium chloride solution*

In the air the optimum material for weight saving, in this example, is the titanium alloy followed by the steel at 1.25 times the weight of titanium followed by the aluminium and beryllium alloys at 1.44 and 2.25 times the weight. In sodium chloride solution the titanium is still marginally the best followed by aluminium and steel at 1.03 and 1.12 times the weight of titanium. The change in the order of merit arises from the differing degrees of susceptibility to stress corrosion cracking of the alloys as shown by the relative weight changes of the three alloys; the weight of titanium must be increased by 1.48, aluminium by 1.06 and the steel by 1.33 times to tolerate the crack in the wet condition. The toughness of the aluminium alloys is least affected by the presence of salt solution and the toughness of the titanium alloy is most affected.

Selection of material for a leak before break situation

The working stress $f\sigma_y$ can be calculated for a thin wall cylindrical vessel with a wall thickness t, working pressure, P, and radius R from the expression

$$f\sigma_y = \frac{PR}{t} \tag{19}$$

The mass of unit area of the vessel, W, is given by

$$W = dt = d\ \frac{PR}{f\sigma_y} \tag{20}$$

where d is the density of the material. The usual method of minimizing W would be to select a material of high yield strength and work at this stress with $f = 1$. However, in this example there is an additional condition that the vessel should leak before breaking and the minimum critical flaw size $a_c \geq t$. The condition for failure of the vessel is

$$K_{1c} = f\sigma_y \sqrt{Ma_c} \qquad (21)$$

so that

$$\frac{K^2{}_{1c}}{Mf^2 \sigma_{y^2}} = a_c \geq t = \frac{PR}{f\sigma_y}$$

and

$$\frac{K^2{}_{1c}}{\sigma_y} = f\ MPR \qquad (22)$$

Equation (22) has a function of the material properties on the left-hand side and design parameters on the right-hand side.

As an example, let the working pressure be 70 MPa, the radius R equal to 0.1 m and take $M = 4$ so that

$$\frac{K^2{}_{1c}}{\sigma_y} = f \times 4 \times 70 \times 0.1 = 28f \qquad (23)$$

Selecting a high strength steel of σ_y = 2336 MPa with a K_{1c} of 38 MN/m$^{3/2}$ gives $f = \dfrac{38 \times 38}{28 \times 2336} = 0.023$ so that the working stress would be 54 MPa and the wall thickness from eqn (19) is $t = \dfrac{7 \times 10^3}{54} = 129$ mm. The mass of unit area of the vessel would then be

$$W = dt = 7.88 \times 10^3 \times 129 \times 10^{-3} = 1010 \text{ kg}$$

The steel selected for this example does not have the highest value of $\dfrac{K^2{}_{1c}}{\sigma_y}$ and does not give the highest value of f from eqn (23). The highest values of $\dfrac{K^2{}_{1c}}{\sigma_y}$ for aluminium and titanium alloys and steel are given in Table 1.2.7 with the results of the previous example reworked using these values of $\dfrac{K^2{}_{1c}}{\sigma_y}$:

Table 1.2.7 *The mass of unit area of pressure vessels working at a pressure of 70 MPa with a radius of 0.1 m which will leak before breaking*

It should be noted that the calculated wall thickness for steel 1 and the aluminium alloy is greater than the radius of the vessel so that these examples cannot be considered as thin-walled vessels. The vessel made from steel 2 has a working stress greater than the titanium alloy vessel and has a thinner wall thickness. The titanium vessel is only 58% of the mass of the steel one and on mass considerations alone the titanium alloy would be selected. It should also be noted that the calculated wall thicknesses for steel 2 and the titanium alloy are less than the minimum thicknesses of 25 and 100 mm respectively for plane strain conditions so that the estimates of toughness would be conservative.

As a second example we consider a vessel of the same dimensions as in the previous example but with a working pressure of 7 MPa. From eqn (22) for the same alloys as used in the previous example we find that f = 5.25, 1.43 and 10 for the steel 2, the aluminium and titanium alloys, respectively. If the working stress is not to exceed the yield stress the maximum value of $f = 1$ must be taken. Hence, from eqn (22)

$$\frac{K^2_{1c}}{\sigma_y} = 1 \times 4 \times 7 \times 0.1 = 2.8$$

and the materials with $\frac{K^2_{1c}}{\sigma_y} = 2.8$ must be selected. To facilitate this choice we can plot the the line $K_{1c} = \sqrt{2.8\sigma_y}$ on Fig 1.2.19; alloys above this line will have adequate toughness when used at σ_y. In order to get the maximum working stress we take the highest strength alloy above the line for the three alloy systems. The previous example has been reworked for these alloys with the results shown in Table 1.2.8:

Table 1.2.8 *The mass of unit area of pressure vessels working at a pressure of 7 MPa with a radius of 0.1 m which will leak before breaking*

Note that in this example although the titanium vessel is still the lightest it is now 90% of the mass of that of the steel vessel.

1.2.2.6 Factors which influence fracture toughness

Several factors, of which composition, environment and strength have already been mentioned, are recognized to influence the value of the fracture toughness of a material. These factors may be divided into two general classes, intrinsic properties of the material such as composition and strength and extrinsic factors which depend on the service conditions such as environment and temperature.

INTRINSIC FACTORS

The fracture toughness of a material depends on the intrinsic properties such as strength, composition, type of heat treatment and final thickness of material.

Effect of strength

There is a general tendency for the higher strength materials to have lower toughness values than the lower strength materials as shown in Fig. 1.2.19. However, for a given strength level there is also a wide range of toughness values. Much of this variability can be attributed to the factors discussed below.

Effect of composition

The toughness of various types of alloy can be roughly divided into strength and toughness combinations on the basis of the major alloying element. For example, the aluminium alloys are quite separate from the steels in Fig. 1.2.19. Within these regions a further classification can be made on the basis of alloying additions as shown by Fig. 1.2.20 which is a plot of K_{1c} vs σ_y values for aluminium copper alloys tested at 20°C:

Fig. 1.2.20 *K_{1c} vs σ_y data for aluminium–copper alloys tested at 20°C*

Minor alloying additions or impurities which have a deleterious effect on ductility also reduce the fracture toughness. Some examples of these are interstitials such as oxygen and hydrogen in titanium alloys, phosphorus and sulphur in steels and the inclusions which occur in most alloys unless special processing is adopted to minimise the inclusion content.

Effect of heat treatment

Most alloys are heat treated to provide a given level of strength. However, the optimum heat treatment for strength is not generally the one for optimum toughness. As can be seen from Fig. 1.2.21, which shows the effect of differing tempering treatments on the strength and toughness of a 0.5Ni–1Cr–1Mo–0.5C steel, it is possible to get widely different toughness values at the same strength level:

Fig. 1.2.21 *The effect of tempering on the strength and toughness of a 0.5Ni–1Cr–1Mo–0.5 C steel*

Effect of material thickness

The section thickness of a material can affect the hardenability of some alloys and, although it may be possible to modify the heat treatment to obtain similar strengths in different section thicknesses the toughness values resulting from the different heat treatments can vary. In the same way the amount of working required to produce wrought products of different section thickness can influence the fracture toughness. Figure 1.2.22 shows the fracture toughness of an Al–6.3 Cu alloy as a function of testing temperature, for different plate thickness. It should be noted that this is the thickness of the material, not the thickness of the test specimen, and all the tests gave K_{1c} values satisfying the $2.5\left[\frac{K_{1c}}{\sigma_y}\right]^2$ criterion:

Fig. 1.2.22 *Effect of plate thickness on K_{1c} of Al–6.3Cu (2219) as a function of testing temperature*

EXTRINSIC FACTORS

The fracture toughness of a material is also influenced by extrinsic factors such as the service conditions and the orientation of the crack.

Effect of environment

The effect of service environment upon K_{1c} has already been discussed. In air the toughness is denoted by K_{1c} but in other environments, such as 3% NaCl, the toughness may be reduced to a value K_{1scc}.

Effect of temperature

The data summary in Fig. 1.2.19 was obtained from tests performed at 20°C although service applications may call for other temperatures. It is well known that some steels undergo a ductile–brittle transition as temperature and rate of loading vary and it is likely that fracture toughness values will also be sensitive to temperature and loading rate.

Figure 1.2.22 shows that the toughness of the Al–6.3 Cu alloy tends to increase as the temperature of testing is reduced. Similar behaviour is also exhibited by most of the aluminium alloys shown in Fig. 1.2.23. Aluminium alloys do not undergo a ductile–brittle transition and are suitable for low temperature applications.

Fig. 1.2.23 *Effect of testing temperature on K_{1c} of aluminium alloys*

Figures 1.2.24 and 1.2.25 show the effect of temperature on Ti–6Al–4V and four steels, respectively. The toughness of the titanium alloy decreases steadily with decreasing temperature and there is a transition in toughness of the steels, similar to the impact transition curves associated with the ductile–brittle transition of steels.

Fig. 1.2.24 *Effect of testing temperature on K_{1c} of Ti–6Al–4V alloy for various crack orientations*

Fig. 1.2.25 *Effect of testing temperature on K_{1c} of steels*

Effect of crack orientation

The orientation of the crack plane and the direction of crack propagation within a material has a marked effect on the toughness. The system of axes used to identify the orientation of a crack is illustrated in Fig. 1.2.26, In a rectangular section or plate the rolling direction or length is designated L, the thickness or short transverse direction is S and the width or long transverse direction is T. In a cylindrical bar the length is still L, a radial direction is R and a chordal direction, normal to a radius, is C. The crack plane is identified by the direction of the normal to that plane and the direction of propagation of the crack corresponds to the appropriate axis. An L–T orientation is thus a crack, the plane of which is normal to the rolling direction propagating in the long transverse direction.

Fig. 1.2.26 *Coding system for specimen orientation and crack propagation direction for specimens from (a) plate and rectangular section and (b) round bars*

When an ingot containing inclusions is worked to produce a plate, for example, the deformable inclusions will form stringers aligned in the rolling direction and a crack with the L–T orientation will have a fracture plane cutting across the stringers. A crack with a T–L orientation will have the inclusions lying in the fracture plane and as a result the toughness will be reduced. Figure 1.2.20 shows that the T–L orientation tends to give lower toughness values than the L–T orientation in Al–Cu alloys. The effect of orientation of the crack in Al–4.5Cu–1.5Mg alloys is shown in Fig. 1.2.23. The same order of ranking of the toughness in terms of crack orientation, L–S, L–T and T–L is also found in the Ti–6Al–4V alloy shown in Fig 1.2.24.

1.2.2.7 Fracture toughness of particular alloys

The following summary indicates the compositions of the particular alloys for which fracture toughness values have been used in the fracture toughness and fatigue section of the *Materials Selector*, together with the locations, in the *Selector*, of the relevant data.

ALUMINIUM ALLOYS

K_{1c} vs σ_y for wrought aluminium alloys at 20°C.

Fig. 1.2.27a	Al–6.3 Cu	(2219)
	Al–5.3 Cu	(2021)
	Al–4.4 Cu–0.8Si	(2014)
	Al–4.5 Cu	(2020)
Fig. 1.2.27b	Al–1.6 Cu–2.5 Mg–5.6 Zn	(7075)
	Al–2 Cu–2.7 Mg–6.8 Zn	(7178)
Fig. 1.2.27c	Al–4.5 Cu–1.5 Mg	(2024)
	Al–0.6 Cu–3.3 Mg–4.3 Zn	(7079)

K_{1c} vs K_{1scc} for wrought aluminium alloys: Fig. 1.2.11.
K_{1scc} vs σ_y for Al 2000 and 7000 series aluminium alloys: Fig. 1.2.28

Susceptibility of aluminium alloys to stress corrosion cracking: Fig. 1.2.11.
Effect of test temperature on K_{1c}: Fig. 1.2.23
Effect of plate thickness on K_{1c} of Al–6.3 Cu (2219) alloy as a function of testing temperature: Fig. 1.2.22

TITANIUM ALLOYS

K_{1c} vs σ_y for titanium alloys at 20°C. } Fig. 1.2.29
Various wrought titanium alloys.
Wrought 8Al–1Mo–1V and 6Al–6V–2.5Sn.
Wrought 7Al–2Nb–1V and 6Al–6V–2Sn alloys and wrought and cast 6Al–4V.

K_{1c} vs K_{1scc} for titanium alloys: section 1.2.2.4.
Various wrought alloys and wrought 7Al–2Nb–1Ta, 8Al–1Mo–1V, 6Al–6V–2.5Sn and 6Al–4V: Fig 1.2.12.
Variation of K_{1c} with test temperature: section 1.2.2.6.
Variation of K_{1c} with temperature of Ti–6Al–4V for various crack orientations: Fig. 1.2.24.

STEELS

K_{1c} vs σ_y for wrought alloys at 20°C.

Fig. 1.2.30a	14Cr	13Co	5Mo					(AFC 77)
	15Cr	13Co	4Mo		2Ni			(AFC 260)
Fig. 1.2.30b	9 Ni	4Co	1Cr	1Mo			0.2C	(HP94)
	2Ni	1Cr	0.5Mo			0.3C		(AISI 4330)
	1.75Ni	0.75Cr	0.5Mo	1.75Si			0.4C	(300M)
Fig. 1.2.30c	13Cr	8Ni	2Mo					(PH 13–8–Mo)
	0.5 Ni	1Cr	1Mo			0.5C		(D6AC)
Fig. 1.2.30d	18 Ni	5Mo	7Co (marageing)					
	1.75 Ni	0.8 Cr	0.25 Mo			0.4C		(AISI 4340)

Hard high-speed steels, tool steels, white cast irons and cemented carbides: Fig. 1.2.31.
Table 1.2.9 *Composition of materials shown in Fig. 1.2.31*
K_{1c} vs K_{1scc}.
18Ni–7Co–5Mo (marageing): Fig. 1.2.13.
Low alloy steels: Fig. 1.2.14.
Martensitic stainless and precipitation-hardening steels: Fig. 1.2.15.
K_{1scc} vs σ_y for some stainless and other high strength steels: Fig. 1.2.32.
Variation of K_{1c} with temperature: Section 1.2.2.6.
Low alloy steels and AFC 77 (14Cr–13Co–5Mo–0.1C): Fig. 1.2.25
Effect of tempering temperature: Section 1.2.2.6.
Effect of tempering on strength and toughness of a 0.5Ni–1Cr–1Mo–0.5C steel (D6AC): Fig. 1.2.21.

1.2.2.8 Elastic plastic fracture mechanics

INTRODUCTION

The linear-elastic fracture mechanics, LEFM, approach outlined in the previous sections is applicable to plane strain conditions where the thickness of material $B > 2.5\left[\frac{K_{1c}}{\sigma_y}\right]^2$ and when the applied stress is less than the yield stress.

According to LEFM the stress normal to the plane of the crack varies inversely as the square root of the distance from the tip of the crack and becomes infinitely large at the crack tip. In elastic plastic material a distance r_y ahead of the crack tip experiences a stress greater than the flow stress σ_y, giving a plastic zone of radius r_y centred on the crack tip. Irvine has suggested the expression $a_{eff} = a_{phys} + r_y$ to allow for the local plasticity where

$$r_y = \frac{1}{6\pi}\left[\frac{K_{Ic}}{\sigma_y}\right]^2 \quad \text{for plane strain} \tag{25}$$

and

$$r_y = \frac{1}{2\pi}\left[\frac{K_{Ic}}{\sigma_y}\right]^2 \quad \text{for plane stress} \tag{26}$$

For plane strain $\left[\frac{K_{Ic}}{\sigma_y}\right]^2 \leq B/2.5$ and from eqn (25)

$$r_y = B/15\pi$$

i.e. $r_y \sim 0.02B$ for plane strain.

Fully plane stress occurs when the plastic zone radius equals the material thickness. Since the plastic zone in plane stress is three times the zone in plane strain the thickness of a material which assumes the fully plane stress condition is 0.06 of the thickness of the same material that would be in plane strain. The plane stress toughness K_c is 2–3 times the plane strain toughness K_{Ic} so that as the material thickness decreases from the minimum for plane strain to 0.06 of this thickness the effective toughness increases from K_{Ic} to 2–3 times K_{Ic}. The use of K_{Ic} for thickness less than 2.5 $\left[\frac{K_{Ic}}{\sigma_y}\right]^2$ is safe but overly conservative and alternative methods of treatment are desirable for non-plane strain conditions.

When the applied stress equals the flow stress of the material the whole section becomes plastic. At lower applied stresses the stress intensification ahead of the crack causes a large plastic zone which may spread throughout the section thickness, although the applied stress is less than the flow stress and general yielding does not occur. The Irvine correction is intended to apply to smallscale yielding at the crack tip in plane strain conditions; larger scale plasticity is treated by the Dugdale model for crack tip plasticity described below.

DUGDALE MODEL

This model proposed by Dugdale, is sometimes referred to as 'The strip yield model'. It is supposed that the plastic strain is concentrated into an infinitely thin strip on the prolongation of the crack; the remainder of the body behaving elastically. Figure 1.2.33 shows a crack of length $2a$ in an infinite body loaded by a remote stress σ, in a direction normal to the crack plane:

Fig. 1.2.33 *Schematic representation of the plastic strip yield (Dugdale) model*

A notional crack of length $2c$ is considered by applying a compressive stress equal to the flow stress σ_y on the edges of the crack over the distance $2c$–$2a$ from the tips. At the tips of the fictitious crack the remote loading produces a positive stress intensity factor and the crack edge loading a negative stress intensity factor. By equating the resulting stress intensity to zero the extent of the plastic strip is defined by:

$$\frac{c}{a} = \sec\left(\frac{\pi a}{2\sigma_y}\right) \tag{27}$$

The model has also been used to find the crack opening displacement, COD†, at the

†The term CTOD, crack tip opening displacement, has become the currently preferred terminology. The original term COD is used here in conformity with BS 5762. Noting that COD refers to the opening at the crack tip, the two terms are synonymous.

physical crack tip, i.e. the distance 'δ' in the figure as

$$\delta = \frac{8\sigma_y a}{\pi E} \log \sec \left(\frac{\pi\sigma}{2\sigma_y}\right) \tag{28}$$

where E is Young's modulus. More generally E can be replaced by E^1 where,

$E^1 = E$ for plane stress

and $E^1 = E/(1 - v^2)$ for plane strain and

v = Poisson's ratio

When $\sigma/\sigma_y \rightarrow 0$ in eqn (28)

$$\delta = \frac{\pi\sigma^2 a}{E\sigma_y} \tag{29}$$

and rearranging eqn (29)

$$E\delta\sigma_y = \pi\sigma^2 a = K^2 \tag{30}$$

for the elastic case. From eqn (30)

$$\delta\sigma_y = K^2/E = G = J \tag{31}$$

where G is the crack extension force and J is Rice's J integral. Equation (31) demonstrates that the COD and J integral approaches are equivalent to the LEFM approach for smallscale plasticity; COD and J integral approaches are used for elastic plastic fracture mechanics where plasticity is not smallscale, i.e. as $\sigma \rightarrow \sigma_y$.

By equating δ from eqn (28) to δ from eqn (29) with the notional length c

$$\frac{8\sigma_y a}{\pi E} \log \sec \left(\frac{\pi\sigma}{2\sigma_y}\right) = \frac{\pi\sigma^2 c}{E\sigma_y}$$

or

$$\phi = \frac{c}{a} \frac{8\sigma_y^2}{\pi^2 S^2} \log \sec \left(\frac{\pi\sigma}{2\sigma_y}\right) \tag{32}$$

Equations (28) and (32) provide the basis for two design criteria or failure assessment procedures, developed from the Dugdale model of the plastic zone for large-scale yielding when LEFM is not appropriate. Equation (28) leads to the COD approach pioneered by The Welding Institute and eqn (32) leads to the two-criteria approach developed by the CEGB.

CRACK OPENING DISPLACEMENT; COD APPROACH

The design curve

The design curve is obtained from equation (28) by defining Φ a dimensionless crack opening displacement, as:

$$\Phi = \frac{\delta E}{2\pi\sigma_y \bar{a}} \tag{33}$$

then substituting for δ from equation (28)

$$\Phi = \frac{4}{\pi^2} \log \sec \left(\frac{\pi\sigma}{2\sigma_y}\right) \tag{34}$$

and replacing stress by strain in equation (34)

$$\Phi = \frac{4}{\pi^2} \log \sec \left(\frac{\pi \varepsilon}{2 \varepsilon_y} \right) \tag{35}$$

where ε is the local strain which would exist in the region of the crack if the latter were not there. Experimental determinations of Φ as a function of $\varepsilon/\varepsilon_y$ using values a/y between 0.025 and 0.5, where y is half the gauge length over which ε is measured, showed that Φ is independent of a/y. On the basis of this work the following design curve was proposed:

$$\Phi = \frac{4}{\pi^2} \log \sec \left(\frac{\pi \varepsilon}{2 \varepsilon_y} \right) \quad \text{for} \quad \frac{\varepsilon}{\varepsilon_y} \leq 0.86 \tag{36}$$

or

$$\Phi = \frac{\varepsilon}{\varepsilon_y} - 0.25 \qquad \text{for} \quad \frac{\varepsilon}{\varepsilon_y} \geq 0.86 \tag{37}$$

The toe of the curve was subsequently revised to a simpler and more conservative expression. The design curve currently used is:

$$\Phi = (\varepsilon/\varepsilon_y)^2 \qquad \text{for} \quad \varepsilon/\varepsilon_y \leq 0.5 \tag{38}$$

or

$$\Phi = (\varepsilon/\varepsilon_y) - 0.25 \qquad \text{for} \quad \varepsilon/\varepsilon_y \geq 0.5 \tag{39}$$

The first term in the expansion of equation (36) is $1/2\ (\varepsilon/\varepsilon_y)^2$ for smallscale yielding, i.e. small values of $\varepsilon/\varepsilon_y$. At these small values the design curve has a safety factor of 2.0 on the strip yielding model represented by equation (36). The two forms of the design curve are shown in Fig 1.2.34.

Fig. 1.2.34 *Current design curve compared with the earlier version*

Method of application

For effective ratios of defect size:width less than 0.1, and when the nominal design stress is below yield, it has been proposed that equations (38) and (39) can be rewritten in terms of stress and the critical COD, δ_c, to give:

$$\bar{a}_{\max} = \frac{\delta_c E \sigma_y}{2 \pi \sigma_1^2} \qquad \text{for} \quad \frac{\sigma_1}{\sigma_y} \leq 0.5 \tag{40}$$

or

$$\bar{a}_{\max} = \frac{\delta_c E}{2 \pi (\sigma_1 - 0.25 \sigma_y)} \qquad \text{for} \quad \frac{\sigma_1}{\sigma_y} \geq 0.5 \tag{41}$$

σ_1 is the total pseudo-elastic stress in the region of the crack. Whilst σ_1 can exceed σ_y the structure may still behave in a predominantly elastic fashion, since the yielded zone is contained by the surrounding elastic material. In welded structures, contained yield occurs by virtue of residual stresses, which may themselves be equal to the yield stress and may be additive to the applied stress. Contained yielding can also occur at stress concentrations.

For general application the values of σ_1 given below have been suggested.

Crack location	*Weld conditions*	σ_1
Remote from stress concentrations	Stress relieved	σ
	As welded	$\sigma + \sigma_y$
Adjacent to stress concentrations	Stress relieved	SCF $\times \sigma$
	As welded	(SCF $\times \sigma$) + σ_y

σ is the nominal design stress and SCF is the elastic stress concentration factor associated with the stress concentration. It is assumed that stress relieving reduced the residual stress to zero and that the residual stress was as high as σ_y

The design curve represented by equations (40) and (41) was derived from a model for through thickness flaws. Part through flaws are treated by assuming that, for contained yielding, the parameters governing flaw shape effects would be the same as those for elastic conditions. The elastic stress intensity factor for a semi-elliptical surface crack is

$$K_1 = \sigma\sqrt{Ma}$$

For an equivalent through thickness crack of length $2\,\bar{a}$.

$$K_1 = \sigma\sqrt{\pi\bar{a}}$$

For the same stress and stress intensity factor in the two cases

$$\frac{a}{B} = \frac{\bar{a}}{B}\frac{\pi}{M} \tag{42}$$

Equation (42) is plotted as a/B vs. $\bar{a}/B$ in Fig 1.2.35. The value of $\bar{a}_{max}$ is calculated from equation (40) or (41) and the figure entered with $\bar{a}_{max}/B$ as abscissa to obtain the maximum allowable surface flaw, $\bar{a}_{max}$ as ordinate from the appropriate $a/2c$ curve.

Fig 1.2.35 *Relation between actual surface defect dimensions and the parameter* $\bar{a}$

Experimental justification of the design curve

A survey conducted by The Welding Institute permitted a comparison to be made of the allowable crack size, predicted by smallscale COD tests, with the critical crack size, measured in wide plate tests, for through thickness and surface flaws in plain and welded materials for 73 sets of tests. The comparison showed that on average the measured critical flaw size was 2.5 times the predicted one and the predicted flaw size represented a 95% confidence limit against failure. It was also found in analysing the wide plate tests for weldments that a residual stress level of σ_y was required to bring the predicted allowable flaw size below the measured critical flaw size.

Critical COD

In the standard test method for determining COD, BS 5762, the crack-opening displacement at the crack tip, δ, is derived from the crack-opening displacement measured at the crack mouth. The load–COD relation derived from the test can be one of three general types, shown schematically in Fig. 1.2.36. The types of relation are associated with different crack propagation behaviours, which are determined by the material and test temperature.

Fig. 1.2.36 *Schematic representation of types of load–COD relation illustrating critical values of* δ *defined in BS 5762*

The crack can suddenly propagate with no prior stable crack growth and the critical COD is defined as δ_c, (Case A). Alternatively the crack may propagate rapidly after some prior stable growth. The COD for the onset of stable crack growth is defined as δ_i (Case B). When the crack propagates rapidly after prior stable growth, the critical COD at the onset of rapid propagation is defined as δ_u (Case B). In the third case there may be no onset of rapid propagation and the load–displacement record reaches a plateau, (Case C). Here the critical COD is defined as δ_m the value of δ at the first attainment of maximum load.

It is important to distinguish between the critical COD for the onset of stable crack growth, δ_i which is associated with ductile tearing, and the critical COD for unstable crack growth, δ_c or δ_u. In yielding fracture mechanics unstable fracture refers to quasi-brittle fracture which, in common structural steels, is usually associated with considerable amounts of cleavage fracture. Experimental evidence of the behaviour of structures indicates that the use of δ_i is over-conservative and justifies the use of δ_c, δ_u or δ_m, as appropriate, in the design curve.

Under the conditions relevant to COD testing, the minimum δ_c values for through thickness, surface and embedded flaws in tension plates are underestimated by the three point single-edge-notched-bend specimen of BS 5762. Similarly, the full thickness square-section three point single-edge-notched bend specimen will underestimate δ_c values for surface flaws in tension, provided that the bend specimen crack length and width match the crack depth and section thickness for the plate tests.

Critical COD is regarded as a 'material property' but the value determined depends not only on the material and its condition but the section thickness. The following values are quoted as examples of the order of magnitude.

Material	*COD*
Low alloy structural steel 141 mm thick, YS 376 Pa as stress relieved	0.25 mm
BS 4360 Grade 50D, 100 mm thick, YS 600 Pa	δ_m = 0.49 mm
Best of 17 matching weld metals	δ_c = 0.12 mm
After post-weld heat treatment five weld metals	δ_m = 0.49 mm

Scope of design curve

The COD test is a useful method of studying fracture toughness in the transition region between linear elastic behaviour, where K_{Ic} should be used, and fully ductile behaviour, where a limit load approach is appropriate. Experience and the semi-empirical theory of the design curve indicate there is an inherent safety factor of 2.5. The design curve is applicable to the selection of material during initial design, the specification of maximum allowable flaw size at the design stage or after fabrication to establish the necessity for repairs, and failure analysis. The design curve has been incorporated in a BSI Public Document[†] on assessing flaw sizes in fusion welds. Reviews of the COD approach and design curve have been published by Harrison[‡] and Dawes[§].

†BSI. PD6493 (1980) 'Guidelines on some methods for the derivation of acceptance levels for defects in fusion welded joints'.

‡Harrison, J. D., Dawes, M. G., Archer, G. L., and Kamath, M. S. (1979). ASTM STP 668. *Elastic-Plastic Fracture*, p. 606.

§Dawes, M. G., (1979). *Advances in Elasto-Plastic Fracture Mechanics.* Applied Science Publishers, Barking, p. 179.

TWO-CRITERIA APPROACH

The failure assessment curve

Taking the LEFM expression for failure in the form:

$$K_{1c} = \sigma_c \sqrt{Ma\,\phi} \tag{43}$$

where ϕ is the plastic zone correction factor and σ_c the failure stress.

From eqn (32), is given by:

$$\frac{8\sigma_y{}^2}{\pi^2\sigma_c{}^2} \log\sec\left(\frac{\sigma_c}{2\sigma_y}\right)$$

and $$K_{1c} = \sigma_c\left[\frac{Ma\,8\sigma_y{}^2}{\pi^2\sigma_c{}^2}\log\sec\left(\frac{\pi}{2}\frac{\sigma_c}{\sigma_y}\right)\right]^{1/2}$$

Rearranging eqn (44):

$$\frac{K_{1c}}{\sigma_c\sqrt{Ma}} = \frac{\sigma_y}{\sigma_c}\left[\frac{8}{\pi^2}\log\sec\left(\frac{\pi}{2}\frac{\sigma_c}{\sigma_y}\right)\right]^{1/2} \tag{45}$$

Then defining $K_R = \dfrac{\sigma_c\sqrt{Ma}}{K_{1c}} = \dfrac{\text{applied stress intensity from LEFM}}{\text{fracture toughness}}$

and $S_R = \dfrac{\sigma_c}{\sigma_m} = \dfrac{\text{applied loading}}{\text{limit loading}}$

eqn (45) can be rewritten as:

$$K_R = S_R\left[\frac{8}{\pi^2}\log\sec\left(\frac{\pi}{2}S_R\right)\right]^{-1/2} \tag{46}$$

Equation (46) represents the failure assessment curve shown in Fig. 1.2.37, the failure assessment diagram. As S_R tends to zero, when σ_c is much less than σ_y, K_R tends to 1 and LEFM is applicable, i.e. $K_{1c} \approx \sigma\epsilon\sqrt{Ma}$. When S_R tends to 1, where σ_c becomes equal to σ_y, K_R tends to zero and K_{1c} is much greater than $\sigma_c\sqrt{Ma}$. Thus at the extremes *either* LEFM is applicable and the failure is brittle when σ is much less than σ_y *or* the failure occurs by plastic collapse or general yielding where the material is tough. The intermediate elastic plastic range is represented by the 'failure assessment curve' in Fig. 1.2.37. Points in the failure assessment diagram with coordinates K_R, S_R which lie inside the curve will be safe from failure.

Fig. 1.2.37 *The failure assessment diagram*

Method of application

The general principle of the method is straightforward; evaluate the parameters K_R and S_R, which are then entered in the failure assessment diagram, for example, the point A in Fig. 1.2.37. The position of this point relative to the failure assessment curve defines the safety of the structure.

The locus of the point A as a function of load is a straight line through the origin, at zero load, to the failure assessment curve, since S_R and K_R are directly proportional to σ. The reserve factor on load is F_L = OF/OA and failure occurs for $F_L = 1$, when A and F are coincident. Other loci, as a function of K_{1c} or collapse stress σ_m are also straight lines parallel to the K_R and S_R axes, respectively. The locus as a function of crack size is generally curved, as shown in Fig. 1.2.37.

The simple properties of the loci permit a sensitivity analysis to study the effects of varying levels of confidence in the input data, regardless of the degree of plasticity. The diagram can also be used to assess the effects of stable crack growth, by fatigue cracking for example.

Input parameters

K_{1c} – plain strain fracture toughness

K_{1c} values can be difficult to measure if the material is very tough because of the size limitation on test specimens, or if the material under consideration has an inadequate section thickness. In such cases it is necessary to resort to estimates of K_{Ic} obtained from full thickness test or from correlations between Charpy impact tests and K_{Ic} (see Section 1.2.2.9).

a and *M* – flaw shape and size

The flaw or crack is treated as an elliptical or semi-elliptical flaw by the method outlined in Section 1.2.2.3 to obtain values for *M*.

σ – primary stress

This is the uniform applied stress or design stress.

σ_m – the collapse stress

This is the nominal stress in the structure when the net-section stress in the ligament at the flaw equals the flow stress of the material σ_f, which is taken as $\frac{1}{2}(\sigma_y + \sigma_u)$ where σ_y is the yield stress and σ_u the tensile strength of the material.
$\sigma_m = \sigma_f(1 - c/w)$ where c/w is a ligament correction parameter dependent on crack shape; for long shallow defects $c/w = a/w$ the fractional reduction in section of the defect.

Secondary stresses

These are local stresses which may be residual stresses, arising from fabrication of the structure, or thermal stresses caused by temperature gradients. These stresses are self-equilibrating and play no part in the regime of plastic collapse under tensile stress. The calculation of S_R involves only the primary stresses.

Locally, secondary stresses are algebraically additive to the primary stress and are involved in the calculation of K_R. For the CEGB procedure, the coordinates of the assessment point, in the presence of secondary loading are evaluated as:

$$S_R = S_R{}^P$$

and $$K_R = K_R{}^P + K_R{}^S + \rho$$

where superscripts p and s refer to primary and secondary loading, respectively and ρ is the necessary plasticity corrections for secondary loading to $K_R{}^P$.

In order to evaluate ρ it is necessary to estimate $K_P{}^S$, the plasticity-corrected value of $K_1{}^S$ due to secondary loading. A simple first-order plasticity correction is taken to be adequate providing that pessimistic plastic zone size corrections, η, are made. The suggested values for η are:

$$\frac{\pi}{16}\left[\frac{K_{1c}}{\sigma_y}\right]^2 \quad \text{for plane strain}$$

and

$$\frac{\pi}{8}\left[\frac{K_{1c}}{\sigma_y}\right]^2 \quad \text{for plane stress}$$

The value of ρ can now be obtained from Fig. 1.2.38 for the appropriate ratio of $K_1{}^S/K_P{}^S$ where $K_1{}^S$ is evaluated as a function of (a) and $K_P{}^S$ as a function of $(a + \eta)$:
Fig. 1.2.38 *Plasticity correction to $K_R{}^S$ for the two-criteria approach*

Justification of the approach

Because of the ease of performing the assessment procedure, reserve factors are not specifically defined. The user is encouraged to determine these factors taking account of where the major uncertainties occur in the input data and the consequences of failure. Thus, reserve factors are not hidden in design curves or formulae and the user can tailor the factors to his knowledge and requirements.

The major criticism of this procedure is centred on the detail of the failure assessment curve in the elastic plastic region. The basic assumption of the procedure is that the failure assessment curve provides a realistic lower bound failure locus which is independent of geometry. Regarded as an empirical interpolation curve between $K_R = 1$ and $S_R = 1$, it can be justified by some 150 assessment points. It has also been demonstrated for a variety of geometries of practical interest that the failure curve can be regarded as a definitive elastic plastic analysis with a theoretical foundation.

The assessment procedure is given in the CEGB report R/H/R6† and further discussion of the procedure, with illustrative examples is available in papers by Milne‡ and Darleston§, for example.

1.2.2.9 Estimation of K_{1c} from Charpy impact tests

When the measurement of K_{1c} is difficult or impossible because of inadequate section size of material or inadequate testing capacity, resort to empirical correlations between K_{1c} and Charpy impact energies (CVN) can be made. Several such correlations are available.

At temperatures above the transition temperature where the impact energies lie on the upper shelf Rolfe and Novak* established the following correlation for steels with yield strength in the range 758–1696 MPa and impact energies 22–121 J:

$$\left[\frac{K_{1c}}{\sigma_y}\right]^2 = 0.64\left[\frac{(\text{CVN})}{\sigma_y} - 0.01\right] \quad \text{Rolfe and Novak}$$

Server¶ has produced data which extend the correlation to lower strength steels and weldments, having yield stress in the range 447–566 MPa and impact energies of 60–192. These data are shown as points in Fig. 1.2.39 with the Rolfe and Novak line:
Fig. 1.2.39 *Upper shelf impact energy–fracture toughness correlation*

†Harrison, R.P., Milne, I. and Loosemore, K. (1977), CEGB Report R/H/R6–Rev. 1.
‡Milne, I. (1979) *Developments in Fracture Mechanics – 1.* Applied Science Publishers, Barking, p. 259.
§Darlaston, B. J. L. (1979). *Advances in Elasto-Plastic Fracture Mechanics.* Applied Science Publishers, Barking, p. 319.
*Rolfe, St. and Novak, S. R. (1970) ASTM STP, 463, 424.
¶Server, W.L. (1979) ASTM STP, 668, 493

The BSI PD 6493[‡] gives a conservative estimate of the upper shelf correlation for ferritic steels of yield stress less than 480 MPa as

$$K_{1c} = 0.53\ (\mathrm{CVN}) + 57.6$$

which is shown in the figure for comparison.

In the transition range between the upper and lower shelf impact energies, different correlations have been established. Rolfe and Novak[§] give the following expression for steels having a yield stress in the range 270–1700 MPa with impact energies in the range 4–82 J:

$$K_{1c}{}^{2}/E = 0.22\ (\mathrm{CVN})^{3/2}$$

Whilst for steels with yield strengths in the range 410–480 MPa and impact energies of 7–68 J, Sailors and Corten[¶] have proposed the relation:

$$K_{1c} = 14.6\ (\mathrm{CVN})^{1/2}$$

These two correlations are compared with the BSI correlation in Fig. 1.2.40:

Fig. 1.2.40 *Correlations between impact energies and fracture toughness in the transition region*

‡BSI PD 6493, 1980.

§Rolfe, S. T. and Novak, S. R. (1970) ASTM STP, **463**, 424.

¶Sailors, R. H. and Corten, H. T. (1972). ASTM STP, **514**, part 1, 164.

Table 1.2.1 Strength and toughness data for selected Al, Ti and ferrous alloys

	σ_y (MPa)	K_{1c} ($MN/m^{3/2}$)	C ($m/(MN^4/m^6)^1$)	B min[a] mm $2.5\left[\frac{K_{1c}}{\sigma_y}\right]^2$
Steels (S1) Fe–9Ni–4Co–1Mo–2C	1350	160	1×10^{-12}	35
Fe–18Ni–5Mo–7Co (S2) Marageing	2300	45	1×10^{-12}	0.96
Titanium (T) Ti–6Al–6V–2.5Sn	1320	60	2×10^{-12}	5.2
Aluminium (A) Al–2Cu–2.7Mg–6.8 Zn	620	20	4×10^{-11}	2.60

[a] B = minimum thickness for constant fracture toughness (see Section 1.2.2.5 – eqn (13)).

Table 1.2.1

Table 1.2.2 Threshold stress intensity factor for fatigue crack growth, ΔK_{TH} of some ferrous materials

Material	***Tensile strength (MPa)***	***R ratio***	ΔK_{TH} ***(MN/m$^{3/2}$)***
Mild steel at 20°C	430	−1.0	6.4
		0.13	6.0
		0.35	5.2
		0.49	4.3
		0.64	3.2
		0.75	3.8
Mild steel at 300°C	480	−1.0	7.1
		0.23	6.0
		0.33	5.8
Low alloy steel at 20°C	835	−1.0	7.1
	680	0.0	6.6
		0.33	5.1
		0.50	4.4
		0.64	3.3
		0.75	2.5
Ni–Cr–Mo–V steel at 300°C	560	−1.0	7.1
		0.23	5.0
		0.33	5.4
		0.64	4.9
Marageing steel at 20°C	2010	0.67	2.7
A533B	700	0.1	8.0
		0.3	5.7
		0.5	4.8
		0.7	3.1
		0.8	3.0
A508	570	0.1	6.7
		0.5	5.6
		0.7	3.1
T–1	850	0.2	~5.5
		0.4	~4.4
		0.9	~3.3
300–M		0.05	8.5
650°C temper, oil quench		0.7	3.7
300–M		0.05	6.2
650°C temper step cool		0.7	2.7
18–8 Austenitic at 20°C	685	−1.0	6.0
	665	0.0	6.0
		0.33	5.9
		0.62	4.6
		0.74	4.1
Grey cast iron	255	0.0	7.0
		0.5	4.5

Table 1.2.3 Threshold stress intensity factor for fatigue crack growth, ΔK_{TH}, of some non-ferrous materials

Material	***Tensile strength (MPa)***	***R ratio***	***ΔK_{TH} ($MN/m^{3/2}$)***
Aluminium	77	−1.0	1.0
		0.0	1.7
		0.33	1.4
		0.52	1.2
A356 Cast Al alloy	200	0.1	6.1
		0.8	2.4
AF42 Cast Al alloy	279	0.0	4.8
		0.5	3.4
		0.8	1.7
L65 Al–4.5 Cu cast alloy	450	−1.0	2.1
	495	0.0	2.1
		0.33	1.7
		0.5	1.5
		0.67	1.2
2024–T3 Wrought Al alloy	483	0.8	1.7
2219–T851 Wrought Al alloy	420	0.1	5.0
		0.5	1.7
Al–Zn Super plastic alloy	250	0.0	1.4
		0.5	0.8
Copper	225	−1.0	2.7
	215	0.0	2.5
		0.33	1.8
		0.56	1.5
		0.8	1.3
60–40 Brass	325	0.0	3.5
		0.33	3.1
		0.51	2.6
		0.72	2.6
Phosphor-bronze	455	−1.0	5.9
ZW1 Magnesium alloy 0.6% Zr	250	0.0	0.83
		0.67	0.66
AM 503 Magnesium alloy 1.6% Mn	165	0.0	0.99
		0.67	0.77
Nickel	430	0.0	7.9
		0.33	6.5
		0.57	5.2
		0.71	3.6
Monel	525	−1.0	5.6
		0.0	7.0
		0.33	6.5
		0.57	5.2
		0.71	3.6
Inconel	655	−1.0	6.4
	650	0.0	7.1
		0.57	4.7
		0.71	4.0
Titanium 6Al–4V	540	0.15	~6.6
		0.6	2.2

Table 1.2.3

Table 1.2.4 Comparison of $\left[\frac{K_{1c}}{\sigma_y}\right]^2$ and $\left[\frac{K_{1scc}}{\sigma_y}\right]^2$ ratios for high strength and high toughness, beryllium, aluminium, titanium and iron base alloys

	High toughness		*High strength*	
	$\left[\frac{K_{1c}}{\sigma_y}\right]^2$ *mm*	$\left[\frac{K_{1scc}}{\sigma_y}\right]^2$ *mm*	$\left[\frac{K_{1c}}{\sigma_y}\right]^2$ *mm*	$\left[\frac{K_{1scc}}{\sigma_y}\right]^2$ *mm*
Beryllium	6.4	–	2.5	–
Aluminium	13.4	7.2	3.9	0.06–2.7
Titanium	44.5	5.2	2.1	0.2
Steel	16.4	5.2	0.3	0.005

Table 1.2.5 Optimum materials to be fracture safe with a 1 mm flaw stressed to σ_y in air

	σ_y *(MPa)*	K_{1c} *(MN/m$^{3/2}$)*	d/K_{1c} *(kg/MN per m$^{3/2}$)*	*W* *(kg/m^2)*
Beryllium	224	15	124	8.1
Aluminium	520	34	79.5	5.2
Titanium	1232	82	55	3.6
Steel	1728	115	68.6	4.5

Table 1.2.6 Optimum materials to be fracture safe with a 1 mm flaw stressed to σ_y in sodium chloride solution

	σ_y *(MPa)*	K_{1c} *(MN/m$^{3/2}$)*	d/K_{1scc} *(kg/MN per m$^{3/2}$)*	*W* *(kg/m^2)*
Aluminium	480	32	84.5	5.5
Titanium	816	55	82	5.35
Steel	1296	86	92	6.0

Table 1.2.7 The mass of unit area of pressure vessels working at a pressure of 70 MPa with a radius of 0.1 m which will leak before breaking

Alloy	$\frac{K^2_{1c}}{\sigma_y}$ *(MN/m)*	*f*	*Stress* $f\sigma_y$ *(MPa)*	*Wall thickness* t *(mm)*	*Mass/m²* *W* *(kg)*	*Relative % mass*
Steel 1	0.63	0.023	54	129	1010	100
Steel 2	15	0.525	740	9.65	76	7.5
Aluminium	4	0.143	50	140	392	38.8
Titanium	28	1.0	720	9.75	44	4.35

Table 1.2.8 The mass of unit area of pressure vessels working at a pressure of 7 MPa with a radius of 0.1 m which will leak before breaking

Alloy	$\frac{K^2_{1c}}{\sigma_y}$ *(MN/m)*	*f*	*Stress* σ_y *(MPa)*	*Wall thickness* t *(mm)*	*Mass/m²* *W (kg)*	*Relative % mass*
Steel	2.8	1	2120	0.33	2.6	62
Aluminium	2.8	1	463	1.51	4.2	100
Titanium	2.8	1	1330	0.525	2.36	56

Table 1.2.7 and Table 1.2.8

Table 1.2.9 Composition of materials shown in Fig. 1.2.31

	High speed tool steels and white cast irons[a]									*Almost stoichiometric WC cemented carbide*					
Material	*C* (%)	*Si* (%)	*Mn* (%)	*Cr* (%)	*Ni* (%)	*Mo* (%)	*W* (%)	*Co* (%)	*V* (%)	*Laboratory material*[b]	*Co* (%)	*Grain size* (μ m)	*Commercial material*[c]	*Co* (%)	*Grain size* (μ m)
1. DIN50 NiCr13	0.55	0.30	0.40	1.0	3.0	0.3	–	–	–	12	3	1.5	21	6	1
2. AISI A2	1.00	0.30	0.60	5.2	–	1.1	–	–	0.2	13	3	2.5–3	22	6	1.5
3. AISI M2	0.86	0.24	0.14	4.4	–	5.2	6.7	–	2.1	14	6	7–8	23	8	1.3
4. AISI M42	1.08	0.20	0.30	3.6	–	9.5	1.7	8.3	1.1	15	9	1.5	24	8	3.2
5. STORA 30	1.20	0.25	0.29	4.02	–	4.8	6.0	10.3	3.18	16	9	2.5–3	25	11	1.9
6. ASP 30	1.23	0.45	0.37	4.1	–	5.0	6.3	8.2	3.14	17	9	7–8	26	11	3.3
7. AISI 01	0.90	–	1.16	0.5	–	–	0.49	–	0.10	18	15	1.5	27	15	1.9
8. AISI D2	1.48	0.26	0.27	12.7	–	0.82	–	–	0.94	19	15	2.5–3	28	15	3.3
9. H-IC-C-525	3.35	0.97	0.30	1.31	4.56	0.35	–	–	–	20	15	7–8	29	20	1.8
10. H-IC-C-625	3.52	0.80	0.33	1.93	4.95	0.30	–	–	–	–	–	–	30	25	3
11. AISI M7	1.0	0.30	0.30	4.0	0.07	8.7	1.7	0.25	2.0	–	–	–	–	–	–

[a] Erikson, K. (1973). Fracture toughness of hard high-speed steels, tool steels and white cast irons. *Scandinavian Journal of Metallurgy*, **2**, 197–203.
[b] Lueth, R. C. (1972). Determination of fracture toughness parameters for tungsten carbide–cobalt alloys. In *Fracture Mechanics of Ceramics*, Vol. 2. Plenum Press, London, p. 791.
[c] Inglestrom, N. and Nordberg, H. (1974). The fracture toughness of cemented tungsten carbide. *Engineering Fracture Mechanics*, **6**, 597–607.

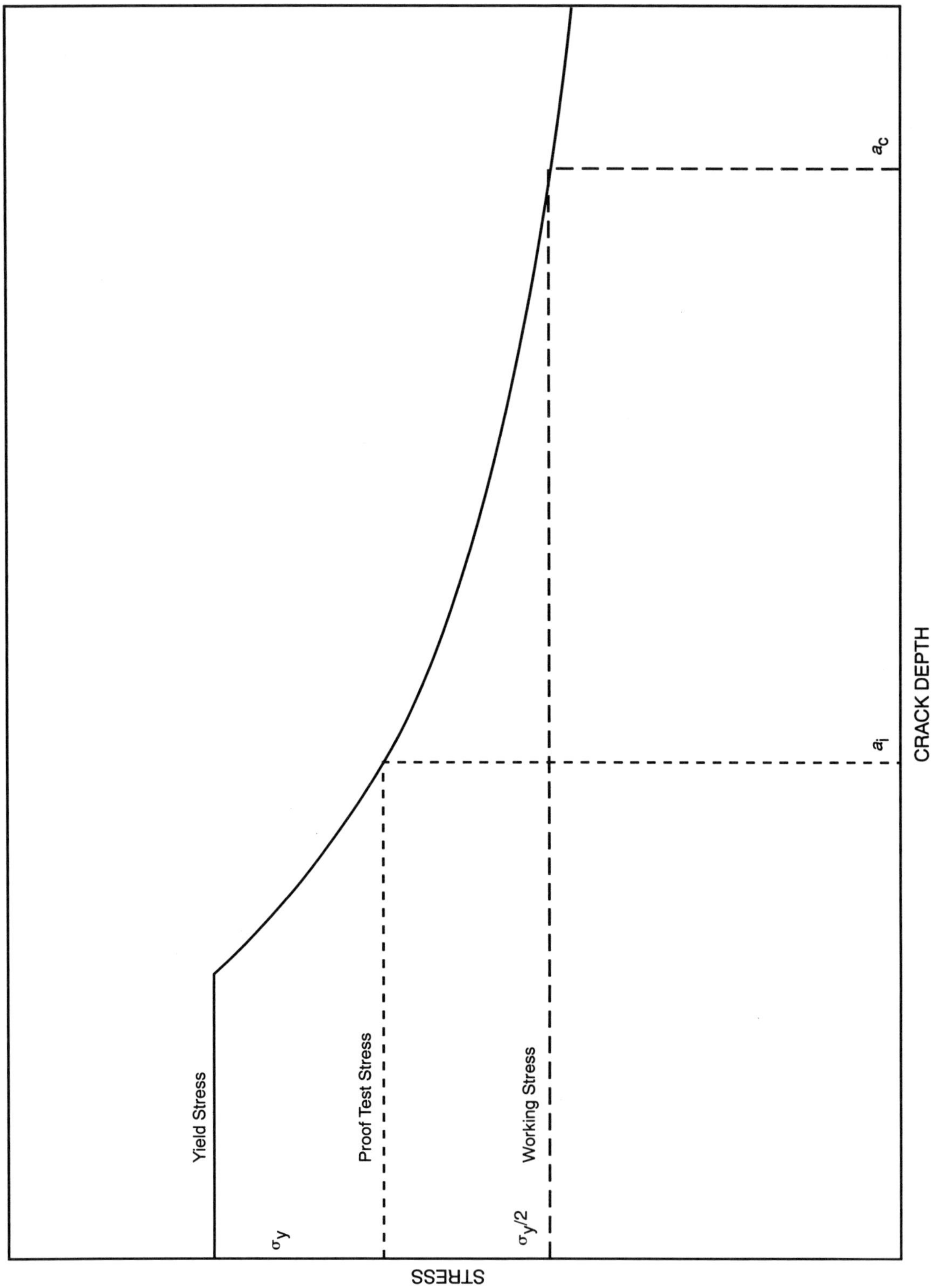

Fig. 1.2.7 Applied stress–crack depth relationship

Fig. 1.2.7

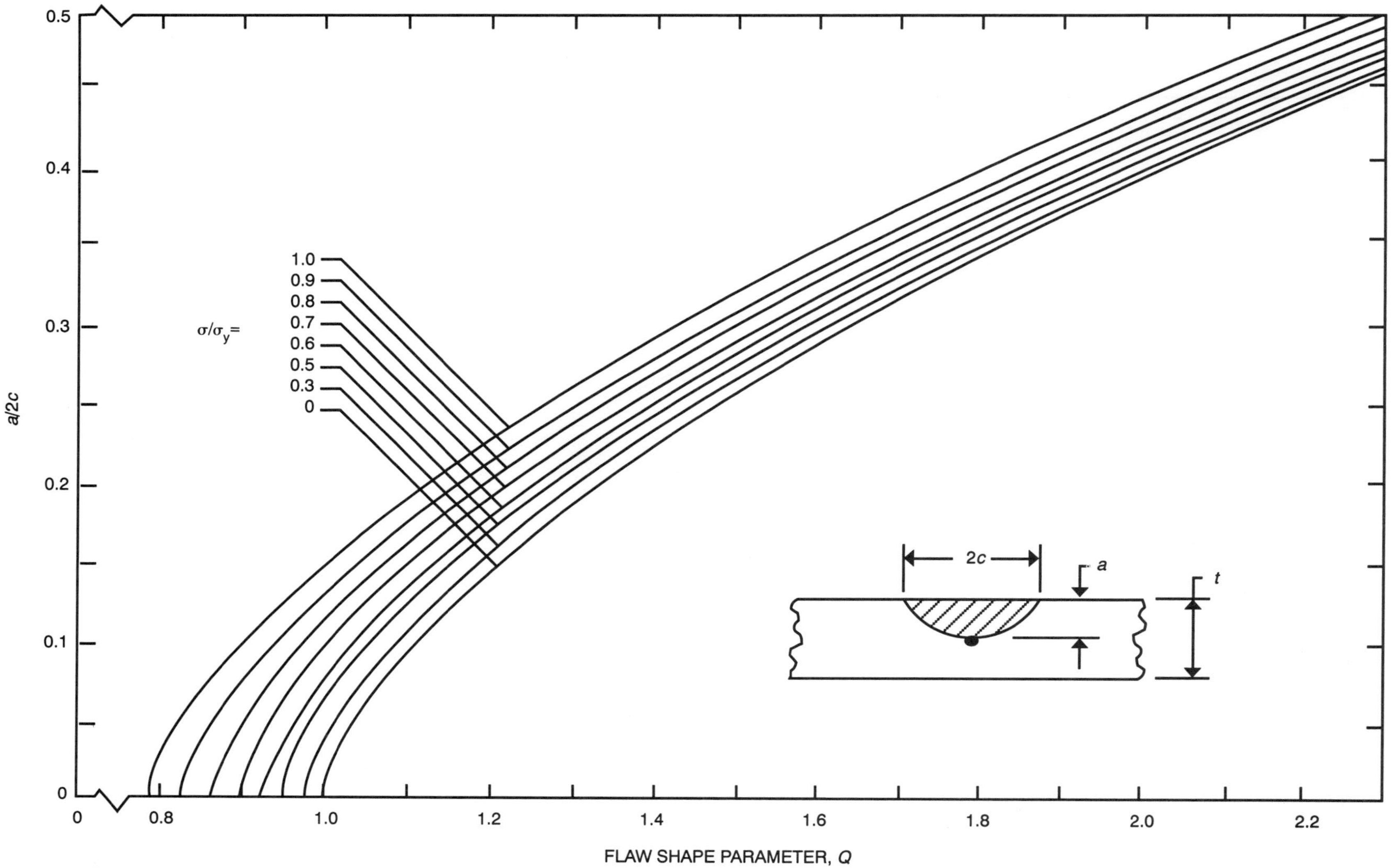

Fig. 1.2.8 Flaw shape parameter for surface and internal flaws ($Q = [\phi^2 - 0.212(\sigma/\sigma_y)^2]$, where $\phi = \int_0^{\pi/2} [\sin^2\theta + (a/c)^2 \cos^2\theta]^{1/2} d\theta$)

Fig. 1.2.8

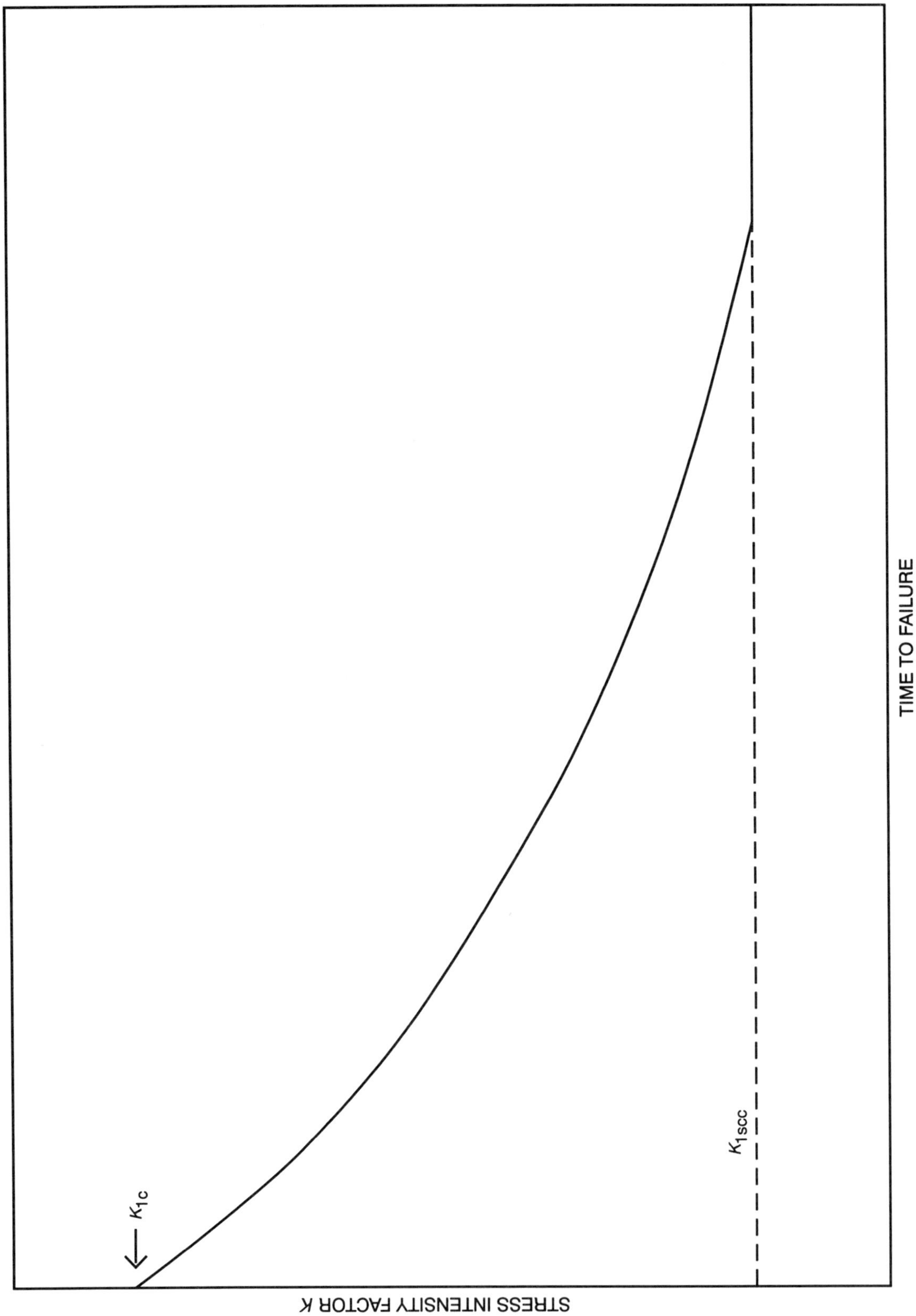

Fig. 1.2.9 Effect of stress intensity factor on time to failure in an aggressive environment

Fig. 1.2.9

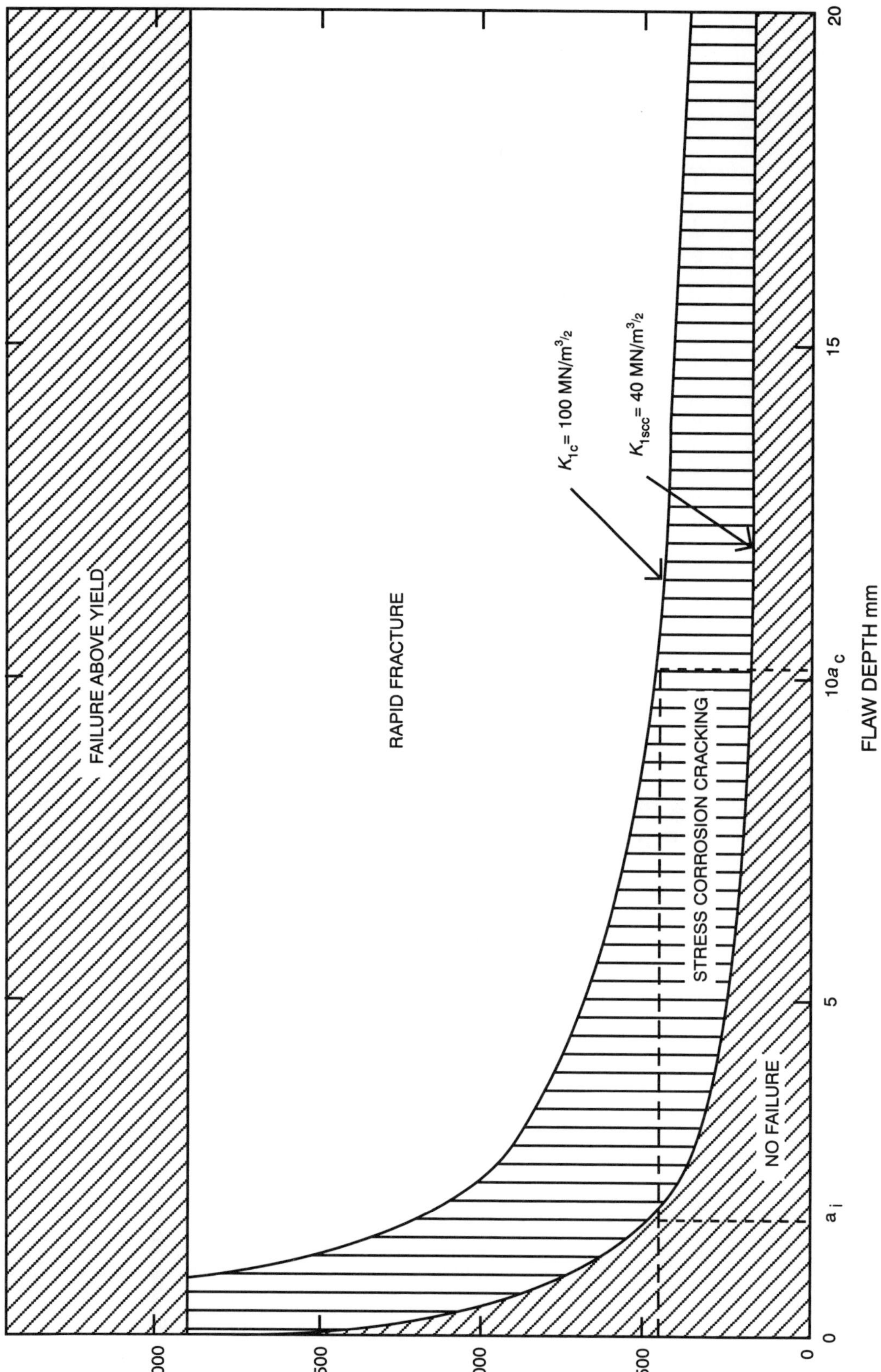

Fig. 1.2.10 Relationship between K_{1c}, K_{1scc} stress and flaw size

Fig. 1.2.10

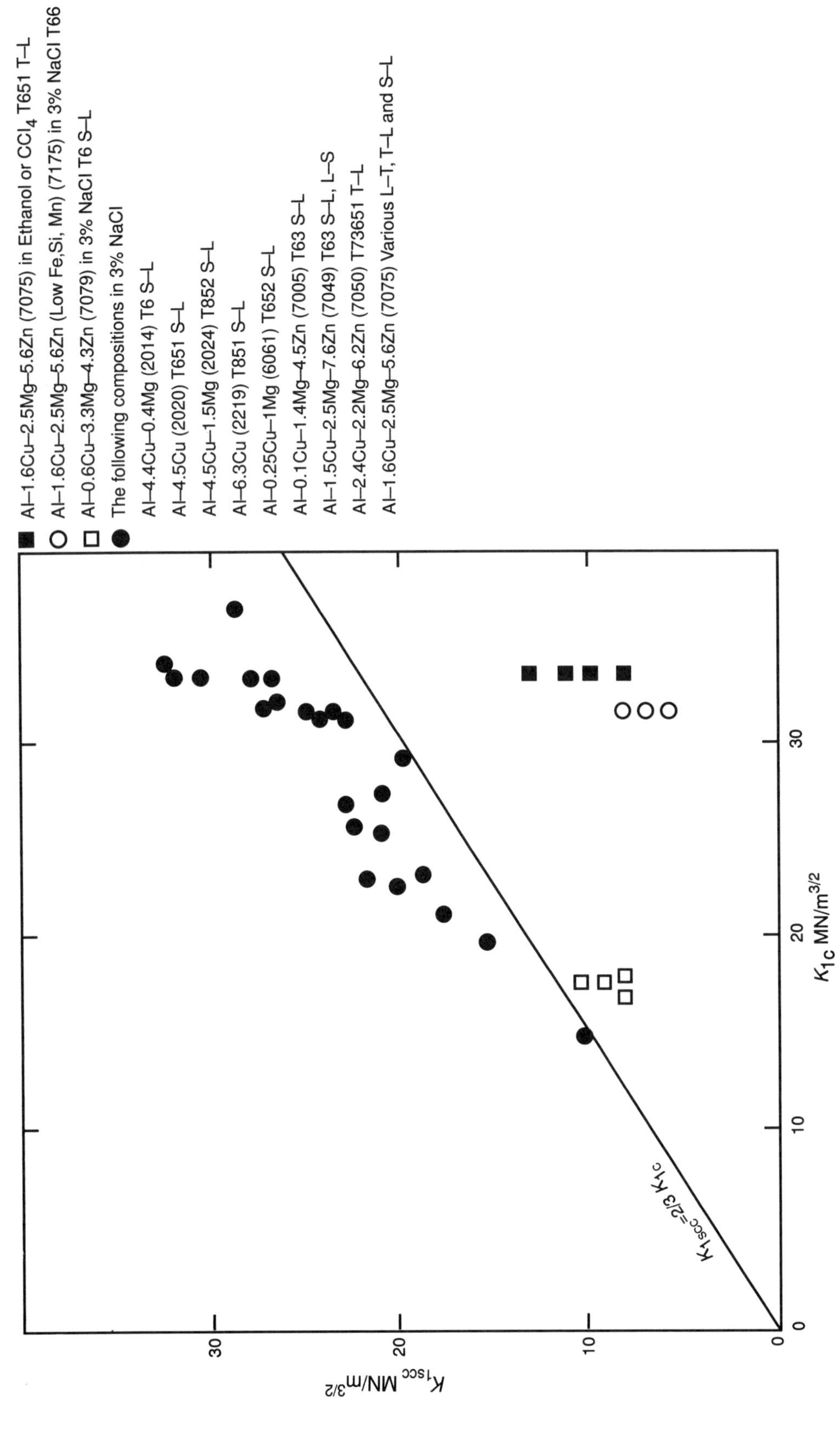

Fig. 1.2.11 Susceptibility of aluminium alloys to stress corrosion cracking in 3% NaCl

Fig. 1.2.11

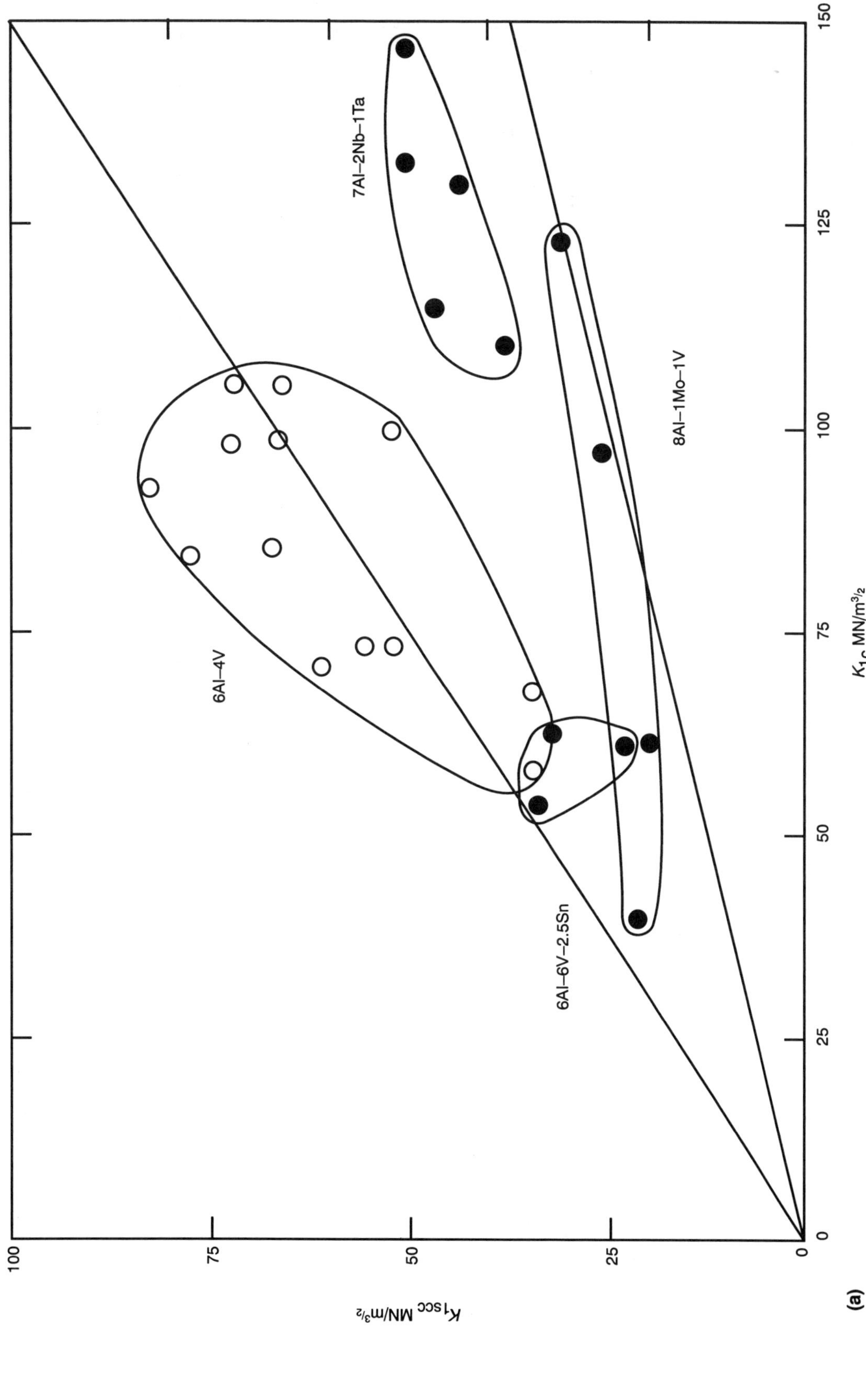

Fig. 1.2.12 Susceptibility of titanium alloys to stress corrosion cracking in NaCl

Fig. 1.2.12

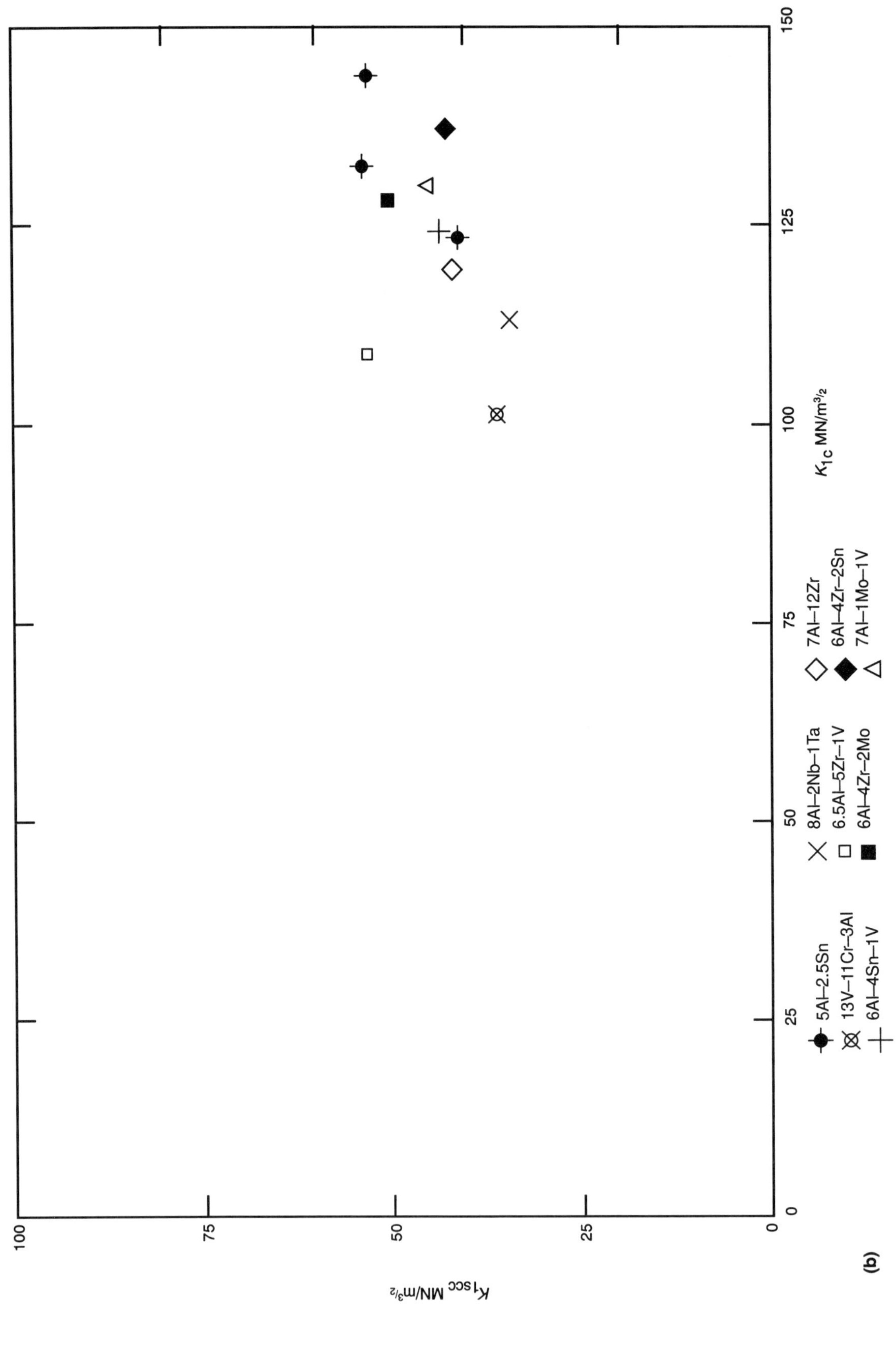

Fig. 1.2.12 Susceptibility of titanium alloys to stress corrosion cracking in NaCl – *continued*

Fig. 1.2.12 – *continued*

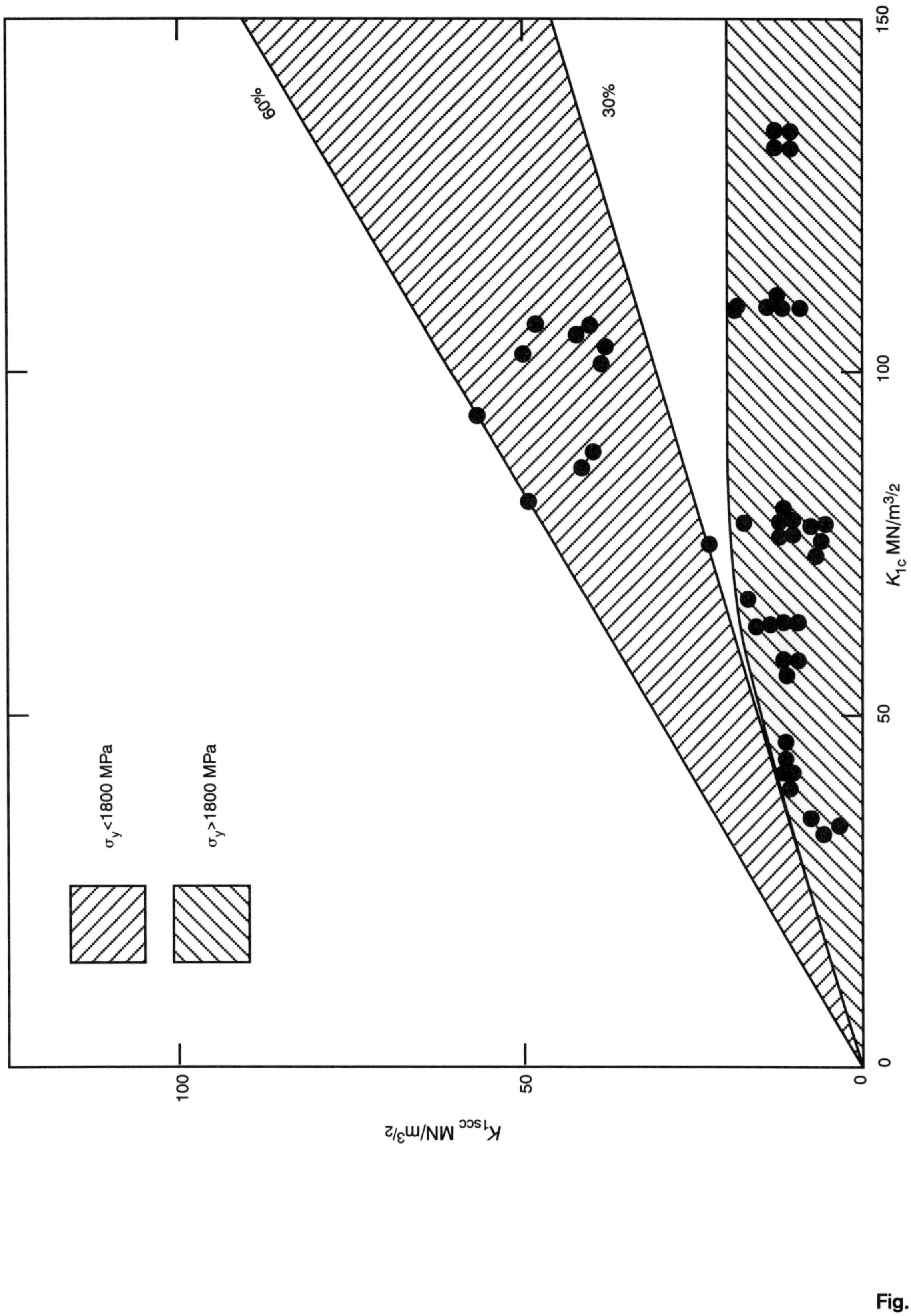

Fig. 1.2.13 Susceptibility of marageing steel (18Ni–5Mo–7Co) to stress corrosion cracking in 3% NaCl

Fig. 1.2.13

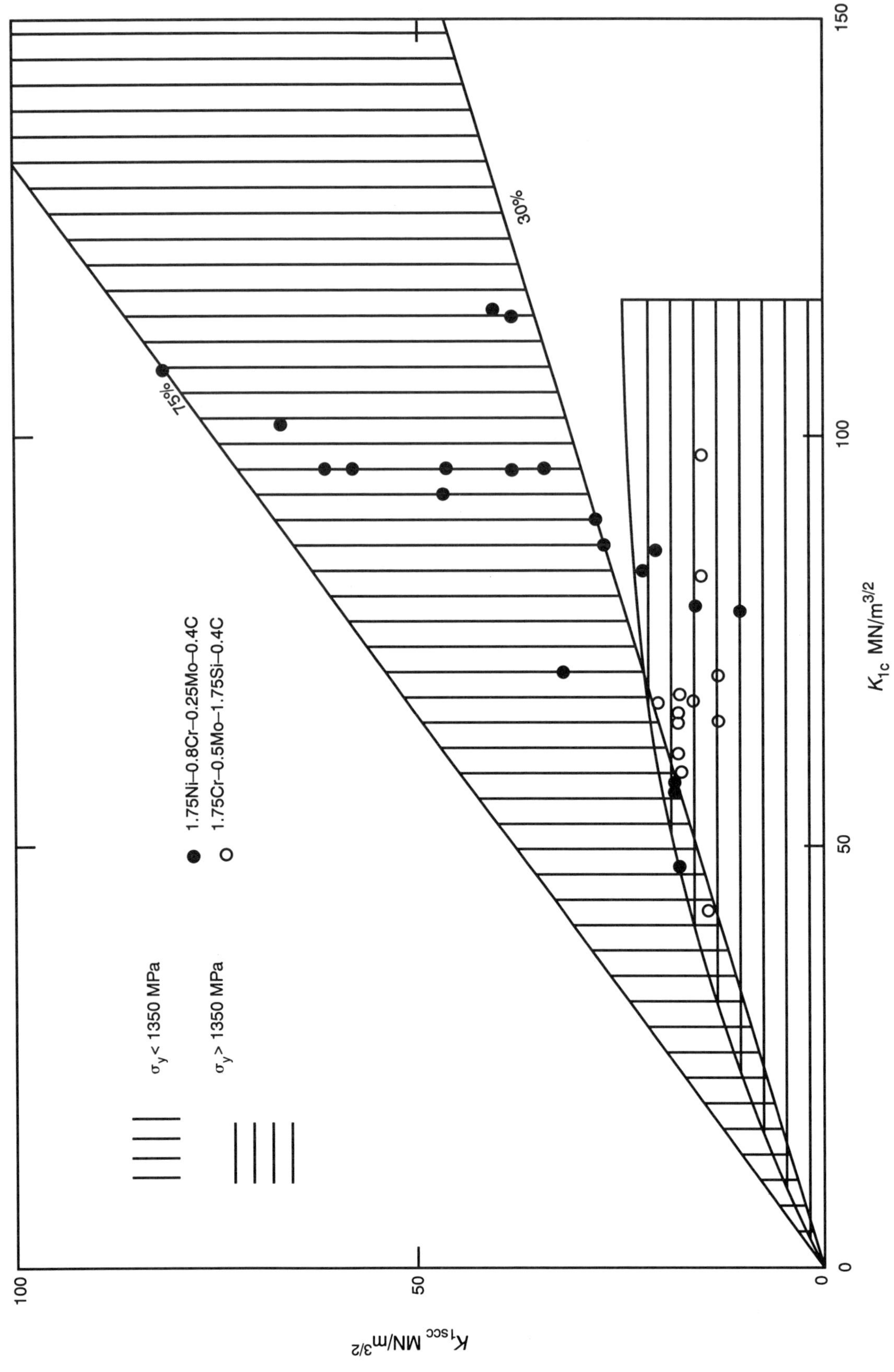

Fig. 1.2.14 Susceptibility of low alloy steels to stress corrosion cracking in 3% NaCl

Fig. 1.2.14

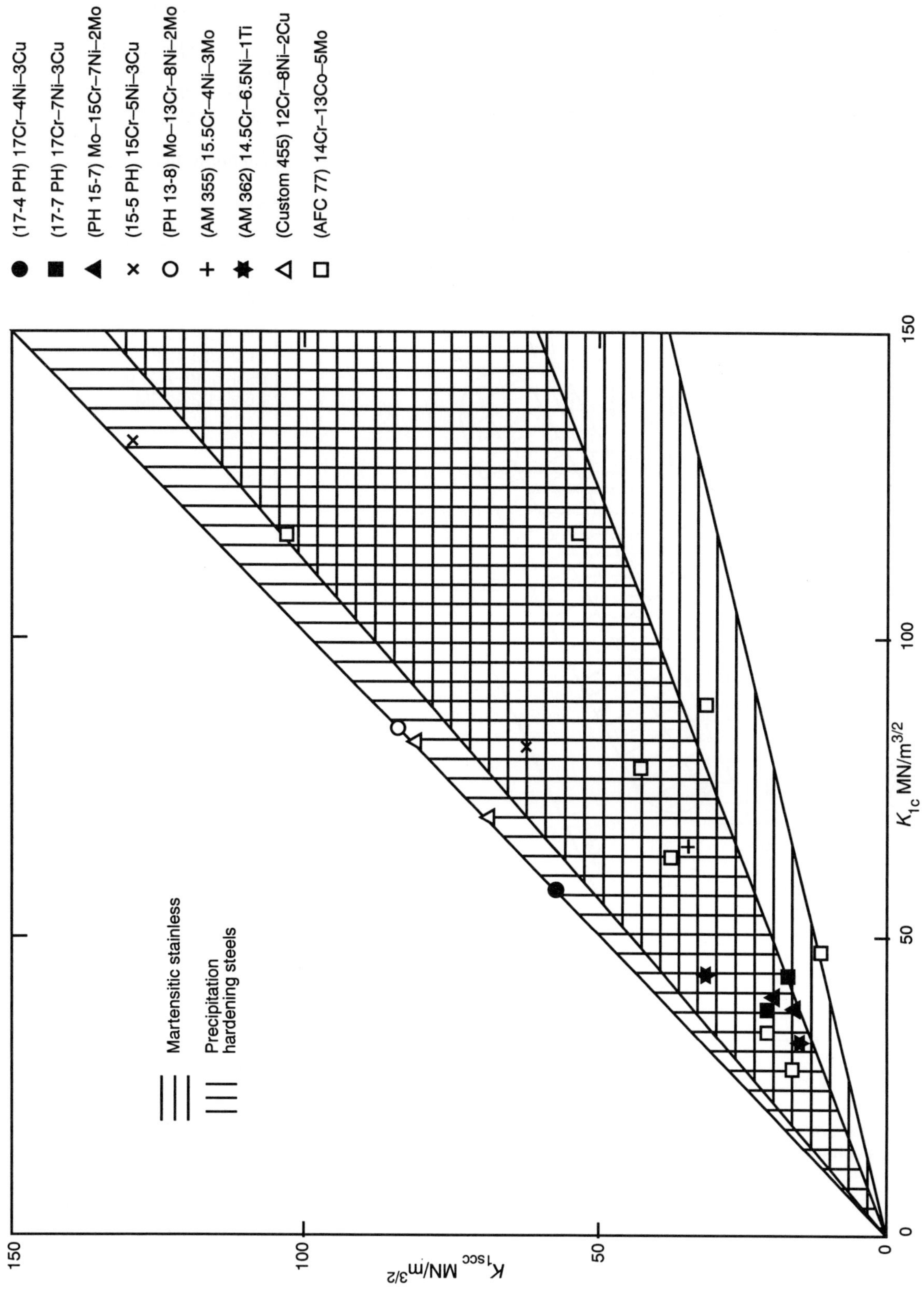

Fig. 1.2.15 Susceptibility of martensitic and precipitation hardening stainless steels to stress corrosion cracking in 3% NaCl

Fig. 1.2.15

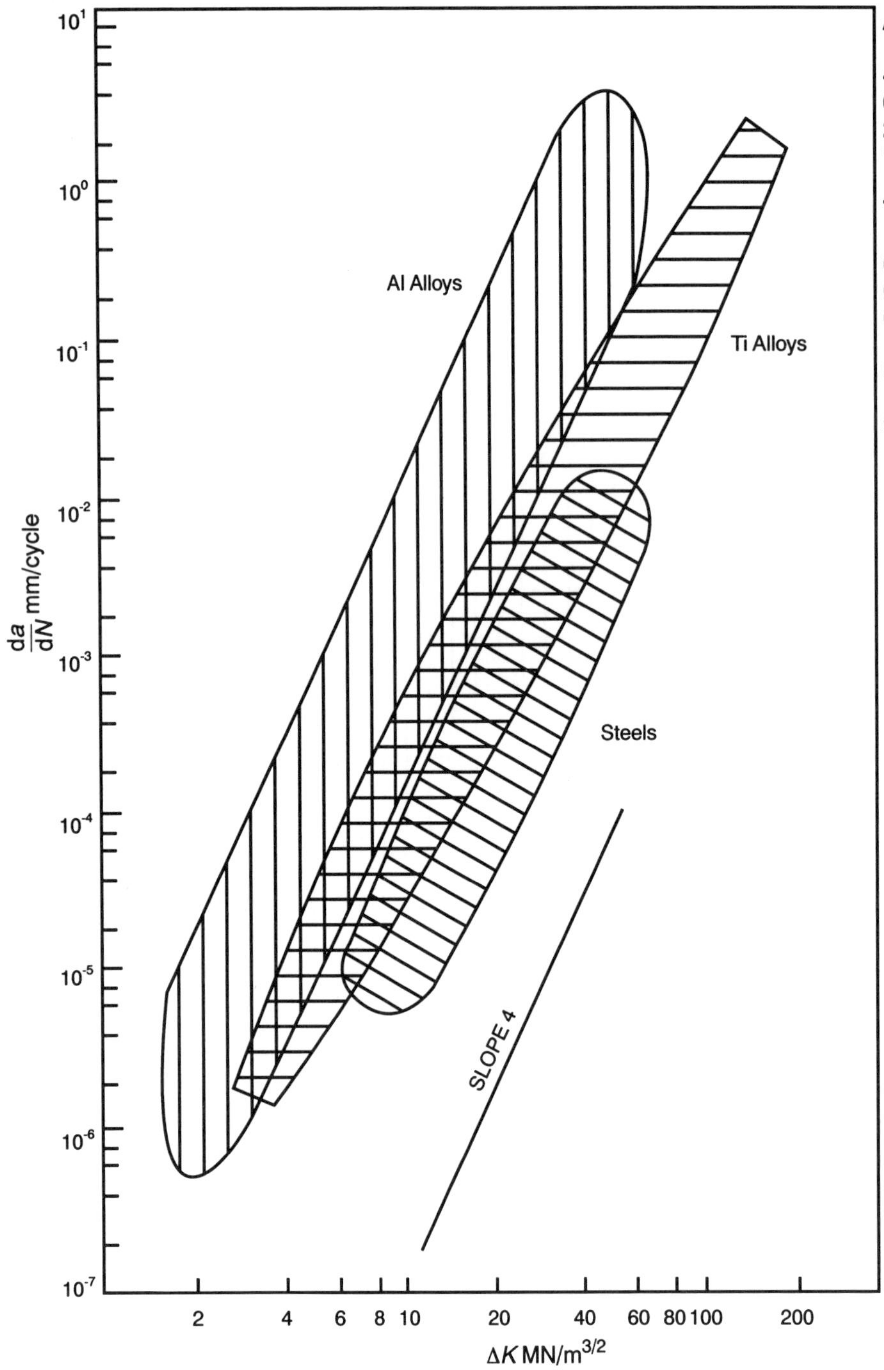

Aluminium Alloys

4.5Cu–1.5Mg (2024)
6.3Cu (2219)
2.4Cu–2.3Mo–6.2Zn (7050)
1.6Cu–2.5Mg–5.6Zn (7075)

Titanium Alloys

6Al–4V
6Al–6V–2Sn
6Al–2Sn–4Zr–6Mo

Steels

9Ni–4Co–0.2C (HP94)
13Cr–8Ni–2Mo (PH13-8Mo)
18Ni–5Mo–7Co (marageing)
1.6Ni–1.2Cr–3Mo–3C (EN24)

Fig. 1.2.16 The effect of stress intensity factor, ΔK, on the rate of fatigue crack propagation, d*a*/d*N*, for Al, Ti and Fe base alloys

Fig. 1.2.16

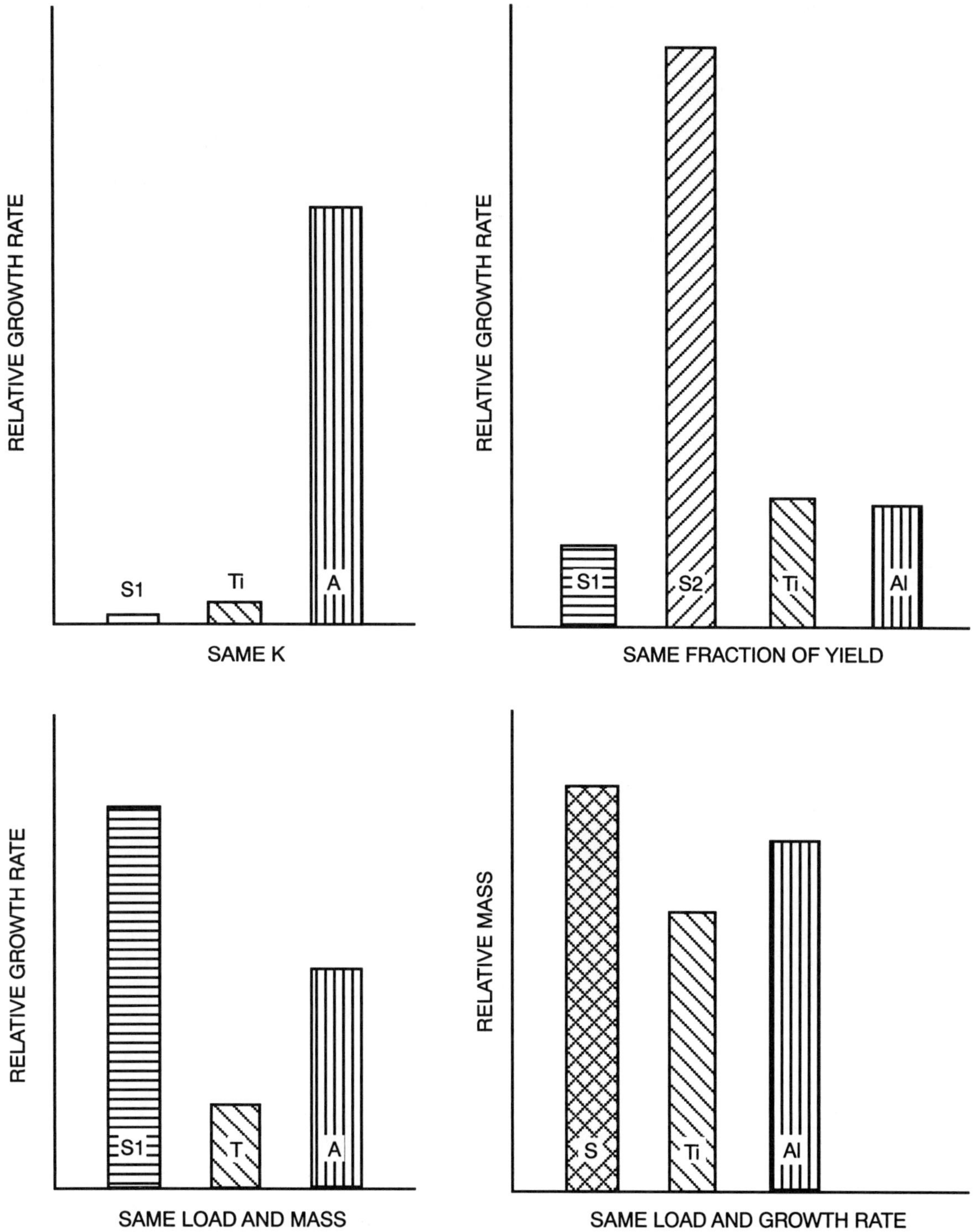

Fig. 1.2.17 Relative growth rates for different design criteria

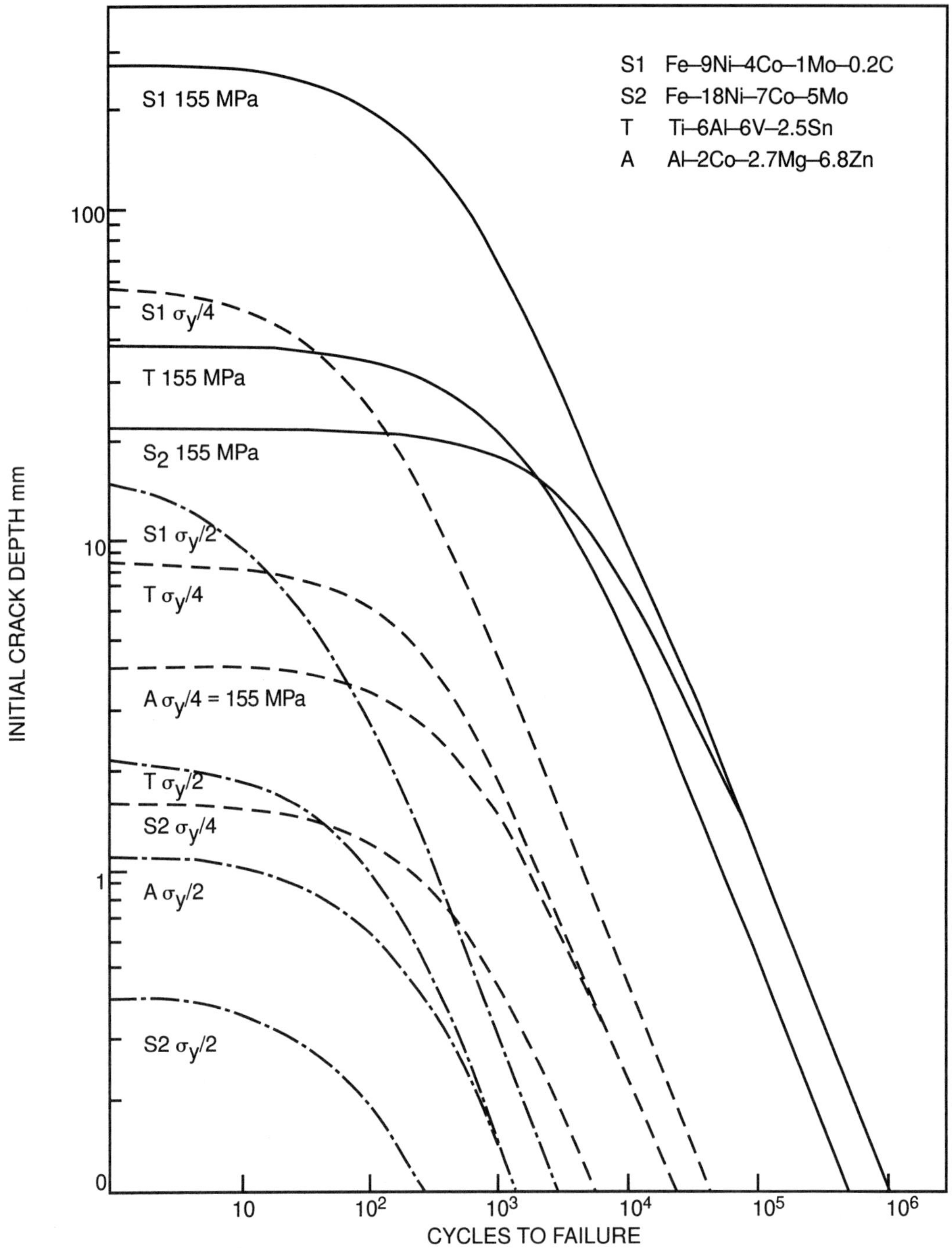

Fig. 1.2.18 A log-log plot of cycles to failure, N, for various initial crack lengths, a_i, calculated from $N = 1 / (CM^2(\Delta\sigma)^4)[1/a_i - 1/a_c$ for $\Delta\sigma = \sigma_y/2$, $\sigma_y/4$ and 155 MPa

Fig. 1.2.18

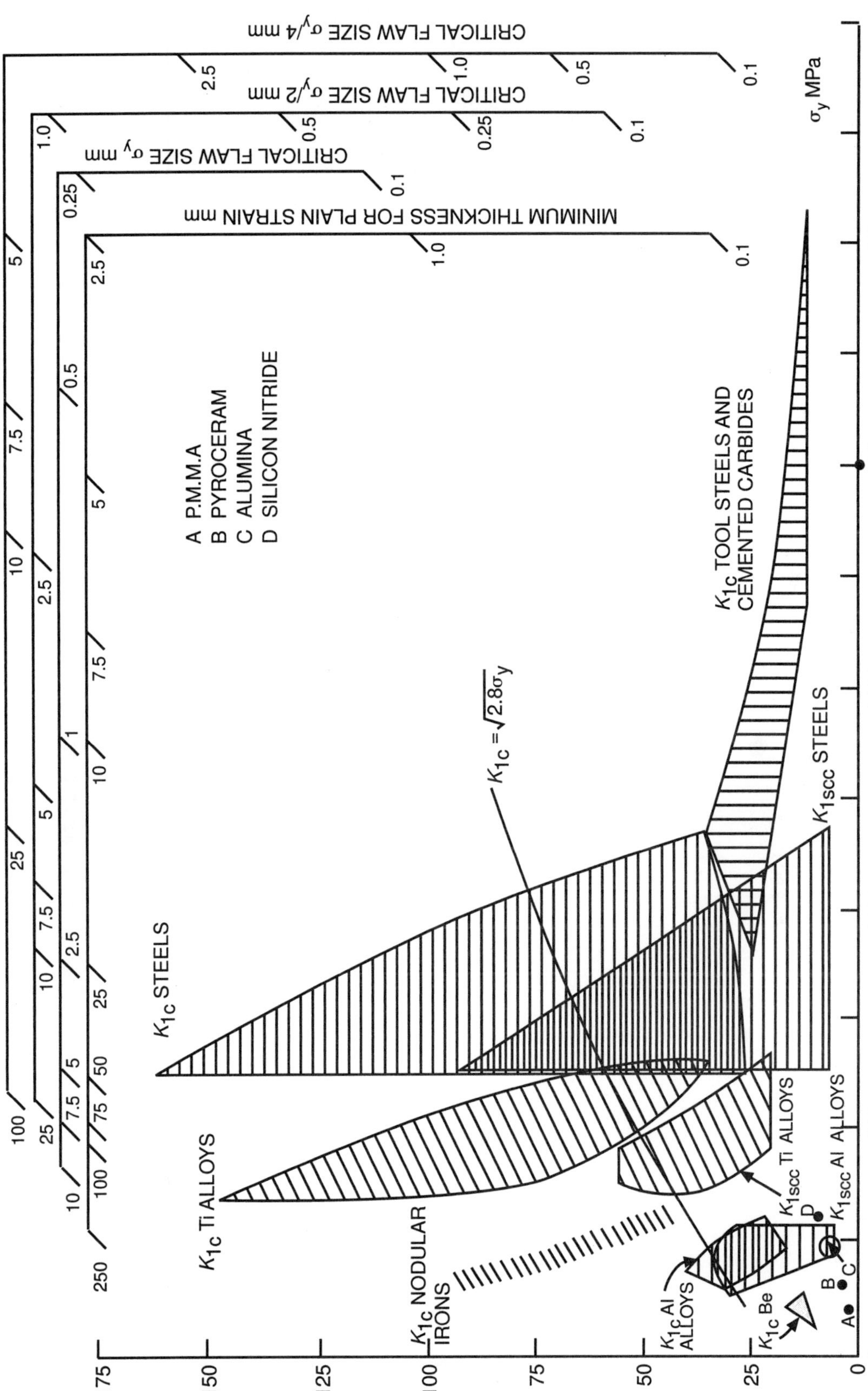

Fig. 1.2.19 Failure analysis diagram showing summary of yield strength and fracture toughness data for several alloy systems and selected non-metallics, tested at 20°C

Fig. 1.2.19

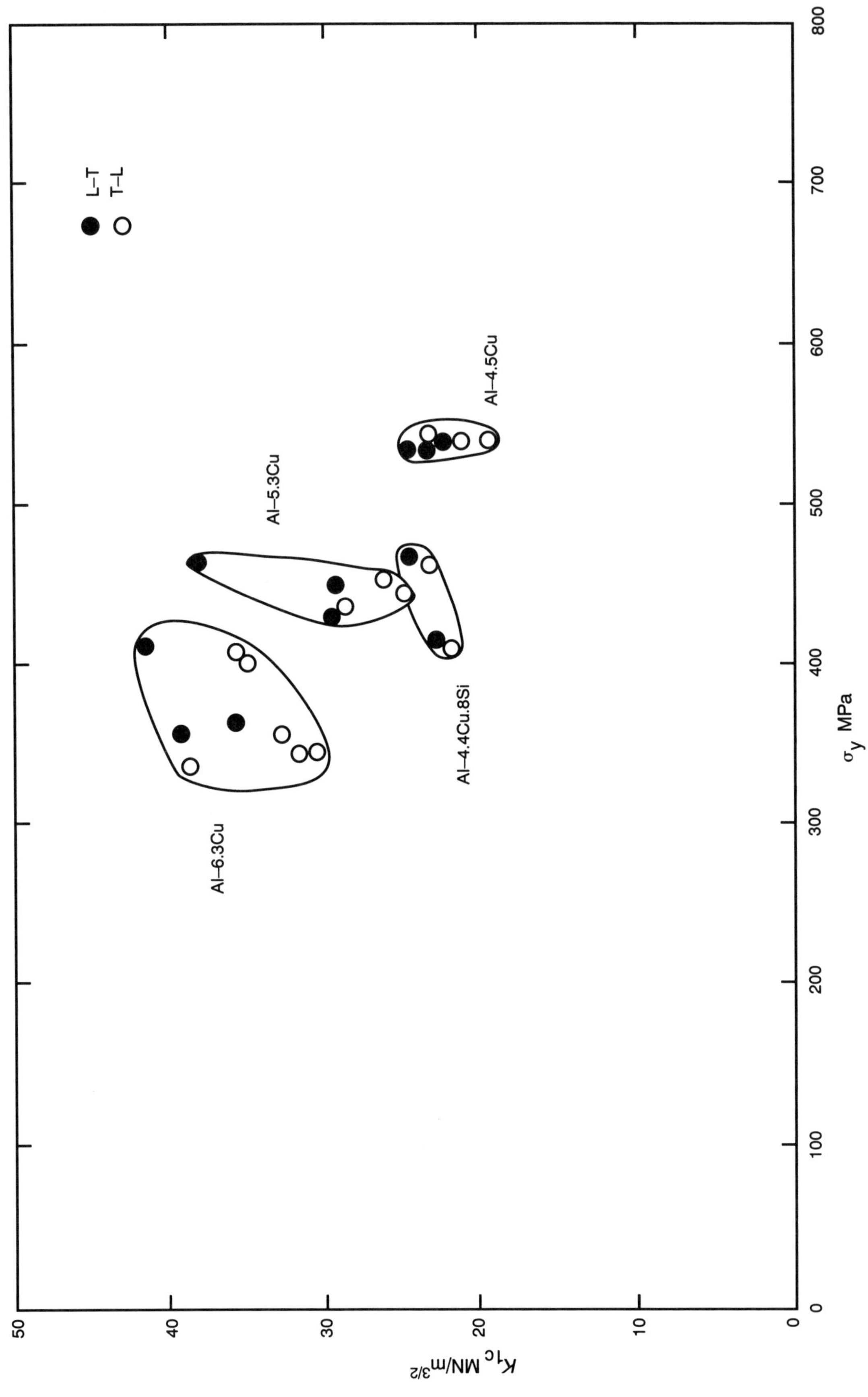

Fig. 1.2.20 K_{1c} vs σ_y data for aluminium–copper alloys tested at 20°C

Fig. 1.2.20

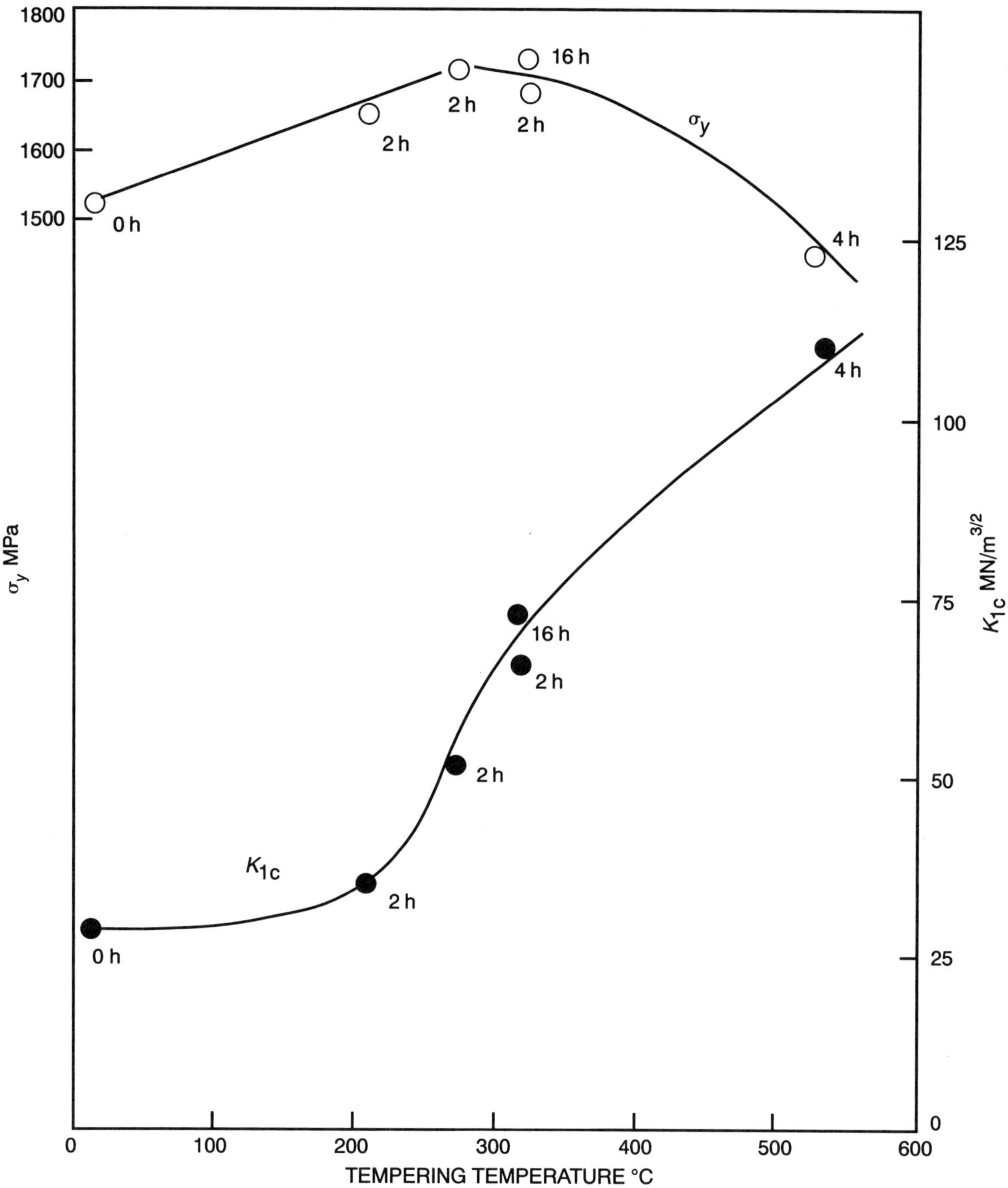

Fig. 1.2.21 The effect of tempering on the strength and toughness of a 0.5Ni–1Cr–1Mo–0.5C steel

Fig. 1.2.21

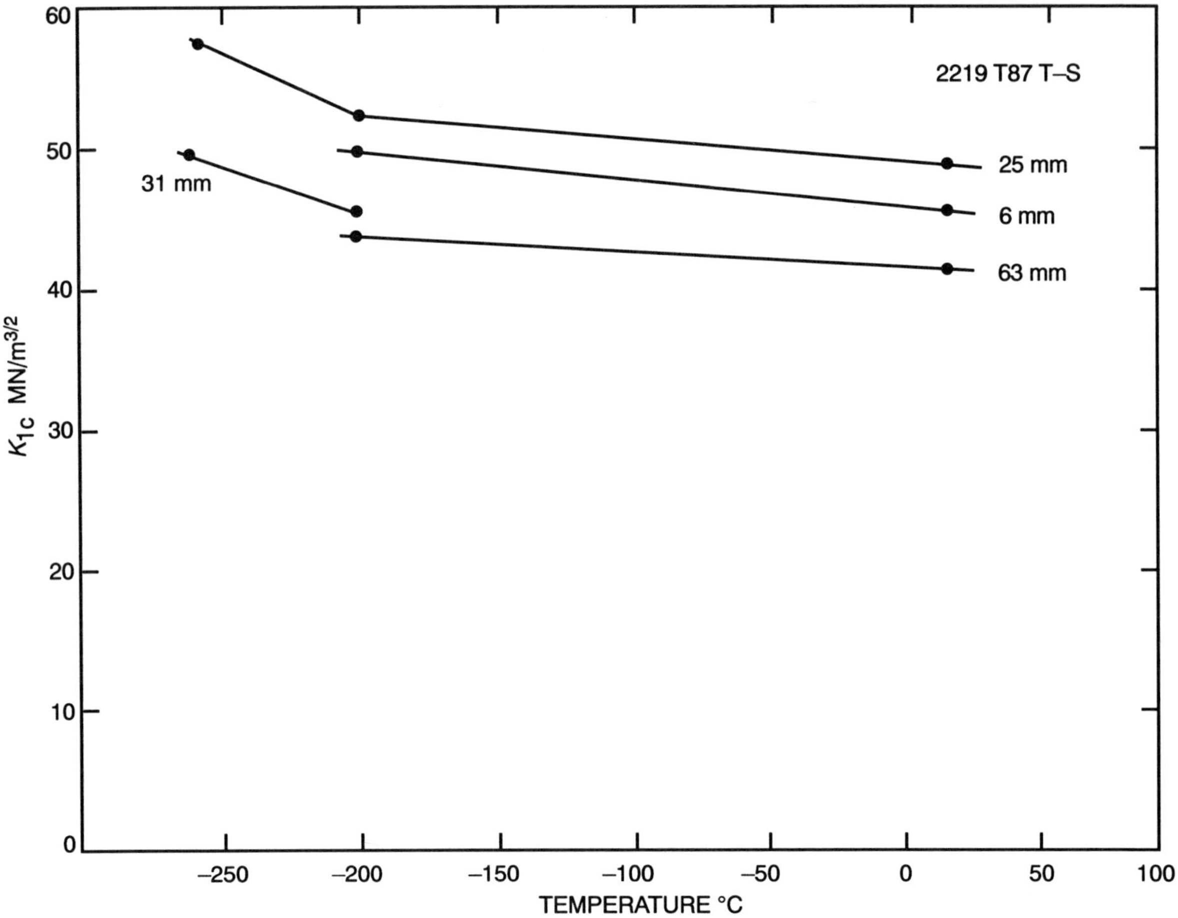

Fig. 1.2.22 Effect of plate thickness on K_{1c} of Al-6.3Cu (2219) as a function of testing temperature

Fig. 1.2.22

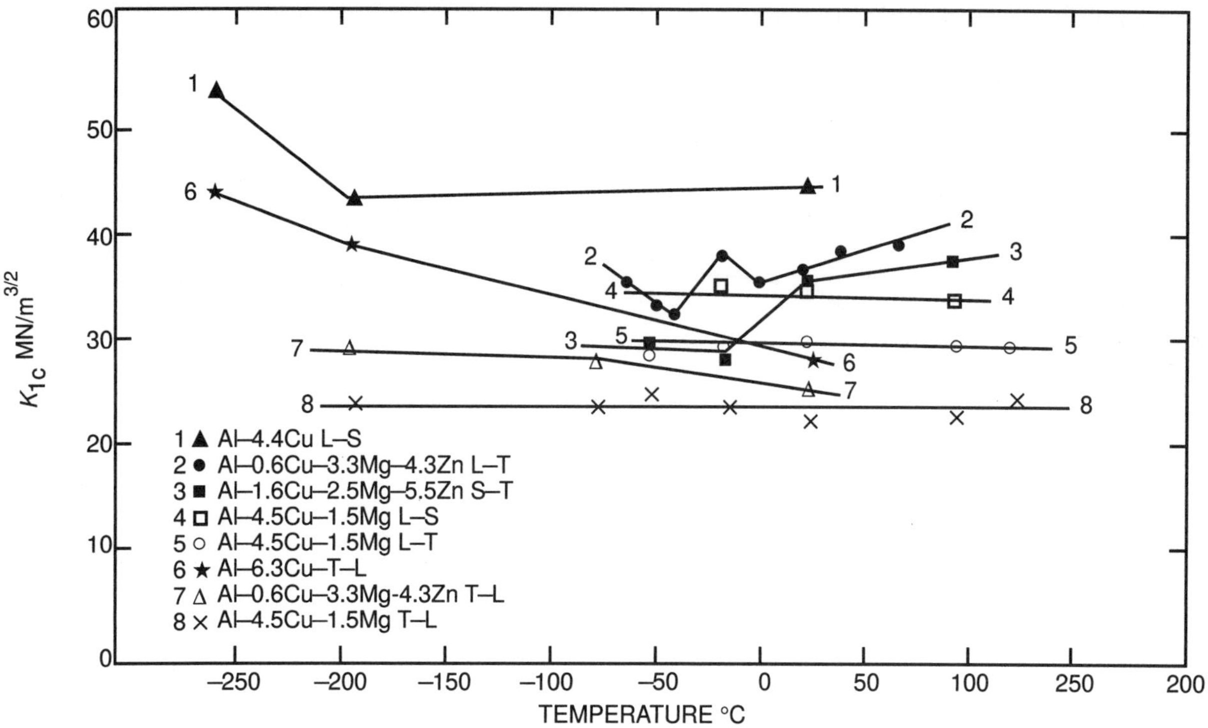

Fig. 1.2.23 Effect of testing temperature on K_{1c} of aluminium alloys

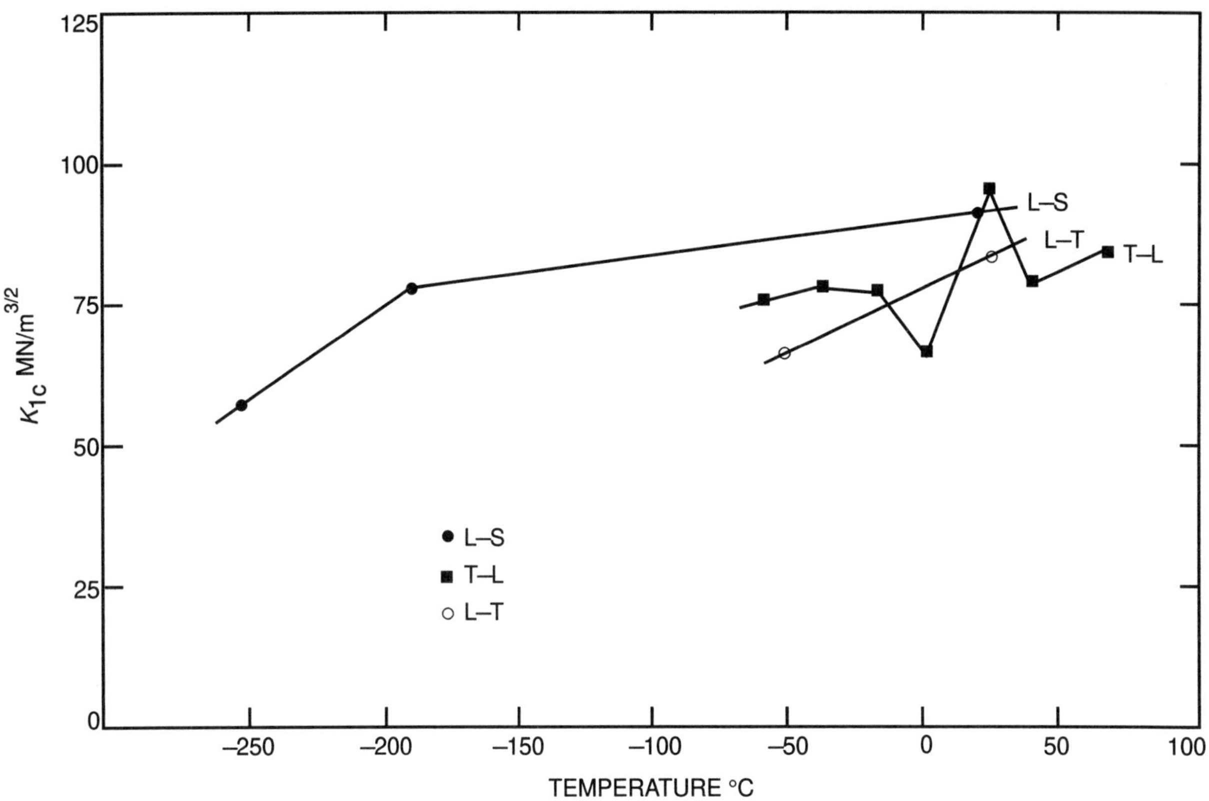

Fig. 1.2.24 Effect of testing temperature on K_{1c} of Ti–6A1–4V alloy for various crack orientations

Fig. 1.2.24

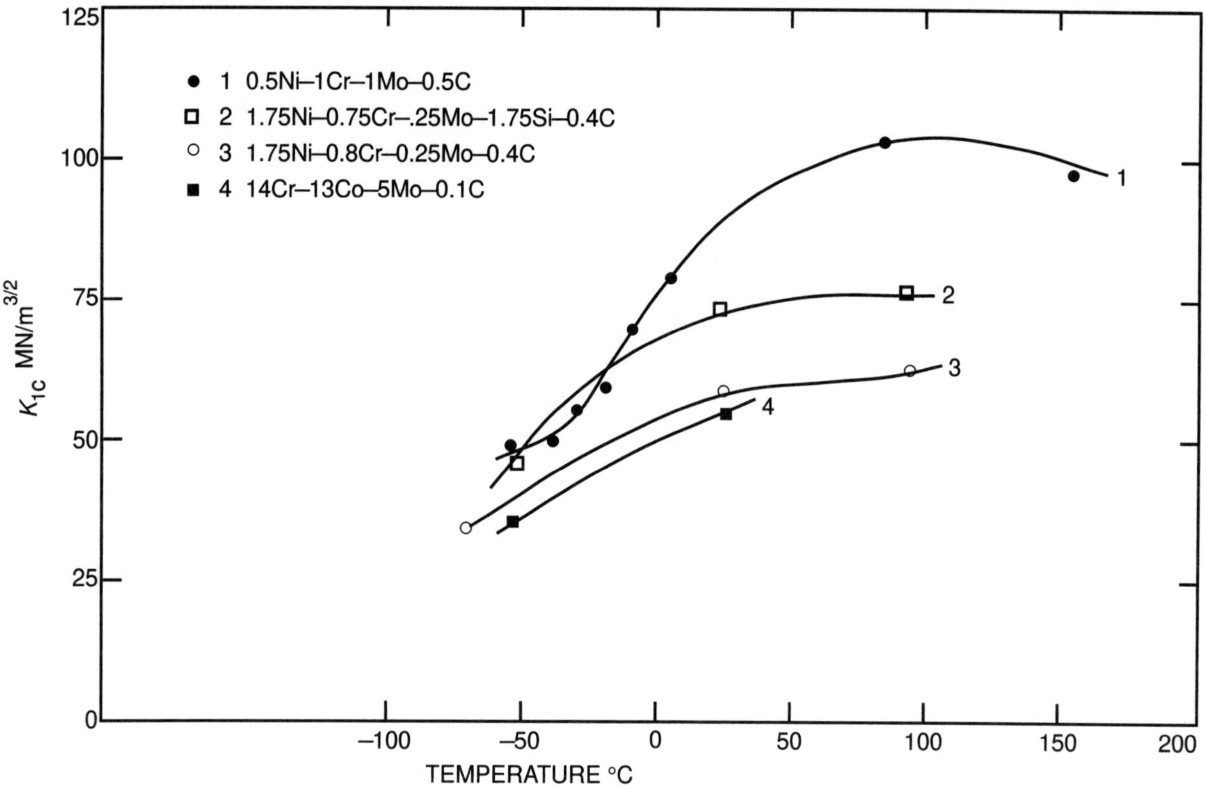

Fig. 1.2.25 Effect of testing temperature on K_{1c} of steels

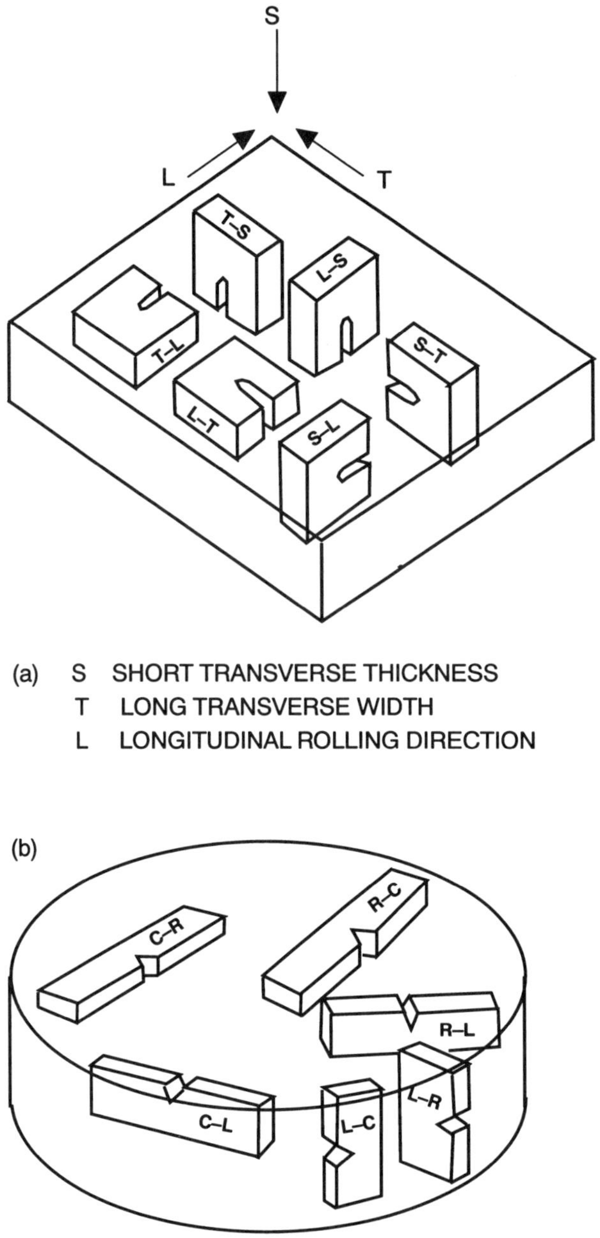

Fig. 1.2.26 Coding system for specimen orientation and crack propagation direction for specimens from (a) plate and rectangular section and (b) round bars

Fig. 1.2.26

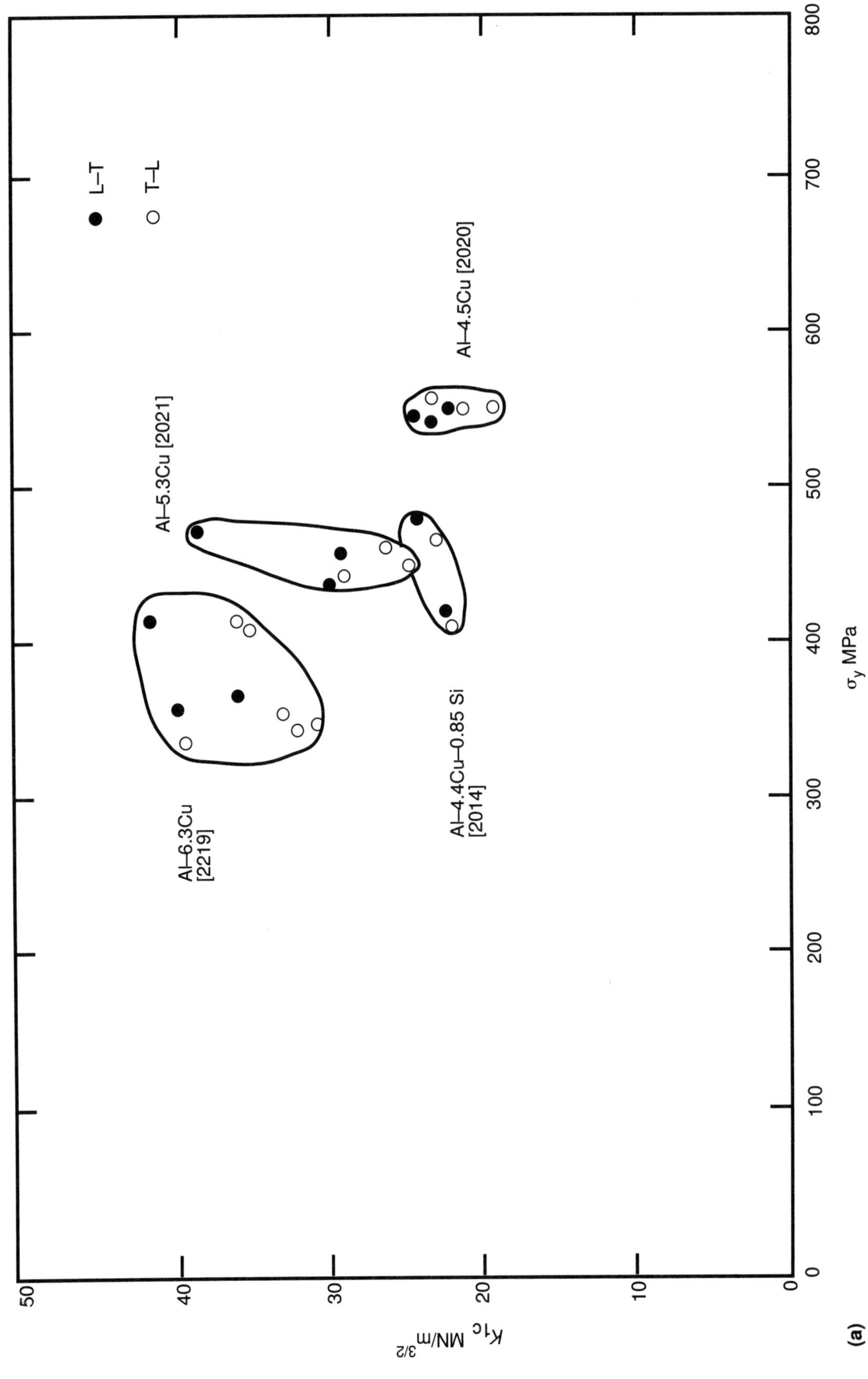

Fig 1.2.27 Fracture toughness vs yield strength for wrought aluminium alloys at 20°C

Fig. 1.2.27

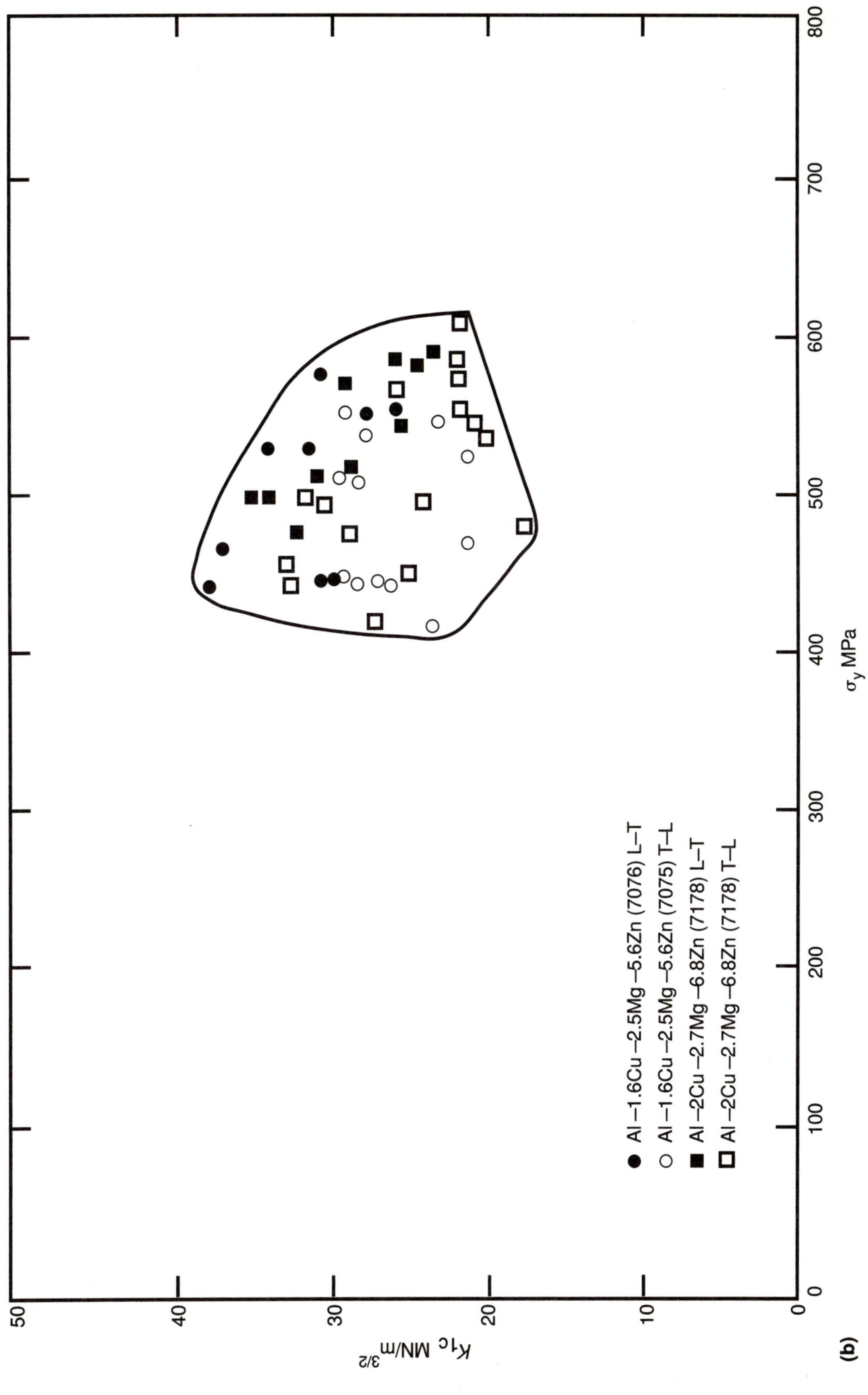

(b)

Fig. 1.2.27 Fracture toughness vs yield strength for wrought aluminium alloys at 20°C – *continued*

Fig. 1.2.27 – *continued*

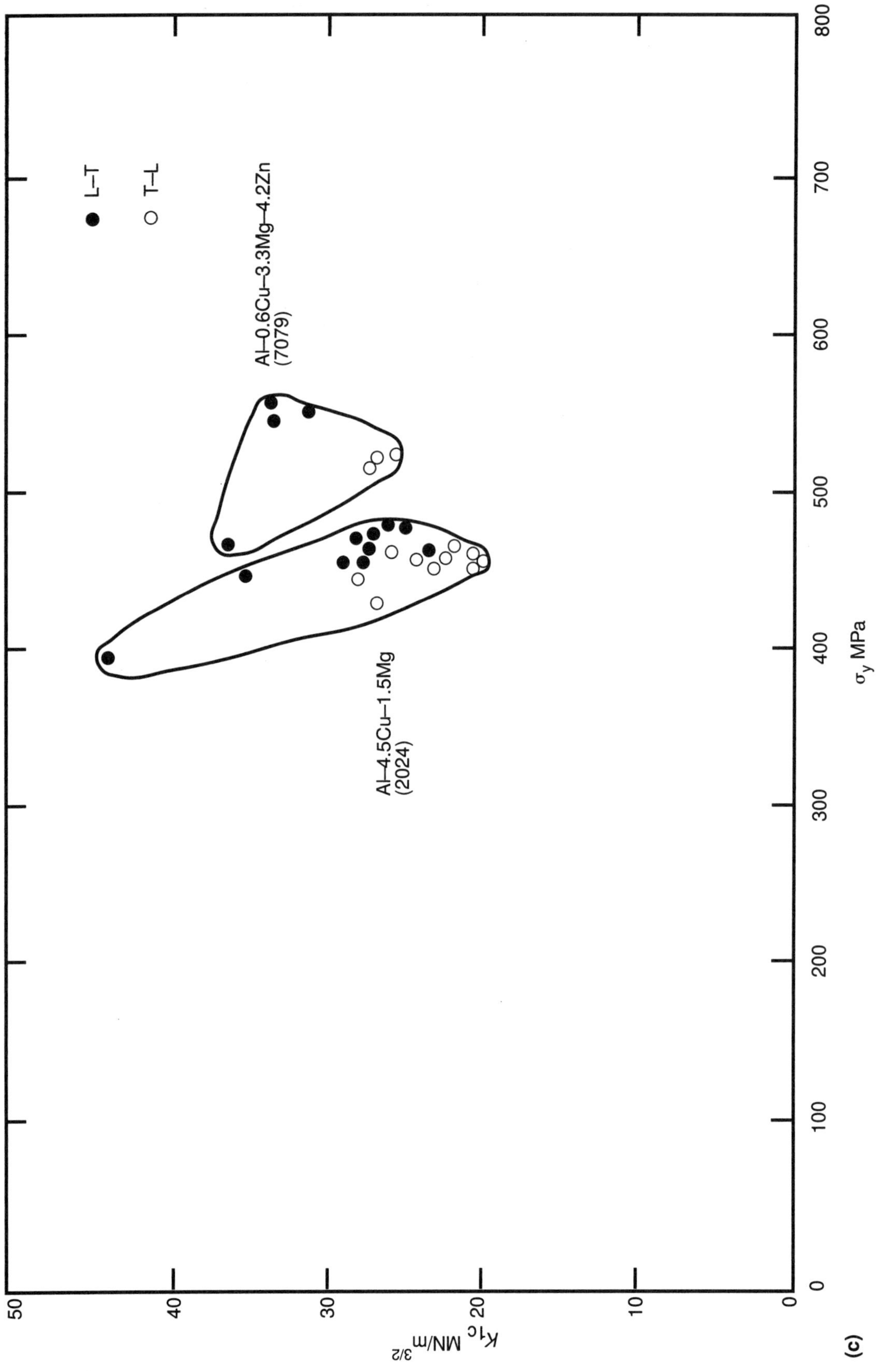

(c)

Fig. 1.2.27 Fracture toughness vs yield strength for wrought aluminium alloys at 20°C – *continued*

Fig. 1.2.27 – *continued*

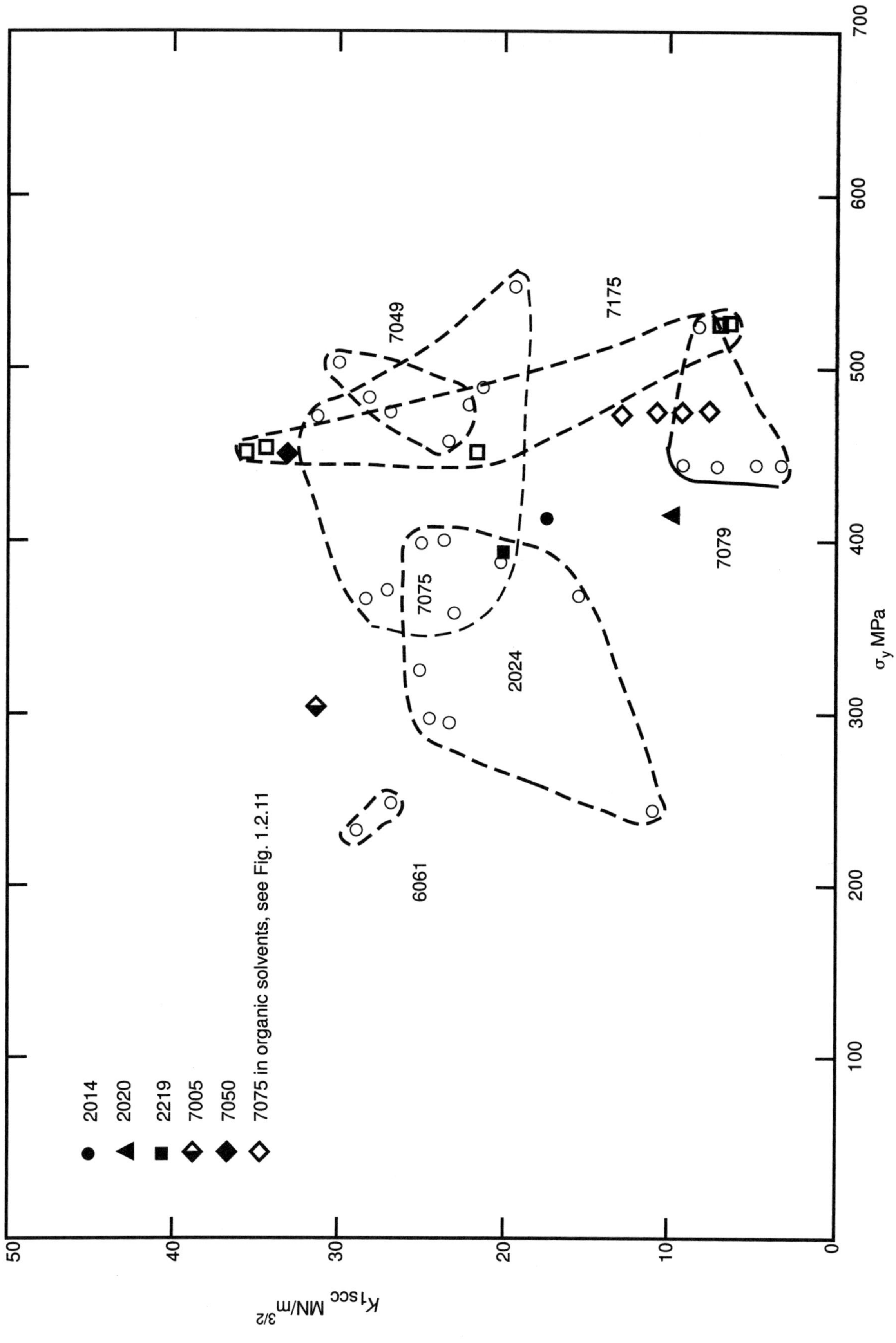

Fig. 1.2.28 K_{1scc} **vs** σ_y **for aluminium alloys**

Fig. 1.2.28

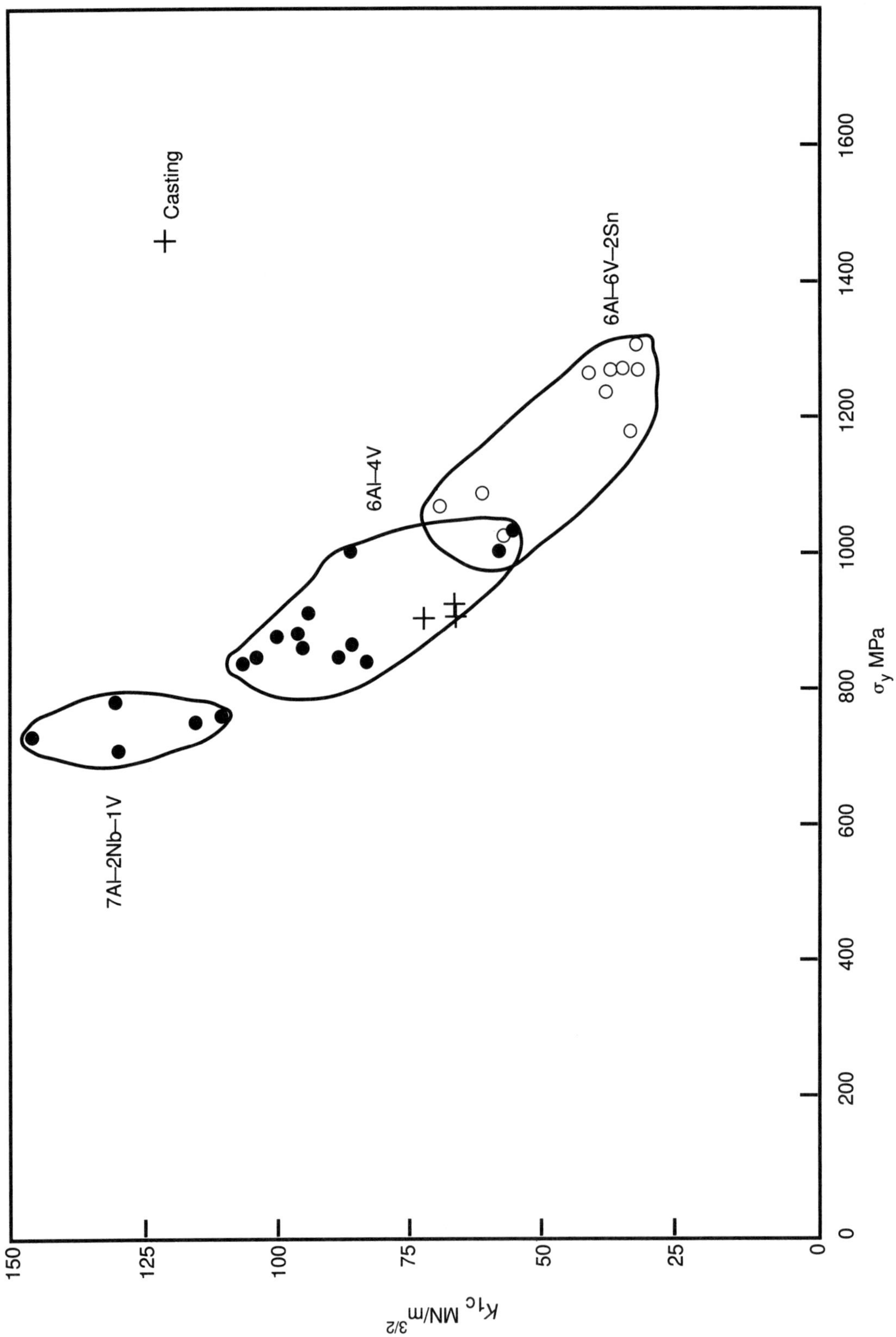

(a)

Fig. 1.2.29 Fracture toughness vs yield strength for wrought titanium alloys at 20°C

Fig. 1.2.29

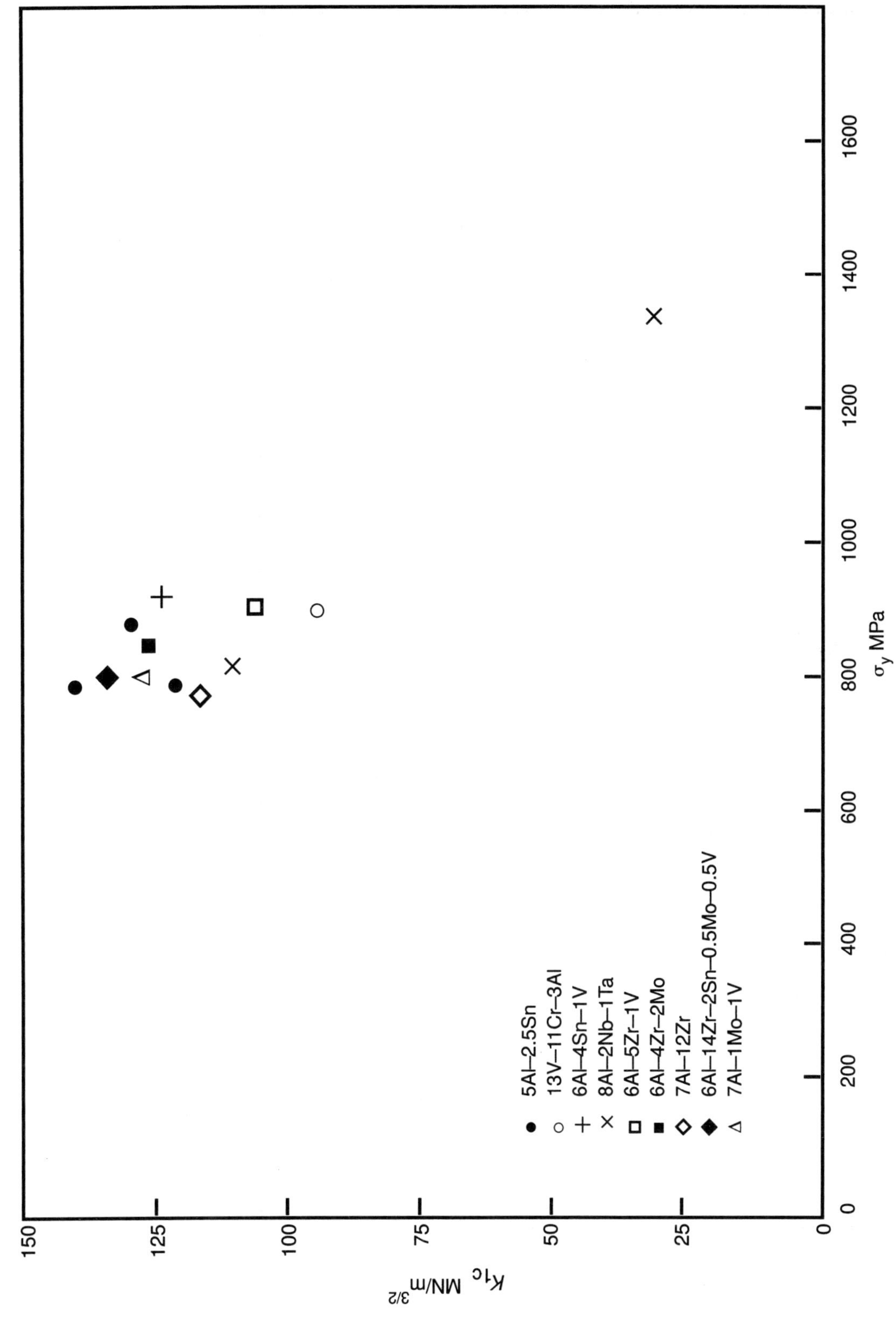

(b)

Fig. 1.2.29 Fracture toughness vs yield strength for wrought titanium alloys at 20°C – *continued*

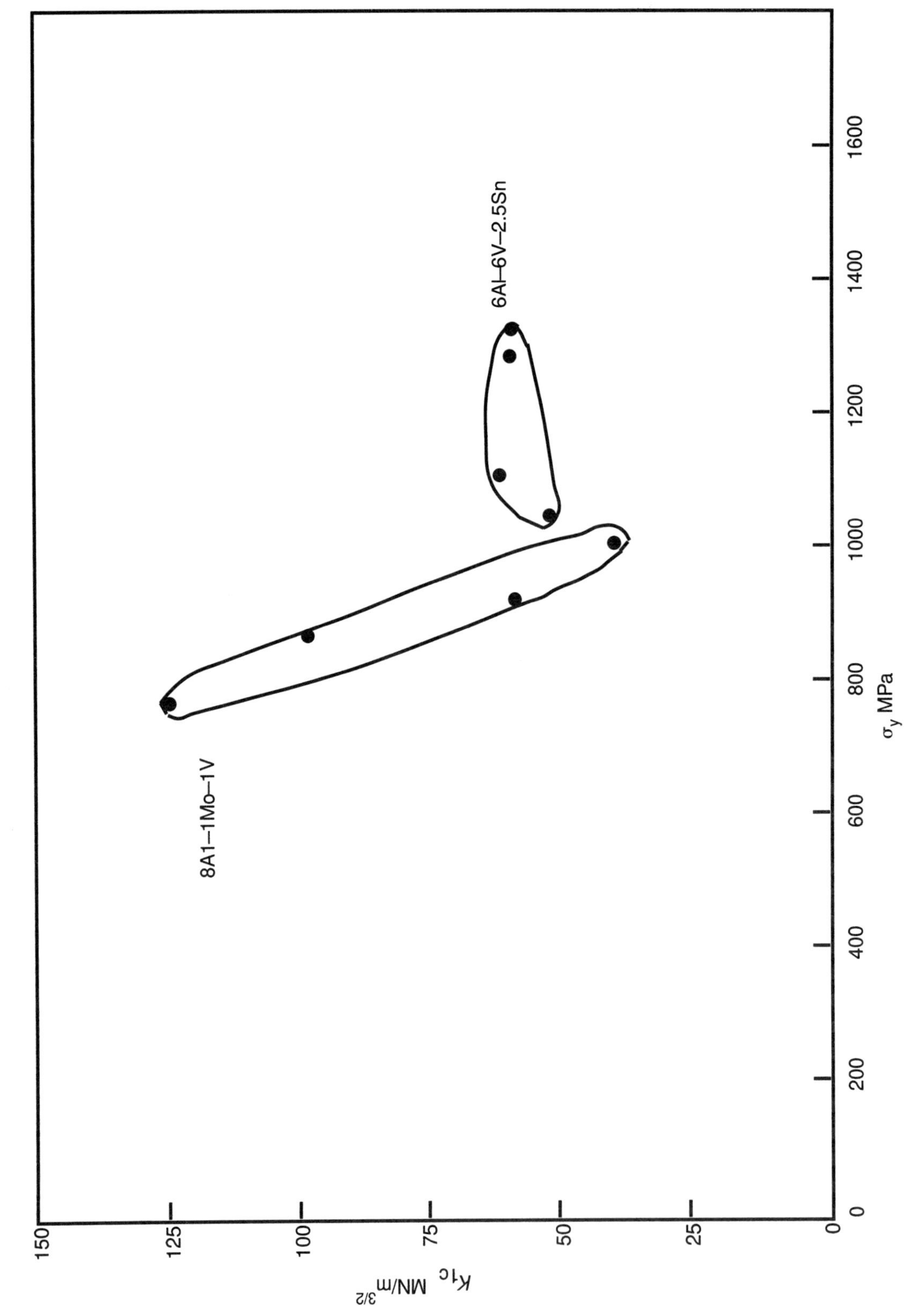

(c)

Fig. 1.2.29 Fracture toughness vs yield strength for wrought titanium alloys at 20°C – *continued*

Fig. 1.2.29 – *continued*

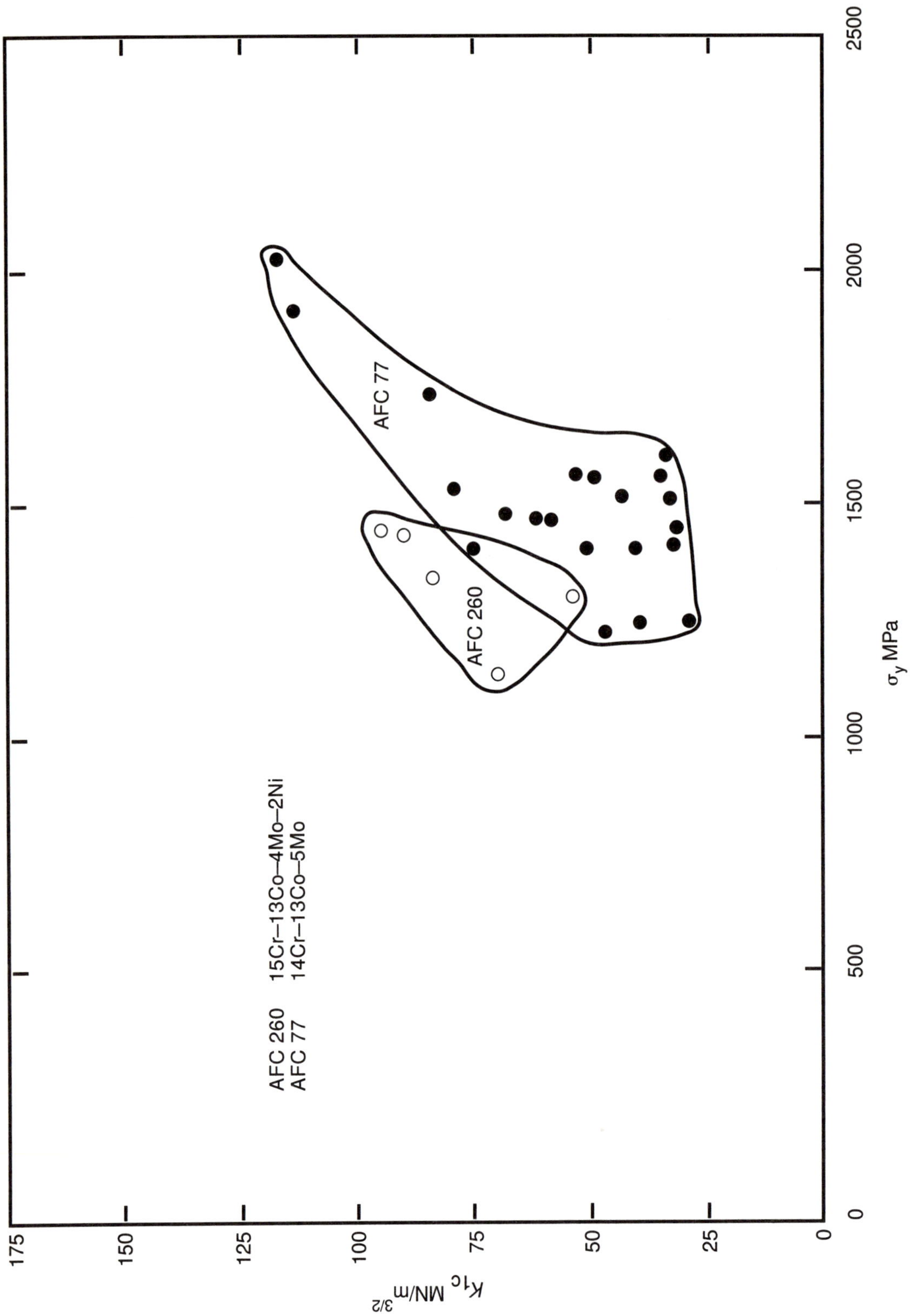

(a)

Fig. 1.2.30 Fracture toughness vs yield strength for wrought steels at 20°C

Fig. 1.2.30

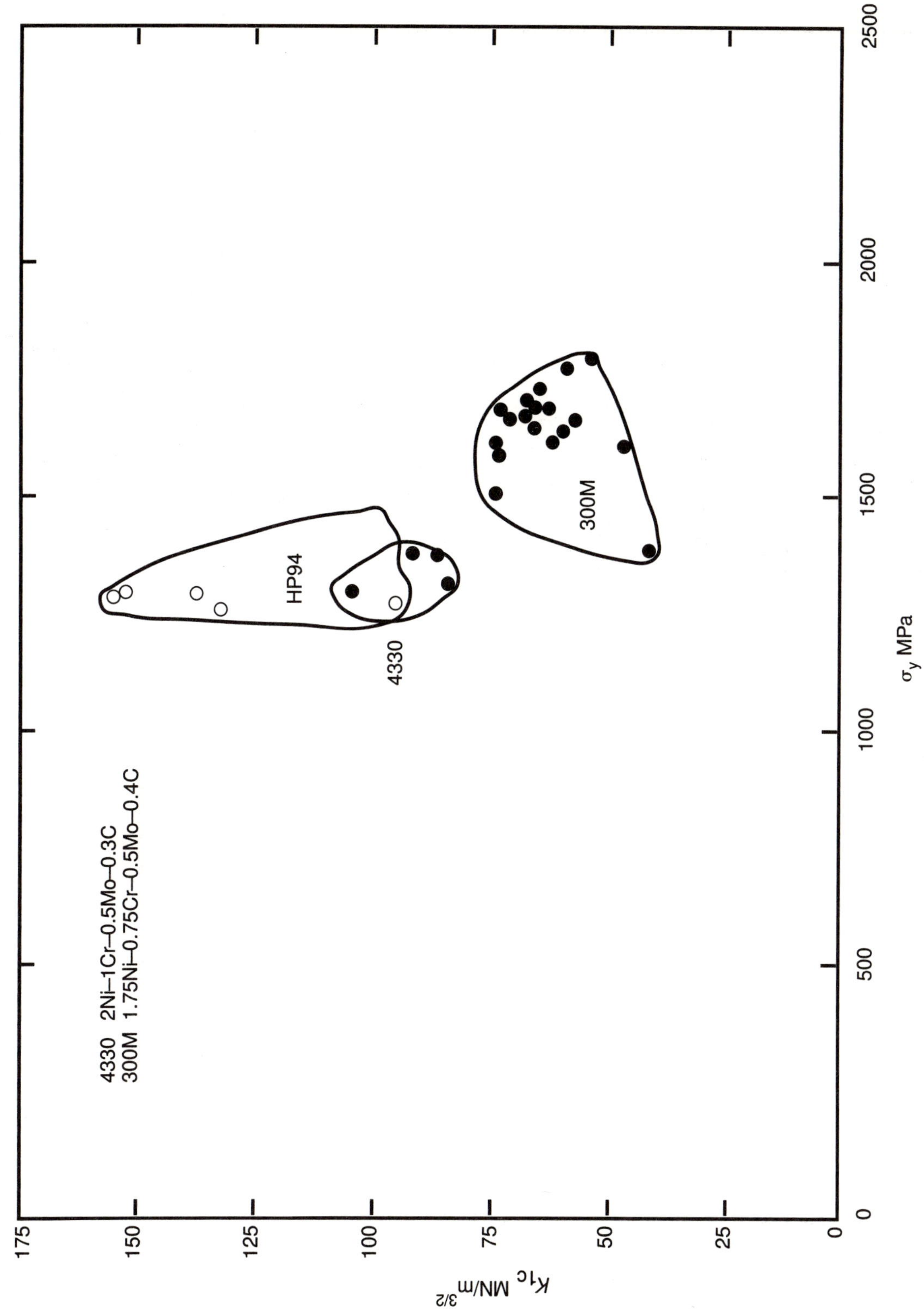

(b)

Fig. 1.2.30 Fracture toughness vs yield strength for wrought steels at 20°C – *continued*

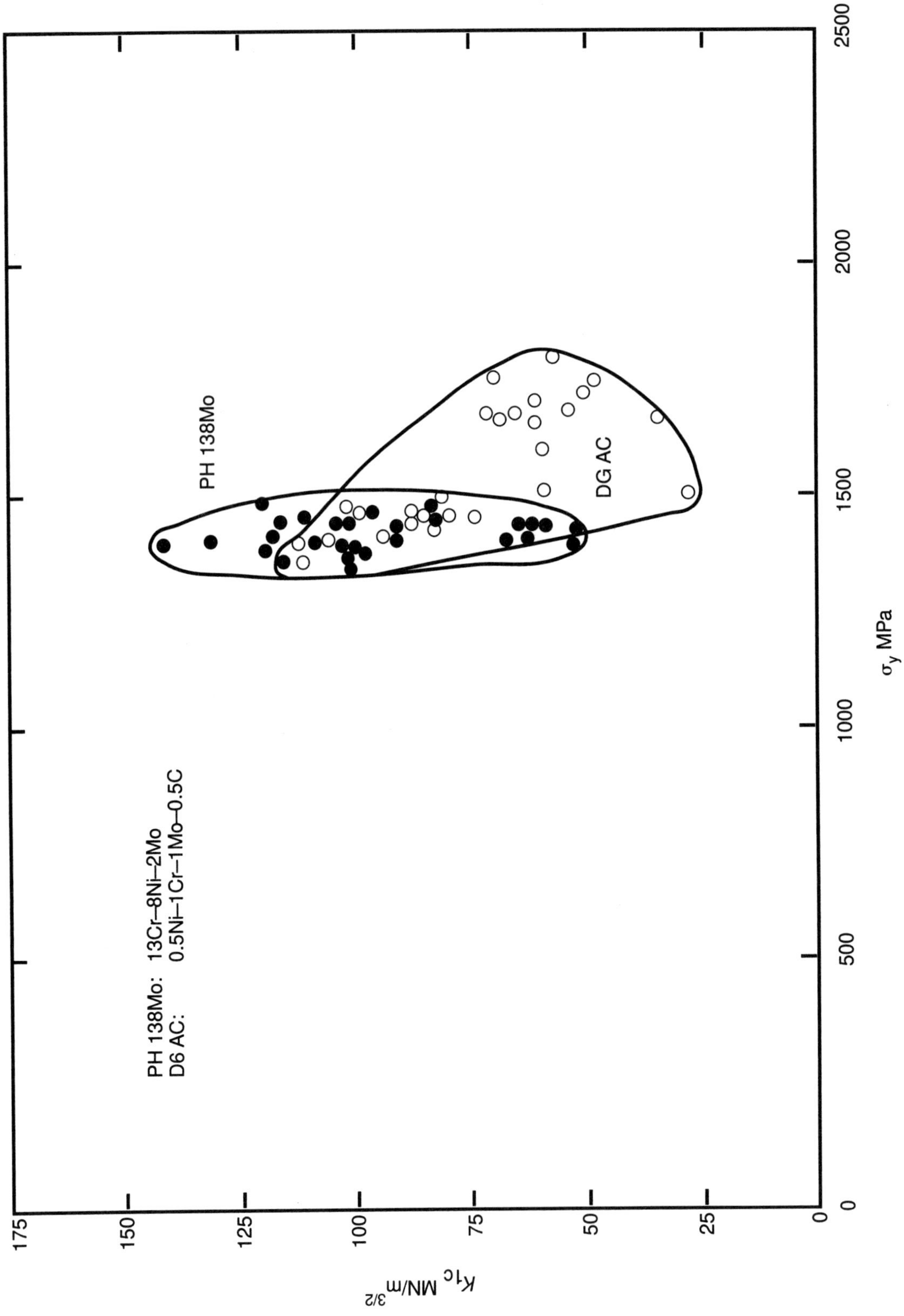

(c)

Fig. 1.2.30 Fracture toughness vs yield strength for wrought steels at 20°C – *continued*

Fig. 1.2.30 – *continued*

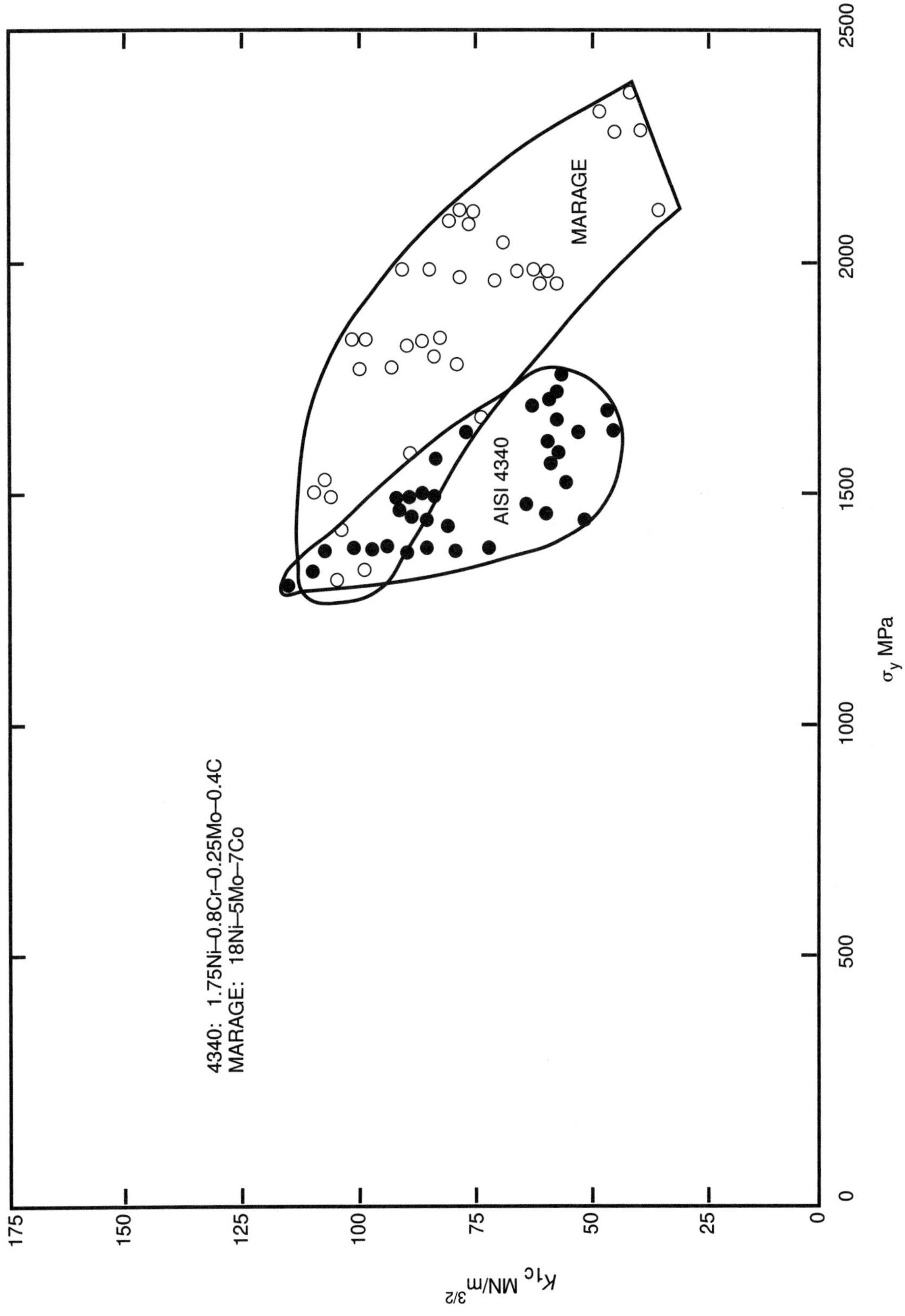

(d)

Fig. 1.2.30 Fracture toughness vs yield strength for wrought steels at 20°C – *continued*

Fig. 1.2.30 – *continued*

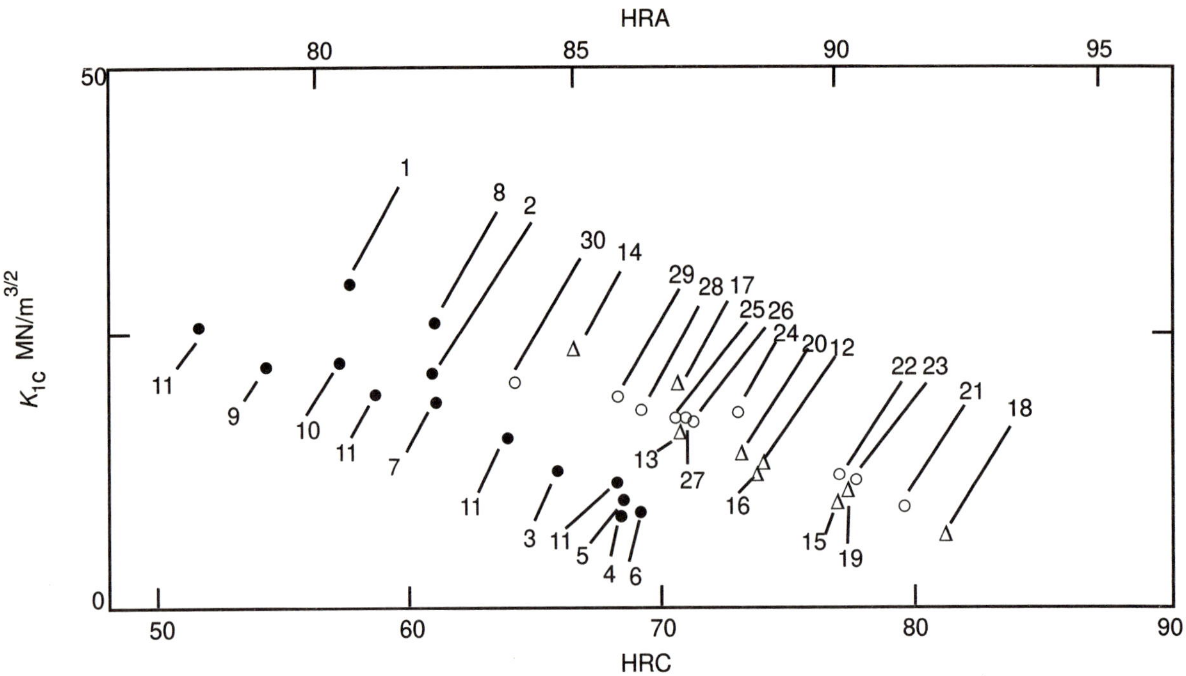

Fig. 1.2.31 Fracture toughness as a function of Rockwell hardness (A and C scales) for cemented carbides, high speed steels, tool steels and white cast irons. (See Table 1.2.9 for compositions)

Fig. 1.2.31

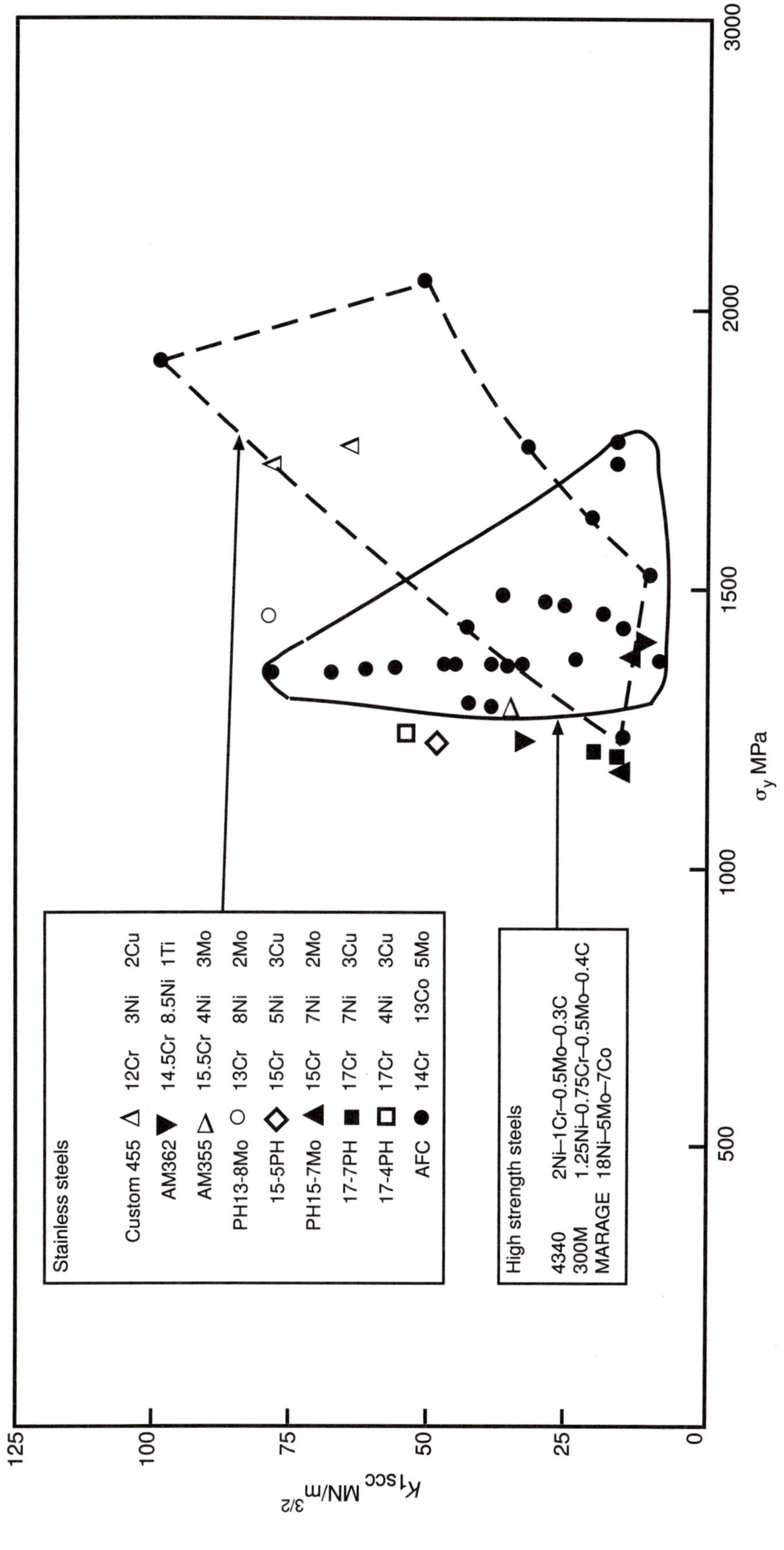

Fig. 1.2.32 K_{1scc} **vs** σ_y **for some stainless and other high strength steels**

Fig. 1.2.32

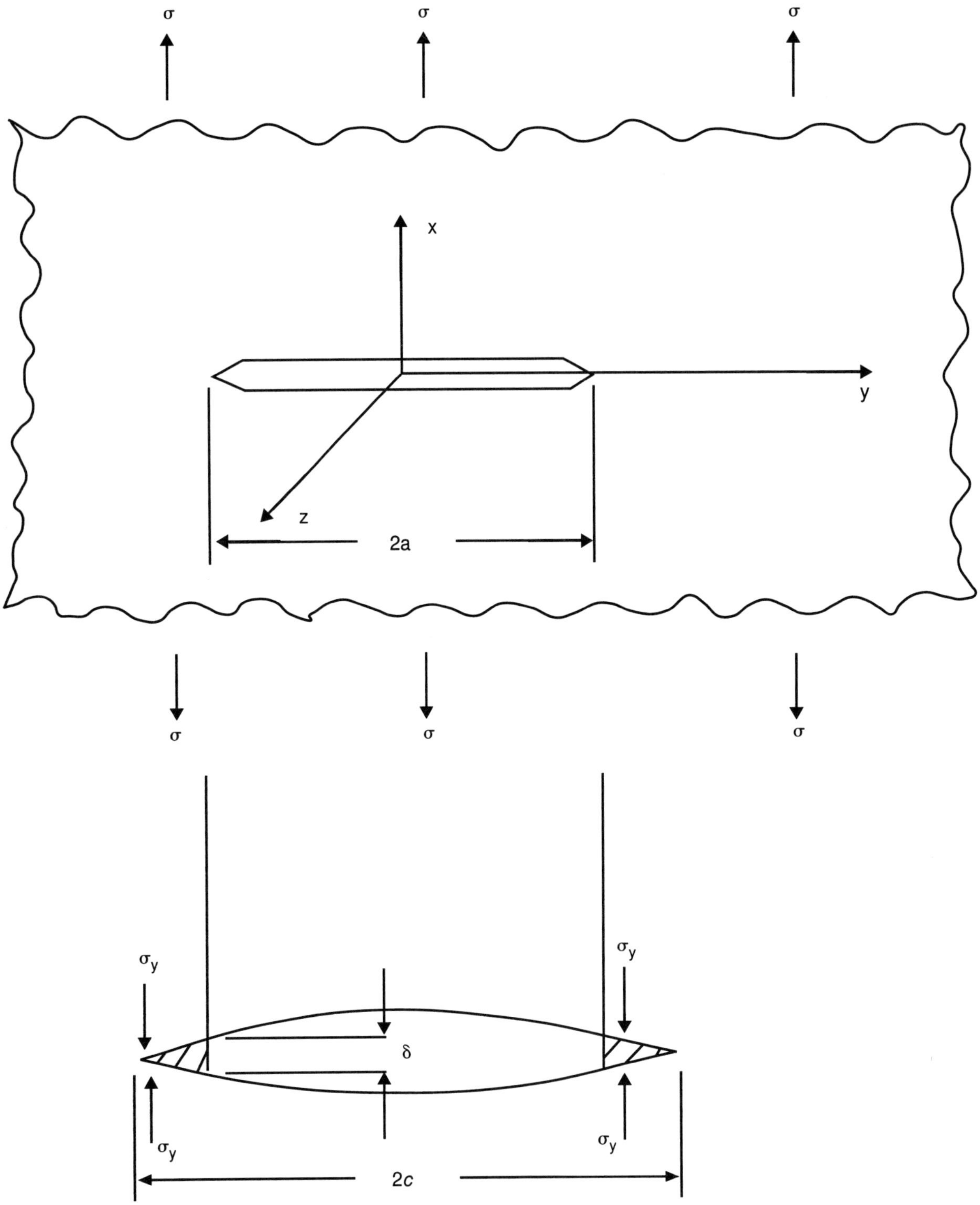

Fig. 1.2.33 Schematic representation of the plastic strip yield (Dugdale) model

Fig. 1.2.33

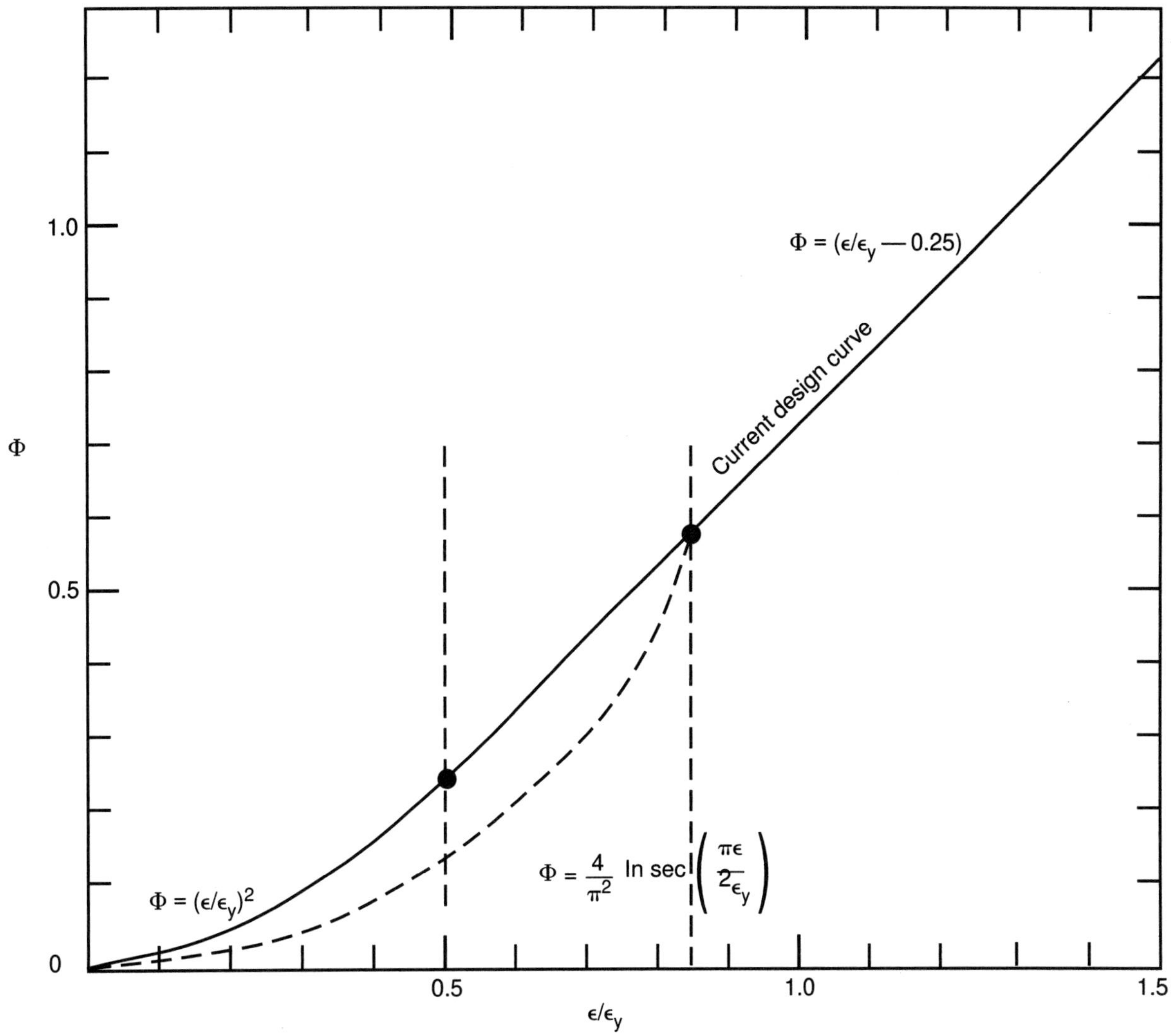

Fig. 1.2.34 Current design curve compared with the earlier version

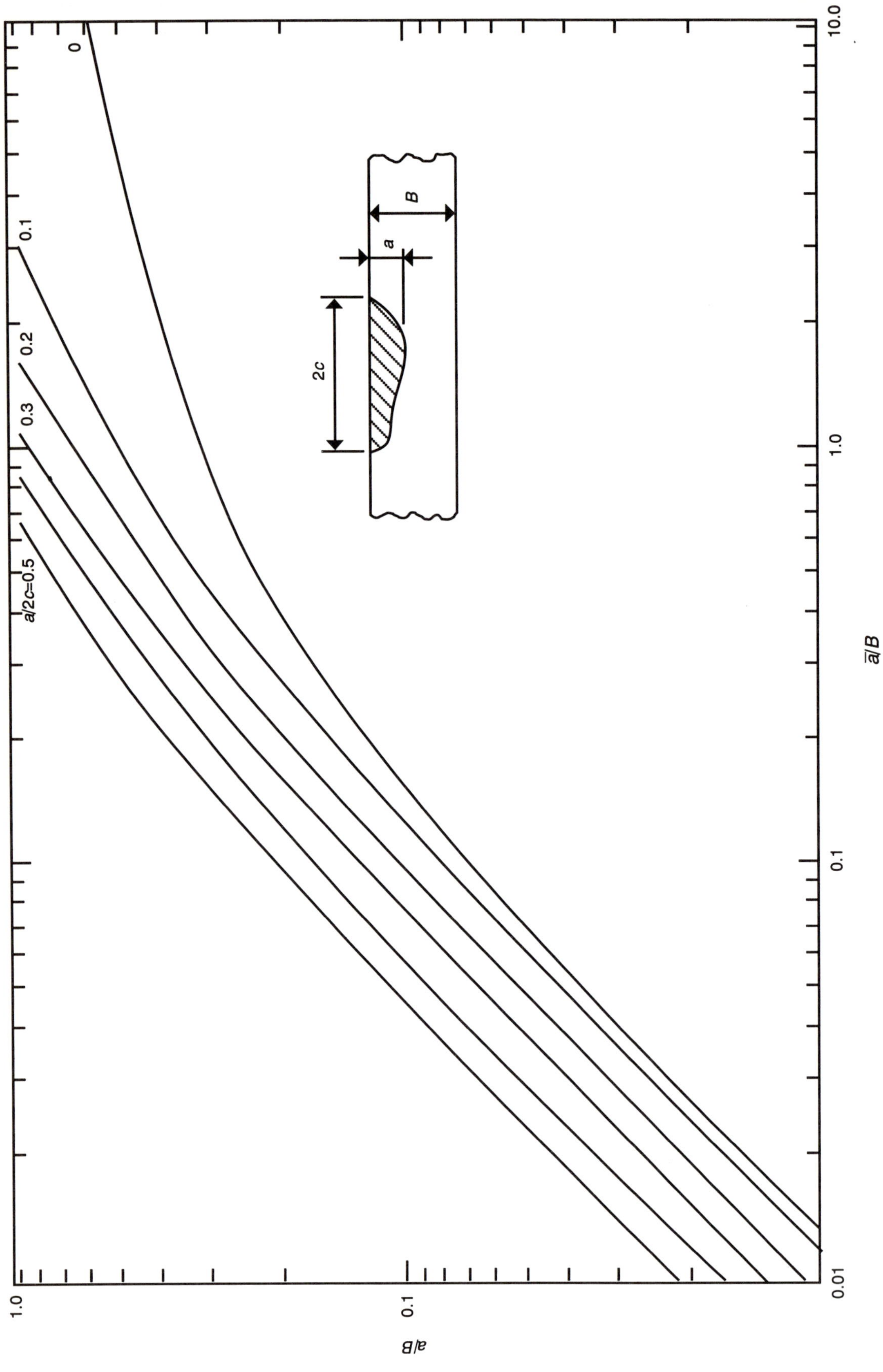

Fig. 1.2.35 Relation between actual surface defect dimensions and the parameter $\bar{a}$

Fig. 1.2.35

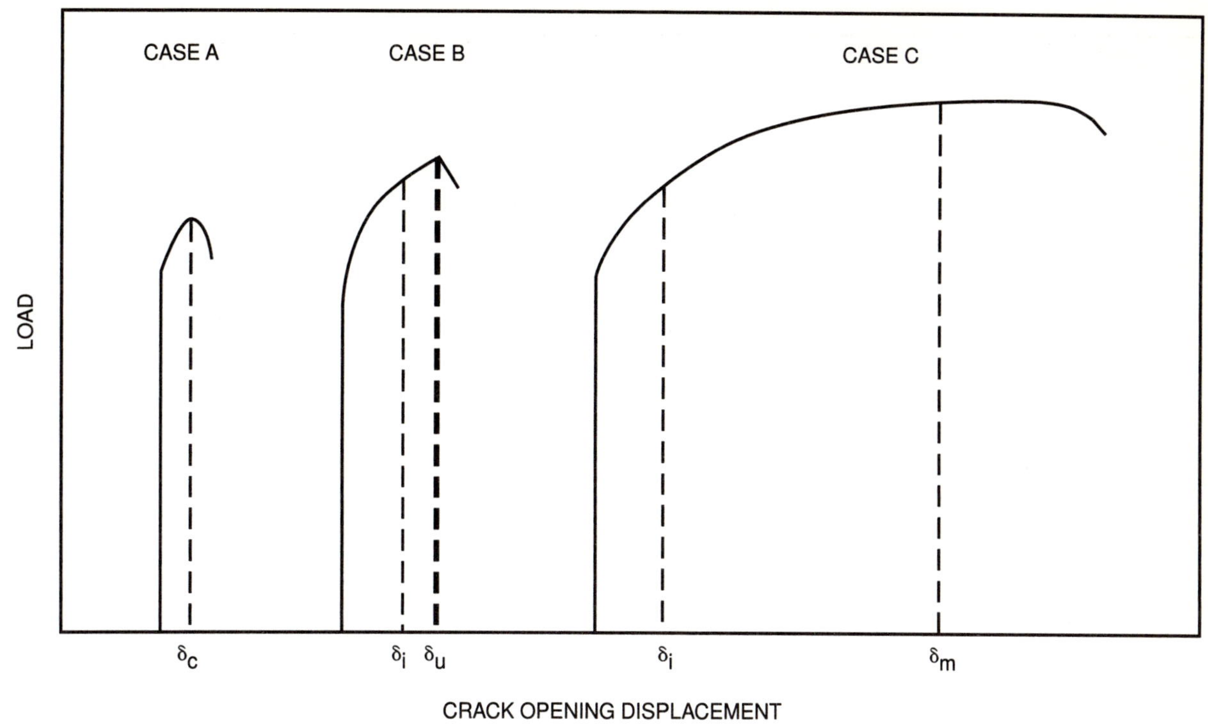

Fig. 1.2.36 Schematic representation of types of load-COD relation illustrating critical values of δ defined in BS 5762

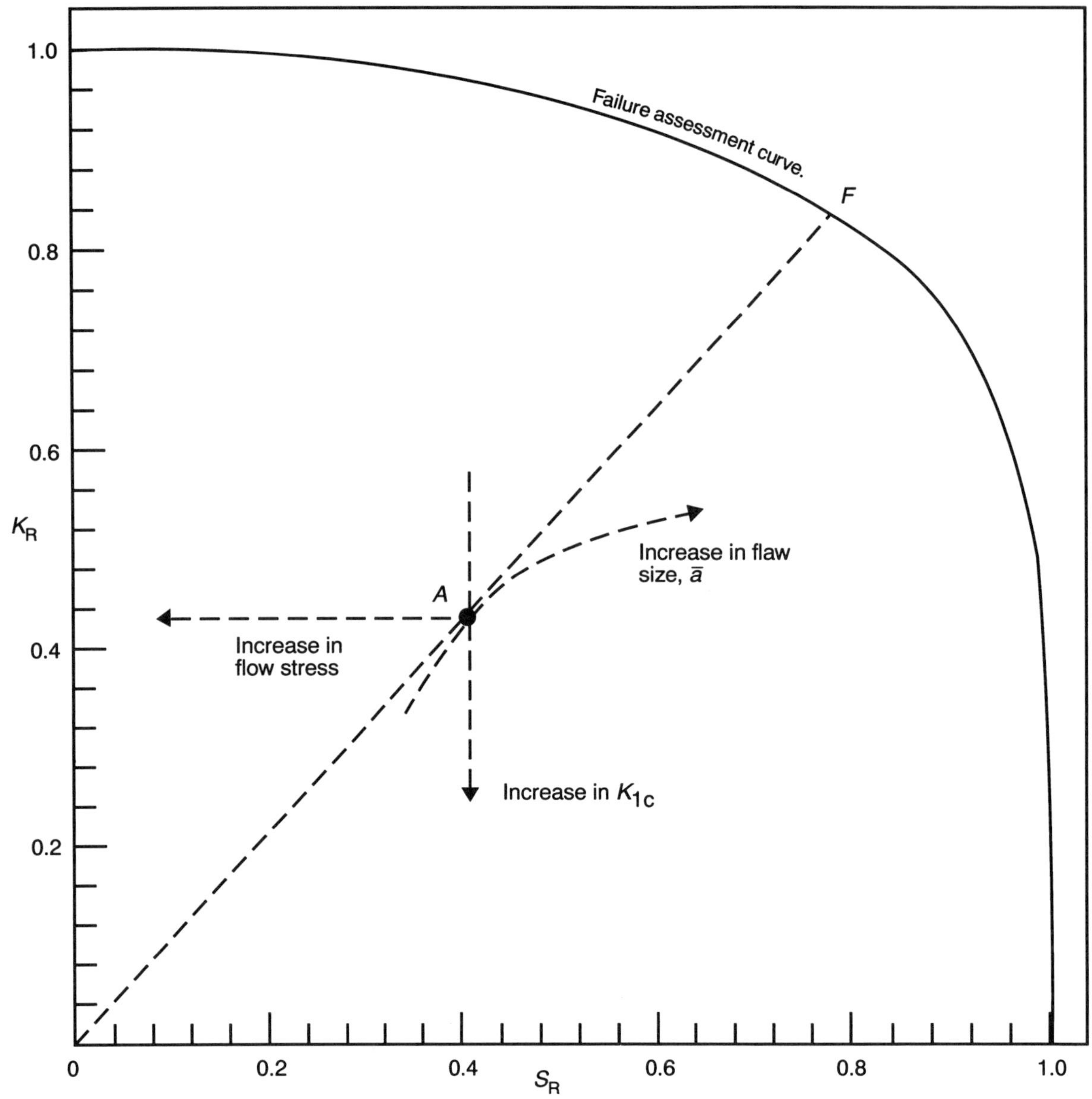

Fig. 1.2.37 The failure assessment diagram

Fig. 1.2.37

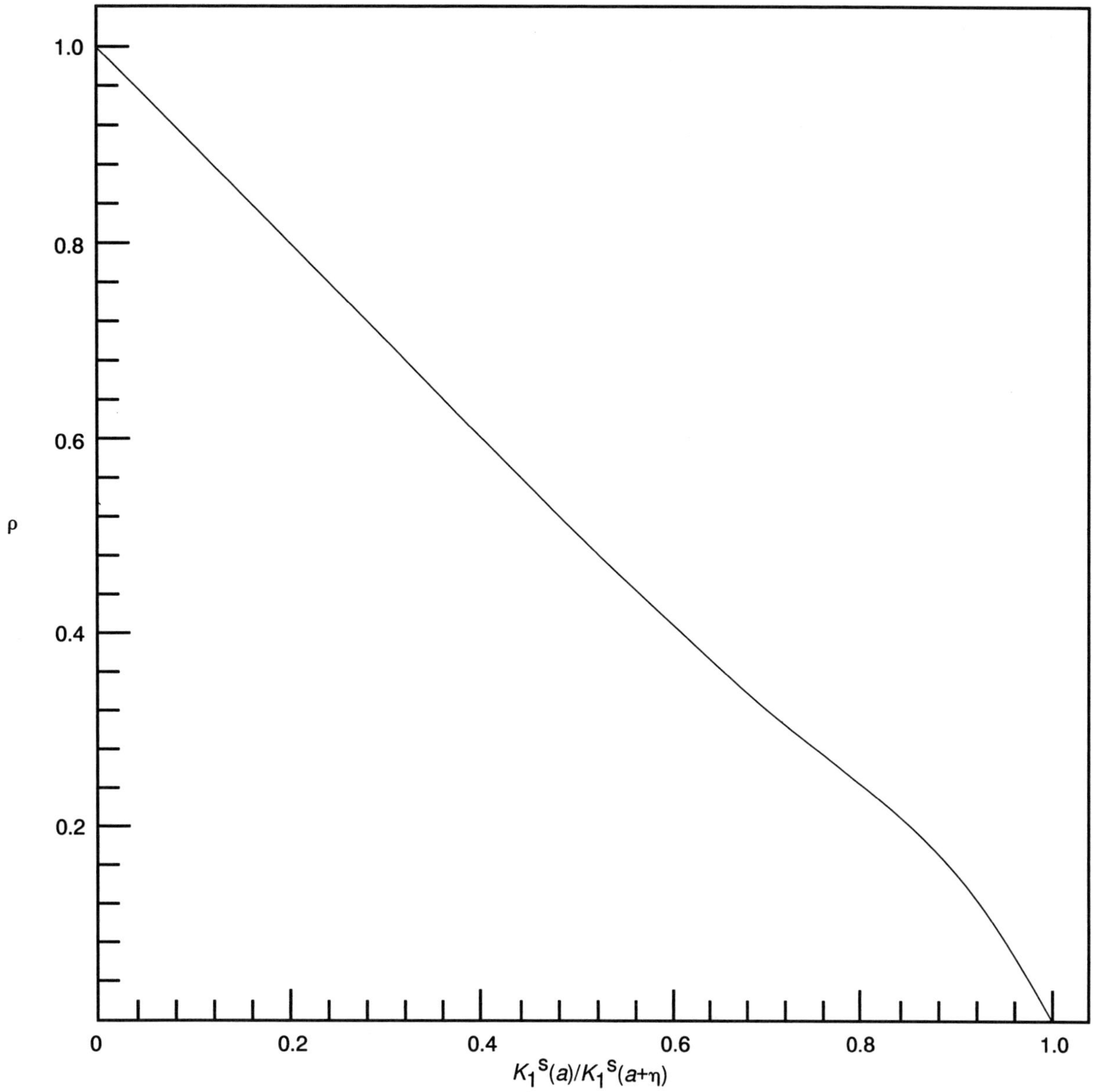

Fig. 1.2.38 Plasticity correction to $K_R{}^S$ for the two-criteria approach

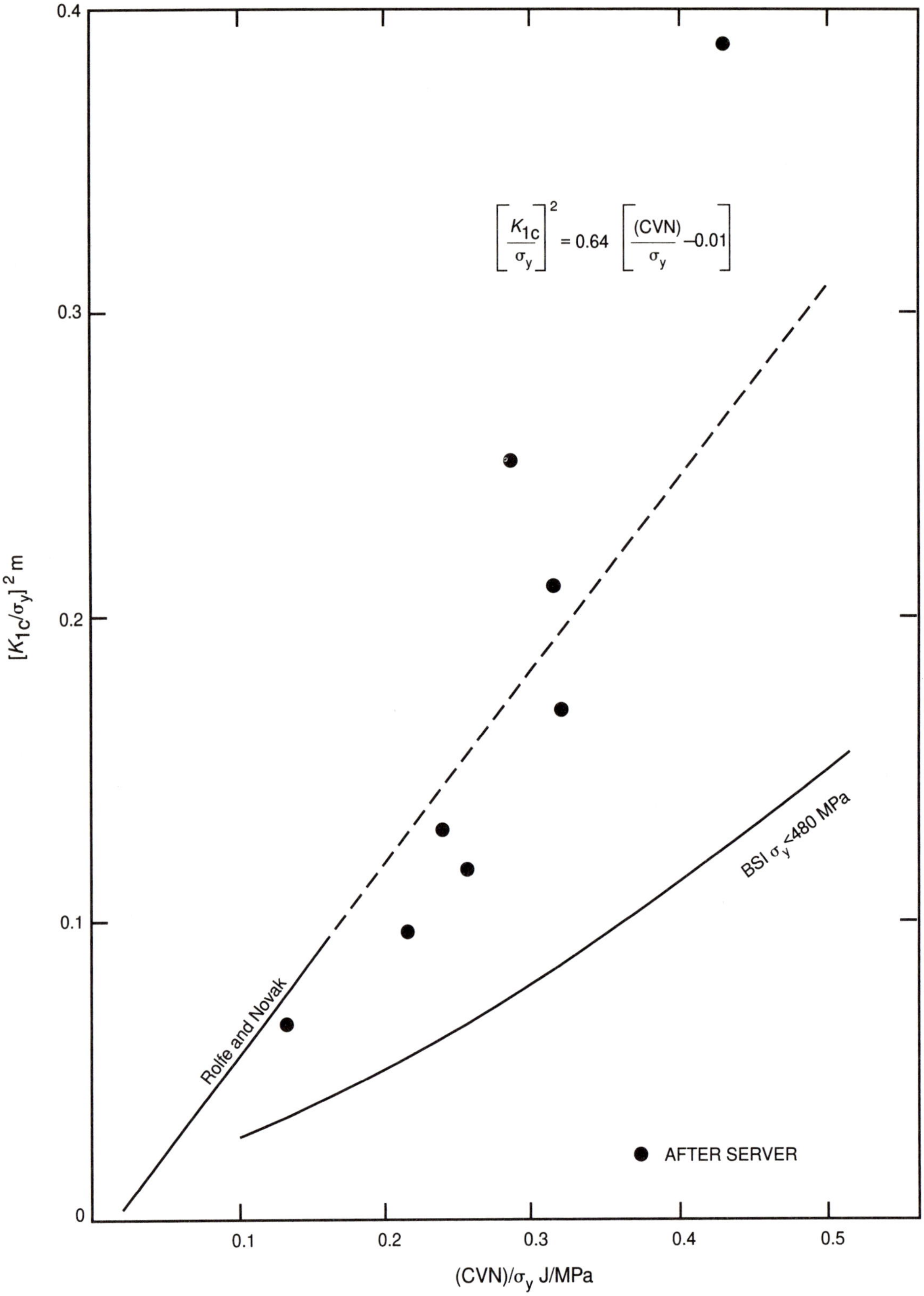

Fig. 1.2.39 Upper shelf impact energy–fracture toughness correlation

Fig. 1.2.39

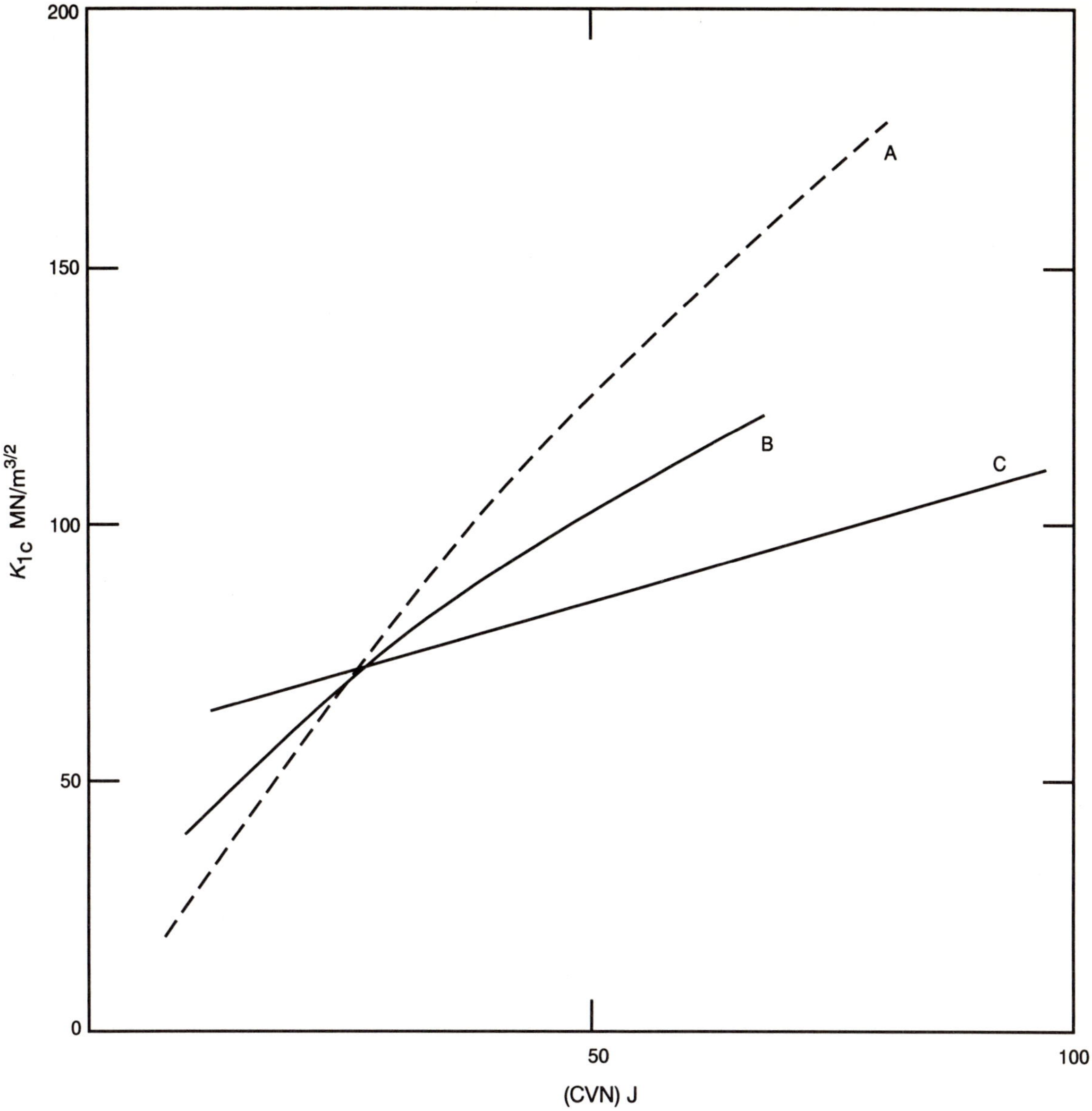

Fig. 1.2.40 Correlations between impact energies and feature toughness in the transition region (A) Rolfe and Novak; (B) Sailors and Corten; (C) BS. PD6493

1.2.3 High temperature mechanical properties

1.2.3.1 Introduction

Increase of temperature above ambient is accompanied by changes in mechanical properties, slow initially but accelerating as the melting point is approached. The elastic modulus also decreases and some typical recorded examples are listed in:
Table 1.2.10 *Typical changes in elastic modulus with temperature*

1.2.3.2 Creep and high temperature strength of metals and alloys

The strength of metals and alloys is generally temperature- and strain rate-sensitive; higher temperatures and slower strain rates decrease the strength.

CREEP

Creep is the time dependent strain which occurs under sustained loading of a material. In the case of an externally loaded component, creep becomes apparent as a change in shape of the component or as a fracture. In self-loaded devices, such as bolts, creep is apparent as stress relaxation and is unlikely to cause fracture.

Creep of metal and alloys can occur at all temperatures. At relatively low temperatures, i.e. at temperatures below half of the melting point in K ($<T_m/2$) the creep strain for constant stress and temperature varies as the logarithm of time and an increase in stress or temperature will enhance the creep rate. Since the creep rate decreases with time, the total strain will tend to a limiting value for a given set of temperature and stress conditions. Failure of the material is unlikely to occur under this type of creep condition.

At high temperatures, in excess of $T_m/2$, there are three well defined stages of creep: primary, in which the creep rate decreases with time, followed by a secondary stage in which the rate is constant, and a final tertiary stage in which the rate increases until the material finally ruptures.

Approximate values of $T_m/2$ are 200°C for aluminium, 650°C for iron, 720°C for titanium and 60°C for zinc. These temperatures are similar to the recrystallization temperatures of the metals and correspond to the temperatures used for process annealing. Where stress relieving heat treatments of lower temperature are employed the internal stresses are reduced by logarithmic creep.

The creep life of a component in service may be limited by the rupture time (time to failure) or the amount of tolerable strain (dimensional change). Creep data can be expressed as the stress required to produce a fixed strain in a fixed time at the service temperature.

Useful creep lives may be of the order of 100 000 h and the experimental determination of data spanning this time is both costly and time consuming. Numerous methods have been proposed to permit short-term test data to be extrapolated in order to predict long-term creep properties. Most of the methods, however, take no account of the metallurgical changes which may take place during long service lives and thereby change the creep properties of the material. The use of such methods is therefore better suited to interpolation between existing data than to extrapolation beyond the range of time or temperature.

SELECTION FOR HIGH TEMPERATURE STRENGTH

Design stresses for components intended for use at ambient temperatures are based on the 0.2% proof stress or the ultimate tensile strength; at higher service temperatures the likelihood of creep deformation has to be considered. As an illustration, consider the following data for carbon steels, which gives a detailed examination of various steels.

			Stress for rupture life of		*Creep strength for*
Temperature	*0.2% PS*	*UTS*	*100 000 h*	*10 000 h*	*1% strain in 100 000 h*
(°C)	*(MPa)*	*(MPa)*	*(MPa)*	*(MPa)*	*(MPa)*
400	150	360	165	220	120
450	130	320	85	130	48

At 400°C the 0.2% proof stress is less than the 10 000 h rupture stress and could, therefore, be used as a safe working stress. (Note that the 100 000 h rupture stress is less than the UTS.) The creep strength for 1% strain in 100 000 h is, however, less than the 0.2% proof stress and creep deformation in excess of 1% would occur during 100 000 h at a stress equal to the proof stress. If the application of the material is one in which a 1% dimensional change cannot be tolerated, the working stress must be limited to a level which produces tolerable creep deformation, i.e. the working stress must be limited to the creep strength for a tolerable strain over the service life. At 450°C, the proof stress equals the 10 000 h rupture stress and working stresses for lives greater than 10 000 h must be based on rupture stresses, or, if only limited strains can be tolerated, creep strengths. Thus short time tensile test data can only be used for design purposes when creep deformation is negligible, otherwise the appropriate creep property, rupture stress or creep strength, must be used. The following method of data presentation has been adopted to facilitate the choice of appropriate property and to compare the high temperature strength of various alloys systems.

By using a time-temperature parameter, such as the Larson–Miller parameter, short term tensile test and creep data can be reduced to a single curve. As shown in Fig 1.2.41 short term tensile strengths and creep rupture strengths, as well as 0.2% proof stress and 0.2% creep strengths, can be represented by single master curves, when plotted in the form of log stress against the Larson–Miller parameter, $P = T\,(20 + \log t) \times 10^{-3}$ where T is the temperature in K and t the time in hours. Some data for Nimonic 90 are first plotted in the form of log stress versus temperature in degrees centigrade to give curves for the 0.2% proof stress and UTS and families of curves for the creep rupture stresses and the 0.2% creep stresses. The same data are then replotted against the Larson-Miller parameter to give two master curves.

Fig. 1.2.41a *Tensile and creep data for Nimonic 90 as a function of temperature*
Fig. 1.2.41b *Data from Fig. 1.2.41 a plotted as a function of the Larson–Miller parameter, P*

When the master curve varies rapidly with the value of P it is readily apparent that the strength is time and temperature sensitive and design stresses must always be based on the appropriate creep property. When the strength is independent of the value of P there will be many values of time and temperature for which the strength is insensitive to time and temperature. Thus, for values of P less than, say, 15, both rupture and creep strengths are independent of P in Fig 1.2.41. Considering the stress required to obtain 0.2% strain in 10^{-1}, i.e. the 0.2% proof stress, then if $P = 19T \times 10^{-3} < 15$ the stress will be independent of temperature, i.e. $T < 789$ K or 516°C, which temperature corresponds to the rapid decline of the 0.2% proof stress with temperature shown in the first plot of Fig 1.2.41.

Considering the 10 000 h creep strength for 0.2% strain, the limiting temperature is now given by $24T \times 10^{-3} < 15$, i.e. 667 K or 395°C, as shown by the back extrapolation of the 10^4 h creep strength in the first plot of Fig. 1.2.41. Thus, although the 0.2% proof stress does not vary with temperature below 500°C, creep does occur at this stress in

10^4h when the temperature is greater than 400°C.

1.2.3.3 Precious and refractory metals

There are a number of metals with very high melting points and these, the refractory metals, tungsten, molybdenum, tantalum and niobium and their alloys have outstanding mechanical properties up to 1200°C, temperatures well in excess of those which can be tolerated by the more common engineering metals, as shown in:

Fig. 1.2.42 *Comparison of the elevated temperature tensile strengths of tungsten, molybdenum, molybdenum alloys, tantalum, niobium, Nimonic 95 and a platinum alloy*

All of the refractory metals are subject to rapid attack by oxidising atmospheres but, where hardness and creep resistance at very high temperatures are paramount and the environment is neutral or reducing, or the exposure so limited in time that attack can be neglected, one of the refractory metals must be used.

Molybdenum has the optimum combination of properties but, where its low ductility is unimportant, or can be overcome by techniques of manufacture, the superior temperature resistance of tungsten may require its use.

Coatings which prevent or delay oxidation are under development but where resistance to oxidation is essential in addition to high temperature mechanical properties a precious metal must be used. Typical alloys are Pt 10% Rh, Pt 5% Au or Pt 4% Ru. Their major drawback is their very high cost.

1.2.3.4 Ceramics

TEMPERATURE LIMITING PARAMETERS

As is the case with metals the high temperature use of ceramics is limited by the onset of stress-induced deformation and environmental attack. An additional requirement is the need to resist thermal shock, a function of heating rate rather than temperature. Many ceramics are stable oxides and therefore resistant to heating in unpolluted air. Environmental attack is not, therefore, a factor unless some constituent, which may in a few instances be water, attacks the ceramic either chemically or by reducing the melting point by a 'fluxing' action. Such breakdown processes as melting vaporisation or dissociating usually initiate well above normal operating temperature.

The operating temperature in air of a non-oxide ceramic is usually much lower than the temperature limited by melting, vaporizing or dissociation in inert conditions, and may well be lower than the limit imposed by deformation.

As with metals, resistance to deformation of a ceramic is a function of the melting point and the limit due to creep may be three or four hundred degrees lower than the temperature at which the material will withstand short-term stress application.

Ceramics with the highest creep resistance are essentially high refractory single phase materials of comparatively large grain size, but many commercial materials have segregated impurity phases or even glasses at grain boundaries and their resistance to creep is therefore significantly reduced.

The short-and long-term service temperatures of typical ceramics are shown in:

Fig. 1.2.43 *Maximum service temperatures of ceramics*

THERMAL SHOCK

Suitability for service at elevated temperature carries with it the risk of exposure to high thermal gradients and when a material of low ductility (even in the creep range the ductility of a ceramic is an order of magnitude lower than that of a metal) is subjected to thermal shock the risk of failure is high.

Resistance to thermal shock is a function of thermal conductivity, thermal diffusivity, fracture strain $\sigma_{f/\epsilon}$ fracture toughness K_{1c} and an inverse function of thermal expansion α.

Silicon carbide offers the best combination of properties. Form of material is important in the performance of a material after the fracturing process is initiated. Weak friable materials or those with high densities of defects retain a high proportion of their initial strength, whereas glass normally fails catastrophically.

Therefore, porous refractories, fire bricks and internally micro-cracked structures such as partially stabilized zirconia will usually outperform a fully dense engineering ceramic.

1.2.3.5 Comparative maximum service temperatures

Most engineering applications lie in the temperature range −20 to 600°C and are met by metals, glasses, ceramics, polymers and natural products (wood, leather and fibres). These materials are adequately strong, cheap, withstand normal environments and most are ductile (or their shortfall in ductility may be compensated by design).

The most important effect of the requirement to increase temperature capability of a material component is to increase its cost. High temperature steels are more than 10 times the price of carbon steels, nickel alloys twice the price of high temperature steels, ceramics (except graphite) more expensive still and refractory and precious metals even more costly.

While the temperature requirement can be satisfied by metal properties, materials will deform plastically and conventional design processes may be used. Ceramic (which may in some cases such as sialon have fracture stresses equivalent to those of steel) are, however, non-ductile and even at temperatures at which deformation occurs plastically it is inadequate for conventional design techniques, which therefore must be modified, possibly increasing cost and almost certainly restricting design.

Typical materials for particular temperature ranges and their critical properties are listed in:

Table 1.2.11 *Materials which may be used for service over specific temperature ranges and their more critical properties*

Table 1.2.10 Typical changes in elastic modulus with temperature

Material	***Condition***	***Temperature change***	***% Decrease in elastic modulus***
300 Series stainless steel	Hard	0–650°C	25
300 Series stainless steel	Soft	0–300°C	25
17.4 PH stainless steel		0–500°C	7
Aluminium bearing alloy		20–120°C	7

Table 1.2.11 Materials which may be used for service over specific temperature ranges and their more critical properties

Material		*Temperature range (°C)*	*Cost*	*Resistance to environment*	*Specific strength*	*Ductility*	*Special properties*
Wood		−40 to 50	Low	Good	Good	Low	Anisotropic
Natural fibres		−40 to 50	Low	Good	Good	Fair	Electrical insulator
Polymers		−40 to 200	Low	Good	Good	Good	Electrical insulator
Polymers		−200 to 300	High		Low	Good	Electrical insulator
Aluminium alloys		−150 to 250	Low	Good	Good	Good	Electrical conductor
Steels	Carbon	−100 to 400	Low	Adequate	Low	Excellent	Electrical conductor
Copper alloys		0 to 300	Moderate	Excellent	Very low	Excellent	Electrical conductor
Glasses		0 to 400	Low	Excellent	Very low	Zero	Electrical insulator
Steel	Low alloy	400 to 500	Moderate	Adequate[a]	Low	Good	Electrical conductor
Titanium		0 to 500	High	Good	Excellent	Good	Electrical conductor
Steel	Austenitic	500 to 600	Moderate/high	Good	Low	Good	Electrical conductor
Steel	HK 40	600 to 800	High	Good	Very low	Good	Electrical conductor
Nickel alloys		0 to 800	Very high	Very good	Low	Good	Electrical conductor
Refractory metals		0 to 1200	Very high	Poor[a]	Low	Moderate	Electrical conductor
Precious metal (platinum)		0 to 1200	Highest	Excellent	Low	Good	Electrical conductor
Glass	Silica	0 to 1400	Moderate/high	Excellent	Moderate/low	Zero	Electrical insulator
Ceramic	Porcelain Silicon nitride	0 to 1600 0 to 1800	Moderate High	Excellent Excellent	Moderate/low Moderate/low	Zero Zero	Electrical insulator Electrical insulator
Ceramic	Alumina	0 to 1800	High	Excellent	Moderate/low	Zero	Electrical insulator
Ceramic	Silicon carbide	0 to 2100	High	Excellent	Moderate/low	Zero	Electrical resistor
Ceramic	Graphite	0 to 2400	Low	Poor[a]	Moderate/low	Zero	Electrical resistor

[a]Air must be excluded at elevated temperature.

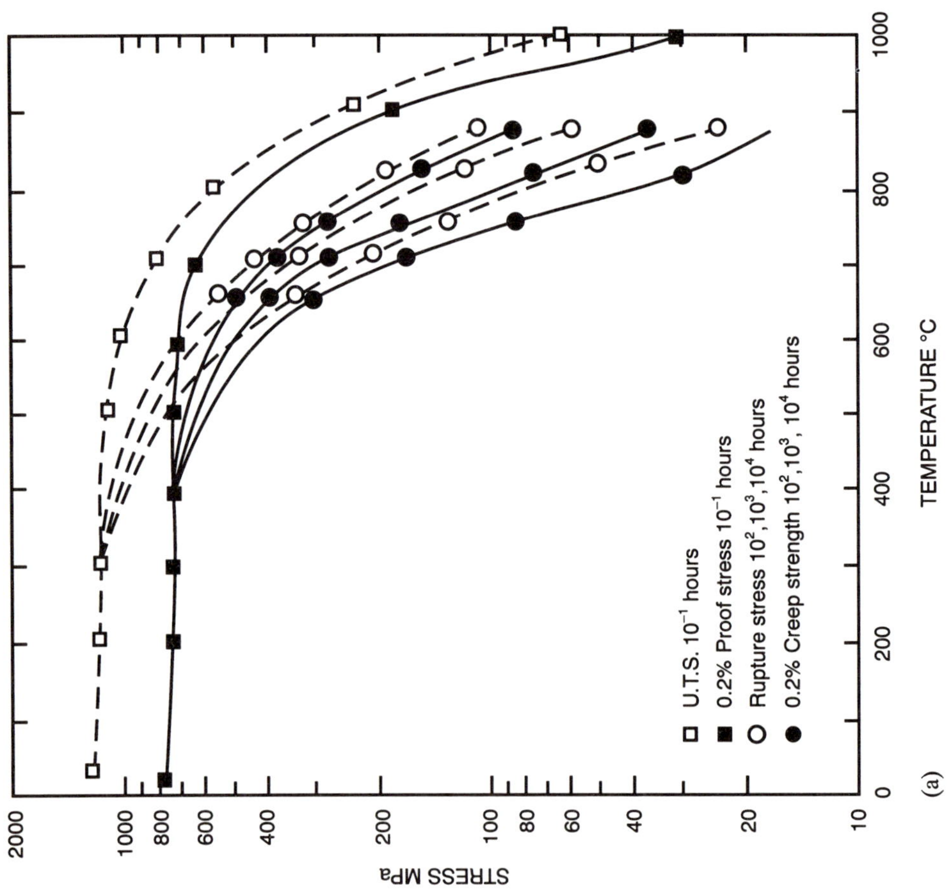

Fig. 1.2.41a Tensile and creep data for Nimonic 90 as a function of temperature

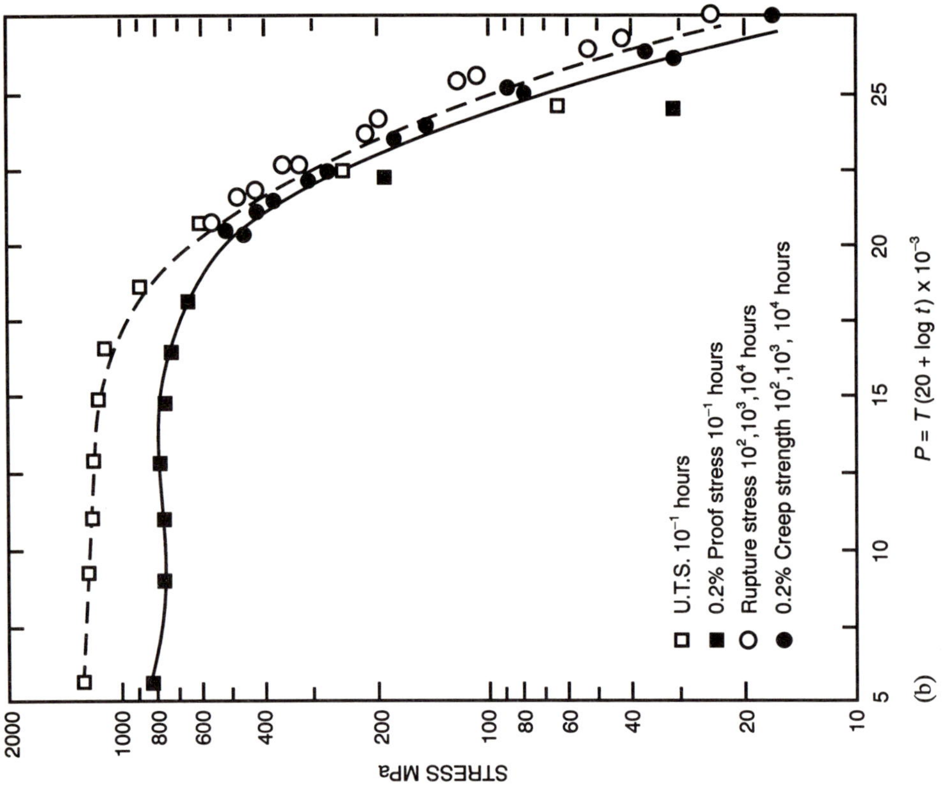

Fig. 1.2.41b Data from Fig. 1.2.41a plotted as a function of the Larson–Miller parameter, *P*

Fig. 1.2.41a and Fig. 1.2.41b

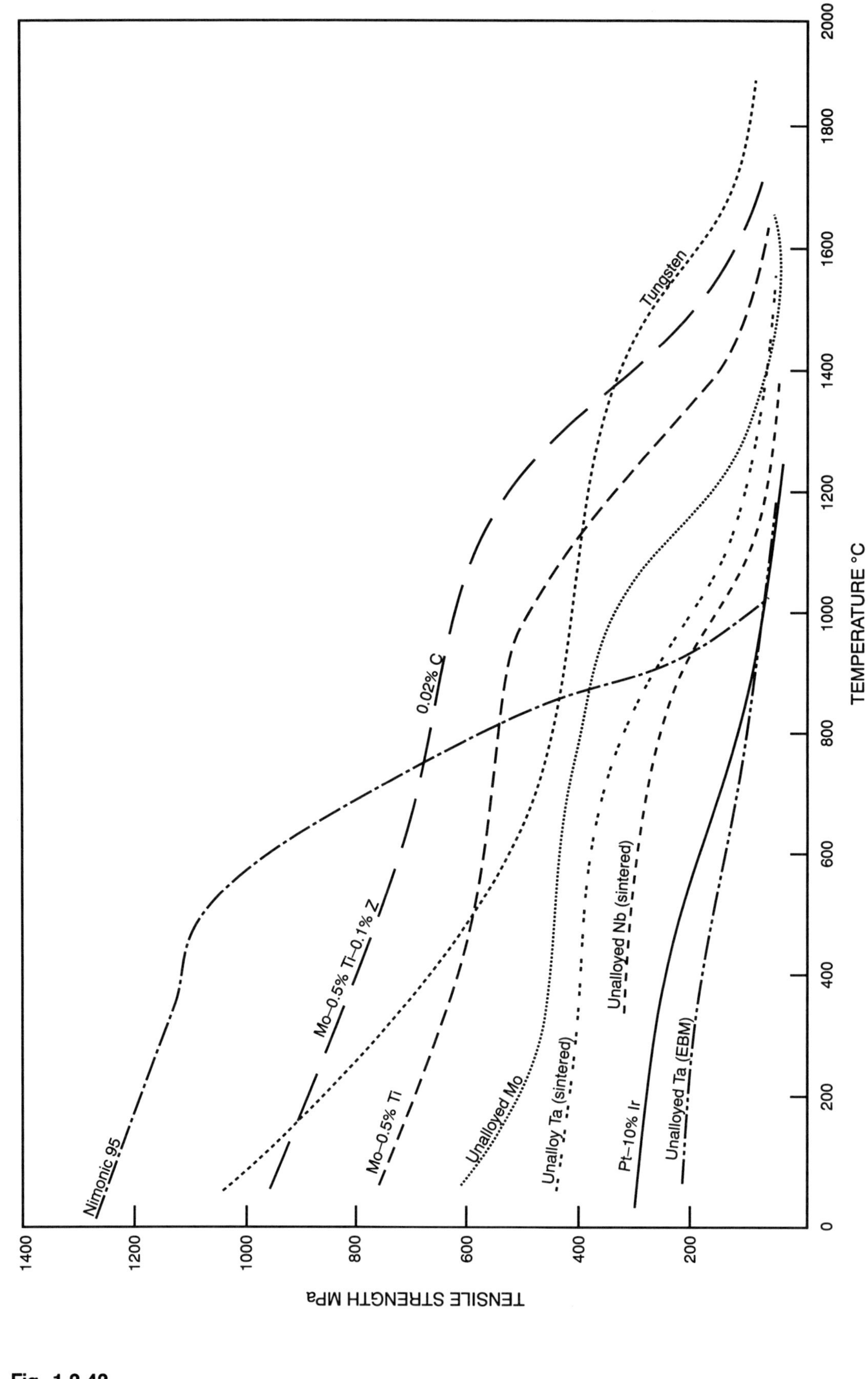

Fig. 1.2.42 Comparison of the elevated temperature tensile strengths of tungsten, molybdenum, molybdenum alloys, tantalum, niobium, Nimonic 95 and a platinum alloy

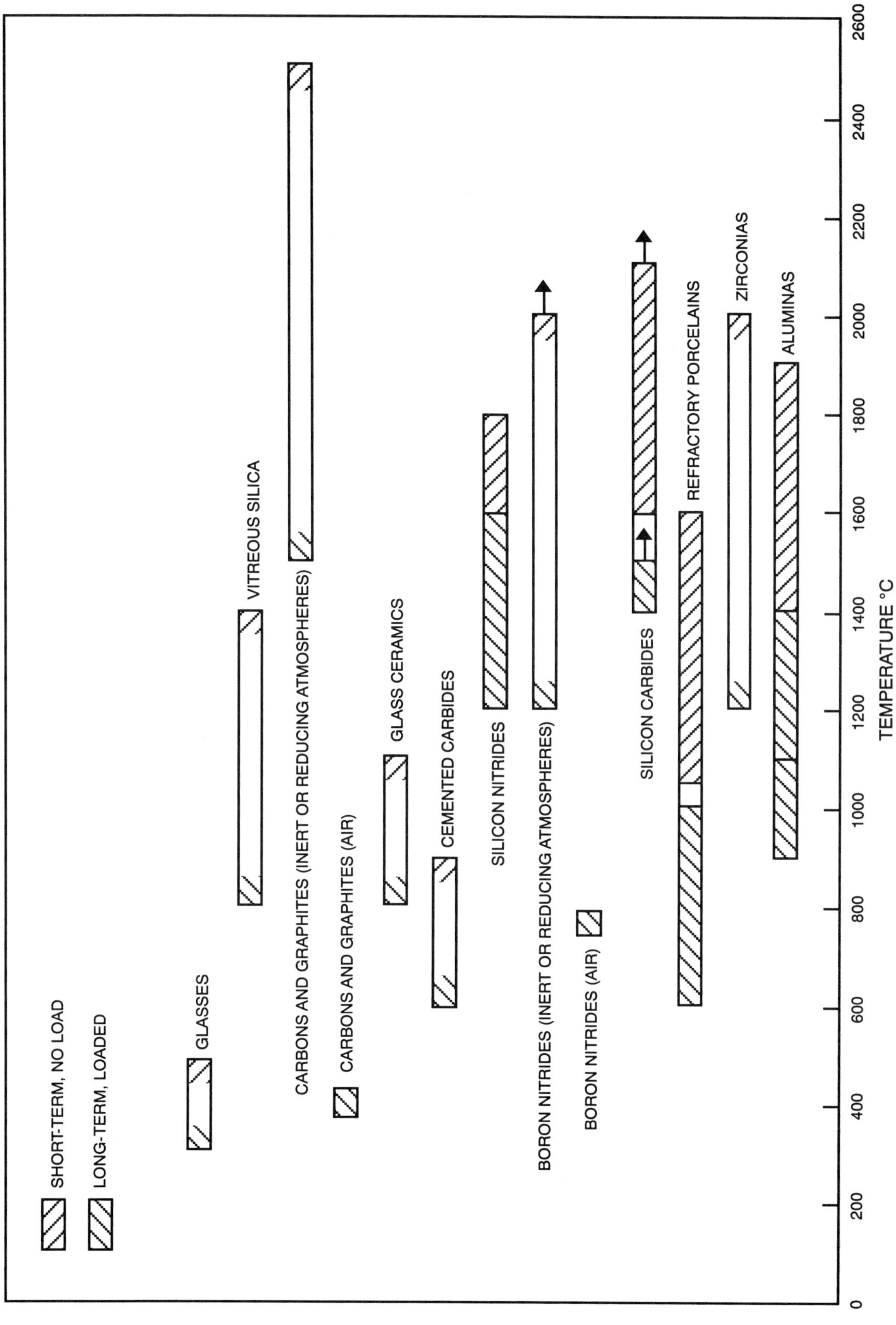

Fig. 1.2.43 Maximum service temperatures of ceramics

Fig. 1.2.43

1.2.4 Low temperature properties of materials

The information in this section has been prepared by A. M. Pye from original material by D. A. Wigley. Permission to reproduce information from the publication *Low Temperature Properties of Materials* by D. A. Wigley (Oxford University Press in collaboration with the Design Council) is gratefully acknowledged.

1.2.4.1 Introduction

This section is an introduction to some of the problems which may be encountered when designing equipment for use below room temperature (300 K). Figure 1.2.44 indicates some of the more significant fixed points as well as typical operating temperature ranges for various structures, and a very general indication of the materials used for their construction:

Fig. 1.2.44 *Some significant cryogenic temperatures and typical operating temperature ranges for various structures and constructional materials*

Many load-bearing structures have to operate satisfactorily over a wide range of climatic temperatures, especially those liable to see service in arctic regions, and it is important that the materials from which they are made perform safely and satisfactorily over this temperature range. As can be seen from Fig. 1.2.44 there are many such applications whose specifications call for operation down to about −40 to −50°C (230 K).

A second major group deals with the handling and storage of liquefied petroleum gases and the operation of low temperature petrochemical process plant, and these call for minimum service temperatures of about −100 to −110°C (170K).

One of the most important temperature regions from an economic point of view is that from −150 to −200°C (125–75 K) which covers the large-scale separation and storage of liquefied atmospheric gases such as oxygen and nitrogen as well as the more recently developed area of liquid natural gas technology.

The last, and most specialized, temperature range is that below −250°C (25 K) where hydrogen and helium can be liquefied and used for sophisticated applications such as rocket fuels and superconducting magnets as well as for a host of smaller scale laboratory research projects.

Properties of and requirements for materials operating at low temperature are listed in:

Table 1.2.12 *Important low temperature materials properties and requirements*

The kind of materials used and their major characteristics are listed in:

Table 1.2.13 *Characteristics of material types employed at low temperatures*

This section is confined to mechanical properties. Low temperature physical properties of materials are considered in Section 1.3.

1.2.4.2 Low temperature properties of metals

Alloys employed at low temperature are listed and their characteristics compared in:

Table 1.2.14 *Properties of FCC structure metals employed at low temperatures*

Table 1.2.15 *Composition and basic mechanical properties of some constructional and pressure vessel steels (BCC structure) used below room temperature (plates)*

Table 1.2.16 *Properties of HCP structure metals at low temperatures*

PROPERTIES OF BCC METALS EMPLOYED AT LOW TEMPERATURE

Metals with BCC crystal structure undergo a ductile/brittle transition, which may be at some temperature above or below ambient.

Exceptions: alkali metals, β brass.

Although this severely limits their use, the whole range of plain carbon, low and medium alloy steels is included in this category. The basic philosophy behind their successful application lies in ensuring that they have adequate toughness at their minimum operating temperature for them to withstand not only their design stresses, but also accidental impact and other overloads without failing in a brittle and catastrophic manner.

Design Codes such as BS 1500, BS 1515 and ASME VIII lay down the maximum allowable stresses at the lowest operating temperature for each grade, section, thickness and condition in which the steel is available. Further details on recommended practice can be obtained from these sources.

NB. Impact tests are a valuable quality control technique, but do not give rise to data which have any intrinsic meaning. To establish the conditions under which a steel may be safely used, more elaborate tests must be conducted.

Table 1.2.15 shows the composition and basic mechanical properties of some constructional and pressure vessel steels employed below room temperature. The beneficial effect of nickel on Charpy test values should be noted.

FRACTURE TOUGHNESS

Fracture toughness theory provides the only sound method of designing with high strength, low toughness materials. This theory is given more detailed consideration in Section 1.2.2.

Materials may be approximately classified into three categories outlined in:

Table 1.2.17 *Low temperature fracture behaviour of engineering alloys*

For details of experimental techniques based on fracture toughness theory for assessment of ferritic structural and pressure vessel steels, Section 4 of the Engineering Design Guide *Low Temperature Properties of Materials* by D. A. Wigley (Oxford University Press) should be consulted.

1.2.4.3 Low temperature properties of ceramics, glasses, concrete and cement

Mechanical properties:	Very little plastic deformation. Fail at low stresses in tension at all temperatures.
Typical applications:	Glass Dewar flasks for cryogenic liquid storage. Glass windows as viewing parts in cryogenic bubble chambers. Large liquid natural gas storage tanks have been constructed from concrete with the concrete maintained in compression by steel reinforcing rods placed in tension around the warmer exterior. For smaller scale applications (floors of loading bays, road storage vessels) high alumina cements have been used.
Remarks:	Careful thermal annealing to remove surface cracks and residual stresses from glass is essential. Extremely slow cooling of thick glass sections is recommended to avoid failure due to differential contraction or thermal shock. Residual compressive stresses can be induced in the surface of glass plates for toughening.

1.2.4.4 Low temperature properties of plastics and composites

THERMOPLASTICS

Desirable properties:	Low density.
	Good electrical and thermal insulating characteristics.
Limitations:	Very high thermal expansion coefficients relative to metals.
	Very prone to thermal shock, particularly below T_g (glass transition temperature).

The glass transition temperatures and LOX (liquid oxygen) compatibilities of a number of thermoplastics and elastomers are listed in:
Table 1.2.18 *Glass transition temperatures and liquid oxygen (LOX) compatibilities of some polymers*

Depending on the type of loading cycle, the lowest working temperature of a thermoplastic or elastomeric material is about 10–30K above its glass transition temperature. Brittle failure is in general encouraged by low temperatures, high strain rates and large stress concentrations. Abrupt section changes and sharp corners should be avoided in design.

Impact toughnesses over the working temperature range of a number of thermoplastics are compared in:
Table 1.2.19 *Comparative impact toughness of some thermoplastics*

THERMOSETS

Inherently brittle materials, similar to thermoplastics below T_g.
Poor thermal shock resistance.
Liable to crazing on rapid cooling.

Thermosets are almost invariably used in conjunction with fillers such as powders or fibres. These generally reduce the thermal expansion coefficient of the material and enable matching. Cloth fabrics, paper and other fibrous materials can be resinated and cured to form composites such as Tufnol which are strong, dimensionally stable, easily machined, available in several forms and with good electrical insulation properties.

COMPOSITES

Glass fibre- and carbon fibre-reinforced plastics have properties which make them suitable for cryogenic applications. Some properties are shown in:
Table 1.2.20 *Low temperature properties of some high performance composites*

LOW TEMPERATURE APPLICATIONS OF POLYMERIC MATERIALS

The use of fluorocarbons as seals

For sealing applications such as O rings, gaskets and bearings fluorocarbons are principally employed. This family (PTFE, TFE, Teflon, FEP, PCTFE, Kel-F) have the following properties:

Excellent electrical insulation
Inert to most chemicals and solvents
Non-stick and low friction characteristics
Useable for long periods at high temperatures
Measurable ductility down to 4K (1%)

Filled PTFE compositions are available (glass fibre, graphite or bronze) to improve the tensile and compressive properties of the polymer.

Adhesives

For cryogenic joining, adhesives are preferred to other conventional techniques. Fillers are used to match the expansion coefficient of the structural adhesive as closely as possible to those of the substrate and adherent. The glass line should be kept as thin as possible. 'Structures' or 'carriers' are sometimes used between the substrate and adherent – usually glass fibre mat. This allows an even bond line and reduces differential contraction and creep.

Common adhesives for cryogenic applications: Epoxy-nylons
Phenolic-nitrile
Epoxy-phenolic
Fluorocarbon-epoxy-polyamides

Superinsulants

Very thin films of polymers (eg. Mylar (Terylene)) are used for electrically insulating wires and thermally insulating flexible cryogenic pipelines. Although below T_g and potentially brittle, the extreme thinness allows them to be bent around very small radii without exceeding the elastic limit.

Examples: Vapour barriers for filament-wound glass fibre fuel tanks and foam insulation. Aluminized grades used to lag liquid helium and hydrogen storage vessels.

Expanded plastic foams for thermal insulation

Advantages: cheap
convenient to apply
lightweight
relatively efficient compared to mineral insulants, but inferior to vac-powder and superinsulation.

Materials: polystyrene, polyurethane
(thermal conductivity depends upon gas used to blow foam, rather than polymer)

Applications: rocket fuel tanks
large storage vessels
transfer lines

Generally used in conjunction with vapour barriers.
See also Vol. 3, Section 3.6.

Lubricants

PTFE filled with bronze, graphite or glass fibre are the only materials suitable for cryogenic operation because organic fluids freeze or are LOX incompatible and MoS_2 cannot operate without moisture and oxygen.

Table 1.2.12 Important low temperature materials properties and requirements

Materials property	***Materials requirement***
Strength	Withstand design stresses, including accidental impact at low temperature
Toughness	Failure must not be catastrophic and brittle.
Compatibility	Resistance to environments which may include liquefied gases including liquid oxygen
Structural and dimensional stability	Thermal cycling
Heat capacity	Minimize initial evaporation during cooldown
Thermal conductivity	**High** to minimize temperature gradients **Low** to minimize heat loss by evaporation
Electrical and magnetic properties	Ferromagnetism can be a disadvantage when dealing with magnetic currents or materials
Fabrication	Joints must withstand thermal cycling and not impair structural integrity

Table 1.2.13 Characteristics of material types employed at low temperatures

Material class	*Examples*	*Advantages*	*Limitations*
Face centred cubic metals (FCC)	Cu, Al, Ni alloys, austenitic steels	Strength, ductility and toughness all improve as temperature falls.	High cost for large structures
Body centred cubic metals (BCC)	Ferritic and martensitic steels	Low-cost methods exist to determine lowest safe operating temperatures (design codes such as BS1500, BS1515 and ASME VIII lay down the maximum allowable stresses at the lowest operating temperature for each grade, section, thickness and condition in available steels)	Materials (except alkali metals and brasses) undergo ductile/brittle transition which may be at temperatures above or below ambient dependent on composition, grain size, notches and defects
Hexagonal close packed metals (HCP)	Ti, Mg, Zr, Be alloys	Properties intermediate between those of FCC and BCC.	High cost compared with BCC metals.
Ceramics, glasses and concretes	–	Strong in compression, e.g. concrete maintained in compression by steel reinforcing bars.	Brittle, low thermal shock resistance and conductivity. Careful annealing to remove surface cracks and residual stress from glasses.
Thermoplastics	Nylons, acetals, polyethylenes, PTFE etc.	Low density, good insulators, fluorocarbons (PTFE, etc.) have exceptional properties.	High thermal expansion compared with metals. Prone to thermal shock below glass transition temperature T_g. Depending upon the type of loading cycle, the lowest working temperature of an elastomeric material is about 10–30 K above its glass transition temperature. Brittle failure in general is encouraged by low temperatures, high strain rates and large stress concentrations. Abrupt section changes and sharp corners should be avoided in design
Thermosets	Epoxies, unsaturated polyesters	Toughness improved by fibre reinforcement which also reduces thermal expansion	More brittle than thermoplastics of equivalent strength and stiffness. Poor thermal shock resistance and liable to craze on rapid cooling
High performance composites	Carbon fibre- and carbon–glass fibre-reinforced epoxies and polyesters	Strong, lightweight, highest strength/thermal conductivity of any materials useful for load-bearing insulated members	High cost.

Table 1.2.14

Table 1.2.14 Properties of FCC structure metals employed at low temperatures

Material			Advantages	Limitations	Typical applications	Remarks
Copper base	OFHC copper		Metal becomes stronger and more ductile at low temperatures. High electrical and thermal conductivity. Ease of joint formation.	High cost. Low-alloyed coppers used for additional strength.	Used where high conductivity, ease of joint formation are important.	
	α brass < 36% Zn		Strength and ductility increases with zinc content and decrease in temperature. Easy to machine. Solderable, brazeable.		Paired tube and high-pressure heat exchangers. Small cryostats.	α brasses generally preferred to α–β, β brasses because their strength can be increased by cold working (although at the expense of ductility).
	Cupro-nickel		Improved resistance to wear and corrosion. Low thermal and electrical conductivity.	Not competitive with stainless steels for corrosion resistance.	Heater windings, thermocouples condenser tubes.	
Nickel base	Nickel				Rarely used in practice, although perfectly suitable.	Nickel alloys can be controlled to give high or low expansion coefficients, hard or soft magnetic properties, constant elastic modulus, controlled electrical resistance
	Monel Inconel		Strength. Excellent corrosion resistance. Low thermal conductivity. Weldable with correct fillers.		Used for severely corrosive environments.	
Aluminium base	Aluminium		Can replace copper for high conductivity applications.	Soft.		Not suitable for use when alkalis or alkaline cleaners are used. LOX Compatible. Low density.
	Solution-hardened alloys	Al–Mn	Very ductile, easy to form. Dip brazeable. Extrudable and cuttable.	Only moderate strengths.	Tubes, bends, junctions, plate fin heat exchangers, tubes, plates, trays in distillation columns, etc.	
		Al–Mg	Higher strength than Al–Mn. Weld strengths approach parent metal.	Cold working severely impairs ductility.	Storage tanks. Road/rail transporters.	
	Heat-treatable alloys	Al–Mg–Si	Weldable.	Lowest strength of heat-treatable types but post-weld heat treatment necessary to maintain weld strength above those of non-heat treatable types.	Only heat-treatable Al alloy commonly used outside the aerospace industry.	
		Al–Cu	High strength/weight ratio.	Notch toughness falls seriously at low temperatures. Not easy to weld reliably.	Not widely used at low temperatures.	
		Al–Zn–Mg	Very high strength.	Very low notch toughness below 77 K.	Rarely, if ever, used.	

Table 1.2.14 Properties of FCC structure metals employed at low temperatures – continued

Iron-based alloys with stabilized austenitic structure	Invar 36% Ni		Completely stable. Strong and tough. Readily weldable, very low thermal expansion coefficient	High cost.	Transfer lines, pipelines, storage tank liners, etc., where contraction stresses have to be minimized.	Additional cost of Invar can be recovered by the smaller numbers of expansion joints that its use permits
	Austenitic stainless steels	Stable '25/20s', e.g. 310 CK20 Kromarc 55	High nickel content ensures stability of structure. Stable non-transforming steels. Reasonable strength and high toughness make suitable for long-term use.	310 becomes super paramagnetic at liquid helium temperatures.	Very important materials with applications ranging from small cryostats, precision components, to large storage and transport vessels and cryogenic fuel tanks.	
		Metastable '18/8s' e.g. 304 316 321 347	Suitable with reservations as outlined in limitations. Welding can usually be accomplished using suitable materials and processes. High N content improves yield strength and stability against martensite transformation.	Effects of martensitic transformation are: reduced toughness, dimensional variation, induced ferromagnetism. For precision equipment thermal cycling between 77 and 300 K should be effected prior to machining to complete transformation. Weld decay unless stabilized by Ti/Nb. TiC streaking can lead to porosity		

Table 1.2.14 – *continued*

Table 1.2.15

Table 1.2.15 Composition and basic mechanical properties of some constructional and pressure vessel steels (BCC structure) used below room temperature (plates)

Type	Condition	Designation	Composition (%)						Mechanical properties at 20°C			[a]Charpy V impact test temperature (°C)
			C	Mn	Si	Nb	Ni	Cr	Yield strength (MPa)	Tensile strength (MPa)	Elongation (%) $5.65\sqrt{S_o}$	
C–Mn semi-killed	As-rolled	BS 4360 Grade 40C Grade 43C	0.22 0.22	1.6 1.6					260	400–480 430–510	25 22	0 0
C–Mn–Nb semi-killed	Normalized	BS 4360 Grade 40D Grade 43D Grade 50C	0.19 0.19 0.24	1.6 1.6 1.6	 0–0.55	0.10 0.10 0.10			260 280 355	400–480 430–510 490–620	25 22 20	−20 −20 −15
C–Mn–Nb silicon-killed	Normalized	BS 4360 Grade 50D	0.22	1.6	0.1–0.55	0.10			355	490–620	20	−30
C–Mn silicon-killed + fine grain	Normalized	BS 4360 Grade 40E Grade 43E Grade 55E	0.19 0.19 0.26	1.6 1.6 1.7	0.1–0.55 0.1–0.55 0–0.65	0.10			260 280 450	400–480 430–510 550–700	25 22 19	−50 −50 −50
Ni–Cr–Mo	Quenched and tempered	BSC QT 445 (USS TI) ASTM A517 Gd F	0.10–0.20	0.60–1.00	0.15–0.35	Mo 0.40–0.60	0.7–1.0	0.4–0.65	690	795–930	18	−45
$2^1/_4$% Ni	Normalized	ASTM A203 Gd A	0.17–0.23	0.70–0.80	0.15–0.30		2.1–2.5		255	450	25	−60
$3^1/_2$% Ni	Normalized and tempered	BS1501/503 ASTM A203 Gd A	0.17–0.20	0.70–0.80	0.15–0.30		3.25–3.75		255	450	22	−100
9% Ni	Double normalized and tempered. As-welded with Ni–Cr–Fe (Inconel 92) electrodes	BS 1501/509 ASTM Code Case 1308 and A353	0.13	0.90	0.13–0.32		8.40–9.60		515 480	690 655	22 –	−196 −196

[a]N.B. In the case of structural steels covered by BS4360, this is the temperature at which the energy absorbed in a Charpy V notch impact test falls to 27 J (20 ft lb). This does not guarantee that the steel, especially if used in thick sections, will be completely satisfactory for use at low temperatures.

Table 1.2.16 Properties of HCP structure metals at low temperatures

Material	*Advantages*	*Limitations*	*Typical applications*	*Remarks*
Titanium alloys	Very high strength/weight ratio. Yield and tensile stresses rise rapidly at low temperatures. Extra-low interstitial grades available for cryogenic applications.	Ductilities strongly dependent upon interstitial impurity concentration. Oxygen must be kept below 0.12%. Welding is liable to cause recontamination and embrittlement.	Pressure vessels. Very low temperature areospace applications. (Down to 4 K).	BCC (β) grades (e.g. Ti–3V–11Cr–3Al) are too brittle. α–β types are only safe to 77 K (e.g. Ti–6Al–4V); because of low notch toughness.
Magnesium alloys	Low density. Good strength/weight ratio.	Notch sensitive. Low ductility. Poor impact strength.	Used to moderately low temperatures in aircraft components, but in construction applications can only be used where the limitations can be tolerated.	
Beryllium		Health hazard.	Has been considered for electrical conductors in cryogenic power cables.	
Zirconium	Alloys used in cryogenic applications.			
Cadmium Indium	Very high ductility at 4 K.		Pure indium wire used as a gasket to seal flanged and bolted joints at low temperatures.	
Zinc		Embrittled completely at temperatures just below ambient.	Not used.	

Table 1.2.17 Low temperature fracture behaviour of engineering alloys

Classification	***Approximate relationships***	***Steels***	***Aluminium alloys***	***Titanium alloys***	***Remarks***
High strength	$\sigma_y > E/150$	$\sigma_y > 1250$	$\sigma_y > 420$	$\sigma_y > 750$	Fail in low energy mode at all temperatures. Fracture toughness criteria provide the only safe basis for design.
Medium strength	$E/150 > \sigma_y < E/300$	$1250 > \sigma_y > 625$	$420 > \sigma_y < 210$	$760 > \sigma_y > 380$	Fail in a high-energy absorbing shear mode above room temperature. **Notch brittle** at low temperatures. Fracture toughness analysis should be used to check likelihood of failure in a low-energy mode at low temperature.
Low strength (including austenitic steels)	$\sigma_y < E/300$	$\sigma_y < 625$	$\sigma_y < 210$	$\sigma_y < 380$	Fail in a high-energy absorbing shear mode.
Ferritic steels	$\sigma_y < E/300$	$\sigma_y < 625$			**Notch brittle** below shear to cleavage transition temperature which is to some extent controlled by composition and structure.

Stress in MPa.
For details of experimental techniques based on fracture toughness theory for assessment of ferritic structural and pressure vessels steels, Section 4 of the Engineering Design Guide *Low Temperature Properties of Materials* by D. A. Wigley (Oxford University Press) should be consulted.

Table 1.2.18 Glass transition temperatures and liquid oxygen (LOX) compatibilities of some polymers

Common name	***Synonyms***	***State at room temp.***[a]	***Glass transition temp. T_g (K)***	***LOX compatibility***	***Comments***
Natural rubber	*cis* 1–4 Polyisoprene, NR.	E	203	No	Crystallizes rapidly below room temperature.
Butadiene rubber	'High-*cis*' polybutadiene, BR.	E	173–203	No	Low temp. resilience of solution and emulsion polymerized BR better than NR.
Styrene/butadiene rubber	SBR	E	215–235	No	Low temp. resilience falls off with increase in styrene content, not as good as NR.
Nitrile rubbers	Acrylonitrile/butadiene copolymer, NBR.	E	221–248	No	Low temp. resilience falls off with increase in acrylonitrile content, not as good as NR.
Carboxylated nitrile latex		E	<173	No	Terpolymer with improved low temp. properties.
Chloroprene rubbers	CR	E	230–243	No	Excellent resistance to most oils, greases and halogenated hydrocarbon refrigerants. Cloth-inserted sheeting best for gaskets.
Ethylene/propylene rubber		E	215–221	No	Good low temp. flexibility and abrasion resistance.
Butyl rubber	Isobutylene/isoprene copolymer, IIR	E	210	No	Low temp. resilience inferior to NR. Cloth-inserted sheeting best for gaskets.
Polysulphide rubber		E	223	Moderate	Outstanding solvent and oil resistance but poor strength.
Polyurethane rubber	AU, EU	E	203–238	Moderate	Can be formulated as an elastomer, coating or adhesive.
Silicone rubbers	Polydimethyl siloxane, MQ, PMQ, VMQ, PVMQ	E	150–160	Moderate	Good chemical and oxidation resistance but high cost: most suitable true elastomer for low temp. use.
Fluoro-rubbers (i) vinylidene fluoride/ hexafluoropropylene copolymers	FKM	E	220–250	Good	Oil resistant, chemically inert, good oxidation resistance.
(ii) Nitroso rubber		E	222	Good	Incombustible in gaseous oxygen.

[a]E = elastomer; TP = thermoplastic.

Table 1.2.18 Glass transition temperatures and liquid oxygen (LOX) compatibilities of some polymers – *continued*

Polytetrafluoroethylene	PTFE	TP	400 (T_g) 166 T_{gg}	Complete	Expensive but irreplaceable for low temperature service. Fabricated by cold pressing and sintering.
Fluorinated ethylene/propylene copolymers	FEP	TP	–	Complete	Can be moulded and extruded, better impact strength than PTFE.
Polychlorotrifluoroethylene	PCTFE	TP	~490	Complete	Less ductile but tougher than PTFE at very low temperatures.
Polyvinylidene fluoride	PVDF	TP	~230	Good	Stronger and tougher than PFTE or PCTFE at low temperatures, allows thinner and more flexible seals.
Polyvinyl fluoride	PVF	TP	–	Good	Widely used as barrier layer film. Good flexibility at low temperatures.
Polyethylene terephthalate		TP	340	Moderate	Widely used as film or fibre. Strong and tough down to very low temperatures.
Polyimide		TP	–	Moderate	Originally developed for superior high temperature performance, good flexibility at low temperatures.
Polyamide	Nylons (hexamethylene adipamide = nylon 66)	TP	323	No	Widely used as engineering plastic at room temperature, embrittles at lower temperatures.
Polyethylene	(Low density, LDPE)	TP	153–173		Retains excellent toughness to below 200 K, but very high thermal contraction.
Ethylene/vinylacetate copolymers	EVA	TP	<200	No	Retain flexibility and toughness down to ~200 K.
Ionomers		TP	<180	No	Cross-linked by ionic bonds to increase toughness, flexible down to <200 K.
Polypropylene	PP	TP	255	No	Stronger than polyethylene at room temperature but embrittles at moderately low temperatures.
Polyvinylchloride	PVC	TP	~360	No	Rigid PVC embrittles at ~240 K but plasticized PVC flexible down to ~220 K.

[a]E = elastomer; TP = thermoplastic.

Table 1.2.18 – *continued*

Table 1.2.19 Comparative impact toughness of some thermoplastics

Material (recognized abbreviation)	***Temperature (°C)***							
	−20	***−10***	***0***	***10***	***20***	***30***	***40***	***50***
Polystyrene (PS)	A	A	A	A	A	A	A	A
Polymethylmethacrylate (PMMA)	A	A	A	A	A	A	A	A
Glass-filled nylon (dry) (PA)	A	A	A	A	A	A	A	B
Methylpentene polymer	A	A	A	A	A	A	A	AB
Polypropylene (PP)	A	A	A	A	B	B	B	B
Craze resistant acrylic (PMMA)	A	A	A	A	B	B	B	B
Polyethylene terephthalate (PET)	B	B	B	B	B	B	B	B
Polyacetal (POM)	B	B	B	B	B	B	C	C
Rigid polyvinylchloride (PVC)	B	B	C	C	C	C	D	D
Cellulose acetate-butyrate (CAB)	B	B	B	C	C	C	C	C
Nylon (dry) (PA)	C	C	C	C	C	C	C	C
Polysulphone	C	C	C	C	C	C	C	C
High density polyethylene (HDPE)	C	C	C	C	C	C	C	C
Polyphenylene oxide (PPO)	C	C	C	C	C	CD	D	D
Propylene–ethylene copolymers	B	B	B	C	D	D	D	D
Acrylonitrile–butadiene– styrene (ABS)	B	D	D	CD	CD	CD	CD	D
Polycarbonate (PC)	C	C	C	C	D	D	D	D
Nylon (wet) (PA)	C	C	C	D	D	D	D	D
Polytetrafluoroethylene (PTFE)	BC	D	D	D	D	D	D	D
Low density polyethylene (LDPE)	D	D	D	D	D	D	D	D

Key: A: Brittle, specimens break even when unnotched.
B: Notch brittle, specimens brittle when bluntly notched but do not break when unnotched.
C: Notch-sensitive, specimens brittle when sharply notched.
D: Tough, specimens do not break even when sharply notched.

(Data by courtesy of ICI Plastics Division)

Table 1.2.19

Table 1.2.20 Low temperature properties of some high performance composites

Reinforcing agent	***Material***	***Advantages***	***Limitations***	***Remarks***
Glass fibre	GRP (thermosetting resin matrix)	Enhanced toughness. Quite high strengths. Extremely high specific strengths. Highest strength/ thermal conductivity ratio of any materials used in cryogenic equipment.	Static fatigue, when moisture penetrates fibre/ matrix interface. Anisotropic mechanical properties. Material becomes porous long before full potential strength is developed (serious for GRP filament wound fuel tanks, overcome by electroplating metal liner inside the tank).	GRPs used in fuel tank construction and aerospace applications. Use for load-bearing, thermally insulating supports is increasing both for tensile and compressive loading configurations.
Carbon fibre	CFRP	Much higher stiffness and specific stiffness than GRP.	Much higher thermal and electrical conductivities than GRP. CFRP can be stiffened by special design modifications.	Only employed where higher modulus is advantageous, e.g. highly stressed fuel tanks wherein the vapour barrier problem would be minimized.

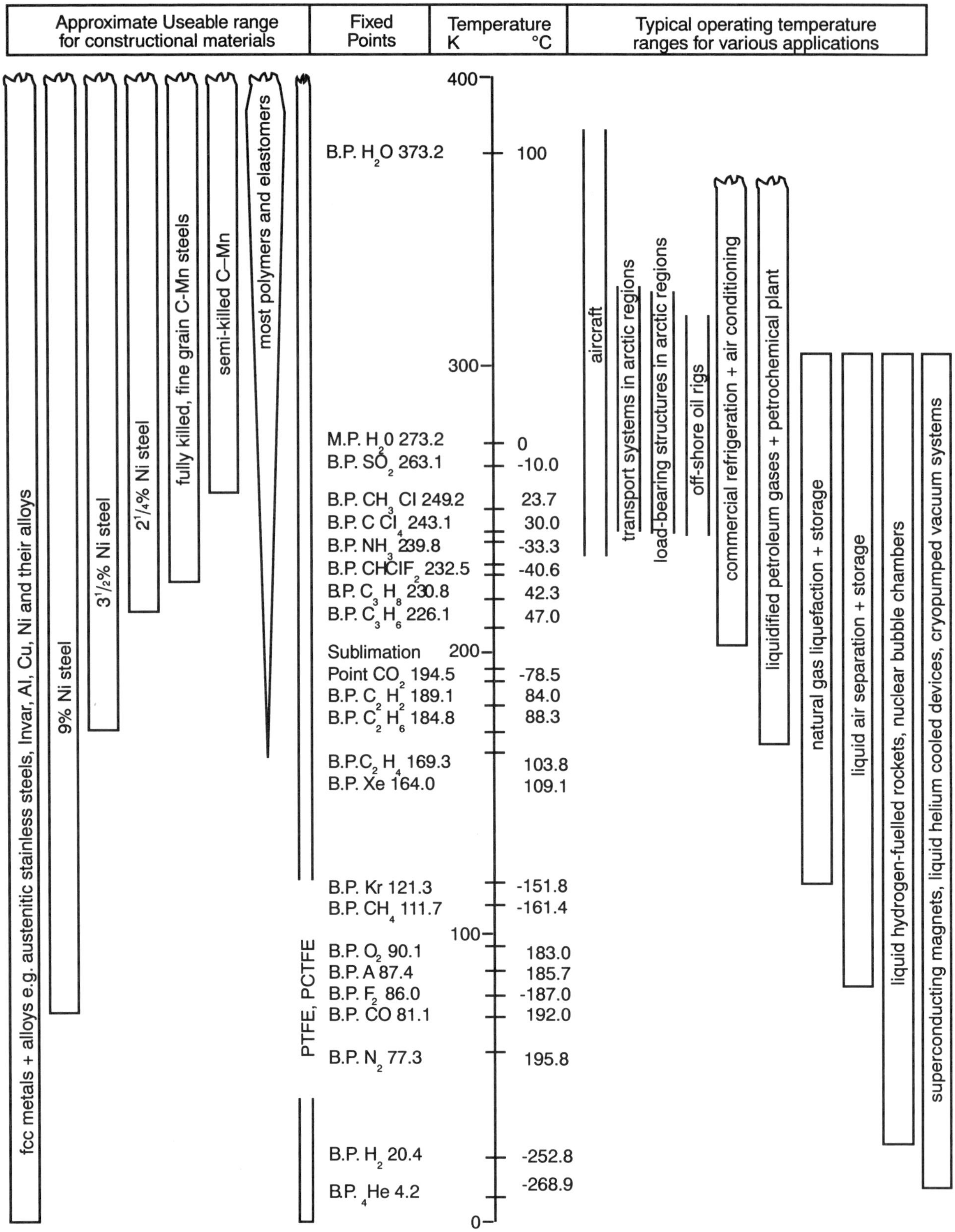

Fig. 1.2.44 Some significant cryogenic temperatures and typical operating temperature ranges for various structures and constructional materials

Fig. 1.2.44

1–3

Physical properties

Contents

List of tables

List of figures

1.3.1 Density (specific gravity, ρ)

Materials classified according to their specific gravity are given in:
Fig. 1.3.1 *Materials classified according to specific gravity and its influence on selection*

The relationship between density and specific gravity can cause some confusion. The specific gravity of a substance is given by its density normalized by the density for water. Hence, at 23 °C:

$$\text{density (g/cm}^3) = \text{specific gravity} \times 0.9975$$

where the density for water at 23°C is 0.9975 g/cm^3.

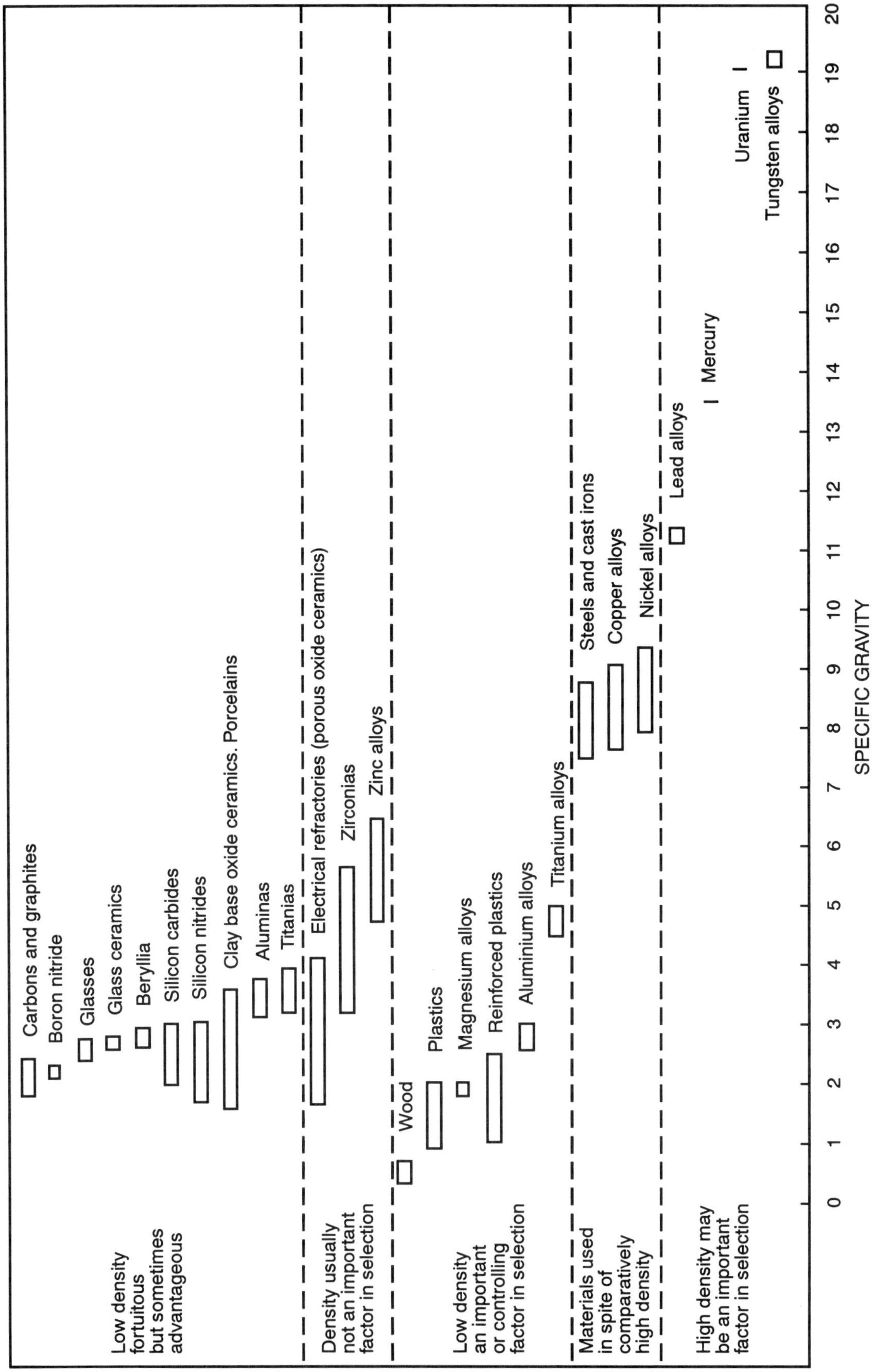

Fig. 1.3.1 Materials classified according to specific gravity and its influence on selection

Fig. 1.3.1

1.3.2 Thermal properties

1.3.2.1 Thermal expansion, α

The thermal expansion coefficient of a solid may vary from zero (or even a negative value) over certain ranges of temperature to a value of 300×10^{-6}/K.

A high coefficient is usually associated with a low melting point and *vice versa*. A very high (or very low) coefficient over a limited temperature range is usually associated with a change in phase or atomic rearrangements.

A high thermal expansion coefficient is usually a disadvantage in design and, other considerations being equal, a material with low thermal expansion coefficient is to be preferred to a material with a high coefficient. The most critical effects of thermal expansion arise when it results in a change in moment of inertia in a timing mechanism.

More usually, however, the influence of thermal expansion becomes apparent when combining different materials, composite materials (e.g. carbon fibre-reinforced epoxy or silicon carbide particulate-reinforced aluminium) being a case in point. For components using different materials, fits and clearances must be precisely evaluated or leaks, oxidation or bursting stresses may result from temperature variations in a single material.

The harmful effects can be minimized by a careful choice of material. Indeed, the use of a material of high expansion coefficient of shorter length or heated to a lower temperature in a two-component system can neutralize the effects of high expansion.

Thermal expansion interacts with thermal conductivity and specific heat to generate thermal and residual stresses when a component is heated non-uniformly. Here a low coefficient is advantageous but a high thermal conductivity and adequate ductility are even more so. For example, copper is superior to silicon carbide, which is itself superior to vitreous silica in spite of the relative thermal expansions of these three materials.

For composite materials, the internal stresses that can occur during thermal cycling can have a dramatic effect on mechanical behaviour, depending on the available mechanisms to relieve the induced stresses, (e.g. diffusion, plastic deformation, fracture, etc). Thermal expansion is of use mainly in assembly, including hot shrinking, and in control mechanisms where thermal displacements can be made substantial or eliminated entirely by the interaction of materials with high and low coefficients of expansion.

The thermal expansion coefficients of typical engineering materials are shown diagrammatically in:

Fig. 1.3.2 *Thermal expansion coefficients of materials*

Notable are the very high coefficients of some plastics and the near to zero expansion coefficients of vitreous silica, boron nitrides and, over a limited temperature range, some iron–nickel alloys.

Some crystalline materials have anisotropic thermal expansions, boron nitride among ceramics and uranium among metals.

Some materials show quite sharp dimensional changes over short ranges of temperature, usually associated with phase changes. Frequent excursions through these ranges should be avoided.

1.3.2.2 Thermal conductivity, λ

The value of thermal conductivity is of design importance in three ways: where heat conduction is part of a component's function; where heat insulation is part of a component's function; and where neither is the case but it is essential (or desirable) to avoid hot spots and consequently material deterioration through distortion, thermal fatigue, melting or environmental attack.

The thermal conductivity of typical engineering materials is illustrated diagrammatically in:

Fig. 1.3.3 *Thermal conductivity of engineering materials*

The thermal conductivity of metals depends on the amount of alloying, pure metals having high conductivities while alloying agents, particularly those in solid solution, reduce conductivity. Thus, very pure copper has a very high conductivity, but that of brasses and bronzes is much lower. Of the steels, carbon steels have the highest conductivities and highly alloyed steels, including stainless steel, the lowest.

Most oxide ceramics (except beryllia) have very low thermal conductivities but other ceramics may be good conductors.

Materials used specifically as heat conductors include copper (locomotive fire boxes and condenser tubes), aluminium (power sinks for transistors) and graphite/clay mixtures for crucibles. Boron nitride is also used as a crucible material, advantage being taken of its high 'C' direction conductivity to even out 'hot spots' in crucibles. Beryllia (and boron nitride) are also used as heat-conducting electrical insulators.

Conductors need not necessarily be solid metals. Liquid sodium, which has a high thermal conductivity, is used to conduct heat from the hot disc of an aero engine valve to the cooler stem.

Materials used specifically as thermal insulators are mostly oxide ceramics. The material itself functions mainly as a refractory, often in the form of fibre, the insulation being provided by the air entrapped in the porosity.

When a material is employed in a position where thermal gradients are likely to occur it is preferable, other considerations being equal, to use the material with the highest thermal conductivity. This is the case with gas turbine blades, where a high thermal conductivity limits the formation of hot spots, and power plant materials where high thermal conductivity of low alloy steels reduces thermal fatigue compared with austenitic steels.

1.3.2.3 Specific heat, *C*, and thermal diffusivity, *D*

Specific heat is defined as the heat required to raise the temperature of a specific weight of material by a specific increment of temperature (S.I. units J/kg per K). As may be seen in Table 1.3.1, it varies very considerably between materials:

Table 1.3.1 *Physical properties of metals and ceramics*

Because specific heat is a function of movement or vibration of individual atoms its value is readily calculable for simple materials over temperature ranges remote from phase changes.

For a pure solid element, specific heat is, to an accuracy adequate for most practical purposes, 25 000 divided by the atomic weight. Ceramics and plastics, because of complications introduced by the differing masses and varying nature of bonding between the constituents, usually have lower average atomic specific heats than elements.

It also follows, since variations in density usually follow those in atomic weight, that the specific heat by volume (J/m^3 per K) shows less variation than the specific heat by weight, lying for most materials, between 2.0 and 3.5 J/cm^3 per K.

Specific heat is significant in design because it directly influences the rate at which the input of heat will raise the temperature of a material, and therefore the cost of heating or heat retention by thermal reservoirs.

It also influences thermal diffusivity D where

$$D = \frac{\lambda}{\rho C}$$

where λ = thermal conductivity, ρ = density, and C = specific heat.

Thermal diffusivity is an inverse measure of the ability of heat flow to induce thermal stress in a material, but because of the comparative invariance of specific heat, thermal conductivity has a much greater influence.

Thermal diffusivity is relatively easy to measure and its measurement is therefore used to determine the value of thermal conductivity of a material.

Table 1.3.1 Physical properties of metals and ceramics

	Density ρ (g/cm^3)	***Specific heat (C) (J/kg per K)***	***Thermal conductivity λ (W/m per K)***	***Thermal diffusivity (λ/ρ per C)***	***Thermal expansion coefficient ($10^{-6}/K$)***	***Electrical resistivity (μΩcm)***
Aluminium Aluminium alloys	2.70 2.46–2.93	900	239 88–201	0.098	23.5 16.5–25	2.68 3.5–8.6
Copper Copper alloys	8.94 7.57–8.94	385	399 21–397	0.12	17.7 16–21.2	1.7 7–35
Magnesium Magnesium alloys	1.74 1.75–1.87	1050 960–1050	167 79–146	0.091	27 26–27.3	3.9 5–14.3
Nickel Nickel alloys	8.89 7.85–9.22	456 373–544	74.9 9.1–21.7	0.019	13.3 7.6–14.9	9.5 51–139
Titanium Titanium alloys	4.51 4.42–4.86	528 400–610	16 4.8–16	0.0067	7.6 6.7–9.8	48.2 70–170
Zinc Zinc alloys	7.13 5.0–6.7	389 109–123	113 26–28	0.04	39.7 27.4	6
Steels Carbon steels Low alloy Stainless	 7.83–7.87 7.83–7.87 7.42–8.69	 435–494 452–494 402–519	 45.2–65.3 33.1–48.6 12.1–26.8	 0.015 0.010 0.005	 10.6–12.62 10.55–12.8 9.3–18	 12–19.7 20–37 48.6–122
Aluminas	3.45–3.99	730–1100	13.8–43.2	0.008	4.5–8	
Beryllias	2.8–2.93	1020–1090	270–300	0.09	5–8	
Titanias	3.5–4.13	690	2.5–5	0.001	5–8.5	
Zirconias	3.5–5.9	450–750	0.9–2	0.0005	7–9	
Porcelains (oxide compounds clay based)	1.8–3.8	850–1100	1.1–6.2	0.0013	1–11	
Porous oxide ceramics (electrical refractories)	1.9–4.4	1100	0.3–3.7	0.0006	-0.05–7	
Silicon carbides	2.2–3.2	600–700	12.6–200	0.06	2.8–4.2	10^5–10^{11}[a]
Boron nitrides	1.9–2.1	780	1.5–250[b]	0.0007–0.16[b]	–2–4.1	
Silicon nitrides	1.9–3.3	700–1100	7–43	0.01	1.5–3.6	
Sialons	3.3	600	21.3	0.01	1.5–1.7	
Cemented carbides	10–15.3		25–120		5–7	
Glass ceramics	2.4–2.6	500–900	1.3–3.6	0.0013	–0.25–9.7	
Carbons, graphites	1.6–2.2	700–800	5–121[b]	0.003–0.08[b]	1–8.3	90–450
Glasses	2.18–2.50	710–830	1.1–1.2	0.0006	0.013–7.8	10^{14}–10^{16}

[a] Other ceramics have resistivity 10^9–10^{17} Ωcm. Plastics 10^6–10^{18} Ωcm.
[b] Depending on direction of heat flow.

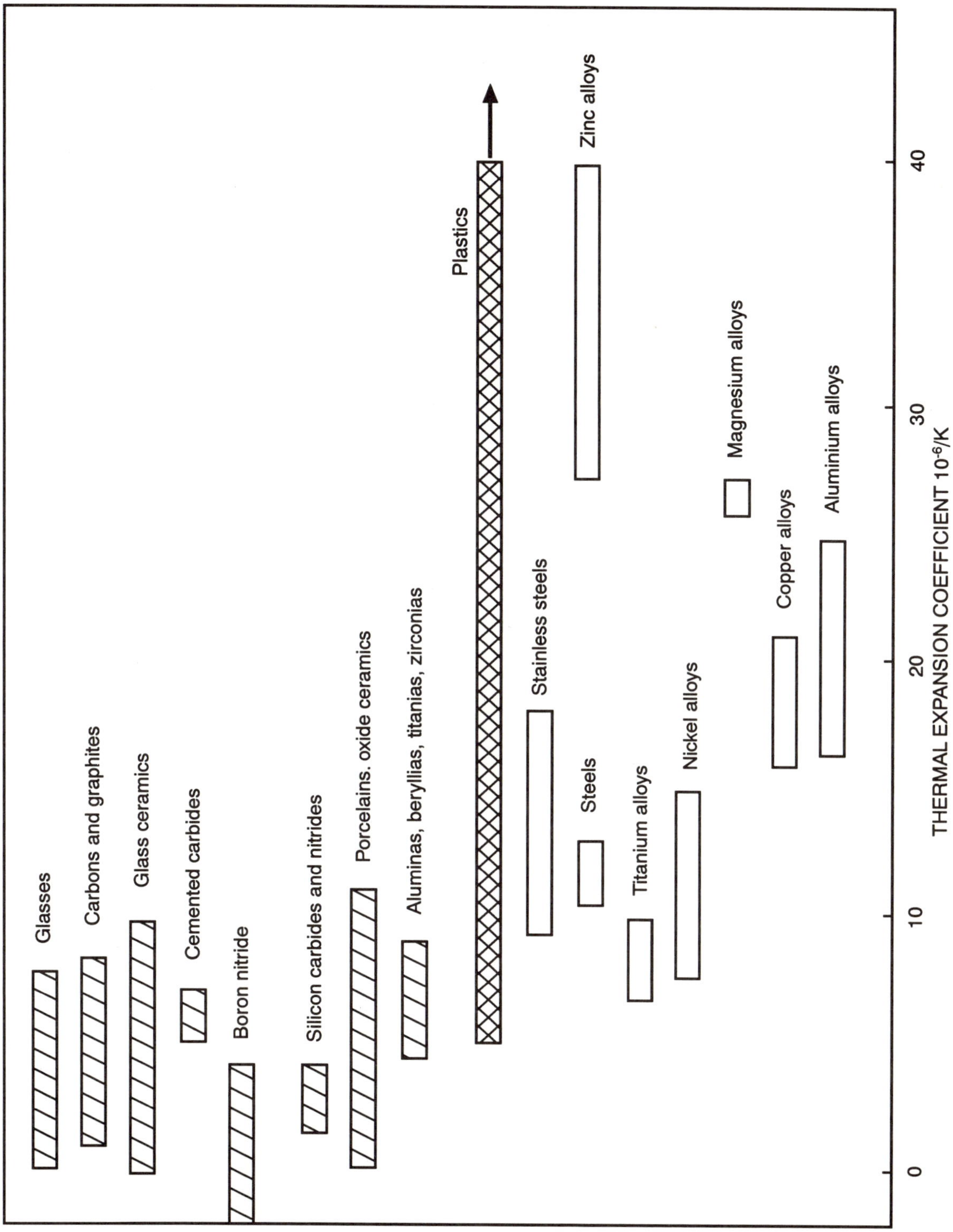

Fig. 1.3.2 Thermal expansion coefficients of materials

Fig. 1.3.2

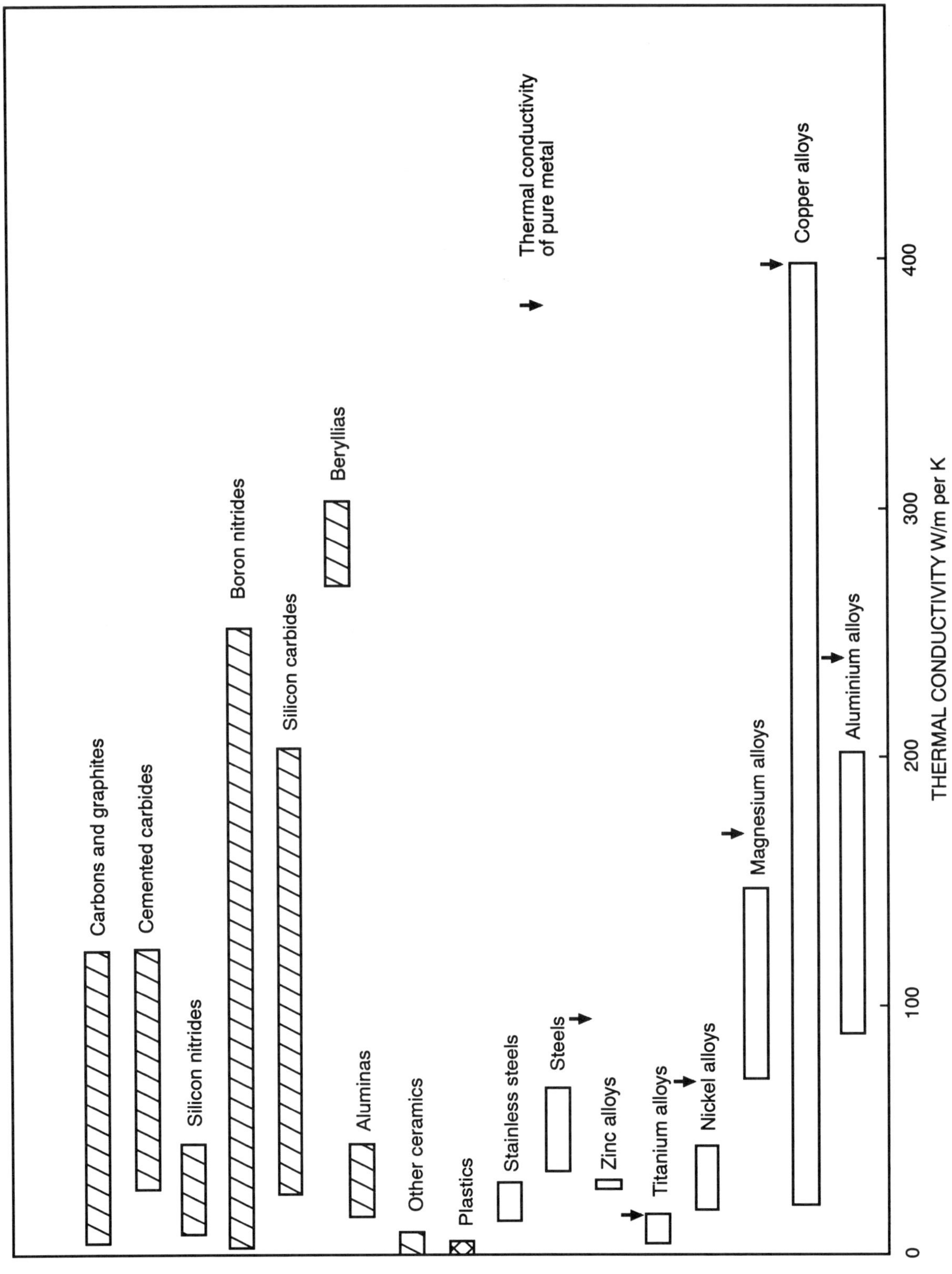

Fig. 1.3.3 Thermal conductivity of engineering materials

Fig. 1.3.3

1.3.3 Electrical properties

1.3.3.1 Electrical conductivity (resistivity, Ω)

Electrical conductivity is usually specified by its inverse parameter, resistivity, in ohms, Ω. It is an important property in conductors, contacts, resistors, semiconductors and insulators.

While relative conductivity influences choice for certain applications, resistance to environment is the most important parameter for many applications. However good a material is as a conductor it is useless if severely corroded.

Conductivity of a metal is normally a function of purity. Except for semiconductors, the higher the purity the higher the conductivity because conductivity is reduced by the presence of alloying elements, particularly in solid solution.

CONDUCTORS

Materials with a very wide range of electrical conductivity may find application as conductors, as shown in:

Fig. 1.3.4 *Electrical resistivity of materials grouped according to function*

Highly pure copper and aluminium are almost exclusively employed for conventional wiring. A small addition of silver marginally improves the temperature resistance of copper without impairing conductivity, strength and resistance to deformation, and wear resistance is conferred by the addition of cadmium. Zirconium– and chromium–copper have significantly improved creep strength, albeit at the expense of conductivity.

Carbon and graphite are used for conduction at very high temperatures including electrodes for arcs, and conducting films.

Superconductivity, the complete absence of electrical resistance, was first discovered in 1911 in mercury at 4.2 K (–269°C), achieved by cooling in liquid helium. The production of liquid helium is complicated and expensive, making materials that could superconduct at 'high' temperatures, for example, above the boiling point of nitrogen (77 K or –196°C), extremely attractive. With zero resistance, superconductors could improve the efficiency of motors, generators and power transmission dramatically.

Until 1986, the highest temperature at which superconductivity had been observed was still only 23 K (–250°C), achieved with the compound Nb_3Ge, which is still the highest temperature known for a metal alloy superconductor. The interest in so-called high temperature superconductors took off in 1986 with the discovery of layered cuprate compounds, superconducting at 40 K. Since then, the highest confirmed temperature at which superconductivity has been observed is about 165 K, with a mercury-based cuprate compound under pressure.

Despite the early high temperature superconductors being incapable of handling the large currents required for power engineering applications, it is still hoped that they will find application in large-scale power transmission systems as well as electronic components and superconducting magnets. There also remains interest in compounds of the type Nb_3Ge and Nb_3Sn, superconducting at around 20 K, as these have high current-carrying capacity.

Gold is used in the form of thin films, for interconnecting transistor circuitry.

A special case is the compensating leads from a thermocouple to a cold junction or a compensating pyrometer. These must be of materials which, over the ambient temperature range, generate the same EMF as the thermocouple, but, since a balanced potentiometer takes no current, need not necessarily be of low resistance.

Conducting polymer composites have resistivities ranging between 10^{-2} and 10^{12} Ω. They are described in Section 3.8.7.

CONTACTORS

There are two types of contactor: contact points which impinge and open directly and brush-type contacts where one member slides over another.

The point types may be made, for progressively severe duty, of copper, tinned copper, silver plated copper, tungsten and platinum. The harder, higher melting point materials resist erosion by sparking. Sulphur hexafluoride atmospheres may impart protection.

Brush-type contacts are normally of carbon, or a copper–carbon cermet. These materials will not produce adhesive wear on the traversed copper contact.

RESISTORS

Resistors are normally either wound from a resistance wire of some alloy with a fairly high resistance and good environmental resistance such as nichrome, nickel silver, or, for temperature stable low duty resistors, constantan, or made by depositing on a carbon film on a ceramic.

High-temperature furnace resistor elements may be made from silicon carbide or platinum. Silicon carbide is a semiconductor whose resistance decreases sharply with increases in temperature and also with increase in direct current.

Very high temperature vacuum furnace resistor elements are spirals machined from graphite.

SEMICONDUCTORS

Semiconductors used extensively in transistors are made from silicon (or germanium) doped with elements such as phosphorus (P), antimony (Sb) or arsenic (As) to make n-type semiconductors or boron (B) to make p-type semiconductors.

Junctions between p-and n-type semiconductors may be used as rectifiers or, by utilizing the increasing current voltage ratio with increasing voltage, as amplifiers.

INSULATORS

Insulators are usually oxide ceramics or plastics. They normally have resistivities in excess of 10^{11} Ω cm tested cold and dry.

Conditions of use may, however, reduce their insulating properties. Increase in temperature reduces resistivity roughly 100 times for 200°C for the purer oxide-type ceramics and by higher factors for other materials. This deterioration is sometimes specified in the case of ceramics as the T_e value, the temperature at which the resistivity has fallen to 10^4 Ωm.

High potential gradients cause breakdown in a material. The capacity to resist this is described as 'dielectric strength' and the method of measurement specified in ASTM D149. Both plastics and ceramics deteriorate because of arcing along the surface. 'Tracking resistance' measurements are specified in ASTM D495.

The properties of individual materials are listed for ceramics in Vol. 2, Chapter 15 and for plastics in Vol. 3.

1.3.3.2 Thermoelectricity

When two electrical conducting materials are connected in series, heating one of the two junctions generates an electrical potential roughly proportional to the temperature difference between them. The ratio of the potential difference to the temperature difference is termed the 'thermoelectric power'. This may be used to generate electricity, or by using carefully calibrated materials for the two limbs of the thermocouple, to measure temperature.

Typical thermocouple materials are:

iron:constantan for low temperatures;
chromel:alumel for intermediate temperatures;
platinum:platinum–rhodium for high temperatures.

1.3.3.3 Enhancement charge storing capacity or permittivity

When conductors are separated by an insulator, a potential difference between the conductors will cause a current to flow. The magnitude of this current depends on the electric polarization of the insulator.

The relative permittivity (or dielectric constant) of a material is the ratio of the electric charge required to raise the potential difference a specific amount to that which would be required in empty space.

The relative permittivity of most plastics lies between 2 and 8 and of most ceramics between 6 and 9 (see details of specific materials in Vols 2 and 3) but certain titanates specially formulated as dielectrics may have values ranging up to 5500.

1.3.3.4 Loss tangent and dielectric power loss

When an alternating voltage is applied to a dielectric, electric power is lost and appears as heat. This is normally described by the phase angle δ introduced by the time taken for polarization to occur after the field is applied.

The power loss per cycle per unit volume of material, P, is:

$$P = \pi \epsilon_0 K' V_0^2 \tan\delta$$

where ϵ_0 is the permittivity of free space ($8.8\,54 \times 10^{12}$ F/m), V_0 is the maximum of the sinusoidal voltage, and K' is the relative permittivity of the material.

Tan δ varies very considerably with frequency but normally at low frequencies, even though tan δ may be quite high, losses are comparatively low and choice of insulator not restricted. However, at high frequencies ceramics containing proportions of glass cause unacceptable loss and more expensive low alkali materials must be used.

Values of relative permittivity and tan δ are listed under specific materials in Vol. 2 Chapter 2.15 for ceramics and Vol. 3 for plastics. ASTMD150 and BS4612 Parts 2.1 and 2.2 specify procedures for measuring relative permittivity and tan δ.

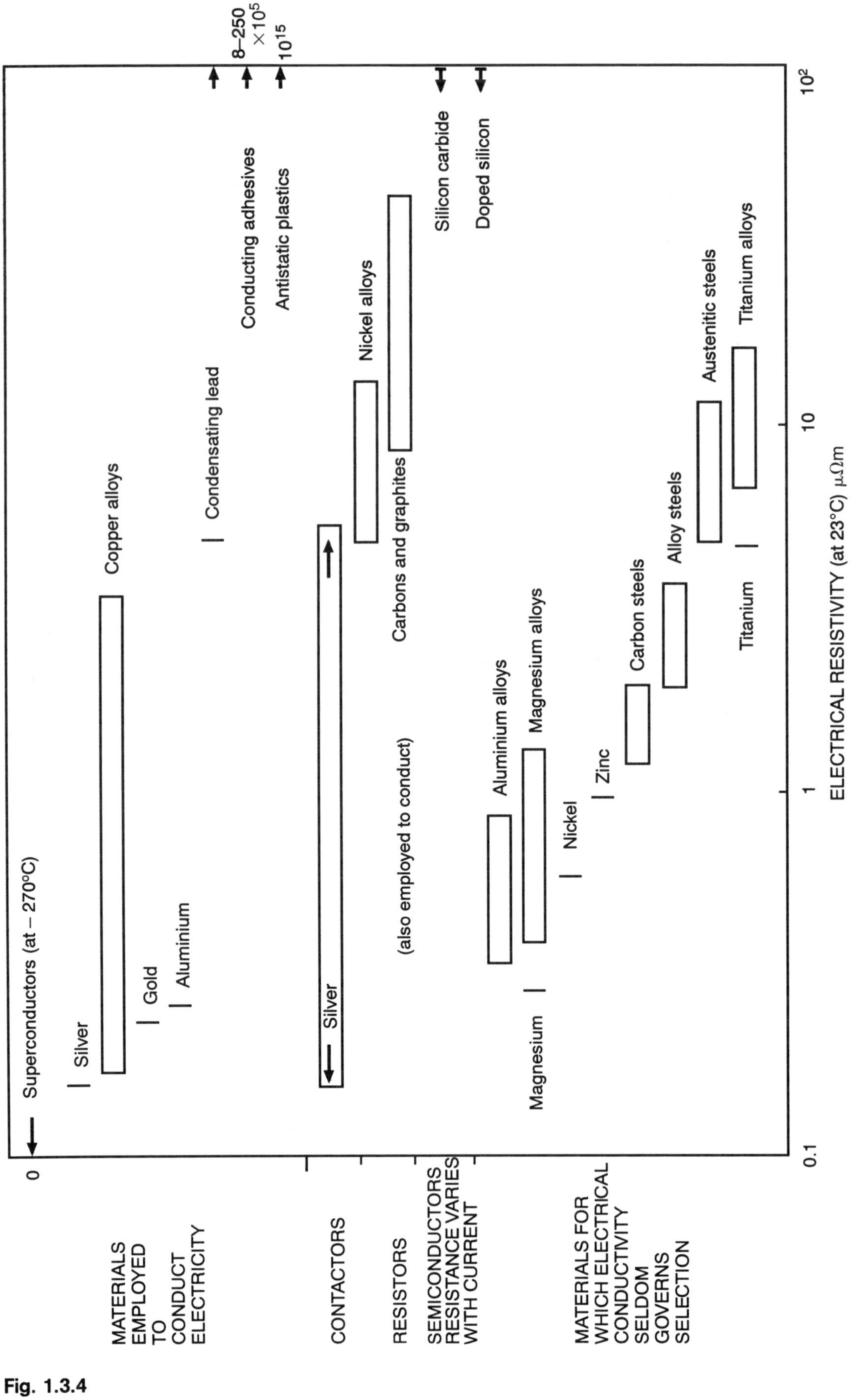

Fig. 1.3.4 Electrical resistivity of materials grouped according to function

Fig. 1.3.4

1.3.4 Magnetic properties

1.3.4.1 Introduction

Some materials show ferromagnetism, becoming strongly magnetized in weak fields. The elements iron, nickel and cobalt are ferromagnetic at room temperature and are the bases of most alloys used in magnetic applications. Some oxides of iron, manganese, nickel, etc., are ferrimagnetic. This property differs in terms of the electron behaviour from ferromagnetism but its manifestations are identical to ferromagnetism, and some useful magnetic materials are mixtures of these magnetic oxides.

On heating a critical temperature, called the Curie temperature or Curie point θ_c, is reached and the material is no longer magnetic, but becomes paramagnetic.

Ferro- or ferrimagnetic materials are used in engineering applications because they provide large values of magnetic flux density B which is of paramount importance in electrical engineering and can be used statically and dynamically.

For any material:

$$B = \mu_0 \ (H \ + M)$$

where B is the flux density, μ_0 the space permeability (or magnetic field constant), H the magnetizing force or field and M the magnetization.

In a ferromagnetic material most of the flux comes from the material rather than the applied field. Without it, B and $\mu_0 H$ would be the same. In a ferromagnet M reaches a constant maximum value, the saturation magnetization M_S, in quite moderate fields. The saturation induction is often called $J_S = \mu_0 . M_S$.

The way in which the high flux density is employed depends critically on the application but in general, it may be used in one of two ways. Firstly to induce a voltage in a coil linked with the magnetic circuit, e.g. in a transformer or a choke. Secondly to generate a useful force between the magnetized material and another magnetic component, e.g. in a motor.

Both these functions, and combinations of them, can be used in a great variety of applications and devices.

Type of interaction	*Role of B*	*Typical device*
Electrical	Induce a voltage proportional to dB/dt	Transformer, choke
Electromechanical	Generate a force on a current carrying conductor proportional to B	Electric motor, loudspeaker
Mechanical	Create an attractive force proportional to B^2	Holding magnet, relay, magnetic separator

Magnetic materials may be 'soft', in which the magnetizing force is only maintained when the magnetizing field is present, or 'hard' which once magnetized stay magnetic. Typical magnetic materials are mapped on a B, H diagram in:

Fig. 1.3.5 *Magnetizing force and flux density attained in fields of up to 10* H_c

SELECTION CRITERIA AND CLASSIFICATION OF MAGNETIC MATERIALS BY MANUFACTURING TECHNIQUES

The broad principles outlined in the previous section can be used for the selection of materials for any given use. But for this purpose, the technique by which the materials are made and the form in which they are available is also relevant. Guidelines for selection will be discussed in this chapter. Figure 1.3.5 gives a diagram of coercive force and flux density attained in fields of up to $10H_c$. It can be seen that for lowest and highest H_c materials, B is limited. The cobalt–iron alloys have the highest values of flux density (2.43 T for Fe–35% Co); the ferrites have the lowest values (0.2–0.4 T). In general, the cost of materials increases as the H_c values deviate from the centre value of $H_c = 10^3$ A/m. For a general index of critical application and choice of most suitable materials, with indication of tables in which further details can be found, see:
Table 1.3.2 *Critical application parameters and choice of magnetic materials*

1.3.4.2 'Soft' magnetic materials

In equipment for the general distribution and use of electrical energy the most important properties are low hysteresis losses at high flux densities. These properties are found in electrical steels, which must often be available in the form of thin sheet, or laminations with insulated surfaces. Because of the tonnage quantity use, cost is a prime consideration and hence materials consisting mainly of iron are used. Alloying fairly pure (carbon-free) iron with up to 4% silicon results in steels with high permeability and fairly low loss whilst retaining high values of B_s. The addition of Si increases the electrical resistivity of iron and in the moderate percentage used does not make it impossible to hot roll.

Bar, rod, slab plate sheet and strip can be made. Materials are used in sheets or strip form at thicknesses between 2 and ~0.1 mm.

Cold rolling and appropriate heat treatment improves the permeability in the rolling direction by orienting the crystal texture. Oriented materials are of higher magnetic quality and only available as sheet or strip. They are more costly than the non-oriented steels but their improved performance may justify their application. Sheets or strip are available in thickness varying from 0.4 to 0.05 mm. The oriented iron–silicon steels contain about 3–3.25% Si.

In relays, cores for magnets, etc., solid pieces are required and hence pure iron or low Si-steel bar, rod, slab or sheet are used, because of the high B_s value. In some cases, e.g. for relays, the retentivity B_η must be very low. For special purposes iron–cobalt alloys with the highest B_s value of all materials known, are used. If the above uses are under AC conditions, laminations are used.

In general, the electrical steels are confined to applications at mains frequencies (except for some high grade Si-steels). However, many applications in electronics, communication, radio and TV require magnetic components which are used at audio and radio frequencies up to 50 or 100 MHz. Here the losses of lower grade materials would be intolerable and essentially two classes of materials are available, the iron–nickel alloys with 45–80% Ni known as permalloys and the ferrites.

The iron–nickel alloys with very high permeabilities fall essentially into two classes, the 45–50% Ni alloys and the 70–80% Ni alloys. These are known under various trade names. The 50% Ni range have fairly high B_s values (~1.5 T), quite high values of μ_i and μ_{max}. The 80% Ni range have lower B_s values (~0.8 T) and very low H_c values. Hence their losses are very low indeed. These alloys are available as thin rolled sheet and strip in thicknesses from 0.35 to 0.003 mm, the preferred thickness being about 0.2 mm. The alloys are available in the form of hot or cold rolled, annealed sheet and bar rod and wire. The main classes of application for these alloys are for audio transformer coils, relays and magnetic screening, etc.

Where the highest quality is needed, e.g. in high efficiency coils and inductors, pulse transformers, magnetic amplifiers and high quality screens the 70–80% grades are preferred. The 50% alloy can be oriented by rolling and heat treatment, for use in applications where rectangular loops are desired, such as amplifiers and contact rectifiers, storage coils, harmonic generators, etc. All the alloys are stress-sensitive and manufacturers instructions with regard to mechanical and thermal treatments vary and are very critical. Selection criteria and properties of these alloys are tabulated in:

Table 1.3.3 *Selection guide for solid soft magnetic materials for DC or very low frequency AC conditions* and

Table 1.3.4 *Selection guide for sheet materials and lamination for AC applications up to audio frequencies (20 kHz)*

However, iron–boron based amorphous alloys, prepared by rapid solidification from the melt, could replace crystalline materials in several areas. Some display properties comparable to the best grades of the permalloys and others exhibit core losses up to a factor of 20 times lower than oriented silicon steels. These alloys, developed in the early 1970s, can be divided into three main classes, listed in:

Table 1.3.5 *Properties of some amorphous magnetic alloys*

The iron–based alloys were originally developed as low-loss substitutes for the grain-oriented silicon steels for power distribution applications, but they also possess high frequency (up to 100 kHz) characteristics, similar to the nickel–iron alloys.

Iron–nickel based alloys are also similar to the crystalline nickel–iron alloys, but with better corrosion resistance. The cobalt-based alloys have the highest permeabilities and the lowest losses, but they are also the most expensive.

Hysteresis losses increase steeply with B and eddy current losses are high for materials with good electrical conductivity and even moderate lamination thickness. At radio frequencies, therefore, laminations must give place to powder cores (cores of compressed high grade metal powder) but even these are often not suitable. Here the ferrites, mixed oxides of Fe_2O_3 with NiO, MnO, MgO, CuO and other oxides which are electrical insulators are preferred. They have fairly low values of B_s and for this reason their use in power applications at low frequencies is limited and they also have fairly low Curie temperatures and therefore high temperature coefficients of magnetic properties. The manganese–zinc ferrites have, compared to nickel–zinc ferrites, higher saturation but lower resistivity and are used up to about 1 MHz. Beyond 1 MHz the nickel–zinc ferrites with higher resistivity are generally used. These materials are produced by ceramic techniques, i.e. firing of compressed powders at high temperatures in controlled atmospheres. They are hard and brittle, and cannot be formed after manufacture. Hence they are available in standard shapes and sizes as transformer cores and pot cores for coils, and the manufacturer's list must be consulted for range of available components.

At the very high frequencies at which the ferrites are used, the total losses as the material is cyclically magnetized become very important. In general, the materials being insulators, the eddy current losses are negligible, but there are two factors contributing to the losses, even a very low values of B. The hysteresis losses, which can be characterized by a hysteresis coefficient, η_B, and the so-called residual losses, which arise from atomic and microscopic magnetization losses in the material, characterized by the residual loss factor $(\tan \delta_r)/\mu$.

The total losses, neglecting the negligible eddy current contribution are proportional to $B\, \eta_B\, (\tan\delta_r)/\mu$.

Only very approximate guidelines for the choice of a specific ferrite material can be provided here. The materials are known under manufacturers' grades (see Section 1.3.4.3 below) and are best classified by their initial permeability. The extensive literature on design data supplied by manufacturers for their materials should be consulted.

The reader is also guided to further, more detailed tabulation:

Table 1.3.6 *Typical physical and magnetic properties of iron and iron-rich alloys*

In addition to the broad ranges of materials mentioned above, there are a number of special materials, which are summarized in Section 1.3.4.4.

1.3.4.3 Permanent magnet materials

The bulk of permanent ('hard') magnet materials now used fall into one of three classes. The iron–nickel–aluminium-based alloys (alnicos), rare-earth magnets and the barium ferrites.

The alnico alloys are usually cast and heat treated, are brittle and hard and have a rough surface. They can only be machined by grinding. The barium ferrites are ceramics, also hard and brittle. Sintered alnicos are also available, usually with slightly reduced properties and higher cost, but they have smoother surfaces, and are somewhat less brittle. The alnicos have fairly high values of B_r (0.7–1.35 T) and moderate values of H_c (40–160 kA/m), although there is a fairly wide range available. The majority of the alloys are heat treated in a magnetic field during manufacture. In some cases the crystal structure is oriented during casting (columnar crystal or oriented grain material). In these cases the direction of the flux in the magnet must be confined to the 'preferred direction', the properties in other direction being very inferior. The alnicos are used for magnets in electrical instruments, microphones, loudspeakers, magnetrons, holding magnets, etc.

The cheaper barium ferrites have low values of B_r (0.42 T) but very high H_c (200–300 kA/m). Hence they are very suitable for application where large demagnetizing fields are encountered, e.g. in multiple alternators and dynamos, or in cheap electric motors for vehicles (windscreen wiper motors, etc.) and toys. They have a higher temperature coefficient of magnetization above room temperature and hence their use in instruments and meters is limited. They can be produced in the isotropic form (in this case especially the demagnetization curve is almost a straight line, very suitable for assembly after magnetizing) or for improved performance they can be produced with a preferred direction of magnetization (anisotropic). All these materials are supplied in the final form and shape. Strontium ferrites have greater coercivity than barium ferrites and are increasing their market share.

The rare-earth magnets were developed in the 1960s and 70s and the first alloys were based on the intermetallic compounds $SmCo_5$ and Sm_2Co_{17}.

The magnets based on $SmCo_5$ have an energy product $(BH)_{max}$ typically in the range 130–160 kJ/m^3, while those based on Sm_2Co_{17} have typical values of $B_r = 1.10\ T, H_c = 520$ kA/m and $(BH)_{max} = 240$ kJ/m^3.

The newest rare-earth magnets, based on $Nd_2Fe_{14}B$, were developed in the 1980s. Apart from their relatively low Curie temperature (310°C), the Nd–Fe–B magnets are the strongest available magnets (see Fig. 1.3.5). They are prepared using powder metallurgy techniques, with two basic methods in common usage: crushing of castings to produce powder or crushing of melt-spun ribbon.

Despite their high cost, rare-earth magnets have found many applications where their high coercivities and energy products can result in design advantages, for example, earphones, magnetic bearings and microwave power devices. Because of their higher energy products and because iron and neodymium are more abundant than cobalt and samarium, the Nd–Fe–B magnets are likely to replace $SmCo_5$ magnets in some applications.

In addition to these three groups there are a number of special materials. Magnet steels can be supplied as rolled bar, and can be machined before the hardening heat treatment needed to develop the permanent magnet properties. Some alloys can be punched or cut (e.g. chromium steels and the alloy of V–Co–Fe known as Vicalloy) and are used as stamped laminations for hysteresis motors.

Other hard magnet materials include Fe–Cr–Co alloys, Pt–Co alloys and elongated single domain (ESD) magnets, as well as the magnet steels fabricated for over 50 years as permanent magnet materials.

Permanent magnet materials and their selection rules can be found in Tables 1.3.14 to 1.3.17:

Table 1.3.14 *Main selection list for permanent magnet materials*
Table 1.3.15 *Typical properties of rolled, drawn strip and wire permanent magnet materials*
Table 1.3.16 *Typical properties of cast permanent magnet alloys (Alnico class)*
Table 1.3.17 *Typical properties of permanent magnet materials based on barium ferrite*

1.3.4.4 Special materials classed with magnetic materials

Apart from the materials mentioned, there are a number of special alloys classed with magnetic materials, these have been developed for a number of special applications, often in magnetic circuits or in conjunction with magnetic materials.

TEMPERATURE COMPENSATION MATERIALS

In some applications, e.g. electricity meters, engines and temperature sensitive relays, the flux in a gap or component has to be kept constant as the temperature varies. The compensator is usually used to shunt the useful flux, thus since the temperature coefficient of the magnet supplying the flux is negative, a shunt with a strong negative temperature coefficient of B_s or μ can be used to compensate for the magnet variations.

The alloys have about 30% Ni and θ_c can vary between 40 and 120°C, (note at θ_c, $B_s = 0$).

Example:

T (°C)	–40	–20	0	+20	+40	+60	+80	+100
B_s (T)	1.17	1.12	1.06	1.0	.91	.84	.74	.64

INVARS (EXTREMELY LOW COEFFICIENT OF THERMAL EXPANSION)

In some precision components the dimensional tolerance is very critical. Invars have exceptional dimensional stability and very good corrosion resistance. They are iron–nickel alloys (~40% Ni) and the expansion coefficient at room temperature is $<2 \times 10^{-6}$/°C between 20 and 100°C. Some types of invars have a constant low expansion coefficient of $\sim 3 \times 10^{-6}$/°C between 20 and 250°C. The alloys have to be heat treated carefully, but are machinable.

GLASS TO METAL SEAL ALLOYS

The invar type of low expansional alloy is also suitable for glass to metal seals. Chromium–iron alloys are also used for this application.

NANOCRYSTALLINE SOFT MAGNETIC MATERIALS

These materials are produced by hyperfine powder compaction and sintering, vapour deposition of thin films or by crystallization by heat treatment from the amorphous state. The crystallite size will be typically in the range 5–50 nm.

The Fe–Cu–M–Si–B-based nanocrystalline soft magnetic FINEMET materials (where M = Nb, Ta, Zr, Hf, Ti, Mo and W) were developed in 1988 by Hitachi. These materials feature high saturation magnetic flux density (1.2–1.6 T), comparable to that of iron-based amorphous materials, and high initial permeability (5000–100 000), comparable to those of the cobalt-based amorphous alloys. In addition, core losses are low, the magnetic distortion is minimal and hardly any thermal instability is evident.

Possible applications include high frequency transformers and magnetic heads.

Table 1.3.2 Critical application parameters and choice of magnetic materials

Critical application parameter	***Most suitable group of material***
High flux density, low frequencies or DC	Iron, low-Si iron, solid or thick lamination.
High flux density mains frequency (50/60 Hz) low losses	Iron–silicon laminations (non-oriented).
Lowest core losses at mains or audio frequencies good flux density	Oriented iron–silicon lamination, 50% nickel alloys. Iron–boron based amorphous alloys
Lowest core losses, high permeability, flux density not of prime importance for high frequencies	50% nickel alloys 70–80% nickel alloys (up to 20 kHz). Soft ferrites (especially RF and HF coils and transformers). Iron–boron based amorphous alloys
Low remanence or very low coercive force	Iron–nickel and special iron–silicon relay materials.
Highest frequency operation	Very thin iron–nickel laminations and spiral cores, soft ferrites.
High temperature applications	Cobalt–iron alloys
Permanent magnets	Alnicos, barium ferrites, rare-earth magnets

Table 1.3.3 Selection guide for solid soft magnetic materials for DC or very low frequency AC conditions

Material	*Range of main properties*	*Application*
Cast, forged, rolled solid iron and low Si-iron	B_s = 2.0–2.2 T μ_{max} = up to 5000 H_c = 7–100 A/m $\rho = 1\text{–}10 \times 10^{-7}\ \Omega m$ SG = 7.8–7.6 Can be machined, forged, etc., anneal after machining	Magnetic circuit components core components, relays, magnetic pole pieces
Solid chromium iron	B_s = 1.3–1.4 T μ_{max} = 1000–2000 H_c = 100–400 A/m $\rho = 6\text{–}8 \times 10^{-7}\ \Omega m$ Can be machined, requires heat treatment after machining	Solenoid cores and housings and core components in corrosive environments
Cobalt–iron alloys	B_s = 0.8–2.3 T μ_{max} = 100 000–450 000 H_c = 0.4–16 A/m $\rho = 4\text{–}6.5 \times 10^{-7}\ \Omega m$ SG = 8.2 Requires annealing	Highest flux density applications, pole tips of DC electromagnets
~50% Iron–nickel alloys	B_s = 1.55 T μ_{max} = 80 000 H_c = 2 A/m $\rho = 4.5 \times 10^{-7}\ \Omega m$ SG = 8.2 Requires annealing	Low retentivity relay parts or armatures
~70–80% Iron–nickel alloys	B_s = 0.8 T μ_{max} = 450 000 H_c = 0.4 A/m $\rho = 6 \times 10^{-7}\ \Omega m$ SG = 8.7 Requires careful annealing	Very low retentivity and high performance relay and armature use where high flux density not needed

Table 1.3.4 Selection guide for sheet materials and lamination for AC applications up to audio frequencies (20 kHz)

Material	*Range of main properties*	*Application*
Iron and low Si-iron (up to ~ 0.8% Si)	Main thickness available 0.50–0.65 mm 0.65 mm thickness grade: typical losses at 50 Hz 4–8 W/kg at 1.0 T 5–15 W/kg at 1.5 T	Magnetic screening fractional horse power motors.
Non-oriented Si-iron 1–3% Si	Main thicknesses available 0.35, 0.5 and 0.65 mm Typical losses 0.50 mm thickness at 50 Hz 1.5–5 W/kg at 1.0 T 3.5–6 W/kg at 1.5 T	Rotating machines, stators of induction motors. Transformers.
Oriented 3–3.25% Si-iron	Main thicknesses available 0.35, 0.30 and 0.28 mm (also thin strip 0.05–0.15 mm) Typical losses for 0.3 mm thickness at 50 Hz 0.9 W/kg at 1.5 T 1.3 W/kg at 1.7 T For 0.1 mm thickness at 400 Hz 14 W/kg at 1.5 T	Quality power transformers and in thinner grades audio frequency transformers, aircraft electrical equipment.
Cobalt alloys	Main thicknesses available 0.05–0.35 mm tapes down to 0.03 mm typical loss for 0.15 mm thickness at 400 Hz 22 W/kg at 1.0 T 75 W/kg at 2.0 T For 0.015 mm thickness at 400 Hz 14.5 W/kg at 1.0 T 39 W/kg at 2.0 T	Telephone earpiece membranes. Magnetostriction transducers. Magnetic lenses (can be annealed to have rectangular hysteresis loop for magnetic amplifiers). High Curie point for higher temperature applications.

Table 1.3.4 Selection guide for sheet materials and lamination for AC applications up to audio frequencies (20 kHz) – *continued*

Material	***Range of main properties***	***Application***
50% nickel–iron alloys	Range of thickness available 0.35–0.003 mm. Also as strip for spiral cores μ_{max} = 70 000 (250 000 with rectangular loop) Typical core losses: W/kg 0.2 mm thick 50 Hz 400 Hz 2 kHz 0.1 T 0.003 0.1 1.0 1.0 T 0.3 6.0 – Requires heat treatment. Can be supplied as complete spiral core	Current transformers, audio transformers, magnetostriction transducers. Can be obtained with rectangular hysteresis loop especially for magnetic amplifiers and storage chokes.
70–80% nickel alloys	Range of thickness available. 0.35–0.072 mm (also thicker for low frequency applications and screening). Foils down to 0.003 mm μ_{max} up to 350 000 Typical core losses: mW/kg 0.2 mm thick 50Hz 400 Hz 5 kHz 0.03 T 0.1 2.5 – 0.1 T 1.15 25 – 0.3 T 13 200 – 0.05 mm thick 0.03 T 0.046 0.8 34 0.1 T 0.6 8.0 400 0.3 T 6 70 4000 Requires heat treatment. Can be supplied as complete spiral core.	Current transformers, high frequency transformers, magnetic recording heads, chokes and inductors. High grade relays. Can be obtained with rectangular loop for magnetic amplifiers. Limit of frequency 20 kHz These are for standard grade. For special quality grades these figures are at least three times better.
Amorphous alloys	Range of thickness available 0.025–0.017 mm μ_{max} = 1 000 000 (Annealed) Typical core losses: W/kg 60 Hz 400 Hz 5 kHz 0.05 T 0.002 0.02 0.5 0.1 T 0.006 0.05 1.3 1.0 T 0.15 2.0 100.0	Pulse power cores, anti-theft security devices, EMI shielding and magnetic amplifiers

Table 1.3.4 – *continued*

Table 1.3.5 Properties of some amorphous magnetic alloys

Alloy	*Density (g/cm^3)*	B_s *(T)*	H_c *(A/m)*	*Maximum permeability* μ_{max}		*Electrical resistivity ($\mu\Omega cm$)*	*Curie temperature (°C)*	*Core losses (W/kg)*	
				Annealed	*As cast*			*60 Hz, 1.0 T*	*25 kHz, 0.2 T*
Iron based									
$Fe_{81}B_{13.5}Si_{3.5}Co_2$	7.32	1.61	3.2	300 000	>4000	135	370	0.14	50
$Fe_{78}B_{13}Si_9$	–	1.56	2.4	600 000	45 000	137	415	–	–
$Fe_{67}Co_{18}B_{14}Si$	7.56	1.80	4.0	400 000	120 000	123	415	0.25	–
Iron-nickel based									
$Fe_{40}Ni_{38}Mo_4B_{18}$	7.90	0.88	1.2	800 000	>50 000	138	353	0.12	50
Cobalt-based									
$Co_{67}Ni_3Fe_4Mo_2B_{12}Si_{12}$	7.80	0.77	0.4	600 000	290 000	136	365	–	7.0
Metglas 2714A	7.59	0.57	–	1 000 000	>80 000	142	225	0.02	4.2

Table 1.3.6 Typical physical and magnetic properties of iron and iron-rich alloys

Material	*Specific gravity SG*	*Si content % (approx.)*	*Resistivity (Ω m)*	B_s (T)	B_r (T)	H_c (A/m)	μ_i	μ_{max}
Electrical iron	7.86	0	10^{-7}	2.15	0.6–0.8	~100	10–200	1000–7000
Low silicon armature iron	7.75	0.5–1.0	2.8×10^{-7}	2.05	0.6–0.8	~70–90	300	3000–4500
Low silicon electrical iron	7.7–7.8	1.6–2.2	$3.7–4.0 \times 10^{-7}$	2.0	0.6	~80	300	5000
Dynamo steel	7.65	3.0	4.7×10^{-7}	1.9	0.6	~50–60	290	7000–9000
Transformer steel (non-oriented)	7.65	3.0–3.2	$4.1–5.0 \times 10^{-7}$	1.9	0.6	~45	~300	9000
Oriented silicon steel	7.65	3.0	5×10^{-7}	1.97	1.1–1.3	~30–40	1000	10 000–50 000
Chromium iron	7.59–7.62	Cr content (approx.) 17	$6–7.6 \times 10^{-7}$	1.4	0.25–0.8	~80–400	–	300–2500

Table 1.3.7 Typical loss data for laminated Fe–Si materials at 50 Hz

Material	*Thickness (mm)*	*Losses (W/kg) at B*		
		1.0 T	*1.5 T*	*1.7 T*
Electrical iron and low silicon armature iron	0.65 0.50	3.5–4.4 2.9–3.9	8.0–10.0 6.7–8.7	
Low silicon electrical iron	0.50	1.8–2.2	5.0	
Dynamo steel	0.50 0.35	1.6 1.4	3.55 3.35	
Transformer steel (non-oriented)	0.35 0.175	1.1–1.2 17.5 (at 400 Hz)	2.8–3.0	
Oriented silicon steel	0.35 0.30 0.28 0.10		1.02 0.90 0.85 1.5 (at 400 Hz)	1.28–1.44 1.17–1.32 1.26

Table 1.3.6 and Table 1.3.7

Table 1.3.8 Typical physical and magnetic properties of iron–nickel alloys

Material	*SG*	*Electrical resistivity (Ω m)*	B_s *(T)*	H_c *(A/m)*	μ_i	μ_{max}	θ_c (°C)	λ_s ($\Delta l/l$)
36% Ni (lamination)	8.15	7.5×10^{-7}	1.3	15–40	2000–3000	8000–20 000	250	22×10^{-6}
36% Ni (solid)	8.15	7.5×10^{-7}	1.3	20	3000	15 000	250	22×10^{-6}
36% Ni (sintered)	8.10	7.0×10^{-7}	1.3	16–24	1500–2500	4500–7000	250	22×10^{-6}
50% Ni (lamination)	8.25	$4–5 \times 10^{-7}$	1.55	1.6–8	3500–5000	40 000–130 000	470	25×10^{-6}
50 % Ni (square loop)	8.25	$4–4.5 \times 10^{-7}$	1.6	8–12		50 000–70 000	470	25×10^{-6}
(sintered and solid)	8.25	4.5×10^{-7}	1.55	8	5000	35 000	470	25×10^{-6}
70–80% Ni (lamination)	8.5–8.8	$5.5–6 \times 10^{-7}$	0.77–0.8	0.55–1.0	60 000–140 000	120 000–350 000	400	–
(sintered and solid)	8.6	6×10^{-7}	0.8	1.2	30 000–40 000	10 000–150 000	400	–

The ranges of iron–nickel alloys are manufactured in various grades and the data given in the tables refer to a medium grade. In some cases the range of properties available is given.

Table 1.3.9 Typical losses for iron–nickel laminated cores

Material	*Thickness (mm)*	*W/kg losses at*											
		50 Hz		*400 Hz*			*2 kHz*			*10 kHz*			
		***B* = 0.5 T**	***1.0***	***0.1***	***0.3***	***1.0***	***0.03***	***0.10***	***0.3***	***0.03***	***0.1***		
36% Ni alloy	0.3		1.1										
50% Ni alloys													
Normal loop	0.3 0.2 0.1	0.06	0.25	0.16 0.1 0.07	1.2 0.7 0.5	9.5 6.0 4.2	0.22 0.13 0.09	1.9 1.1 0.85	14 8 5.0	– 1.5 1.05	– 13 9		
Normal loop special grade	0.1		0.1										
Square loop	0.1	0.2	0.4–0.6										
Square loop special grade	0.05	0.02	0.07										
	Thickness (mm)	*Losses in mW/kg*											
		50 Hz		*400 Hz*			*2.4 kHz*			*5 kHz*			
60–80% Ni		***B* = 0.5 T**	***0.1***	***0.5***	***0.1***	***0.2***	***0.1***	***0.02***	***0.004***	***0.1***	***0.02***	***0.004***	
Standard grade	0.35 0.2 0.05	8.0 40 20	2.7 1.2 0.6	1200 500 190	64 25 7.5	2.5 1.1 0.36	– – 130	– – 5	– – 0.18	– – 400	– – 14	– – 0.6	
Better grade	0.2 0.1 0.05	22 18 –	1.0 0.55 –	700 250 110	38 13 6	2 0.6 0.25	– 25 13	– 9 5	– 0.45 0.25	– 750 320	– 34 14	– 1.6 0.65	
Best grade	0.1 0.05	12 2.5	0.32 0.075	250 35	9 1.5	0.35 0.06							

Table 1.3.9

Table 1.3.10 Typical properties of high B_s iron–cobalt alloys (permendur) (49% Fe–49% Co–2% V nominal composition)

	SG	*Resistivity* *(Ωm)*	B_s *(T)*	θ_c *(°C)*	H_c *(A/m)*	λ_s *(Δl/l)*	μ_{max}
Permendur	8.1	3.5×10^{-7}	2.3–2.35	980	30–120	60×10^{-6}	7000–1200
Permendur 24 (a special ductile grade)	7.95	2×10^{-7}	2.35	925	–	–	–
Sintered permendur	8.1	1.5×10^{-7}	2.35	950	65	60×10^{-6}	10 000

Permendur sheet, thickness 0.05–0.5 mm $H_c > 40$ A/m
50 Hz core losses W/kg at B

thickness	0.3 T	1.0 T	1.7 T
0.3 mm	0.32	1.9	4.0
0.2 mm[a]	0.12	0.52	1.4

[a]Sintered and hot rolled, special quality.

Table 1.3.11 Applications of ferrites (see Tables 1.3.12 and 1.3.13 for further details)

Type: Mainly manganese–zinc based						
Main application	Inductors	Inductors and aerial rods	Wide band and pulse transformers	High B_s application TV and power transformers		
Approx. range of frequency	<200 kHz	100 kHz–2 MHz	up to 200 MHz	100 kHz–300 MHz		
Initial permeability μ_i	800–2500	500–1000	1500–10 000	1000–3000		
Type: Nickel–zinc based						
Main application	Antenna rods HF wide band and power transformers	Antenna rods HF power transformers	Antenna rods HF power transformers Inductors	Inductors		
Approx. frequency range	100 kHz–300 MHz	500 kHz–5 MHz	2–30 MHz	10–40 MHz	20–60 MHz	>30 MHz
Initial permeability μ_i	500–1000	160–490	70–150	36–65	12–30	<10

Table 1.3.12 Typical values of magnetic properties of ferrites – permeability classification (see Table 1.3.11)

Type: Manganese–zinc based						
Property						
μ_i	800–2500	500–1000	1500–10 000	1000–3000		
B_s (T)	0.35–0.5	~0.4	0.3–0.5	0.32–0.52		
B_r (T)	0.08–0.14	0.15–0.20	0.07–0.14	0.1–0.25		
H_c (A/m)	10–30	40–100	2.8–24	10–30		
Power loss density (f = 10 kHz, B 0.2 T)	45–130	250	50–150	50–120		
Curie point (°C)	140–210	200–280	90–200	180–280		
Resistivity (Ωm)	0.5–7	1–70	0.02–0.5	0.2–1.0		
Type: Nickel–zinc based						
Property						
μ_i	500–1000	160–490	70–150	36–65	12–30	10
B_s (T)	0.28–0.34	0.3–0.36	0.25–0.42	0.24–0.28	0.15–0.26	0.1–0.2
B_r (T)	0.15–0.19	0.12–0.16	0.24–0.34	0.15–0.20	0.08–0.15	0.05–0.1
H_c (A/m)	16–50	80–160	160–500	300–500	500–1600	800–1600
Curie temperature (°C)	90–200	200–370	350–490	300–500	250–510	250–510
Resistivity (Ωm)	$10–10^7$	$>10^3$	$>10^3$	$>10^3$	$>10^3$	$>10^3$

Table 1.3.13 Typical values of residual loss factor $(\tan\delta_r)/\mu \times 10^6$ for soft ferrites (at <0.1 mT), and hysteresis coefficient, η_B, in m/T × 10^6, type of Mr–Zn ferrite (μ_i range)

Residual loss factors $(\tan\delta_r)/\mu \times 10^6$			
Frequency (kHz)	800–2500	500–1000	1500–10 000
10	0.8–1.8		1–10
30	1.3–3.0		2–20
100	2.0–10	5–15	4–60
1000		10–40	
Hysteresis coefficient, η_B, at 10 kHz			
B1–3 mT	0.3–1.3	0.48–1.9	0.1–1.3

Table 1.3.14 Main selection list for permanent magnet materials

Form and type	*Materials group*	*Range of main properties*			*Principal applications*	*Table details*
		BH_{max} *(kJ/m³)*	*B (T)*	H_c *(kA/m)*		
Rolled, drawn, bar, strip, wire	Cr–steels Co–Cr steels	42	1.35	48	Where machinability or thin strip (laminations) are required, thin-walled tubular magnets clock motors, compass needles, hysteresis motors.	15
	Cu–Ni–Fe and Cu–Ni–Co	7–14	0.35–0.55	35–50		
	Fe, Co, V (Vicalloy)	9–20	0.9–1.2	22–30		
Cast alloys (Alnico type)	Isotropic	10–14	0.6–0.8	40–50	Magnets in block or ring form. Loudspeakers. Magnets with more than 2 poles. Horseshoe magnets.	16
	Anisotropic	26–48	0.8–1.25	52–145	Magnetrons, centrepole loudspeakers, instruments, magnetos, large scientific instruments.	16
	Anisotropic grains	47–60	1.3–1.35	55–60	Where weight and size is important. Only blades and cylinders.	16
Barium ferrite	Isotropic	8–10	0.22–0.28	135–200	Low cost toys, two-pole motors, alternators, where low electrical conductivity is needed (in AC fields).	17
	Anisotropic	25–30	0.35–0.4	150–240	As above, loudspeakers, also holding devices, where assembly after magnetization matters. Filters, torque drives.	17
	Resin bonded	3.2	0.15	80	Clocks, tachometers, motors, toys.	17
Rare earth–cobalt (RE–Co)	Sintered	120–200	0.6	420–500	Miniaturization, microwave devices, slipping motors for watches, medical implants, magnetic bearings.	18
	Resin bonded	60–80	0.9	560–700		
$Nd_2Fe_{14}B$	Sintered	290	1.23	880	As above	
77% Pt–23% Co		73	0.65	360	Timing motors, relays, thermostats	

Table 1.3.15 Typical properties of rolled, drawn strip and wire permanent magnet materials

Material	*SG*	B_r (T)	H_c (kA/m)	BH_{max} (kJ/m^3)	B_D (T)	H_D (kA/m)	*Vickers diamond hardness before heat treatment*
Fe–Cr	7.7	1.35	48.00	42.0	1.12	38.0	286
Cu–Ni–Co (US)	8.6	0.53	35–50	6.8–8	–	–	–
Cu–Ni–Fe (US)	8.6	0.54–0.6	47	8–14	–	–	–
Vicalloy[a]	8.0	0.9–1.2	23–30	9–20	0.65–0.9	17–22	400[b]

[a] Can be punched, in the 'as supplied condition', heat treatment to produce permanent magnet properties: 2 h. at 600°C.
[b] After heat treatment hardness 840–950 Vickers.

Table 1.3.16 Typical properties of cast permanent magnet alloys (Alnico class)

Material	B_r (T)	H_c (kA/m)	BH_{max} (kJ/m^3)	B_D (T)	H_D (kA/m)	*Recoil permeability*
Isotropic	0.6–0.77	40–50	10–14	0.39–0.46	26–32	5–5.5
Anisotropic normal H_c	1.27–1.32	48–52	42–44	1.0–1.1	41–42	3.5–4.0
Anisotropic high H_c	0.80–0.85	96–144	32–48	0.5–0.55	60–96	1.9–2.5
Anisotropic grains	1.29–1.35	56–60	47–60	1.05–1.2	45–51	2.5–3.0

These alloys consist of Fe, Ni, Al, Co and Ti (sometimes with small other additions). In the anisotropic alloys a specially square demagnetization curve can be developed by directional cooling from the melt. The highest H_c-alloys contain more Ti. Vickers hardness 450–700. Specific gravity 6.9–7.3.

Table 1.3.17 Typical properties of permanent magnet materials based on barium ferrite

Material	B_r (T)	H_c (kA/m)	$_iH_c$[a] (kA/m)	BH_{max} (kJ/m^3)	B_D (T)	H_D (kA/m)	*SG*	*Recoil permeability*
Sintered isotropic	0.2–0.28	135–200	220–320	8–10	0.12–0.14	70–100	4.3–4.9	1.2
Anisotropic high B_r	0.4	140–160	150–180	26–28	0.22	140	4.9	1.1
Anisotropic high H_c	0.37	240	250	26	0.18	145	4.7	1.05
Resin bonded[b] isotropic	0.15	80	160	3.2	0.07	40	3.6	1.2

[a] $_iH_c$ the 'intrinsic' coercive force gives the demagnetization field which would reduce the magnetization M to zero. It is a good quality parameter for magnets subject to a very strong demagnetizing field.
[b] The ferrite powder is also available, which can be used by the user to manufacture his own magnets with Bakelite resin (instructions from manufacturers!).

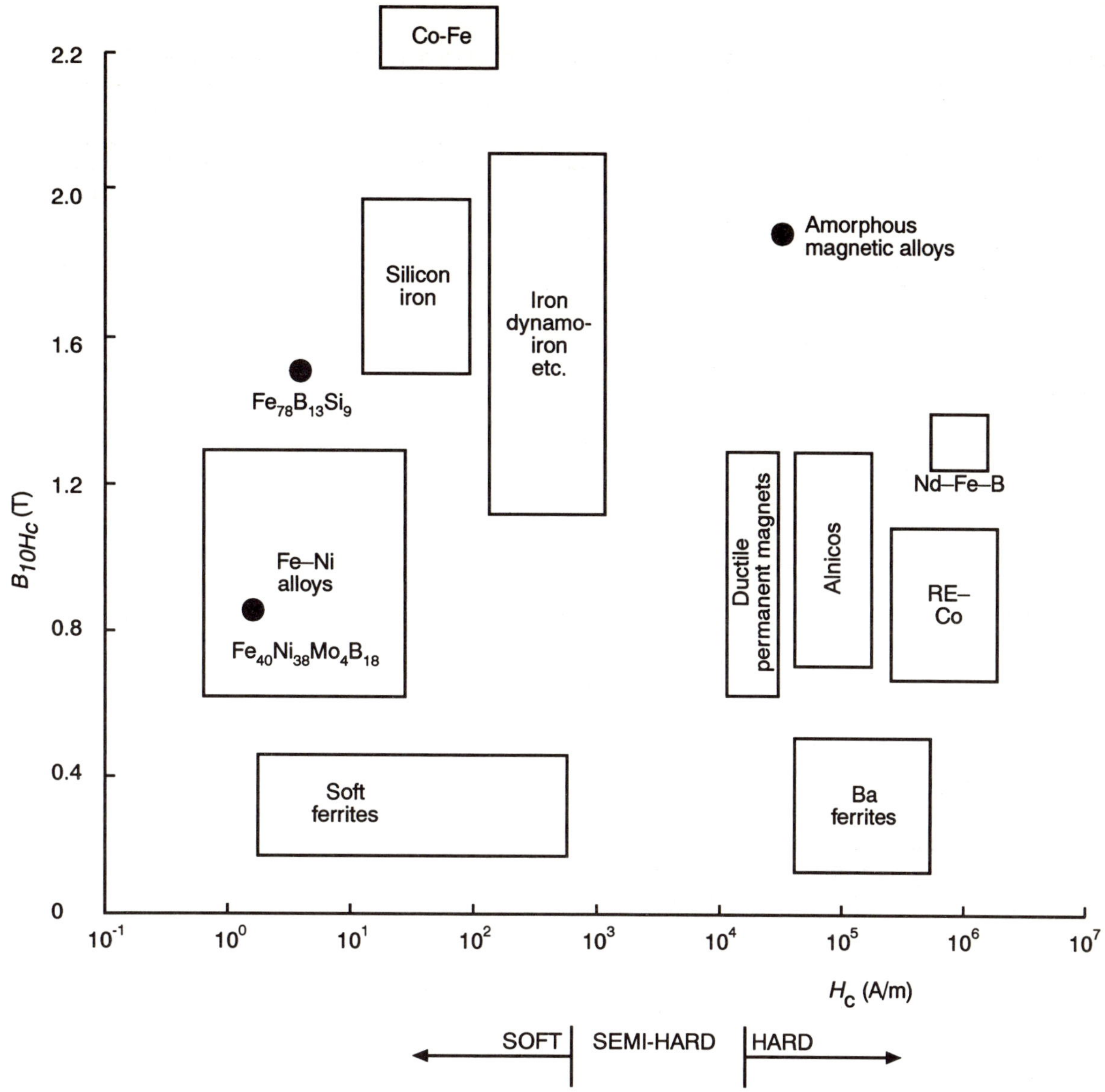

Fig. 1.3.5 Magnetizing force and flux density attained in fields of up to $10H_c$

1.3.5 Optical properties

For the majority of engineering applications, optical properties have no significance. Most ceramics and all metals are opaque in bulk. Glasses, many polymers and a few ceramics are, however, sufficiently transparent for optical use. The main parameters governing suitability are transparency, refractive index, gloss and emissivity, described below.

1.3.5.1 Transparency

Transparency, which is a general quality depending largely upon bulk and superficial homogenity, is measured by loss of clarity and loss of contrast. These two effects are closely interconnected and constitute the two different aspects of one phenomenon.

For most purposes, transparency can be described by the following characteristics. A certain amount ϕ_{sc} (scattered) of the monochromatic luminous flux ϕ (parallel beam) falling perpendicularly on a translucent film or sheet, will be scattered in all directions. If ϕ_A is the flux absorbed by the material (ignoring re-emission with changed wavelength, known as fluorescence) the undeviated flux, $\phi_{undeviated}$, is given by:

$$\phi - \phi_A - \phi_{sc} = \phi_{undeviated} ,$$

The direct transmission factor, T, is defined by $T = \phi_{undeviated} / \phi$. For weak scattering

$$T = e^{-(\sigma+K)l}$$

where l is the material thickness, σ is a measure of the amount of light scattered (scattering coefficient or turbidity) and K is a measure of the amount of light absorbed (absorbtion coefficient). (σ + K) is known as the alternation or extinction coefficient.

Transparency is a function of wavelength of the incident light. Typically, a fused quartz may show 90% transmission for radiation at wavelengths between 0.2 and 3 μm.

The total transmission factor (transmittance) is defined by the ratio of the total transmitted flux to the incident flux.

The total reflection factor (reflectance) is defined (for normal incidence) as the ratio between the backward scattered flux and the incident flux.

Haze characterizes the loss of contrast which results when objects are seen through a scattering medium. Deterioration of contrast is mainly due to the light scattered forward at high angles to the undeviated transmitted beam. The 'milkiness' of translucent samples when viewed from the side on which light is incident is largely due to the backward scattering.

Clarity is a measure of the capacity of the sample for allowing details in the object to be resolved in the image which it forms. It is strongly dependent upon angular distribution of scattering intensity and distance between object and sample.

1.3.5.2 Refractive index

The refractive index (n) of a material for a given wavelength, in vacuum, of electromagnetic radiation is defined by the ratio of the velocity of the radiation in vacuum and the velocity in the material. The refractive index is a function of wavelength and temperature.

Dispersion is a measure of the change in refractive index between wavelengths at the extremes of the visible spectrum.

Birefringence is a measure of optical anisotropy. Some materials are isotropic unless stressed elastically; permanent birefringence may be introduced by processing, or by dispersions of one isotropic material within another.

Refractive indices of most optical polymers are tabulated according to the individual material in Vol. 3.

Glasses are available with a very wide range of refractive indices and dispersions. Manufacturers should be consulted for materials with specific properties.

1.3.5.3 Gloss

Gloss is a property of the surface of a plastic specimen. Its magnitude depends on the refractive index of the plastic, the smoothness of the surface and the occurrence of sub-surface optical features. The relative importance of these factors depends upon the angle of incidence of light falling on the surface.

1.3.5.4 Colour

Colour in plastics may be produced by the use of specific fillers. It is produced in specific oxide ceramics and glasses by impurity atoms in the lattice. 'Whiteness' is often preferred, because products tend to sell better, but requires more expensive processing, perhaps purer raw materials and sometimes reduction firing.

1.3.5.5 Emissivity

Emissivity of ceramics is usually specified at 1000°C. It may vary from above 0.9 for carbons, graphites and silicon nitride to below 0.3 for pure white oxide ceramics.

1.3.6 Physical properties of materials at low temperatures

The physical properties specific heat, thermal expansion, and thermal and electrical conductivity are important in low temperature applications.

1.3.6.1 Specific heat

Specific heat approaches zero at absolute zero temperature as shown in:
Fig. 1.3.6 *Specific heats of materials at low temperatures*
Knowledge of this is important because:

(i) Materials with high heat capacity boil off more cryogenic liquid during cool-down.
(ii) In soldering, the high heat capacity of lead can result in joints having higher heat capacity than the parts being joined (this can be used to advantage, e.g. in regenerators).

1.3.6.2 Thermal expansion

Thermal contraction of materials as temperature is lowered, in conjunction with temperature gradients and thermal expansion mismatch, can produce very substantial displacements and, if these are not allowed for in design, very high thermal stresses.
The relative changes in length that can occur are illustrated in:
Fig. 1.3.7 *Total linear contraction of materials*

1.3.6.3 Thermal (and electrical) conductivity

Thermal conductivity behaviour varies very considerably with composition and, to a lesser extent, mechanical history. Most materials show a progressive decrease in thermal conductivity as the temperature is lowered but in pure annealed elements (and alumina) the thermal conductivity rises to roughly a hundredfold at between 10 and 20 K, as shown in:
Fig. 1.3.8 *Thermal conductivity of materials at low temperatures*
Great care must be taken in selection of materials to avoid heat losses through materials with enhanced conductivities.
Electrical conductivity shows even more variation, some materials showing very low or even zero resistivity near 0 K.

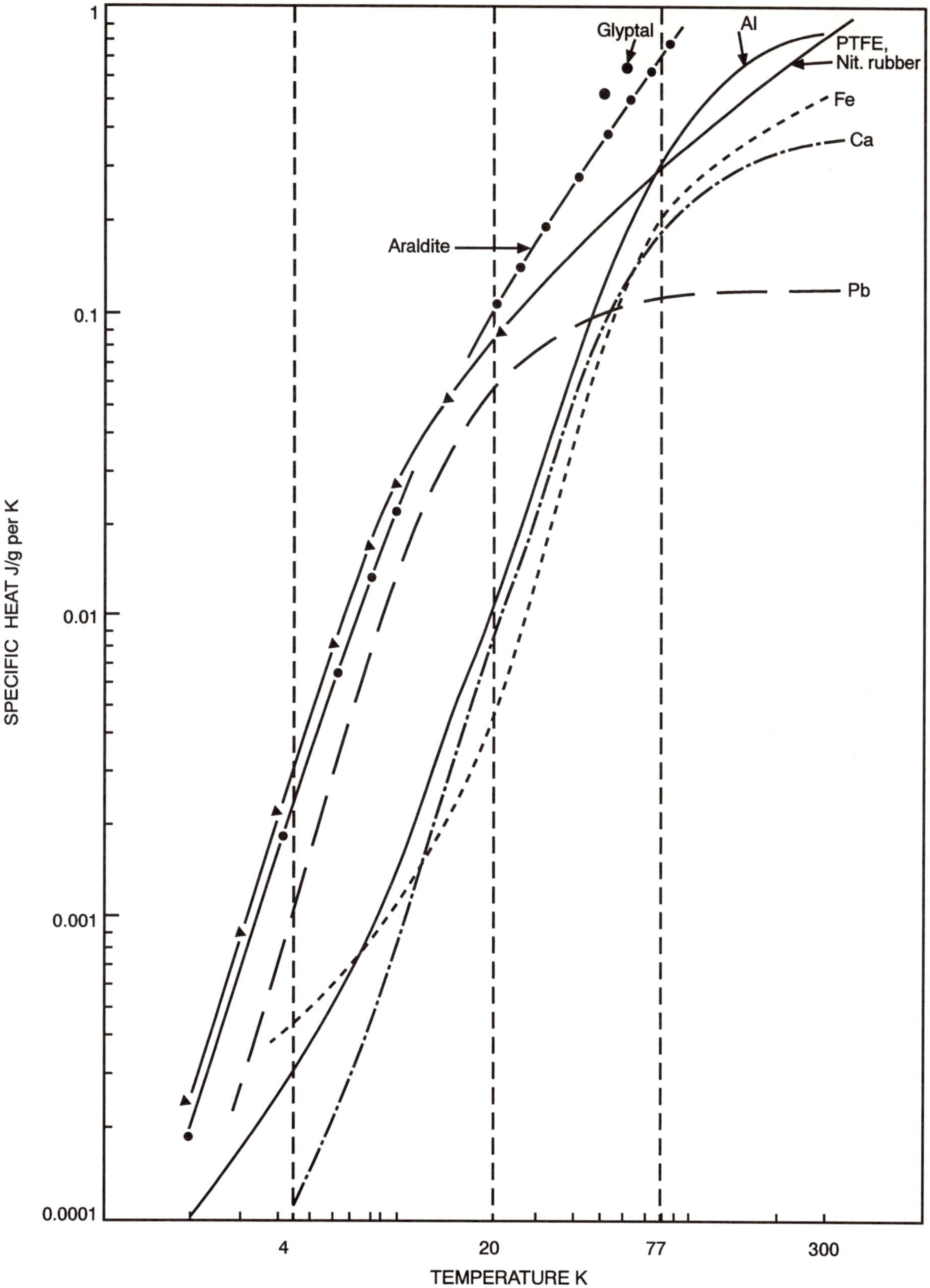

Fig. 1.3.6 Specific heats of materials at low temperatures

Fig. 1.3.6

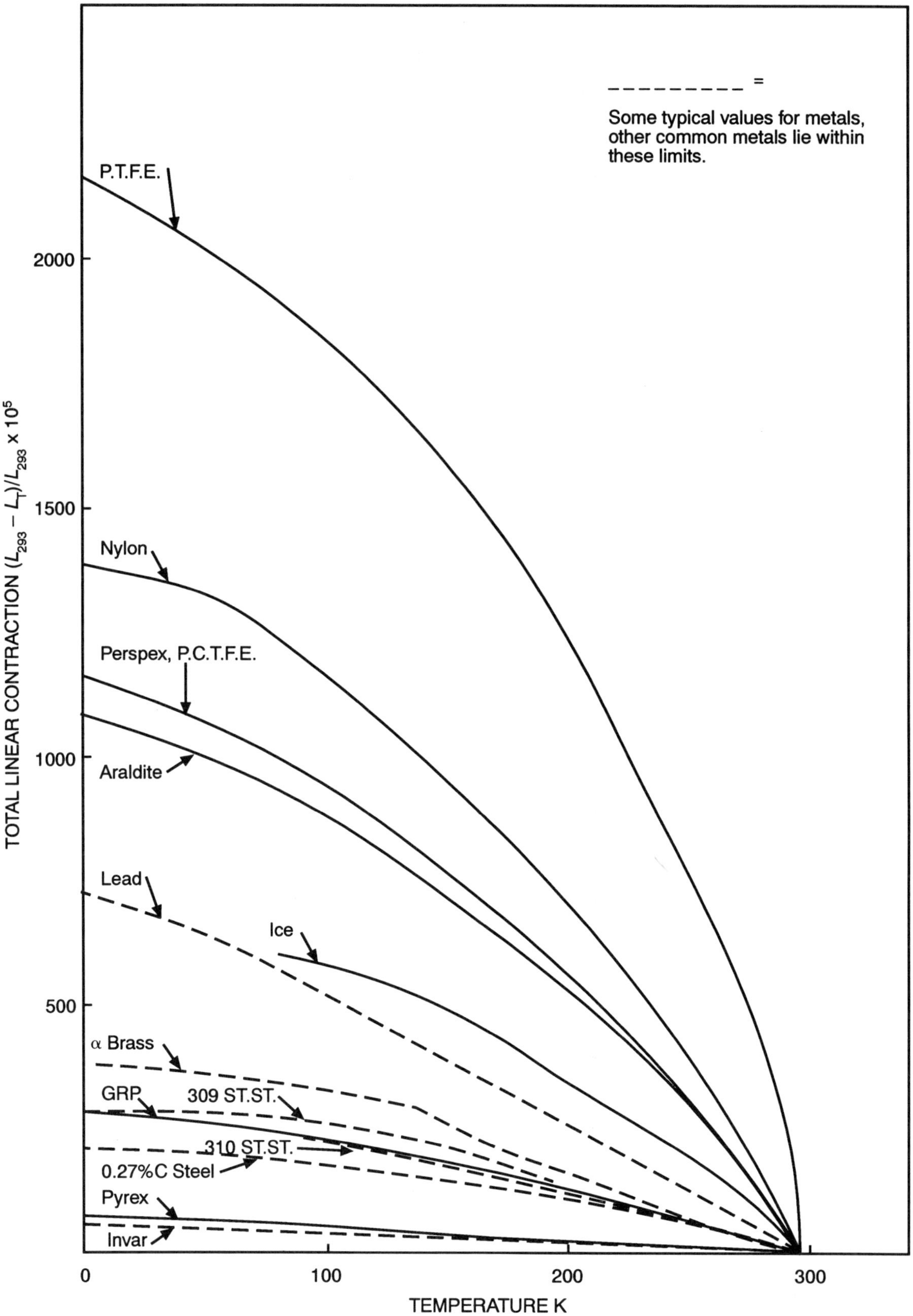

Fig. 1.3.7 Total linear contraction of materials

Fig. 1.3.7

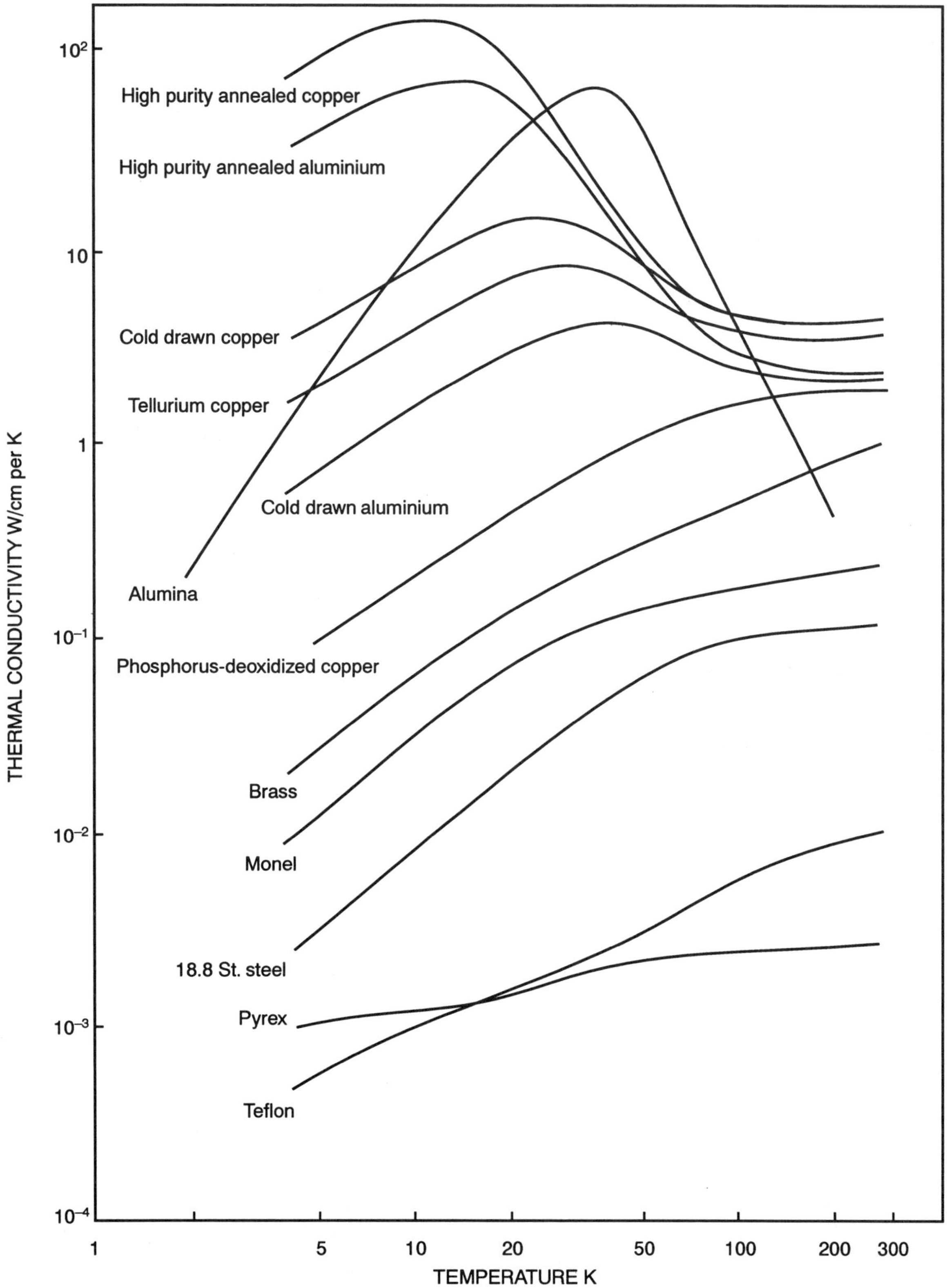

Fig. 1.3.8 Thermal conductivity of materials at low temperatures

1–4

Corrosion

Contents

List of tables

List of figures

1.4.1 Corrosion behaviour of metals

1.4.1.1 Introduction to corrosion of metals

CORROSION AND ITS CONSEQUENCES

Corrosion is the degradation of the surface, and possibly a near-surface zone, of a metal or alloy as a result of chemical reaction with its environment. The degradation is caused by an oxidation reaction, where 'oxidation' is used in the general electrochemical sense of a metal atom M losing electrons (not necessarily to oxygen):

$$M \rightarrow M^{Z+} + Ze^-.$$

This is a half-reaction. The other half-reaction required for electrical neutrality affects the rate of corrosion and the nature of the reaction product. In dry corrosion, charged metal ions move by solid state diffusion in the reaction product and this usually requires high temperatures to give significant metal loss. In aqueous corrosion, charged ions move in aqueous solution, permitting rapid attack at low temperatures unless blocked by an impervious surface film.

The corrosion of metals is a complex phenomenon influenced by many parameters. Even with full information about all the necessary variables it is not always possible to predict exactly how any given material will behave in a specific environment. In these situations it becomes necessary to carry out corrosion tests comparing possible materials.

A metal may simply dissolve into, say, a slightly acidic medium. Even a small amount of dissolution could contaminate a food or pharmaceutical product, or discolour a paint or dye.

Whether or not a surface layer of corrosion product is formed, more substantial attack can weaken a component by:

- **general corrosion** reducing the load-bearing cross section and in some cases,
- **demetallification** (also called dealloying or selective leaching), which is the selective removal of an alloying element from below the surface, or
- **intergranular attack**, where corrosion proceeds more rapidly down grain boundaries.

The rate and nature of attack can be influenced by the application of mechanical or thermal stress, heat flux and cyclic stress. A corrosion scale can sometimes be caused to grow at a faster rate, or to spall off, both creating debris and exposing fresh surface to attack.

There are also two special phenomena in aqueous electrolytes: stress corrosion cracking, where cracks propagate more readily under constant stress as a result of corrosion at the crack tip and corrosion fatigue, where the electrolyte similarly reduces fatigue resistance.

If the corroding component is pressurized, or is simply required to remain leak free, another phenomenon found with some metal/environment combinations can be important: the occurrence of localized pitting attack which can cause complete penetration even though general corrosion remains acceptable.

Corrosion within pipework can give rise to a variety of additional problems that need to be considered by a designer, even when the pipe retains its mechanical integrity. These can be local blockage by corrosion product formed *in situ*, or formed elsewhere and redeposited, blockage by spalled corrosion product, and reduced heat transfer leading to elevated temperatures in heat exchangers.

Corrosion products frequently occupy a larger volume than the metal from which they are formed. If they are also porous and can continue to grow in a confined space, they can exert large stresses. This has led to phenomena as diverse as cracking of concrete bridges as a result of rusting of steel reinforcing bars and straining of bolts in gas-cooled nuclear reactors by oxidation of the steel plates and washers being held together.

Generally, 'rusting' of external metal surfaces is regarded as unsightly.

These are all adverse effects of corrosion. With the right materials/environment/temperature combination it is often possible to produce 'protective' corrosion films on a metal surface that reduce the rate of transfer of metal ions to the reactive medium to a low level.

In contrast to unsightly rusting, protective corrosion films may have aesthetic appeal. Very thin films on metals such as titanium produce decorative interference colours. The green patina on copper roofs is thought by many to add to their attraction.

MATERIALS SELECTION

Materials selection for a particular service environment clearly involves the avoidance of combinations that might lead to any of the more catastrophic consequences described above. The corrosion rate must be kept acceptable, but very low corrosion rates achieved via very expensive materials may not always be necessary. Thus selection of the most suitable material in many situations is essentially an economic decision.

However, it should be noted in this regard that the cost of a fabricated and installed component in a particular material is merely the most obvious cost. Costs incurred in inspecting, servicing or replacing components also require assessing. Thirdly, the cost arising due to plant shut-down during these maintenance operations can often be enormous. Additionally the safety aspect of a component failure has to be assessed in determining the risks of component failure by corrosion. It follows that the selection of the best material for a particular application is a complicated decision. It may, on occasions, be cheaper to produce a component of very expensive material rather than incur the costs arising from frequent replacing of the same component made of cheaper material.

DESIGN ASPECTS, CONSTRUCTION, AND MAINTENANCE

Appropriate materials selection for the service environment is the first essential in the avoidance of corrosion problems, but care is required with other plant or component design aspects.

For dry atmospheres, metal surface temperature is important. The design should take account of any increase in average temperature over service lifespan as a result of deteriorating performance, such as fouling of heat transfer surfaces.

For low temperature aqueous corrosion, the possibility of an increase in electrolyte concentration under fault conditions (e.g. seawater ingress into freshwater pipes) or altered service conditions (e.g. use of de-icing salts) must be considered.

Certain features can cause local acceleration of attack, particularly in aqueous environments. These are:

- crevices, where corrosive electrolytes may become concentrated,
- pipe bends and the like, where change in flow direction can sometimes lead to local erosion-corrosion or impingement attack,
- galvanic coupling of dissimilar metals (that is, electrical connection, commonly via direct contact, in the presence of an electrolyte).

Galvanic coupling can lead to very rapid attack of the metal that is 'less noble' in the galvanic series (described below). This has sometimes been the inadvertent consequence of replacing a worn component with one fabricated from a different alloy, or even the use of a graphitic grease with the graphite acting as a more noble 'metal'.

Situations where a porous scale can continue to grow in a confined space, interface, or crevice and cause mechanical damage by expansive reaction are to be avoided.

The shut-down environment during plant maintenance and not just the normal operating environment needs to be considered by designers and operators, especially if condensation can produce corrosive films on metal surfaces.

Access to areas where corrosion protection may need to be renewed and avoidance of moisture traps, especially concealed moisture traps, should be considered at the design stage.

If a component is to be pre-treated or coated to confer corrosion resistance, the designer should consider how to minimize the possibility of damage to the coating during assembly.

Finally, training and/or supervision of construction workers must be adequate to prevent use of 'inappropriate' materials, particularly ones producing galvanic couples, especially for safety-critical items.

CORROSION IN DRY GASEOUS ENVIRONMENTS

Whether or not an oxide scale can form on a metal surface in a hot dry gas is determined by the thermodynamics of the reaction:

$$4M + ZO_2 = 2M_2O_Z .$$

If the oxygen partial pressure of the gas is greater than the dissociation pressure of the oxide, it can form. If it is less, oxidation cannot take place and a pre-existing oxide film will be reduced. It is useful to consider the oxygen partial pressure of a gas mixture in the initial screening of candidate materials. However, there may remain considerable uncertainty that can be resolved only by testing under service conditions if the gas mixture is a non-equilibrium mixture and the actual oxygen partial pressure at the metal surface is uncertain.

The greater the difference between the oxygen partial pressure and the oxide dissociation pressure, the greater is the free energy change or driving force for oxidation. Even so, the largest free energy changes do not always give the highest corrosion rate. The initial reaction of a base metal surface with oxygen is extremely rapid even at room temperature. However, once an initial adherent oxide layer has formed, subsequent growth depends on oxygen passing through the oxide to react with the metal or metal atoms passing through the oxide to react with oxygen. The rate of growth is thus likely to be dependent upon the diffusion coefficient of metal or oxygen in the oxide and therefore to increase with increasing temperature, possibly according to an Arrhenius relationship. Growth stresses, thermal cycling stresses, or concurrent deformation of the underlying metal may produce defects in the oxide scale that provide more rapid diffusion paths.

In some systems, once an initial thin layer of oxide has formed this prevents further rapid reaction; the metal or alloy has good corrosion resistance. In other cases the reaction rate, though it decreases with time, may remain unacceptably high. Or an initially satisfactory protective oxide film may break down, leading to rapid 'breakaway' corrosion.

Similar considerations apply to sulphidation rather than oxidation that occurs in some environments. Many real situations are more complicated, however, with both sulphidation and oxidation, and with many alloying additions involved in the overall corrosion process.

Certain alloying elements such as chromium and aluminium confer good corrosion resistance because they produce very adherent compact oxide films with low ionic diffusion rates. Other alloying additions act in a more subtle way at low concentration, increasing scale–metal adhesion, increasing scale plasticity, or altering diffusion rates in the scales.

Some gaseous constituents accelerate corrosion of particular metals and alloys. Hydrogen chloride is one example. Another is any atmosphere with a high carbon potential and which can 'dump' carbon in a growing oxide scale.

Rapid attack can proceed under molten salts deposited on a metal surface.

ELECTROCHEMICAL CORROSION

Anode–cathode reactions

Metals corrode in aqueous environments or other electrolytes by an electrochemical mechanism. It is thus necessary for there to be both anodic and cathodic sites. At an anodic site an oxidation process takes place (i.e. loss of electrons). Typically this could be the passage into solution of metal as ions, e.g.

$$\underset{\text{(metal)}}{Ni} - 2e^- \longrightarrow \underset{\text{(ion in solution)}}{Ni^{2+}}$$

At a cathodic site a reduction process takes place (i.e. gain of electrons). Two of the most common cathodic reactions are the reduction of dissolved oxygen and the liberation of hydrogen gas:

$$O_2 + 4e^- + 2H_2O \longrightarrow 4OH^- \quad \text{reduction of oxygen in solution}$$

$$2H^+ + 2e^- \longrightarrow H_2 \quad \text{liberation of hydrogen gas}$$

The anodic and cathodic sites or areas can either be permanently separated from each other or be constantly moving about the metal surface.

In either situation it is necessary for there to be a complete electrical circuit established to permit the flow of electrons. One path is through the electrolyte whilst the other will be through metal contacting the anode and cathode regions.

The process will be dependent upon the change in free energy and the characteristics of the product of reaction. Thus if the metal forms an insoluble layer the reaction is likely to be slowed or terminated. Conversely, if the reaction products are soluble the metal can dissolve in an unimpeded fashion.

Galvanic series

If dissimilar metals are immersed in an electrolyte and connected externally, it can be expected that a potential difference will develop between them, giving rise to a current. One metal will act as an anode while the other as a cathode. This potential difference arises from the different potentials that metals establish in solution.

The metal that is the anode will undergo an oxidation process (i.e. loss of electrons),

e.g.
$$\underset{\text{(metal)}}{Cu} - 2e^- = \underset{\text{(ion in solution)}}{Cu^{2+}}$$

Similarly at the metal that is the cathode a reduction process (i.e. gain of electrons) will occur.

e.g.
$$\underset{\text{(in solution)}}{2H^+ + 2e^-} \longrightarrow \underset{\text{(evolved as gas bubbles)}}{H_2}$$

It can be seen that the metal that is the anode is passing into solution and is thus corroding away.

Which metal of a pair will corrode away can be seen from the galvanic series:

Table 1.4.1 *Typical galvanic series of metals and alloys*

The more noble metals will act as cathodes when coupled to the more base metals which act as anodes and therefore corrode. The position of the metals and alloys in this series do differ in different environments and are influenced by temperature and oxygen content.

Area effects

The anode and cathode must have two conducting paths, normally one through the electrolyte and the other usually by metal connecting the two electrodes.

The current that flows means that to maintain a balance the number of electrons lost at the anode must be the same as the number of electrons gained at the cathode.

It follows that if a small anode is coupled to a large cathode the rate of anode reaction will be faster than if a large anode is coupled to a small cathode.

Compositional differences

(a) Within the metal.
An electrochemical cell can also arise in just one metal due to compositional differences within the metal. Thus, for example, if one phase within a two phase material is anodic with respect to the other then the anodic one will corrode away.

(b) Within the solution.
Difference in concentration within a solution can also produce electrolyte cells with one region of an immersed metal acting as a cathode and the other acting as a corroding anode.

(c) Temperature differences.
The chemical reactivity of a metal in solution is temperature dependent. Temperature differences can thus give rise to an actively corroding anode region usually in the hot zone, the cooler parts then being cathodic.

Electrolyte additions

In electrochemical corrosion two paths must exist for the flow of current (movement of electrons) between the anode and cathode. Since the anode and cathode are usually connected by conducting metal the conductivity of the electrolyte is important.

Additions to the electrolyte which increase its conductivity can therefore accelerate the corrosion rate.

Chloride ions frequently accelerate corrosion by increasing the solution conductivity. Other species which increase the conductivity and therefore corrosion include CO, CO_2, H_2S and SO_2.

TYPES OF CORROSION

It is clear from the foregoing that corrosion can occur in many forms for which there are different terms. The following is a simplified and brief explanation of the more common types.

General corrosion

This is the general wastage of a metal by reaction with its environment to produce a chemical compound of the metal. General corrosion proceeds evenly over the surface of a metal and the rate of reduction in metal thickness is often predictable.

Breakaway corrosion

There are some combinations of materials, environment, and/or operating conditions where a slowly growing protective corrosion scale breaks down and much more rapid 'breakaway' attack ensues. Where this has been identified as a possibility, it may be possible to delay the onset or restrict the rate of attack by control of materials variables and adjustments of operating conditions. The phenomenon is known in both aqueous and dry environments, and may arise from mechanical or chemical breakdown of the protective scale.

Galvanic corrosion

If dissimilar metals are connected in the presence of a conducting solution an electrolytic cell can be produced. The most active metal will then become an anode and be corroded away at a greater rate than if not coupled to the dissimilar metal. Conversely the less active metal will be protected.

Demetallification

The corrosion action on an alloy can be highly selective, and effectively remove only one metal, for instance zinc, from a copper–zinc alloy. The metal remaining after this form of attack generally has similar dimensions to the original component, but is porous and lacking in mechanical strength.

Intergranular corrosion

This is a highly selective form of attack, in which the material between the grains of metal is either corroded away or weakened to the extent that the individual metal crystals can separate. It is when the gaps between the crystals are opened up by corrosion processes that deterioration of the metal can occur. If the metal is now stressed, it literally tears, using the corroded hairline cracks as starting tears.

Pitting corrosion

This term is used to describe all those forms of corrosion where the attack is localized into small but possibly deep holes. The total amount of metal removed may be small but a metal section thickness may be perforated. The pits can be either randomly distributed or associated with changes in section, chemical composition or microstructure of the alloy.

Crevice corrosion

The differences in chemical composition between the environment inside a crevice and that outside a crevice can stimulate pitting from the crevice. The crevice may, for instance, be a sharp corner where debris can accumulate or a gap at the edge of a gasket or packing material.

Deposit attack

This is a form of pitting where the different chemical environment under a scale or deposit gives rise to pits.

Waterline attack

Pitting or grooving can occur at the waterline of a container due to the higher oxygen concentration in the surface layers of solution.

Impingement attack

Where a flowing stream of fluid strikes a plate or pipe bend a combination of the greater supply of reactive species, mechanical abrasion by suspended particles and the moving fluid can give localized attack.

Cavitation attack (see also Chapter 1-5)

Sudden changes in pressure in a liquid can produce vapour-filled cavities, which subsequently collapse. The collapsing cavities release energy, producing mechanical damage. Propellers and pumps suffer from this form of attack which is mainly mechanical, but otherwise has similarities to impingement.

Stress corrosion

Metals supporting stresses well within their normal capabilities can crack in some specific environments. Where the environments that produce stress corrosion cracking in a particular alloy are known they can be avoided. Environments which are likely to cause stress corrosion can sometimes be predicted from existing information.

Corrosion fatigue

Metals subjected to cyclic loading are prone to failure by the gradual growth of cracks. This process of metal fatigue is frequently accelerated in corrosive environments and the term 'corrosion fatigue' is then used.

Liquid metal embrittlement

Some metals undergo cracking in the presence of liquid metals and care is therefore required in handling liquid metals.

Comparative corrosion performance of metals

A summary of the susceptibility of metals and alloys to the different types of corrosion is given in:

Table 1.4.2 *Comparative susceptibility of metallic materials to different types of corrosion*

The performance of metals in common environments is given in:

Table 1.4.3 *General environmental performance of metals*

Further information on the corrosion behaviour of metals and alloys is given in the following section (1.4.1.2). Detailed information on corrosion in specific environments is to be found in the relevant chapters of Vol. 2.

1.4.1.2 Corrosion susceptibility of metals

The corrosion susceptibility of common metals and alloys is given in:

Table 1.4.4 *Corrosion susceptibility of iron and steel (except stainless steel)*
Table 1.4.5 *Corrosion susceptibility of stainless steels*
Table 1.4.6 *Corrosion susceptibility of copper and its alloys*
Table 1.4.7 *Corrosion susceptibility of aluminium and its alloys*
Table 1.4.8 *Corrosion susceptibility of titanium and its alloys*
Table 1.4.9 *Corrosion susceptibility of nickel and cobalt and their alloys*
Table 1.4.10 *Corrosion susceptibility of zinc and cadmium and their alloys*
Table 1.4.11 *Corrosion susceptibility of magnesium and its alloys*
Table 1.4.12 *Corrosion susceptibility of tin and lead and their alloys*
Table 1.4.13 *Corrosion susceptibility of precious metals*

Table 1.4.1 Typical galvanic series of metals and alloys

increasing reactivity ↓	**Protected end (cathodic or most noble)**
	Platinum
	Gold
	Graphite
	Titanium
	Silver
	Hastelloy C
	18-8-Mo Stainless steel
	18-8 Stainless steel[a]
	Chromium irons
	Inconel
	Nickel
	Silver solder
	Monel
	Copper–nickel alloys
	Copper
	Aluminium bronze
	Brasses
	Stainless steels[a]
	Tin
	Lead
	Lead–tin solders
	Nickel cast irons
	Cast iron
	Carbon steel
	Aluminium alloys (2017, 2024)
	Cadmium
	Aluminium alloy 1100
	Galvanized steel
	Zinc
	Magnesium alloys
	Magnesium
	Corroded end (anodic or least noble)

[a] Stainless steels are shown as occupying two positions because they exhibit erratic potential depending upon the incidence of pitting. The two positions are intended to represent possible extremes of behaviour.

Table 1.4.2

Table 1.4.2 Comparative susceptibility of metallic materials to different types of corrosion

Alloys	General	Demetallification	Galvanic corrosion	Intergranular	Pitting	Stress corrosion
Silver and precious metals	Only in very aggressive acidic/alkaline environments	Occurs to least noble constituent.	Does not occur with metals below.	Not generally applicable.	Not generally applicable.	Alloys with base metals are susceptible.
Titanium	As above but less resistant	Not known.	Titanium, copper, stainless steel and nickel can safely be used together in many cases but not with large areas of silver and other precious metals.	Occurs infrequently.	Occurs mainly at high temperatures in chlorine – containing environments.	Generally only found in service at higher temperatures on specimens precracked at low temperatures or in fretting situation.
Copper	Less resistant than above but better in alkalis.	Occurs in Cu/Zn, Cu/Al, Cu/Sn and to lesser extent in Cu/Ni.		Occurs during demetallification.	Not generally susceptible but may occur in tubes.	Wrought Cu/Zn most susceptible. Other alloys less so.
Stainless steels	More resistant than above in oxidizing environments	Not known.		Can occur in sensitized material.	As for titanium but more susceptible.	More susceptible than Ti and Cu/Ni alloys.
Nickel	Similar to above but less attack in HCl and H_2SO_4.	Not known.		Generally only occurs in aggressive chloride-containing environments.		As for stainless steel but less susceptible.
Lead and tin	Less resistant, particularly in alkalis.	Not known.	May be safely connected with Cu, Ni, Fe, Zn, Cd, Al, and stainless steel.	Not applicable.	Not applicable.	Not generally applicable.
Iron	Readily corroded in many environments.	Prolonged exposure to sea can reduce cast irons to a graphite skeleton.	Corrosion promoted by the alloys above.	Not generally applicable in low alloy steels.	Occurs more often than in stainless steels in less aggresive solutions.	Mainly in higher strength alloys but also mild steels in nitrates and caustic media.
Zinc and cadmium	Less resistant than above in acids and alkalis but better in atmospheres.	Not applicable.	These alloys are only purposely connected to those above to protect them.	In zinc alloys, impurities such as cadmium, tin and lead segregate at the grain boundaries as the alloy solidifies. The grain boundaries are anodic compared to the remainder of the grains, thus corrosion of the grain boundary metal occurs.	Not applicable since these are used as coatings.	Not generally applicable.
Aluminium	As above but more resistant in strongly oxidizing environments.	Not known.		Can be susceptible depending on heat treatment.	Occurs more often than in stainless steels and in less aggressive solutions.	Higher strength alloys generally more susceptible.
Magnesium	As for Zn and Cd but resistant in acids.	Not known.		Not generally found	Occurs more frequently than in aluminium alloys.	More susceptible than aluminium alloys.

Table 1.4.3 General environmental performance of metals

Environment / Metal group	*Inorganic acids*[a]	*Organic acids*[b]	*Oxidizing acids*[c]	*Alkalis*[d]	*Seawater and neutral chloride solutions*[e]	*Freshwater and atmospheric exposure*[f]
Noble metals	Good	Excellent/ Good	Good/Poor	Excellent/ Good	Excellent	Excellent
Titanium	Good	Excellent	Excellent/ Good	Good	Excellent	Excellent
Nickel and alloys	Good/Fair	Good	Fair	Good[g]	Excellent/ Good	Excellent
Stainless steels	Fair	Good	Excellent/ Good	Good/Fair	Good[h]	Excellent/ Good[h]
Copper and alloys	Fair/Poor[i]	Fair[i]	Poor[i]	Fair/Poor[j]	Good[k]	Good
Tin	Good/Fair	Good	Fair/Poor	Fair	Good/Fair	Good
Lead	Good/Poor[l]	Fair/Poor	Good/Poor[l]	Fair	Good	Good[m]
Carbon steel and iron	Poor	Fair	Poor	Good	Fair	Fair
Aluminium and alloys	Poor	Fair	Poor	Poor	Fair	Good/Fair
Cadmium	Poor	Poor	Poor	Poor	Fair	Good
Zinc	Poor	Poor	Poor	Poor	Fair	Good/Fair

[a] Sulphuric and hydrochloric acids most important.
[b] Acetic acid most important.
[c] Includes nitric acid and non-oxidizing acids containing oxidizing agents such as oxygen or sodium nitrate.
[d] Includes sodium and potassium hydroxides and ammonia.
[e] Seawater contains ~3.4% NaCl, pH ~8.
[f] In unpolluted areas, rainwater is slightly acidic. In polluted areas pH may be as low as 4. Chloride content of mains water varies from a few ppm to over 100 ppm in British Isles. Hardness salts reduce corrosion rates.
[g] Not resistant to ammonia.
[h] Corrosion tends to be pitting or crevice corrosion.
[i] Cupro-nickels are resistant to acids.
[j] Satisfactory resistance to potassium and sodium hydroxides but rapid attack and stress corrosion in ammonia.
[k] High corrosion rate if flow rate exceeds 1 m/s.
[l] Lead is attacked by nitric and hydrochloric acids but is resistant to sulphuric, chromic, hydrofluoric and phosphoric acids.
[m] Lead mains pipes corrode slowly in soft water.

Table 1.4.4 Corrosion susceptibility of iron and steel (except stainless steel)

<table>
<tr><th>Corrosion type</th><th>Low alloy cast irons</th><th>High alloy cast irons</th><th>Mild steel</th><th>Carbon steel</th><th>Low alloy steels</th><th>High alloy steels</th></tr>
<tr><td rowspan="2">General corrosion</td><td rowspan="2">Atmospheric corrosion resistance superior to mild steel but not resistant to strong acids except for sulphuric above 65%. Good resistance to alkalis. Grey iron is susceptible to cracking in fuming sulphuric acid.</td><td rowspan="2">High alloy cast irons in general are more resistant to corrosion than the low irons. Austenitic cast irons are resistant to sulphuric and hydrochloric acids at room temperature and to most alkalis and salts. They are attacked by oxidizing acids such as nitric and chromic. High silicon irons are resistant to most acids except for HF, sulphuric acid and fuming sulphuric acid. Molybdenum bearing grades are particularly resistant to HCl. They are attacked by strong alkalis fluorides and sulphides.</td><td colspan="4">Steels generally exhibit relatively poor resistance to atmospheric corrosion and to acidic aqueous systems. In neutral waters the corrosion rate increases as the oxygen content of the water increases. Alkaline solutions (pH>11) lead to passivity and a very low corrosion rate. While steels are generally attacked by aqueous solutions of acids, some strong acids, such as chromic, concentrated nitric and sulphuric and commercially pure phosphoric, can be successfully handled in mild steel equipment, although it is essential that the range of concentrations handled should be carefully controlled. Steels are resistant to commercial solutions of alkalis, although corrosion rates become relatively high in hot concentrated solutions of strong alkalis such as NaOH (see also stress corrosion below). Generally, the rate of rusting will be similar for all mild steels and low-alloy steels. The presence of up to approximately 0.2% copper in such steels has a marked improving effect. Amounts between 0.1%–0.2% are common in steels, arising from recycled scrap steel, with the amount tending to build up, since copper is not removed in the process of steelmaking.</td></tr>
<tr><td></td><td>Essentially the same as mild steel.</td><td>Frequently more resistant to atmospheric corrosion than mild steel. In particular nickel and copper additions give improved corrosion resistance. Chromium may also be beneficial although it can lead to severe pitting problems.</td><td>Steels with above 3% nickel have particularly good corrosion resistance in atmospheric and marine exposure. 5% Ni steels give good service in concentrated alkalis. Chromium additions – as low alloy steels.</td></tr>
<tr><td rowspan="2">High temperature oxidation and corrosion</td><td rowspan="2">Satisfactory up to 320°C in steam and 430°C in air, above which oxidation and dimensional changes may lead to problems. Ductile iron has markedly better dimensional stability than grey or white irons.</td><td rowspan="2">High alloy cast irons are satisfactory in air up to 820°C, while 15% Cr irons may be used up to 900°C and 35% Cr up to 1000°C. Also more resistant in steam than low alloy cast irons.</td><td rowspan="2">Useful up to 500 and 600°C and may be satisfactory for some applications up to 700°C. Oxidation in steam becomes serious at 540 to 600°C.</td><td rowspan="2">Similar to mild steel.</td><td colspan="2">Chromium additions generally increase resistance to high temperature oxidation (see also stainless steels).</td></tr>
<tr><td>A 2 1/2% Si 1% Al alloy has been develloped which is useable at up to 1000°C in dry conditions.</td><td></td></tr>
<tr><td>Breakaway corrosion</td><td colspan="6">Alloys with less than 12% Cr are known to be susceptible to breakaway corrosion in high pressure CO_2. Susceptibility and corrosion rate decrease with increasing Cr and Si. Ferritic steels may suffer severe attack in sulphidizing environments.</td></tr>
<tr><td>Liquid metals</td><td>Grey iron may be used for handling liquid sodium, sodium potassium alloys, magnesium, cadmium, aluminium, zinc, lead and tin, although attack may become unacceptable at higher temperatures.</td><td>Little better than grey iron in most cases.</td><td colspan="2">Suitable for most of the metals that can be handled by cast iron, but susceptible to stress corrosion by liquid copper and brazing alloys.</td><td colspan="2">No data</td></tr>
<tr><td>Aqueous galvanic corrosion</td><td colspan="6">Cast irons and steels may be safely coupled with aluminium, magnesium, zinc and cadmium, and in most cases these metals will provide cathodic protection of iron and steel. One exception to this occurs with zinc-iron couples in neutral waters, when the steel will be corroded at temperatures above about 70°C owing to a change in the structure of the oxide film on the zinc. Some caution should also be employed with the above metals when using high strength steels in corrosion fatigue conditions as there are circumstances in which these metals can increase the rate of fatigue crack growth. Coupling iron and steel with stainless steel, copper, nickel and their alloys will increase the corrosion rate, although the effect will be of minor importance in de- aerated, relatively pure neutral solutions such as are found in recirculating heating systems. Tin and lead are very close to steel in the galvanic series and the galvanic effect of these metals will be of minor importance in most cases. Significant galvanic corrosion may occur between different grades of steel, and owing to the area effect, care should be taken to ensure that fasteners are cathodic to the larger area of metal to be fastened, otherwise they may be rapidly corroded. Hence, bolts for mild steel should be made of high Ni, low Cr alloy which will be galvanically protected by the steel structure.</td></tr>
<tr><td>Localized corrosion (i) Demetallification</td><td>Graphitization of grey and ductile irons (the dissolution of the iron leaving a soft porous graphite structure) frequently occurs in soils, some waters and mildly acidic solutions.</td><td>Normally immune to graphitization.</td><td colspan="4">No data</td></tr>
<tr><td>(ii) Intergranular corrosion</td><td colspan="6">No data</td></tr>
</table>

Table 1.4.4 Corrosion susceptibility of iron and steel (except stainless steel) – *continued*

Corrosion type	***Low alloy cast irons***	***High alloy cast irons***	***Mild steel***	***Carbon steel***	***Low alloy steels***	***High alloy steels***
(iii) Pitting corrosion	Frequently occurs in soils	Superior to low alloy cast irons.	Pitting frequently occurs in the corrosion of steels, especially in neutral waters and in seawater. Surface deposits of millscale significantly increase severity of pitting.			
				Similar to mild steel.	Depends on alloy composition. Alloying with chromium tends to increase pitting.	High chromium steels are particularly susceptible to pitting in chloride-containing solutions.
(iv)	Impingement, erosion corrosion and cavitation	Generally suffer from erosion corrosion and impingement. Grey iron is very sensitive to cavitation damage, but hardened ductile iron can have good resistance.	Generally better than low alloy cast irons.	In general increasing hardness and corrosion resistance will increase impingement, erosion corrosion and cavitation resistance.		
(v) Crevice corrosion and deposit attack	Less susceptible than steels.		Steels are generally susceptible to crevice corrosion in aerated aqueous environments.			
Corrosion fatigue	No data		Steels are susceptible to corrosion fatigue in most environments, although inhibitors may significantly improve their performance. There is no fatigue limit for steels in the majority of corrosive environments. High-strength steels will frequently show some benefit over low strength steels for only a few ($<10^4$) high stress cycles, but for operation to a large number of cycles ($>10^6$) there is little difference between different steels. For steels to be used in thick sections crack growth rates in corrosion fatigue may be very high if the steel is susceptible to stress corrosion. Very limited data are available in this area, and for critical components experimental determination of crack growth behaviour is recommended.			
Stress corrosion and hydrogen embrittlement		No data		Stress corrosion cracking of steels occurs in caustic solutions above about 40°C; in certain compositions of commercial liquid ammonia, in nitrate solutions and in solutions containing hydrogen sulphide. Many higher strength steels are also susceptible to stress corrosion in seawater and other chloride solutions. High strength steels are also susceptible to hydrogen embrittlement resulting from hydrogen pickup during plating and pickling operations or from excessive cathodic protection.		
Corrosion prevention and control	Cast irons and steels can be protected against corrosion by three methods, coating (e.g. painting, electroplating, metal spraying), cathodic protection (sacrificial anodes, impressed current) and inhibition (e.g. chromate, polyphosphate or silicate in waters, and a wide range of organic materials for atmospheric corrosion prevention). Some protective systems may combine two of these methods, such as inhibitive paints and zinc and cadmium coatings which provide galvanic protection at breaks in the coating.					

Further information on iron and steel is given in Vol. 2, Chapters 2-1–2-3.

Table 1.4.4 – *continued*

Table 1.4.5 Corrosion susceptibility of stainless steels

	Ferritic stainless steels	***Martensitic stainless steels***	***Cr–Ni austenitic stainless steels***	***Cr–Mn austenitic stainless steels***	***Precipitation-hardening stainless steels***
General corrosion	All grades exhibit good corrosion resistance, with austenitic steels being better than ferritic and martensitic grades. Are attacked by non-oxidizing acids such as HCl and H_2SO_4. In particular chlorides and other halides tend to make the steel active, leading to rapid corrosion.			Essentially the same as nickel austentic steels, although fewer data are available.	Generally better than martensitic steels, and some may be superior to austenitic steels.
		Martensitic stainless steels have less corrosion resistance than other stainless steels due to the low chromium content. However it is the combination of strength, hardness and corrosion resistance which makes the alloy attractive for many applications, including high-quality knives, ball bearing and values.			
High temperature oxidation	The high temperature oxidation resistance of stainless steel is primarily controlled by the chromium content. 12 Cr steels are usable up to 760°C. 18 Cr to 930°C and 25 Cr up to 1100°C. Corrosion is significantly increased by contaminants such as sulphur dioxide, carbon dioxide and water. In sulphidizing atmospheres, at least 20% Cr may be needed. Steam is initially more corrosive than air at the same temperature, but the long-term behaviour appears to be very similar once fairly thick oxide films have formed. Stainless steels are susceptible to severe corrosion from the combustion products of fuels containing vanadium or chlorides.				
Liquid metals	Generally attacked by the majority of liquid metals.				
			May be used for Na–K alloys up to 870°C, but very sensitive to oxygen contamination.		
Molten salts	For service in chlorides high nickel and lower chromium give the best resistance, cast 15 Cr–35 Ni and 12 Cr–60Ni (ACI HT and HW) being generally used. Most austenitic steels are suitable for use in fused potassium/sodium nitrate mixtures. Molten cyanides are particularly aggressive, 15 Cr–27 Ni (AISI 330) or 15 Cr–60 Ni giving the best resistance.				
Aqueous galvanic corrosion	Galvanic attack on stainless steels is complicated by the possibility of a transition between the active and passive states. Passive steel will not normally corrode, and is cathodic to most materials. If, on the other hand, the oxide film is damaged and as a result of the composition of the environment it cannot reform, this leads to an area of active material with behaviour very similar to alloy steels. Consequently, severe galvanic corrosion can occur between active and passive regions on one piece of steel. Since the stainless steel can itself act as cathode, there is no need to worry about the effect on stainless steel of coupling with other metals. The effect on the other metals will generally be more important.				
Localized corrosion (i) Demetallification	No data				
(ii) Intergranular corrosion	If stainless steels are 'sensitized' by incorrect heat treatment or by service temperature conditions, intergranular corrosion generally occurs in relatively aggressive media such as nitric acid or nitric–hydrofluoric acid mixtures, although it may occur in mild environments, e.g. seawater.				
	Sensitized after rapid cooling from above 930°C. May be desensitized by short annealing treatment at 650– 820°C or avoided by slow cooling.	May occur – maximum resistance is in the hardened condition.	Will be sensitized to intergranular corrosion by heat treatment at 430– 900°C. May be reduced by using extra low carbon grades (e.g. AISI 304L) or by using grades stabilized with titanium or niobium (columbium) (e.g. AISI 321 and 347).	Similar behaviour to nickel austenitic stainless steels. May generally be reduced by keeping carbon below 0.06%.	No data.
(iii) Pitting	When corrosion of stainless steel does occur it will frequently be pitting corrosion. Molybedenum is markedly beneficial in protecting against pitting while the stabilizing elements, titanium and niobium, tend to increase pitting. Pitting is most frequently found in chloride-containing environments, in which the austenitic molybdenum-containing grades (e.g. AISI 316) offer the best resistance.				
(iv) Impingement, erosion corrosion and cavitation	Generally fairly resistant to impingement, erosion corrosion and cavitation.				
(v) Crevice corrosion, deposit attack	Active–passive corrosion can occur if the low oxygen content of the solution in the crevice cannot maintain the passivity of the steel. Hence crevice corrosion will readily occur in seawater. Alloys resistant to pitting (e.g. AISI 316 and 317) will tend to be more resistant to crevice corrosion, but for sensitive applications crevices should be avoided.				

Table 1.4.5

Table 1.4.5 Corrosion susceptibility of stainless steels – *continued*

	Ferritic stainless steels	*Martensitic stainless steels*	*Cr–Ni austenitic stainless steels*	*Cr–Mn austenitic stainless steels*	*Precipitation-hardening stainless steels*
Corrosion fatigue	Providing the steel is not sensitive to intergranular corrosion, stainless steels will generally be more resistant to fatigue crack initiation than low alloy steels of similar strength. On the other hand, once a fatigue crack has started to grow the rate of growth will frequently be greater for the stainless steel. Thus, for applications involving carefully finished unwelded components, stainless steels may give a significant improvement in life, whereas the use of stainless steel in welded load-bearing structures using thick plate is inadvisable (unless, general corrosion conditions require it).				
Stress corrosion and hydrogen embrittlement	Generally resistant, although cracking has been reported in waters at elevated temperatures.	Stress corrosion of hardenable martensitic steels has been reported in acid sulphide, chlorine and marine and industrial atmospheres.	Stress corrosion occurs predominantly in chloride solutions, although it has been reported in the absence of chloride (e.g. in sulphide and caustic solutions). The problem most frequently occurs at temperatures above ambient, but failures have been observed for work-hardened material at ambient temperatures.		Susceptible in chlorides and sulphide solutions, especially in highest strength heat treatments.
Corrosion prevention	In general, corrosion prevention for stainless steels is directed at ensuring that the steel remains in the passive condition. This can involve design to avoid crevices and regions likely to accumulate deposits and the use of anodic protection to maintain the passive film in reducing environments.				

Further information on stainless steels is given in Vol. 2, Chapter 2-1.

Table 1.4.5 – *continued*

Table 1.4.6 Corrosion susceptibility of copper and its alloys

	Copper	*Brasses*	*Al–bronzes*	*Sn–bronzes (and phosphor bronzes)*	*Si–bronzes*	*Cupro-nickels*	*Cu–beryllium alloys*
General corrosion	Copper and its alloys can be used in NaOH, KOH but not in aqueous ammonia. Copper and its alloys are attacked by sulphide in solution and by sulphide and sulphur vapours. High copper alloys are more resistant to sulphide and oxygen attack than copper if phosphorus and arsenic are not added. High copper alloys are more resistant than copper to acids and are used for handling HF.						
	Copper corrodes in: (i) Oxidizing acids or acid solutions containing oxidizing agents, e.g. HNO_3, aerated H_2SO_4, H_3PO_4, $FeCl_3$; (ii) HCl is most aggressive non-oxidizing acid.	High zinc brasses (>15% Zn) are less resistant than copper in both oxidizing and non-oxidizing acids.	Generally more resistant than brasses although influenced by heat treatment.	More resistant to sulphuric acid	Si improves the resistance to acids and ammonia salts.	The most corrosion resistant alloys in acids, alkalis or salt solutions.	Slightly better than copper in most environments.
High temperature oxidation	Copper and its alloys are oxidized fairly rapidly at elevated temperatures. High copper alloys are generally more resistant than pure copper. Additions of Al, Be, Mg are particularly beneficial.						
Aqueous galvanic corrosion	Generally copper and its alloys may be safely used in contact with stainless steels, lead, nickel and tin. Corrosion will occur if copper and its alloys are connected to more noble elements (platinum group metals, carbon etc.). Copper alloys should not be used with small areas of less noble metals (e.g. iron, steel, zinc, aluminium) which will corrode preferentially.						
Localized corrosion (1) Demetallification		Dezincification occurs at low and high pH typically <4 and >9, particularly if oxygen supply restricted. Surface flaws, crevices, scale and high chloride levels increase risk. Dezincification less likely in alloys with $<15\%$ Zn. Fe and Mn increase rate of dezincification. Sn and especially As, Sb inhi- bit attack. With α/β alloys (e.g. 60/40 brass) dezincification of β phase not inhibited by As, and Sb.	Less likely in alloys with 8% Al. Dealuminification reduced by As additions. Occurs more at low pH and is influenced by the other factors noted for brasses.	Can occur but less likely than in other alloys.	No data.	Denickelification may occur but infrequently.	No data.
(ii) Intergranular corrosion	Occurs predominantly by contact with liquid metals, mercury salt solutions and in aqueous solutions often during demetalification.						
(iii) Pitting corrosion	Copper and its alloys may be susceptible to pitting in some soils and in seawater especially where organic sulphur compounds have been produced by microbiological action. Pitting of copper tubes may occur in some mains waters of low pH and if tubes imperfectly processed during manufacture.						

Table 1.4.6 Corrosion susceptibility of copper and its alloys – *continued*

(iv) Impingement and cavitation	Not normally used under impingement conditions.	Brasses are the least resistant of copper alloys (except for Al–brass which with As addition is similar to Al–bronze).	Al–and Mn–bronzes more resistant than other bronzes and brasses.	Slightly less resistant than Al–bronzes.		Most resistant alloys, particularly 70/30 and 90/10 alloys with Fe addition.	
Corrosion fatigue	All copper alloys show a reduction in fatigue strength in aqueous solutions (e.g. seawater) and contaminated atmospheres.						
Stress corrosion cracking	Stress corrosion cracking of copper alloys occurs predominantly in moist ammonia, ammonium salts, amines, moist sulphur dioxide and some nitrate and chloride solutions, as well as during exposure to industrial atmospheres due to presence of SO_2, NH_3, or nitrogen oxides.						
	Tough pitch copper very resistant.	If >20% Zn then least resistant of all copper alloys.	Intermediate resistance.	Intermediate resistance.	High resistance.	High resistance	Intermediate resistance.

Further information on copper and its alloys is given in Vol.2, Chapter 2-4.

Table 1.4.7

Table 1.4.7 Corrosion susceptibility of aluminium and its alloys

Corrosion type	*Commercial purity aluminium*	*Al–Cu*	*Al–Mn*	*Al–Si*	*Al–Mg*	*Al–Mg–Si*	*Al–Zn–Mg*
General corrosion	Pure aluminium has a good corrosion resistance which is increasingly impaired by alloying.						
	V. GOOD Except in acids (HCl, HNO_3, H_2SO_4) and alkalis (NaOH, KOH).	FAIR Corrodes in acids, alkalis and additionally in industrial and marine environments. Corrosion resistance improved by 0.5% Mg.	GOOD	V. GOOD Somewhat worse than pure aluminium.	V. GOOD Worse than Al/Si.	GOOD Slightly worse than Al/Mg.	GOOD
Aqueous galvanic corrosion	Zinc and cadmium can be safely coupled to aluminium. In decreasing order of safety, titanium, stainless steel, chromium plate, steel and lead can be coupled to aluminium except in marine environments. The metals copper and nickel should, together with their alloys, be avoided.						
Localized corrosion (i) Demetallification	Not applicable.						
(ii) Intergranular corrosion		All the aluminium alloys can undergo intergranular corrosion, the incidence and severity being dependent upon heat treatment. The solution treated condition is generally less susceptible than fully heat treated, whilst overageing is usually beneficial.					
	Not usually susceptible.	Properly heat treated these are only susceptible in severe environments: polluted industrial, severe marine, seawater immersion.	Not usually susceptible.	Not usually susceptible.	Not usually susceptible if properly treated except in unusual environments (e.g. NH_4NO_3 solution).	If Mg and Si are in correct proportions to form Mg_2Si only slightly susceptible.	Susceptible if improperly heat treated.
(iii) Pitting corrosion	Pitting is common in aluminium alloys, where the normally protective film of oxide is not quite fully protective. Generally the purest metal is most resistant, whilst increasing alloy content increases the attack. Pitting is promoted by Cu, Ni, Fe, carbon and chloride present in solution or deposited on the aluminium surface.						
	V. GOOD	POOR	GOOD	FAIR	GOOD	Good if Mg_2Si precipitates. See intergranular corrosion.	GOOD
(iv) Impingement and cavitation	Aluminium has a lower resistance to impingement and cavitation attack than other common constructional metals. The attack is increased by heavy metal ions and chloride.						
(v) Exfoliation	This form of corrosion, also termed layer of lamellar corrosion, produces a splitting of the metal into layers, the resulting corroded material resembles the leaves of a partially opened book. Attack is usually along grain boundaries aligned parallel to the surface but can also be transgranular. Occurs in atmospheric exposure and in solution. In the case of less corrosion-resistant alloys, a sprayed metal coating of pure aluminium obtains good corrosion resistance and the alloy structural member is completely protected against exfoliation or stress corrosion for long periods.						
	No data.	Slow in normal atmospheres. Very common in Al–Cu–Mg alloys.	Not data.	No data.	Observed.	Observed.	Observed.

Table 1.4.7 Corrosion susceptibility of aluminium and its alloys – *continued*

Corrosion fatigue	A reduction in fatigue properties in aggressive environments (e.g. seawater) is common to aluminium alloys.						
Stress corrosion cracking	Not susceptible	In aluminium alloys stress corrosion cracking tends to be most marked for stresses applied in the short transverse direction, and is markedly affected by heat treatment.					
		Susceptible.	Not susceptible.	Not susceptible.	Not very susceptible at low Mg (below 3%).	Not susceptible with normal heat treatment.	Increasingly susceptible as alloy content and strength increase.

Aluminium alloys are frequently clad with a layer of another alloy having better corrosion properties (for example commercial purity aluminium). The surface will then have the properties of the cladding alloy but precautions must be taken to avoid edge attack.
Further information on aluminium and its alloys is given in Vol. 2, Chapter 2-5.

Table 1.4.8 Corrosion susceptibility of titanium and its alloys

Corrosion type	***Commercial purity titanium, Ti–Al–Sn, Ti–Cu, Ti–Pd, Ti–Al–V (Mo), Ti–Al–V–Cr, Ti–Mn, Ti–Mo***
General corrosion	Titanium and its alloys have excellent resistance to many acids, bases and neutral environments. Commercially pure titanium is relatively weak, loses its strength at elevated temperatures, but has the best corrosion resistance. Collectively, titanium and its alloys should not be used in hydrofluoric acid or mixtures thereof, red fuming nitric acid (<2% water or N_2O_4>6%), boiling NaOH or KOH. High corrosion rates can also occur in H_2SO_4 (40 and 80%) (minor amounts of Cu^{2+}, Fe^{3+} inhibit attack). Some organic acids (oxalic, formic and trichloroacetic) give rapid corrosion. High concentrations of HCl and H_3PO_4 (30%) may cause unacceptable rates of corrosion (5 mpy). Dry Cl_2 (gas or liquid with 0.3% H_2O), Br_2, hot fluorine, liquid O_2 and hot gaseous O_2 (>500°C) cannot be safely handled by titanium alloys. Halides, particularly $AlCl_3$, can rapidly attack Ti. Oxidation and nitrogen pick-up increase greatly above approx. 500 °C, Ti is generally unsuitable for use in hydrogen-containing atmospheres.
Aqueous galvanic corrosion	In the passive condition titanium alloys are more noble than most other common alloys (Al, carbon steel, Zn, Mg) but they can be used with stainless steel, Monel, Cu and high Cu alloys. Because of the high electrical resistance of the oxide film it may be permissible to use large areas of steel with small areas of titanium, although it would in any case be advisable to insulate the contact area. Under reducing conditions, e.g. low oxygen concentration, titanium may become active (see crevice corrosion) in which case accelerated corrosion of the Ti can occur.
Localized corrosion (i) Demetallification	At present there is no evidence of demetallification in Ti alloys.
(ii) Intergranular corrosion	Not generally encountered although liquid metals, e.g. Al, Pb, Hg, Cd, Ga, Li, can rapidly attack Ti alloys.
(iii) Pitting corrosion	Pitting of titanium alloys can occur rapidly in hot concentrated halide solutions. In particular, $CaCl_2$, $MgCl_2$ cause pitting. in all cases there must be localized breakdown of passivity and this is often induced by crevice effects (see crevice corrosion).
(iv) Impingement, erosion and cavitation	Titanium alloys have good resistance but there is a critical liquid velocity which depends on the environment. In oxidizing solutions this velocity will be higher than in reducing environments. It will also depend on surface finish, solids content and temperature. In seawater velocities up to approx. 8 m/s can be tolerated. The higher strength alloys generally have better resistance.
(v) Exfoliation	At present there is no evidence of exfoliation in Ti alloys.
(vi) Crevice corrosion	Occurs normally in halide solutions and is more prevalent above room temperature, e.g. 100 °C. It is usually associated with gasketed joints, scale deposits, overlaps. Teflon and fluorocarbon adhesives in joints accelerate Ti corrosion. The Ti–0.2 Pd alloy is most resistant to crevice corrosion.
Corrosion fatigue	Most information is concerned with the influence of chloride solutions (seawater, salt solutions). A reduction in the fatigue limit can occur e.g. in the α–β alloys particularly if heat treated in the α–β temperature range. Fatigue crack growth rates are increased in aqueous environments, the actual rate depending on temperature, amplitude, waveform, frequency, pH and concentration. Data for all the alloys is incomplete.
Stress corrosion	Occurs in alloys (not commercially pure Ti) in hot chlorides above about 250 °C and in some chlorinated and brominated hydrocarbons (e.g. $CHCl_3$) particularly at boiling point. Also occurs in hot Cl_2 and Br_2, red fuming HNO_3, N_2O_4 and some alcohols (methanol and ethanol), particularly if they contain chloride. In aqueous halides (salt solution, seawater) many Ti alloys exhibit reduced fracture toughness, e.g. commercially pure Ti, Ti–5Al–2.5 Sn, Ti–8Al–1 Mo–1V although the susceptibility is reduced by β-stabilizing elements and heat treatment in β field.

Further information on titanium and its alloy is given in Vol. 2, Chapter 2-6.

Table 1.4.9 Corrosion susceptibility of nickel and cobalt and their alloys

Corrosion type	*Nickel and alloys*	*Cobalt and alloys*
General corrosion	Nickel has good corrosion resistance generally, forming a layer of corrosion resistant passive oxide film. Nickel is, however, attacked by oxidizing acid salts and will not resist many halides. Nickel is attacked by mineral acids, particularly if oxidizing salts are present. However, nickel can be used to handle cold sulphuric acid or cold hydrochloric acid. Oxidizing acids rapidly attack nickel. Nickel based alloys have good general corrosion resistance in seawater but can exhibit severe crevice corrosion.	Scaling and oxidation rates of unalloyed cobalt in air can be 25 times those of nickel. However in many respects cobalt has a similar corrosion resistance to nickel. Cobalt is attacked by mineral acids, halogens and halides.
Aqueous galvanic corrosion	Nickel coupled to aluminium will produce rapid corrosion of the aluminium. Coupling to nickel will also produce accelerated corrosion of mild steel, iron and cast iron. Nickel can generally be coupled to stainless steels in neutral and alkaline solutions but the nickel will be attacked in oxidizing acid or oxidizing salt solutions. The potential difference between nickel and nickel copper, copper, brasses or bronzes is generally small and coupling can therefore be safely carried out. In neutral solutions nickel can be coupled to tin, lead and tin and their solders.	Cobalt is more active than nickel but otherwise the same remarks apply.
Localized corrosion (i) Demetallification	Monel and cupro-nickels undergo denickelification in aqueous hydrofluoric acid.	Stellite No. 1 undergoes decobaltification in waste sulphuric acid containing ferrous sulphate.
(ii) Intergranular corrosion	Commercial wrought nickel can suffer from intergranular weakening due to graphite precipitation as a result of prolonged heating above about 400°C. The use of a low carbon nickel is recommended for high temperatures.	No data.
(iii) Pitting	Nickel and its alloys are susceptible to pitting by seawater which is stagnant or slow moving. Nitric acid causes pitting.	Cobalt can undergo pitting attack in seawater. Nitric acid pits cobalt more quickly than nickel.
(iv) Impingement, erosion and cavitation	Nickel and those alloys which age-harden are highly resistant to impingement attack and erosion.	Impingement tests on a 65/30/5 cobalt chromium tungsten alloy have shown that its behaviour resembles a 12% Cr steel.
(v) Crevice corrosion and deposit attack	Nickel and alloys will pit due to oxygen concentration cells which can be formed under fouling organisms or other deposits.	No data.
Corrosion fatigue	The fatigue properties of nickel, cobalt and the alloys of either are reduced when in corrosive environments.	
Stress corrosion cracking	Commercial nickel can stress corrosion crack in caustic soda. To resist chloride environments which will SCC austenitic stainless steel, Inconel (77/16/7 Ni Cr Fe) can be used. Nickel is in general relatively free from SCC.	No data.

Further information on nickel and cobalt is given in Vol. 2, Chapters 2-7 and 2-11 respectively.

Table 1.4.10 Corrosion susceptibility of zinc and cadmium and their alloys

Corrosion type	***Zinc and alloys***	***Cadmium and alloys***
General corrosion	Zinc has reasonably good corrosion resistance in natural atmospheres and neutral aqueous solutions, but corrodes rapidly in both acidic and alkaline media. Compared with steel, for which it is frequently used as a protective coating, zinc corrodes much more slowly.	Cadmium's corrosion behaviour is very similar to that of Zn, but it is stable in alkalis and slightly more stable in acid and neutral solutions.
Aqueous galvanic corrosion	Zinc will corrode if connected to steel whilst in an aqueous electrolyte, and for this reason is frequently used as a protective coating for steel. Similarly zinc will preferentially corrode if connected to most common metals other than aluminium and magnesium.	Cadmium is not as protective to steel as zinc. Cadmium will corrode if connected to most common metals other than aluminium, magnesium and zinc whilst in an aqueous electrolyte.
Localized corrosion (i) Demetallification	Since zinc is not commonly used with any significant alloy content demetallification is not expected.	No data.
(ii) Intergranular corrosion	In zinc alloys containing aluminium, particularly if tin and lead are also present, intergranular attack can occur. Additions of magnesium, and to some extent copper, can reduce the severity. A steam environment will suffice to produce this attack. In zinc alloys, for example, impurities such as cadmium, tin, and lead segregate at the grain boundaries during solidification. The grain boundaries are anodic compared to the remainder of the grains and corrosion of the grain boundary metal occurs.	No data.
(iii) Pitting corrosion	Zinc and cadmium are not usually regarded as susceptible to pitting.	
(iv) Impingement and cavitation	Zinc and cadmium have poor resistance to impingement and cavitation attack.	
Corrosion fatigue	No data.	
Stress corrosion	The intergranular attack occurring in Zn–Al alloys may be due to stress corrosion.	No data.

Further information on zinc and its alloys is given in Vol. 2, Chapter 2-8.

Table 1.4.11 Corrosion susceptibility of magnesium and its alloys

<table>
<tr><th>Corrosion type</th><th>Mg</th><th>Mg–Al</th><th>Mg–Al–Zn</th><th>Mg–Zn</th><th>Mg–Zn–Zr</th><th>Mg–Zr</th></tr>
<tr><td>General corrosion</td><td colspan="6">Magnesium and its alloys are resistant to atmospheric corrosion providing high concentrations of CO_2, SO_2/SO_3 and chloride are not present. Mg–Al alloys are slightly better than others. With the exception of HF and chromic acids, corrosion in acids occurs rapidly. In caustic solution the corrosion resistance is greater than for Al alloys. Heavy metal impurities in the alloy (e.g. Fe, Ni, Cu) cause large increases in corrosion rate; the maximum advised levels are approx. 0.017% Fe, approx. 0.0005% Ni and 0.1% Cu in the Mg. For Mg/Al alloys the limit for Fe is reduced considerably; this is offset by the addition of approx. 1% Zn or Mn. The corrosion rate of Mg is little affected by small additions of Pb, Sn, Al, Na, Si, Mn, Th, Zr, Be, Ce, Pr, Nd. The corrosion rate increases with more than 2% Zn, 0.3% Ca, 0.5% Ag. Magnesium and its alloys are resistant to fluoride solutions, and fluoride and chromate additions to neutral solutions inhibit corrosion.</td></tr>
<tr><td>Galvanic</td><td colspan="6">Galvanic corrosion occurs if Mg or its alloys are coupled to metals other than Al, in particular Fe, Ni, Cu and Ti and to a lesser extent Cd and Zn. Zn and Cr plates have been used to reduce the effects of the Fe/Mg couple. Chromates, vanadates and sulphides have been used to reduce galvanic effects, although problems in providing complete and continuing protection still exist. Some Al alloys will, particularly if they contain Fe or Cu, cause corrosion of Mg alloys in salt solution.</td></tr>
<tr><td>Localized corrosion
(i) Intergranular</td><td colspan="4">Intergranular corrosion is rarely observed in Mg alloys.</td><td>Has been found in Mg–2Zn–1Zr and Mg–5.5 Zn–1Zr–2Th, both in the aged condition.</td><td></td></tr>
<tr><td>(ii) Pitting</td><td colspan="6">Pitting corrosion of Mg and its alloys in humid (~90% RH) atmospheres and aqueous solutions is usually associated with foreign particles which are cathodic to Mg. This is particularly true for cast alloys because it is more difficult to avoid contamination. Mechanical damage to the oxide film can also initiate pitting attack. Concentration cells also cause pitting.</td></tr>
<tr><td>(iii) Crevice corrosion</td><td colspan="6">Crevice corrosion can occur in Mg and its alloys and good design should avoid lapped joints, etc., in any aqueous solutions, e.g. NaCl, and chlorides in general.</td></tr>
<tr><td rowspan="2">Stress corrosion</td><td rowspan="2">Not reported in service but SCC produced in lab. KHF_2 solutions.</td><td colspan="5">Cast alloys are generally more resistant than wrought alloys of similar composition. Welds are less likely to suffer SCC, if they are stress relieved.
Mg alloys containing 1.5% Al are generally sufficiently resistant to SCC not to warrant special treatments. In the laboratory SCC has been produced in KHF_2, HF, HNO_3, NaOH, $NaCl/H_2O_2$, $NaCl/H_2CrO_4$, Air + H_2O + SO_2 + CO_2.</td></tr>
<tr><td colspan="2">Atmospheric exposure can cause failures if Al > 2.5%, particularly if welded.</td><td colspan="3">These alloys very much less susceptible to SCC by atmospheric exposure. SCC resistance improved by rare earth additions.</td></tr>
<tr><td>Corrosion fatigue</td><td colspan="6">The corrosion fatigue limit of Mg and its alloys in condensing conditions or aqueous solutions is approx. 30% of the normal air fatigue limit. Higher limits may be obtained if the solution restricts CO_2 and O_2 access. Some coatings do help to substantially retain the air fatigue limit, e.g. petroleum jelly, paint systems, epoxy resin seals. Chromate and chemical etches have a slightly adverse effect as can some anodic oxide films. Fluoride anodizing and surface sealing have a beneficial effect.</td></tr>
</table>

Further information on magnesium and its alloys is given in Vol. 2, Chapter 2-9.

Table 1.4.12 Corrosion susceptibility of tin and lead and their alloys

Corrosion type	***Tin and alloys***	***Lead and alloys***
General corrosion	Tin has good corrosion resistance and is relatively free from attack by weak acids and alkalis. It is, however, attacked by strong acids, alkalis and many halides. Tin is very readily applied as a corrosion resistant coating to metals such as copper and steel.	Lead has good corrosion resistance. It is, however, attacked by many alkalis and acids with the notable exceptions of sulphuric, chromic, sulphurous and phosphoric acids.
	ALLOYS OF TIN AND LEAD: *Solders* A particular problem associated with solders is enhanced corrosion pitting etc. from residual fluxes (frequently chloride based). This is particularly likely where aggressive fluxes (containing fluoride) have to be used, as in the soldering of aluminium alloys. *Bearing metals* These are at risk to corrosion by lubricating oils and greases and information should be available from the suppliers of lubricants. Acid constituents which can occur by contamination with combustion products give corrosion attack. In this respect tin-based bearings are considered to be superior to lead-based alloys. The corrosion resistance of lead-based alloys increases with tin content and it is considered that at least 5% tin is necessary in lead-based alloys for adequate corrosion resistance.	
Galvanic corrosion	Tin is more noble than Cd, Fe, Cr, Zn, Al and Mg and is therefore not corroded by coupling to these metals.	Pb is more noble and therefore is not normally corroded by coupling to Cd, Fe, Cr, Zn, Al and Mg. Cu, Ni and Fe in alkaline solution can accelerate the corrosion of Pb and solders.
Localized corrosion (i) Pitting and deposit attack	Tin can undergo waterline pitting associated with deposits from hard waters. Tin can also undergo pitting attack in seawaters.	Lead can undergo pitting in soils, sands giving least attack, clays the most. Nitrates, chlorides alkalis and organic acids are the corrosive constituents. The attack can be by the formation of differential aeration cells, microbial attack or stray current corrosion.
(ii) Intergranular corrosion	No data.	Lead with greater than 0.05% Mg can undergo intercrystalline attack by decomposition of Mg_2Pb precipitates. Commercial lead and lead with calcium/lithium additions can undergo intercrystalline attack in acetic/nitric acid solutions. Large grained lead can undergo grooving in sulphuric acid. Eutectic structures are to be avoided.
(iii) Cavitation, impingement, erosion	Being soft metals neither lead nor tin can be regarded as suitable for exposure to impingement cavitation or erosion attack.	
Corrosion fatigue	No data.	Lead has a better fatigue life in oil, vacuum and acetic acid than in air.
Stress corrosion cracking	No data.	Stress can give an increased incidence of pitting but since lead is not used as a load-bearing material the absence of data is to be expected.

Further information on lead and tin–lead alloys is given in Vol. 2, Chapters 2-10 and 2-14 respectively.

Table 1.4.13 Corrosion susceptibility of precious metals

Corrosion type	*Ag*	*Au*	*Pd*	*Pt*	*Rh*	*Ir*	*Ru*	*Os*
General corrosion	The corrosion behaviour of the precious metals and alloys is characterized by their excellent resistance in many aggressive environments. Only Ag and its alloys suffer surface tarnishing in the atmosphere.							
(i) Acids	Not suitable for use in most hot conc. acids. Suitable in HF at below boiling point. Aeration increases corrosion. Suitable for dilute acids except HNO_3.	Suitable for use in all common single acids except *aqua regia*, hot HBr.	Not suitable in conc. HI, HBr, selenic acid (H_2SeO_4) HNO_3, hot $HClO_4$, *aqua regia* and conc. H_2SO_4.	Not suitable in *aqua regia* or in hot conc. HI, HBr or H_2SeO_4. Pt/Ir or Pt/Rh more resistant.	Not suitable for hot HBr. Suitable in *aqua regia*.	Most resistant. Suitable in all common acids and *aqua regia*. No data on $HClO_4$ and H_2SeO_4.	Suitable for all common acids and *aqua regia*. No data on H_2SeO_4.	Not suitable for conc. *aqua regia*, HNO_3, H_3PO_4, or hot conc. HI, HBr, HCl.
	All of the above comments apply to concentrated acids and in several cases equally well to dilute solutions. The resistance of the metals and alloys in hot HCl is detrimentally influenced by oxidizing agents, e.g. ClO_3, H_2O_2, MNO_2. Data on hot HF are not readily available. Halide contamination of HNO_3 can greatly increase the corrosion rate. Ag is usually sufficiently resistant to organic acids and therefore precludes the necessity for using the more noble metals. Acetic, benzoic, boric, carbolic, citric, formic, lactic, propionic, salicylic, tartaric and fatty acids can all be used with Ag. Ag/Pd may be used in boiling acetic acid.							
(ii) Alkalis	Solutions of NH_3, cyanide and alkali sulphides attack Ag. Hot NaOH and KOH solutions can be handled by Ag. Fused alkalis attack at liquid interface.	All the more noble metals are more resistant to NaOH and KOH than Ag. Oxidizing agents, if present, greatly increase attack on Ru, e.g. $NaOH/H_2O_2$. Ru and Os attacked in 10% $NaOH/Cl_2$ at RT, while Pd, and to a lesser extent Rh, attacked in hot solution. Cyanide solutions attack Au and other more noble metals if hot.						
(iii) Halogens	The presence of moisture increases attack by halogen gases.							No data.
	Ag resistant only to Cl_2 at room temperature.	Au not suitable in wet Cl_2 or Br_2 or in dry Br_2.	Pd not suitable in Cl_2 or Br_2.	Pt not suitable in Br_2.	Rh, Ir and Ru suitable in all halogens although some attack in I_2 for Rh, saturated Cl_2/H_2O, Br_2/H_2O and I/alcohol for Ir.			
(iv) Salts	Precious metals, except Ag, are not affected by many salt solutions. Ag is attacked by several persulphates, alums, thiosulphates, polysulphides, ferric and ferrous salts, CuCl, $CuCl_2$, $HgCl_2$, alkali halides, cyanides. Hot $FeCl_3$, and $CuCl_2$ solutions attack Au, Pd, Pt, Rh/Pt and Ir/Pt alloys.							

Table 1.4.13 Corrosion susceptibility of precious metals – *continued*

High temperature corrosion (i) Oxidation	Only Au and Pt do not form oxides up to melting point. Oxides of Rh, Ir and Os are volatile.
(ii) Other gases	Dry Cl_2 attacks most noble metals although Pt is used in Cl_2 up to approx. 250°C. Dry HCl can be handled by Pt and Au up to approx. 1200°C and 700°C respectively. Pt is most suitable for sulphur gases.
(iii) Fused chemicals	Generally suitable for NaOH, KOH and fused salts, Pt is most suitable under oxidizing conditions. Pt and Pt/Rh alloys are suitable for handling mixed silicates and lead components.
Galvanic corrosion	Couplings between precious metals are best avoided, although in many instances corrosion is trivial. With precious metal/base metal couples rapid attack of the base metal occurs.
Stress corrosion	SCC of precious metal alloys can occur in chloride solutions at room temperature, e.g. $FeCl_3$, HCl/HNO_3. Information is only available on Ag and Au alloys including Ag/Au, Ag/Pt, Au/Cu. Low carat golds are generally more susceptible and should be stress relieved to avoid SCC.

Further information on precious metals is given in Vol. 2, Chapter 2-12.

Table 1.4.14 Corrosion susceptibility of tungsten and molybdenum

Corrosion type	*Tungsten*	*Molybdenum*
High temperature corrosion	Tungsten starts to oxidize at a linear rate in air from 400°C, in steam from 700°C and in CO_2 from 1200°C. It is more resistant to H_2S than to SO_2 at elevated temperatures (~700°C). Carbon and hydrocarbon give carburization above 1200°C, CO above 1700°C. CS_2, NO_2 and NO also attack tungsten. Hydrogen does not react with tungsten at any temperature. Nitriding occurs at 1500°C in N_2. Tungsten forms a fluorine at 20°C and a chloride with chlorine above 250°C. Bromides and iodides will form at red heat. A silicide is formed by reaction with silicon above 1000°C. Sulphur and H_2S will both give slow corrosion of tungsten at red heat.	In air the oxidation rate is extremely high above about 700°C. Where molybdenum is used for furnace heaters, a vacuum or reducing atmosphere (typically 95% N_2, 5% H_2) is used to overcome the oxidation problem. In non-oxidizing gases, e.g. dry oxygen-deficient combustion gases, CO_2, H_2S, it is rapidly attacked by hot NO, N_2O, SO_2, F, Cl_2, Br_2. Fused alkalis (NaOH) also cause severe attack.
Aqueous corrosion	Tungsten is resistant to cold dilute HNO_3 and H_2SO_4 and cold conc. HCl but suffers some attack from hot dilute H_2SO_4, conc. H_2SO_4 (hot or cold), hot conc. HCl, hot conc. HNO_3. *Aqua regia* and HF/HNO_3 mixture both give very rapid attack. Tungsten is attacked slightly by hot alkaline solution (KOH, NaOH, NH_4OH). If used as an anode in acid or alkali solutions more rapid attack occurs.	Molybdenum is extremely resistant to attack by hot and cold solutions of non-oxidizing acids. In particular HCl, HF, H_3PO_4, H_2SO_4 cause little corrosion except at high temperature in concentrated H_2SO_4 (>50%). However, oxidizing agents (air, H_2O_2, NO_3 and ClO_3 etc.) cause rapid attack in the above acids. Similarly, oxidizing acids (HNO_3) can rapidly corrode Mo. Cold concentrated HNO_3 passivates Mo. Alkaline solutions (NaOH, KOH, etc.) are corrosive only in the presence of oxygen and oxidizing agents.
Liquid metal corrosion	Tungsten shows slight amalgamation with mercury. Tungsten is not affected at 600°C by sodium, sodium–potassium, gallium or magnesium.	Molybdenum has good resistance to Bi (up to 1560°C), Bi/Pb (<1100°C), Bi/Pb/Sn (at 800°C), Ca (<550°C), Pb (at <1100°C), Li (up to 1000°C) Hg (<550°C), K (up to 900°C), Na (probably suitable up to 1500°C) and Sn (<550°C). Liquid Zn causes rapid corrosion at and above the melting point.
Fused salts	Molten alkalis (NaOH at ~500°C), nitrates and peroxides rapidly attack tungsten. Other oxidizing salt or salts containing oxygen also cause rapid attack at elevated temperatures. Resistant to molten $AlCl_3$ and $BeCl_2$ when used as an anode. Several other compounds cause somewhat less attack, e.g. S, P, B and the refractory oxides MgO, Al_2O_3, ThO_2 and ZnO_2.	

Further information on the refractory metals is given in Vol. 2, Chapter 2-13.

Table 1.4.15 Corrosion susceptibility of niobium and tantalum

<table>
<tr><th>Corrosion type</th><th>Niobium</th><th>Tantalum</th></tr>
<tr><td>General corrosion</td><td>Niobium, though having good corrosion resistance compared to most metals, is not as resistant as tantalum.

Niobium has good corrosion resistance to acids except HF, H_2SO_4 (hot conc.), HCl (hot conc.) and a mixture of HF/HNO_3.</td><td>Pure tantalum has particularly good resistance to corrosion. There is very little attack by mineral acids except HF and H_2SO_4 (hot conc.). Tantalum is resistant to Cl_2, Br_2, I_2 (wet or dry) up to 150°C but is attacked by F_2.

Tantalum is attacked by strong alkalis but resists most organic compounds except those containing fluorine, free sulphur trioxide or strong alkalis.</td></tr>
<tr><td>High temperature oxidation</td><td colspan="2">Above 500°C niobium and tantalum react rapidly with air. The surface oxide dissolves in the metal leading to eventual complete oxidation. Niobium oxidizes more slowly than tantalum.

Oxygen and nitrogen dissolve in the metal giving a reduction in corrosion resistance to other environments.

Various coating techniques have been attempted (mainly with niobium) to counter the oxidation problems at high temperatures.</td></tr>
<tr><td rowspan="2">Liquid metal attack</td><td colspan="2">Niobium and tantalum show relatively good resistance to lithium at 1000°C and to liquid sodium up to 870°C unless oxygen is present. The behaviour in potassium and potassium–sodium alloys is similar to that in solution.

Niobium and tantalum may possibly be resistant to thallium and also to tin, but both dissolve slowly in uranium with the probable upper temperature limits for container usage being 1400°C and 1450°C respectively. Both niobium and tantalum are corroded in molten zinc at 440°C.</td></tr>
<tr><td>Niobium has a good resistance to bismuth up to about 500°C but above that temperature is increasingly attacked.

Niobium has a good resistance to lead at 1000°C but is rapidly dissolved by lead–bismuth eutectic alloy at 1095°C.</td><td></td></tr>
<tr><td>Hydrogen embrittlement</td><td colspan="2">The ductility of niobium and tantalum is reduced by absorption of hydrogen gas.</td></tr>
</table>

Further information on refractory metals is given in Vol. 2, Chapter 2-13.

Table 1.4.16 Corrosion susceptibility of beryllium and its alloys

Corrosion type	*Beryllium and alloys (Be–BeO, Be–Ca)*
	Available information on corrosion of Be is almost completely concerned with its application in nuclear reactors and missile/satellite systems.
High temperature corrosion	In O_2 rapid oxidation can occur at 700°C depending on processing, impurity content and H_2O content in the O_2. In N_2, the weight increase is parabolic at 1100°C m.pt. Exfoliation and blistering may occur in air if the Be contains approx. 300 ppm Cl. The H_2O content is important above approx. 600°C. In CO_2 at 650°C breakaway corrosion occurs after a certain time depending on the H_2O content. The deleterious effect of water increases with increasing temperature. In CO the oxide formed is not protective above 550°C. The Be–BeO and Be–Ca alloys were developed for improved resistance to wet CO_2. Severe corrosion occurs in steam above 325°C.
Aqueous corrosion	Most of the data on aqueous corrosion are from tests in H_2O and H_2O_2 simulating water cooling systems. At approx. 850°C attack is about 1 mpy after 12 months in demineralized water; but pitting occurs and is also influenced by pH, dissolved O_2 and foreign ions. Sodium dichromate and nitrate inhibit pitting. At higher temperatures (approx. 350°C) oxidation and blistering occur readily. Galvanic corrosion does not appear to occur with stainless steel, Pt, Zr and Zircalloy-2 in demineralized water at about 350°C. Cu can have a detrimental effect.

Table 1.4.17 Corrosion susceptibility of zirconium and its alloys

Corrosion type	*Zirconium and alloys*
Corrosion in water and steam	Good corrosion resistance to boiling water. Small amounts of nitrogen (more than 0.005%) give rapid breakaway of the oxide. Alloying additions can counteract the deleterious effect of nitrogen.
General corrosion	Zirconium has excellent corrosion resistance to most acids both cold and hot (HCl, H_2SO_3, H_2SO_4 up to 80%, HNO_3, CH_3COOH, tannic, formic, lactic, chromic, citric, monochloroacetic), with the exceptions of trichloracetic (100%, 100°C, poor) dichloroacetic (100%, 100°C, fair to poor) *aqua regia* (poor), phosphoric (65 to 85%, poor). Zirconium resists most chlorides with the exception of ferric chloride and chlorine gas saturated with water. Zirconium has excellent resistance to alkalis, KOH, NaOH, NH_4OH. Breakaway corrosion of zinconium in high-pressure high-temperature water can be reduced or prevented by alloying additions.
Stress corrosion	Zirconium will undergo stress corrosion in aqueous $FeCl_3$, fluoride-bearing solutions and iodine vapour.

1.4.2 Methods of corrosion prevention of metals

1.4.2.1 Introduction

Since many metals do not have adequate corrosion resistance to many environments, protective coatings are often applied. Coatings may be polymeric, ceramic or metallic. Metallic coatings can be applied by electrolytic or chemical processes and by hot-dip, diffusion and spray processes. Corrosion of metals can also be prevented by cathodic protection in some circumstances.

The advantages of metallic and polymeric (paint) coatings are compared in:

Table 1.4.18 *Characteristic advantages of metallic and paint coatings for the protection of metal components*

The various processes are described and their advantages, limitations and applications listed in:

Table 1.4.19 *Comparison of surface protection methods for metal components*

Steel corrodes rapidly in normal environments and there is generally a requirement for a corrosion-protective coating. Zinc is anodic to steel and is commonly used as a protective coating. It may be applied by galvanizing (hot-dip), sheradizing (diffusion), spraying, electroplating and painting (zinc-rich paints). Further details of the processes are given in sections 1.4.2.2–1.4.2.9.

The coating processes are compared in:

Table 1.4.20 *Comparison of zinc coatings on steel for corrosion prevention*

The relative lifetime costs of the various coating systems must take into account maintenance costs as well as the initial cost of the coating process (see Table 1.4.18 for a qualitative guide). Although paint coatings may initially be cheaper than metallic coatings, their lifetime costs can be much higher.

1.4.2.2 Paints and polymeric coatings

INTRODUCTION

In general, the corrosion protection offered by a paint or polymer-based coating is due to its barrier properties with respect to oxygen and vapour permeability, which are necessary for the metal corrosion mechanism. The coating must therefore be adequately thick and free from defects such as pin-holes. In some cases protection is enhanced by other forms of protection (e.g. phosphate pre-treatment, corrosion inhibiting pigmentation, etc.).

Paints and polymer-based coatings can be classified according to coating thickness, although there is considerable overlap and such distinction is blurred.

(a) Paints – coating thickness typically 20–50 μm per coat. However high-build paints and solventless paints are applicable at thicknesses of 200 μm per coat or more.

(b) Polymer-based protective coatings – intermediate coating thickness. This class includes organosols, plastisols and powder coatings.

(c) Plastic (organic) lining – coating thickness greater than 0.5 mm obtained either by multiple applications of organic linings such as coal tar or bituminous coatings, or in one operation by using prefabricated sheet called 'in-mould coating'. In many cases the coatings are polymer-based and similar to the powder coatings described in (b).

The decision whether to coat or to apply a lining material is frequently made on the basis of ease of application. Very large items or processing equipment, storage tanks, etc., are usually 'lined' rather than coated, since coating would involve a multicoat application (to build the required film thickness). However, with the development of new spraying techniques, e.g. hot spray, dual gun, electrostatic powder spray, airless spray, etc., and new 'high-build' coatings, the distinction between the two becomes even more blurred, particularly when chemically the same material is used.

A general guide to the advantages and limitations of paint systems, plastics coatings and lining materials is given in:
Table 1.4.21 *Comparative properties of paints and polymeric coating systems*

SELECTION OF PAINT AND POLYMERIC COATINGS

The selection of the optimum organic coating for a given component will only be achieved after consideration of several factors. These are:

(a) The environment which the component will encounter. Typical corrosive environments and their critical factors are:
Air – humidity, atmospheric pollutants, seawater environment
Soil – oxygen supply, water supply, resistivity, composition
Water – oxygen supply, electrical conductivity
Chemicals – pH, moisture content, chemical nature, concentration, bacteria, fungi, etc.
The temperature of the environment is an overriding critical factor and in addition abrasion and scratch resistance will often be required.
(b) The lifespan of the component to be produced and whether maintenance is a practical possibility.
(c) The possible coating methods and the feasibility of the necessary pre-treatments. Evidently these considerations are not independent as the successful performance of a coating for the requisite lifespan will depend on the correct method of application of the coating.
(d) Cost. Where more than one coating and/or coating process is technically possible and adequate, cost will dictate the best choice. The cost of a coating or lining is dependent upon the following interrelated factors:
    (i) Surface preparation – this cost will depend on the type and standard of surface preparation which will in turn depend on:
    – the type of component involved;
    – the accessibility of the component;
    – the properties of the coating or lining to be applied.
    (ii) Coating process – the cost will depend on the method of application adopted and the number of coats applied which will in turn depend on:
    – the size and shape of component;
    – the accessibility of the component;
    – the coating employed – its covering power, thickness per coat and coating material utilization.

The total cost of a coating or lining for a given component is given by:
total cost = cost of surface preparation + cost of coating process + cost of maintenance measured over the lifetime of the component.

PAINTS

Paints are polymeric coatings which employ pigments to give opacity, to absorb UV light, to colour and to act as corrosion inhibitors. The pigments used can be divided into two main categories: prime pigments which confer optical properties and comprise inorganic and organic pigments and extender pigments which confer mechanical and rheological properties. Both categories may comprise naturally occurring or synthetic materials.

Inorganic pigments (naturally occurring and synthetic)

Typical examples of these materials are: rutile titanium dioxide (the most widely used pigment), red or yellow iron oxides, and earth pigments (ochres).

Organic pigments

These include pigment dyestuffs (e.g. phthalocyanine blue or green), toner pigments (insolubilized dyes such as rubine toner) and lake pigments (absorbed dyes on mordents, such as crimson lake) used to give vivid colours in paints.

Extender pigments

These 'extenders' are used to modify the mechanical or rheological properties of a coating and to give film build. Certain extenders can be used to control pH (carbonates). They have negligible opacifying or colourant properties. Examples include barytes, whiting, chinaclay, talc, mica and silica as quartz or as fumed silica.

Effect pigments

Active anti-corrosion pigments that have leachable anti-corrosion salts, such as red lead, lead cyanamide, lead silicochromate, zinc chromate, zinc tetraoxichromate, dibasic lead phosphite, zinc phosphate, calcium phosphosilicate and basic calcium molybdate are included among effect pigments. The last three pigments are considered non-toxic; the other leachable pigments are toxic. Passive anti-corrosion pigments which improve barrier properties of a film are flake pigments such as micaceous iron oxide and aluminium flake. Zinc powders can provide cathodic protection in zinc-rich coatings or they are used as reactive pigments in zinc silicate coatings. 'Glamour' pigments such as pearlescent pigments and metallic pigments (aluminium or bronze flake for instance) are also included in this category, since they produce special optical effects, although aluminium flake is also used in reflective and anti-corrosion coatings.

The generic paint types are compared in:

Table 1.4.22 *Comparative properties of generic paint types*

The characteristics of non-convertible and convertible paints are given in:

Table 1.4.23 *Properties of non-convertible paints*, and

Table 1.4.24 *Properties of convertible paints*

The application methods and pre-treatment processes are outlined in:

Table 1.4.25 *Application methods for paints*

Table 1.4.26 *Painting pre-treatment processes and primers for metal*

POWDER COATINGS

Thermoplastic powder or plastic coatings are thermoplastic materials applied by a fluidized bed process, dipping or electrostatic spray. Their characteristics, applications and properties are given in:

Table 1.4.27 *Characteristics and applications of thermoplastic powder coatings and linings* and

Table 1.4.28 *Properties of plastic powder coatings*

Thermosetting powder coatings are thermosetting materials. They are applied by electrostatic spray on to cold components, followed by curing which melts and cross-links the thermoset. Average coating thickness is about 50 μm, compared to an average

25–30 μm thickness for paint films (except high- build types). The minimum thickness is 40 μm, which is attractive for some applications.

The characteristics, applications and properties of thermosetting powder coatings are given in:

Table 1.4.29 *Characteristics and applications of thermosetting powder coatings*, and
Table 1.4.30 *Properties of thermosetting powder coatings*

1.4.2.3 Vitreous enamels

Vitreous enamelling consists of the coating of a surface with a thin unbroken layer of glass. From the point of view of corrosion protection, it is imperative that the glass layer remains continuous. Any slight thermal or mechanical shock can result in hair cracks in the glass, and corrosion then proceeds rapidly. Other terms often used are porcelain enamelling, glass lining and glass coating. High temperature coatings and ceramic coatings are vitreous enamels only if they are bonded by a glass. Steels, cast irons, aluminium, copper, bronze, gold and silver may be enamelled.

ENAMELLING STEEL

Not all steels are suitable as enamelling stock. Low carbon content is preferred in order to prevent interaction between carbon and enamel which may lead to formation of gas blisters.

Suitable steel stock choices, in order of desirability, are:

1. Special enamelling steel with low carbon and manganese and with titanium (0.3%) and silicon (0.12%) present to combine with the carbon.
2. Very low carbon steel – carbon 0.002% max.
3. Enamelling iron – moderately low carbon, manganese 0.04% max., phosphorous 0.01% max. and sulphur 0.03% max.
4. Cold rolled rimming steel which, owing to its method of manufacture, has an outer skin purer than its centre. This property enhances enamelling quality.

Steel thicknesses for typical enamel ware applications are given in:
Table 1.4.31 *Steel thicknesses for enamel ware*

The recommended maximum area for a given gauge and width is given in:
Table 1.4.32 *Maximum area for given width of enamel ware*

ENAMELLING CAST IRON

Consideration of the many factors involved in producing defect-free enamels on cast iron indicates that the following iron composition is likely to be most suitable:
Total carbon 3.4% max., Si 2.6%, S 0.08% max., Mn 0.5–0.6%, P dependent upon pig iron used. Changes in silicon can be made if the carbon is altered to preserve the required structure. (Decreasing carbon by 0.1% requires increase of silicon by 0.3%.) Iron castings must be clean and heavily shot blasted before enamelling. If a white iron shot of high combined carbon is used, there is a risk of the enamel blistering if any of the shot becomes embedded. This can occur in both steel sheet and castings.

A thin film of rust following cleaning or the phosphate film left behind as a result of phosphoric acid pickling, are both considered to be advantageous in promoting adhesion.

ENAMELLING ALUMINIUM ALLOYS

The vitreous enamelling of aluminium alloys requires the use of special low softening point enamels. Although lead enamels were originally used, these have been replaced by alkali and alkaline-earth silicate enamels to avoid the toxicity associated with lead. Since enamelling is a relatively high temperature process, account of this must be taken in consideration of the mechanical properties of the aluminium alloy after enamelling. Enamelling is therefore difficult for heat treatable alloys or for those required in the cold worked condition. Aluminium alloys must have good, clean surfaces before enamelling.

ENAMELLING COPPER

Copper requires special enamels based mainly on lead silicate glasses. Undercoats are required so that colour which may be caused by reaction with the metal does not show. Copper should be chemically cleaned before enamelling.

PHYSICAL AND CHEMICAL PROPERTIES OF VITREOUS ENAMELLED METALS

General corrosion resistance

The resistance of good quality vitreous enamels on iron or steel is excellent. Enamelled signs withstand severe weathering over very long periods. Vitreous enamel parts are also resistant to corrosion when buried in soil.

Acid resistance and toxicity

Most vitreous enamels are resistant to the common mineral acids, but not to hydrofluoric acid which dissolves glass. They may also be attacked by citric acid (lemon juice). For use in contact with foods it is important that poisonous constituents like lead, cadmium and antimony should not be present in the enamel. A standard acid resistance test for vitreous enamels is ASTM C282–53.

Alkali resistance

Vitreous enamels on iron and steel are usually very resistant to alkalis. A standard test consists of exposure to 5% hydrated tetrasodium pyrophosphate at 90°C for 2 h. Loss of weight in mg from a 3.5 inch square panel is called the alkali index.

Electrical resistance

Vitreous enamels are good electrical insulators at room temperature. They find application in sealing resistor wires. For high quality electrical resistors, enamels of low alkali content are preferred. Enamel on steel will resist electrical breakdown up to 14–19 V/μm. Maximum service temperature is 315–540°C depending upon enamel type.

Colour

Almost any colour can be given to vitreous enamels. The incorporation of insoluble mill additions of high refractive index like titania gives opacity to the enamel. In this way good whites can be obtained with diffuse reflectance properties similar to those of magnesia.

Emissivity

By the incorporation of ceramics like fused silica or magnesia low emissivity coatings can be applied to metals. By incorporating green chromic oxide as a mill addition very high emissivity coatings can be applied. Enamels thus provide a method of controlling total emissivity for operation up to about 900°C.

Oxidation resistance

Mild steel can be protected from oxidation in air at 700°C by vitreous enamelling. A very effective enamel has the following frit composition:
Quartz 31.4%, zinc oxide 4.1%, barium carbonate 47.0%, calcium carbonate 5.9%, boric acid 9.5%, beryllium oxide 2.1%.

Thermal expansion

The coefficient of expansion of enamels varies widely according to composition. Enamels must be formulated so that the expansion is a good fit to the substrate being enamelled.

Thermal conductivity

Varies with composition. Typical ranges are 0.71–1.17 kW/m per K at 0°C and 0.76–1.36 kW/m per K at 100°C.

Thermal shock

Vitreous enamels on steel have good thermal shock resistance. This property has promoted the application of enamelled cooking utensils.

A test often applied is to heat to about 250°C and quench in water, and then to increase the heating temperature by some 15°C in successive cycles until failure occurs. The thermal shock resistance is given as the number of these cycles required to cause failure.

The oxidation resistance imparted by enamelling is compared with other methods in:
Fig. 1.4.1 *Oxidation of protected mild steel at 700°C in air*

MECHANICAL PROPERTIES OF VITREOUS ENAMELS ON METALS

Hardness

Vitreous enamels fall between 5 and 6 on the Mohs scale, that is harder than bottle glass but not as hard as stoneware or quartz. On the Brinell scale they fall between 600 and 700.

Impact resistance

Usually measured by allowing a heavy ball to fall a standard distance down a guide tube onto the surface. Values (averaged from a number of tests) are comparative and should be checked against acceptable standards.

Rigidity

The modulus of elasticity of most enamels is about 27% of the modulus of elasticity of steel. If vitreous enamel is coated on the compression face of a steel, the steel gauge can be lessened to give the same deflection for the same load as was present in the uncoated steel. However the combined thickness of enamel and steel will be greater than that of the unenamelled steel for these constant conditions. Using 175 μm of enamel in a design needing 437 μm thick steel uncoated, the steel can be thinned by 14% for enamelling to give the same rigidity. Again with 175 μm of enamel thickness but with a thicker requirement of 1.3 mm steel the saving on steel thickness is only 4%

Fatigue strength

Fatigue strength of steel is increased by vitreous enamelling, probably owing to pre-stressing in the application of the enamel and perhaps also to the bridging of stress raisers.

1.4.2.4 Electroplating

Almost all commercial electroplating processes at present are carried out in aqueous solutions of a suitable metal salt by passing a direct current through the solution with the article to be treated cathodic to an anode of the same metal or of an inert metal. The depositing metal forms on the substrate article by cathodic reduction of its ions in the electrolyte.

The more common processes used to deposit the metals, the ease and relative cost of producing the coating, the corrosion resistant behaviour of the deposit, and other relevant engineering properties are given in:

Table 1.4.33 *Common corrosion resistant electroplates, their processing baths, properties and applications*

Pre-treatment of the substrate is essential for successful electroplating. The ease of substrate preparation is shown in:

Table 1.4.34 *Ease of pre-treatment and preparation for plating of various base metals*

PROCESSING COST

This parameter is a complex one depending on a number of factors, e.g. basic cost of the deposit metal, thickness of deposit and density of metal (i.e. cost per μm of deposit metal), ease of handling and stability of electrolyte, size and shape of substrate articles, possibility of barrel plating or large batch vat plating for low unit plating costs, surface finish of base metal required before plating and ease of pre-treatment of substrate material, (see Table 1.4.33).

If a decorative finish is required for a mild environment several possibilities are available (bright zinc, nickel plus chromium, gold, satin nickel, etc.), and relative costs can be calculated on the above factors. However, should the application demand the unique properties of gold then cost estimation is a much more limited exercise, the objective of which is how economically can the necessary gold plate be produced.

Often overlooked, one way of reducing cost is to produce parts from plated stock where available, instead of plating a fabricated part.

CORROSION RESISTANCE BEHAVIOUR OF PROTECTIVE ELECTROPLATES

Generally electroplates protect the base metal in one of two ways: sacrificial corrosion or barrier protection.

Sacrificial corrosion

The electroplate is more susceptible to corrosion than the base metal so that when environmental contact is made between the two metals via pores and other defects in the plate the galvanic couple set up accelerates the corrosion of the coating rather than the base. In this case, the effective life of the protective system is dependent on the rate of corrosion of the coating and its thickness.

Barrier protection

The electroplate is fairly resistant to the expected environment and protects by shielding the base metal from corrosion. Since the coating is more noble than the base metal the success of the system depends upon complete continuity of the coating to avoid accelerated corrosion of the substrate. Sufficient thickness is therefore the important criterion for corrosion protection by this mode. The deposit must also be free from stress-cracking which might allow atmospheric contact with the base metal and severe local corrosion. In some cases a deliberately produced micro-cracked or microporous deposit (e.g. Cr) to reduce localized attack on the underlying metal is used.

SELECTION OF ELECTROLYTE

Once the choice has been made between the various candidate metals for a protective electroplate a second choice is often presented to the processor; that of the electrolyte to be used. Some metals, e.g. copper, cadmium, tin, gold, nickel can be plated from two or more commercially available processes which may have quite different solution parameters (pH, valency of the metal in solution, etc.), operating parameters (usable range of current density, temperature, throwing power) and deposit properties.

Solution pH

Several substrates, e.g. ferrous and zinc alloys, are readily attacked in acid solutions of more noble metals causing replacement reactions in which loosely adherent, powdery immersion deposits of the more noble metals are formed on the substrates. Electroplating on this surface provides poor quality coatings. It is necessary in these cases to plate initially from complexed solutions (usually alkaline, e.g. cyanide, pyrophosphate, stannate) of the deposit metal in which the concentration of free metal ion is very low so that the replacement reaction takes place only very slowly or not at all. When such a continuous plate has been formed, further plating can be carried out in a faster, acid bath if required. Some acid zinc and acid alloy zinc electrolytes are very corrosive, and are unsuitable for electroplating hollow tubes.

Chemical valency of the deposit metal in the electrolyte

The amount of electrical charge required to produce a given amount of deposit is directly proportional to the valency state in which the metal exists in solution. Thus, twice as much current, or twice the deposition time, is required to produce the same weight of tin plate from an alkaline stannate bath as from an acid fluoroborate bath. The reverse is true for alkaline cyanide and acid sulphate copper electrolytes.

Practical current densities and temperatures

Costs of plating are adversely affected by low practical current densities and high operating temperatures, often with special temperature control and agitation of the electrolyte. Where the effect of these parameters on overall cost is significant, the differences exhibited by different electrolytes should be carefully considered.

Throwing power

This property of electroplating baths is all important when coating intricate shapes. Whilst a plater can go a long way towards ensuring uniform current distribution on a cathode by careful jigging and matching of cathode and anode shapes, the eventual deposit thickness distribution over the cathode is determined by the solution throwing power. This parameter is a measure of the ease with which a uniformly thick deposit can be obtained on all parts of a substrate surface, and varies markedly from bath to bath. As a general guide, when processing complex shaped articles only electroplating baths of good throwing power should be used. The throwing power of common electrolytes is shown in:

Table 1.4.35 *Throwing power of common electrolytes*

1.4.2.5 Chemical deposition coatings

Chemical deposition coatings, applied without the use of applied current, are of two types, immersion replacement and autocatalytic (electroless). Their characteristics are shown in:
Table 1.4.36 *Characteristics of chemical plating processes*

IMMERSION REPLACEMENT

The substrate metal atoms at the surface are replaced by deposited ions from solution, which are reduced at the substrate surface due to the difference in electrode potential of the two metal/ion systems.

The electrochemical series of metal/ion systems is a useful guide to the immersion behaviour of various metals in different electrolytes and shows that the electrode potential decreases and the ease of formation of ions of metals increases as follows:

Au, Hg, Ag, Cu, Pb, Sn, Ni, Cd, Fe, Zn, Al

In immersion plating, deposition continues until an effectively continuous coating has formed so that no further ion displacement can occur. Deposits are therefore thin (<0.1 μm), and are normally only of use in mildly aggressive environments for mainly decorative purposes.

AUTOCATALYTIC (ELECTROLESS)

In this process metal ions are reduced at a catalytic substrate surface by the action of a reducing agent present in the solution. Once started the reaction can continue (autocatalysis) and there is no theoretical limit to the thickness of coating, since the continuance of deposition is not dependent on the participation of the substrate in the reaction. The main systems in commercial use at present are nickel and copper.

Pure gold is claimed to have been auto-catalytically deposited from a bath using sodium borohydride as the reducing agent. There are now some commercial baths available, but limited success has been reported in use.

Ferromagnetic alloys Co–P, Ni–Co–P, Ni–Fe–P have also been auto-catalytically plated on magnetic tapes.

1.4.2.6 Chemical conversion coatings

The three main processes used to produce corrosion protective conversion films on metals are:

- anodizing
- phosphating
- chromating

The essential difference between this class of coating and electrodeposited and chemically deposited metal coatings is that in a conversion coating the substrate is an integral part of the protective non-metallic film which is formed by a chemical (electrochemical in the case of anodizing) reaction of the substrate metal with the process solution. The characteristics of chemical conversion coatings are given in:
Table 1.4.37 *Characteristics of chemical conversion coatings for metals*

ANODIZING

This is a protective system used primarily for aluminium and its alloys, although zinc and magnesium are also known to form satisfactory anodized films. Anodizing is also referred to as the passivation of aluminium, in which a thick oxide coating is produced. The coating remains passive, or more cathodic, after the voltage has been removed. However, there is the danger that the metal may enter the transpassive region, if the potential of the anode is too high. In this instance the passive film breaks down and the high corrosion current causes rapid pitting.

Decorative anodizing is very dependent on the aluminium alloy being processed and close attention should be paid to the alloy composition and its suitability for protective decorative or bright anodizing. An approximate guide is given in:
Table 1.4.38 *Suitability of aluminium alloys for sulphuric acid anodizing*

PHOSPHATING

Corrosion resistant applications of phosphating are restricted largely to iron and steel. When used as an undercoat for paint to improve bonding, phosphating is extended to other metals such as aluminium, cadmium and zinc. The phosphate films are less effective for corrosion protection than chromating or anodizing and are used primarily as temporary protectives. Unsealed phosphate coatings (i.e. no oil or paint) do not provide substantial corrosion resistance.

CHROMATING

The corrosion resistant character of a chromate film is due to (a) a non-porous structure acting as a moisture barrier and (b) the inhibiting properties of chromate on corrosion of substrate metal should penetration of the film occur. The corrosion protection afforded by chromating is superior to that obtained with phosphating, and, for aluminium, can approach that achieved with some anodic finishes.

Metals protected by chromating include aluminium, cadmium, copper, iron, magnesium, steel, tin and zinc. Although substrate dissolution in a soluble chromate solution is the first stage in each case, the character of the coatings varies from metal to metal and the films on aluminium, for instance, contain considerably more base metal than chromate films on zinc. Steel workpieces are normally zinc or cadmium plated before chromating.

1.4.2.7 Hot dip coatings (including galvanizing)

Hot dipping is a process for coating substrates, mainly ferrous metals, with low melting point metals (zinc, tin, aluminium and to a limited extent lead) from the molten state by immersion of the substrates in the liquid metal in either batch or continuous mode. Good adhesion is achieved by the formation of an intermediate alloy layer between substrate and coating metal. This alloy layer must be carefully controlled so that its brittle intermetallic constituents do not have a significantly deterimental effect on the mechanical properties of the coating. In the process of hot dipping, the pure coating metals are softer than the substrate metals. This explains why such a coating method is rarely suitable for applications in which resistance to wear is required. Despite the pure coating metals being soft, alloying with the substrate metals increases hardness and reduces ductility. The alloy layers are known to exhibit corrosion resistances different from those of pure metals.

The structure of the hot dipped coating is dependent on:

- immersion time. Long times increase the alloy thickness because of heavier attack on the steel.
- rate of withdrawal of the workpiece. The faster the workpiece leaves the bath, the thicker is the final pure metal coating.
- heat capacity of the workpiece. The slower the post-withdrawal cooling rate the

more chance there is of further alloying occurring. Heavy articles are sometimes quenched immediately after hot dipping to minimize this tendency.

Typical processing parameters, properties and applications of hot dipped coatings are given in:

Table 1.4.39 *Hot dip processes, coating properties and applications*

GALVANIZING

Galvanizing is the term applied to coating steel with zinc by a hot dipping process. Galvanized steel is used for many structural applications. Galvanizing can protect steel from corrosion for up to 40 years in non-aggressive environments. The life of galvanized steel depends on the coating thickness and the environment, shown in:

Fig. 1.4.3 *Typical lives of galvanized zinc coatings*

A recommended paint coating adds about 8–15 years to the total life of the coating. For decorative purposes repainting is needed after 8–12 years. A guide to painting galvanized steel is given in:

Fig. 1.4.3 *Guide to painting galvanized steel*

1.4.2.8 Diffusion coatings

The coating element is in the form of, or is converted to, a volatile chemical compound (usually a halide) which on heating reacts with the surface of the substrate metal. In this reaction the volatile halide dissociates and deposits metal which alloys with base metal with inward diffusion. The process rate is mainly diffusion controlled. Thus, similar to hot dipped coatings, a typical diffusion coating consists of an outer pure or highly concentrated coating metal with the interior alloyed to a varying degree with the substrate, and integrally bonded to it.

The main members of the diffusion coating family – apart from the special case of carburizing of steel – are aluminizing (calorizing), siliconizing, sherardizing and chromizing. Other diffusion coatings produced include iron–aluminium, nickel–aluminium on copper and steel, nickel and cobalt alloys, chromium–aluminium and chromium–silicon coatings for sulphidation resistance and good mechanical properties, metal silicide coatings on refractory metals, titanium carbide and silicon carbide on steel alloys, and boride coatings. These coatings are more widely used on ferrous materials.

The characteristics of the main processes are given in:

Table 1.4.40 *Characteristics of some important diffusion coating processes*

1.4.2.9 Sprayed coatings

Corrosion protective coatings can be applied by spraying molten metal (usually aluminium or zinc) on to most metal substrates. The coating is always porous and therefore must be applied in conjunction with primers and sealants. Zinc spraying is competitive with galvanizing for one-off applications, specialized areas and thicker coatings for longterm applications. The recommended zinc and aluminium sprayed coatings for several corrosion environments are given in:

Table 1.4.41 *Sprayed coatings for corrosion resistance*

Sprayed aluminium and nickel–chromium alloys are also used to protect steel in elevated temperature environments. A guide is given in:

Table 1.4.42 *Sprayed coatings for the protection of iron and steel in elevated temperature environments (BS 2569)*

1.4.2.10 Cathodic protection

The corrosion reaction which leads to metal loss in electrolytes is the discharge of positively charged metal ions into solution. By making the metal electrically negative (cathodic) with respect to the solution this reaction can be slowed down, and in some cases stopped altogether. Cathodic protection consists of using this effect to protect metal structures, the metal protected almost invariably being iron or steel. This is the only method of corrosion control which can prevent corrosion completely or can be controlled to achieve a required level of protection. The two main types of cathodic protection are impressed current and sacrificial anodes. Their characteristics are compared in:

Table 1.4.43 *Relative merits of cathodic protection systems*

In order to reduce the current drawn from an impressed current system or to reduce the rate of consumption of sacrificial anodes, it is generally worthwhile coating the objects to be protected, using the cathodic protection system to maintain the corrosion resistance at breaks in the coating. An elegant example of this technique is galvanized steel where the zinc acts both as a protective coating and as a sacrificial anode to protect the steel at breaks in the coating.

IMPRESSED CURRENT

Impressed current systems obtain the current from a power supply operating off mains electricity. The positive connection is made to the solution (or soil) via an 'anode' of a relatively inert material such as graphite (the positive voltage on the anode would give rapid attack on most metals, since the discharge of positive metal ions would be accelerated). Platinum, platinum-sheathed titanium, tantalum, niobium, lead–silver alloys and silicon–iron alloys are also employed in this capacity. Frequently anodes are surrounded with protective shields to avoid too high cathodic current densities in the immediate vicinity of the anodes; high current densities damage usual protective coatings, particularly in marine service (e.g. shipbottom paint systems). A common shield is made from cold bonded neoprene rubber.

Comparative properties of anodes are summarized in:

Table 1.4.44 *Comparison of anode materials for impressed current systems*

SACRIFICIAL ANODES

When materials such as zinc, aluminium and magnesium are immersed in water in electrical contact with iron they are preferentially corroded. The iron is protected at the expense of increased corrosion of the sacrificial element.

Economic considerations in choosing the metal for a galvanic anode are based on:

(i) Theoretical amount of current generated per unit weight of anode corroded.
(ii) Efficiency, i.e. useful current generated in external circuit.

Properties of galvanic anodes are summarized in:

Table 1.4.45 *Comparison of sacrificial anodes*

1.4.2.11 Protection of steel by wrapping tapes and sleeves

Cathodic protection is often combined with protective wrappings. In this way, the protective current can concentrate on any defects in the coating, existing or at an early stage of development and treat such areas with a small amount of current. This provides a more economical approach to protection.

Wrapping with adhesive tapes protects ferrous metals, particularly pipelines, joints, valves and other fittings by excluding the environment. For further protection against accidental damage and to promote adhesion, thorough cleaning and inhibitive priming is recommended.

Three types of wrapping are commonly available:

(i) Petroleum jelly tapes:	Natural or synthetic fibre fabric or glass cloth impregnated with petroleum jelly and neutral mineral filler. Coating is permantly plastic, suitable for application to irregular profiles. If used above ground and subject to abrasive wear, an overwrap of bituminous tape should be used. Also suitable as insulation to avoid bimetallic contacts.
(ii) Synthetic resin or plastic tapes:	Polyethylene or PVC strips of thickness 150–250 μm, coated with synthetic rubber adhesive on one side. Used in damp or dirty conditions as insulation to avoid bimetallic contacts. PVC is usually supplied black to minimize UV degradation.
(iii) Coal tar and bitumen tapes:	Used for buried pipelines. High moisture resistance and good adhesion to steel. Fabric reinforcement normally glass fibre. Various different grades are available to suit differing environmental service conditions.

1.4.2.12 Protection of steel by cement and allied products

Cement mortar linings are widely used for the internal protection of water mains. Special formulations and coating procedures are used. They have limited impact resistance but may be repaired on site by fresh applications.

Exposed steel may be covered with gypsum plaster and magnesium oxychloride cements, but it should first be coated with a suitable bitumen coating that is resistant to water penetration while the plaster or cement is curing.

It is important to note that it can be dangerous to add calcium chloride to accelerate cement hardening, as the chloride ions initiate corrosion. In order to protect the steel, galvanizing coupled with an adequate covering of dense concrete should give the performance required.

Table 1.4.18 Characteristic advantages of metallic and paint coatings for the protection of metal

Metallic coatings	***Paint/polymer coatings***
Predictable life.	Ease of application in factory or on site.[a,b]
Single application system (unless over-painting specified).	Wide availability of painting facilities.[a,b]
No drying time needed.	No effect on mechanical properties of steel substrate.[c]
Protection of damaged areas by sacrificial (cathodic) protection – also true of zinc rich paints.	Easy repair of damage to coatings.[a]
	Wide range of colours available for aesthetic purposes.[d]
Good abrasion resistance.	Thermally insulating.
Additional benefits of hot dip coatings	Can be used to protect wood, stone and concrete installations.
Good adhesion.	
Coating thickness unaffected by contours.	
Major faults easily visible.	
Additional benefits of sprayed coatings	
Can be applied on site. No size limit.	
Thickness of coating built up as desired.	

[a] Not true for high temperature cure paints and polymeric coatings.
[b] Not true for polymeric coatings.
[c] Not necessarily true for high temperature cure paints and polymeric coatings.
[d] Some polymeric coatings very limited on colour range.

Table 1.4.19

Table 1.4.19 Comparison of surface protection methods for metal components

Surface protection method		Description of process	Section	Materials used	Advantages	Limitations	Typical application areas	Relative costs
Paints and polymerized coatings.		Inert coating preventing contact of substrate with environment. Thicknesses range from 5 μm (paint) to 0.5 mm (powder).	1.4.2.2	Non-convertible coatings. Convertible coatings. Thermoplastic powders. Thermosetting powders.	Ease of application. Paints easy to maintain. Wide range of materials to suit many situations. Pitches and bituminous coatings are cheap and can provide effective protection.	Relatively soft and thus susceptible to mechanical damage. Many require curing.	Exterior structures and decorative applications. Convertible paints used for continuous exposure to corrosive fluids.	Range from low to moderate.
Vitreous enamelling		Application of ceramic coating to metal surface subsequently fired on to form glassy fused coating.	1.4.2.3	Glasses on steel, cast iron, copper, aluminium.	Complete protection of substrate. Easy to clean, good appearance. Heat and acid resistant. Electrically insulating.	Substrate must be resistant to fusing temperature. Cracking and chipping under abrasion or heavy deformation. Avoid sharp edges in the design for enamelling otherwise there can be difficulty in obtaining a uniform coating.	Industrial, e.g. some chemical plant. Domestic and decorative items.	High.
Electroplating		Electrolytic reduction of metal ions at the cathode in an aqueous solution by the passage of current through the solution from an outside source.	1.4.2.4, 1.4.2.5	Principally Zn, Cd, Cr, Ni, Cu, Sn, Ag, Au.	Many variations of metal coatings and electrolytes available. Low heating costs. Batch or continuous application.	Local coating thickness depends on plating bath and component design. Bonding not as strong as in hot dip or diffusion coatings.	Exterior use. Marine environments. Limited use in corrosive media. Limited use on Al for electrical purposes.	Moderate depending on metal cost.
Chemical deposition (Electroless auto-catalytic)		Autocatalytic chemical reaction of metal ions at a catalytic surface from an aqueous solution. The deposited metal is also catalytic, hence the process is self-perpetuating.	1.4.2.6	Principally Ni, also Cu, Au, Co, others.	Excellent throwing power gives uniform coatings on complex shapes.	Limited range of materials.	Aggressive environments in conjunction with wear.	High.
Conversion coatings	Anodizing	Anodic oxidation of substrate in aqueous acid medium under prescribed voltage/current density conditions.	1.4.2.7	Used on Al (mainly); also Mg, Zn, Ti.	Wear resistant and durable. Can be coloured. Thick coatings up to 25 μm possible. Room temperature process. Very adherent (integral).	Close attention needs to be paid to alloy composition. Limited range of substrates.	Aluminium alloy cladding for buildings, decorative items, hard coatings and electrical insulation.	Medium to high anodizing > chromating > phosphating.
	Phosphating	Chemical interaction with phosphate-containing aqueous solution to form mixed phosphate conversion coating.		Steel, Zn, Cd, Al substrates.	Excellent bond to subsequent paints, etc. Batch or continuous process.	Requires very clean, oil-free substrate prior to coating, then protection of coat from mechanical damage by painting.	Temporary corrosion protection only; topcoat necessary. Also post-treatment after Zn, Cd, electroplating.	Medium to low.
	Chromating	Chemical interaction of substrate with chromate-containing aqueous solution to form mixed chromate conversion coating.		Steel, Zn, Cd, Al substrates.	Excellent bond to subsequent paints, etc. Good corrosion resistance.		Temporary corrosion protection but excellent pre-painting treatment for Al.	Low.

Table 1.4.19 Comparison of surface protection methods for metal components – *continued*

Surface protection method	*Description of process*	*Section*	*Materials used*	*Advantages*	*Limitations*	*Typical application areas*	*Relative costs*
Hot-dip coatings	Immersion of workpiece in molten coating metal. Interfacial alloying occurs and the coating is solidified in a cold air stream after removal from bath.	1.4.2.8	Zn (galvanizing), Sn, Al.	Good metallurgical bond. Rapid. Batch or continuous.	High energy costs. Less uniform than electroplating. Limited to low melting point metals. Poor control of thickness; minimum thickness less with electroplating.	Exterior use. Al on steel for high temperature uses.	Low.
Diffusion coatings	Alteration of the chemical composition and properties of the substrate surface by diffusion of elements from gaseous, liquid or solid medium.	1.4.2.9	Al (aluminizing) on steels, nickel and cobalt alloys. Cr (chromatizing) on steels. Si (siliconizing) on steels. Zn (sherardizing) on steels. C (carburizing) on steels.	Good metallurgical bond.	Slow. High energy costs. Cannot be used for continuous processing.	Industrial, marine and hot corrosive environments.	Moderate to high.
Sprayed coatings	Molten metals sprayed on to substrate.	1.4.2.10	Most metals and alloys. Zn or Al or Al–Zn alloy on steel.	Used in conjunction with primer and sealant. Zn spraying competitive with galvanizing for one-off applications with thicker coatings attainable. Single articles.	One-off, labour intensive process. Complex items difficult to coat as 'line of sight' process.	Specialized areas, often for longterm protection, e.g. large structures.	Low but higher than galvanizing (Zn) particularly in volume applications.

Table 1.4.19 – *continued*

Table 1.4.20 Comparison of zinc coatings on steel for corrosion prevention

Coating type	*Application areas*	*Adhesion*	*Continuity and uniformity*	*Thickness*	*Mechanical properties*	*After treatments*	*Other considerations*	*Section*
Galvanizing (hot dip)	Suitable for fabricated sheet metalwork, relatively large structures and small articles where the finish or machining is not to close limits.	Good. The zinc coating is alloyed with the basis metal.	Generally uniform. Some excess zinc at drainage points.	Generally: 45–125 μm on products; 25 μm on sheet. Thicker coatings possible.	Conventional coatings applied to finished articles, not formable. Alloy layer is abrasion-resistant but brittle on bending. Galvanized sheet has little or no alloy layer and is readily formable. Higher process temperatures than sherardizing can result in loss of mechanical properties.	Etching treatment or weathering required before painting. Some chlorinated rubbers, polyurethane, isocyanate and polyamide–epoxy paints may be directly applied. Chromate conversion coatings used to prevent wet storage stain.	Inexpensive Not always satisfactory for machined, threaded or mating parts because of the thick and uneven coating which has to be removed from the threads by retapping. Thin gauge material liable to distort during processing. Limited by bath size – specialized works can handle parts up to 30 m long and fabrications 18×2×5 m. Continuous galvanized wire and strip available.	1.4.2.7
Sherardizing	Suitable for bolts, nuts, threaded articles, small pressings, machine castings, forgings, springs, moulded sections and tubing. NOT suitable for soldered assemblies, although welded and brazed assemblies can be sherardized.	Good. The zinc coating is alloyed with the basis metal.	Continuous and very uniform even on threaded and irregular shaped parts.	Variable at will, 15–40 μm. Thicker coatings possible (up to 75 μm for very severe exposure).	Usually applied to finished articles. The thinner coatings can be formed but heavy coatings can be brittle on bending. Excellent abrasion resistance, hardest of all zinc coatings.	Good key for paint. Phosphate and chromate conversion coatings used to prevent wet storage stain.	Inexpensive, even for thick deposits. No danger of heat distortion. Lower temperatures than galvanizing. Limited by size of containers – generally used for fairly small complex components. Useful when close control of tolerance important. Uneconomical for large bulky items, tanks, etc. Narrow parts up to 6–7 m long can be treated. Lacks eye appeal of electroplate (matt light grey).	1.4.2.8
Spraying	Suitable for large units and assemblies, either in component or assembled form. No size limit. Not suitable for intricate shapes or thin sheet. (Also fasteners when done after fabrication). An extremely good base for final paint coatings.	Good mechanical interlocking provided the abrasive grit blasting pre-treatment is done correctly.	Depends on operator skill.	Variable, coating thickness is built up by controlling the speed of movement of the spray gun and its distance relative to the surface to be coated, also by the number of successive spraying passes. Generally 100–200 μ m but can be applied thicker.	Usually applied to finished articles. Not highly abrasion resistant (soft coating). Thin sheet buckles during preparatory shot- blasting.		Cheaper than galvanizing or sherardizing for heavy compact parts. Attractive light grey.	1.4.2.9

Table 1.4.20

Table 1.4.20 Comparison of zinc coatings on steel for corrosion prevention – *continued*

Coating type	*Application areas*	*Adhesion*	*Continuity and uniformity*	*Thickness*	*Mechanical properties*	*After treatments*	*Other considerations*	*Section*
Electroplating	Suitable for small pressings, threaded and machined parts. Also for continuous sheet or wire.	Good. Comparable with other electroplated coatings.	Uniform within limitations of 'throwing power' of plating bath.	Variable at will generally 2–25 μm. Thicker coatings possible but generally uneconomic.	Can be formed if required but small parts usually finished before plating.	Treatment required before painting. Chromate conversion coatings used to prevent wet storage stain.	Not suitable for tubular interiors or deeply recessed articles. Barrelled coatings (2.5– 5 μm) cheap, but offer little protection to outdoor exposure, abrasion or handling. Pre-treatment required before painting. Attractive, bright appearance. Alloy zinc coatings have been developed which can give a greatly enhanced corrosion performance over pure zinc, e.g. zinc cobalt, zinc iron, zinc nickel.	1.4.2.4
Painting (zinc-rich paints)	Suitable for erected structures, site jobs and patch repair of galvanized sheet. No size limitations. Dip/spin/dip/spin coatings suitable for small components.	Good. Abrasive grit-blasting of the steel gives best results.	Depends on operator skill. Reasonably uniform.	Up to 40 μm per coat; more with special formulations. Over-painting easy.	Painted sheet can be formed. Harder than simple paint film, reasonably abrasion and chip resistant. Unsatisfactory for normal wear and tear.	Can be used alone, or as a primer under conventional paints.	Can be applied by brush, spray or dipping without technical experience. Very useful for articles too large for galvanizing, sherardizing or electroplating. No distortion. Expensive as normal production finish.	1.4.2.2

Table 1.4.21 Comparative properties of paints and polymeric coating systems

Coating type	*Sub-group*	*Characteristics*	*Advantages*	*Limitations*	*Examples*
Paints	Non-convertible	Dry by release of solvent. No other change occurs in chemical structure.	Quick-drying	Short-term solution to corrosion problems. Solvent non- resistant. Normally applied to weathered or wire-brushed surfaces. Can be applied to wire-blasted surfaces.	Bituminous vinyl resins Cellulosics Acrylic lacquers Latex paints
	Convertible	Change chemically after release of solvent by reacting with air (autoxidation) or by reaction with a chemical initiator such as an organic peroxide. They are nearly always applied to a blast-cleaned surface. Most of the coatings are based on thermosetting resins.	Most high performance coatings fall into this group. Greater solvent resistance. Better recoatability and higher film build. Much longer life than non-convertibles and are frequently left maintenance-free for long periods. Consequently the cost is nearly always less over a period of time than with a non-convertible coating.	High cost. Two-pack systems can be wasteful as limited pot life once mixed. Intolerable to dirt and millscale.	Alkyds Modified alkyds (e.g. vinyl toluene, styrenated, silicone- modified) Epoxy resins Amino resins (e.g. melamine, urea and benzoguanamine resins) Isocyanates Blocked isocyanates (polyurethanes) Unsaturated polyesters Phenolics Silicones Siloxanes Alkyl silicates Thermosetting acrylics
	Zinc-rich paints	Zinc can be in an organic (e.g. silicone) or inorganic (e.g. silicate) binder.	Good corrosion resistance at high zinc loadings. Good temperature resistance.	Limited colour range (sealer coats available to improve appearance). Exposure of freshly applied zinc silicate can be deleterious.	Organic (e.g. two-pack epoxy or silicone resin binder) Inorganic (e.g. silicate binder)
	Grease paints	Coatings based upon greases have two uses: (i) permanent non-curing coatings for application on the inside of box sections; (ii) temporary protectives.	High corrosion resistance and self-healing properties.	Coatings very fragile and should only be used where no mechanical damage can occur.	Mixed oil and wax compositions containing nitrohydrocarbons for anti-corrosion properties.
	Water-based paints	Solutions and water dilutable resins, emulsions and latex dispersions in water.	Lower VOC emissions, more environmentally friendly products. Lower viscosity materials enables use of high molecular weight polymers.	Difficult to produce high gloss finishes. Require biocides in formulations. Susceptible to freezing.	Aqueous dispersion paints based on various types of latex polymers including halogenated materials. Emulsions of epoxy or acrylic based paints. Industrial two-pack polyisocyanate paints. Polyester emulsion based paints.
	Non-aqueous dispersions (NADs)	Polymer particles dispersed in aliphatic hydrocarbon non-solvent (non- aqueous analogue to latexes)	Ability to have very high molecular weight polymeric film-forming material without paying penality of very high viscosity.	Formulations restricted by high non-solvent content. Materials must be baked to film-form.	Acrylic copolymer. High gloss and durability acrylic lacquer car finishes.
Powder coatings	Thermoplastic[a] ('plastic coatings')	Very much thicker coatings than paints (500 μm powder, 6.5 mm plastisol). Relative cost to paints is dependent upon material. Dry application.	Better product utilization. Environmentally friendly. Relatively tough coating finish. Good corrosion protection. Coloured thick films readily obtainable. Overall production costs can be low. Coating plant expenditure up to 40% less than with paint.	Material is applied to all surfaces of the article which can be wasteful. Problems in post-coating fabrication will be encountered. Fairly limited range of materials. New equipment required if changing from paint. Heat-curing necessary (constraints on section changes). Difficult to coat large heavy articles.	PVC Nylon 11 or 12 Chlorinated polyether Polyethylene Fluoroplastics (Polypropylene)
	Thermosetting	Applied by electrostatic spraying or fludized bed coating. Average thickness 50 μm. Dry application. Very good materials utilization.	Intricate shapes can be coated at low cost. Good corrosion protection. Better finish than thermoplastic. Greater film strength in relatively thin coatings.	Films thinner than about 25–30 μm unobtainable. Films generally have some 'orange peel' effect and not quite up to 'solution' paint standard. Faraday-cage problems with electrostatic spray application of hollow components.	Epoxy Polyester Acrylic Polyurethanes Hybrids

[a] the dividing line between paints and plastic coatings is blurred, but generally thermoplastics materials applied by dipping or electrostatic spray are referred to as 'plastic coatings'. They are also applied as linings for sheets, etc.

Table 1.4.22 Comparative properties of generic paint types

Generic type	Coating material	Properties							
		Mechanical Properties	Resistance to:						Recoating
			Water	Acids	Alkalis	Solvents	Temperature	Weathering	
Alkyd	Short oil alkyd[a]	Hard	Good	Best in group	Fair	Good	Good	Fair	Easy
	Long oil alkyd	Flexible	Fair	Poor	Poor	Poor	Good	Good	Easy
	Silicone alkyd	Tough	Good	Fair	Poor	Fair	Best in group	Very good	Fair
	Vinyl alkyd	Tough	Good	Good	Poor	Fair	Fair	Very good	Difficult
Vinyl	Polyvinyl chloride Acetate copolymers	Tough	Very good	Excellent	Excellent	See note 1	65°C	Very good	Easy
	Vinylacrylic copolymers	Tough	Good	Very good	Very good	Good	65°C	Excellent	Easy
Epoxy	Epoxy amine	Hard	Good	Good	Good	Very good	Very good	Fair (some chalking)	Difficult
	Epoxy polyamide	Tough	Very good	Fair	Excellent	Fair	Good	Good (some chalking)	Difficult
	Epoxy coal tar	Hard	Excellent	Good	Good	Poor	Good	Poor	Difficult
	Epoxy ester	Flexible	Good	Fair	Poor	Fair	Good	Good (some chalking)	Reasonable
Chlorinated rubber	Resin modified	Hard	Very good	Very good	Very good	See note 2	Fair	Good	Easy
	Alkyd modified	Tough	Good	Fair	Fair	See note 2	Fair	Very good	Easy
Polyurethane	Air dry varnish	Very tough	Fair	Fair	Fair	Fair	Good	Some yellowing and chalking	Care
	Two-pack	Tough, hard	Good	Fair	Fair	Good – hard better	Good	Some yellowing and chalking	Difficult
	Moisture cure	Very tough, abrasion resistant	Fair	Fair	Fair	Good	Good	Fades in light, yellows in shade	Difficult
	Non-yellowing	Hard to rubbery	Good	Fair	Fair	Good – hard better	Good	Very good	Difficult
Water base	Polyvinyl acetate	Scrub resistant	Poor	Poor	Poor	Poor	Fair	Very good	Easy
	Acrylic	Scrub resistant	Poor	Poor	Poor	Poor	Fair	Excellent	Easy
	Epoxy	Tough	Good	Good	Good	Good	Good	Fair	Fair
Inorganic zinc	Water base	Tough, abrasion resistant	Good	Poor	Poor	Excellent	Excellent	Excellent	Easy
	Organic base	Tough, hard	Good	Poor	Poor	Good	Excellent	Excellent	Easy

NOTES: 1. Aliphatic hydrocarbon = good. Aromatic or oxygenated hydrocarbon = poor.
2. Aliphatic hydrocarbon = fair. Aromatic hydrocarbon = poor.
[a] as an alkyd–melamine formaldehyde system (short oil alkyds are very rarely used on their own).

Table 1.4.23

Table 1.4.23 Properties of non-convertible paints

Paint	Typical main constituents	Drying		Advantages	Limitations	Typical applications	Method of application
		Air	Stove				
Coal-tar pitches Bitumens (i) Solution (ii) Emulsions	Heavy coal tar and petroleum fractions.	X	X	Fair resistance to weak acids and alkalis. Low cost. Air-drying possible. Will adhere to poorly prepared surfaces. Non-toxic grades of bitumens available for use with potable water.	No resistance to solvents. Upper temperature limit 65°C. Dark colour. Soft. No inhibitive effect – rusting often continues under these paints. Thick layers may be required for sensible effect. May be brittle in cold weather.	Protection of underground pipes (may be wrapped with cotton or hessian).	Brush or spray. Coal tar pitches and asphalts may require a stiff brush and travelling, so that considerable skill is required.
Vinyl resins (i) Primers (ii) Topcoats	PVC, polyvinyl butyral or PVC/acetate copolymers; vinylidene chloride and vinylidene flouride; organic solution lacquers based on PVC/acetate; latex materials (aqueous dispersion of vinyl, acrylic or styrene polymers or copolymers).	X		Fair air-drying. Resistance to weak alkalis dilute non-oxidizing organic acids and salts is good. Alternative to chlorinated rubber. Inert to foodstuffs. Tough and flexible. Very durable coil coatings produced from vinylidene flouride-based coatings.	PVA polymers have poor resistance to solvents, while $PVCl_2$ polymers are good and PVF_2 even better. PVCs frequently present adhesion problems. Low thermal resistance (65°C).	Solvent-based: metal, concrete vessels, fume protection. Organosol: steel strip, fume extractors, domestic appliances.	Solvent: brush-spray, roller, knife, dipping. Organosol: spray, brush, spread, knife, dipping.
Chlorinated rubber	Chlorinated rubber plus plasticizer; low build coatings; high build coatings up to 120 μm thickness	X		Excellent adhesion. Cheaper than epoxies. Can be pigmented with micaceous iron oxide or leafing aluminium for protection against atmospheric corrosion and low temperature immersion. Good resistance to mineral oils, acids and alkalis.	Poor solvent resistance	Marine coatings. Underseals for automotive vehicles.	Airless spray. Low build coatings normally brush applied.
Cellulosics	Nitrocellulose Cellulose acetate butyrate	X	X	Fast-drying. Hard, high gloss lacquer finishes.	Low build finishes. Require strong solvents with high VOC emission. Great care required in handling and storing nitrocellulose as a raw material.	Nitrocellulose used in high gloss automotive topcoats (now superseded by acrylics). Both nitrocellulose and CAB used as lacquers in furniture industry.	Spray applied.

X = technique in use.

Table 1.4.24 Properties of convertible paints

Paint	Typical main constituents	Drying		Advantages	Limitations	Typical applications	Method of application
		Air	Stove				
Oil paints (drying-oil type)	Linseed oil	X		Cheap. Easy to apply. Air cure. Red lead/linseed oil paint provides good corrosion protection.	Very slow drying (days). Poor weather resistance (except red lead/linseed oil paint). Very soft films easily damaged. Largely obsolescent.		Brushing
Oleoresinous paints (drying-oil type)	Naturally occuring resins such as rosin, fossil gums or phenolics which are cooked with drying oils	X		Cheap. Easy to apply. Air cure. Flexible. Can be coated.	Inadequate corrosion resistance. Superseded by the alkyds.	Cheap primer/ undercoats, drum finishes.	Brush or spray
Silicone paints (drying-oil type)	Silicones Silicone alkyds	 X	X X	Keep cleaner and retain colour and gloss better than most coatings. Heat and weather resistant (up to 550°C). High chemical resistance.	More expensive than oil or alkyd paints. Poor resistance to solvents. Surface pre-treatment required.	Used where better durability and appearance is required. Steel structures and marine finishes, for example.	Brush or spray
Alkyds	Oil-modified polyesters. 20–40% oil content, 'short oil' alkyd; 40–60% oil content, 'medium oil' alkyd; >60% oil content, 'long oil' alkyd. Constitute more than half of all resins used.	X	X[a]	Good performance for atmospheric exposure. Very versatile. Properties range from stoving, hard durable finishes to air-drying soft and flexible wood finishes. One of the cheaper synthetic resins.	Long oil alkyds used in gloss finishes tend to yellow in darkened surroundings. Generally not as durable as vinyls or acrylics.	Wide range of applications from car finishes to decorative coatings.	Most methods
Polyurethanes	Very wide range, from oil-modified polyurethanes, to one- and two-pack isocyanates, and blocked isocyanates.	X	X[b]	Ambient temperature cure. Two-pack finishes have high corrosion resistance (with primer). Can have excellent chemical and solvent resistance and high temperature immersion (130°C).	Aromatic isocyanates have a tendency to yellow. Those based on toluene diisocyanate (TDI) are cheaper than aliphatic types. Isocyanates are becoming less acceptable due to toxicity concerns.	Finishes for aircraft, agricultural and heavy duty equipment. Cross-linking resins for durable industrial products. Automotive refinishes. Floor finishes.	Airless or electrostatic spray or brush
Polyesters	Saturated or unsaturated polyesters based on polybasic acids reacted with polyols.	X	X	High film build. Relatively versatile materials with good durability. Two-pack coatings can produce very tough, flexible finishes.	Short pot life of two-pack systems can lead to wastage.	Large variety of coatings ranging from primer/ surfacers for the automotive and industrial markets, thick clear finishes for wood, marine finishes, GRP mouldings and heavy duty machinery finishes.	Hand lay-up. Special two- component spray techniques. Curtain coating. Electrostatic spray.

Table 1.4.24 Properties of convertible paints – *continued*

Epoxies	Most commonly used are reaction product of bisphenol A with epichlorohydrin.	X[c]	X[d]	Excellent chemical resistance. Working temperature limited to about 70°C unless modified. Good strength and adhesion. Harder and more abrasion resistant than chlorinated rubber.	Poor weathering properties as topcoat. High coat. Strong acid resistance inferior to chlorinated rubber. Critical application parameters.	Tank and pipe lining. Adhesives.	Various
	Water-based. 25–100 μm per coat.		X	No solvent fumes or fire hazard. Produces good anti-corrosion electrocoat primers. Good adhesion.	Susceptible to UV degradation. Less chemically resistant than organic solvent-base or solvent-free types.	Priming of concrete surfaces prior to application of an epoxy topcoat.	Various, including electrodeposition.
	Solvent-based. 25–75 μm per coat.	X	X	Very good chemical resistance (two-pack forms have best corrosion resistance). Grades available for use with potable water and foodstuffs. Ambient curing possible with two-pack.	Thin coatings dictate the use of several coats to cover blast profile. Intercoat adhesion, contamination and entrapment problems may develop.	Fuel and petrol tanks. Equipment exposed to corrosive environments. Consumer durables (white goods).	Brush. Spray.
	Coal tar epoxy	X		Lower cost. Higher film builds. Improved water resistance. Decreased temperature sensitivity.	Black or dark brown only. Some solvents may leach out coal tar, unless special sealants used.	Sewage outfall pipes. Sewage treatment plants below waterline. Cheap, heavy duty coatings on steel structures and marine finishes below waterline.	Brush. Spray.
	Urethane tar epoxy (as above but with isocyanate curing agent)	X		Can be cured at room temperature or below. Immersion resistance extended to higher temperatures. Very desirable abrasion resistant coating possible with special fillers. Improved solvent resistance.	Shorter pot life.		
	High build solventless epoxy resins. Liquid resins incorporating 10% solvent for easy spray. 100 μm per coat.	X	X	Reduced application costs due to thicker films.	Chemical resistance slightly inferior to solvent- based forms. Blistering due to solvent entrapment can result on immersion. Adhesion can be a problem.		Airless or air atomized spray.

Table 1.4.24 – *continued*

Table 1.4.24 Properties of convertible paints – *continued*

Paint	Typical main constituents	Drying		Advantages	Limitations	Typical applications	Method of application
		Air	Stove				
Epoxies (cntd.)	Solvent-free			Can be applied to any thickness. None of the problems associated with solvents. Excellent adhesion. Used in powder coatings. May be extended with coal tar to provide a cheaper, more flexible coating.	380 μm required to give real long- term protection to blasted steel. Very short pot life.	Concrete. Effluent treatment plants.	Sprayed at high temperature. Special machinery may be used, which enables the application of 650 μm coatings with high abrasion and chemical resistance, superior to solventbearing resins.
Phenolics	Phenol formaldehyde resins (novolacs and resoles)		X	Excellent exterior durability and chemical resistance. Service temperature 200°C (dry) 100°C (wet sterilizable). Phenolic/silicone also available for use in steam to 180°C.	Tendency to discolour. Poor adhesion to metal. Relatively brittle. Not suitable for use with alkalis above pH 10. Stoving necessary.	Tanks, vessels, hoppers, filter plates, valves, pumps. Heat exchanger tube linings (interior and exterior – heat transfer properties little affected). Widely used in can coatings, wire enamels, industrial coatings, marine finishes.	Spray.
Epoxy-phenolics			X	Very durable, chemically resistant (esp. strong alkalis). Similar to epoxy. Slightly less hard, but more flexible than phenolic.	Stoving necessary (150–200°C).	Chemically aggressive environments such as linings for storage of solvents.	Spray
Acrylic	Methyl methacrylate copolymer Thermosetting acrylic	X	X	Good weathering properties. Very durable, tough finishes.	Poor chemical resistance.	Aircraft finishes. Clear hard wearing finishes for floors. Thermosetting automotive finishes.	Various
Amino resin	Condensation products of formaldehyde with urea, melamine, benzoguanamine and glycoluril. Used to cross-link other functional resins.	X	X	Wide range of solubility, ranging from water to aliphatic hydrocarbon soluble.	Stoving required for all except urea resins, which are the least durable. Resins that release formaldehyde are becoming less acceptable.	Wide range of durable and hard wearing industrial finishes available. Automotive finishes. Coil coatings.	Various

NB Various industrial paint systems are based upon combinations of the above (e.g. epoxy-phenolics, urethane oils, urethane alkyds, epoxy esters, etc.).
X = technique in use
[a] short and medium oil length alkyds are used in stoving finishes
[b] blocked isocyanate
[c] polyamine curing agents
[d] polyol or carboxyl-containing resin curing agents

Table 1.4.25 Application methods for paints

Method	*Description*	*Advantages*	*Limitations*
Spray	Paint ejected under pressure as fine droplets from spray gun.	Rapid coverage of complex shapes. Can be hand-held (low equipment costs) or automated.	Difficult to obtain uniform, reproducible coverage. Overspray wastage.
Airless spray	Paint atomized under higher pressure (1.7–35 MPa) in the absence of air. Used for thixotropic paints, chlorinated rubbers, etc.	Better surface finish (dense, smooth). Less overspray wastage. Normally used on large flat expanses. Rapid application.	More expensive equipment than spray. Tip blocking can be frequent.
Air-atomized spray	Paint spraying using suction or gravity feed cups.	Simple, effective, cheap equipment	Slow. Wasteful due to overspray. Not suitable for thick coatings.
	Paint fed from pressure pot (200–550kPa)	Most paints using thinners can be spray applied to give smooth finish.	Slower than airless spray. Air entrapment likely with thick films. Wastage high due to overspray. Requires skilled operators.
Two-component hot spray	Either airless or air-atomized, used for the application of solvent-free epoxy resin paint films, polyurethane foam extrusion, etc. Base and hardener heated separately and pumped through heated lines by metering units and mixed.	Saving on manpower can be considerable. Hot application (100°C) reduces risk of air entrapment. Flat tile-like finish. Solvent entrapment and intercoat adhesion problems eliminated. High-density, thick (500 μm) films obtained.	High capital cost of machinery.
Dip (paint)	Components immersed in bath of paint, removed and drained.	Good coverage. Suited to simple shapes and high volume production.	Paint bath needs careful monitoring and operation. Components must give even drainage. Effluent disposal necessary.
Electrophoretic (more commonly known as electrodeposition)	Variant of dip process. Metal components act as electrodes. An electrocoagulation process which can either be anodic or cathodic.	Even, pore-free coatings on complex shapes. Suited to large volume production. Very efficient fully automated painting process with little wastage.	Components must be electrically conducting. Requires deionized water and very careful control of bath temperature and maintenance of paint deposition control factors.
Roller	Application by paint loaded roller.	Rapid, even thickness coating of large areas by continuous process.	Flat sheet or coil stock only. Large volume production necessary.
Brush, knife, trowel	Application of paint resin or plastisol by one of these implements.	Very low equipment cost. Versatility.	Labour intensive. Too slow and expensive for large volume production. Difficult to achieve even coating thickness.
Electrostatic spray	Paint or powder thrown as charged droplets or particles on to earthed component – adherent film is then stoved or fused in an oven.	Can be hand-held or automated. Rapid coverage with low overspray wastage. No solvent disposal problems.	High equipment cost. Specially formulated coatings necessary. Labour intensive if not automated.

Table 1.4.26 Painting pre-treatment processes and primers for metal[a]

Requirement	***Treatment***	***Industrial processes***	***Remarks***
Metal cleaning	Degreasing Chemical cleaning Descaling Blast cleaning Derusting	(i) Solvent (trichloroethylene)[b] (ii) Detergent (iii) Alkali (pickling) (iv) Acid (pickling) (v) Ultrasonic (vi) Vapour blasting (vii) Grit blasting (viii) Shot blasting (ix) Vibratory finishing (x) Wire brushing (xi) Hot water blasting (xii) Grit sand blasting	(i), (ii), (v) and (vi) are used industrially (mainly for steel) to remove grime and grease while (iii), (iv), (vii), (viii), (ix), (x), (xi) and (xii) remove also rust, millscale, etc. Painting or priming should be carried out soon after. Blasting is preferable to wire-brushing or pickling. Manual blast cleaning on site is very expensive compared with automated blast cleaning at steelworks or fabrication workshop and pre-treated materials should be used wherever possible.
Pre-treatment	Pickling Anodizing	(i) Acid pickling (ii) Alkali pickling (iii) Phosphating (iv) Chromic acid anodizing (v) Chromate passivation	(i) and (ii) are similar to (iii) and (iv) above. (iii) deposits a phosphate film while (iv) and (v) deposit also a chromate film, both electrically insulating and localizing corrosion. (iv) and (v) used for aluminium.
Priming	Wash primers Anti-corrosion priming Paint primers Shop primers	(i) Wash primer (ii) Etch primer (iii) Electrophoretic (iv) Water-based (v) Zinc rich (vi) Red oxide (vii) Lead primers (viii) Metallic (ix) Epoxy	(i) and (ii) are frequently synonymous and contain corrosion inhibiting pigment (zinc tetroxychromate). (iii) is used for autopriming. (v) (vi) and (vii) contain rust and corrosion inhibiting pigments. (viii) is based on corrosion resisting metallic powder (e.g. bronze, aluminium, stainless steel). Some primers develop good adhesion to metal on stoving (e.g. vinyl primers). Red lead primers still most effective for wire brushed steel surfaces. Metallic lead primers less effective but still have inhibitive capabilities and zinc-rich primers are very effective when applied to blast cleaned surfaces. Zinc/epoxy resin formulations are quick drying. Zinc content should be not less than 85% in dry film. Zinc chrome primers should be used on aluminium surfaces.

[a]The correct surface preparation, pre-treatment and primer are essential if the chosen paint is to perform adequately for the desired time span. Suppliers advice should always be sought on the paint system employed.
[b]Trichloroethylene is becoming environmentally unacceptable and is being replaced with other solvents.

Table 1.4.27 Characteristics and applications of thermoplastic powder coatings and linings

Material	*Characteristics*	*Applications*
PVC organosol or plastisol (properties very dependent upon pre-treatment, processing and quality of polymer.)	Heated component treated with adhesive primer dipped in cold PVC dispersion and oven cured, giving rubbery coating of 0.25–1.25 mm thickness. Proven weathering resistance. Resists acids, inhibits marine growth. Abrasion resistant, tough, strongly adhering cushion coating. Electrically and thermally insulating. Does not oxidize like rubber. Low temperature grades available. Not suitable for use with solvents due to extraction of plasticizers. Low maximum continuous service temperature of 60°C.	Heavy duty coatings for industrial plant and marine applications. Cushion coatings for storage, transportation and corrosion protection of accurately machined metal components, floor gratings, ducting and pipework.
PVC powder	Applied by fluidized bed (preferably) or electrostatic spray with subsequent oven cure. Weathering resistance. Resists corrosion by seeping water, vehicle exhaust gases, and hot detergents. Thermally and electrically insulating. Grades available with similar hardness to nylons but with better edge coverage. Seam free homogeneous coatings. Non-brittle in winter temperature. Limitations as for PVC plastisol.	Wirework, dishwasher baskets, etc. Tractor radiator grilles. Rainwater goods. Safety handrails. Tunnel lining supports.
Cold applied PVC	Permanent or strippable coatings for on-site application; wide range of formulations for most substrates including foams, walls and steelwork. Application by spraying, brushing or dipping.	Sound proofing, insulation, protection of automobile electrical parts against wet conditions, coatings for long-term outdoor storage of machinery.
Nylon 11 and 12	Applied by fluidized bed or spray. Good abrasion and impact resistance. Low coefficient of friction, improved by MoS_2 impregnation. Resistant to oils, alkalis, seawater and most solvents (e.g. trichlorethylene). Non-toxic, can be steam sterilized. High upper service temperature (120°C). Lower water absorption. Can be applied over relatively rough surface, e.g. castings. Not resistant to strong mineral acids. Difficult to repair if damaged. Special primers needed if Cu alloys are to be coated. More expensive than PVC.	Demineralized water containers. Window frames. Balustrading. Components in hospitals and food industry. Castings and presswork. Electrical insulation.
Polyethylene	Applied by fluidized bed or spraying. Cheapest and most easily applied of all thermoplastic coatings. Good resistance to acids and alkalis. Can be used up to 100°C. Prone to environmental stress-cracking. *High density grades* – higher melting point, tougher, better natural adhesion to substrate metals and can withstand sterilization temperatures. Semi-matt finish. *Low density grades* – more easily applied, gloss finish, thicker coatings.	*High density* – electrical applications. Heavier duty applications. Chemical plant tanks and pipework. *Low density* – non-drip decorative coating for wirework – refrigerator shelves, racks, clothes, airers, etc.
Cellulose acetate butyrate (CAB)	Applied by fluidized bed or spraying. Good physical properties, low thermal conductivity. Gives pleasant feel. Poor chemical resistance.	Steering wheels and totally enveloping coatings. Door handles. Tool handles.

Table 1.4.27

Table 1.4.27 Characteristics and applications of themoplastic powder coatings and linings – *continued*

Material	***Characteristics***	***Applications***
Fluoroplastics	Application methods include dipping, spraying and multicoat dispersions, but not all fluoroplastics can be applied by each method (e.g. PTFE and PTFCE can only be applied by flame spraying and while coating itself has exceptional corrosion resistance, porosity in coating can limit usefulness). Pore-free coating of exceptional chemical resistance possible if flame spraying is avoidable as application technique. High upper service temperature (180–225°C). Low friction. Better strength and thermal shock resistance than glass. Electrically insulating. Expensive.	Reaction vessels and pump impellors for chemicals and slurries. Non-stick kitchen utensils.
Chlorinated polyether (Penton)	Applied by fluidized bed, flame spray, as multicoat dispersion, or prefabricated lining. Chemical resistance between that of PVC or polyethylene and fluorinated polymers. Upper service temperature of 120°C. Expensive.	Chemical process plant linings.

Table 1.4.27 – *continued*

Table 1.4.28

Table 1.4.28 Properties of plastic powder coatings

Coating	Normal thickness range (mm)	Shore hardness	Safe working temperature (°C)		Toxicity	Burning rate	Chemical resistance			Abrasion resistance	Volume resitivity at 20°C(Ωcm)	Process temperature[b] (°C)	Thermal conductivity (W/m per K)	Colour	Surface finish	Applied by
			Minimum	Maximum continuous, maximum intermittent			Acids	Akalis	Solvents							
PVC plastisol	0.63 12.7	55–65	–35	60, 80	Very slight	Slow to self ext.[a]	Good	Good	Poor	Very good	3×10^9	170	0.15	Full range[a]	Gloss	Dipping or spray
Low temperature PVC plastisol	0.76 11.4	45–55	–50	60	Very slight	Slow to self ext.[a]	Fair	Good	Poor	Good	3×10^9	170	0.12	Green	Gloss	Dipping
PVC powder	0.25 0.76	80–98	–10	60, 80	Very slight	Slow to self ext.[a]	Good	Good	Poor	Very good–excel.	4×10^{12}	260	0.17	Full range	Gloss	Dipping
PVC cold applied	0.13 2.5	85–95	–25	60, 80	Very slight	Slow to self ext.[a]	Fair	Good	Poor	Good	4×10^9	Ambient	0.17	Full range	Gloss	Spraying or brushing
Polyurethane	0.25	95	–70	100, 120	Non-toxic	Medium	Poor	Fair	Fair	Out-standing	2×10^{11}	300	0.21	Limited range	Gloss or matt	Dipping or spraying
Neoprene (E)	0.4	75–80	–50	110, 130	Very slight	Self ext.	Fair	Good	Good	Very good	10^{12} variable	Ambient	0.12	Grey	Matt	Spraying or brushing
Epoxide (TS)	0.05 0.2	98	–50	70 140	Non-toxic[a]	Non-flammable	Good	Good	Fair	Good	5×10^{16}	150	0.17	Full range	Gloss	Dipping or spraying
Nylon 11 and 12	0.25 0.76	95	–50	100, 120	Non-toxic	Resistant to flame	Poor	Good	Good	Excel.	6×10^{13}	300	0.29	Full range	Gloss or matt	Dipping or spraying
Cellulose acetate butyrate, CAB	0.38 1.3	100	–40	85, 100	Non-toxic	Slow	Poor	Poor	Poor	Good	10^{13}	325	0.21	Full range	Gloss	Dipping
Low density polyethylene	0.38 1.9	80	–70	70 105	Non-toxic	Medium	Very good	Good	Poor	Fair	3×10^{17}	200	0.29	Full range	Gloss	Dipping or spraying
High density polyethylene	0.38 1.7	90	–70	70, 115	Non-toxic	Medium	Very good	Good	Good	Good	10^{16}	250	0.04	Limited range	Matt	Dipping or spraying
Fluoroplastics for corrosion resistance	0.25 1.0	Varies	–70	180, 225	Non-toxic	Non-flammable	Excel.	Excel.	Excel.	Good to excel. varies with grade	10^{15} -10^{18}	270–450 depends on grade	0.12 to 0.25	Limited	Gloss or matt	Dipping or spraying

Details marked
[a] are dependent on various factors.
[b] depending on size and thermal capacity.
Data supplied by Plastic Coatings Ltd.

TS = Thermoset E = Elastomeric

Table 1.4.29 Characteristics and applications of thermosetting powder coatings

Type	*Characteristics*	*Typical applications*
Epoxies	Most common thermoset powder coating. Electrically insulating. Non-toxic. Abrasion resistant. Can withstand boiling fat. Wide range of colours. Good chemical resistance. Poor weathering resistance Good wetting of clean metal substrate. No volatiles evolved during cure. Hollow articles and 'inaccessible' areas can be coated.	Electrical components Kitchen utensils Fire extinguisher linings Pipe linings
Polyester	Intermediate between epoxies and acrylics on performance and price. Better (fair) weathering resistance than epoxies, but inferior to acrylics. Alkyd/melamine powders less weather resistant than other types (saturated or unsaturated). Better resistance to chalking and yellowing than epoxies on heating. Lower storage life than standard epoxy types. Worse chemical resistance than epoxy.	
Polyurethane	Very good exterior exposure resistance. Better chalking, but inferior yellowing resistance than polyesters. Unpleasant odour. Pinholing at 70 μm thickness. Excellent gloss flow and mechanical properties. Reduced edge coverage. Highest costs.	
Acrylic	Good heat resistance. Outstanding weathering. Lower temperature cure than polyesters. Excellent gloss and colour retention. Very poor mechanical properties. Chemical resistance fair, worse than epoxy. Limited storage life.	
Hybrid (or epoxy polyesters)	Developed post-1974 and can contain up to 80% polyester. They have the general characteristics of epoxies, but are non-yellowing by heat and UV. They are designed for indoor and general industrial use.	

Table 1.4.30 Properties of thermosetting powder coatings

Property / *Material*	*Epoxy*	*Epoxy-anhydride*	*Epoxy-polyester*	*Polyester*	*Polyurethane*	*Acrylic*
Finish	Good	Good	Good	Fair-good	Very good	Good
Stoving	180°C 10 min	180°C 10 min	180°C 15 min	200°C 10 min	190°C 10 min	160°C 20 min
Mechanical properties	Good	Good	Good	Good	Fair–good	Poor–moderate
Heat resistance (100°C)	Limited	Fair–good	Fair–good	Very good	Varies	Good
Weathering	Chalking	Chalking	Chalking	Good	Varies	Good
Solvent resistance	Fair–good	Good	Fair–good	Fair–good	Fair	Fair
Alkali resistance	Fair	Fair	Fair	Fair	Good	Good
Acid resistance	Good	Good	Good	Good	Good	Good

Table 1.4.30

Table 1.4.31 Steel thicknesses for enamel ware

Application	*Gauge no.*	*mm*
Kitchen ware	31–24	0.3–0.6
Store parts, signs	24	0.6
Table tops, refrigerators, heavy signs	18	1.2
Washing machine tubs	18	1.2
Chemical tanks, food	4–10	3.3–5.9

Table 1.4.32 Maximum area for given width of enamel ware

	Maximum area for width:		
Gauge	*15.2 cm*	*30.5 cm*	*45.7 cm*
24 (0.6 mm)	0.046 m^2	0.279 m^2	0.465 m^2
22 (0.71 mm)	0.093 m^2	0.325 m^2	0.557 m^2
20 (0.91 mm)	0.139 m^2	0.465 m^2	0.743 m^2

Table 1.4.33 Common corrosion resistant electroplates, their processing baths, properties and applications

Deposit metal	Protected substrate	Common electrolytes	Processing Ease	Cost	Corrosion resistance behaviour	Typical thickness (μm)	Hardness (kg/mm^2)	Wear resistance	Typical applications
Cadmium BS1706	Ferrous and copper alloys	a. Alkaline cyanide.	H	H-M	Sacrificial. Better resistance to marine atmospheres than zinc.	2.5–20 depending on severity of conditions.	40	M	Severely restricted because of health and safety, and environmental considerations. Effectively for military and some aerospace use only.
		b. Acid sulphate	M	H-M					
Zinc BS 1706	Ferrous	As above	H	L	Sacrificial	2.5–50 depending on severity of conditions.	40–50	M-H	Many domestic fittings.
Zinc/ Cobalt		Alkaline cyanide or acid chloride.	M	M	Sacrificial	5–30	110–210	M	Fasteners, cadmium replacements.
Zinc/ Iron		Alkaline	M	M	Sacrificial	2.5–30	100–125	M	Fasteners, can be blackened easily.
Zinc/ Nickel		Acid chloride complex alkaline	M	M	Sacrificial	5–30	Acid 140–160 Alkaline 200–250 Acid (Boeing) 250–310	M	Motor industry, aerospace, cadmium replacement.
Chromium BS 1224 BS 4641	Ferrous, copper, zinc alloys. Aluminium.	Chromic acid with sulphate or fluoride catalysts.	H	H	Barrier, used over nickel undercoat, often itself over copper.	0.2–1 0.1–100 for hard, wear-resistant coats.	900	H	Bright decorative finish over Ni and Cu plates on automobile trim and domestic fittings.
Nickel BS 4758 BS 1224 BS 4601	Ferrous, zinc, copper, aluminium alloys.	a. Watts – sulphate, chloride, boric acid. b. Sulphamate c. Chloride.	H	M	Barrier to substrate. Used as an integral part of Cu–Ni–Cr protective scheme on corrodible substrates. Sacrificial to Cr. Bright and dull Ni coats protective in own right.	2.5–40. Up to 500 for hard, wear resistant coats.	150–500	H M-L on softer Al alloys which may distort under load.	Cu–Ni–Cr protective scheme on automobile trim and domestic fittings, or can be used on own when bright plated.
Lead	Ferrous, copper alloys	Acid fluoroborate	H	L	Barrier.	2–5	5	L	Storate coating.
Tin BS 1872	Ferrous, copper alloys.	a. Acid fluoroborate	H	M	Barrier to substrates.	0.4–15	5–10	L	Steel strip continuous plating for containers. Tarnish resistant, solderable storage coat for copper wire.
		b. Stannous sulphate	H	M					
		c. Sodium stannate	H	H					
Lead–tin alloys	Ferrous, copper alloys	Fluoroborate	M	L	Barrier.	0.4–5	5–10	L	Tarnish resistant, solderable coatings. Useful for heavy duty bearings.

Table 1.4.33

Table 1.4.33 Common corrosion resistant electroplates, their processing baths, properties and applications – *continued*

Copper	Ferrous, zinc, aluminium alloys.	a. Acid sulphate b. Alkaline cyanide. c. Mildly alkaline pyrophosphate.	H	M	Barrier to substrates. Usually used as undercoat for Ni and Cr.	10–50	60–150	L	Base coat for Ni and Cr. Heat conducting layer on cooking utensils. Electroforming. Drawing lubricant.
Copper–zinc alloys	Ferrous	Cyanide	H	M	Barrier.	5–20	100–200	M	Mainly decorative.
Gold and low alloys BS 4292 BS 3315	Copper alloys. Silver.	a. Alkaline cyanide b. Acid and neutral complex cyanide c. Neutral–alkaline complex sulphite.	M	H	Barrier to all substrates. Totally tarnish resistant in all environments.	0.5–5	70–250 controllable by choice of alloy and conditions of operation of bath.	L–M	Electrical and electronic contact and connector surfaces in low duty service conditions. Jewellery and cosmetic articles.
Silver BS 4290 BS 2816	Ferrous, non-ferrous alloys	Alkaline cyanide.	H	H	Barrier. Tarnishes in moist and sulphurous atmospheres but attack is superficial.	2.5–25	50–150	M	Heavy duty conductor surfaces. Jewellery domestic, cosmetic articles
	Aluminium				Mating surfaces on some aluminium conductors.	2–4.	50	L	Electrical items.
Rhodium	Silver	Sulphate acidified with sulphuric or phosphoric acid.	L	M	Barrier. Tarnish resistant in most environments. Porous when thin, cracked at >10 μm.	0.1–0.5 (decorative) 1–5 (industrial).	700	H	Heavy duty contact and connector surfaces. Decorative tarnish free surface on jewellery, watch cases, light, etc.

H = high; M = moderate; L = low

Table 1.4.33 – *continued*

Table 1.4.34 Ease of pre-treatment and preparation for plating of various base metals

Substrate base metal	***Ease of preparation***	***Comments***
Plain carbon steel	H	All rust must be removed, especially from pits.
Heat treatable steels	M	Scale formation may be considerable.
Stainless steel	M–L	Tenacious oxide film to be removed before adherent coatings can be applied.
Cast iron	M–L	All rust and surface defects must be removed.
High strength steel	L	Susceptibility to hydrogen embrittlement, therefore much care required.
Copper, brass	H	
Zinc die-castings	M–H	Reactive to etches, must use care. Subcutaneous pores may cause staining and blistered deposits.
Nickel	M	Easily passivates.
Aluminium	M–L	Tenacious oxide film and presence of insoluble intermetallic compounds. Requires special treatment.
Magnesium	L	Oxide film.
Titanium	L	Oxide film.

H = high, M = moderate, L = low

Table 1.4.34

Table 1.4.35 Throwing power of common electrolytes

Deposit metal	*Electrolyte type*	*Throwing power*
Cadmium	Cyanide Acid	Good Poor
Zinc	Cyanide Acid	Good Poor
Zinc–cobalt	Alkaline	Fair
Zinc–iron	Alkaline	Fair
Zinc–nickel	Acid	Poor
Zinc–nickel	Alkaline	Fair
Copper	Cyanide Pyrophosphate Acid	Good Fair Poor
Nickel	Acid	Fair
Chromium	Acid	Very poor
Lead	Acid	Poor
Tin	Acid Alkaline	Poor Very good
Rhodium	Acid	Poor
Silver	Cyanide	Very good
Gold	Cyanide Neutral Acid	Good Fair Fair
Brass	Cyanide	Good
Tin–nickel	Acid	Poor

NB When using proprietary electrolytes some modification to the above gradings may be experienced and advice should be sought from the supplier concerned.

Table 1.4.36 Characteristics of chemical plating processes

Process	*Suitable substrate*	*Deposit thickness*	*Properties*	*Applications*
Autocatalytic nickel (electroless nickel)	Most ferrous, non-ferrous metals, plastics.	Theoretically unlimited, practically 125 μm. Plating rate 12–25 μm/ h for Ni/P. Less for Ni/B.	Very hard, 500–800 kg/ mm^2. Excellent throwing power. Wear resistant. Corrosion resistant.	Plating on plastics to give continuous conducting layer for building up electroplate. Building up worn engineering parts. Plating uniform deposits on complex shapes.
Autocatalytic copper	Steels, brass, aluminium, plastics, circuit boards.	1–25 μm. Plating rate 5 μm/h.	Hardness as electroplated Cu. Excellent throwing power. High conductivity.	Plating on plastics. Base coating on printed circuitboards – through-hole-plating.
Immersion gold	Ferrous, non-ferrous, metals plastics.	< 0.5 μm	Hardness as electroplated Au. Bright. Porous, subject to substrate tarnishing. Good conductivity and emissivity.	Inexpensive decorative uses. Printed circuits, transistors, connectors for high conductivity and good solderability. Emissive properties useful in aerospace and missile applications.
Zincate (immersion zinc)	Aluminium alloys.	< 0.5 μm.	Uniformity varies with substrate alloy.	Plating on aluminium. Used as base plate for copper–nickel– chromium.

Table 1.4.36

Table 1.4.37 Characteristics of chemical conversion coatings for metals

Metal	*Process*	*Cost*	*Coating properties*
Aluminium and alloys	**Anodize**: chromic acid 3–10%, 0.4–0.6 A/dm^2, anode voltage cycled with time, 40°C, 30 min. Sealing essential.		Opaque, white grey film. Excellent corrosion resistance. Growth rate less than with sulphuric acid. Thickness up to 15 μm.
	Anodize: sulphuric acid 20%, 1.2–1.5 A/dm^2, 15 V, 18–20 °C, 20 min. Sealing essential.	Cheapest of anodizing processes.	Translucent, porous film. Growth rate ~ 0.5 μm/min. Thickness 1–100 μm. Excellent corrosion resistance.
	Anodize: oxalic acid 5%, 1.5 A/dm^2, 60 V, 35–40°C, 30–60 min. Sealing essential.		Slower growth than sulphuric acid. Similar film. Excellent corrosion resistance. Little dye absorption.
	Chromate: MBV calcined soda, sodium chromate, 95°C, alkaline pH, 15–30 min.	Higher than phosphating, lower than anodizing.	1–2 μm. Good corrosion resistance.
	Chromate: silicate sodium carbonate, sodium chromate, sodium silicate, 95°C, alkaline pH, 8–10 min.	As above.	1 μm. Good corrosion resistance.
	Chromate: fluoride chromic acid, sodium dichromate, sodium fluoride, 60°C, acid pH, 1–5 min.	As above.	0.1–1 μm. Good corrosion resistance.
	Chromate: phosphate chromic acid, sodium dichromate, sodium fluoride, phosphoric acid, 20°C, acid pH, 1–5 min.	As above.	1–5 μm. Good corrosion resistance.
	Phosphate: zinc or manganese phosphates, nitric acid, hydrofluoric acid 60°C, 5–10 min.	Low.	2–5 μm. Temporary protection.
Zinc and cadmium	**Chromate**: sodium dichromate, sulphuric acid, pH 1.2–1.6, 1–10 sec.	Higher than phosphating.	0.3 μm. Yellow-brown film, prevents tarnishing in industrial and heavy oxide formation in marine atmospheres.
	Phosphate: as for aluminium minus the fluoride.	Low.	2–5 μm. Temporary protection only. Main use for paint bonding.
Iron and steel	**Chromate**: most ferrous chromating is carried out on zinc or cadmium coated steel as above.		
	Phosphate: very similar solutions to zinc.	Low.	2–5 μm. Temporary protection.

Table 1.4.38 Suitability of aluminium alloys for sulphuric acid anodizing

	Protective				*Decorative*				*Bright*			
	E	***G***	***M***	***U***	***E***	***G***	***M***	***U***	***E***	***G***	***M***	***U***
Wrought	1 1080 1050 1200 5005 5154 5251 5083 6060 6063	3103 6082 6083 3105 7620	2031 2014 2024 7020 2618		1 1080 1050 1200 N4 5005 N5 5154 5251 5005 5083 6060 6063	3103 6082 6083 3105 7020	2014[a] 2024[a] 2031 2618 7020		1 1080 1050 BTR1 BTR2 BTR3	1200 3105 5154 5251 H9 5083 6060	N3 3103 5454 5005 6061 6082	2014 2024 2031 2618 7020
Cast	LM0 LM5 LM18 LM25	LM4 LM10 LM16 LM21 LM22	LM2 LM6 LM9 LM13 LM20 LM24	LM12	LM0 LM5	LM18[a] LM25	LM4 LM10 LM13[a] LM16[a] LM21[a] LM22 LM24[a]	LM2 LM6 LM9 LM12 LM20	LM0	LM5	LM10	LM2 LM4 LM6 LM9 LM12 LM13 LM16 LM18 LM20 LM21 LM22 LM24 LM25

E, excellent; G, good; M, moderate; U, unsuitable; [a] dark as-anodized film, therefore only suitable for dark colours.

Table 1.4.39 Hot dip processes, coating properties and applications

Process		*Typical coating properties*	*Applications*
Type	*Parameters*		
Zinc (galvanizing)	Temperature 432–466°C, optimum economy and coating quality approx. 450°C. Time of immersion dependent on mass of workpiece, 1–5 min. Typical withdrawal rate 1.5 m/min.	Thickness for corrosion resistance – 30–50 μm depending on environment. Sacrificially protects steel. Tougher than steel substrate.	Sheet for fabricating corrugated roofing, water storage tanks, fasteners, pipe and fittings, exposed structural members, railing, fencing, etc.
Tin (continuous)	Entry temperature 325°C. Operating speeds 2.4–14 m/min depending on size, thickness requirements. Withdrawn through palm oil at a controlled temperature for the product to emerge at 240°C i.e. controlled quenching. The omission of this process may result in oxidation of the tin coating.	Thickness for corrosion resistance – 20–30 μm. Usually flow brightened. Protects steel by barrier protection, therefore must be pinhole-free.	With lacquer for food containers. Good base for paints.
Aluminium (both continuous and batch)	Immersion time critical due to rapid reaction of aluminium and steel. Bath temperature 660–705°C.	Thickness for corrosion resistance – 40–75 μm. Sacrificially protects steel if natural oxide film is penetrated. Decreases hardness and strength of steel substrate if too thick. Cannot draw aluminium dipped steel sheet.	Similar applications to galvanized steel. Aluminium dipped steel has generally better corrosion resistance, especially at raised temperatures such as automobile exhausts, and equipment for handling hot flue gases.

Table 1.4.40 Characteristics of some important diffusion coating processes

Process	*Substrate*	*Process parameters*	*Coating properties*	*Applications*
Aluminizing (calorizing)	Steels, nickel and cobalt alloys	Aluminium or ferro- aluminium powder, volatile halide and ceramic aggregate heated at 815–1200°C for 6–24 h.	After coating, the deposit is ~60% Al at surface and ~150 μm thick. On post-treating in air, diffusion occurs and the surface Al decreases to <25% with corresponding decreases of coating thickness to <125 μm. The coating is resistant to combustion gases and sulphurous combustion products.	High temperature corrosion resistance, chemical and petroleum process equipment and aircraft engine parts exposed to combustion gases, automobile exhausts.
Chromizing	Steels	Chromium powder, halide and inert aggregate heated around work load at 800–1300°C for ~20 hours. The presence of hydrogen is essential for thick coatings.	Surface layer on steel (>12% Cr) ferritic, bulk austenitic. Up to 100 μm possible when hydrogen used as a reducer. Claimed to have better oxidation resistance at high temperatures than many stainless steels and to be cheaper.	High temperature oxidation resistance applications. e.g. long periods of exposure to temperatures of 750°C maximum.
Siliconizing	Steels	Vapour phase process using silicon tetrachloride or solid phase using a powder pack of silicon carbide or ferrosilicon and circulating chloride gas at 900–1100°C using silicon tetrachloride as a carrier.	Coating thickness 350–750 μm, 14% Si at surface, 12% Si in bulk. Unusual combination of corrosion, wear and moderate heat resistance. Brittle case.	Pump shafts, cylinder liners, valves, valve guides and fittings.
Sherardizing (zinc)	Steels	Diffusion zinc coating formed from heating zinc dust and sand diluent mixture around the workpiece at 350–400°C for 3–10 h. No carrier is normally required, the zinc vapour itself providing this service (BS 4291).	Matt, grey appearance. Corrosion resistance proportional to thickness. Coating ~8–9% Fe. Possibly diffusion occurs outwards in this process. Harder than zinc coatings and can be painted without any pre-treatment. Very uniform thickness.	Small pressings, forgings, castings, machined parts, nuts, bolts, washers etc., and longer lengths of rod tube, etc.

Table 1.4.41 Sprayed coatings for corrosion resistance

Corrosion environment	Recommended coatings (In order of decreasing resistance)	Remarks	Typical applications	Typical lifetimes
Industrial atmospheres Encountered in or near large cities and heavily industrialized areas. Coating must protect substrate against smoke and chemical fumes as well as sun, wind and rain.	0.075 mm of aluminium or Al–Zn alloy plus one coat of primer and two coats of aluminium vinyl sealant.	Best protection in most severe condition. Aluminium better than zinc in high sulphur atmospheres. Application requires optimum blast preparation.	All types of structural and fabricated steel. Electrical conduits. Bridges. Power line hardware. Lawn and patio furniture.	10–20 years.
	0.075 mm of zinc plus one coat of primer and one coat of aluminium vinyl sealant.	Lower cost. Zinc bonds better to poorly prepared surfaces. Not as effective as above in severe conditions.		10–20 years.
	0.05–0.12 mm of zinc. No after-spray treatment.	For use instead of or to replace galvanizing. Lowest cost. Easy application.		5–10 years.
Rural atmospheres Encountered inland away from industrial areas or salt spray. Coating must protect substrate from sun, wind and rain. Mildest of atmospheres where many low cost coatings can be employed by other techniques (e.g. by brushing).	0.075 mm of aluminium or A–Zn alloy plus one coat of primer and two coats of aluminium vinyl sealant.	Best protection but primarily too expensive for most applications which must in any case permit perfect blast preparation.		20 years.
	0.075 mm of zinc plus one coat of primer and two coats of aluminium vinyl sealant.	Zinc bonds better to poorly prepared surfaces. Lower cost than above but probably still too expensive for most applications.		10–20 years.
	0.05–0.12 mm of zinc. No after-spray treatment.	Cheapest sprayed metal coating, replaces galvanizing. Probably adequate for most rural atmospheres.		10–20 years.
Marine atmospheres Encountered near sea coast or salt lakes. Coating must protect substrate against humid, heavily laden salt atmospheres and salt spray as well as sun, rain and wind.	0.075 mm of aluminium plus one coat of primer and two coats of aluminium vinyl sealant.	Best protection in most severe conditions. Application requires optimum blast preparation.	Bridges – above water line. Dock structures – above water line. Exterior of storage tanks. Ship superstructures. Trestles. Transformer cases.	10–20 years.
	0.075 mm of zinc plus one coat of primer and one coat of aluminium vinyl sealant.	Lower cost. Zinc bonds better to poorly prepared surfaces. Shorter life than above.		10–15 years.
Potable fresh water Coatings must protect substrate against cold fresh water and not contaminate the water.	0.2 mm of zinc with no subsequent treatment.	Extra thickness ensures longer life and possible total cost savings.	Freshwater storage tanks. Aqueducts. Filter troughs. Conduits.	10–20 years.
Non-potable fresh water Coatings must protect substrate against fresh water (not intended for drinking purposes). Water temperature 25°C, pH 5–10.	0.075 mm of aluminium or Al–Zn alloy plus one coat of primer and two coats of aluminium sealant.	Best protection but most expensive. Requires optimum blast preparation.	Power plant inlets. Structure immersed in water. Boat hulls operating in fresh water.	10–20 years.

Table 1.4.41

Table 1.4.41 Sprayed coatings for corrosion resistance – *continued*

Corrosion environment	*Recommended coatings (In order of decreasing resistance)*	*Remarks*	*Typical applications*	*Typical lifetimes*
Hot fresh water Coatings must protect substrate against hot fresh water. Not intended for drinking purposes. Water temperatures 50–200°C (steam), pH 5–10.	0.15 mm aluminium or Al–Zn alloy plus two coats of 2-pack polyamide cured epoxy (one clear followed by one ivory finish).	Lower total cost treatment.	Heat exchangers. Hot water storage tanks. Steam cleaning equipment. Parts exposed to steam.	
Sea-water immersion Water moving or still. Immersion is considered to extend from just below tide limit to 1 m above high tide limit.	0.075 mm of aluminium plus one coat of primer and two coats of aluminium vinyl sealant.	Best protection for severe conditions. Optimum blast preparation necessary. Not recommended where fouling is a problem.	Marine engine oil pans. Steel piping and piers. Ships hulls. Fish holds and tanks. Below water line where fouling is a problem.	5–10 years.
	0.075 mm of aluminium plus one coat of primer, one coat of aluminium vinyl sealant and two coats of tributyl tin oxide anti-fouling paint.	Useful for seawater and anti-fouling protection.		10–20 years.
Food and organic chemical processing Coatings should resist the action of chemicals such as oils, fuels, solvents and foodstuffs without contaminating or altering the taste.	0.15 mm of aluminium plus two coats of 2-pack polyamide cured epoxy (one clear followed by one of ivory finish).	Lowest total cost treatment.	Storage tanks for gasoline oils, fuels, solvent such as xylene, ethyl alcohol, butyl alcohol, amyl acetate. and toluene. Malt syrup tanks in breweries. Soft drink plants. Dairies. Tanker compartment for edible oils or molasses. Glycerine tank linings.	

Table 1.4.41 – *continued*

Table 1.4.42 Sprayed coatings for the protection of iron and steel in elevated temperature environments (BS 2569)

Class	*Service temperature (°C)*	*Coating metal*	*Thickness (μm)*	*Intermediate treatment*	*Final treatment*
D	550	Aluminium	180	None.	None.
E	900	Aluminium	180	Coat uniformly and completely with coal tar pitch solution.	Heat to 800–900°C.
F	900	Aluminium alloy	180	None.	Heat to 800–900°C, mildly oxidizing atmosphere.
G	1000 No sulphur gases	Nickel–chromium alloy[a]	380	Coat uniformly and completely with coal tar pitch solution with added aluminium pigment.	Heat to 1050–1150°C.
H	1000 Sulphur gases present	Nickel–chromium alloy[a] followed by aluminium.	380 100	As above.	As above.

[a]Typical alloys used are 60% NI–15% Cr–25% Fe or 80% NI–20% Cr.

Table 1.4.43 Relative merits of cathodic protection systems

Cathodic protection system	***Advantages***	***Limitations***
Sacrificial anodes	Can be used where there is no power. No initial outlay for power equipment. Relatively foolproof and little supervision required. Current cannot be supplied in the wrong direction with consequent promotion of corrosion instead of protection. Installation is simple. If protection is inadequate, additional anodes are easily introduced if required. Anode leads are cathodically protected.	The current available depends on the anode area, which for large ships may be cumbersome and heavy. Where DC power is available electrical energy can be obtained more cheaply than from galvanic anodes. In large anode arrays the high current flowing in the leads requires heavy wiring to keep resistance losses low.
Impressed current	If sufficient voltage is available the protecting current can be increased to any desired amount, as long as the anode material remains functional. The leads need not be large since resistance losses can be corrected by an increase in voltage. Large, even uncoated, structures can be protected in high resistivity environments by using a large driving voltage.	Continuous DC power must be available. Care must be taken that the current is never connected in the wrong direction. More technical and trained personnel are necessary for supervision. The anode leads must be insulated and waterproofed since exposure of the lead will allow preferential corrosion at any break in the insulation. Impressed current systems with inert anodes must be provided with current shields. There is greater danger of cathodic corrosion of aluminium and other amphoteric metals, and damage to coatings.

Table 1.4.44 Comparison of anode materials for impressed current systems

Material	*Characteristics and applications*
Graphite	Susceptible to mechanical damage. Lower current density limits for extended life. Used to protect moored, rather than mobile ships, where installation of anode can be protected.
Platinum	Used extensively. No loss of material at high current densities. May be foil, supported by GRP. Corroded by AC – must be rectified.
Titanium, Tantalum, Niobium } Sheated or plated with platinum	More rugged than platinum. Titanium does not undergo anodic dissolution at < 10 V impressed even at high current densities (e.g. at Pt discontinuities). Ta and Nb preferred where impressed voltage > 10 V.
Lead–2% silver	High current density operation.
Silicon iron	Brittle. Not susceptible to anodic dissolution.

Table 1.4.45 Comparison of sacrificial anodes

Material	*Characteristics*	*Efficiency (%)*
Aluminium	Low efficiency due to local action currents. Can be improved by alloying with Hg, but this can promote stress-cracking of nearby copper alloys.	50 90 with 0.04% Hg
Magnesium	Dangerous, explosive except in well vented open spaces clear of sparks or flames. Used as a sacrificial anode on ships where its total immersion in water renders it safe in operation. The only real danger is in installation.	60
Zinc	Must be high purity, free from iron.	95 (in seawater).
Mild steel	Used to protect alloys in saltwater-cooled condensers; also to protect stainless steels to minimize crevice corrosion effects in seawater.	

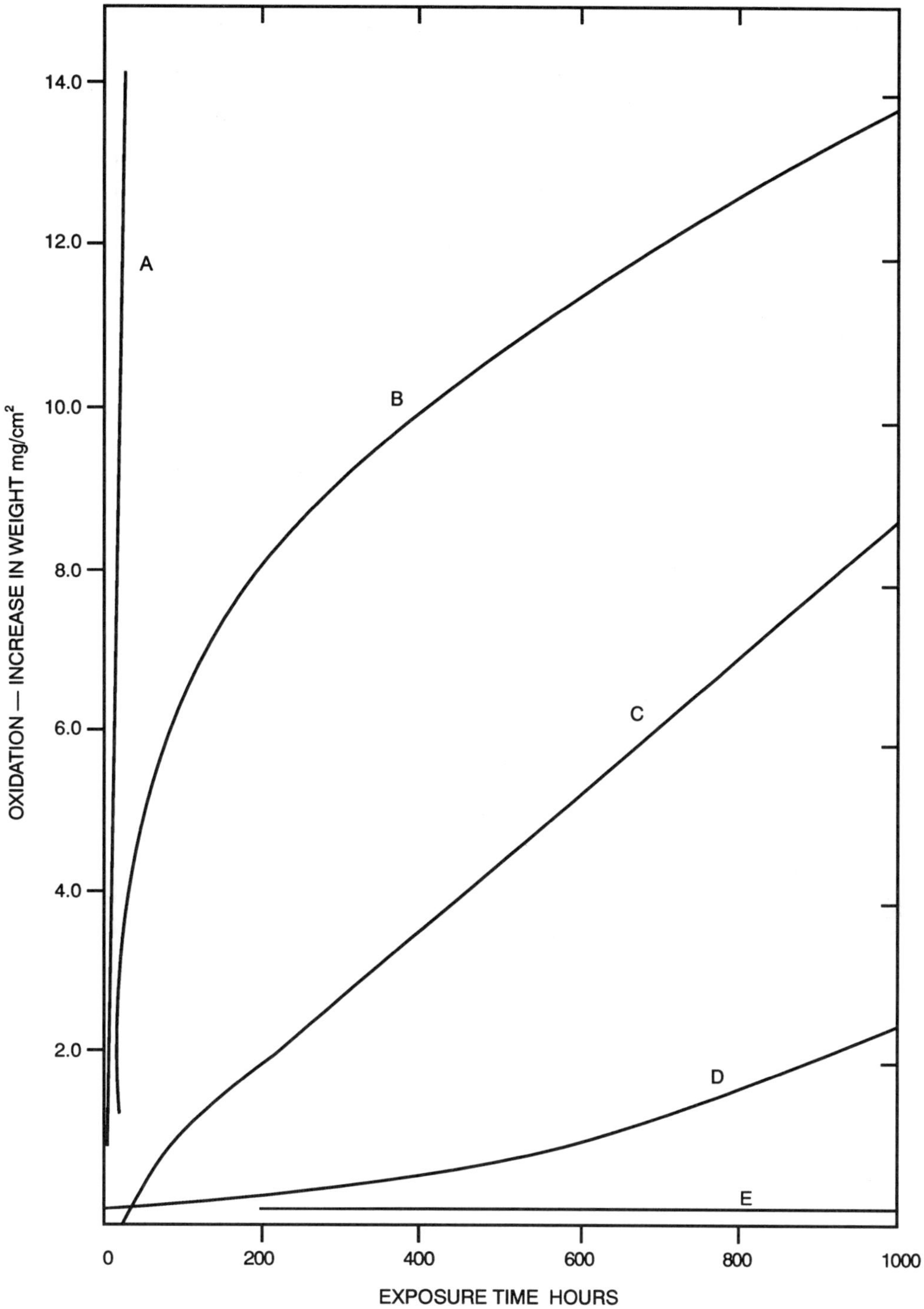

A - Unprotected
B - Aluminium spray coating
C - Commercial vitreous enamel
D - Special vitreous enamel
E - Chromized steel

Fig. 1.4.1 Oxidation of protected mild steel at 700°C in air

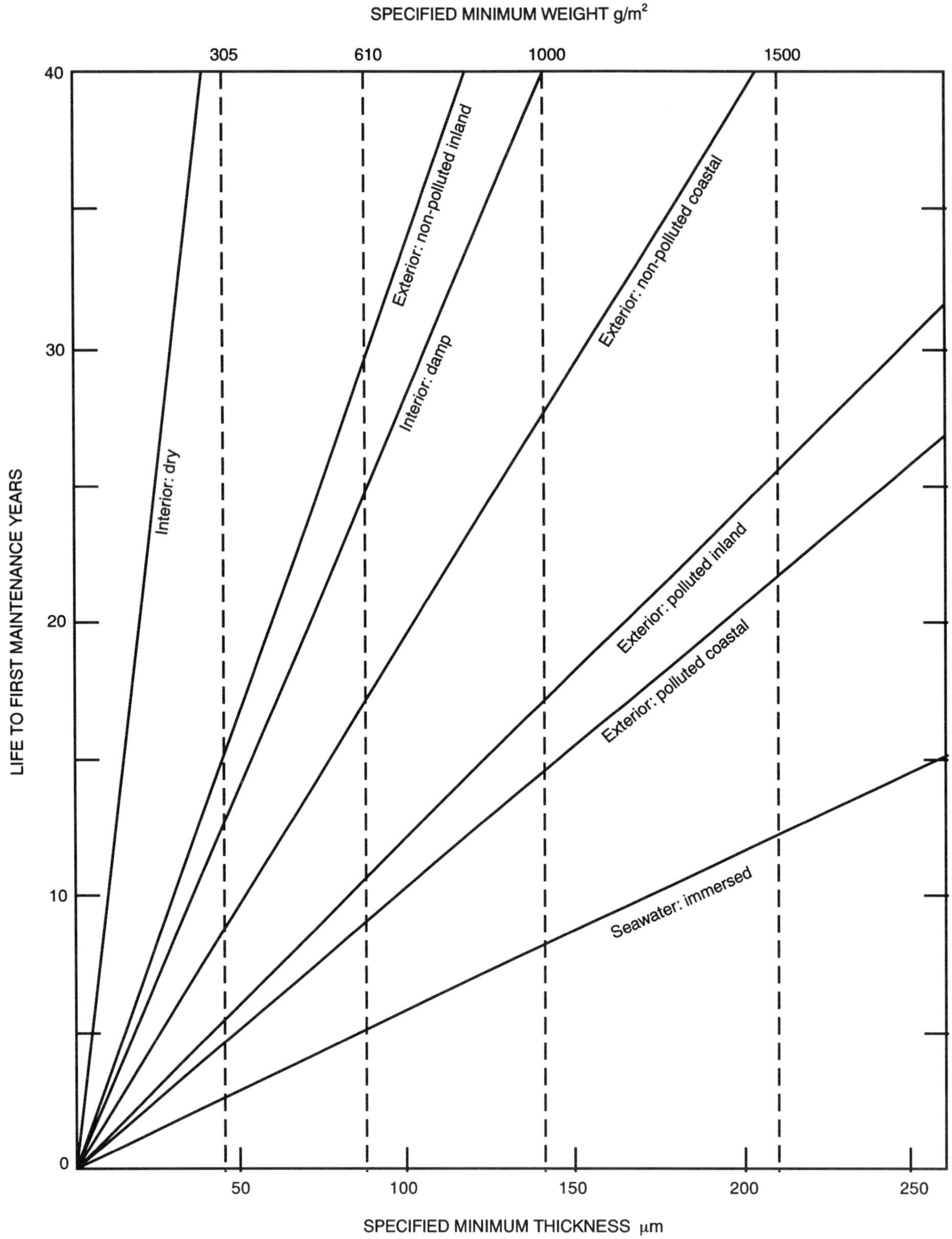

Fig. 1.4.2 Typical lives of galvanized zinc coating

Fig. 1.4.2

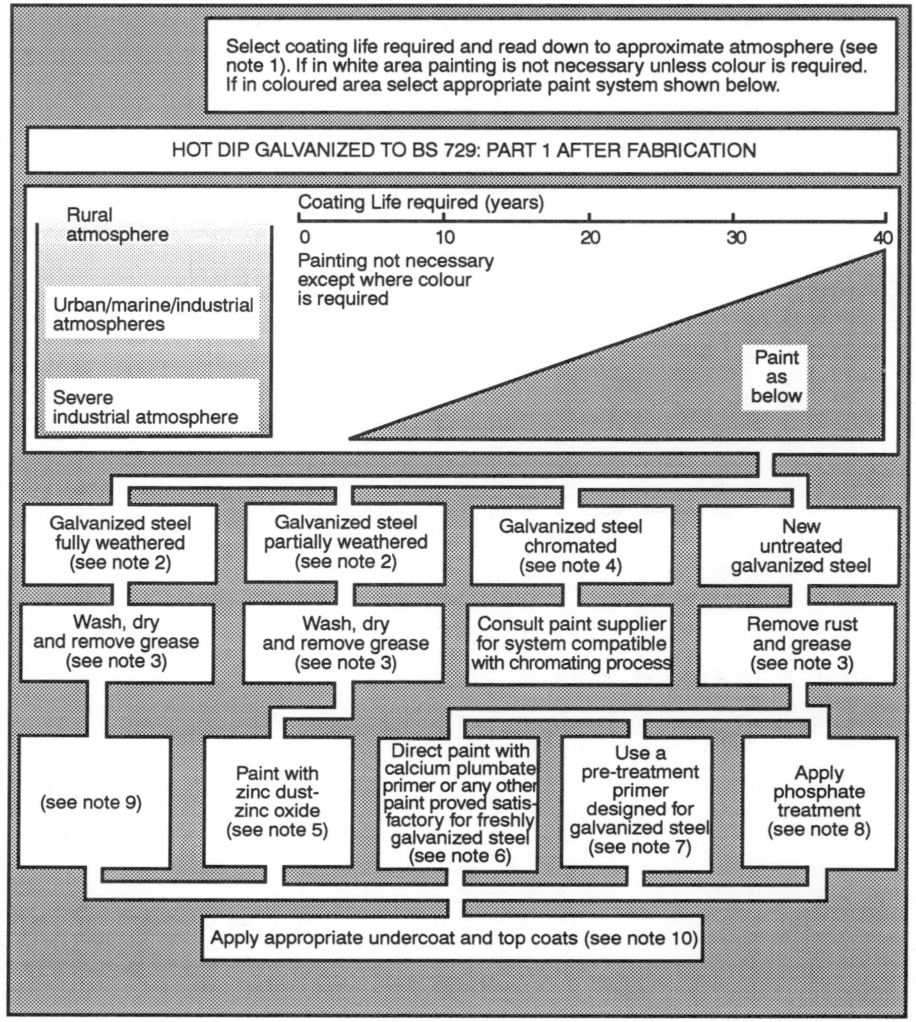

NOTES

1. The life of a zinc coating depends on the precise atmosphere and the amount of zinc present. The figures cover most normal outdoor uses in the United Kingdom and apply to a coating thickness of 610 g/m^2.

2. When galvanized steel is completely weathered and all the bright zinc surface has changed into a dull surface layer, most paints will perform satisfactorily. The time needed for weathering varies with the exposure conditions, and since it is not always easy to clean a weathered surface completely it may be more satisfactory to paint new galvanized steel . Structures that are not painted for some years will of course be fully weathered.

3. Freshly galvanized steel is free from grease, but it is necessary to remove any grease which appears after transport, assembly, erection, or use.
The simplest way to remove grease is to swab generously with white spirit, using several fresh swabs on each area.

4. Chromating is a treatment often applied to new galvanized sheet and wire, and sometimes to fabricated products, to prevent wet storage stain ('white rust').

5. Zinc dust-zinc oxide paints are based on conventional paint media pigmented with zinc dust and zinc oxide, generally in the ratio 80:20.

6. Calcium plumbate primers are pigmented with specially prepared calcium orthoplumbate; some conventional topcoats, notably the simplest alkyd paints, do not remain adherent to calcium plumbate in damp or marine environments. A number of other materials such as certain types of zinc dust paint and some epoxy resin paints are claimed to adhere directly to new galvanized steel; the makers' recommendations for these materials should be followed strictly.

7. Pre-treatment primers (also called etch primers or wash primers) slightly etch the metal surface and provide a thin bonding layer. Almost any paint may be applied over pre-treatment primers.

8. Phosphate treatment is most suitable for factory painting schedules. Long delay in painting after phosphating is unwise.

9. No special adhesive primers or pre-treatments or anti-corrosive paints are needed on fully weathered galvanized steel.

10. Choice of finishing coat is affected by its appearance and durability and its compatibility with treatments already applied. Normal decorative paints (undercoat and topcoat) are suitable from the point of view of appearance, but they may need modifying when applied over calcium plumbate or zinc dust paints. For long life under adverse conditions, a paint based on a synthetic medium may be necessary (e.g. epoxy resin paints, chlorinated rubber, etc.). These paints are best applied over pre-treatment-primed or phosphated surfaces. When overcoating other primers follow the maker's instructions.

Fig. 1.4.3 Guide to painting galvanized steel

1.4.3 Corrosion behaviour of ceramics

1.4.3.1 Introduction

Although engineering ceramics do not corrode in the conventional sense of the word (they are mostly electrical insulators so cannot undergo electrochemical attack), they do have some limitations in their ability to withstand harsh environments. The key to understanding the corrosion resistance of engineering ceramics is to view their performance from several viewpoints simultaneously:

1. resistance to attack of the main phase,
2. resistance to attack of all secondary phases,
3. arrangement and continuity of the secondary phases.

Thus while thermodynamics and practical experience might dictate that the principal crystalline phase should resist corrosion, the Achilles heel of most ceramics is the secondary phase(s). These are usually deliberately present as a result of design of the manufacturing process, but they rarely have the corrosion resistance of the principal phase. As an example, an alumina ceramic may be fabricated with a few per cent of clay and lime to produce a liquid phase during firing. Impurities and some of the alumina dissolve in this liquid, which on cooling forms a glassy phase which is continuous throughout the material. The acid resistance of this glassy phase is rather less good than that of the alumina, so the ceramic component can become penetrated by the acid without actually changing size or appearance. However, it becomes softer, weaker, and eventually will allow leakage. The formulation of secondary phases, or their elimination, is thus a key factor in determining corrosion resistance.

1.4.3.2 Ceramics for harsh environments

The suitability of a given ceramic for a particular application usually has to be assessed under simulated or more severe conditions than that application will impose. Since these vary widely, and the composition of ceramics is usually not published, it is difficult to provide meaningful numerical data for this compilation. Materials recommended for evaluation and those best avoided in selected environments are listed in:
Table 1.4.46 *Ceramic materials for resisting harsh conditions*

Other factors may need to be taken into account, such as required lifetime, concentration of corroding species, size and shape of component, or cost, which might restrict or expand the choice.

It is highly unlikely that individual manufacturers will have documented evidence of the performance of their products in more than a small number of specific conditions, and the user may have to resort to test or trials of candidate materials to provide the evidence required.

Table 1.4.46 Ceramic materials for resisting harsh conditions

Corroding conditions	***Materials recommended***	***Materials to avoid***
Water, room temperature	Most	Beta alumina
Water, 100°C	Most	Lanthanum chromite
Steam, 150°C	Most	Y-TZP
Dilute inorganic acids, room temperature	Most	Reaction-bonded SiC for HF/HNO_3 mixtures. Magnesia
Hot concentrated acids, >100°C	Chemical porcelain Borosilicate glassware High purity aluminas Silicon carbide	Some electrical aluminas Cordierites Beryllia Magnesia. Sintered silicon nitrides
Dilute alkaline solutions	Most silicate-free alumina materials	Glasses
Hot concentrated alkalis	Silicate-free alumina materials Vitreous carbon Boron nitride Magnesia.	Glasses. Glass ceramics. Some electrical aluminas. Titania
Oxidizing conditions, high temperatures	Oxides, such as alumina, mullite, zirconia (Si_3N_4 and SiC based materials may be satisfactory if continuous oxide scale develops)	Carbon Boron nitride Boron carbide Tungsten carbide Titanium carbide Titanium nitride
Reducing conditions, high temperatures	Carbon/graphite Alumina Thoria	Magnesia
Basic slags and glasses	Magnesia	Silica and some aluminosilicates
Fused alkaline salts	Magnesia. Boron nitride	Alumina. Zirconia
Molten oxides	Magnesia for PbO	Most non-oxides unless atmosphere is inert or reducing
Molten metals	Alumina, alumina-based, or zirconia-based for most ferrous metals Silicon nitride, for aluminium, lead, etc. Thoria, for many high melting or reactive metals Vitreous carbon, for gallium, arsenic and phosphorus containing compounds	Carbon, for many ferrous and other heavy metals which form carbides
Alkali metal vapours	Magnesia	Alumina Zirconia
Halogen gases at high temperatures	Alumina Silica, except against fluorine	Magnesia Boron nitride Silica-based materials for fluorine

For further more detailed review of information on corrosion-resisting properties of ceramics, see L A Lay, *Corrosion resistance of technical ceramics*, HMSO, 2nd edition 1991.

1.4.4 Degradation of plastics

1.4.4.1 Introduction

Whilst corrosion of metals generally involves oxidation leading to growth of metal oxide (or other) films or dissolution of metal by an electrochemical reaction, the effect of the environment on plastics generally leads to degradation and a deterioration in physical and mechanical properties. Plastics are prone to attack by weathering, chemicals, water and radiation (particularly ultraviolet) to a greater or lesser extent. The effects of the environment include swelling, chemical breakdown of the plastic, cracking and deterioration in properties. A guide to the environmental degradation of plastics is given in:
Table 1.4.47 *Degradation of plastics*

1.4.4.2 Comparative corrosion susceptibility of plastics

The resistance of plastics to a range of organic chemicals, acids, bases, salts, weathering and water is shown in:
Table 1.4.48 *General environmental performance of plastics*

In general, crystalline resins have much better chemical resistance, and solvent resistance in particular, than amorphous thermoplastics, because of their insoluble nature. However, crystalline polymers are particularly susceptible to environmental stress-cracking (ESC), a time dependent phenomenon. The environment is often a liquid or vapour that would not attack the plastic in the absence of stress. ESC often first makes itself apparent in the form of fine hairline cracks that reflect the light on the polymer surface. These cracks are normally bridged by polymer chains able to transfer some load, but if subjected to an impact load, failure is likely. Over a long period of time, the cracks may develop and promote failure in their own right.

The manufacturing conditions are important in determining degradation behaviour, especially in injection moulding, as they influence the size of any crystalline structures and the internal stress state. For this reason, raw materials manufacturers' guidelines should be followed.

Further comparative information on the environmental performance of plastics (particularly fire and high temperature performance) is given in Section 3.1. Details of particular plastics are given in the relevant sections in Vol. 3.

1.4.4.3 Degradable plastics

While degradation of plastics is generally undesirable, the pollution of the environment resulting from the disposal of non-degradable plastics is of increasing concern. While incineration (with energy generation) and recycling are options, plastics that degrade into harmless end-products have been another possibility since the 1970s.

Degradable plastics can be classified as photodegradable, semi-biodegradable or biodegradable. The photodegradable plastics are based on polyethylene and are not biodegradable, and their rate of degradation is dependent on climatic conditions. They have been used for agricultural mulch films, carrier bags and disposable packaging. The most typical semi-biodegradable plastics are master batches based on LDPE, HDPE and PS filled with starch, typically at 10–15%. The starch component is biodegradable, but not the base polymer. The applications are similar to those of the photodegradable plastics, although starch and the base polymer are basically incompatible and difficult to process, and the biodegradation of starch is slow.

Polyglycolic acid (PGA) and polylactic acid (PLA) are used widely for biodegradable suture materials. There is also interest in the use of PLA for biodegradable packaging. Blends of biodegradable polymers with starch are also promising (eg Mater-Bi^{TM} from

Novamont (IT). A number of other systems have been developed for controlled drug release implant systems (polyamide-enamides, polyhydroxy esters and polyanhydrides). Polyhydroxybutyrate (PHB) is biodegradable and biocompatible and may allow for the design of temporary medical implants. It is, however, expensive and difficult to process, although ICI (UK) have developed Biopol^{TM}, a copolymer with improved processing characteristics.

The market for degradable plastics remains relatively small and is currently estimated at less than 0.5% of all thermoplastics sold. The reasons for this would appear to be the relatively high development costs of the resin systems and their application in the very price sensitive area of packaging. Medical devices, however, look certain to develop into a niche market for certain biodegradable plastics.

1.4.4.4 Paint finishing of plastics

Plastics may be painted to protect them from corrosion, light, weathering and mechanical influences. Although paint finishing is more normally employed to:

- impart surface properties such as electrical conductivity, antistatic, abrasion resistance, smoothness, barrier to diffusion, colour (especially in small batch numbers when colouring of the moulding compound may be uneconomic);
- conceal surface defects (e.g. weld lines).

Certain difficulties may occur in the paint finishing of plastics which have to be taken into consideration in contrast with painting metals. Apparently identical plastics in the same material group, for example PVC, can sometimes differ very considerably. Factors which affect paintability include molecular structure and weight distribution, manufacturing process, material thickness, surface texture and incorporated additives.

PRE-TREATMENT

Plastics frequently have poor adhesion for coatings, and this adhesion principally depends upon the polarity of the plastics molecules. In many instances, pre-treatment of some kind is essential to improve coating adhesion. Physical methods include cleaning with solvents or ionized air, mechanical roughening, tempering and controlled slight swelling in powerful solvents. Chemical methods include flaming, corona discharge, and surface oxidation/etching.

Plastics are insulators and greater care must be taken than in the painting of metal parts, because of electrostatic build-up and possible adherence of dust. This is particularly true with thermoplastics, which have higher dielectric constants than thermosets.

A guide to pre-treatment and painting of some common plastics is given in:
Table 1.4.49 *Paint finishing of some common plastics*

DIRECTIONS FOR PAINTING SOME COMMON PLASTICS

ABS

When ABS is painted, a decrease in impact strength results and stress-cracking may occur. The more flexible the paint film the less the fall in impact strength.

Special grades of ABS can be electroplated.

Phenolics, melamine and urea formaldehydes

Readily coated by most paints, occasionally after slight surface roughening. Because of the high heat resistance of thermosets, stoving finishes are particularly suitable. In the coating of phenolic plastics, traces of free phenol may cause discoloration; with oxidative drying paints, extended drying times may be required.

Polyesters and epoxies

Polyesters must be degreased and freed from adherent dirt. As a rule, a simple wipe-over with a suitable solvent suffices (such as isopropanol). When trichloroethylene, acetone or other powerful solvents are used, a check must be made on the surface of the polymer to determine whether there is any attack on the surface which could give rise to coating defects. Surfaces of outstanding quality can be produced, indistinguishable to the eye from painted metal parts.

Moulding defects may occur on the surface or within the moulded part (blisters or voids) caused by:

air trapped between SMC layers or in the mould;
bubbles of air resulting from badly impregnated glass fibres in the pre-loaded mat;
voids resulting from particles being torn open.

Simple application of a primer coat or a base coat cannot adequately fill and render invisible relatively large surface defects; these have to be filled in with stopper compounds so that they can thereafter be painted over with normal stoving finishing systems.

For hand lay-up and prepreg moulding, a gel coat can be applied initially to the mould. Given the correct processing, 'print through' of any reinforcing fibres in the laminate should be minimized.

Polycarbonate

Unsuitable aggressive solvents and paint systems which are not sufficiently flexible can reduce impact strength significantly.

Special grades of polycarbonate can be electroplated with pre-treatment.

Polyethylene and polypropylene

There is no paint at the present time for non-pre-treated PE and PP. Virtually all paints however have good adhesion to pre- treated PE and PP. The real difficulty lies in the fact that pre-treatment for polyolefins is very costly and time consuming.

Blends based on EPDM copolymers with polypropylene, as used in motor vehicle construction, after suitable degreasing (e.g. solvent/steam) are directly painted with the aid of a special adhesion-promoting primer. Flexible paint systems are used for PP/EPDM externally mounted components.

Special grades of PP can be electroplated.

Polyamides (nylons)

Polyamide moulded parts have good resistance to solvents and show no tendency to stress cracking or crazing. Difficulties in the painting of polyamides are not caused by the plastic itself, but are due to surface adsorbed moisture which can amount to about 2–8%. For this reason paint systems selected are chiefly such as combine water (PUR paints) or are relatively hydrophilic, as for example, acid catalyzed systems.

Alternatively, the adhering moisture film may be removed immediately prior to painting by washing the surface with, for example, ethanol. No pre-treatment is required when stoving paint systems are employed.

Acrylic (polymethyl methacrylate)

PMMA has a pronounced stress-cracking tendency and is unstable towards polar solvents. A tempering process to relieve inner stresses (e.g. 6 h at 60–80°C) is desirable before painting with special materials.

Acetal

The high degree of crystallinity combined with the molecular structure of acetal polymers create severe adhesion problems. Without pre-treatment, painting of acetals with any hitherto known system is not possible. Only after a mechanical pre-treatment (e.g. surface rubdown with abrasive paper) or a chemical pre-treatment (e.g. 5 min/60°C immersion in 85% phosphoric acid) are certain systems effective.

Care should be taken to ensure that no oxidative drying alkyd resins are used on acetals which have been pre-treated with phosphoric acid, since a certain amount of formaldehyde is set free on the surface through the pre-treatment and this can lead to extended drying times.

Polystyrene

Polystyrene shows limited stability towards organic solvents, a tendency to stress-cracking and greater than normal dependence of the visual appearance on the geometry, type and location of the gating and on the processing parameters. Most of these considerations exist to an even greater extent with polystyrene foam, for which the sensitivity increases with the pore size of the foam, however this does make it suitable for solvent bonding.

Polyurethane

Finishing of integral foam moulded elements produced by RIM/RRIM is not straightforward. Blisters and pores in the surface layer are not always avoidable and when present make painting very troublesome. To facilitate the demoulding operation in the manufacture of the parts, effective release agents, usually waxes, are as a rule employed. These frequently interfere with subsequent painting, becoming noticeable through levelling or adhesion failure. After-cleaning of the moulding parts is therefore indispensable. In largescale operations this is most effectively carried out by means of a degreasing plant, operating with, for example, trichloroethylene.

Flexible RIM mouldings for automotive use are given a suitably flexible priming coat onto which a two-coat metallic finish is applied. The topcoat systems are almost invariably highly flexible two-component PUR finish materials, since resistance to fuel, fungicides, fungicide removers, dilute alkali solutions and dilute acids is required as well as flexibility. Light stability, weather resistance and gloss retention are also required.

PVC

When considering the application of a gloss finish for PVC mouldings, a distinction has to be made between unplasticized and plasticized PVC. Depending on the extent and the nature of the plasticizer, exudation and/or migration may occur, particularly under the influence of elevated temperature. With internally plasticized PVC grades these relatively undesirable characteristics do not occur.

Calendered PVC normally contains large amounts of interfering release agents such as stearates, silicone oil, paraffin oils and polyethylene. These must be completely removed before gloss finishing commences; a mixture of toluene and cyclohexanone has proven to be excellent for this purpose.

As far as paint adhesion is concerned, PVC is not a difficult substrate.

OPTIMIZATION OF PAINT FINISHES ON PLASTICS

Improvement of surface

A prerequisite for a perfect paint finish on plastics parts is the minimizing or avoiding of surface defects with the object of producing a substrate surface of the quality which has been usual for many years with metals. Such a surface can then be given a finish coat without excessive effort and with the same certainty. About 70% of the rejection rate of paint finished critical plastics parts is attributable to faults in the workpiece surface and 'only' about 30% to faults in the actual paint finishing plant.

More suitable paint systems

Optimization of the paint finishing of plastics is also assisted by the use of 'more suitable' paint systems which conceal the surface faults, or do not reveal them so clearly. Possible alternatives of this type are textured or delustered paint systems.

Lowering the drying temperature

Many faults in the product finish do not occur until after drying at elevated temperature in the paint dryer. Reducing the paint drying temperature also reduces the defects, although a longer residence time in the dryer results. Best of all would be a suitable paint system which cured at room temperature. The only process of this type which appears promising is curing by means of electron beams, using paint materials still to be developed commercially.

An advantage of this electron beam process is that the radical cure of the paint film applied proceeds with practically no appreciable heating. Further, identifiable advantages are:

the extremely short cure time – a matter of seconds,
the hard, scratch resistance surface,
the excellent adhesion, even on troublesome substrates.

Disadvantages of the electron beam systems at the present time are:

trouble-free cure available only in inert gas atmosphere,
restricted flexibility,
matt finish with spray painting systems is difficult to obtain,
poor overpainting on a similarly applied previous coating,
limited weather resistance,
physiological objections to the monomeric reactive diluent required.

Despite these limitations, interest in electron beam systems is relatively high since this method permits the use of quasi solvent-free paint systems, the energy requirement for curing is small in comparison with other methods, and production capacity of possible plants is relatively very great.

Use of more economical application processes

Plastics materials are usually coated by means of normal high pressure spray guns, since the plastics mouldings as a rule have too small and too fissured a surface for airless spraying. With level plastics surfaces, roller coating and flow coating have been found suitable. Dipping and flow coating are confined chiefly to thermosets because of the long time in contact with solvents. Manual operation with brush or roller may be eliminated from consideration since productivity is too low and attainable quality almost always unacceptable.

Plastics are electrical non-conductors and cannot therefore be electrostatically coated unless an electrically conducting surface layer is provided. This can be done, for example, by immersing the workpiece in a dilute solution of *p*-toluene sulphonic acid, or of quaternary amines, or through addition of these substances in a basecoat. A disadvantage in introducing conductive materials is the fact that these materials are water soluble and will remain so. If exposed to steam there is a risk of blister formation through osmotic action.

By using special conductive basecoats these basic faults can be avoided and plastics parts basecoated in this way can be electrostatically finish coated. The reduction in overspray loss by the use of electrostatic spray equipment is of the order of between 10 and 20%.

Table 1.4.47 Degradation of plastics

Type of degradation	***Description***	***Significance***	***Materials (polymers) particularly prone***
Hydrolysis	Chemical breakdown resulting in the addition of the elements of water. Can be caused by the action of solutions of acids or alkalis; exposure to UV light; exposure to weathering action.	General deterioration of physical properties: strength, stiffness with corresponding embrittlement; loss of clarity and surface finish; surface crazing; deterioration and/or loss of uniformity of colour; deterioration of electrical properties; increase in water (vapour) transmission.	Many thermoplastics but in particular the cellulose esters; some nylons; some polyesters to a limited extent and under certain conditions. Loss of clarity in transparent materials.
Oxidation	Treatment with oxidizing acids; exposure to UV light (photo-oxidation); prolonged application of excessive heat; weathering (see also below).	Deterioration of mechanical properties resulting in embrittlement and in some cases stress-cracking; increase in power factor; loss of uniformity of surface; discoloration and loss of clarity.	Most thermoplastics in varying degrees and in particular the polyolefins, PVC, the nylons, and cellulose derivatives.
Weathering (including water absorption)	Generally results from the combined effect of UV exposure and water absorption. The former causes the breakdown in the bonds in the polymer chain, the latter has a plasticizing action on the polymer which increases material flexibility but ultimately (on the elimination of the water) causes embrittlement.	General deterioration of physical properties, resulting in some cases in environmental stress-cracking; deterioration in colour and/or clarity. Absorption of water reduces the dimensional stability of the plastics sheet or moulded unit.	Most thermoplastics, and particularly cellulose derivatives; polyethylene; PVC; nylons.
Physical attack	Physical attack by ultrasonic or high energy radiation causes rupture of polymer chains with consequent degradation of overall physical and mechanical properties. By contrast high energy radiation (gamma rays) affects cross-linking of polymer chains under controlled conditions, thus converting a thermoplastic into a thermoset (or nearly so).	General deterioration in physical, mechanical, optical and electrical properties either singly or in combination in extreme cases.	All polymers suffer (including synthetic rubbers and polyurethanes); action is less in the case of the thermosetting materials (suitably selected according to the end use), which are less prone to attack.
Chemical attack (including environmental stress-cracking or crazing)	Organic solvents have solvent or swelling action on many thermoplastics (hydrocarbons less so). Chemical attack by strong acids and/or alkalis (see also hydrolysis; oxidation). Residual catalyst or trace metal in the polymer can cause breakdown depending on temperature and/or time of action.		

Corrective action will always be taken by the manufacturer to prevent or minimize degradation. Thus anti-oxidants and UV stabilizers for example are standard additives, which are used as needed depending on the polymer in question and the end use envisaged; in no case can the degradation be completely eliminated but sufficiently so for many practical and industrial purposes.

Table 1.4.47

Table 1.4.47 Degradation of plastics – *continued*

Type of degradation	*Description*	*Significance*	*Materials (polymers) particularly prone*
Mechanical breakdown	Plastics are subject to creep; this applies in particular to thermoplastics, and can take place even at relatively low temperatures, depending on the time the load is applied for. Wear reduces the dimensional accuracy of plastic components, and this also occurs when plastics (e.g. elastomers) are used as tank linings. Fatigue, static or dynamic (the latter due to work heat), tends to result in molecular break down of the polymer chain. Mouldings which have 'built-in' stresses (due frequently to incorrect mould design and inappropriate processing conditions) are prone to breakdown or exhibit surface cracking in ultimate use.	Creep in particular will adversely affect the dimensional accuracy of moulded components, which would otherwise withstand normal wear. The same applies with 'built-in' or developed moulding stresses. Friction (plastic on plastic or plastic/ metal etc.) is an important factor as frictional heating can cause degradation. With metal–plastic combinations some wear, in the form of a transfer film is desirable for low total wear. Reinforced and laminated products should be selected for stressed applications.	All plastics are subject to a greater or lesser degree. Suitable reinforcement can however upgrade most plastics. In the case of laminates the orientation of the reinforcement must be such that applied stress does not induce delamination.
Thermal action	Has relatively small effect on thermosetting materials, but in the case of thermoplastics it increases the distance between the molecular chains resulting in loss of strength and ultimate flow of the material. If the applied temperature is sufficiently high and prolonged, complete decomposition (of thermoset or thermoplastic) will result. In the case of polymethyl methacrylate the polymer is 'cracked', with regeneration of the monomer. The application of moderate heat to thermoplastics is a reversible process. A more specialized case of thermal degradation is that of electrical 'tracking', in which the surface breaks down (carbonizes) due to the actual 'flash-over' between the terminals. Whilst most thermosets char gradually when excessive heat is applied, thermoplastics burn freely, some almost explosively (celluloid), others melt and burn whilst dripping. Some are self-extinguishing.	Thermal action draws the distinction between thermoset and thermoplastic materials. In the case of the latter the design must take note not only of the melting point of the polymer, but the temperature at which it begins to soften, and in particular the point at which it cannot further sustain the required load. Repeated and prolonged application of heat (even below softening point), may reduce mechanical properties, and ultimately result in chemical breakdown. Prolonged exposure even to moderate heat under load can cause creep in many thermoplastics. The thermal cracking of polymethyl methacrylate has no particular significance beyond the economics of regeneration and recovery of scrap material. Particular regard must be paid to the fire hazard in the use of plastics.	Most thermoplastics. Thermosetting composite laminates can be formulated to counter creep and thermal action, to a limited degree. In high voltage electrical apparatus the plastics used must be selected to avoid surface tracking, as once this occurs consequences may be serious. Fire risks need careful attention depending on the application. Hydrocarbon polymers offer the greatest risk, chlorinated polymers are relatively safe. In any case the danger can be reduced by the addition of suitable 'anti-flash' additives.

Table 1.4.47 – *continued*

Table 1.4.48 General environmental performance of plastics

Material		Aliphatic hydro-carbons	Aromatic hydro-carbons	Oils fats waxes	Fully halo-genated hydro-carbons	Partially halo-genated hydro-carbons	Alcohols mono-hydric	Alcohols poly- hydric	Phenols	Ketones	Esters	Ethers
Acrylics		Fair	Poor	Good	Poor	Poor	Poor to fair	Good	Poor	Poor	Poor	Poor
Cellulosics		Fair	Variable	Good	Fair to poor	Fair to poor	Fair to poor	Excellent	Poor	Poor	Poor	Fair
Chlorinated polyether		Good	Good	Good	Fair	Fair	Good	Excellent	Good	Fair/poor	Fair	Good
Fluoroplastics	PTFE	Excellent	Excellent	Excellent	Excellent	Good	Excellent	Excellent	Excellent	Excellent	Excellent	Good
	PEP	Excellent	Excellent	Excellent	Excellent	Good	Excellent	Excellent	Excellent	Excellent	Excellent	Good
	PFA	Excellent	Excellent	Excellent	Excellent	Good	Excellent	Excellent	Excellent	Excellent	Excellent	Good
	PVDF	Excellent	Good	Excellent	Good	Fair/good	Excellent	Excellent	Fair	Fair	Fair	Good
	PTFCE	Excellent	Fair	Good	Good	Good	Excellent	Excellent	Good	Fair	Fair	Poor
Polyacetals[a]		Good	Variable	Good	Fair/good	Good	Good	Good	Poor	Good	Good	Good
Nylons (polyamides)		Good	Fair	Good	Fair	Poor	Good	Good	Poor	Good	Fair	Good
Polyamide-imide		Good	Good	Good	Good	Good	Good	Good	Poor?	Good	Good	Good
Polycarbonate		Fair	Poor	Poor/fair	Poor	Poor	Mainly good	Good	Poor	Poor	Poor	Poor
Thermoplastic polyesters		Good	Poor/good at RT	Good	Poor/good	Poor	Good/ excellent	Excellent	Poor	Good	Good	Good
Polyethylene		Fair	Fair/poor	Excellent	Poor	Poor	Good to fair	Good	Excellent	Good	Good	Good
Polyolefin copolymers		Good/ poor	Poor/fair	Good	Poor	Poor	Good	Good		Poor	Good at RT	Poor
Polyphenylene oxide (mod.)			Poor	Variable	Poor	Poor	Excellent	Excellent		Poor		
Polyphenylene sulphide		Excellent	Good	Good	Good	Good	Excellent	Excellent	Good	Good	Excellent	Good
Polypropylene		Fair	Generally good	Good	Good at room temp.	Good at room temp.	Good	Excellent	Good	Good	Good	Fair
Polysulphones	Polysulphone	Poor	Poor	Good	Poor	Poor	Good to fair	Good	Poor	Poor	Poor	Poor
	Polyethersulphone	Good	Variable	Good	Poor	Poor	Good			Poor	Poor	Poor
	Polyarylsulphone	Fair	Poor	Good	Poor[c]	Poor[c]				Poor[b]	Poor	
	Polyphenylsulphone	Good	Good		Poor	Poor				Poor	Poor	
Polyurethanes		Fair	Fair to poor	Good	Poor	Poor	Fair to poor	Fair to good	Poor	Fair to poor	Poor	Poor
Polyvinyl chloride		Good	Poor	Good	Poor	Poor	Good	Good	Poor	Poor	Poor	Poor
Vinyl esters		Good at RT	Good at RT		Good at RT	Good at RT				Poor	Poor	

Table 1.4.48

1.4.48 General environmental performance of plastics – *continued*

Inorganic acids		*Bases*		*Salts*			*Organic acids*		*Oxidizing acids*		*Sunlight and weathering*	*Fungus and bacteria*	*Compatibility with foodstuffs*	*Hot water*	*Sterilizability (medical applications)*	*Detergents*
Conc.	*Dilute*	*Conc.*	*Dilute*	*Acid*	*Neutral*	*Basic*	*Conc.*	*Dilute*	*Conc.*	*Dilute*						
Poor to good	Good	Good	Good	Good	Exc.	Good	Poor	Fair to poor	Poor	Good	Exc.			Depends upon grade		Excellent
Poor	Fair	Poor	Fair	Good	Good	Fair	Poor	Poor	Poor	Poor	Some grades OK			No	No	Fair to Excellent
Fair	Good	Fair	Good	Good	Good	Good	Fair	Good	Poor	Fair	Fair			Good		Excellent
Exc.	Exc.	Exc.	Exc.	Exc.	Exc.	Exc.	Good/ exc.	Exc.	Exc.	Exc.	Exc.	Exc.		Exc.		Excellent
Exc.	Exc.	Exc.	Exc.	Exc.	Exc.	Exc.	Good/ exc.	Exc.	Exc.	Exc.	Exc.	Exc.		Exc.		Excellent
Exc.	Exc.	Exc.	Exc.	Exc.	Exc.	Exc.	Exc.	Exc.	Exc.	Exc.	Exc.	Exc.		Exc.		Excellent
Exc.	Exc.	Exc.	Exc.	Exc.	Exc.	Exc.	Fair	Exc.	Good	Exc.	Good	Exc.	Exc.	Exc.		Excellent
Exc.	Exc.	Exc.	Exc.	Exc.	Exc.	Exc.	Fair	Exc.	Exc.	Exc.	Exc.	Exc.	Exc.	Exc.		Excellent
Poor	Poor	Fair to poor	Fair to poor	Fair	Good	Good	Poor	Fair	Poor	Poor	Generally good		Not recom.	Good		Fair to excellent
Poor	Good	Good	Exc.	Poor	Good	Fair	Poor	Fair	Poor	Poor	Fair to good	Very resistant	Resistant but stains	Some grades fair	By steam or irridation	Fair to excellent
Good	Good	Poor	Poor	Good	Good	Poor	Generally good	Generally good	Good	Good	Good					
Fair	Good	Poor	Poor	Good	Exc.	Fair	Fair	Fair	Poor	Good	Good		Approved grades	Degrades <65°C	No	Fair
Good/ fair	Good	Poor	Poor/ fair	Good	Good	Poor/ fair	Poor/ fair	Good	Poor	Fair/ poor	Some good			Degrades <70°C	No	Good
Fair to good	Exc.	Good	Good	Exc.	Exc.	Exc.	Exc.	Exc.	Poor	Good	Poor	Resistant	Non-toxic grades	HDPE has been used at 90°C	No	Fair
Fair to poor	Good at RT; Variable at elevated temperatures						Generally good	Generally good	Good	Good	Good (esp. EVA)				By low level radiation or chemicals	Good
Fair	Good	Exc.	Exc.	Good	Good	Good					Poor		OK	Exc.	Exc.	Excellent
Good	Good	Good	Exc.	Good	Good	Good	Good	Good	Poor	Fair	Good		Good	Exc.		
Exc.	Exc.	Exc.	Exc.	Exc.	Exc.	Exc.	Good	Exc.	Poor	Good	Poor		Non-toxic grades	Exc.	Steam sterilized	Excellent
Good	Good	Good	Exc.	Good	Exc.	Good	Fair	Good	Poor	Good	Good to exc.		Approved grades	Good[d]	Repeatedly steam sterilisable	Good
Variable	Good	Good	Exc.	Good	Good	Fair	Fair	Fair	Good	Good	Not rec. without pig.					Good
	Good		Exc.					Exc.[f]								
	Exc.		Exc.											Exc.[d]		
Poor	Fair	Poor	Fair	Good	Exc.	Exc.	Poor	Fair	Poor	Poor	Good		Restricted applic.	Not rec.		Fair to poor
Good	Exc.	Exc.	Exc.	Exc.	Exc.	Exc.	Poor	Good	Good	Exc.	Good	Poor	Special grades. Some concern exists.	Not rec.[e]		Moderate to Exc.
Good at RT	Good at RT	Good at RT	Good at RT	Good at RT	Good at RT	Good at RT										

Table 1.4.48 – *continued*

Table 1.4.48 General environmental performance of plastics – *continued*

Material		Aliphatic hydro-carbons	Aromatic hydro-carbons	Oils fats waxes	Fully halo-genated hydro-carbons	Partially halo genated hydro-carbons	Alcohols mono-hydric	Alcohols poly- hydric	Phenols	Ketones	Esters	Ethers
Styrene-based polymers	ABS	Fair	Poor	Good	Poor	Poor	Fair	Good	Poor	Poor	Poor	Poor
	PS	Fair/poor	Poor	Mainly poor	Poor	Good	Good	Good	Poor	Poor	Poor	Poor
	SAN	Good	Poor	Good	Poor	Poor	Fair	Good	Poor	Poor	Poor	Poor
TPX			Poor	Good	Poor	Poor				Fair to poor		
Polybutylene		Good at RT	Fair/ poor		Poor	Poor				Good at RT	Good at RT	
Polyesters and alkyds		Good	Fair to poor	Good	Poor	Poor	Fair	Fair	Poor	Poor	Poor	Good
Aminos	Melamine formaldehyde	Good	Good/ excellent	Fair to good	Excellent	Excellent	Excellent	Excellent	Good	Good	Good	Exc.
	Urea formaldehyde	Good	Good	Good	Excellent	Excellent	Good to excellent	Excellent	Good	Good	Fair to excellent	Exc.
Diallyl phthalate			Good	Good	Poor	Poor	Fair			Poor		
Epoxy resins		Excellent	Good to excellent	Excellent	Excellent	Good	Excellent	Excellent	Fair	Fair	Good	Fair
Furan-e		Good	Good	Good	Exc.	Exc.	Exc.	Exc.	Good	Good	Good	Exc.
Phenolics		Excellent	Excellent	Excellent	Excellent	Good	Good	Excellent	Excellent	Good	Good	Exc.
Polybutadiene		Good	Good	Good	Good	Good	Excellent	Excellent	Good	Good	Good	Good
Polyimides		Excellent	Excellent	Excellent	Excellent	Excellent	Excellent	Excellent	Good	Excellent	Excellent	Exc.
Silicones		Fair	Fair to poor	Excellent	Poor	Poor	Good	Excellent	Poor	Good	Fair	Fair

[a] Wide variations exist since copopolymers are more inert than homopolymers.
[b] Stress-cracks.
[c] Freons have little effect.
[d] Withstands prolonged immersion in boiling water.
[e] CPVC commonly used for hot water pipes.
[f] Highly swollen by hot acetic acid.

Table 1.4.48 – *continued*

1.4.48 General environmental performance of plastics – *continued*

Inorganic acids		***Bases***		***Salts***			***Organic acids***		***Oxidizing acids***		***Sunlight and weathering***	***Fungus and bacteria***	***Compatibility with foodstuffs***	***Hot water***	***Sterilizability (medical applications)***	***Detergents***
Conc.	***Dilute***	***Conc.***	***Dilute***	***Acid***	***Neutral***	***Basic***	***Conc.***	***Dilute***	***Conc.***	***Dilute***						
Poor	Good	Good	Good	Good	Exc.	Good	Poor	Fair to poor	Poor	Good	Fair to good		Special grades only		No	Excellent
Poor to fair	Good	Fair	Exc.	Good	Exc.	Good	Poor	Fair	Poor	Fair	Fair to good		OK	No	No	Great variation
Good	Good	Good	Exc.	Good	Exc.	Good	Fair	Good	Poor	Good	Fair		OK	No	No	Good to excellent
Good	Good	Good	Good	Good	Good	Good			Poor		Fair		Yes	Yes	Yes	Good
	Good at RT		Good at RT	Good at RT	Good at RT	Good at RT			Good at RT	Good at RT						
Good	Poor	Fair	Good	Exc.	Exc.	Good	Good	Good	Poor	Fair	Fair to good			Good		Good
Poor	Good to fair	Poor	Good	Exc.	Exc.	Good	Good	Good	Poor	Fair	Good		OK	Some effect on properties		Good
Poor	Fair	Poor	Fair	Good	Good	Good	Good	Good	Poor	Fair	Good		OK	Some effect on properties		Very Good
Fair	Good	Fair	Good			Poor	Poor to fair				Good					Good
Fair	Exc.	Exc.	Exc.	Exc.	Exc.	Exc.	Fair	Fair to good	Poor	Fair	Good		Fair	Fair		Good
Good	Good	Good	Good	Exc.	Exc.	Good	Good	Exc.	Poor	Fair	Good					
Fair to poor	Fair to good	Poor	Poor	Exc.	Exc.	Fair	Good	Fair	Poor	Poor	Good					Fair to good
Fair	Exc.	Good	Good	Exc.	Exc.	Good	Fair/ good	Exc.	Fair	Good	Very good					
Poor	Fair	Poor	Fair	Good	Good	Poor	Good	Exc.	Poor	Good	Exc.	Very resistant		Exc.		Good
Depends on acid	Good	Good	Good	Exc.	Exc.	Exc.	Fair	Good	Poor	Exc.	Exc.					

Table 1.4.48 – *continued*

Table 1.4.49 Paint finishing of some common plastics

Plastic		Paint finishes									
		Pre-treatment[a]	Single coat		Multicoat						
					Primer			Topcoat			
			Components		Components		Special primer	Special basecoat	Components		Stoving finish
			1	2	1	2			1	2	
Thermoplastics											
PE/PP	polyethylene/polypropylene	4		X		X				X	
PP-EPDM	polypropylene blends with varying amounts of ethylene, propylene, diene and synthetic rubber	1,2					X			X	
POM	polyacetal	4		X		X				X	
PPO-PS	polyphenylene oxide–polystyrene	1	X	X				X°		X°	
PS	polystyrene	1	X		X					X	
ABS	acrylonitrile–butadiene–styrene	1	X	X		X				X	
PVC	polyvinyl chloride	1	X	X				X°		X°	
PMMA	polymethyl methacrylate	1,4[b]	X	X							
PA	polyamide (nylon)	1	X	X		X				X	X
PC	polycarbonate	1		X	X					X	X
CAB	cellulose acetate butyrate	1	X	X				X°		X°	
Thermosets											
PUR rigid	polyurethane rigid foam	1,2		X	X	X				X	X
PUR flexible	polyurethane flexible foam	1,2	X				X	X°		X	
GF–UP	glassfibre-reinforced unsaturated polyester resin	2,3		X	X	X				X	X
GF–EP	glassfibre-reinforced epoxy resin	2,3		X	X	X				X	X

[a]Pre-treatment: 1 Solvent degreasing
2 Solvent/steam degreasing
3 Aqueous cleaning
4 Protracted special method
[b]With parts of complicated design tempering is necessary.
X° 2-coat metallic finishing application

Table 1.4.49

1.4.5 Biodeterioration

1.4.5.1 Introduction

The performance of a material in an environment may be adversely influenced by biological action. Certain examples of industrially important mechanisms of attack, which indicate where the possibility of biodeterioration may have to be taken into account, and the countermeasures which can be used, are included in this section.

Action may take the form of a direct attack by a living organism, for example the consumption of cellulose by a mould, or by beetle larvae, or may entail the conversion of a non-aggressive to an aggressive environment, for example the attack of metals by sulphuric acid produced from sulphides by thiobacilli. Most of these forms of attack are inhibited by maintaining a dry environment. Where this is impracticable or ineffective a biocide may be incorporated in the material or the environment.

Other, larger scale, forms of biodeterioration include the destruction of aerospace components by the impact of birds, and comminution by rodents of relatively soft materials.

The number of aggressive organisms and of materials subject to attack is very large and the method of conferring resistance may be specific to a given combination.

An important form of biodeterioration which does not necessarily consume structural material is marine fouling. Any non-toxic surface immersed in seawater will accumulate a substantial organic growth; vegetable such as algae, and animal such as barnacles. These incrustations greatly increase a ship's frictional resistance and may provide sites for organisms that cause pitting corrosion. This was reduced or prevented in wooded ships by copper sheathing which cannot, however, be used for iron ships which must be protected by anti-fouling paints containing cuprous oxide and mercury compounds, or other anti-fouling components.

Fouling is not restricted to ships. Serious problems can arise in industrial seawater intakes, shipboard and dockside fire mains and condenser systems. The most effective means of prevention in these locations is chlorination of the water if this can be tolerated ecologically.

1.4.5.2 Biodeterioration of metals

No direct attack by bacteria on metals has been observed, but the electrochemical processes fundamental to the aqueous corrosion of the metals can be initiated or augmented or in certain cases inhibited by their metabolic action. Some bacteria promote the formation of surface biofilms, creating an anaerobic environment where other damaging bacteria may grow. Some of the metabolic activities which can modify the environmental resistance of the most commonly used metals are as follows. These processes are generally seen at temperatures in the range 4–60°C.

ABSORPTION OF NUTRIENTS (INCLUDING OXYGEN) BY MICROBIAL GROWTHS

All types of bacteria that can grow in a coherent colony or clump are potentially capable of producing a differential aeration cell. The process has been mainly observed in heat exchangers incorporating a wet cooling tower which concentrates organic matter. Metal loss and pitting corrosion have been observed on cast iron, carbon steel, low alloy steel and copper alloys. The rate of attack is reduced by plating with zinc or cadmium, and still further reduced in fresh water by the use of stainless instead of carbon steel. No instances of attack on titanium have been observed.

Attack may be reduced or prevented by lowering the biostatic concentration in the system by more frequent changes of water. Infected systems must be disinfected by the addition of biocides.

PRODUCTION OF ORGANIC ACIDS AS END-PRODUCTS OF FERMENTATIVE GROWTH

Organic acid attack may affect most non-ferrous metals or aerial systems. The best known case is the attack of aluminium aircraft fuel tanks by the kerosene metabolizing mould *Cladosporium resinae* which grows at kerosene water interfaces. The attack can be reduced by protective coatings or biocides, but both these remedies have drawbacks.

PRODUCTION OF SULPHURIC ACID FROM LOWER VALENCY SULPHUR COMPOUNDS

Sulphuric acid attack can take place in buried metals and in sewers. In mines, one of the causative organisms, *Thiobacillus ferroxidans*, may be used in the recovery of metals from low grade ores.

The cure in the case of buried metals is to use lime as an inhibitor. Sulphide formation in sewers should be prevented by efficient aeration. In mines, pumping and other machinery should be constructed of acid resistant stainless steel.

INTERFERENCE WITH THE CATHODIC (PROTECTION) PROCESS IN OXYGEN-FREE CONDITIONS BY SULPHIDES PRODUCED BY SULPHATE REDUCING BACTERIA

Sulphate reducing bacteria occur in wet clay soil and produce reducing sulphides from sulphate. The reducing sulphide reacts with iron to produce a loose scale which can easily be prized off leaving a light metallic surface. The most serious problem is the corrosion of pipelines, but piers, sheet piles and the hulls of berthed ships in contact with estuarine mud suffer corrosion which is almost impossible to prevent. Food cans may also be attacked and the contents spoiled.

The cure so far as pipe lines is concerned may lie in the provision of a non-aggressive surround, consisting of a backfill of sand or chalk (possibly with a biocide incorporated), the use of cathodic protection or the provision of a barrier coating, but this latter must be completely impervious which is, in the case of a pipeline, difficult to ensure.

MISCELLANEOUS FORMS OF ATTACK

Other biological attack on metals includes corrosion by bird droppings which may attack old metal roof structures and cavitation caused by the presence of shellfish in heat exchangers. This can be controlled by regular chlorination.

1.4.5.3 Biodeterioration of plastics

The number of polymers and of aggressive micro-organisms (bacteria and moulds) is too great for the production of a detailed schedule of specific resistance.

There are, however, principles which may be used to guide the choice of materials which will resist most environments met with in practice.

Parameters which confer resistance or immunity to pure polymeric materials are:

- Increased halide content.
- Increased chain length.
- Increased cross-linking.
- Toxicity of breakdown products to bacteria.
- Presence of sulphur within the molecular structure.
- Absence from the molecular structure of naturally occurring groups on to which enzymes can 'key' or which may be broken down into products upon which the micro-organisms can feed.

Most polymers, including fluoroplastics, vinyls, epoxies and polystyrenes are, so far as is known, immune, whereas natural products such as animal and vegetable glues and melamine formaldehyde, cellulose nitrate, polyvinyl acetates and polyester type polyurethanes are attacked by micro-organisms. Natural rubbers are attacked, particularly when degraded by UV light or by heat.

Most commercially used plastics are, however, formulations of a polymer with a filler and/or a plasticizer or stabilizer. Most stabilizers are inorganic salts of heavy metals and might be expected to increase resistance to biodeterioration. A filler, such as saw-dust or starch is attacked unless the polymer forms linkages with at least some of the OH groupings of the carbohydrate.

Attack of the plasticizer in industrial vinyls is the major cause of breakdown in properties of plastic materials. Consumption of the plasticizer (most of which are long-chain organic acids or esters, able to support the growth of micro- organisms) seldom causes the plastic to disintegrate, but can embrittle and cause changes in shape and texture.

Attack of the plasticizer (or the polymer) may be controlled by the incorporation of small concentrations of a biocide in the formulation (or on the surface) of the plastics. Biocides include a variety of compounds including those of tin, copper and sulphur (or sulphur/zinc). There is evidence that they are not really satisfactory for plasticized vinyls, mainly because access of the biocide to the plasticizer may be impeded by the polymer.

Internally plasticized thermoplastic resins such as monochlorotrifluoroethylene, Nylon 66, polyethylene, vinylidene chloride/vinyl chloride, Teflon, Tygon, Velon, Vinyon, and vinylstearate/vinyl chloride mixtures have been shown to resist micro-biological attack.

The resistance of polymers is strongly influenced by environmental conditions, and the exclusion of moisture inhibits microbiological attack on both stored ingredients and finished products. Because of their reduced hardness relative to most metals, polymers are subject to mechanical attack by insects (termites, cockroaches, or beetle larvae) or rodents. The plastic materials do not appear to be used as food by the organisms but are removed to facilitate access to food, or reproductive activity. Almost all plastics have been shown to be attacked by some organisms. This is unlikely to cause severe widespread damage except perhaps in tropical regions. It may be prevented by increasing the hardness of the plastic or by incorporating a suitable insecticide or rodenticide.

One form of mechanical attack is the destruction of plastic aircraft components by bird strikes which impinge harmlessly on metals. These may be reduced or eliminated by acoustic broadcasting of distress calls which disperse the birds. See also Section 1.4.4.3.

1.4.5.4 Biodeterioration of stone and concrete

The sulphuric acid production by thiobacilli described in relation to metallic corrosion above may have a dramatic effect on concrete. Concrete sewers and even cooling towers have been known to collapse from acid attack. When this occurs the sulphur source should be eliminated. Efficient aeration is the most effective cure.

Autotrophic sulphur bacteria, autotrophic nitrifying bacteria, and heterotrophic bacteria have all been implicated in decay of limestone and calcareous sandstone (sand particles bound together by a calcium carbonate cement). The main causes of stone decay appear to be natural weathering by CO_2-acidified rain and attack by atmospheric sulphur dioxide. However, large colonies of the aforementioned bacteria have been found in pores and crevices where organic material may accumulate, and they may contribute to damage at these sites by the production of sulphuric, nitric or organic acids which attack calcium carbonate.

Other micro-organisms such as lichens and fungi may also grow on porous stone surfaces and contribute to disfigurement or decay. Lichens are seen especially in areas where atmospheric pollution is low.

1.4.5.5 Biodeterioration of textiles

WOOL

Wool may be damaged by bacteria or mould or may be consumed by moth larvae. Mildew is promoted by high humidity (95% RH), warmth (25–40°C) and a neutral pH (6.5–8.5). Well ventilated, clean goods, stored under cool dry conditions are unlikely to be damaged. Mildew can be prevented and its growth terminated by the application of products based on dichlorophene and pentachlorophenol. Pentachlorophenol also inhibits attack by moths.

COTTON

Cotton is attacked by both bacteria and moulds if stored under damp conditions. These deterioration mechanisms are prevented by the short time wear and wash cycle to which cotton articles of clothing are normally subjected.

However, material for heavy outdoor use, tarpaulins, tents, ropes and twine are subject to attack, particularly damage and soiling by mould which must be prevented by the addition during manufacture, and the renewal during waterproofing, of a biocide. Effective biocides are chlorinated phenol derivatives and tributyl tin compounds.

1.4.5.6 Biodeterioration of wood

There are many varieties of wood and even more varieties of attacking organisms. These include moulds and bacteria which attack wood by producing enzymes which convert it into food, insects (termites, beetles and wood wasps) and also marine molluscs and crustaceans.

The most important means of protection (after ensuring that the wood chosen is the most appropriate for the purpose) is to keep the wood dry (fungi and most beetle larvae need a moisture content of 20% to survive).

1–5

Wear

Contents

List of tables

List of figures

1.5.1 Introduction to the various types of wear

Wear occurs at the surface of components which are subject to relative movement under load. The movement may occur in contact with another component or with fluid or particulate solids. Wear involves the removal of material from the surface and may lead to the degradation or ultimate failure of components.

All the moving parts of machinery may be subject to wear. The rate of wear is generally low on the internal components of machines, which move relative to each other, such as bearings and shafts, piston rings and cylinder liners, cams and tappets, and gear teeth.

Higher rates of wear can occur on the components of machines which handle hard particulate materials, such as earth moving machines and mining or mineral processing equipment. Fluids flowing rapidly through pipes and ducts or over turbine blades and propellors, can also erode these components particularly if solid particles are entrained in the fluid.

The types of wear encountered in typical practical situations are given in:

Table 1.5.1 *Practical wear situations*

The characteristics of the different types of wear are summarized in:

Table 1.5.2 *Basic wear mechanisms*

Table 1.5.1 Practical wear situations

General situation	*Type of contact*	*Practical examples*	*Type of wear*	*Remarks*	*Typical type of materials used*
Movement between two components in contact.	Conformal or area contact.	Plain bearings, pistons in cylinders.	Adhesive wear. (some surface fatigue abrasive wear and fretting).	Surfaces usually need some running in.	Soft materials operating with hard materials, and with a hardness ratio of at least 3.
	Counterformal or concentrated contact.	Rolling bearings, gears, cams and tappets.	Surface fatigue. (some adhesive wear and abrasive wear.)	Adhesive wear only becomes a factor if there is a substantial amount of sliding.	Hard with hard materials.
	Both the above types when the components are new.	As above.	Running-in.	Involves mutual adjustment of the surfaces of the two components.	Surface treatments and coatings or special surface finishes can help.
Movement between a component and a mass of material.	Particulate solids rubbing over a surface.	Excavator buckets, sand-blast apparatus and other particulate materials handling plant.	Abrasive wear.	The type of abrasive wear experienced depends mainly on the contact pressure.	Hard materials for mainly sliding contact. Elastic materials for mainly bouncing contact.
	Particulate solids or liquids in a fluid, flowing over a surface.	Equipment pumping abrasive slurries, turbine blades in wet steam, aircraft leading edges in rain.	Erosion.	Only becomes noticeable at high impact velocities and/or high particle densities.	Hard materials for low impact angles. Elastic and ductile materials for high impact angles.
	Stream of fluid flowing over a surface.	Ships' propellers, pipes and ducts.	Cavitation erosion.	Requires low local pressure to initiate the process. Corrosion can accelerate the wear.	Materials with high ultimate resilience.

Table 1.5.2 Basic wear mechanisms

Type of wear	***Physical mechanism and performance relationship***	***Material factors***	***Remarks***
Adhesive wear	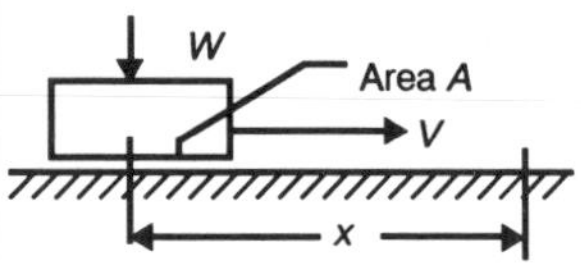 Volume wear = $\frac{kWx}{\text{in time } t}$ Rate of depth wear = $\frac{kWx}{At}$ = kPV When one surface slides over the other interaction between the high spots produces occasional particles of wear debris.	The constant k varies over a factor of about 1000 depending on the material combination. See later Tables and Graphs.	This type of wear occurs when two nominally smooth surfaces are in sliding contact.
	Metals on metals 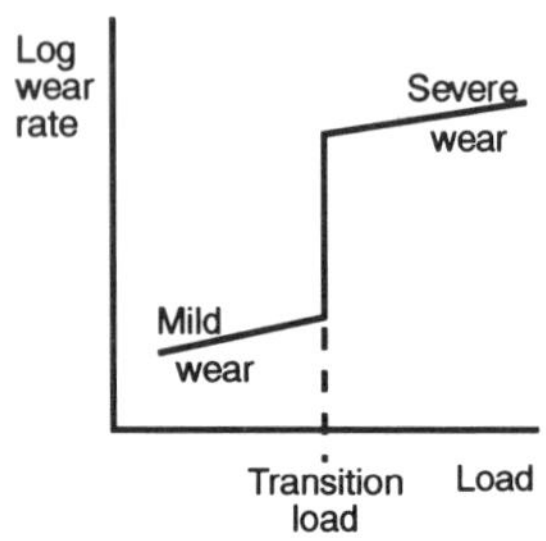 The presence or otherwise of adherent oxide films has a major effect on the wear rate. At low loads and speeds the wear is mild. At high loads and speeds the wear is severe.	The ability to form a low shear strength surface film such as an oxide or a smeared soft constituent is important. The ability to accept a surface treatment giving similar effects is also useful. (See Table 1.5.45.)	
	Metals on plastics The topography of the metal surface has a major effect. Transferred plastic films on the metal and fatigue of high spots on the plastic are major factors.	Plastic materials with filling or reinforcement can carry higher loads. The filler reduces the wear rate, often by conditioning the mating surface.	
Scuffing	Scuffing occurs particularly when a high degree of sliding occurs under poor lubrication conditions, and is particularly likely to start from local high spots due to poor surface finish. It is mainly a running-in problem, and it occurs especially if full load and speed are applied too rapidly. Scuffing can also occur after longer running periods if there is a deterioration in the lubrication conditions. Scuffing occurs by the local welding of two heavily loaded surfaces followed by the tearing away of the welded material. This gives very high rates of material removal. Scuffed surfaces have a roughened and sheared appearance.	To reduce the likelihood of scuffing, material pairs should be chosen having different mechanical and preferably chemical properties. Also materials having a heterogeneous structure have a reduced tendency to scuffing. Thermochemical surface treatments can be used to increase scuffing resistance. Using EP additive in oil or increased oil viscosity tends to increase resistance to scuffing.	If at all possible, avoid running-in at full load and speed. See section on running-in.

Table 1.5.2 Basic wear mechanisms – *continued*

Type of wear	*Physical mechanism and performance relationship*	*Material factors*	*Remarks*
	Surfaces can wear by fatigue when they are subject to fluctuating loads. High surface stresses cause cracks to spread into the material, and when two or more of these cracks become joined together large loose particles are formed. There are two basic types of surface fatigue:		
Surface fatigue	(1) That which occurs on hard components in counter-formal contact where the loose particles are usually triangular in section, and fall away from the surface leaving pits. The allowable contact stresses corresponding to a given life before failure by this mechanism depend on the materials and their application. 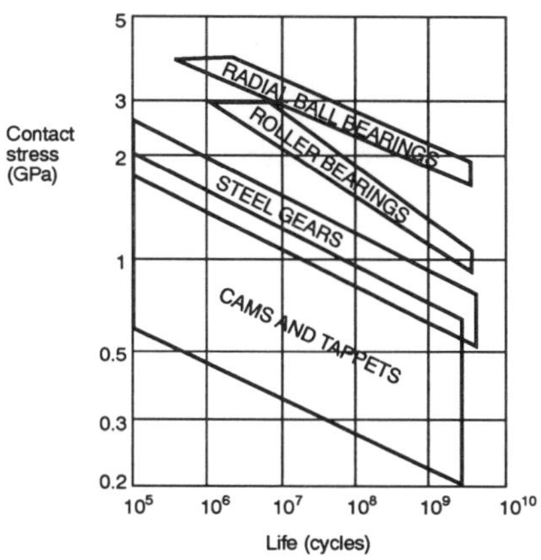	In general, increasing hardness increases resistance to pitting. However, increased hardness reduces the tendency for an imperfect surface to adjust itself by wear or surface flow. To enable this effect to take place, one component is often made slightly softer to assist running-in. Residual compressive stresses in the surface due to, say, carburizing can greatly improve the resistance to pitting. Surface and material defects reduce pitting resistance. Smooth polished surfaces and homogeneous materi-als are therefore desirable.	In some components, such as gears, very mild pitting may occur as part of the running-in process. This normally ceases as soon as the surfaces of the teeth have bedded in together.
	(2) That which occurs in plain bearings where the loose particles are usually rectangular in section, and generally remain trapped by the shaft in approximately their original position. This type of failure is particularly sensitive to bearing temperature. Adequate oil flow through the bearing is therefore important.	Harder materials have higher fatigue strengths, but are less conformable and more dirt-sensitive. A material which is only just strong enough should therefore generally be chosen. Higher fatigue strengths can be obtained by using materials in thinner layers. If the layers are too thin, however, the embeddability of dirt will be reduced, and increased shaft wear may occur.	

Table 1.5.2 – *continued*

Table 1.5.2 Basic wear mechanisms – *continued*

<table>
<tr><th>Type of wear</th><th>Physical mechanism and performance relationship</th><th>Material factors</th><th>Remarks</th></tr>
<tr><td>Running-in wear</td><td>Running-in wear involves at least two separate processes:
(1) The roughness peaks and surface irregularities are worn away or smoothed over, so that the surfaces conform more closely. This alone does not ensure a low wear rate after running-in, unless it causes a transition to fluid film lubrication.

(2) The build up of work-hardened surface layers on the mating surfaces. This increased hardness will in most cases lead to a lower wear rate after running in.</td><td>Materials requirements for running-in:
Low coefficient of friction.
High conformability.
Appreciable running-in wear rate (reduces tendency to scuffing). This effect can be produced by using abrasive running-in accelerators.</td><td>If temperatures attained during sliding become too high, scuffing – a catastrophic form of wear – becomes a problem.</td></tr>
<tr><td>Abrasive wear</td><td>This involves the removal of material from a surface by the mechanical action of an abrasive. Abrasives are substances which are usually harder than the abraded surface and have an angular profile. Abrasive wear is a complex physical process, but may be considered to consist of three main types.

(1) <u>Deformation wear</u>
This wear occurs when coarse abrasive material tears off sizeable particles from the wearing surface. Substantial plastic deformation of the surface occurs due to the very high specific contact pressures involved.

(2) <u>Cutting wear</u>
This less severe form of wear occurs by the cutting action of abrasive particles. The mechanism is most easily thought of as similar to a machining operation with each abrasive particle cutting a wear chip from the surface. This type of wear only occurs if the abrasive particles are sharp.

(3) <u>Rubbing wear</u>
The abrasive particles in this type of wear merely rub a groove in the surface of the material. Most of the displaced material forms ridges on either side of the groove. A small amount of material is scooped out at the beginning of each rubbing action, and eventually breaks off to give a wear particle. The amount of material removed in this way, each time, is quite small compared with the other two processes.</td><td>In a situation where the abrasion is due to abrasive being trapped between two mating surfaces, e.g. in a bearing, mating surfaces of equal hardness should be avoided
because these give high wear rates.
A material must be tough to resist deformation wear.
Hardness becomes the most important materials factor with the less severe wear mechanisms, so for cutting and rubbing wear, hardness affects wear resistance as follows.
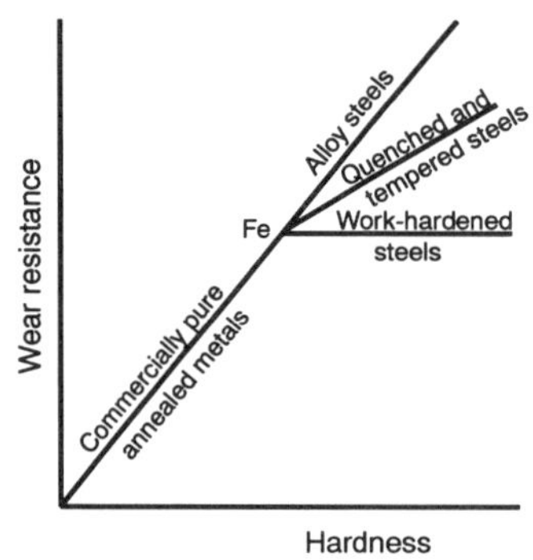

Work-hardenable materials only have useful abrasive wear resistance if the wear situation is severe.</td><td>The volume wear,
$V \propto W \times L$
where W is the load and L is the distance slid.
However, in most practical situations both the load and the speed are impossible to quantify.</td></tr>
</table>

Table 1.5.2 – *continued*

Table 1.5.2 Basic wear mechanisms – *continued*

Type of wear	***Physical mechanism and performance relationship***	***Material factors***	***Remarks***
Fretting	Fretting occurs where two contacting surfaces, nominally at rest, undergo minute oscillatory tangential relative motion. Small particles of metal are removed from the surfaces and then oxidized. It may manifest itself by debris oozing from the contact, particularly if the contact is lubricated with oil. Colour of debris is red on iron and steel, black on aluminium and its alloys. On inspection the fretted surfaces show shallow pits filled and surrounded with debris. If the debris is trapped, seizure can occur. If the debris can escape from the contact, loss of fit may result. Fretting can reduce fatigue strength by 70–80%. This effect reaches a maximum at an amplitude of movement of about 8 μm. At higher amplitudes of relative movement there is less reduction in strength, but the amount of material abraded away increases. Typical situations, where fretting occurs are: press fits on shafts, riveted, bolted and pinned joints, stationary bearings under vibration, splined couplings, etc. Fretting wear damage increases roughly linearly with normal load, amplitude of relative movement and number of cycles. Damage rate on mild steel, – approx. 0.1 mg per 10^6 cycles per MPa normal load, per μm amplitude of relative movement. Increasing the pressure can, in some instances, reduce or prevent the relative movement, and hence reduce fretting damage. However, if the load is increased and movement still occurs the fretting will be worse.	To reduce fretting damage, unlike metals in contact are recommended – preferably a soft metal with low work hardenability and which anneals at a low temperature (such as Cu) in contact with a hard surface, e.g. carburized steel. Phosphate and sulphide coatings on steel and anodized coatings on aluminium prevent metal to metal contact. Their performance may be improved by impregnating them with lubricants, particularly oil-in-water emulsions. Baked-on molybdenum disulphide can be effective but liquid MoS_2 is not recommended. Electrodeposited coatings of soft metals, e.g. Cu, Ag or sprayed coatings of Al allow the relative movement to be taken up within the coating. Chromium plating is not generally recommended. Inserts of rubber, PTFE or rubber cements can sometimes be used to separate the surfaces and take up the relative movement.	Although a fretting problem can be eased by using appropriate materials combinations, it is better to eliminate the problem by making design changes such as: (1) Elimination of stresses and vibrations which cause the relative movement. (2) Separating the surfaces at which fretting is occurring. (3) Increasing the pressure by reducing the area of contact provided this eliminates the relative movement.

Table 1.5.2 – *continued*

Table 1.5.2 Basic wear mechanisms – *continued*

Type of wear	***Physical mechanism and performance relationship***	***Material factors***	***Remarks***
Particle impact erosion	Erosion is the wear of a surface due to the flow across it of a suspension of small solid particles in a fluid, or a suspension of liquid particles in a gas. The contact pressure between the particles and the surface arises from their kinetic energy as they encounter the surface. This type of attack has been shown to comprise two forms of wear: (1) Deformation wear: This occurs if particles strike the surface at a large impact angle, i.e. with a flow velocity which has a large component at right angles to the surface. The repeated deformation of the surface leads to the formation of wear particles. (2) Cutting wear: This occurs if particles strike the surface at a more acute angle, scratching out some material from the surface by the cutting action of the free moving particles. In practice these two types of wear occur simultaneously. The amount of wear varies roughly linearly with the concentration of particles in the carrier fluid.	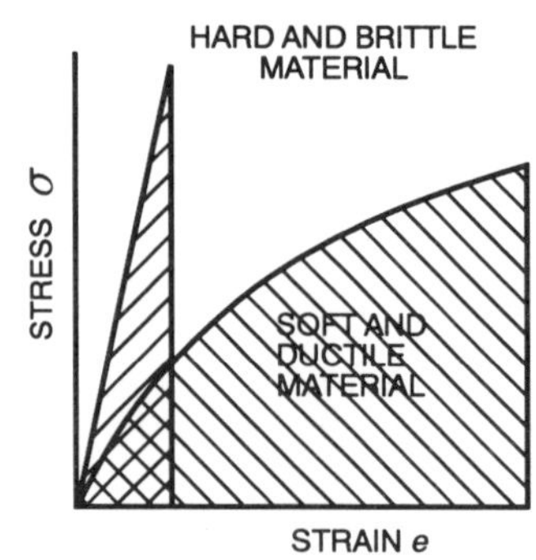 In deformation wear the material absorbs the energy of impact by deformation either elastically (brittle substances) or elastically and plastically (ductile substances). Thus, the deformation wear resistance of a material depends on the amount of energy that the material can absorb without fracture, which is proportional to the area under the stress–strain graph. From the graph it can be seen that ductile materials tend to have a better resistance to deformation wear. Resistance to cutting wear is roughly proportional to the strength of the material in the final work hardened state, and so a harder, less ductile, material will have a better resistance to cutting wear.	The impact angle and impact velocity have a large effect on the amount of wear experienced in particle impact erosion. It is useful to correlate these and the particle concentration in the practical situation to indicate the likely severity of the wear, and the most suitable materials to combat the wear.
Cavitation erosion	Hydrodynamic cavitation is the formation of bubbles or cavities in an otherwise homogeneous fluid flow when the local pressure falls to the vapour pressure of the liquid at the ambient temperature. The cavities become unstable when carried by the flow into regions of higher pressure and consequently collapse. Continual bombardment of a surface by imploding cavities can lead to cavitation erosion of the material. The cavities may be vaporous, or gaseous if the liquid contains a lot of gas, and the damage caused by the latter will be less than that of the former. The physical instability of the bubbles depends on such factors as surface tension, fluid vapour pressure and viscosity at the ambient temperature. Liquid density and bulk modulus, as well as the amount of dissolved gas, may be significant in cavitation, but since many of these factors are inter-related it is difficult to assess their individual importance. Under cavitation erosion conditions, materials in contact with a fluid can also be susceptible to corrosive action. This is intensified when the fluid is flowing over the component, and if damage to any protective film occurs. If corrosive action is likely in a particular situation, reference should be made to the data in Chapter 1-4.	No simple, exact relationship exists between cavitation erosion and materials properties. At high rates of cavitation erosion, however, materials with a high ultimate resilience characteristic tend to have better resistance to cavitation erosion. The ultimate resilience characteristic $= \frac{1}{2}\,\frac{(\textit{tensile strength})^2}{\textit{elastic modulus}}$	As well as reducing the damage to a surface by choosing suitable materials, the phenomenon of cavitation can sometimes be avoided altogether by design changes aimed at avoiding regions of low pressure or arranging for the bubbles to collapse in some region away from the surface. Reducing vibrations and pressure pulsations can also reduce or eliminate cavitation erosion.

Table 1.5.2 – *continued*

1.5.2 Conformal contact

1.5.2.1 Introduction to plain bearings and running-in

Components have a conformal contact when they conform with each other over a substantial area. Typical examples are plain bearings supporting a shaft, or a piston sliding in a cylinder.

The substantial contact areas give relatively low contact pressures, and enable fluid films to be formed between the surface at higher sliding speeds.

A guide to the selection of materials for conformal contact is given below and in Fig 1.5.1:

Fig. 1.5.1 *A guide to the maximum steady load capacity of various plain bearing materials*

Operating characteristics required	*General material combinations which meet the requirements*	*Remarks*
The ability to withstand the inevitable misalignments which will occur in practice, without excessive edge loading The ability to accept particles of dirt in the contact area between the surfaces.	The two working surfaces are made from materials with a hardness difference ratio of at least 3:1. The softer one can conform to the harder one and maintain an adequate contact area. The softer material can also embed dirt particles and prevent a local high load rubbing contact from developing into a general seizure. The softer material must, however, have adequate strength and typical materials used are: <u>For low loads</u> A soft material reinforced with a suitable filler or to fibrous structure, e.g. a filled polymer. <u>For high loads</u> A harder material permeated with a very soft material, improve its surface rubbing properties e.g. lead bronze	The component which extends outside the contact area is made hard, and its mating component is coated or lined with the soft bearing material. Guidance on a suitable bearing material is given by the maximum bearing pressure and the sliding speed, which indicate the conditions of operation, as shown in Fig. 1.5.1.

Bearings may operate dry, with occasional lubrication, with oil lubrication or with lubrication by water or some other process fluid. A guide to selection of materials for bearings with the various lubrication methods is given in:

Table 1.5.3 *General guide to the selection of plain bearing materials*

Whilst one component of a couple in relative motion may be manufactured in a bearing material, this may be expensive and the component may not be strong enough. There are various methods of providing a bearing surface on a component, including lining, coating and providing inserts or shells. Methods of incorporating bearing materials are given in:

Table 1.5.4 *Methods of incorporating bearing materials into an assembly*

RUNNING-IN

The achievement of a low wear rate in plain bearings depends on conformity of the mating surfaces (at both the macro- and micro-level), cleanliness and the generation and maintenance of protective surface layers. These conditions are generally not met in new assemblies and running-in procedures are designed to generate the appropriate

conditions for low longterm wear. The beneficial effects of running-in are outlined in:
Table 1.5.5 *Effects of running-in procedures*

During running-in, continuous high speed, high load conditions should be avoided otherwise high temperatures may develop giving rise to scuffing. Continuous low speed, high load combinations are also undesirable as plastic flow of material may occur and lubricating films may not form. High speed and shock loading during running-in may also give rise to scuffing.

Whilst running-in is often a requirement for components in conformal contact such as plain bearings, it is often also necessary for components in counterformal contact such as gears, cams and tappets. The criticality of running-in for various components is given in:
Table 1.5.6 *Relative importance of running-in for various components*

Surfaces operating under a combination of high contact pressure and high sliding velocity may tend to scuff in the early stages of running-in. Various surface treatments can prevent or reduce this. Characteristics of typical surface treatments are given in:
Table 1.5.7 *Surface treatments to aid running-in at high load and high speed*

1.5.2.2 Dry rubbing bearings

Dry rubbing bearings operate with rubbing contact between two components, usually a shaft and bearing. Because of the absence of lubrication, friction and wear can be high compared to other types of bearing. Important factors in selection of dry rubbing bearing materials are therefore low coefficient of friction and high wear resistance.

Some typical applications of dry rubbing bearings, together with suitable materials, are given in:
Table 1.5.8 *Typical rubbing bearing applications and suitable materials*

Details of some typical bearing materials are given in:
Table 1.5.9 *Typical rubbing bearing materials and their relative costs*

Selection of the most appropriate bearing material must take into account the coefficient of friction of the material, the bearing pressure required, the wear resistance of the material and the temperature. Coefficients of friction of typical bearing materials are given in:
Table 1.5.10 *Coefficients of friction of typical bearing materials*

The maximum allowable bearing pressure as a function of rubbing speed is given for a number of bearing materials in:
Fig. 1.5.2 *Bearing pressures and rubbing speeds for typical bearing materials operating unlubricated against steel*

The variation in maximum allowable bearing pressure with temperature is given in:
Fig. 1.5.3 *Pressure and temperature limits for rubbing bearing materials against steel*

The wear rate of materials is a function of the bearing pressure and the rubbing velocity and can be expressed as:

$$W = k\,PV$$

where W = depth wear rate,
k = the wear factor,
P = bearing pressure,
V = rubbing velocity.

The wear factor, k, can be used as an indicator of wear resistance. For any particular material the value of k varies with temperature and pressure. A low value, k_o, applies at moderate temperatures and pressures, and the value of k rises rapidly as either the temperature limit or the pressure limit is approached. This is illustrated in:
Fig. 1.5.4 *Typical wear factor variation with temperature and pressure*

Values for wear factors at moderate temperatures over a range of bearing pressures are given in:

Fig. 1.5.5 *Wear factors at room temperature of typical rubbing bearing materials as a function of bearing pressure*

More precise data, and the way in which wear factors vary with temperature, can be found in ESDU data item 87007, Design and material selection for dry rubbing bearings, Engineering Sciences Data Unit, 1988.

The performance of a rubbing bearing is affected by a number of design and operational factors. These are outlined in:

Table 1.5.11 *Factors to be considered in designing rubbing bearings*

1.5.2.3 Bearings with occasional lubrication

At surface speeds of less than about 0.5 m/s plain bearings may be used with occasional lubrication. This lubrication may be supplied by hand with an oil can or grease gun, and in such a situation wear is inevitable. An improvement in wear life is possible with automatic lubrication, but bearings of this type are never likely to give lives which are comparable with those obtained from full fluid film lubricated plain bearings, or properly lubricated and sealed rolling element bearings. Bearing life can be considerable extended by surface treatment of metals and use of filled or reinforced plastics.

Some guidance on the selection of materials for bearings of this type is given in:

Table 1.5.12 *Selection of materials for plain bearings with occasional lubrication*

1.5.2.4 Liquid lubricant impregnated bearings

Bearings can be made from porous materials by sintering, and they can then be vacuum impregnated with liquid lubricants. In operation the bearing temperature rises and oil exudes from the pores and lubricates the surface.

During running-in, some wiping of the bearing material occurs in the area under the load, and the resulting pore closure gives increased load capacity. The allowable performance is limited by the temperature rise, since this reduces the life of the lubricant.

A guide to the selection of suitable materials is given in:

Table 1.5.13 *Selection of materials for porous metal bearings*

A guide to the allowable bearing pressure and sliding speed is given in:

Fig. 1.5.6 *Limits of bearing pressure and sliding speed for porous metal bearings, with suggested grades of mineral oil for impregnation*

1.5.2.5 Fluid film lubricated bearings

If a bearing is kept flooded with a viscous lubricant, such as mineral oil, the rotating shaft is capable of generating a pressure in those regions of the clearance space which have a converging taper. This converging clearance can arise from tilting or tapered pads in a thrust bearing, or in journal bearings from the shaft adopting an eccentric position. This enables these bearings to carry steady loads of the order of 1–10 MPa without any solid contact except when starting or stopping.

Bearings of this type can also carry very high loads of the order of 10–100 MPa provided that these are of very short duration.

The characteristics of typical materials used for fluid film lubricated bearings are outlined in:

Table 1.5.14 *Characteristics of oil lubricated bearing materials*

The properties, forms available and applications of oil lubricated bearing materials are given in:

Table 1.5.15 *Properties and applications of materials for oil lubricated plain bearings*

Table 1.5.3 General guide to the selection of plain bearing materials

Type of bearing	*Typical plain bearing material*	*Remarks*
Dry rubbing bearings	Non-metallic materials which, when operating against metals, give greater resistance to seizure and wear than bearing metals.	Reinforced plastics are usually best at low temperatures, while graphites and ceramics are usually best at high temperatures.
Bearings which are required to operate with only occasional manual lubrication.	Bronze, cast iron and plastics are generally the most suitable materials. At very low rubbing speeds hardened steels may be satisfactory.	Generally only suitable for crude applications where a short life is acceptable. However, surface treatment of metals and use of filled or reinforced plastics can increase lifetime considerably.
Oil impregnated porous metal bearings.	Porous bronzes and irons are commonly used. High porosity bronzes are suitable for higher speeds, and lower porosity irons for higher loads.	The main limitation on life is deterioration of the oil, which is accelerated by higher temperatures.
Bearings operating with fluid films and using oil as the lubricant.	Bearing metals such as whitemetal, bronze and aluminium bearing alloys. Some cast irons and some plastics might also be used.	Whitemetal is usually the preferred material for steadily loaded bearings. Other stronger bearing metals may be required for high dynamic loads as in piston engine crankshafts.
Bearings operating with fluid films and using water or some other process fluid as the lubricant.	Generally the same materials that are suitable for rubbing bearings, to assist operation during starting and stopping. Phenolic resins and graphites have been commonly used. Rubbers are also a possibility for liquid lubricants, and PTFE based materials for gaseous lubricants.	Some materials show high dimensional changes in contact with water and other process fluids. This can be a major limitation on material choice.

Table 1.5.3

Table 1.5.4 Methods of incorporating plain bearing materials into an assembly

Typical method	*Materials which can be incorporated in this way*	*Remarks*
Making one component from the bearing material.	Cast irons. High strength bronzes and aluminium alloys.	Can be expensive in the case of bronzes. Bearing material may not have enough structural strength at high loads.
Direct lining or coating one component with bearing material.	Low melting point materials such as whitemetals or plastics.	These are applied by methods such as static or centrifugal casting, or by hot dipping. Low melting point materials reduce any distortion of the coated component.
	Thin coatings of higher melting point bearing materials.	These can be applied by spraying or electroplating.
Inserting or attaching a component made entirely from the bearing material.	Plastics, graphites, ceramics, cast irons, bronzes and aluminium bearing materials.	Interference fits can easily be lost due to heat generated in operation. Large amounts of bearing material are used, and this can increase costs.
Inserting a thick steel shell lined with bearing material or attaching a steel thrust washer or pad lined with bearing material.	All whitemetals and bronzes and most aluminium bearing materials. Could also be used with plastics.	Shells can be machined on standard machine tools. Housings do not need to be very precise, but shells usually need individual fitting. Clamping loads on shells need not be very high. Relatively high cost compared with thin shells.
Inserting a thin steel shell lined with bearing material.	All plain bearing metals, and some plastics.	Shells can only be made on special machines. Very precise housings and high clamping loads are needed. Shells are readily replaceable without individual fitting. Relatively low cost.

Table 1.5.5 Effects of running-in procedures

Condition required	*Reasons why requirements not met by new assemblies*	*Condition produced by running-in*
Macro-conformity	Latitudes in fits and tolerances, misalignment on assembly, deflections of shafts and gear teeth, etc., under load, and thermal distortions. The shaping of piston rings to a suitable profile on the running face.	Increase in apparent area of contact and the relief of contact stresses. Reduction in friction and temperature.
Micro-conformity	Initial composite surface roughness exceeds oil film thickness obtainable with smooth surfaces.	Decrease in roughness. Increase in effective bearing area, coherence of oil film and, ideally, full film lubrication.
Cleanliness	Casting sand, machining swarf, wear products of running-in, etc.	Safe expulsion of contaminants to filters.
Protective layers	Scuff and wear resistant low friction surfaces are not naturally present on freshly machined surfaces.	Safe formation of work-hardened layers, low friction oxide films, oil additive reaction films.

Table 1.5.6 Relative importance of running-in for various components

Critically	***Component***	***Remarks***
Most critical	Piston rings and liners.	Highly critical, especially in high speed IC engines.
	Gears.	Especially hypoid gears.
	Cams and tappets.	In vehicle engines, these are usually covered by the running-in procedures used for the piston rings or gears.
	Rubbing plain bearings.	Running-in desirable for successful transfer of protective layer, e.g. PTFE to countersurface.
	Porous plain bearings.	Running-in desirable to achieve conformity and pore closure in the loaded area without overheating impregnant oil.
	Fluid film plain bearings.	Usually none unless marginally lubricated – mainly a precaution against minor assembly errors.
Least critical	Hydrodynamic thrust bearings. Rolling element bearings. Externally pressurized bearings. Gas bearings.	Assuming correct design, finish and assembly, no running-in required. For certain high precision rolling element bearings (e.g. gyro bearings) and for operation in reactive gas environment, consult makers.

The running in requirement of assembled machinery is that of the most critical part.

Table 1.5.7 Surface treatments to aid running-in at high load and high speed

Surface treatment	*Suitable base materials*	*Description of treatment*	*Typical size effect*	*Primary advantages*	*Primary disadvantages*
Phosphating	All ferrous metals.	Inorganic film produced on surface by chemical or electrochemical surface modifications at 40–100°C.	Adds 5 μm to each surface.	Resists scuffing. Porous nature helps retain lubricant. Little heating effect to cause distortion or tempering or softening of the metal being treated. Inexpensive.	Less effective than Sulfinuz, Tufftride, Noskuff or Sulf BT in reducing general wear rate and scuffing tendency.
Tufftride	Low C, Med C, C–Mn, Cr, Cr–Mo–V, and stainless steels. Cast irons.	Salt bath treatment adding carbon and nitrogen. Nitriding action accelerated by blowing air through solution. Temperature 570–590°C.	Adds 5–10μm to each surface. Depth of treatment up to 1.0 mm. Top 10–16 μm most effective.	Improves resistance to wear and fatigue, and improves corrosion resistance except on stainless steels. Surface is not brittle, although Tuffride gives some increase in surface hardness. Better corrosion resistance than Sulfinuz for non-stainless steels.	Process operating temperatures sufficiently high to temper or anneal many ferrous metals. Also, slight distortion problem as process normally last one in production line.
Sulfinuz	All ferrous metals, including stainless steels. Nimonic alloys.	Salt bath treatment at 540–600°C introducing carbon, nitrogen and sulphur into ferrous metals. Sulphur accelerates nitriding process.	Size change from loss of 5 to gain of 3 μm per surface. Depth of treatment up to 0.5 mm. Top 4–25 μm most effective.	Gives good scuffing resistance with some reduction in coefficient of friction. Also resists wear and fatigue. More resistant than Tufftride to adhesive wear.	Reduction in corrosion resistance of stainless steels. High process temperature may temper base metal and cause distortion.
Noskuff	Case-hardening and direct hardening steels.	Adds nitrogen and carbon in salt bath at 700–760°C.	Adds 5–10 μm per surface. Depth of treatment 45–50 μm – supported by carburized case.	Good anti-scuffing properties to a large depth without reducing fatigue strength and hardness obtained previously by, say, carburizing.	More distortion than Sulfinuz or Tufftride due to high process temperatures. Lapping can be used to correct this due to larger case depth. May cause tempering of base metal.
Sulf BT	All ferrous metals (except stainless steels and 13% Cr).	Sulphur added to metal surface by low temperature (180–200°C) salt bath. Does not add nitrogen or carbon.	Removes 2–5 μm per surface.	Low process temperature minimizes distortion. Iron sulphide gives good anti-scuffing properties. Unlike higher temperature salt baths, no cyanide problems.	No increase in fatigue strength.
Tin plating	All ferrous metals.	Plated by dipping in molten tin or by electroplating.		Good control of thickness. Soft plate easily flows under load to aid conformity of mating surfaces. Low melting point so minimum thermal distortion by dip process.	With sintered materials only the high density type should be plated. Low melting point may give liquid metal embrittlement in some situations.
Silver		Electroplating.		Corrosion resistant except with sulphur-containing oil additives. Good control of thickness.	Expensive – not normally used in mass production.
Copper		Electroplating.		Good adhesion to base steel. Good control of thickness. Often used as a scuff resistant coating on gears and piston rings.	

Table 1.5.8 Typical rubbing bearing applications and suitable materials

Typical application	*Important material property*	*Suitable materials*
Load-carrying linkages with oscillating motion such as vehicle suspensions and mechanical handling linkages.	Maximum allowable bearing pressure.	The choice is limited to the stronger materials such as reinforced thermosetting plastics, PTFE fibre materials and plastic impregnated metals.
Control linkages with oscillating motion.	Wear resistance. Acceptable friction.	Filled PTFE materials and PTFE fibre materials or impregnated metals.
Low speed continuously rotating components, particularly for operation in difficult environments such as high temperatures.	Wear resistance. Acceptable friction. Ability to withstand the environment.	For high temperature applications, ceramics, graphites, polyimides and PTFE materials are possible choices.

Table 1.5.8

Table 1.5.9 Typical rubbing bearing materials and their relative costs

Material	*Common types*	*Typical trade names*	*Cost in bearing form relative to nylon*[a]
Thermoplastics	Nylon/polyamide: monocast extruded plain reinforced solid lubricant filled (graphite/MoS_2)	Nylon 66 Nylatron Zytel Plaslubes Maranyl Ertalon Grafil	1–3
	Polyacetal: plain, solid lubricant filled (PTFE)	Delrin Glacetal Railko Pv80 Fulton Kemetal	1–3
Metal backed thermoplastic	Nylon on steel. Polyacetal + additive – impregnated in porous bronze on steel.	Nyliner Glacier DX Glycodur A	2–5
Reinforced thermosetting plastics	Fabric reinforced thermosetting resins (some with solid lubricant additions). Resins include phenolic, polyester, epoxy and others.	Tufnol Feroform Railko Xylok Orkot	2–4
PTFE impregnated metal	PTFE – lead impregnated in porous bronze on steel.	Glacier DU Glycodur F	3–10
Filled PTFE	Glass-filled Mica-filled Bronze–graphite-filled Graphite-filled Bronze–lead oxide-filled Ceramic-filled	Chemloy Fluorosint Glacier DQ Graflon Permaflon Rulon	3–8
Woven PTFE fibre	PTFE cotton weave – thermoset reinforced + graphite, PTFE glass weave – thermoset reinforced PTFE wire weave – thermoset reinforced	Fiberglide Fiberslip Uniflon Unimesh Fabrilube	6–20
High temperature thermoplastics	Polyimide, PEEK solid lubricant-filled (graphite/PTFE/MoS_2)	Vespel Feuralon Victrex Kinel	6–12
Graphites	High carbon, low carbon/electrographite: with copper and lead with white metal impregnated with thermoset	Morganite carbon Nobrac carbon Purebon	10–15
Graphite impregnated metal	Graphite-filled irons + MoS_2 Graphite-filled bronzes + MoS_2	Deva Lubrite	12

[a]Based on 1000 components.

Table 1.5.10 Coefficients of friction of typical bearing materials

Bearing material	*Typical dry coefficient of friction**	*Typical mating materials*
Nylons Polyacetals Thermosets	0.1–0.4	Hard chromium plate. Stainless steel Mild steel – only satisfactory if the environment is non-corrosive.
PTFE materials	0.05–0.2	
Graphite materials	0.1–0.2	
Ceramics Cermets	0.2–0.8	Material of the same or similar composition as the bearing material. If this is not possible, hardened stainless steels may be satisfactory.

* Note that in general the lower values of coefficient of friction occur with high bearing contact pressure. Allowable pressure (and temperature) limits are shown in Figures 1.5.2 and 1.5.3.

Table 1.5.11 Factors to be considered in choosing materials for rubbing bearings

Design consideration	***Effect***	***Remarks***
Surface finish of mating material	In general, the finer the surface finish the lower the bearing wear. However, for economic reasons 0.2–0.4 μm CLA is normally used. Grinding, or grinding and lapping, are generally satisfactory finishing operations.	The provision of a very smooth mating surface is particularly important for plain unfilled materials, since these do not have the ability to improve the surface in operation by the polishing action of a filler. With ceramic mating materials it is particularly important to avoid severe grinding operations which can cause local stress concentrations and hence rapid wear.
Bearing material thickness	For plastic materials, and particularly for high speed applications, it is usual to use materials which can be manufactured in a thin form to reduce the effect of dimensional changes and to maximize heat dissipation. Some materials, notably graphite, are usually made thick enough to give adequate strength on assembly.	The bonding of a plastic to a metal back often allows very thin plastic layers to be used. Ceramic coating thicknesses up to 0.5 mm on metal substrates can offer the high wear resistance of these materials at low processing cost, if an adequate bond can be obtained between the coating and the substrate.
Bearing temperature	This should generally be kept as low as possible, and an estimate of the equilibrium bearing temperature can give a useful indication of rubbing bearing material suitability, especially for continuously rotating and rapidly oscillating components.	An estimate of equilibrium temperature can be obtained by considering the balance of frictional heating and thermal dissipation of the bearing. For operation at very low temperatures, the bearing material manufacturers must usually be consulted.
Running clearances	Plastic-based materials have large coefficients of thermal expansion and some can absorb certain liquids (e.g. water) so that large clearances of about 0.005 m/m of shaft diameter are typically used with a minimum of about 100 μm. Materials with lower coefficients of thermal expansion, e.g. graphites, graphite impregnated metals and ceramics, are capable of operating with lower running clearances, and 0.002 m/m of shaft diameter with a minimum of 75 μm is typical in order to allow wear debris to escape.	Under exceptional conditions of high temperature, high humidity, etc., greater clearances may be required, particularly with bearings which are thick-walled and made from materials with poor dimensional stability. Typical factors which can affect dimensional stability, and hence clearance are: Metallurgical stability; Corrosion build-up Creep or centrifugal growth; Wear or wear debris Differential thermal expansion; Erosion from process fluid borne particles.
Corrosive environments	It is essential that the mating material is non-corrodable in its working environment, otherwise excessive wear of the bearing material will result. Mild steel is only satisfactory if the environment is completely non-corrosive.	Absorption of moisture by the bearing material (especially graphites) can result in excessive mating material corrosion.
Dynamic load effects	Very high loads, and particularly shock loads, must be avoided if very hard rubbing bearing materials, e.g. ceramics and hard cermets, are used. High toughness cermets which are less susceptible to shock are available, but tend to have less wear resistance than the harder cermets.	The wear data given in this section is for unidirectional loads, i.e. loads that do not rotate relative to the bearing. A rotating or reversing load may have an adverse effect on materials which are subject to fatigue.
Process fluid lubrication (The bulk of applications use, water, liquid metals, air, steam or inert gas.)	The selection of bearing materials is critical, as such fluids have a low viscosity giving small film thicknesses, and also have poor boundary lubrication characteristics. In general the material requirements are: 1. Corrosion resistance in working environment. 2. Ease of fabrication. 3. Dimensional stability – critical as bearing clearances are often small (e.g. gas bearing typically 0.001 m/m of shaft diameter). 4. Ability to withstand sliding contact: (i) At low speeds, at startup and shutdown. (ii) At high speeds.	The lower the process fluid viscosity and the poorer its boundary lubrication characteristics, the more essential it is that the materials are capable of withstanding rubbing contact and so, in general, materials which are good dry rubbing materials are good from this point of view.

Table 1.5.12 Selection of materials for plain bearings with occasional lubrication

Possible material	*Advantages*	*Disadvantages*	*Remarks*
Bronze	Grades with lead and graphite minimize the effects of lubricant starvation. Excellent machinability. High thermal conductivity.	High cost.	At least 10% lead is desirable to minimize lubricant starvation effects. For increased life, shaft should be hardened and given smooth finish.
Cast irons	Bearing surface can be integral part of cast iron component. Particularly suitable for large bearings where cost of bronze becomes prohibitive.	Ferrite content must be controlled or severe galling and seizure can occur. Lower thermal conductivity than bronze.	Flake cast iron gives best results. Performance can be improved by phosphating which helps retain lubricant or surface treatments which add sulphur and/or nitrogen.
Plastics	Will work with less lubrication than metals and as a result often give an increased life. PTFE-filled grades give exceptionally good friction and wear characteristics. Further addition of fibre reinforcement improves load-bearing performance. Corrosion resistant. Thermoplastics have ability to absorb abrasive particles.	Compared with metals, high thermal expansion and lower thermal conductivity, since plastics, notably nylons, absorb moisture with consequent change of dimensions and properties.	Fibre reinforced thermosetting plastics are generally best for high loads. Shafts must be smooth and must not corrode, otherwise rapid wear of bearing will occur.
Hardened steels	Better fatigue life than above materials.	Lowest machinability rating. Greater tendency to scuffing if surfaces are not treated.	Surface treatments which add nitrogen and/or sulphur improve performance. Generally satisfactory at speeds up to 0.05 m/s.

Table 1.5.13 Selection of materials for porous metal bearings

Maximum bearing pressure (MPa)	*Suitable material composition*	*Typical range of porosity*	*Remarks*
100	90% 0.7 C steel, 10% Cu	15–20%	Hardened steel shafts are required.
70	98% 0.7 C steel, 2% Cu	20–25%	Hardened steel shafts are required.
60	75% Fe, 25% Cu	16–18%	Hardened steel shafts are required.
50	98% Fe, 2% Cu	17–20%	Hardened steel shafts are required.
40	88% Fe, 10% Cu, 2% graphite	20–22%	Hardened steel shafts are required.
20–40	89% Cu, 10% Sn, 1% graphite	15–25%	Hardened steel shafts are required.
7–20	85% Cu, 10% Sn, 5% graphite	18–27%	Can operate with unhardened shafts or accept some oil starvation.

Table 1.5.12 and Table 1.5.13

Table 1.5.14 Characteristics of oil lubricated bearing materials

Property	*Significance of property in service*	*Characteristics of widely used materials*			
		White metals	*Copper base alloys*	*Aluminium base alloys*	*Plastic-based*
Compressive strength	To support the bearing load without extruding or dimensional change.	Adequate for many applications, but falls rapidly with rise of temperature.	Wide range of strength available by selection of composition.	Similar to copper base alloys by appropriate selection of composition.	Plastic materials adequate for many applications at low temperatures. See thermal expansion.
Fatigue strength	To sustain imposed dynamic loadings at operating temperature.	as above.	as above.	as above.	as above.
Corrosion resistance	To resist attack by acidic oil oxidation products or contaminants in the lubricant.	Tin base white metals excellent in absence of seawater. Lead base white metals attacked by acidic products.	Lead constituent, if present, susceptible to attack. Resistance enhanced by lead alloy overlay.	Good, but in cold-running petrol engines can be attacked by halides derived from fuel additives.	Dependent on polymer.
Embeddability	To tolerate and embed foreign matter in lubricant, so minimizing journal wear.		Inferior to white metals. Softer weaker alloys with a low melting point constituent (e.g. lead) superior to harder stronger alloys in this category. These properties can be enhanced by provision of overlay, e.g. lead–tin or lead–indium, on bearing surface, where appropriate.	Inferior to white metals. Alloys with high content of low melting point constituent (e.g. tin) superior in these properties to copper base alloys of equivalent strength. Overlays may be provided in appropriate cases to enhance these properties.	Good
Conformability	To tolerate some misalignment or journal deflection under load.	Excellent – unequalled by any other bearing material.			Very good
Compatibility	To tolerate momentary boundary lubrication or metal-to-metal contact at operating speed without seizure (touchdown).				Excellent
Start/stop capability	To tolerate some sliding without seizure when velocity is too low to support the load hydrodynamically.				Excellent
Thermal expansion	Can affect operating clearance or interference fit of lining at operating temperature.	Not usually a problem as white metals are invariably used as thin layers on a steel shell or bush.	Can be a problem when the materials are used as inserted bushes or shells without a steel backing. Temperature changes then cause loss of interference fit in ferrous housings or large changes in operating clearance when in non-ferrous housings.		Can be necess clearan steel b effect.

Table 1.5.15

Table 1.5.15 Properties and applications of materials for oil lubricated plain bearings

Typical commercially available material	*Form*	*Maximum dynamic loading in (MPa)*[a]	*Relative cost*	*Seizure load relative to 80/10/10 lead bronze*[b]	*Material hardness at 20°C (HV)*	*Recommended journal hardness at 20°C (HV)*	*Melting range (°C)*	*Coefficient of thermal expansion ($\times 10^{-6}$/°C)*	*Forms available*[e]	*Applications*
Tin base white metal 88% Sn 8% Sb 4% Cu		10.3–13.7 (~0.5 mm thick); >17.2 (~0.1 mm thick)	1.0	>5	23–32	Soft journal (~140) satisfactory	240–340	~23	Lining of thin, medium and thick walled steel-backed half bearings, split bushes and thrust washers. Lining of bronze backed components, unsplit bushes. Lining of direct-lined housings and connecting rods.	Crankshaft and cross-head bearings of IC engines and reciprocating compressors within fatigue range. FHP motor bushes, gas turbine bearings (cool end); camshaft bushes; marine gearbox bearings; large plant and machinery bearings; general lubricated applications.
Lead base white metal 83% Pb 6% Sn 10% Sb-1% Cu		As above	0.35	>5	~26	As above	245–260	~28	Lining of steel, cast iron, and bronze components.	General plant and machinery bearings operating at lower loads and temperatures.
Silver		20	>9	–	30–60	As above	960	~20	Normally used as an electroplated coating, but can also be cast to form a lining on components.	Applications requiring a soft conformable material, but where high temperatures reduce the effective strength of whitemetals.
Acetal copolymer on steel		28	0.5	>5	19	As above	165	~97[d]	As lining of steel-backed slides, bushes, thrust washers and half-bearings.	General plant and machinery operating in non-acidic oil below 110°C where excellent anti-seizure and low friction properties are required.
70/30 Copper/lead on steel 70 % Cu 30 % Pb	Unplated Plated	24–27.5 on steel 27.5–31	0.35	4–5 –(3)	35–45	~250 ~230	Matrix ~1050 Lead constituent ~237	~16	As lining of thin-. medium- or thick walled half bearings bushes and thrust washers.	Crankshaft bearings for high speed and medium speed petrol and diesel engines; gas turbine and turbocharger bearings; compressor bearings; camshaft and rocker bushes.

[a] Maximum dynamic load ratings determined on single cylinder test rig. Only rough guide for engine design purposes.
[b] Seizure load determined by stop/start tests on bushes. Maximum load on rig equivalent to relative load of 5.
[c] Overlay does not seize, but wears away. Seizure then occurs between interlayer and journal at loading, depending upon thickness of overlay, i.e. rate of wear. The overlay thickness on aluminium–tin is usually less than that on copper–lead and lead–bronze, hence the slightly higher fatigue rating.
[d] Applies to plastic lining only.
[e] All materials are available as a lining to steel backing. The first five materials can also be applied direct. The remaining materials are available as solid inserts.

Table 1.5.15 Properties and applications of materials for oil lubricated plain bearings – *continued*

Lead bronze on steel 20–25% Pb 3–5% Sn remainder Cu	Unplated Plated	35–42 42–52	0.40	2–4 –(3)	45–70	~500 ~230	Matrix ~900 Lead constituent ~327	~18	Machined cast components. As lining of thin, medium or thick walled steel-backed half-bearings, bushes and thrust washers.	As above, for more heavily loaded applications. Usually overlay plated for crankshaft bearing applications.
80/10/10 lead bronze 80 % Cu 10% Sn 10% Pb	Unplated	>48	0.45	1	65–90	~500	Matrix 820–920 Lead constituent ~327	~18	Machined cast components tubes, bars. As lining of steel-backed components. Hard, strong bronze.	Machined bush and thrust washer applications, as lining of thin walled split bushes for small-ends, camshafts gearboxes, linkages, etc. Wide range of applications.
Phosphor bronze, e.g. BS1400–PB1C min. 10% Sn min. 0.5% P remainder Cu.		~62	0.5	–	70–150	~500	~800	~18	Machined cast components bushes, bars, tubes, thrust washers, slides, etc.	Heavy load, high-temperature bush and slide applications, e.g. crank-press bushes, rolling mill bearings, gudgeon pin bushes, etc.
Low tin aluminium 6% Sn 1% Cu 1% Ni remainder Al	Unplated	~42 45–42	0.06	2–5[c]	45–60	~500 ~280	Matrix ~650 Tin eutectic ~228	~24	Cast or rolled machined components. As lining of steel-backed components.	Unsplit bushes for small-ends, rockers linkages, gearboxes. Crankshaft half-bearings for diesel engines and linings of thin and medium walled steel-backed crankshaft bearings for heavily loaded diesels and compressors; also as split bushes for gearboxes, rockers, small-ends, automatic transmissions, etc. Usually overlay plated.
High tin aluminium 20% Sn 1% Cu remainder Al	Unplated	~35	0.12	>5	~40	~200	Matrix ~650 Tin eutectic ~228	~24	As lining of thin and medium walled half-bearings, split bushes and thrust washers.	Heavily loaded crankshaft bearings for high speed petrol and diesel engines, usually without overlay. Small-end, rocker, camshaft, gearbox and linkage bushes; thrust washers.
High tin aluminium 40% Sn remainder Al	Plated and unplated	20	0.23	>5	20	Soft Journal (~140) satisfactory	Matrix ~650 Eutectic ~228	~23		Cross-head bearings of IC engines. Surface properties approaching those of white metal but fatigue strength is retained at high temperatures. Can be used with an overlay plate in applications requiring high conformability.
Aluminium silicon 11% Si 1% Cu remainder Al	Plated	~50	0.04	–	50–60	~500	~600	~20	As lining of thin walled half-bearings, split bushes and thrust washers.	Heavily loaded crankshaft bearings, particularly in turbocharged diesel engines. Always used with an overlay plate.

Table 1.5.15 – *continued*

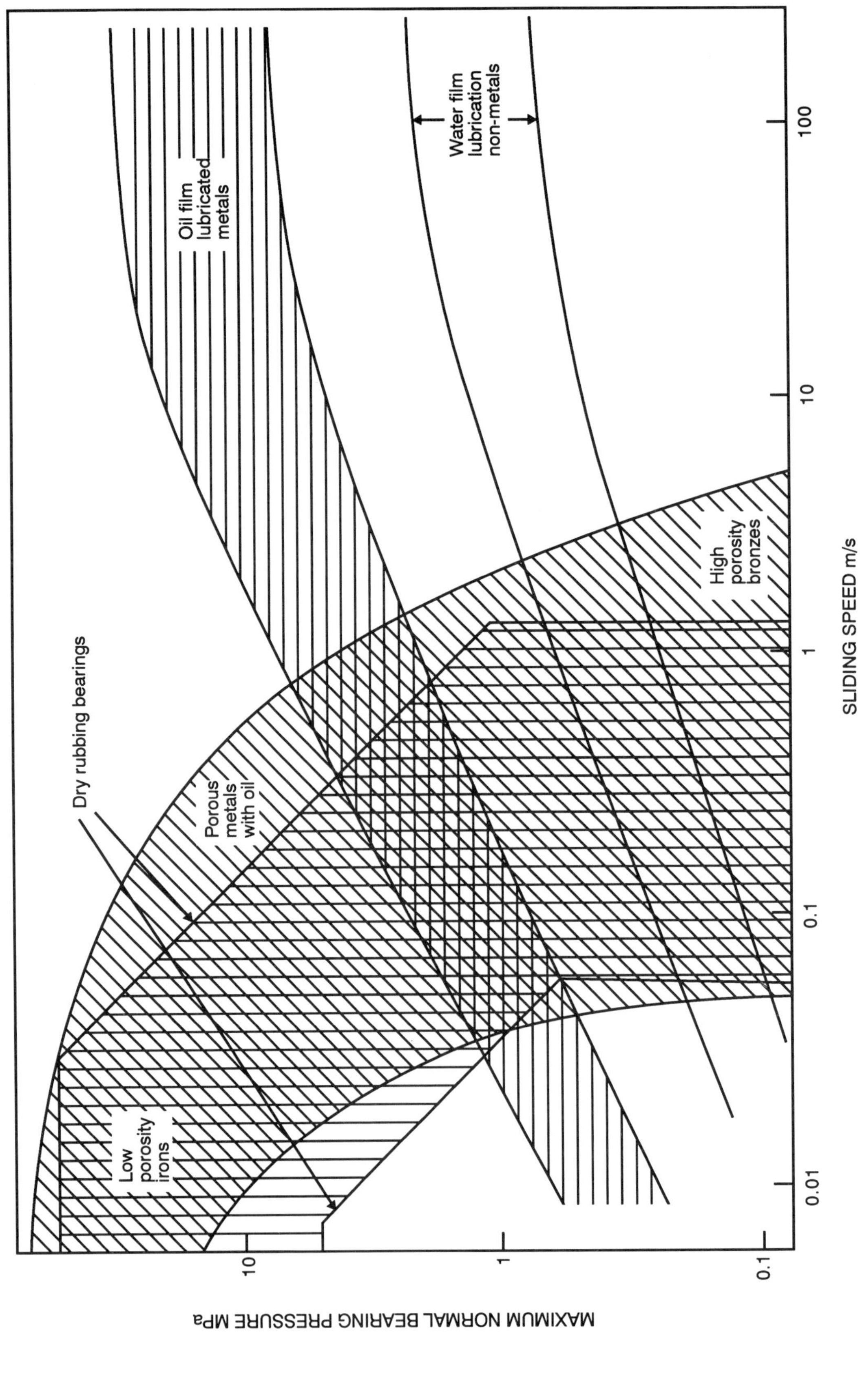

Fig. 1.5.1 A guide to the maximum steady load capacity of various plain bearing materials

Fig. 1.5.1

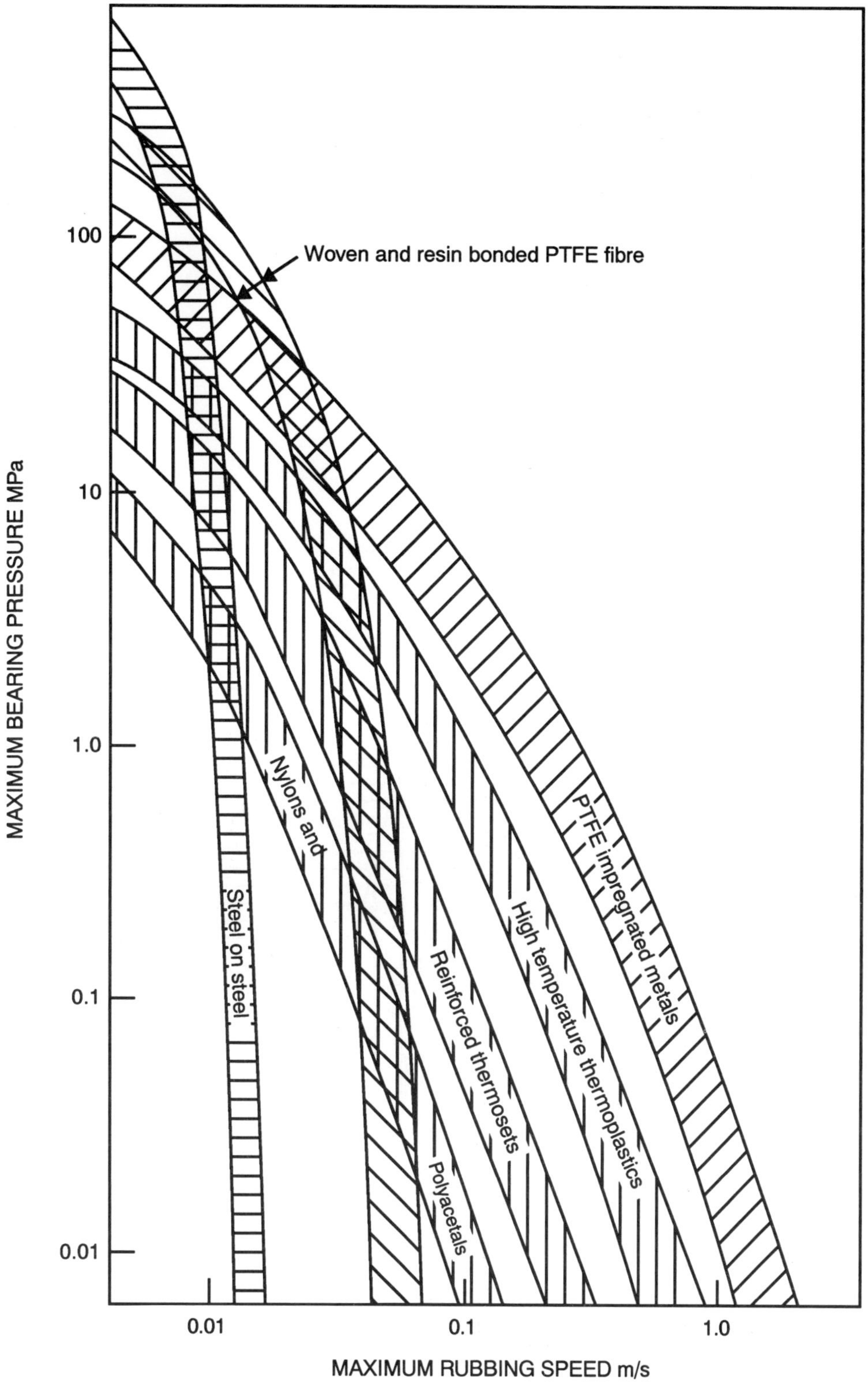

Fig. 1.5.2 Bearing pressures and rubbing speeds for typical bearing materials operating unlubricated against steel

Fig. 1.5.2

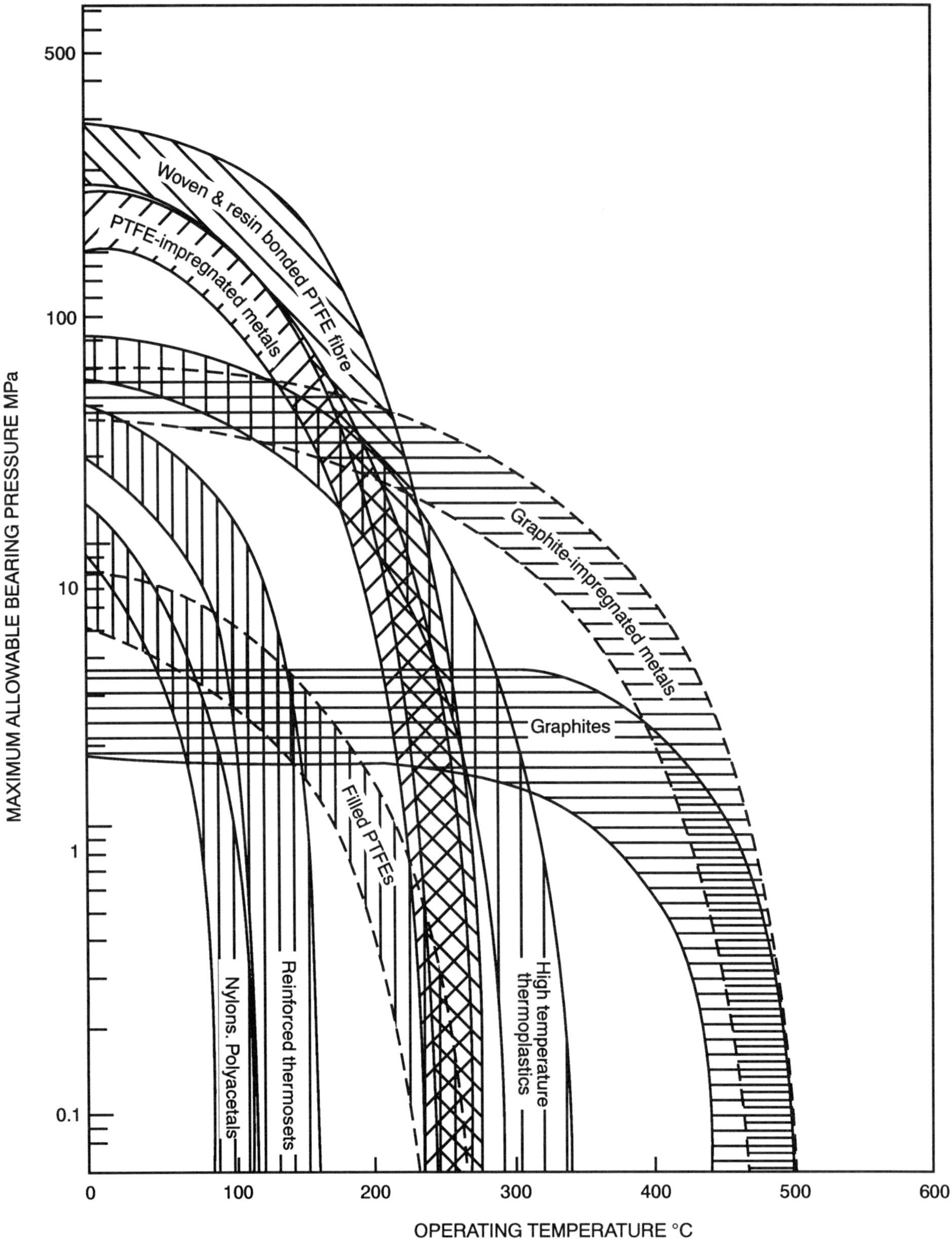

Fig. 1.5.3 Pressure and temperature limits for bearing materials rubbing against steel

Fig. 1.5.3

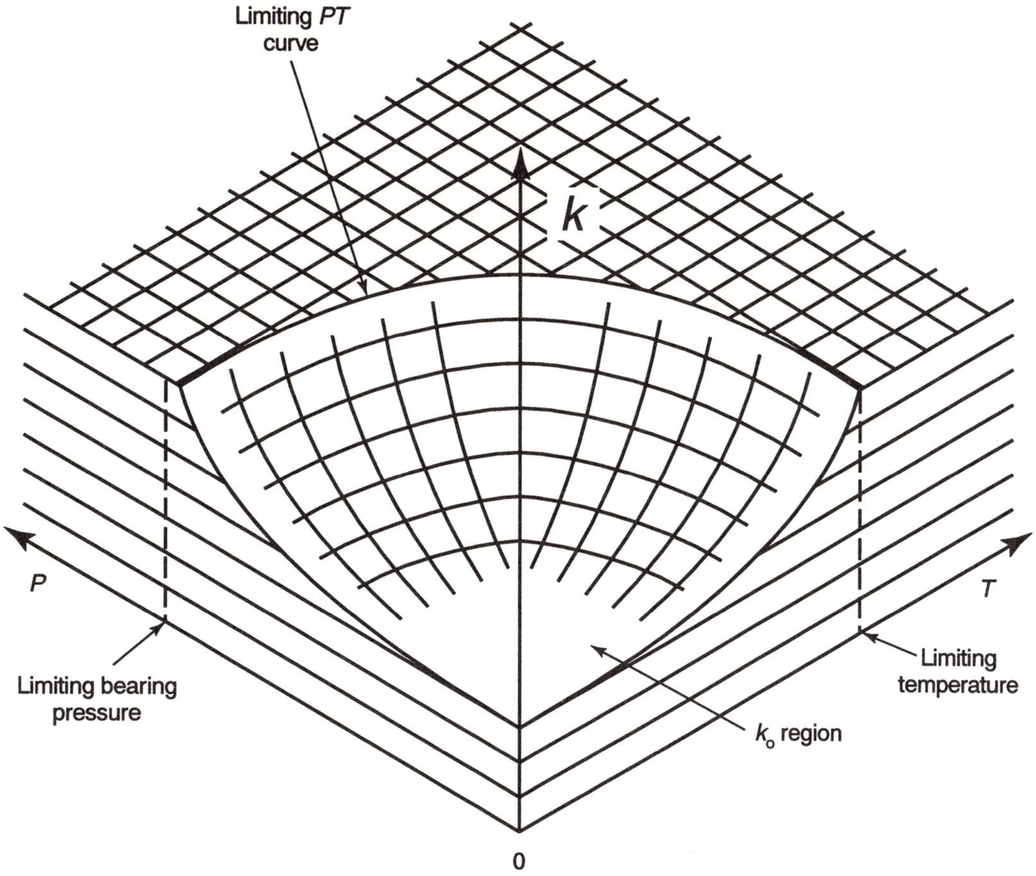

Provided that the operating conditions are moderate, the wear factor *k* is in fact constant, but as operation becomes more severe, the conditions at the surface change and the value of *k* increases. The actual relationship is a constant plateau value of k_o at moderate coditions and rising values at higher combinations of pressure and temperature. Values of *k* for various rubbing bearing materials at different conditions have been measured and can be used for the estimation of likely wear rates in practical applications.

Fig. 1.5.4 Typical wear factor variation with temperature and pressure

Fig. 1.5.4

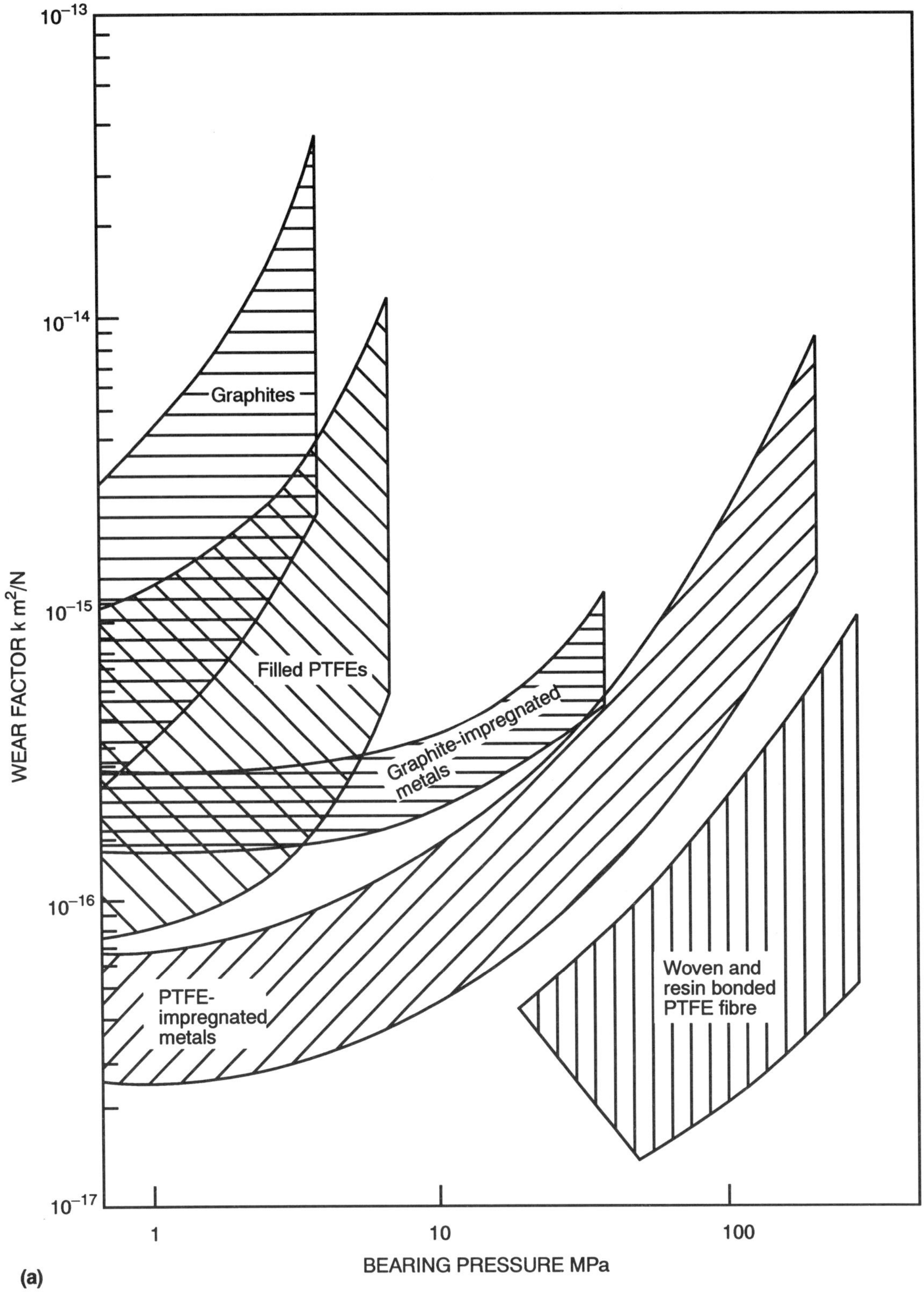

(a)

Fig. 1.5.5 Wear factors at room temperature of typical rubbing bearing materials as a function of bearing pressure

Fig. 1.5.5

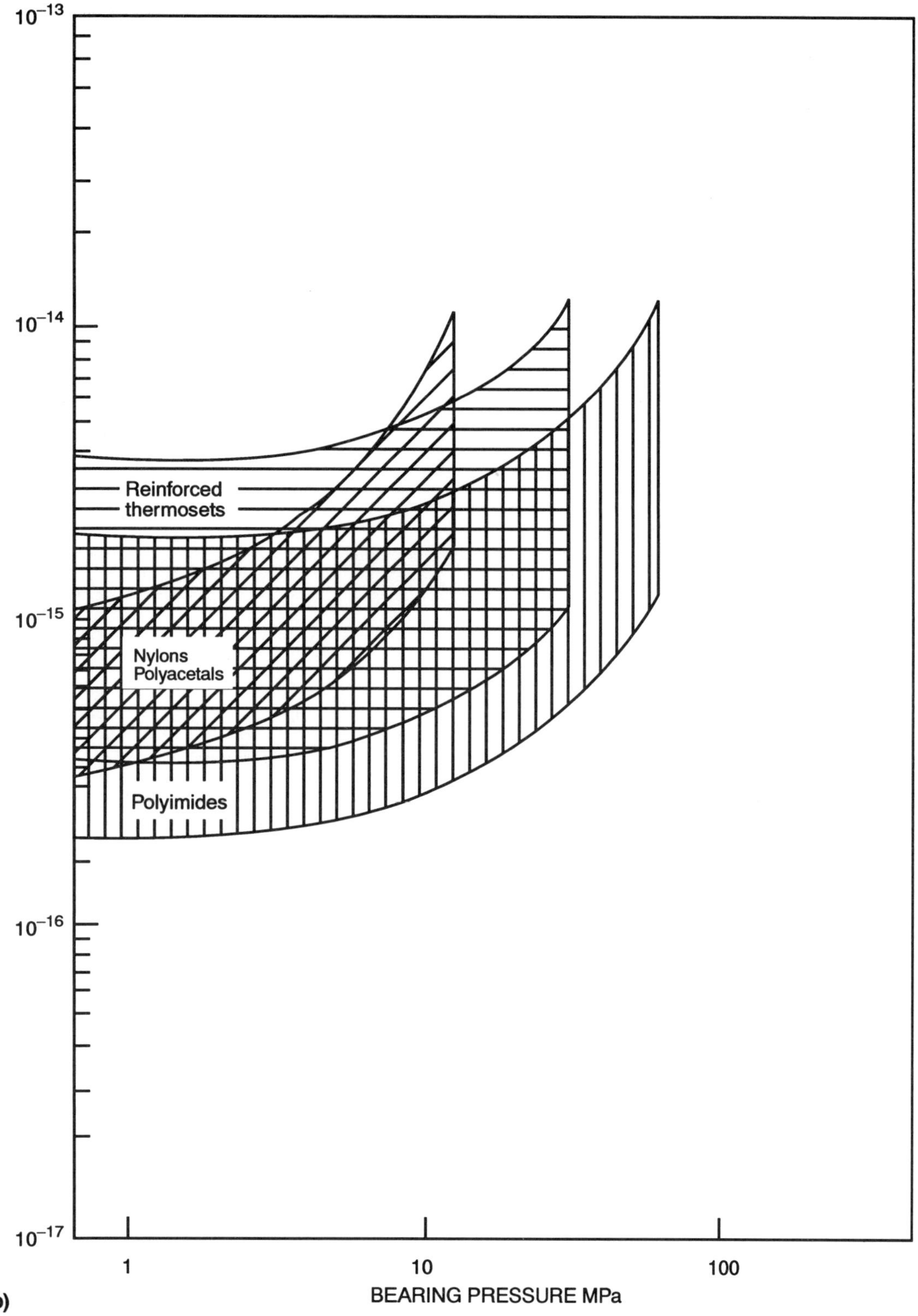

Fig. 1.5.5 Wear factors at room temperature of typical rubbing bearing materials as a function of bearing pressure – *continued*

Fig. 1.5.5 – *continued*

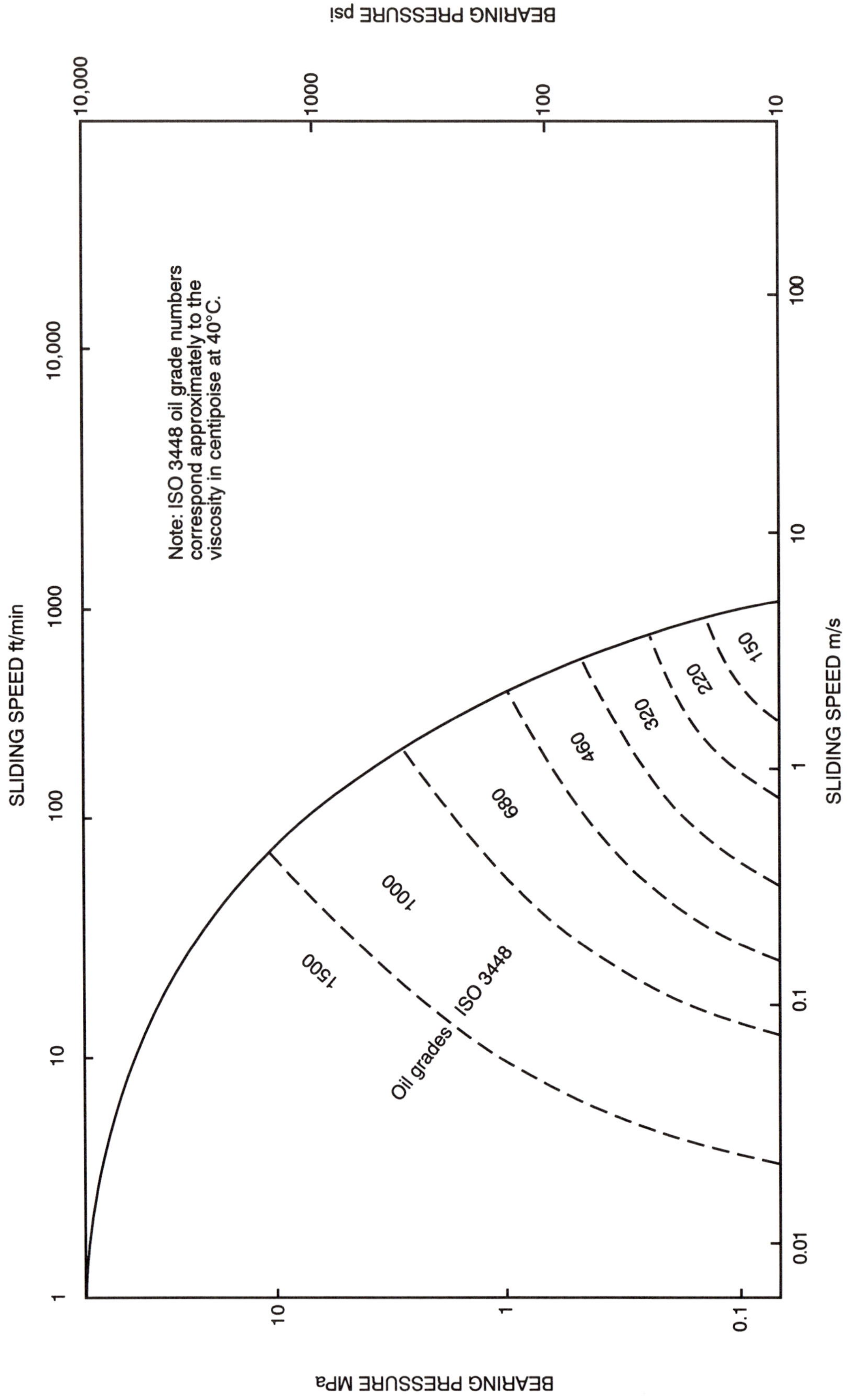

Fig. 1.5.6 Limits of bearing pressure and sliding speed for porous metal bearings, with suggested grades of mineral oil for impregnation

Fig. 1.5.6

1.5.3 Counterformal contact

1.5.3.1 Introduction

Components have a counterformal contact when this contact is concentrated in a small area. Typical examples of applications with counterformal contacts are rolling bearings, gears and cams and followers.

For useful load capacity, very high pressures have to be carried on the small contact areas, and so relatively hard materials are generally required to carry such pressures without surface deformation or fatigue.

In counterformal contacts with predominantly rolling motion, the materials selection requirements are basically similar to those for rolling bearings and for spur, helical and bevel gears. The data given in the sections dealing with these components may therefore be used for the choice of materials for mechanisms which have limited relative sliding.

In counterformal contacts with appreciable amounts of relative sliding, the allowable contact stresses must be reduced, particularly for very hard material combinations. The data on material selection for crossed axis gearing and for cams and tappets provides a guide to the selection of materials for mechanisms with these contact conditions.

A guide to the materials requirements is given in:

Table 1.5.16 *Guide to materials requirements for counterformal contacts*

Components with counterformal contact may be subject to running-in wear and running-in procedures may be necessary. For further details see Section 1.5.2.1 and Tables 1.5.5–1.5.7.

1.5.3.2 Rolling element bearings

In rolling element bearings the components must be hard in order to carry the high contact stresses without surface fatigue. The most commonly used materials are through-hardened steels, and case-hardened steels, though ceramic materials for balls are appearing for use in very high speed applications. Characteristics are shown in:

Table 1.5.17 *Characteristics of commonly used rolling element bearing materials*

At high temperatures and in corrosive environments, tool steels, martensitic stainless steels, refractory alloys, cermets and ceramics may be used. A guide is given in:

Table 1.5.18 *Rolling element bearing materials for use in hostile environments*

The variation of load capacity with temperature is shown in:

Fig. 1.5.7 *Effect of temperature on load capacity of rolling element bearing materials*

To obtain maximum life, the hardness of both the races and the rolling elements should be within an optimum hardness range and the balls or rollers should be up to 10% harder than the race. Cages are used to guide and separate the rolling elements, and are available in a wide range of materials. A guide to the selection of materials for cages is given in:

Table 1.5.19 *Characteristics of rolling bearing cage materials*

1.5.3.3 Gears

Gears have two main material requirements: high fatigue strength in bending, and high resistance to surface fatigue.

Typical gear material combinations are given in:

Table 1.5.20 *Guide to the selection of gear material combinations*

The advantages, limitations, mechanical properties and fatigue properties of common gear materials are given in:

Table 1.5.21 *Characteristics of commonly used gear materials*

The data in the table can be used in designing spur, helical and bevel gearing, but for gears with greater amounts of sliding motion such as worm and crossed helical gears, the surface load capacity factors will be reduced and the order of merit of materials may be affected. Note that the data should *not* be used for detailed design calculations, but an appropriate gear design standard should be used (eg. BS, ISO, DIN, AGMA, etc.).

A gear material must have adequate fatigue strength in bending to carry the required loading on the roots of the teeth for the design lifetime. The bending stress endurance limit for a life of 10^7 cycles is correlated to the ultimate tensile strength of various commonly used gear materials in:

Fig. 1.5.8 *Bending stress endurance limits for commonly used gear materials*

The gear material must also have adequate resistance to surface fatigue to be able to carry the tooth surface loads for the design lifetime. The surface load capacity factor C is a measure of the resistance to surface fatigue of a material, and is shown for gear materials of various hardnesses in:

Fig. 1.5.9 *Guide to resistance to surface fatigue for commonly used gear materials*

The surface load capacity data in Fig. 1.5.9 assumes that the mating gear is steel. If the mating gear is some other material, e.g. brass, then the lower Young's modulus results in an increased allowable load. The effect of the mating material on load factor is given in:

Fig. 1.5.10 *Effect of mating material on surface load capacity factor C*

1.5.3.4 Cams and followers

Many cams and followers operate at very high contact pressures, so surface fatigue (pitting) is a common mode of failure. Scuffing and polishing wear can also occur with plain follower systems where there is a large degree of sliding. Rolling followers have nominally zero sliding, so do not usually suffer from scuffing or polishing wear.

In order to make best use of the fatigue strengths of the materials, various factors need to be taken into account when designing cam and follower systems. These are given in:

Table 1.5.22 *Important working conditions for cams and followers*

The performance of commonly used cam and follower materials is given in:

Table 1.5.23 *Performance of various material combinations for cams and followers*

Typical allowable contact stresses for cam and follower materials are shown in:

Fig. 1.5.11 *Maximum allowable contact stresses for cam and follower materials*

1.5.3.5 Knife edges and pivots

Knife edges and pivots are bearings in which two members are loaded together in nominal line or point contact respectively and can tilt relative to one another through a limited angle by rotation about the contact. A pivot can also rotate freely about the load axis. Examples of knife edges and pivots are illustrated in:

Fig. 1.5.12 *Illustration of knife edges and pivots*

The main requirement of materials for knife edges and pivots is high hardness, giving a high load capacity while keeping the width of the contact area small for low frictional torque and high positional accuracy. Application requirements and desired material properties are given in:

Table 1.5.24 *Important material characteristics for various applications of knife edges and pivots*

Relevant properties are given in:

Table 1.5.25 *Relevant properties of some knife edge and pivot materials*

1.5.3.6 Flexural bearings

Flexural hinges and torsion suspensions are devices which connect or transmit load between two components while allowing limited relative movement between them by deflecting elastically. Examples are given in:

Fig. 1.5.13 *Illustration of a flexure hinge and torsion suspension*

Selection of the most suitable material from which to make the elastic member will depend on the various requirements of the application and their relative importance. Common application requirements and the corresponding desired properties of the elastic member are listed in:

Table 1.5.26 *Important material properties for various applications of flexure hinges and torsion suspensions*

Typical properties which are relevant to applications involving flexure hinges and torsion suspensions are given in:

Table 1.5.27 *Typical properties of some flexure materials*

Table 1.5.16 Guide to materials requirements for counterformal contacts

Essential design requirements	***Material requirements***			***Remarks***
	Rolling bearings	***Gears***	***Cams and tappets***	
Ability to carry load satisfactorily.	Dimensional stability under load as well as absolute load capacity. High yield strength and high creep strength.	Adequate yield strength in bending and compression.	Adequate compressive yield strength to accommodate loading.	Not normally a problem if the materials are strong and hard.
Ability to withstand the impact loads and shocks imposed in use.	Adequate toughness or ductility.	Adequate toughness or ductility.	Adequate toughness or ductility.	Case-hardened materials may provide a solution in difficult cases.
Ability to carry load for the design lifetime.	Adequate resistance to surface fatigue. In operation the surfaces of the components are subject to fluctuating loads.	Adequate bending fatigue strength. Adequate resistance to surface fatigue, which is a common failure mode on gears which have limited amounts of sliding, e.g. spur, helical and bevel gears.	Adequate resistance to surface fatigue.	A material with high resistance to surface fatigue normally has a high elastic limit, which usually means a material of high hardness. It is essential to have a large depth of hardness with surface-hardened materials. The components should have a defect-free structure and good surface finish as both of these qualities significantly increase fatigue life.
Ability to run-in without scuffing.	Not normally a problem.	Scuffing can be a severe limitation, especially on gears with large amounts of relative sliding, e.g. hypoids.	Scuffing can be a problem, particularly when a high degree of relative sliding occurs under poor lubricating conditions.	See Section 1.5.2.1 on running in. Note especially the use of various surface treatments. (Table 1.5.7.)
Wear resistance (usually adhesive and/or abrasive in nature).	Not normally a problem with good sealing and lubrication.	Not normally a problem with good sealing and lubrication. If abrasives are present the use of unlike materials of different hardnesses is recommended. Good lubrication with clean oil reduces the problem by flushing out any abrasive present and reducing the temperature of the contacting surfaces.		Wear rates due to adhesive/abrasive wear are normally low with correct design, finish and assembly, and so are apparent only after a long period of operation.
Ease of manufacture.		If the component can be designed to a reduced level of contact stress, it may be possible to use softer materials which are easier to manufacture.		

Table 1.5.17 Characteristics of commonly used rolling element bearing materials

Type of material	*Structure*	*Ease of heat treatment*	*Environmental limitations*		*Additional treatments to improve performance*	*Remarks*
			High temperature	*Corrosive situations*		
Through-hardened steel	Heat treatment is very important to give a satisfactory carbide structure of optimum hardness in order to achieve the maximum rolling contact fatigue life.	Small components are easy; larger sizes and sections are more difficult.	535A99 (En 31) is not satisfactory at elevated temperatures due to loss of hardness and fatigue resistance. High speed tool steels containing Cr, W and Mo are better at higher temperatures.	Poor resistance to corrosion. Stainless steels can be used, but give reduced load capacity.	Vacuum melted steels improve mechanical properties and fatigue life.	Materials must be of good quality and of uniform structure. Through-hardening requires less process control than case-hardening. 535A99 (EN 31) is the most commonly used steel for conventional bearings.
Case-hardened steel	Carburizing is an effective method of obtaining a deep case with correct structure, supported on a satisfactory transition zone. Can be as effective as a through-hardened structure.	Carburizing is used for convenience in the manufacture of large rolling elements and larger sizes of races. Problems of excessive surface brittleness can arise, especially on small components of alloy steel, particularly Cr steels	Not satisfactory at elevated temperatures (above 180 °C).	Poor resistance to corrosion.	Increase Cr and/ or Ni content to increase hardenability on heavier sections. For very high shock resistance and core hardness, nickel contents up to 4% may be used.	Very important to use sufficiently deep case to avoid subsurface fatigue in the transition zone. Due to the high cost for a deep case, nitriding is seldom used. The toughened core reduces risk of failure by shattering.
Ceramics (silicon nitride)	Hot isostatically pressed.		Not known	Good		Ceramic balls with steel races give enhanced high speed capability and lower heat generation.

Table 1.5.17

Table 1.5.18 Rolling element bearing materials for use in hostile environments

Hostile environment	*Materials requirements*	*Typical types of materials used*	*Cost*	*Ease of processing*	*Remarks*
High temperature	Retention of high hardness and good fatigue performance at elevated temperature (see Fig. 1.5.7).	Through-hardened high speed tool steels (to 550°C)	Low	Easy – except with large components.	Simplest development from standard bearing steels. Often used when corrosion is not a problem.
		Martensitic stainless steels (to 400°C).	Medium	Easy to moderate	More expensive than above. Hardness at elevated temperatures can be increased by increasing the Mo, V, and Co content.
		Refractory alloys (to 800°C).	High	Often difficult	Generally their use has not been widespread due to their limited application, high cost and processing difficulties, and with some, their brittle nature. User must develop and evaluate own material.
		Cermets and ceramics[a] (to 900°C).	High	Difficult	
Corrosive	Good corrosion resistance. See Chapter 1.4.	Martensitic stainless steels.	Medium	Easy to moderate	Most used material.
		Refractory alloys.	High	Often difficult	Generally their use has not been widespread due to their limited application, high cost and processing difficulties, and with some, their brittle nature. User must develop and evaluate his own material.
		Cermets and ceramics[a]	High	Difficult	

[a] Avoid ceramic-on-ceramic combination as this will generally give a very short life because of the very high contact stresses produced, whereas ceramic rolling elements used in conjunction with steel races can give a considerably increased life.

Table 1.5.19 Characteristics of rolling bearing cage materials

Operating conditions		***Typically used cage materials***	***Temperature limitations (°C)***	***Oxidation resistance***	***Wear resistance***	***Method of manufacture***	***Remarks***
Good Lubrication	Low speed or non-critical	Low carbon	260	Poor	Fair	Made from riveted strips.	Standard material
	Low speed or non-critical	Thermoplastic (filled or unfilled)	100	Excellent	Fair	Moulded	–
	Low speed and corrosive environment.	AISI 430 stainless steel. Hardened and tempered.	Limited by elevated temperature performance of lubricant.	Excellent	Poor	Made from riveted strips	Standard material for 440°C stainless steel bearings.
	Medium speed and medium temperature.	Iron silicon bronze.		Excellent	Good at 150°C. Excellent at 260°C.	Machined	e.g. jet engine applications.
	High speed applications.	Non-metallic retainers, fabric base phenolic laminates.	135	–	Excellent	Machined	
	High temperatures.	S. Monel.	Limited by elevated temperature performance of lubricant.	Excellent	Fair	Machined	Excellent high temperature strength.
		17-4-PH stainless steel		Excellent	Poor in air	Made from riveted strips.	
		AISI 430 stainless steel. Hardened and tempered.		Excellent	Poor	Made from riveted strips.	Low speed applications.
Marginal lubrication	Low temperature at any speed.	Silver plated metal.	Possibly 180.	–	Excellent to 150°C.	Plate pre-formed component.	Can be simple remedy if problem of lubrication not very severe.
No lubrication	Preferably low loads.	Composite of PTFE/MoS_2/ glass fibre	~185 to + 300 (life decreased by a factor of 10 at + 150°).	–	Moderate	Hot pressing; machining.	Can often work where no other material will, e.g. vacuum.

Table 1.5.20 Guide to the selection of gear material combinations

Gear type	***Gear duty***	***Material combinations***	
		Pinion (preferably of harder material)	***Wheel (preferably of softer material)***
Spur, helical and bevel gears[b]	Light duty ↓ High duty[c]	Plastics, brass, mild steel and stainless steel in any combination.	
		Malleable cast iron	Plastics, brass. Phosphor-bronze[a]. Malleable cast iron
		Cast iron	Cast iron
		Medium carbon steel normalized.	Cast steel. Medium carbon steel, normalized. Low carbon low alloy steel, normalized.
		Medium carbon high strength low alloy steel, hardened and tempered.	Medium carbon steel, normalized.
		Medium carbon high strength low alloy steel, nitrided.	Medium carbon low alloy steel, nitrided.
		Medium carbon low alloy steel, induction hardened.	Medium carbon low alloy steel, induction hardened.
		Low carbon high strength low alloy steel, carburized.	Low carbon high strength low alloy steel, carburized or hardened and tempered.
Worm gears		Cast iron	Phosphor-bronze
		Through-hardened steel	Soft bronze
		Case-hardened steel	Phosphor-bronze
Crossed helical gears[c]		Case-hardened steel	Case-hardened steel
		Hardenable cast iron (hardness 260–340 VPN)	Hardenable cast iron (hardness 260–340 VPN)
		Low alloy case-hardening steel (hardness 640–670 VPN)	Complex aluminium bronze (hardness 200–260 VPN)

[a] Phosphor bronze should not be run against soft steel pinions.
[b] The pinion material on a given line is suitable for engaging with wheels in materials on and above that line.
[c] At high duty, improving the surface finish increases the load capacity.

Table 1.5.21 Characteristics of commonly used gear materials

Type of material	*Typical operating conditions*	*Primary advantages*	*Primary disadvantages*	*Typical materials*	*Surface hardness VPN (m = minimum)*[a]	*Minimum ultimate tensile strength (MPa)*	*Max. Hertz contact stress for life of 10^7 cycles (MPa)*[e]	*Surface load capacity factor C for life of 10^7 cycles (MPa) (mating with steel)*[e]	*Max. bending stress for life of 10^7 cycles (MPa)*[e]	*Remarks*
Plastics	Motion only. Lightly loaded.	Low cost for large numbers due to ease of manufacture (moulding). Low friction. Silent operation. Will operate with minimum of lubrication. Good corrosion resistance.	Low duty, limited by bending strength. Poor elevated temperature resistance.	Nylon 66 Polyacetal	– –	70 70	– –	– –	40[c] 40[c]	Wear is not generally critical in lubricated plastic gears.
Reinforced plastics	Slight increase in load capacity (normally only used at higher ambient temperatures).		Can improve the poor elevated temperature resistance, but has poor impact loading performance.	Reinforced phenolic laminates. Glass-filled Nylon 12.	– –	– 70	– –	– –	40[c] 40[c]	Reinforcing improves dimensional stability, but has been mainly superseded by use of improved plastics.
Zinc die-casting	Motion only, lightly loaded.	Easy to make large numbers of components with close reproduction in detail and dimensions, and with smooth surface finish.	Zinc has no true elastic limit; creep will occur at room temperature at rates depending on the stress. Hence tensile strength data has no real meaning in terms of design.		90	280	260	3	35	Strength depends on copper content – highest with about 3 % Cu but usually more expensive.
Brass	Very light duty only.	Ease of manufacture (extrusion or machining).	Expensive. Poor wear resistance.	61–64% Cu–Zn Yellow high tensile brass.	190/240	–	690	20	–	
Phosphor-bronze	Wormwheels	Low friction. Material work hardens. Compatible with hardened steel.	Unable to withstand high loading.	BS1400 PB2C: Sand cast Chill cast Centrifugally cast	70 m 80 m 90 m	185 230 260	430 500 560	8 11 14	100 130 145	Best results with fine grained structure and density developed by casting blanks centrifugally.
Brasses or bronzes treated with Delsun	Wormwheels and drives. Increases load capacity, especially in well lubricated conditions.	Increases surface hardness and improves corrosion resistance in saline environment. Improves surface fatigue resistance. Improves running- in in heavily loaded/marginal or no lubrication conditions. Improves compatibility with other mating surfaces.	Treatment can cause thermal distortion as it occurs at about 400°C. Extra process in production line therefore more expensive.	BS265 and BS1400 treated with Delsun.						Often used in seawater applications.

Table 1.5.21

Table 1.5.21 Characteristics of commonly used gear materials – *continued*

Through-hardened alloy steels.	Light power to high duty. Worms and pinions.	Through-hardening increases load capacity but machining becomes difficult above ultimate tensile strengths of 1000 MPa. Hardness is controlled mainly by carbon content (0.3–0.6%). Plain carbon manganese steels have low hardenability and for section sizes greater than 15 mm, additions of Mn, Cr, Mo, Ni are required to achieve through-hardening. The amount of alloying additions required is dependent on section size. Increased hardenability is associated with increased strength, toughness and impact resistance. Molybdenum, vanadium and to a lesser extent chromium, increase strength and wear resistance in the tempered condition. Molybdenum reduces the susceptibility to temper embrittlement. Alloying additions can be combined to give the required materials properties for a given application; however, as alloy content increases the steel becomes more expensive and more difficult to machine.	Steels with good pitting resistance have greater tendency to scuffing, often require surface treatments; see section on running in, 1.5.2.1. Distortion on heat treatment.	Through-hardened and tempered steels.						Wide variety of through-hardening steels to give desired properties.
				Tensile range R e.g. 150M28, 150M36	200 m	695	825	19	235	
				Tensile range S e.g. 708M40, 817M40	225 m	770	895	23	255	
				Tensile range T e.g. 830M31	250 m	850	965	26	275	
				Tensile range U e.g. 830M31, 709M40	270 m	925	1030	30	295	
				Tensile range V e.g. 830M31, 709M40	295 m	1000	1100	34	310	

[a] Gear cutting becomes difficult if hardness exceeds 270 VPN.
[b] This factor indicates the maximum load capacity per unit length of contact (for a life of 10^7 cycles) for counterformal line contacts, such as gears, at which rolling predominates. The corresponding maximum load per unit length of contact is equal to $C\left(\frac{R_1 R_2}{R_1 + R_2}\right)$ where R_1 and R_2 are the radii of curvature of the contacting surfaces. C is proportional to the maximum Hertzian stress squared and also depends on the elastic moduli of the mating materials. The values shown in the table assume a steel mating material, and the effect on C of different mating materials moduli is shown in Fig. 1.5.10. The values of C also assume good surface finish, alignment and lubrication, and if these are poor or if there is appreciable sliding (as in cams and followers or crossed axis gearing) at the contacts, the load capacity factors will be reduced and the order of merit may be affected.
[c] At 20°C; reduces greatly with increasing temperature and hence pitch line speed.
[d] Depending on grade.
[e] The data is not to be used for detailed design calculations, and an appropriate gear design standard should be used (e.g. BS, ISO, DIN, AGMA etc.)

Table 1.5.21 – *continued*

Table 1.5.21 Characteristics of commonly used gear materials – *continued*

Type of material	*Typical operating conditions*	*Primary advantages*	*Primary disadvantages*	*Typical materials*	*Surface hardness VPN (m = minimum)*[a]	*Minimum ultimate tensile strength (MPa)*	*Max. Hertz contact stress for life of 10^7 cycles (MPa)*[e]	*Surface load capacity factor C for life of 10^7 cycles (MPa) (mating with steel)*[e]	*Max. bending stress for life of 10^7 cycles (MPa)*[e]	*Remarks*
Flame- or induction-hardened steels	Low to high duty, especially where bending stresses are not severe, e.g. sprockets.	Increase in surface hardness, so increased resistance to surface fatigue. Both case and core properties are easily adjustable to suit particular application. Depth of hardening depends on amount of heating. Selective hardening of the surface can minimize thermal distortion.	No associated increase in bending strength except at large depths of hardening in the root fillets. Overheating renders the case brittle. Hardening must be done carefully, otherwise the hardened layer may contain high residual tensile stresses leading to surface and bending fatigue.	0.4 C Steel e.g. 080A35 Normalized and water quenched.	550 m	540	1380	54	45	Flame hardening: for small components, achieved simply and in small numbers. Induction hardening: easily applicable to large components and very suitable for mass production techniques.
				0.55 C Steel e.g. 070M55 Normalized and air hardened.	580 m	695	1450	59	179	
				1.5% Ni–Cr–Mo, 2.5% Ni–Cr–Mo e.g. 817M40, 826M31	480 m–600 m	850[d]	1510	65	220	
				2.5% Ni–Cr–Mo e.g. 826M40	550 m	925	1580	72	248	
Sintered materials	Low and medium duty spur and straight bevel gearing.	Cheap to make large quantities of simple components. Sintered components afford a dimensional accuracy superior to that achieved by forging if the ultimate tensile strength of the material is below 620 MPa. This reduces or eliminates additional machining. Powder metallurgy enables the manufacture of alloys not obtainable by other means.	Expensive to produce small number of components. Large amounts of distortion occur when sintering high tensile strength materials.	Sintered brass, bronze, iron, iron copper, steel.	Up to 300 (as sintered) Can be hardened to 600.	Up to 620	–	–	40–240	Allows greater flexibility in choice of alloy to suit particular application.
Malleable cast iron	Light power gears.	Rigid and wear resistant. High damping capacity.			140 m	300	480	7	100	

Table 1.5.21 Characteristics of commonly used gear materials – *continued*

Cast iron	Light power gears and wormwheels. Load capacity can be increased by using nodular or spheroidal graphite irons and induction hardening. Not suitable for heavy duty applications.	Good casting properties give low cost, especially for complex component shapes. The graphitic structure gives good machinability if the phosphorus content is low. The graphite also improves the scuff and wear resistance in both lubricated and unlubricated conditions. Coarse graphite flakes increase the wear resistance but decrease the resistance to bending and impact loads. The matrix should be pearlitic, not ferritic. Shell moulding gives iron gears cast to very close tolerances. Very high damping capacity.	Poor shock resistance. Weaker than steels.	Ordinary grade as cast. Medium grade as cast. High grade as cast. High grade heat-treated. Nodular Austempered Austenitic	165 m 210 m 220 m 300 m 205/240 270/290 255/300	185 250 340 340 – – –	430 520 530 690 720 805 805	9 12 12 19 20 27 24	50 65 90 90 – – –	Cast iron can be used as whole component or as centre of large heavy duty component.
Cast steels	Light power gears.	Slight increase in bending load capacity over cast iron.	More expensive than cast iron.	0.35–0.45% C. Cast steel.	150 m	540	660	12	170	
Forged steels	High duty, industrial and marine gears.	Low cost and good mechanical properties including work hardening. Wear resistant. Good compatibility with other mating materials. Easily machinable.	Not suitable for very high duty and critical applications.	0.4 C 080A35 Normalized Hardened and tempered 0.55 C (070H55) Normalized Hardened and tempered	150 m 180 m 200 m 220 m | 250 m	540 615 695 770 | 850	720 760 825 895 | 965	14 16 19 23 | 26	190 210 235 255 | 275	Use for both small and large gears.
Carburized steel	High duty applications especially in automotive field. When ground, used in more critical applications like aircraft.	Carburizing gives a deep tough layer of high strength and hardness with residual compressive stresses – aids resistance to contact and bending fatigue stresses. Machining to good surface finish, though difficult and costly, can greatly improve resistance to contact and bending fatigue.	Distortion caused during quench part of treatment – can be ground off but grinding leaves tensile stresses in the surface which reduce contact and bending fatigue resistance. However, the improved surface finish normally makes the above effect unimportant.	Carbon case-hardening steels, e.g. 080H15	750 m	495	1860	99	385	Special surface treatments can be required to aid running-in, i.e. to reduce tendency to scuffing – see Section on running-in, 1.5.2.1.
				1.7% Ni–Mo 3% Ni–Cr e.g. 665M17, 665M13	750 m	695–850	1930–2000	105–114	460–480	
				4% Ni–Cr e.g. 659M15	710 m	1310	2130	130	515	

Table 1.5.21 – *continued*

Table 1.5.21 – *continued*

Table 1.5.21 Characteristics of commonly used gear materials – *continued*

Type of material	*Typical operating conditions*	*Primary advantages*	*Primary disadvantages*	*Typical materials*	*Surface hardness VPN (m = minimum)*[a]	*Minimum ultimate tensile strength (MPa)*	*Max. Hertz contact stress for life of 10^7 cycles (MPa)*[e]	*Surface load capacity factor C for life of 10^7 cycles (MPa) (mating with steel)*[e]	*Max. bending stress for life of 10^7 cycles (MPa)*[e]	*Remarks*
Nitrided steels	High duty industrial and marine applications.	Higher surface hardness than carburized steels and greatly reduced, but not necessarily zero, distortion, thus reducing amount of machining after treatment.	Expensive: hardening period long and depth of case limited. Case tends to be susceptible to shock. For large gear sizes the nitrided case must be supported by a core of high strength. More difficult to machine than carburized cases.	3% Cr–Mo. e.g. 722M24	850 m	695–925	1450–1580	59–72	190–255	Must be well finished, have all sharp edges removed and have excessive 'white layer' from the process removed. If profile grinding is used for above purpose, minimum amount of material should be removed.
				3% Cr–Mo–V. e.g. 897M39	850 m	1230	1720	83	290	
Carbo-nitrided steels	High duty – seldom used.	Leaves very thin and hard surface layer. Increases resistance to abrasive wear. Causes less distortion than carburizing, but more than nitriding. Resistance to scuffing due to nitrogen present. Corrosion resistance better than by carburizing.	Does not confer resistance to contact stresses associated with the greater case-depths of carburized gears.	Carbon case-hardening steels 1.7% Ni–Mo 3% Ni–Cr e.g. 665M17, 655M13, 659M15						Sometimes used for small components. Higher processing temperature for more severe service conditions.
Sulfinuz on through-hardened steel.	To reduce tendency to scuffing during running-in – see section on running-in 1.5.2.1.	Gives good scuffing resistance and also some reduction in coefficient of friction.	Sulfinuz causes some component distortion which can be a problem if it is the last process in the production line. Process temperature is sufficiently high (540–600°C) to temper or anneal many ferrous metals.	Sulfinuz treated 1.5 Ni–Cr–Mo. (e.g. 817M40) which was previously hardened and tempered.	400 m	850 (tensile range T)	–	49 (about twice as high as without sulfinuz)	275 (about the same as without sulfinuz)	Data is based on limited experimental evidence.

Table 1.5.22 Important working conditions for cams and followers

Factor	***Effect***	***Remarks***
Nominal contact stress	Maximum permitted stress depends on materials, see Fig. 1.5.12	Plain follower systems usually more compact than roller follower systems.
Misalignment	Leads to local high loading and early failure.	Avoided by self-aligning design, or slightly domed or barrelled follower.
Surface finish	Highest contact stresses can only be used with very good surface finishes.	In IC engines, typically 0.15 μmRa (followers), 0.4 μmRa (cams). Phosphating or other surface treatments help.
Sliding velocity (plain follower systems only)	Risk of scuffing or polishing wear, particularly if entrainment velocity is low for part of cycle.	Methods for calculation of entrainment velocity exist[a] Alternatively, avoid problem by use of roller followers.
Lubricant (critical for plain follower systems).	High working oil viscosity is desirable with a copious supply of clean lubricant for best results. Supply oil at lowest practical temperature to reduce tendency to scuffing. Avoid small quantities of water in lubricant as this significantly reduces the allowable contact stresses.	Unfortunately, lubrication conditions are frequently dictated by the needs of other parts of the machine. Oil has useful effect of flushing out any abrasive presents.

[a] M J Neale, *Drives and Seals a Tribology Handbook*, Butterworth-Heinemann, 1994 also ESDU 84033. Material selection and lubrication conditions for cam and follower systems, ESDU Ltd, 1984

Table 1.5.23 Performance of various material combinations for cams and followers

Cam material	*Follower material*	*Allowable contact stress (MPa)*	*Scuffing and polishing resistance*
Steel, hardened steel	Steel, hardened steel	See Figure 1.5.12	Moderate; improves with hardness
Grey cast iron	Grey cast iron		Very good
Hardenable iron	Hardenable iron		Good
Chill cast iron	Chill cast iron		Very good
Nodular (S.G.) iron	Nodular (S.G.) iron		Moderate
Grey cast iron, hardenable iron	Hardened steel		Very good
Chill cast iron	Hardened steel		Very good
Nodular (S.G.) iron	Hardened steel		Good
Phosphor-bronze (SAE 65)	Hardened steel	590	Very good
Zinc (die-cast)	Steel	180	Poor
Phenolic (fibre-reinforced)	Steel	130	Good
Ceramic (e.g. silicon nitride)	Hardened steel	No data	Very good

Table 1.5.23

Table 1.5.24 Important material characteristics for various applications of knife edges and pivots

Application requirement	***Desired material property***
High load capacity for a given bearing geometry.	High $\frac{(\text{hardness})^2}{\text{modulus of elasticity}}$
Ability to tolerate overload, impact or rough treatment generally.	A measure of ductility in compression, so that overload can be accommodated by plastic deformation rather than chipping or fracture.
Preceding two requirements together (for example for weighbridges, strength testing machines, etc.).	High hardness together with some ductility. In practice various metallic materials with hardness greater than 60 Rc (750 VPN) are usually specified.
Very low friction with useful load capacity where freedom from impact and overloading can be expected (for example in sensitive force balances and other delicate equipment).	Very high $\frac{(\text{hardness})^2}{\text{modulus of elasticity}}$ using various brittle materials having exceptionally high hardness.
High wear resistance.	High hardness is generally beneficial.
Little indentation of block by knife edge or pivot.	Hardness of block > hardness of knife edge or pivot. (This is nearly always desirable; the differential should be at least 5%.)
The two members of the bearing have to slide relative to one another at the contact and must be metallic to withstand impact, etc.	Low tendency to adhesion to avoid high sliding friction and wear; in practice it is often sufficient to avoid using identical materials.
Bearing to be used in a sensitive force balance.	Non-magnetic; should not absorb moisture or be subject to any other weight variation. (Agate, for example, is unsatisfactory in the latter respect since it is hygroscopic).
Bearing to be used in a potentially corrosive environment (includes 'normal' atmospheres).	Good corrosion resistance, especially if preceding requirement has to be met.

Table 1.5.25 Relevant properties of some knife edge and pivot materials

Material	*Hardness, H (VPN)*	*Modulus of elasticity, E (GPa)*	*Load capacity factor, H^2/E (arbitrary units)*	*Ductility*	*Approx. maximum operating temp. in air (°C)*	*Corrosion resistance*[a]
High carbon steel	to 750	210	2.3	Some	250	Poor
Tool steels	to 1000	210	3.4	Some	to 650	Poor–good
Stainless steel (440 C)	710	210	2.1	Some	430	Moderate
Agate	820	72	7.4	None	575[b]	Excellent
Synthetic corundum (Al_2O_3)	2100	380	11.6	None	1500	Excellent
Boron carbide	3000	450	17.4	None	540	Excellent
Silicon carbide	2800	410	16.4	None	800	Excellent
Hot pressed silicon nitride	2000	310	13	None	1300	Excellent

[a] Materials with poor corrosion resistance can often be protected by grease, oil bath or surface treatments such as chromizing of steels.
[b] Phase change temperature.

Table 1.5.25

Table 1.5.26 Important material properties for various applications of flexure hinges and torsion suspensions

Application requirement	***Desired material property***
1. Small size.	High M, equal to Y unless fatigue critical in which case $M = F$.
2. Flexure hinge with maximum movement for a given size.	High M/E
3. Flexure hinge with the maximum load capacity for a given size and movement.	High M^3/E^2.
4. Flexure hinge with minimum stiffness (for a given pivot geometry).	High $1/E$. note that stiffness can be made zero or negative by suitable pivot geometry design.
5. Torsion suspension with minimum stiffness for a given suspended load.	High M^2UG. U is not a material property but emphasises the value of being able to manufacture the suspension material as thin flat strip. Stranded silk and other textile fibres are commonly used to give suspensions with minimum torsional stiffness.
6. Elastic component has to carry an electric current.	High electrical conductivity.
7. Elastic component has to provide a heat bath.	High thermal conductivity.
8. Elastic component has to provide the main reactive force in a sensitive measurement or control system.	Negligible hysteresis and elastic after-effect. Materials commonly used to give this effect are: 91.5% Pt–8.5% Ni alloy. Platinum/silver alloy 85/15 to 80/20. Quartz. Non-magnetic.
9. As 8 and may be subject to temperature fluctuations.	Low temperature coefficient of thermal expansion and elastic modulus (E or G).
10. As 6 but current has to be measured accurately by system of which elastic component is a part.	Low thermoelectric emf against copper (or other circuit conductor) and low temperature coefficient of electrical conductivity.
11. Elastic component has to operate at high or low temperature.	As for 1–10 above, but properties, for example strength, must be those at the operating temperature.
12. Elastic component has to operate in a potentially corrosive environment (includes 'normal' atmospheres).	Appropriate, good, corrosion resistance, especially if requirements 8 or 10 have to be met.

Symbols:
M = Maximum permissible stress.
Y = Yield strength.
F = Fatigue strength.
E = Young's modulus.
G = Shear modulus.
U = Aspect ratio (width/thickness) of suspension cross-section.

Table 1.5.27 Typical properties of some flexure materials

Material	*Yield strength[a] (MPa)*	*Fatigue strength[b] (MPa)*	*Young's modulus E[c] (GPa)*	*Thermal conductivity (W/m°C)*	*Electrical conductivity (% IACS)[d]*	*Atmospheric corrosion resistance[e]*	*Approximate max. operating temperature in air (°C)*
Spring steels 0.6–1.0 C 0.3–0.9 Mn	800–2100	400–700	210	45	9.5	P	230
Carbon chromium stainless steel (BS420 S45)	1500	600	210	24	2.8	M	540
High strength alloy steels: nickel marageing steel	2100	660	190	17	4	P	480
Nickel–chromium–molybdenum-vanadium steel (DTD 5192)	2100	800	210	35	6	P	400
Inconel X	1600	650	210	12	1.7	E	650
High strength titanium alloy	950	650	110	9	1.1	G	480
High strength aluminium alloy	500	150	72	120	30	P	200
Beryllium copper	900	380	120	100	25	G	230
Low beryllium copper	650	240	110	170	45	G	200
Phosphor-bronze (8% Sn; hard)	600	200	110	55	12	G	180
Glass fibre-reinforced nylon (40% GF)	200	–	12	0.35	Negligible	E	110
Polypropylene	37[f]	–	1.4[f]	0.17	Negligible	E[g]	50

[a] Very dependent on heat treatment and degree of working. Figures given are typical of fully heat-treated and processed strip material of about 2.4 mm thickness at room temperature. Thinner strip and wire products can have higher yield strengths.
[b] Fatigue strengths are typical for reversed bending of smooth finished specimens subjected to 10^7 cycles. Fatigue strengths are reduced by poor surface finish and corrosion, and may continue to fall with increased cycles above 10^7.
[c] Modulus of rigidity, $G = E/2\,(1 + \upsilon)$ where υ is Poisson's ratio. For many materials $\upsilon \sim 0.3$ and $G \sim E/2.6$.
[d] Percentage of the conductivity of annealed high purity copper at 20°C.
[e] P = poor, M = moderate, G = good, E = excellent. Protection from corrosion can often be given by grease or surface treatment.
[f] At high strain rate. Substantial creep occurs at much reduced stress levels, probably restricting applications to where the steady load is zero or very small, and the deflections are of short duration.
[g] Deteriorates rapidly in direct sunlight.

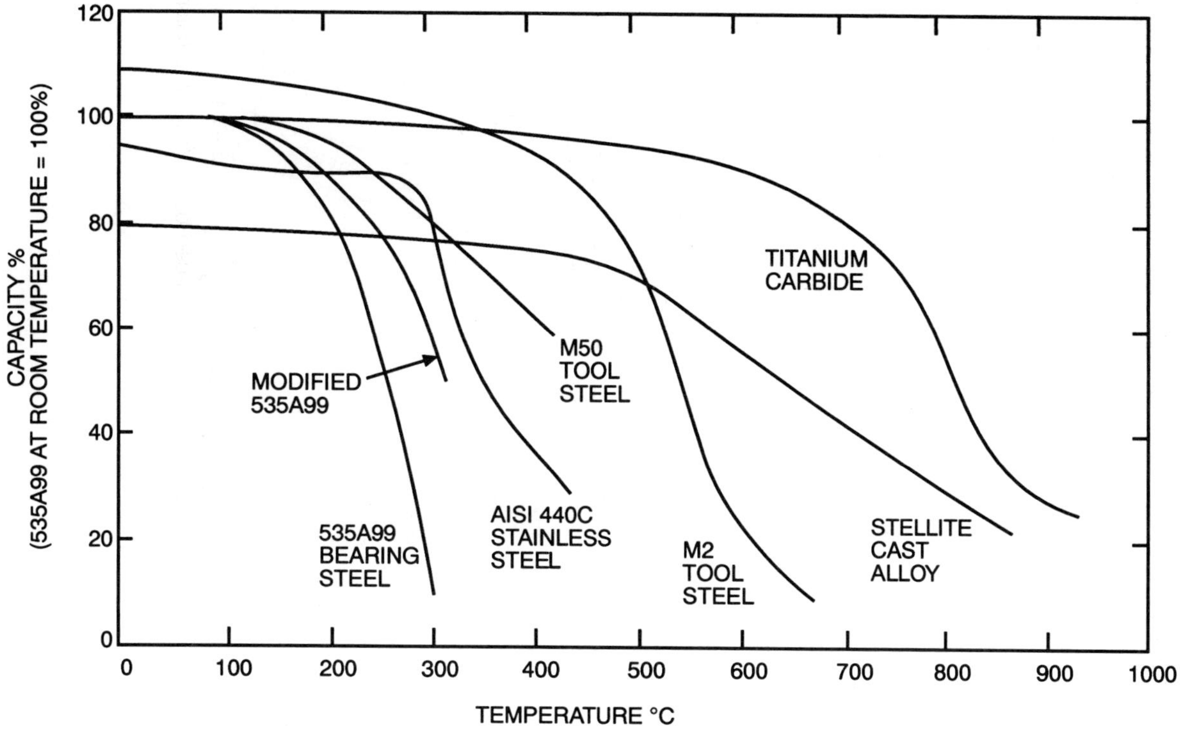

Fig. 1.5.7 Effect of temperature on load capacity of rolling element bearing materials

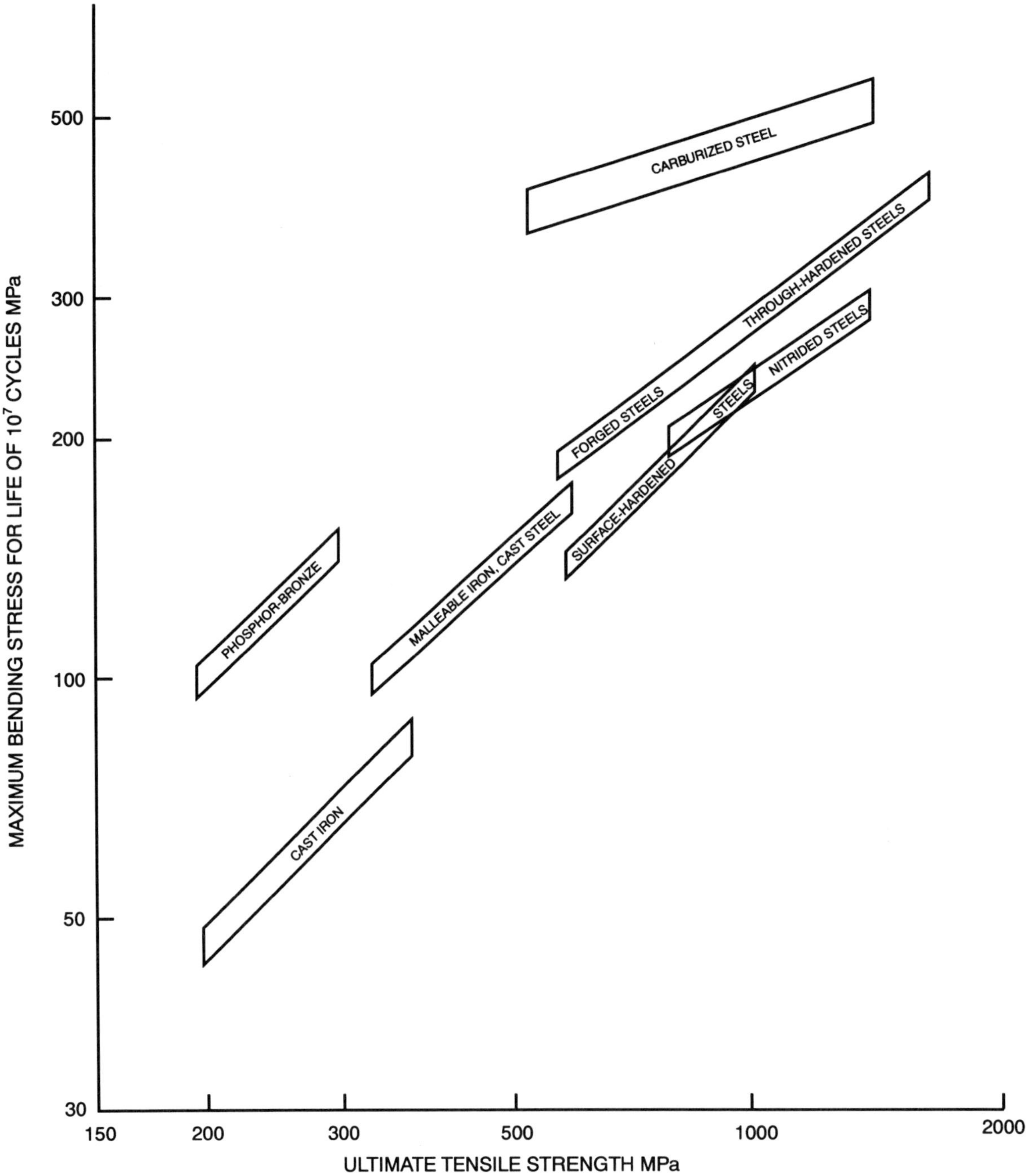

Fig. 1.5.8 Bending stress endurance limits for commonly used gear materials

Fig. 1.5.8

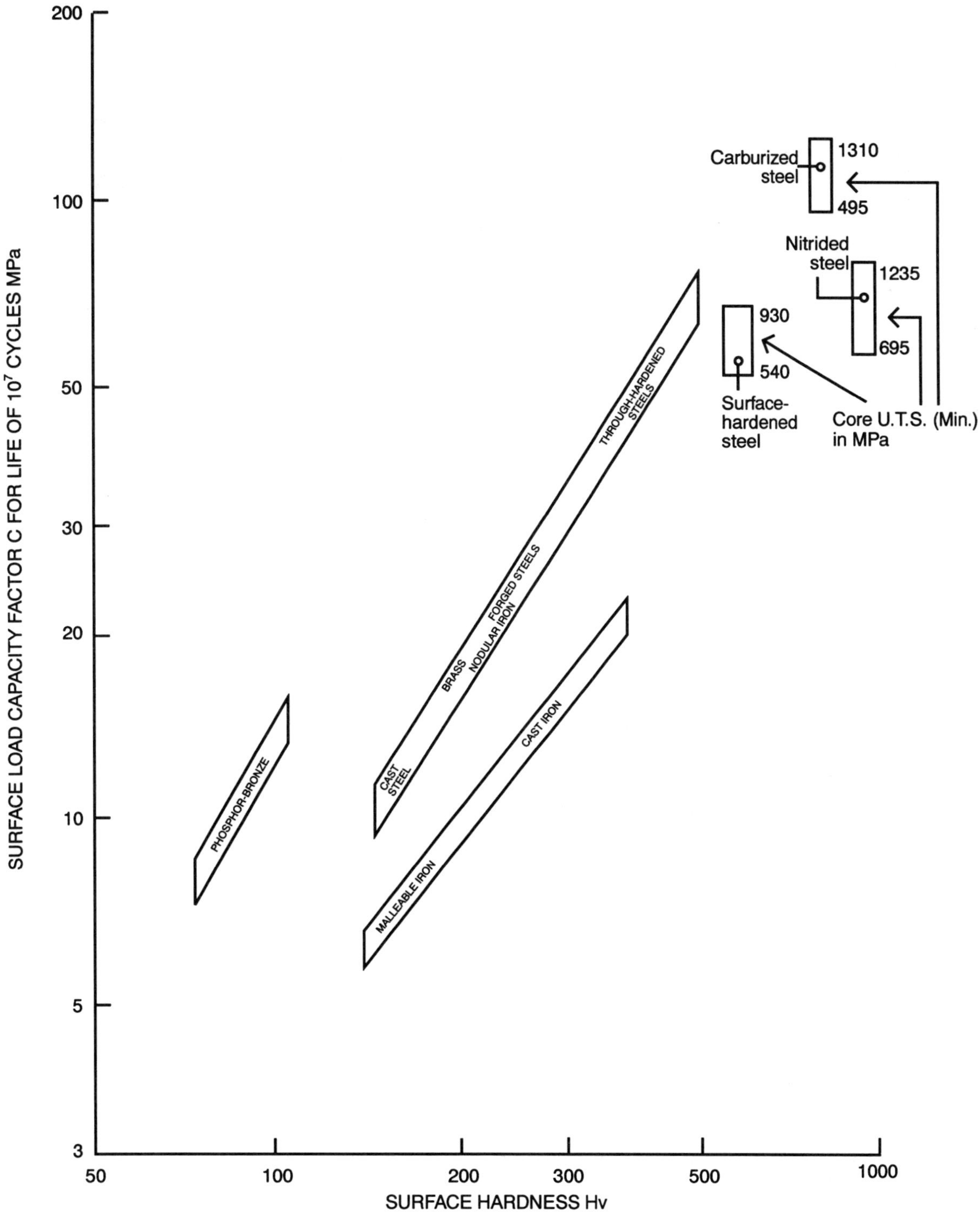

Fig. 1.5.9 Guide to resistance to surface fatigue for commonly used gear materials

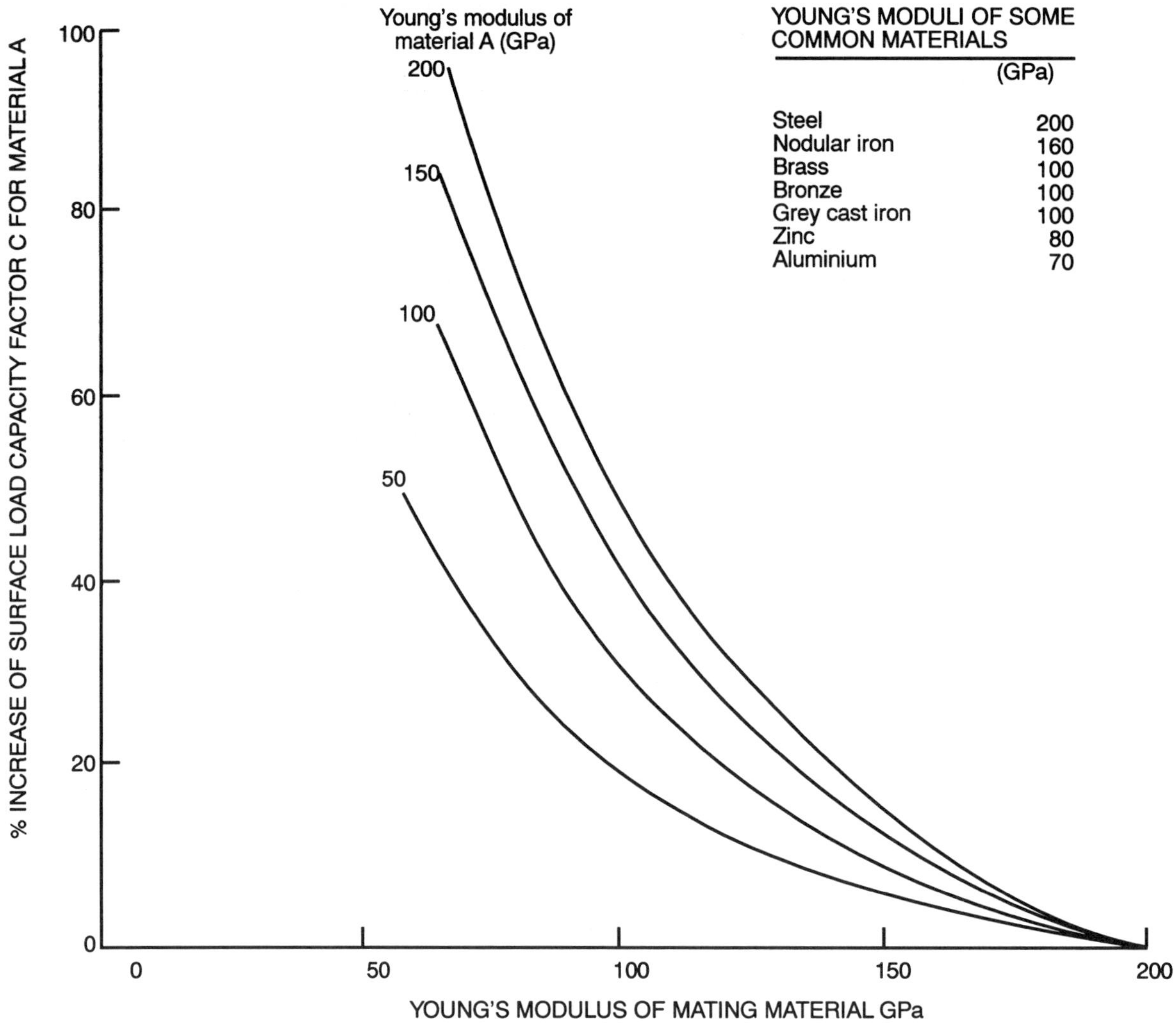

To find surface load capacity factor for material A when mated with a material other than steel:

1. From Table 1.5.21 find the value of the load capacity factor, C, for material A when mated with steel.

2. From the values of the Young's moduli of material A and its mating material use the figure above to read off the amount by which the load capacity factor for material A will be increased.

Note: Both materials of the pair may be treated in a similar way.

Fig. 1.5.10 Effect of mating material on surface load capacity factor, *C*

Fig. 1.5.10

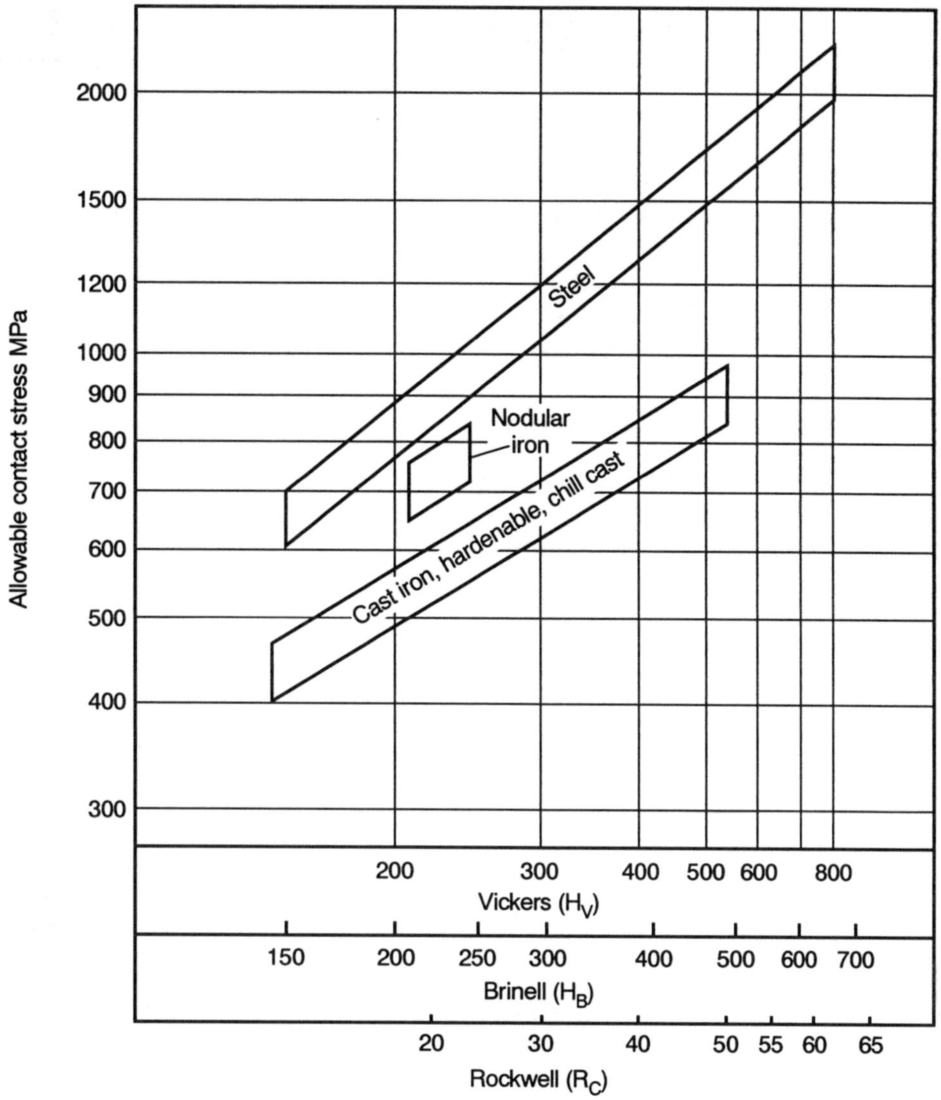

Fig. 1.5.11 Typical allowable contact stresses for cam and follower materials

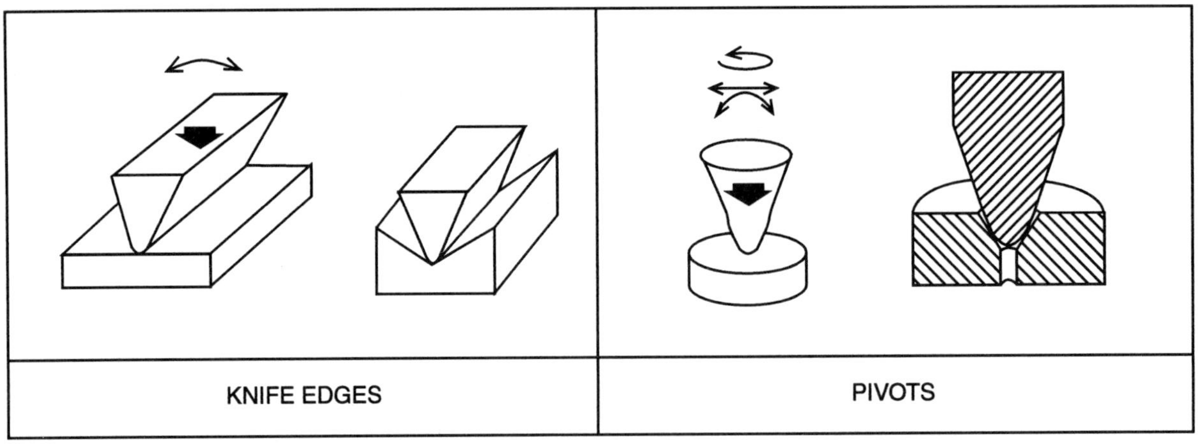

Reproduced with permission from Neale, M.J. (ed.), *Tribology Handbook;* published by Butterworth-Heinemann, 1993.

Fig. 1.5.12 Illustration of knife edges and pivots

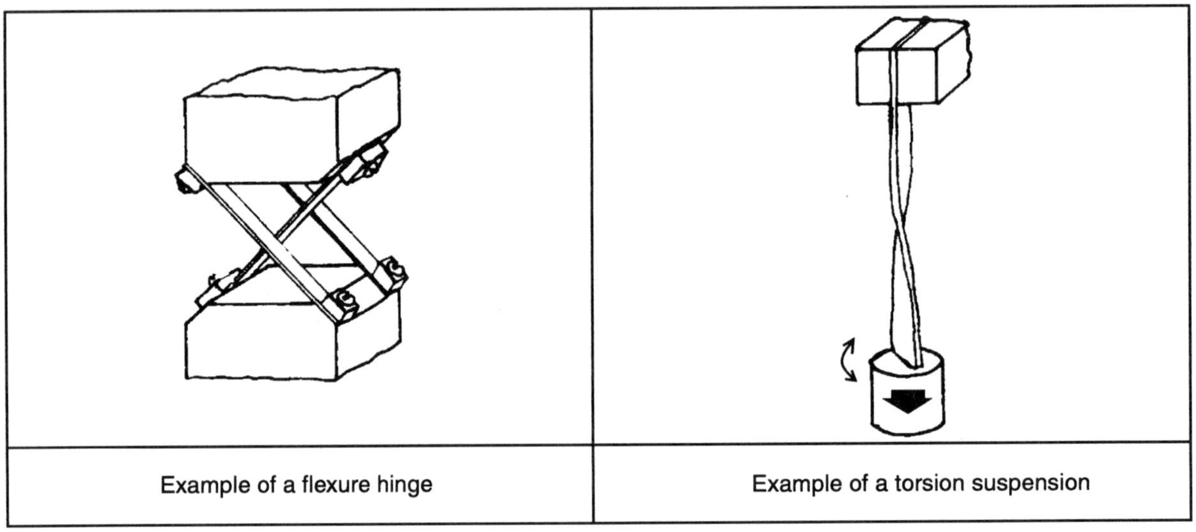

Reproduced with permission from Neale, M.J. (ed.), *Tribology Handbook;* published by Butterworth-Heinemann, 1993.

Fig. 1.5.13 Illustration of a flexure hinge and torsion suspension

Fig. 1.5.12 and Fig 1.5.13

1.5.4 Abrasion

1.5.4.1 Introduction to abrasive wear

The term 'abrasion' refers to wear of materials by hard particulate solids rubbing over a surface. Practical abrasive wear situations can be classified in terms of the contact pressure between the abrasive material and the component. At different pressures, different types of abrasive wear mechanisms predominate although there are no sharp dividing lines. In high stress situations, deformation and cutting mechanisms predominate and toughness is an important material requirement. At lower stresses, rubbing and cutting prevail and hardness is more important than toughness. A general guide is given in:

Table 1.5.28 *General guide to selection of materials for practical types of abrasive wear*

Further considerations, such as the chemical environment and temperature may also need to be taken into account in materials selection, and a guide is given in:

Table 1.5.29 *Secondary considerations in materials selection for abrasive environments*

The materials which are commonly used for abrasive wear resistance include cast irons, steels, hard facings, ceramics, glass, concretes, elastomers and plastics. Some wear resistance data and typical applications for these materials are given in:

Table 1.5.30 *Commonly used materials for abrasive wear resistance*

Wear rates and relative wear performance vary considerably for different wear environments and therefore the wear data given in Table 1.5.30 should be used only as a general guide. Further detailed information on abrasive wear of the materials is given in Sections 1.5.4.2–1.5.4.5.

Wear rates are influenced by several factors, including abrasive hardness, abrasive size, abrasive shape and applied load.

The hardness of common abrasives is given in:

Table 1.5.31 *Hardness of common abrasives and rock forming minerals*

To resist abrasion by these materials a machine component should ideally be about five times as hard as the abrasive. If the component hardness is less than this, the wear rate increases progressively until, if it is of equal hardness, the wear rate may be up to 1000 times greater.

Abrasive particle size also has an effect. For most situations, the wear rate is generally constant for particle sizes above 100 μm. With smaller particle sizes the wear rate reduces and below 25 μm it can reduce substantially.

Particle shape is also important and sharp edged particles can produce wear rates that are up to three times those obtained with more rounded particles.

1.5.4.2 Abrasive wear resistance of ferrous materials

Wear rates of ferrous materials are dependent on structure, hardness and strain hardening properties. For homogeneous materials, and for wear by coarse abrasives at high stress, wear resistance increases approximately linearly with hardness, and at the same hardness, increases as the ratio of UTS to proof stress increases. For materials containing significant volumes of carbide, wear rates decrease rapidly as the carbide content increases (care must be taken when using high carbon white cast irons to avoid high impact environments unless substantial support can be given to the wearing components). High carbon content irons with coarse structures may have very low wear rates on fine abrasives and under low stresses, but generally fine structures are preferable for wear on coarse abrasives and under high stresses.

Wear resistance tends to correlate with bulk hardness and be insensitive to material structure in high stress situations particularly if the abrasive is coarse and hard and there is a low volume fraction of fine carbides in the material. Conversely, wear resistance may not correlate with bulk hardness or be sensitive to structure in low stress situations with fine, soft abrasives or a high volume fraction of coarse carbides.

The effects of structure, heat treatment, alloy content and carbon content on wear of ferrous materials are shown in:

Fig 1.5.14 *Effect of structure, heat treatment and alloy content on wear resistance of steels*

Fig 1.5.15 *Effect of the carbon content on their wear rate under high stress abrasion of ferrous materials in a jaw crusher*

1.5.4.3 Abrasive wear resistance of hard facings

Wear rates of hard facings often vary significantly with different deposition techniques. For most abrasion applications fused deposits are superior to sprayed deposits; the latter rely on mechanical bonding and may have inadequate adhesion. On small components dilution of surfacings by the substrate may cause coating degradation and lead to decreased wear resistance.

1.5.4.4 Abrasive wear resistance of ceramics and cermets

Since many ceramics and cermets are harder than commonly encountered abrasives they have very low wear rates compared to metallic materials. The wear rate of ceramics is strongly dependent on the relative hardness of material and abrasive and tends to decrease with increasing fracture toughness. However, the low fracture toughness of ceramics limits them to application in low stress, fine abrasive environments.

The wear rates of carbide composites (cermets) is also dependent on hardness. These materials have higher fracture toughness than ceramics and are thus more suitable for application in low stress impact and medium stress abrasive environments. The hardness and fracture toughness of these materials is given in:

Table 1.5.32 *Approximate values for the hardness and fracture toughness of ceramics and cermets.*

1.5.4.5 Wear resistance of polymeric materials

Wear rates of polymeric materials are strongly dependent on surface properties of the counterface material and temperature.

When worn against rough metal counterfaces or abrasive grits, polymeric materials often have much greater wear rates than most metallic materials. But polymeric materials have significant advantages over metals when minimum damage to the counterface is important.

GENERAL MATERIALS REQUIREMENTS

To resist fatigue wear a material should have a high fatigue life (i.e. a high value of β in $n = (\sigma_0/\sigma)^\beta$ where n is the number of stress reversals to failure, σ_0 is the uniaxial tensile strength and σ is the amplitude of the stress cycle). Under conditions of fatigue, wear rates increase rapidly with load and are proportional to $(\text{load})^\alpha$, where α is commonly between 1 and 3. To resist cutting wear a material should have high hardness, and a high product of tensile strength and ultimate elongation. Tensile strength and ultimate elongation are more important than hardness.

The types of wear of polymeric materials are classified in:

Table 1.5.33 *Classification of types of abrasive wear of polymeric materials*

Table 1.5.28 General guide to selection of materials for practical types of abrasive wear

Operating condition	*Basic abrasive wear mechanisms*	*Physical description of situation*	*Surface finish after wear*	*Typical practical situations and depth wear rates (μm/h)*	*Material properties required*	*Typical materials*
High stress abrasion.	Involves large amounts of deformation and cutting. Rubbing wear less prevalent at high stresses as abrasive particles fracture and so remain angular.	High specific contact pressures resulting from abrasion at high overall loads and impact of large particles.	Considerably roughened with a combination of large gouges and/ or pits.	Hammers in impact pulverizers 1–25 Shovel digger teeth 0.1–13 Wearing blades in coarse ore scrapers 0.1–2.5 Ball mill scoop lips 0.08–0.4 Crusher liners for crushing siliceous ores 0.05–0.5 Chute liners handling coarse siliceous ores 0.003–0.25	High toughness rather than hardness. Ability to work-harden.	Austenitic manganese steel. Hard-facings. Thick resilient rubber. Cast iron – avoid large particle impact.
Medium stress abrasion.	Considerable amounts of cutting wear prevails with smaller contributions made by the rubbing and deformation processes.	Medium specific contact pressures arising from physical situations which are intermediate between high and low stress abrasion. Most common form of abrasive wear.	Surfaces have a scratched and abraded appearance with some small surface pitting.	Rod and ball mill liners in siliceous ores 0.01–01. Grinding balls in wet grinding siliceous ores 0.004–0.01 Grinding balls in wet grinding raw cement slurries 0.001–0.004 Grinding balls in dry grinding cement clinkers 0.0001–0.0003	High hardness. Toughness less important.	Hard facings. Hardened and/or heat-treated metals, e.g. cast iron, ceramics, quarry tiles, concretes. Cermets, e.g. tungsten carbide for maximum wear resistance if cost is justified.
Low stress abrasion.	Considerable amounts of rubbing wear as well as cutting wear. Wear by deformation does not occur to any appreciable degree.	Low specific contact pressures which are caused by a combination of low overall loads and rounded and smooth particles, or by small abrasive grains.	Smooth polished surfaces with slightly scratched appearance. No large-scale pits or gouges.	Sandblast nozzles 2.5–25 Sandslinger liners 1.3–6.4 Pump runner vanes pumping abrasive mineral slurries 0.003–0.1 Agitator and flotation impellers in abrasive mineral slurries 0.001–0.03 Screw type classifier wear shoes in sand slurries 0.001–0.005	High hardness, becoming less essential at lower stresses. Toughness unimportant at low stresses. Low coefficient of friction.	Tungsten carbide for max. wear resistance if cost justified. Ceramics, smooth metal surfaces. Rubbers, polyurethane and plastics (especially at lowest stresses).

Table 1.5.29 Secondary considerations in material selection for abrasive environments

Operating conditions	*Material properties required*	*Typical materials*
Wet and corrosive conditions.	Corrosion resistance.	Stainless metals, ceramics, rubbers and plastics.
High temperatures.	Resistance to cracking, spalling, thermal shock. General resistance to elevated temperatures.	Chromium-containing alloys of iron and steel; some ceramics.
Minimum periods of plant shutdown.	Ease of replacement.	Any material that can be bolted in position and/or does not require curing. Curved and irregular surfaces may, however, require hard facing weld materials or trowellable materials.
Very arduous and hot conditions.	Any one or combination of the above properties.	Hard facing weld materials.

Table 1.5.29

Table 1.5.30 Commonly used materials for abrasive wear resistance

Type of material	*Typical commercially available materials*	*Wear rates relative to 0.4% C low alloy steel quenched and tempered to about 500 Vickers hardness* — *Sliding wear by coke*	*Wear by blast furnace sinter*[a] — *Sliding*	*Wear by blast furnace sinter*[a] — *Impact*	*Wear of ball mill media grinding quartz ores*	*Wear by flint stone sand loam agricultural soil*
Cast irons	Low alloy 2.5–2.8% C, ~800 VPN	1.2	0.8	1.2	~1.0	0.3
	Heat-treated modular graphite, 700 VPN	1.0	0.15	0.5	0.8	1.5
	15/3 Cr/Mo martensitic	0.55–0.8	0.15	0.3	0.8	0.3
	High Cr, 25–30% martensitic					
	Ni-hard type (3% C, 4% Ni, 2% Cr)	0.6	0.07	0.3	~1.0	0.4
Cast and rolled steels	0.4% C, low alloy, ~500 VPN	1.0	1.0	1.0	1.0	1.0
	0.8% C, ~800 VPN				~1.0	0.5
	0.3% C, 0.6% Mn, 1.5% Cr, 0.75% Ni, 0.4% Mo, ~450 VPN	0.3				
	0.2% C, 1.2% Mn, 1.3% Cr, 0.25% Mo, 350 VPN	1.2	0.85			
	2% C, 12% Cr, ~700 VPN				0.9	0.5
	1% C, 6% Mn, Cr/Mo, austenitic				~1.0	
	1% C, 12–14% Mn, austenitic	0.7–1.1		1.1	1.2	0.9
Hard facings	3–5% C, 20–30% Cr, Co/Mo/V/W/B/Mn/Ni ferrous alloys, manual arc deposited	0.45–0.8	0.09	0.6		0.25–0.4
	Tungsten carbide/ferrous matrix, arc or gas tubular rods		0.2			~0.3
	3.5% C, 33% Cr, 13% W, Co alloy					0.2–0.5
	1% C, 1% Fe, 26% Cr, 4% Si, 3.5% B, Ni alloy					~0.3
Ceramics	Fusion cast 50% Al_2O_3, 32% ZrO_2, 16% SiO_2	0.1–0.2	~0.2	0.6		
	Sintered 95–99% Al_2O_3		~0.2	~0.7		
	Reaction bonded Si C		6.9			
	Cast basalt	0.9	0.9			
	Tungsten carbide / 6% Co	0.07				
Glass	Plate glass	4.5				
Concretes	Aluminous cement based concrete with proprietary aggregates	3.5	15			
	As above with 2% by volume 25 × 0.4 mm dia. wire fibres			2		
	Concrete tile – 6 mm wear-resistant surface		6.7			
Elastomers	Wear resistant rubbers, 55–70° shore hardness	7.8				
	65° shore hardness rubber with saw tooth surface profile	15.1				
Plastics	Polyurethane	18.5	2.7			
	High density polyethylene	15.5–31				
	Epoxy resin based PTFE	40				
	Calcined bauxite filled epoxy resin	11				

[a] The sinter was produced from foreign ore with ASTM 1/4-strength index of about 47.

Table 1.5.30

Table 1.5.30 Commonly used materials for abrasive wear resistance – *continued*

Wear in laboratory jaw crusher siliceous ore	*Wear on commercial bonded 384 μ m flint abrasive*	*Ease and convenience of replacement*	*Typical fields of application*	*Remarks*
 0.04–0.3 0.08–0.6 0.08	0.6 0.7–0.8 ~0.5 0.6	Usually convenient with good design to facilitate replacement.	Cast irons are very suitable materials to resist medium to high stress abrasive wear due to their good wear resistance and reasonable cost. At very severe levels of impact abrasion, however, inadequate toughness can be a problem and only materials of the work-hardening type should be used. Also, cheaper materials may be preferred due to the excessively high wear rates involved.	These materials have the merit that a combination of toughness and hardness may be readily obtained by varying the alloying, method of manufacture, and treatment; thus giving suitable combinations of these properties to suit a particular application and wear situation. Various techniques of surface hardening can also be employed to improve resistance to abrasive types of wear. Other products are sintered metals and metal coatings, e.g. Cr plate and sprayed coatings.
1.0 0.3–1.3 0.35–1.4	1.0 0.55 0.55 0.8	Usually convenient with good design to facilitate replacement.	Due to the very large quantity production involved, steels tend to be comparatively cheap. Thus steels with low wear rates become a competitive materials choice. Their main application lies in hardened steels to resist medium stress abrasion as very low wear rates can be obtained. Austenitic manganese steels can be used in more severe situations due to their work-hardening capability.	
0.7	0.25–0.7 0.35 0.45–0.8 0.85	Replacement can be difficult if applied *in situ.* These materials are often chosen because hard weld may be built up and worn away several times to its total depth under severe wear situations.	For medium and high stress abrasion hardfacings give low wear rates generally, and so are used in many situations to resist abrasive wear, e.g. excavator teeth and other earth moving applications.	
	0.04–0.3 ~0.02 ~3 0.007	Convenient if ceramic is bolted in place. Less convenient if ceramic is fixed by adhesive or cement as long curing times may lead to unacceptably long down-times.	Possible to achieve very high hardness but brittleness tends to be a problem. Most suitable to resist low stress abrasion by low density materials and powders.	
	22	Used in sheet form where transparency is required.	Glass is brittle and so it is only used at the lowest levels of abrasive wear.	
		Long curing times can lead to unacceptably long down-times. Can be messy and difficult under dirty conditions.	Useful to resist wear of irregular shaped components and when abrasion is of low to medium stress.	Also useful in large flat areas, especially when curing time is no real problem, e.g. aircraft hanger flooring, etc. Easily castable.
		Bonded and bolted. Sticking with adhesive can be difficult under dirty conditions.	Very useful to resist impact abrasion – most wear resistant at 90° impact angles. Softer types of rubber are used for low stress impact abrasion. Resilient rubber for more severe impact.	Bonding of rubber to component is a problem in high stress abrasive wear. Good anti-sticking properties and low density.
		Usually used in sheet form. Difficult to bond plastic to component. Solid moulded components are superior but are limited to small sizes.	Low coefficient of friction, good anti-sticking properties. Best for low stress abrasion by fine particles. Resin bonded aggregates are trowellable and so are useful for irregularly shaped components.	Composite plastics are only as tough as their bonding matrix and therefore find more applications where low stress abrasion by powders or small particles takes place.

Table 1.5.30 – *continued*

Table 1.5.31 Hardness of common abrasives and rock forming minerals

Mineral type	***Vickers hardness (kg/mm^2)***
Silicon carbide	2100–3000
Corundum	1900–2100
Topaz	1200
Quartz (flint)	750–1100
Garnet	600–970
Olivene	600–750
Plagioclase feldspar	470–600
Pyrite	470–600
Orthoclase feldspar	470
Magnetite	370–600
Haematite	370–600
Pyroxene	300–470
Hornblende	300–470
Leucite	370–470
Ilmenite	300–470
Limonite	300–370
Apatite	300
Zeolites	180
Flourite	180
Siderite	145–180
Dolomite	145–180
Serpentine	90–180
Calcite	115
Mica	70–115
Chlorite	70–90
Kaolin	70–90
Gypsum	70
Talc	45–56

Table 1.5.31

Table 1.5.32 Approximate values for the hardness and fracture toughness of ceramics and cermets

Ceramic material	***Vickers hardness***	***Fracture toughness ($MN/m^{3/2}$)***
Hot pressed B_4C	3000	8
Reaction bonded SiC	2000	8
Hot pressed Al_2O_3	1700	6
Spinel	1600	2
Hot pressed Si_3N_4	1600	7
ZrO_2	1500	3
Sintered Al_2O_3	1300	6
Reaction bonded Si_3N_4	700	5
Sintered TiO_2	700	3
MgO	700	2

Cermet material	***Vickers hardness***	***Fracture toughness ($MN/m^{3/2}$)***
96%WC 4%Co	1700	8
94%WC 6% Co	1600	12
93%WC 7%Co	1500	12
92%WC 8%Co	1500	12
91%WC 9%Co	1300	13
85%WC 15%Co	1100	18
70%WC 30%Co	850	–

Table 1.5.33 Classification of types of abrasive wear of polymeric materials

Material	*Examples*	*Contact conditions*	*Type of abrasive wear*
Highly elastic	Rubbers	Elastic / partly plastic	Tearing
Elastic	Polyolefins Polyamides	Plastic / partly elastic	Cutting and partly tearing or fatigue
Plastic / elastic	PVC PTFE	Plastic	Cutting
Rigid	Polystyrene PMMA Thermosetting resins	Plastic	Cutting

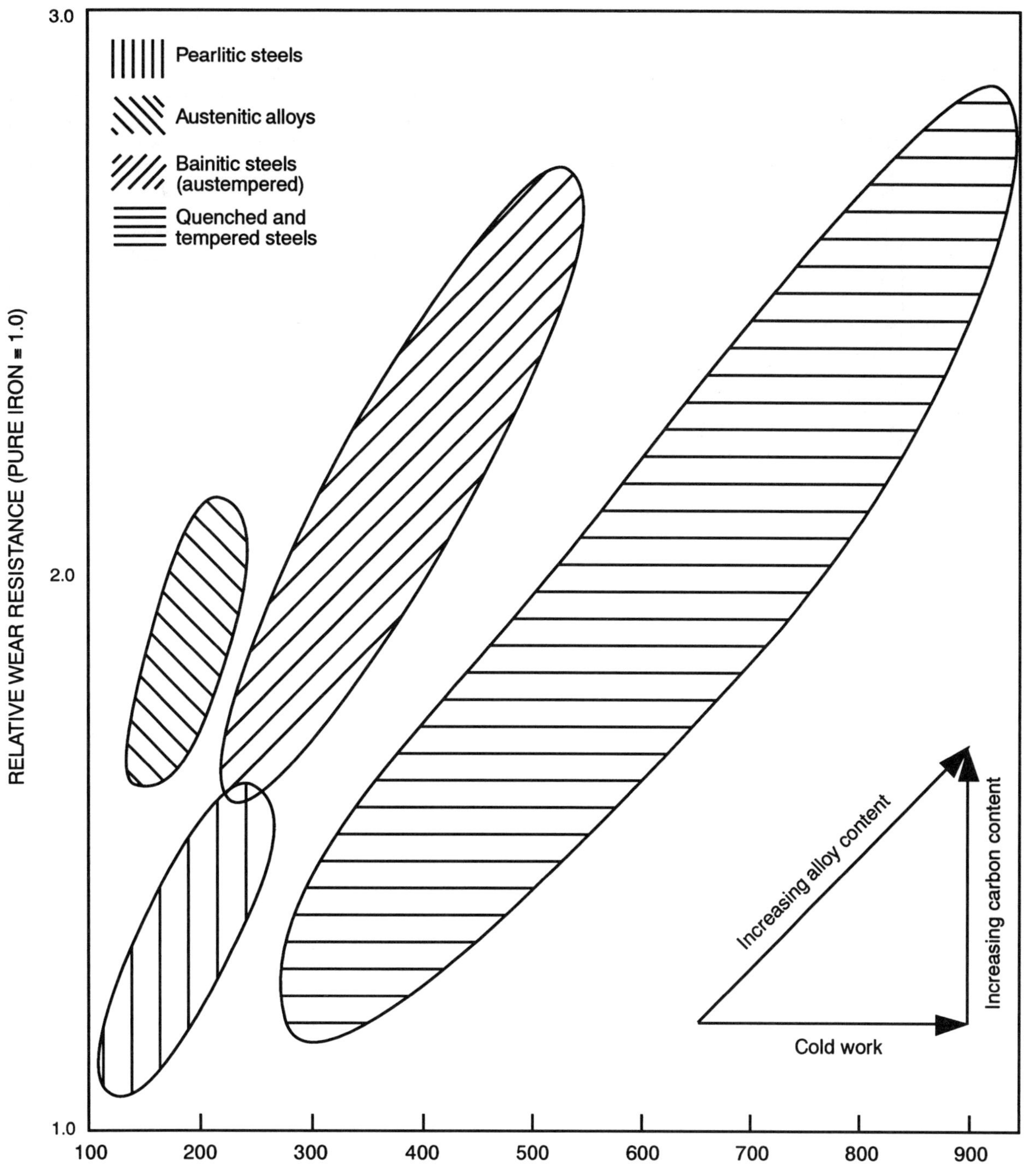

Fig. 1.5.14 Effect of structure, heat treatment and alloy content on wear resistance of steels (worn on 90 μm corundum grit at ~1 MPa applied load/unit area)

Fig. 1.5.14

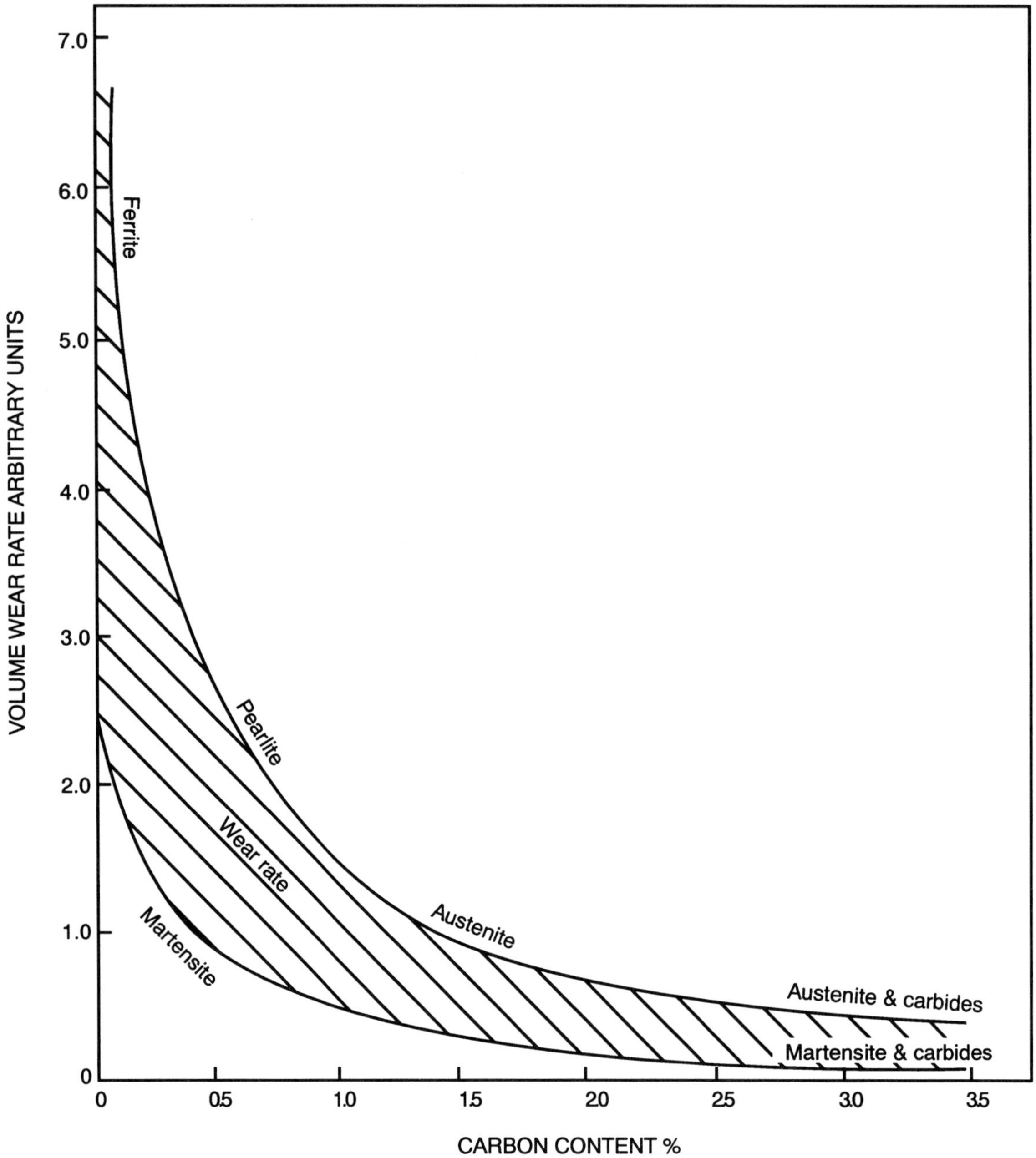

Fig. 1.5.15 Effect of the carbon content of ferrous materials on their wear rate under high stress abrasion in a jaw crusher

Fig. 1.5.15

1.5.5 Erosion

1.5.5.1 Introduction to erosive wear

The term 'erosive wear' refers to wear of a material by particulate solids or liquids in a fluid flowing over a surface. Service tests are the only method to obtain accurate data for a particular application. However, this is very expensive to do, so materials selection must be based on a combination of practical experience coupled with any test data which may be relevant. A general guide to the selection of materials for improved erosive wear resistance in typical applications is given in:

Table 1.5.34 *General guide to the selection of materials to reduce erosive wear by particulate solids and liquids*

The factors which may be controllable in the design and selection of materials are the impact velocity and impact angle. Information on the effect of these is given in:

Fig. 1.5.16 *Effect of impact velocity on erosive wear by 125 μm particles impacting at 90°*

Fig. 1.5.17 *Effect of impact angle on erosive wear of materials impacted with dry 0.2–1.5 mm quartz*

1.5.5.2 Erosive wear resistance of materials

The relative resistance of materials to erosive wear is generally similar to their relative resistance to abrasive wear, and information from Section 1.5.4 can therefore be used as a guide.

Very limited experimental erosive wear data are available. The most comprehensive data is for materials in cylindrical form moving sideways through sandy water, shown in:

Fig. 1.5.18 *Erosive wear resistance of materials in sandy water*

Some data for pipe linings conveying natural mineral substances are given in:

Fig. 1.5.19 *Erosive wear resistance of pipe linings for pneumatic conveying of natural mineral substances*

Table 1.5.34

Table 1.5.34 General guide to the selection of materials to reduce erosive wear by particulate solids and liquids

Erosive situation: Impact velocity	Erosive situation: Impact angle	Typical practical situations	Wear normally experienced	Prevailing basic wear mechanisms	Surface finish after wear	Design advice to reduce wear rate	Materials requirements to reduce wear rate	Typical materials used to reduce wear rate	Remarks
Low velocity (up to 100 m/s)	Low (0–30°)	Low pressure sand-blast nozzles. Wear of pipes handling slurries. Impellors of large extractor fans working in dirty environments.	Rates of wear controllable by suitable design and materials selection if particle size not extremely large.	Cutting wear.	A scratched and lightly polished surface.	Minimize attack angle, i.e. reduce changes in direction of erosive stream.	Use harder material. Reduction of impact angles gives improved wear resistance.	Hardened irons, e.g. Ni-hard – Hardened steels – Chrome plating – Concretes, etc. – Plastics – Ceramics and cermets –	Cheap and castable. More useful strength. Can be hardened very easily. Gives useful wear resistance. Cheap and trowellable. (Cheap and mouldable). Can have enough wear resistance. Best wear resistance. Expensive.
	Medium (30–60°)	Wear by slurries at bends in piping. Ducts of large extraction fans – high wear rates if working in dirty environments.		Cutting and deformation wear.	A flattened and deformed surface with scratches and small-scale surface pitting.	Reduce changes in direction of the erosive stream.	Use harder material. Reduction of impact angle gives improved wear resistance.		
	High (60–90°)	Internal parts of sand-blast apparatus opposite inlet for abrasives. Leading edges of large extractor fans, etc., especially apparent if working in dirty environments.		Deformation wear.	A flattened and deformed surface layer with small-scale surface pitting.	Use highest possible attack angle coupled with ductile material.	Use more elastic material. (Increased hardness often not advantageous).	Soft natural rubber – Neoprene Polyurethane – Plastics Also hard materials as above.	More resilient rubber required as impact velocity increases. Can give acceptable wear resistance.
Medium velocity (between 100 and 250 m/s)	Low (0–30°)	Medium pressure sand-blast nozzles. Runner vanes, etc., of centrifugal pumps handling slurries or solids.	Wear rates are normally high but can be reduced to acceptable levels by suitable design and materials selection. Ease of replacement of materials should be considered.	Cutting wear.	A very scratched and abraded surface.	Reduce changes in direction of erosive stream.	Use harder material if particles are small. Do not use brittle materials if particles are large.	Hardened irons, e.g. Ni-hard – Hardened steels – Chrome plating – Ceramics and cermets –	Small particles. Tougher larger particles. Good wear resistance if base metal preparation good. Best wear resistance. Expensive.
	Medium (30–60°)	Centrifugal pumps handling slurries or solids.		Cutting and deformation wear.	A deformed surface, scratched and abraded with surface pitting.	Reduce changes in direction of erosive stream.	Use tougher and possibly harder materials.	Hard facing welds – Irons and steels of adequate hardness –	Applied to manufactured component. Range of hardnesses and toughnesses readily available.
	High (60–90°)	Leading edges of centrifugal pumps carrying abrasive slurries or solids. Subsonic aircraft leading edges, radomes, windscreens, etc., especially in rain or dirty environments. Leading edges of low speed turbines in wet stream or abrasive situations.		Deformation wear.	A flattened and deformed surface with surface pitting.	Use highest possible attack angle coupled with ductile material.	Use tougher material.	Hard facing welds – Austenitic manganese or similar high toughness steels.	Readily replaceable.

Table 1.5.34 General guide to the selection of materials to reduce erosive wear by particulate solids and liquids – *continued*

High velocity (above 250 m/s)	Low (0–30°)	High pressure sand-blast nozzles. High speed turbine blades in wet steam.	Wear rates normally severe. Optimum design and materials may still give high rates of wear, so easily replaceable, readily available and cheap materials should be considered. Avoidance of situation is another solution, e.g. supersonic aircraft fly at high altitudes, so avoiding rain.	Cutting wear.	A highly scratched and abraded surface with possibly some surface pitting.	Reduce changes in direction of erosive stream.	Use harder material which has not got excessive brittleness.	Ceramics and cermets – Chrome plating – Hardened irons and steels –	Best wear resistance but expensive. Possibly good enough wear resistance. Lower wear resistance offset by cheapness.
	Medium (30–60°)	Supersonic aircraft airframes, windscreens, radomes, etc.		Cutting and deformation wear.	A highly deformed surface, highly scratched with large-scale surface pitting.	Reduce changes of direction of erosive stream.	Use tougher and possibly harder material.	Hard facing welds – Austentic managnese – or similar high toughness steels.	Readily replaceable. Too expensive if wear rate is very high.
	High (60–90°)	Leading edges of turbine blades in wet stream. Supersonic aircraft leading edges, radomes, windscreens, etc., in rain.		Deformation wear.	A flattened and deformed surface with large-scale surface pitting.	Use highest possible attack angle coupled with ductile material.	Use tougher material.	Hard facing welds – Austenitic manganese – or similar high toughness steels.	Readily replaceable. Too expensive if wear rate is very high.

Table 1.5.34 – *continued*

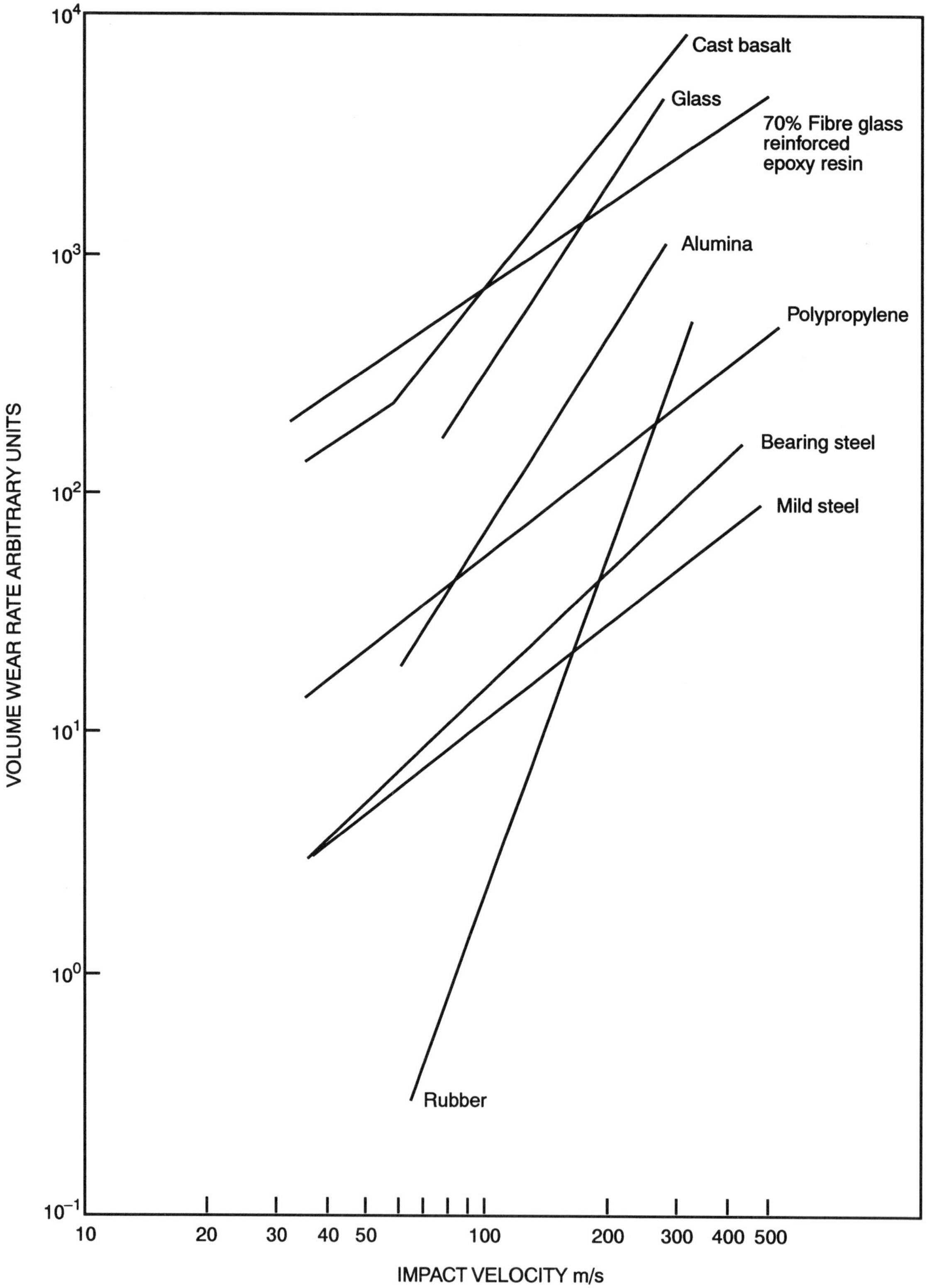

Fig. 1.5.16 Effect of impact velocity on erosive wear by 125 μm particles impacting at 90°

Fig. 1.5.16

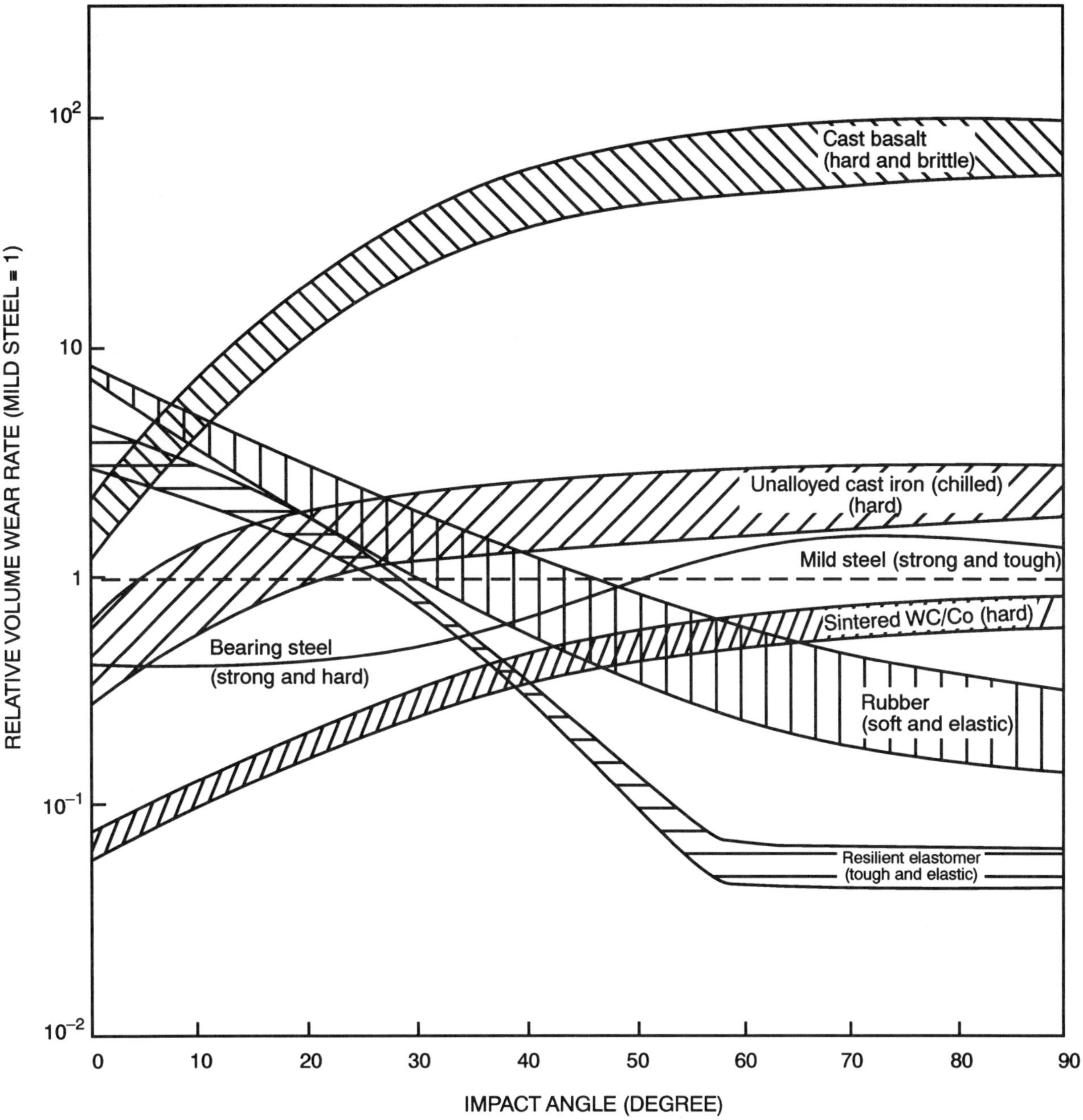

Fig. 1.5.17 Effect of impact angle on erosive wear of materials impacted with dry 0.2–1.5 mm quartz

Fig. 1.5.17

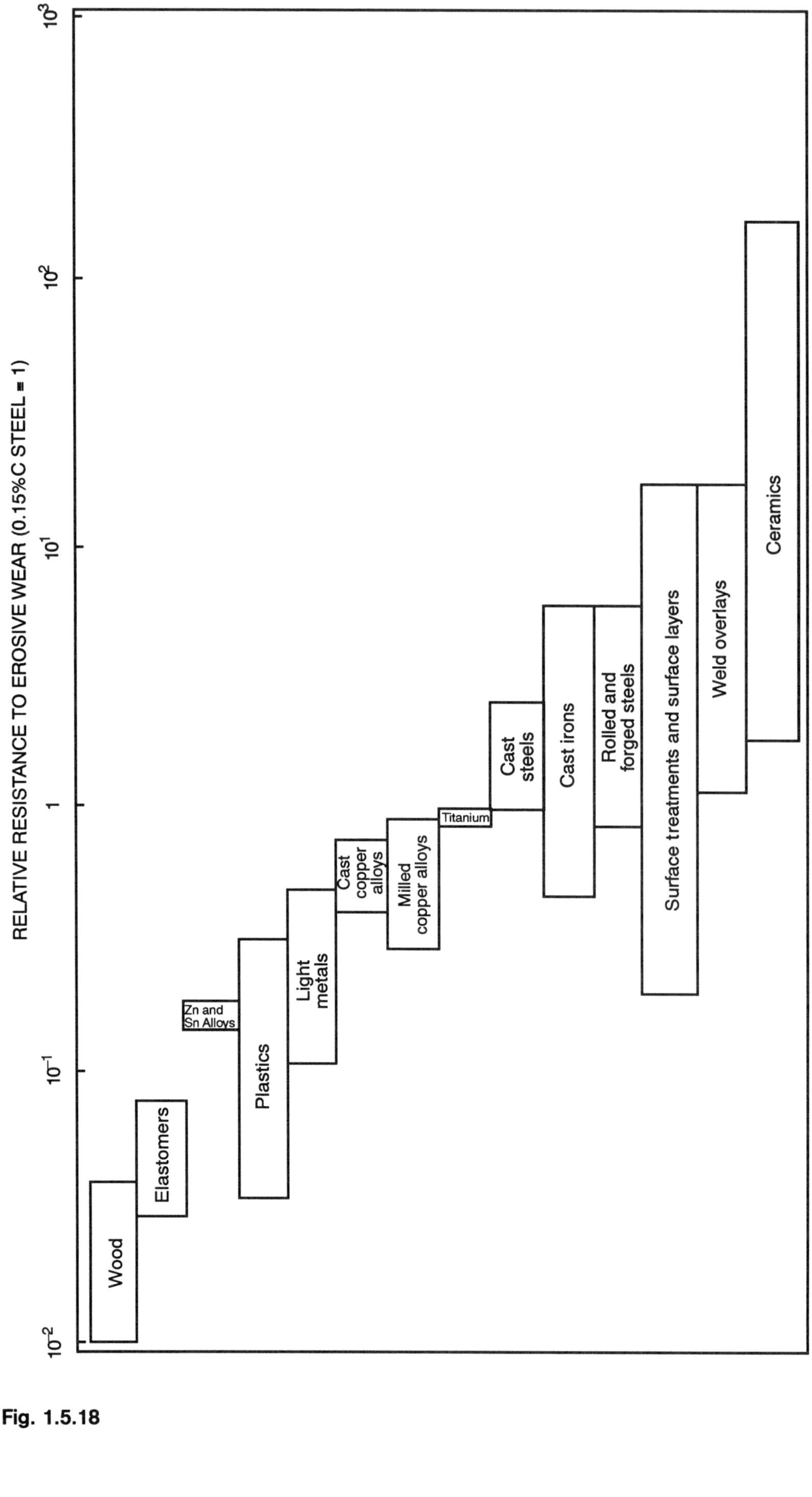

Fig. 1.5.18 Erosive wear resistance of materials in sandy water

Fig. 1.5.18

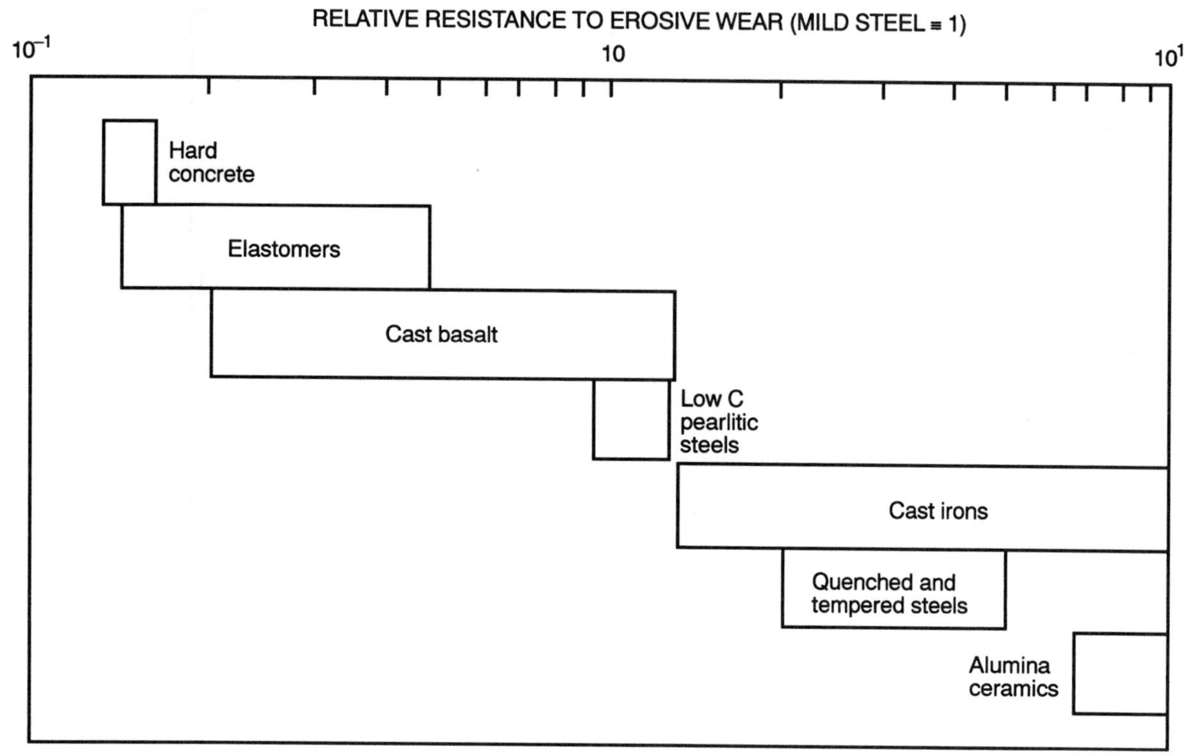

Fig. 1.5.19 Erosive wear resistance of pipe linings for pneumatic conveying of natural mineral substances

1.5.6 Cavitation erosion

1.5.6.1 Introduction to cavitation erosion

Cavitation erosion of surfaces occurs when they are in contact with a disturbed liquid flow. Low pressure regions in the flow create vapour bubbles which collapse on to the downstream surface in regions of higher pressure. The impact of this collapse erodes the surface. Vibrations of the solid surface (e.g. diesel engine wet cylinder liners) can have the same effect, as can pulsations in the fluid flow.

To resist cavitation erosion, a material should have a high ultimate resilience characteristic. This is the mechanical energy per unit volume to take the material to failure.

$$\text{Ultimate resilience characteristic} = {}^1/_2 \; \frac{(\text{tensile strength})^2}{\text{elastic modulus}}$$

Methods of reducing cavitation erosion are given in:

Table 1.5.35 *General guide to the selection of materials to reduce cavitation erosion*

1.5.6.2 Cavitation erosion resistance of materials

Corrosion can be a large component of the wear rate of materials subject to cavitation erosion. The optimum choice of material is therefore affected by its corrosion resistance in the working fluid, as well as by its resistance to cavitation erosion as such.

The resistance to cavitation erosion of some commonly used materials is shown in:

Table 1.5.36 *Resistance of materials to cavitation erosion*

The resistance of metals to pure cavitation erosion generally increases with ultimate resilience characteristic, as shown in:

Fig. 1.5.20 *Cavitation erosion resistance as a function of ultimate resilience characteristic*

Table 1.5.35 General guide to the selection of materials to reduce cavitation erosion

Operating condition	Typical practical situations	Appearance after wear	Amount of wear normally experienced	Corrosive effects	Design advice to reduce wear	Materials requirement to reduce wear	Typically used materials	Remarks
Bulk material moving through liquid	Marine propellers, hydrofoils, pump impellers, marine jet propulsion pumps.	Fine to deep surface pitting, often with a degree of corrosive attack also.	Severe wear rates with poor design and poor materials. Controllable by good design and good materials.	Although not apparent at the high cavitation erosion rates, when these are reduced by appropriate measures it may become necessary to consider corrosion resistance as well.	Use boundary layer devices: (i) supercavitating sections, (ii) air injection. Increase basic working pressure of pump impeller, thus preventing it from reaching cavitation threshold.	High ultimate resilience characteristic. Corrosion resistance if working in saline or other hostile environment.	Rubber coatings – Stellite – Aluminium bronze – Plastics (nylon or polythene) – Stainless steels –	if speed and pressure are low. easily applicable to problem areas. excellent corrosion resistance. with high impact strength. if hardened.
Material vibrating in liquid	Cylinder liners of diesel engines vibrate, causing cavitation by coolant.	Fine to deep surface pitting, normally showing corrosive attack also.	Pitting which can be so deep as to eat completely through component. This normally concentrates in a few spots, so problems arise even with small amounts of material removal.	Very often the cavitation erosion is substantially aided by a corrosion mechanism.	Increase the working pressure of the fluid to prevent it from reaching the cavitation threshold. In the case of a cylinder liner, reduce the vibration by reducing piston impact or increasing stiffness of liner.	Materials normally selected for other design reasons, therefore coatings must be applied, but may be expensive. Wear, therefore, is normally reduced by design innovations and corrosion considerations.	Hard chrome plate	
Movement of liquid past material	Edges of fluid power valves especially spool valves and poppet valves, diesel engines main and big-end bearings, delivery pipes from injector of high output diesel engines.	Pitting or grooving. Material can be removed in apparently random areas – normally associated with sharp changes in direction of flow of liquid.	Although amounts of material removed are normally small, this can have serious consequences when it occurs in mechanisms with small clearances. Also, material choice often determined by other factors, so it is impossible to use a material with good cavitation erosion resistance.	Cavitation erosion normally without undue corrosion problems. Many other factors may be apparent, e.g. material softening at working temperature.	Avoid sharp changes in direction of flow paths by obstructions, etc. Design to prevent the sudden decrease in pressure which causes cavitation – increase escape area. Increase the working pressure, of the fluid to prevent it from reaching the cavitation theshold.	Materials normally selected for other design reasons; therefore, if possible coatings must be applied, but may be expensive. Wear therefore normally reduced by design innovations.		

Table 1.5.36 Resistance of materials to cavitation erosion

Range of Materials	*Typically used materials*		*High intensity relative cavitation erosion resistance (aluminium bronze = 1)*	*Remarks*
Cast copper alloys	Aluminium bronze AB2			1.3 Aluminium bronze has good corrosion resistance in seawater.
	Aluminium bronze AB1		1.0 (standard)	
	Novoston bronze		0.5	
	Gunmetal		0.22	
	High tensile brass		0.18	
	Phosphor-bronze		0.15	
	Nickel gunmetal		0.15	
	Leaded gunmetal		0.14	
Wrought copper alloys	Manganese bronze		0.2	
	Everdur A		0.16	
	Aluminium brass		0.15	
	Free-turning brass		0.14	
	Naval brass		0.12	
	Hot stamping brass		0.11	
	Tellurium copper		0.03	
Nickel alloys	Monel K-500 (aged)		0.7	
	S Monel		0.5	
	Monel K-500 (cold drawn)		0.3	
	Nimonic 75		0.22	
	Nickel (hard)		0.21	
	Nickel (annealed)		0.2	
Wrought aluminium alloys	Al 4% Cu, solution treated and aged		0.08	Low density
	Al 5% Mg, as rolled		0.04	
	Al 3.5% Mg, as rolled		0.03	
Titanium alloys	Ti–2.25 Al–4 Mo–0.255 C–11 Sn		1.0	Excellent corrosion resistance. Low density. High cost.
	Ti C.P. with medium interstitial content		0.7	
Cast irons and cast steels	Martensitic stainless	hardened	1.8	Stainless steels are reasonably corrosion resistant. Martensitic stainless steels are less corrosion resistant than austenitic stainless steels.
	Austenitic stainless	annealed	0.8	
	Ferritic stainless	annealed	0.77	
	SG cast iron	tempered	0.7	
	Martensitic stainless	tempered	0.5	
	Alloy cast iron (D2 Ni-resist)		0.2	
Wrought steels	HT low alloy	hardened	2.5	
	Martensitic stainless	tempered	0.67	
	Martensitic stainless	tempered	0.62	
	Austenitic stainless	annealed	0.5	
	Mild steel		0.2	
Stellites (cobalt alloys)	Stellite 12 (cast) – 29% Cr 9% W 1% C		19.3	Most resistant to severe cavitation erosion.
	Stellite 4 (cast) – 31% Cr 14% W 1% C		15.0	
	Stellite 6 (cast) – 26% Cr 5% W 1% C		5.3	
	Stellite 6 (wrought)		3.4	
	Stellite 7 (cast) – 26% Cr 6% W 0.4% C		2.5	
Plastics	Nylons, acetals; High impact polyethylenes		1.7–3.0	Values of cavitation erosion resistance fall sharply at elevated temperatures. Generally plastics with high impact strengths have good cavitation erosion resistance. Very suitable for mass manufacture of small components, e.g. boat propellers. Corrosion resistant.
	Polyethylenes; Polypropylenes		0.9–1.6	
	PVC; PTFE		0.15–3.5	
	Chlorinated polyether		0.04–0.1	
	Perspex		0.01–0.035	
Rubbers	Bonded rubber coatings; Neoprene		–	Rubber requires careful preparation and application. Rubber gives satisfactory resistance to mild cavitation erosion, and where relative fluid velocities are low. Failure usually occurs at bond to components at higher fluid velocities.

Table 1.5.36

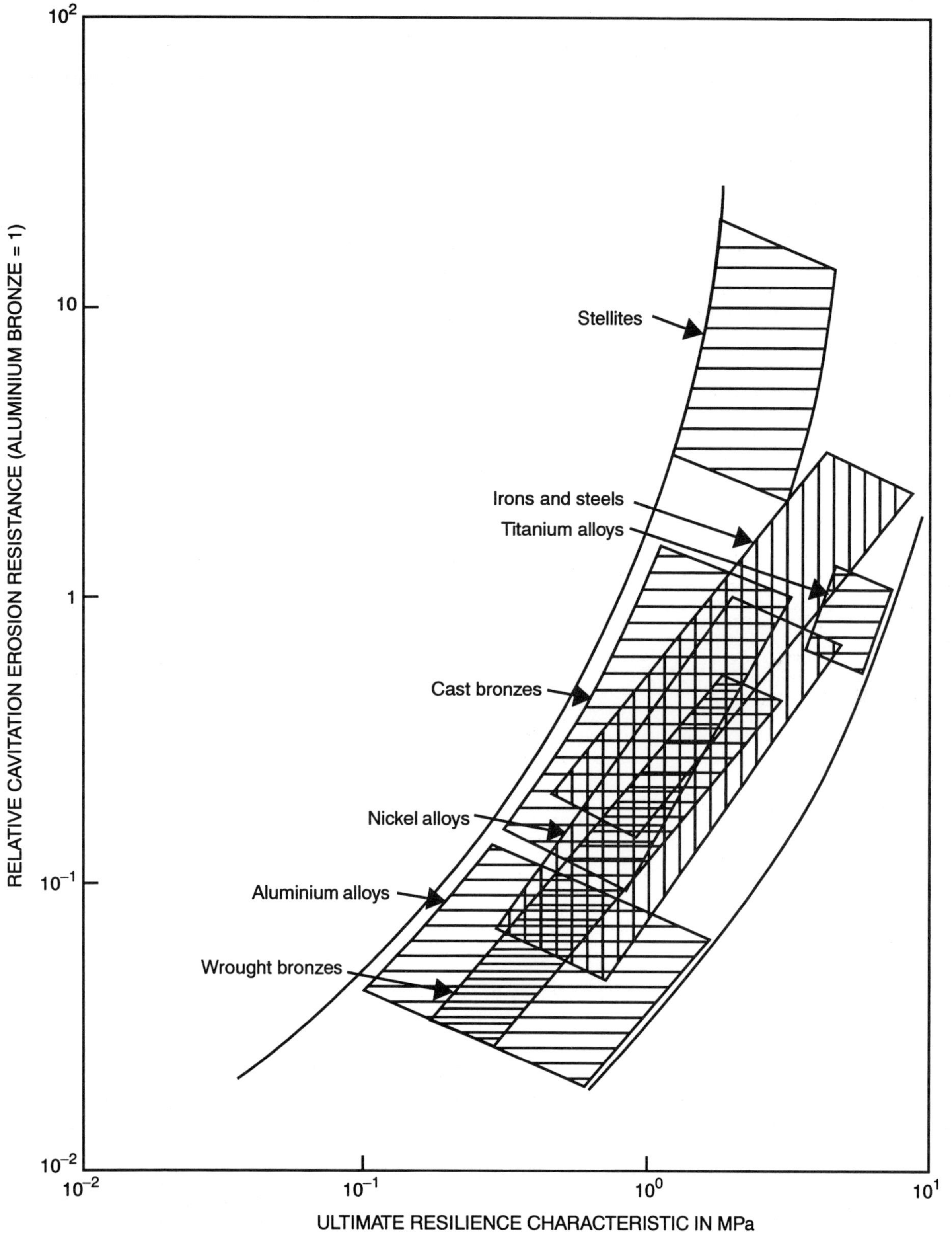

Fig. 1.5.20 Cavitation erosion resistance as a function of ultimate resilience characteristic

Fig. 1.5.20

1.5.7 Surface treatments to improve wear resistance

In many cases, a surface treatment or surface coating may be applied to improve wear performance. The most appropriate treatment or coating depends on the application and the base material. A general guide for bearings and other interacting surfaces is given in:

Table 1.5.37 *General guide to the selection of treatments for bearing surfaces to reduce friction and wear*

A guide to surface treatments to aid running-in is to be found in Fig. 1.5.7 of Section 1.5.2.

Surface coatings can be applied by a number of methods. The techniques and some important factors in their use are outlined in:

Table 1.5.38 *General guide to the selection of coating methods*

In some cases, surface coatings may be required for high temperature environments and guidance on this is given in:

Table 1.5.39 *Guide to the selection of coating materials for sliding conditions*

For conditions where surfaces impact during sliding (hammering), guidance is given in:

Table 1.5.40 *Guide to the selection of coating materials for hammering conditions*

Ceramic based coatings are also useful in some applications and guidance on this is given in:

Table 1.5.41 *Properties of some plasma-sprayed ceramic-based coatings*

Table 1.5.37 General guide to the selection of treatments for bearing surfaces to reduce friction and wear

Treatment	*Suitable base materials*	*Description of treatment*	*Typical hardness (VPN)*	*Typical size effect*	*Primary advantages*	*Primary disadvantages*	*Applications*
Case-hardening (carburizing)	Low C, Med C, C–Mn, Cr–Mo, Ni–Mo and Ni–Cr–Mo steels	Carbon is diffused into low carbon steels by heating the metal in contact with a carbonaceous material, solid, liquid or gas to a temperature above the transformation range (800–950°C) and holding at that temperature to form austenite. Carburizing is generally followed by quenching to form a martensitic hardened case. In any steel, carbon content determines the maximum hardness which may be obtained by a successful quench. The time for a given case depth depends on the temperature, the diffusion rate, the carbon solubility in the austenite, and the rate of reaction with the steel surfaces.	Surface: 630–920 Core: 230 +	Thermal distortion is main effect. Some growth.	Light case depths (up to 0.5 mm): high wear resistance and light loads. Moderate case depths (0.5–1.0 mm): high wear resistance and moderate to heavy loads. Heavy case depths (1.0–1.5 mm): high wear resistance to rolling or abrasive wear and for resistance to crushing or bending loads. Extra heavy case depths (1.5 mm and over): maximum wear and shock resistance.	Treatment causes component distortion. Alloy steels, particularly Cr steels, have a tendency to overcarburize giving brittle surface carbides which are liable to cause spalling under impact loads. Little improvement in adhesive wear resistance.	Light cases: e.g. water pump shafts, sockets and small gears. Moderate cases: e.g. valve rocker arms, bushes and gears. Heavy cases: e.g. transmission gears and king-pins. Extra heavy cases: e.g. camshafts.
Carbonitriding	Low C, Med C, Cr, Cr–Mo, Ni–Mo and Ni–Cr–Mo steels	Both carbon and nitrogen are diffused into the surface by treatment in a molten salt bath or gas atmosphere. Such cases harden by the same mechanism as carburized cases, by transforming austenite to martensite. A carbon-nitrogen austenite is stable at a lower temperature, has greater hardenability, and has a lower temperature range for the formation of martensite. Cyaniding is an example of carbonitriding. Treatment at a temperature of 720–900°C followed by quenching.	Surface: 540–920 Core: 320–350	Depth of treatment 0.075–0.5 mm. Some thermal distortion.	Carbonitriding causes less distortion than carburizing. Wear resistance and corrosion resistance better than by carburizing. Resistance to scuffing due to nitrogen. High processing temperatures are better if service conditions are severe.	Increased cost of atmosphere and slower rate of case formation. Thus, carbonitriding is best for small components and not used if heavy cases are desired.	Uses include ratchets, small gears, hydraulic cylinders and textile machinery.
Nitriding (a) in dissociated ammonia (b) in salt baths (c) ion nitriding	Used on any steel containing Al, Cr, Mo, V or W including austenitic stainless. Avoid free ferrite, decarburized surfaces and cast irons.	Develops high surface hardness caused by precipitation of alloy nitrides at nitriding temperature of 500–570°C without quenching. (a) Nitrogen introduced by soaking in furnace containing 'cracked' ammonia gas. (b) Nitrogen introduced via cyanide or cyanide chloride solutions. (c) Nitrogen introduced by bombardment of nitrogen ions formed by electric discharge in gas mixture (90% N_2, 10% H_2) Thin, brittle outer surface layer needs to be polished off after treatment	Surface: 900–1150 Up to 1250 by ion nitriding.	Adds up to 25 μm to each surface. Depth of treatment: up to 0.75 mm.	High hardness of case which is retained at the range of temperatures in which most low alloy steels soften. Therefore resistance to seizure and galling, and can be used in poorly or non-lubricated conditions. Distortion reduces to a minimum due to low process temperature and lack of quenching. Some fatigue resistance due to the compressive stresses in the case. Ion nitriding considerably reduces treatment time even at the lower nitriding temperatures used which reduce distortion further.	High cost of process due to long treatment time, technical control necessary and extra cost of special steels used for optimum properties. Nitriding is not particularly good for applications involving low stress abrasion or erosion. Uneven hardness, brittleness or spalling will take place if free ferrite or a decarburized surface is nitrided.	Anywhere involving sliding or rolling friction and components requiring high surface hardness, e.g. gears, crankshafts.

Table 1.5.37

Table 1.5.37 General guide to the selection of treatments for bearing surfaces to reduce friction and wear – *continued*

Boriding	Low C, med C low alloy steels	Boron diffuses into steels forming Fe_2B and FeB at temperatures of 900 and 1000°C. The FeB phase is undesirable as this has extremely high hardness and is brittle. The undesired phase can be eliminated by using pack boriding granulates. Electrolytic baths of fused salts are under development.	1800–2200	Adds to the surface by 20–25% of the depth of treatment 25–250 μm.	Gives good resistance to abrasive wear.	Difficult to grind or polish to size. Diamond tipped tools and silicon carbide or aluminium oxide wheels have to be used.	Extrusion dies, drawing dies, forging and pressing dies, gauges, stone crushers piercing tools, rock drills.
Flame and induction hardening	Med C, C–Mn, Cr–Mo and Ni–Cr–Mo steels. Grey cast iron. Pearlitic malleable iron and nodular cast iron.	Heat surface using flame or magnetic induction to above the transformation temperature to give austenite. A rapid quench gives martensite with some untransformed austenite. Stress relief at 175–200°C often desirable.	Flame: Surface: 370–630. Core: 240. Induction: Surface: 740. Core: 240.	Case depth: 0.25–3.2 mm more easily controlled in induction hardening.	Produce high surface hardness and unaffected core. Depth of hardening depends on amount of heating. Both case and core properties are adjustable very easily to suit particular application. Does not require special material as does nitriding. Cheaper than carburizing and nitriding for parts suitable for treatment by both processes.	Overheating must be avoided as it renders the case brittle. No improvement in anti-scuffing qualities of surface.	Flame: sprocket and gear teeth, track rails and runner steels, lathe beds. Induction: camshafts, crankshafts, sprockets and gears, shear blades and bearing surfaces.
Siliconizing	Low C, Med C, unalloyed steels. (Alloying elements affect case formation adversely.)	Iron will dissolve silicon, forming a solid solution containing up to 15%. This silicon is added at 1000°C in a chlorine atmosphere. A siliconizing part may be heat treated to give desired core properties.		Case depth: 0.1–2.5 mm	Low coefficient of friction, non-galling and some lubricant retention so adhesive wear reduced under poorly or non-lubricated conditions. Good in applications where wear and corrosion must be resisted.	Unpopular due to high process temperature. Unsuited for reducing high velocity abrasion, and in applications involving high unit pressures, e.g. ball bearings.	Pump shafts, cylinder liners, valves, valve guides and fittings and conveyor chain links, especially for the chemical industry.
Tufftride	Low C, Med C, C–Mn, Cr, Cr–Mo, Cr–Mo–V. Stainless steels and cast irons	Salt bath treatment adding carbon and nitrogen. Nitriding action accelerated by blowing air through solution which is at 540–600°C.	Surface: 800	Adds 5–10 μm to each surface. Depth of treatment up to 1.0 mm. Top 10–16 μm most effective.	Aids running-in up to mild/severe transitional load. Improves resistance to wear and fatigue and improves corrosion resistance except on stainless steels. Surface is not brittle although Tufftride gives some increase in surface hardness.	Process operating temperatures sufficiently high to temper or anneal many ferrous metals. Also slight distortion problem as process normally last one in production line.	Crankshafts, camshafts, cylinder liners, piston rings, gears, slideways and bearing cages.
Sulfinuz	All ferrous metals including stainless steels. Nimonic alloys. Titanium alloys.	Salt bath treatment at 540–600°C introducing carbon, nitrogen and sulphur into ferrous metals. Sulphur accelerates nitriding process.	Surface: 400	Size change from loss of 5 to gain of 3μm per surface. Depth of treatment up to 0.5 mm. Top 4–25 μm most effective.	Gives good scuffing resistance with some reduction in coefficient of friction. Also resists wear and fatigue. Better than Tufftride for adhesive wear resistance.	Cyanide effluent disposal necessary. Sulfinuz reduces corrosion resistance of stainless steels.	Hydraulic pumps and ball bearing cages.
Noskuff	Case-hardening and direct hardening steels.	Applied after carburizing. Adds nitrogen and carbon in salt bath at 700–760°C.	Surface: 400	Adds 5–10 μm per surface. Depth of treatment 45–50 μm supported by carburized case.	Good anti-scuffing properties to a large case depth without reducing fatigue strength and hardness obtained previously by, say, carburizing.	More distortion than Sulfinuz or Tufftride due to high process temperature. Lapping can be used to correct this since case is thicker.	Heavily loaded gears, idler shafts, rocker shafts, marquette rods.

Table 1.5.37 – *continued*

Table 1.5.37 – *continued*

Table 1.5.37 General guide to the selection of treatments for bearing surfaces to reduce friction and wear – *continued*

Treatment	*Suitable base materials*	*Description of treatment*	*Typical hardness (VPN)*	*Typical size effect*	*Primary advantages*	*Primary disadvantages*	*Applications*
Sulf B. T.	All ferrous metals (except stainless 13% Cr steels).	Sulphur added to metal surface by low temperature (180–200°C) salt bath. Does not add nitrogen.	Depends on prior heat treatment.	Removes 2–5 μm per surface.	Low process temperature minimizes distortion. Iron sulphide gives good anti-scuffing properties. No cyanide disposal problems.	No increase in fatigue strength.	Cylinder liners, heavily loaded gears, hydraulic and pneumatic components. Any close tolerance components.
Conversion coatings (phosphate)	All ferrous metals.	Inorganic film produced on surface by chemical or electrochemical surface modifications at 40–100°C.		Adds 5 μm to each surface.	Resists scuffing and aids running-in. Porous nature helps retain lubricant. Little heating effect to cause distortion or tempering or softening of the metal being treated. Inexpensive.	Less effective than nitriding types of treatment or Sulf B. T. in improving wear resistance.	IC engine for running-in.
Delsun	Brasses and bronzes	Electrolytic deposition of an alloy containing principally tin, antimony and cadmium followed by a diffusion treatment at about 400°C in a molten salt bath during which the deposited material becomes integral with the brass or bronze.	480–600	Adds 7.5 μm to each surface. Depth of treatment about 30 μm.	Increases surface hardness thus improving abrasive and adhesive wear resistance. Good corrosion resistance in saline environment. Improves performance in heavily loaded conditions in oil and in marginal or no lubricant cases. Runs well against conventional structural steels, heat treated or case-hardened steels and on hard chromium plate surfaces. Runs satisfactorily on stainless steels and other copper alloys.		Worms, wormwheels and heavily loaded parts in oil. Pump spindles and other seawater applications.
Trical	Aluminium bronze	Electrolytic deposition of a metallic alloy on the surface followed by a diffusion treatment suited to the alloy used and carried out in a neutral atmosphere or a molten salt bath.	280–850	Adds about 12 μm to each surface.	Produces a hardened zone overlaid by thin soft anti-adhesion layer which gives good resistance to wear. Preservation of excellent corrosion resistance normally given by aluminium bronzes. Mates well with stainless steels.		Pump spindles and seawater applications.
Zinal	Aluminium alloys	Electrolytic deposition of metallic alloys on the surface followed by a diffusion treatment of the deposited alloy at a temperature below 200°C.	200–500	Adds 20 μm to each surface approx.	Increases surface hardness and improves lubricant retention. Protects against seizure and improves wear resistance. Increases ability to distort without breaking. Allows running against steel, chromium, nickel, molybdenum and light alloys themselves.		

Table 1.5.37 General guide to the selection of treatments for bearing surfaces to reduce friction and wear – *continued*

Tiduran	Titanium	Electrolytic deposition followed by diffusion. Adds nitrogen and oxygen			Resists scuffing and reduces coefficient of friction. Increase of about 40% in bending fatigue strength.		
Conversion coatings.	Aluminium titanium, etc.	Inorganic films produced by chemical or electrochemical surface modification.			Wear and corrosion resistance.		

Table 1.5.38 General guide to the selection of coating methods

	Flame deposition	*Spray/fusion deposition*	*Electroplating*[a]	*Weld deposition*
Technique	The deposition of wire or powder material through a plasma arc, oxy-fuel gas flame or oxy-fuel gas detonation on to a grit-blasted component surface, where it keys to form a coating layer.	The deposition of powder alloys, usually through an oxy-fuel gas flame, on to a prepared surface to form a coating which is subsequently fused to the component using either a furnace, a separate oxy-fuel flame, or induction coil.	The electrodeposition of coatings from a chemical solution or suspension.	The deposition of overlay coatings from rod using either argon arc or oxyacetylene heat sources.
Accessibility	Spraying angle and distance affects practicability, e.g. bore diameter relative to depth.	Spraying angle and distance affects practicability, e.g. bore diameter relative to depth.	Generally no limitations, except for internal diameters and corners due to need for internal electrode access.	Generally no limitations, except for internal diameters making rod or torch access difficult.
Distortion	Generally no limitations, except abrasive blast pretreatment (particularly on thin gauge material).	As for flame deposition. Post-spray fusing treatment (1000°C) may cause distortion.	Generally no limitations.	Major limitations due to localized high temperature gradients.
Surface preparation	Attention should be given to the desirability of applying the wear resistant coating without delay after surface preparation. This is necessary to prevent the reformation of oxides that could interfere with adhesion or fusion.			
Application compatibility (thermal expansion)	Generally no limitations.	Compatibility of expansion coefficients important between base material and deposit.	Generally no limitations.	Weld compatibility of base material and weld rod is important, as well as expansion coefficients.
Effect on component	Some deposits can result in some loss in base metal fatigue strength.	Hard, low ductility deposits in some environments, may cause loss in base metal fatigue strength.	Deposits may result in some loss of fatigue strength. Hydrogen embrittlement on high UTS metals must be corrected.	Some loss of fatigue strength possible, but will depend on deposit type and thickness.
Mechanical properties	It is important to note that applied coatings do not usually contribute to the base metal strength and allowances should be made for this, particularly when wear damage is machined out prior to reclamation of the part.			
Deposit integrity	Susceptible to edge chipping (from impact or point loads). Bonding to substrate is by mechanical keying action only.	Metallurgical bond, low ductility deposits susceptible to high point or impact loads.	Care should be taken with edges.	Alloyed interfacial bond. High deposit integrity.
Finishing	Grinding is the only finishing technique; some deposits can be used 'as deposited'.	Grinding is generally employed, but some alloys can be finished by other forms of machining.	Grinding is the only finishing technique.	Grinding is generally employed, but some alloys can be finished by other forms of machining.

[a] When selecting the required coating, care should be exercised to ensure that corrosion due to galvanic action is not likely between the applied and the base material. Substances far apart in the electromotive series are especially susceptible but the severity of attack will depend very much upon the environment.

Table 1.5.38 General guide to the selection of coating methods – *continued*

	Flame deposition	*Spray/fusion deposition*	*Electroplating*[a]	*Weld deposition*
Recom-mended final thickness	Optimum 0.20–0.25 mm Maximum 0.40 mm Minimum 0.08 mm Thickness will depend on deposit type and service environment.	Optimum 0.50–0.75 mm Maximum 1.25 mm Minimum 0.25 mm	Optimum 0.08–0.12 mm Maximum 0.25 mm approx. Minimum 0.01 mm	Generally no thickness limitation, but 1.25 mm normal for light engineering applications. Optimum 0.75 mm Maximum 1.25 mm Minimum 0.50 mm
Economics	For thicker deposits over large areas this technique can be economically attractive.	Spray and fuse operation can make this an expensive technique.	For thin deposits, particularly on small components or areas, large batch processing can make this a very competitive technique.	Disadvantage of individual component application is offset by high integrity and thickness of deposit (assuming distortion not a problem).
Surface quality	Inherent coating porosity can preclude realistic measurement of surface finish. Finishes equivalent to 32 μ in CLA are normally adequate. Finishes better than the equivalent of 16 μ in CLA are normally unnecessary.	Normal engineering finish requirements apply.	Normal engineering finish requirements apply.	Normal engineering finish requirements apply.

[a] When selecting the required coating, care should be exercised to ensure that corrosion due to galvanic action is not likely between the applied and the base material. Substances far apart in the electromotive series are especially susceptible but the severity of attack will depend very much upon the environment.

Table 1.5.38 – *continued*

Table 1.5.39 Guide to the selection of coating materials for sliding conditions (unidirectional or reciprocating sliding (up to 100 cm/s) under steady load)

Suitable coatings	*Temperature of environment (°C)*			
	Under 300°C	*300–500°C*	*500–800°C*	*800–900°C*
Flame deposits	Cobalt base/chromium/ nickel/tungsten powder. Molybdenum powder or wire. Nickel/aluminium composite powder. Tungsten carbide/cobalt composite powder. 13% Chromium steel wire.	Cobalt base/chromium/ nickel/tungsten powder. Nickel/aluminium composite powder. 25% Tungsten carbide/ 5% nickel, plus mixed carbides. Tungsten carbide/cobalt composite powder. 13% Chromium steel wire.	Cobalt base/chromium/ nickel/tungsten powder. Chromium carbide/ nichrome powder. Nickel/aluminium composite powder. 13% Chromium steel wire (to 600°C). 25% Tungsten carbide/ 5% nickel plus mixed carbides.	–
Spray-fusion deposits	Nickel/chromium/boron/ tungsten carbide self-fluxing powder. Nickel/chromium/boron self-fluxing powder.	Nickel/chromium/boron tungsten carbide self-fluxing powder. Nickel/chromium/boron self-fluxing powder.	Nickel/chromium/boron tungsten carbide self-fluxing powder. Nickel/chromium/boron self-fluxing powder.	–
Electroplate	Chromium; rhodium (copper; silver up to 1 cm/s).	Chromium; rhodium silver; cobalt/chromium carbide co-deposit.	Cobalt/chromium carbide co-deposit.	–
Weld deposits	Cobalt base/chromium/ nickel/tungsten rod. Cobalt base/chromium/ tungsten rod.	Cobalt base/chromium/ nickel/tungsten rod. Cobalt base/chromium/ tungsten rod.	Cobalt base/chromium/ nickel/tungsten rod. Cobalt base/chromium/ tungsten rod.	Cobalt base/chromium/ tungsten rod. Cobalt base/chromium/ nickel/tungsten rod.
Others	Anodic films on aluminium alloys	–	–	–

Table 1.5.40 Guide to the selection of coating materials for hammering conditions (where surfaces impact during sliding process)

Suitable coatings	***Temperature of environment***			
	Under 300°C	***300–500°C***	***500–800°C***	***800–900°C***
Flame deposits	Cobalt base/chromium/ nickel/tungsten powder. Nickel/aluminium composite powder. Tungsten carbide/cobalt composite powder. 13% Chromium steel wire.	Cobalt base/chromium/ nickel/tungsten powder. Nickel/aluminium composite powder. 25% Tungsten carbide/ 5% nickel, plus mixed carbides. Tungsten carbide/cobalt composite powder. 13% Chromium steel wire.	Chromium carbide/ nichrome powder. Cobalt base/chromium/ nickel/tungsten powder. Nickel/aluminium composite powder. 25% Tungsten carbide/ 5% nickel, plus mixed carbides.	–
Spray-fusion deposits	–	–	–	–
Electroplate	–	Cobalt/chromium carbide co-deposit	Cobalt/chromium carbide co-deposit	–
Weld deposits	Cobalt base/chromium/ nickel/tungsten rod	Cobalt base/chromium/ nickel/tungsten rod	Cobalt base/chromium/ nickel/tungsten rod	Cobalt base/chromium/ nickel/tungsten rod

Table 1.5.41 Properties of some plasma-sprayed ceramic-based coatings

Coating	*Alumina* Al_2O_3	*75% Cr_2C_3 25% Ni–Cr*	*Chromia* Cr_2O_3	*WC + Fe Cr*	*85% WC 15% Co*
Hardness (VPN)	1100	480	1300	800	1050
Strength of adhesion (MPa)	69	76	83	83	172
Porosity (%)	0.5/1.0	1/3	0.25/1.0	1.0/2.0	0.25/0.5
Density of coating (g/cm^3)	3.45	–	5.03	9.9	13.2
Strength (MPa)	138	–	–	241	689
Coeff. of expansion (10^{-6}/C°)	6.8 (20/980)	–	6.7 (20/540)	6.8 (20/540)	8.4 (20/540)
Max. operating temp. (°C)	980	870	540	540	540
Guidance on applications	Wear and abrasion resistant. Stands high temperature. Electrical insulator.	Good wear properties at high temperature. Smooth surface, corrosion resistant.	Ant-galling. Wear resistant.	Resistant to wet abrasion.	Wear resistant and resistant to mechanical and thermal shock.

1–6

Component manufacturing processes

Contents

List of tables

List of figures

1.6.1 General characteristics of manufacturing processes for metal parts

1.6.1.1 Introduction

Metal parts are manufactured by a wide variety of processing routes which convert the as-bought material into its finished product.

The choice between routes is determined by economics (which will depend on production volume), by the need to produce dimensional accuracy or specific properties, or by the equipment available. A flow chart of the routes available for the production of metal parts is shown in:

Fig. 1.6.1 *Production routes for metal parts*

1.6.1.2 Summary of production routes

A description of the various production routes together with their respective advantages and limitations are summarized in:

Table 1.6.1 *Production routes for metal parts*

The table outlines the major production process types and also many specific processes. Further information on selected processes is given in sections 1.6.2–1.6.7 of this Chapter.

The size and shape limitations associated with the manufacturing process, including the tolerances which can be achieved, are detailed in:

Table 1.6.2 *Comparison of manufacturing processes for metal parts*

The number of components to be manufactured also influences the choice of manufacturing process, as shown in:

Table 1.6.3 *Choice of process for a part depending on quantity required*

Table 1.6.1 Production routes for metal parts

Process	***Advantages***	***Limitations***
Casting *Billet and ingot casting* Almost all metal forming operations have as their origin a metal casting. This has a regular and usually optimized shape, is easy to feed and is cast rapidly and progressively. Ideally continuous casting is employed.	Metal can be refined and alloyed in the molten state. Continuous casting is a very economical process which produces very high quality material.	Some metals cannot be cast. Powder metallurgy may be more economic in specific cases. Electroforming may produce superior material.
Sand processes Wood or metal patterns are prepared, and sand is packed around pattern to form a two-part mould, which is opened, the pattern removed, the mould closed and metal poured.	Almost any metal which can be melted in air can be cast; almost no limit on shape or size – most direct and cheap route from pattern to casting. Rapid production time if patterns available and proved.	Some machining necessary; rough surface finish; tolerances poor; mould parting line introduces further errors. Can be long production time for new complex jobs requiring sampling, or where standards of inspection are high.
Green sand Sand is moist, and bonded with clay and cereals.	Cheap, quick process. Most widely used method. Useful for single castings or mass production.	Sometimes porosity caused by gases evolved from mould; rough surface; tolerance poor.
Dry sand Same as above, but core boxes used instead of patterns. Mould is dried at about 200°C prior to closing and casting.	Mould is stronger, giving ability to tolerate long, thin projections.	More expensive; higher investment in ovens and moulding boxes. More parting line flash than green sand.
Sodium silicate Sand bonded with sodium silicate, which, after moulding, is hardened with CO_2. Additions of, for example, glycerol mono-acetate, cement, ferro silicon and calcium silicate make the mould self-hardening which gives higher productivity and avoids accidents due to pressure of CO_2.	Higher strength than dry sand without need for stoving. Good for cores. Tolerances better than green sand. High production rate.	Some mould burn-on problems in large steel castings. Relatively poor knock-out properties, together with dust problem.
Furane Sand self-sets within minutes by addition of catalyst to resin binder. Sand block can be lifted out and used without moulding box.	Four times strength of CO_2 hardened sand. Handleable without a moulding box. Room temperature process. Excellent moulding qualities requiring no ramming, therefore high permeability requiring no venting. Tolerances best of all sand processes. Less fettling. High production rate. Good breakdown and recycling.	Can give unpleasant fumes. Nitrogen porosity in ferrous castings if high percentage of urea formaldehyde resin used. Requires mould coat for best surface finish.
Full mould expendable pattern Lightweight, foamed plastic patterns are moulded on injection machine or machined and fabricated from blocks. Pattern is surrounded by dry unbonded sand, fluid sand (see below) or other cold setting sand, and is vaporized during pouring of molten metal.	Plastic patterns are easily made; light and easy to handle. Negligible tool costs for prototypes and single castings. No draft required; no parting line mismatch; no flash. Good for very large castings.	Pattern costs can be high for low quantities. Some limitations imposed by low strength of pattern. Can be supply problems for pattern materials. Sometimes incomplete vaporization causes defects in small and medium size castings.
Fluid and castable sand process Sand with sodium silicate and a selfsetting agent, suspended in a detergent foam, is poured around patterns and allowed to set.	Often used with full mould process. No ramming of mould necessary. Low water content of sand needs no drying prior to casting.	Only useful for simple shaped heavy castings. Needs special sand mills and high degree of technical control. Mould requires substantial coating of mould dressing. Process subject to licensing agreements.

Table 1.6.1

Table 1.6.1 Production routes for metal parts – *continued*

Process	***Advantages***	***Limitations***
Shell moulding (croning process) Sand coated with a thermosetting resin is dropped on to a metal pattern heated to approx. 250°C. This melts and cures resin to depth approx. 8 mm forming a shell. Shell halves are stripped from pattern and clipped together to form mould. Shells may be backed by steel shot.	Rapid production rate; exceptionally permeable mould gives reproduction of detail; good dimensional accuracy; uniform grain structure; minimized finishing operations. Suitable for stainless steel castings. Moulds can be stored for long periods.	Some metals cannot be cast. Patterns expensive. Size of part limited. Minimum 200–500 castings to be economic.
Vacuum moulding ('V' process) Plastic film is sucked down onto pattern and mould box filled with sand over pattern. Second plastic sheet then sucked on to top of mould by evacuating air from mould box. Vacuum under pattern released and pattern withdrawn.	Excellent surface finish and exacting dimensional control including possibility of zero draft.	Process subject to patent rights. (Patented and licensed from Japan).
Full mould casting (evaporative pattern lost foam, replicast) Pattern is made in polystyrene foam, coated with special refractory and packed around with sand. Pattern is burnt off by molten metal.	Low cost consumables. Very good surface finish. Best suited to low to medium production runs.	Fumes and carbon deposits from burnt polystyrene can cause problems. Proper venting is essential.
Premium quality castings Aluminium alloy castings in which excellent soundness is achieved by varying the cooling of the metal by controlling the thermal conductivity of the mould locally by varying the material composition, and by optimizing gating and feeding. The rapid solidification produces a fine structure which responds to heat treatment.	When fully heat treated ductility is up to five times that of conventional castings and UTS up to 50% greater. The properties are practically indistinguishable from those of forgings.	Castings are more expensive than conventional sand casting process.
Die-casting processes Mould cavities are machined into metal die blocks designed for repetitive use.	Good surface finish and grain structure; high dimensional accuracy; ability to produce thin sections; rapid production rate because of high mould conductivity; low scrap loss.	High initial mould costs and initial delays for making of die. Shape and size of casting limited.
Slush-casting Process for making hollow articles by decanting excess liquid shortly after pouring. Automated in the Robar Process where liquid is forced into die by piston and sucked out again as piston retracts. Can also be carried out in low pressure die-casting process if conditions right.	Cheap process for models, statuary, teapot spouts and the like. Wall thickness can be accurately controlled.	Usually for low melting point alloys with narrow freezing range. Internal surface rather rough. Slow manual process requiring good temperature control of liquid metal.
Gothias process Process for making hollow articles by partly filling die with liquid metal and displacing it to fill the entire cavity by insertion of a metal core.	Cheap process for models and statuary, etc.	Usually for low melting point metals. Not usually for engineering applications.

Table 1.6.1 – *continued*

Table 1.6.1 Production routes for metal parts – *continued*

Process	*Advantages*	*Limitations*
Gravity die-casting permanent mould process Metal is poured, often from a hand ladle, into a die under gravity only. Mould is hinged and clamped for easy removal of castings. Usually all manual operation. Variants for iron casting are Eaton process (dies arranged on a carousel) and Parlanti process (using anodized aluminium dies).	Good surface finish (Al alloys anodizable). Low internal porosity; good accuracy; useful where capital outlay is limited (die costs approx. 20% of that of high pressure die-casting) but regular cool-off required. Can be mechanized, e.g. 10 000 cylinder heads a day have been produced.	
Squeeze-casting Metal is forced to solidify fully under high hydrostatic pressure after having been poured in a manner that avoids entrainment of gas. Used for thick components where excellent mechanical properties are required.	Void formation is precluded. Dissimilar metal bonding is enhanced. Higher die wall–metal contact produces a faster cooling rate which leads to a finer grain size and improved mechanical properties. Heat treatment is facilitated. Surface finish better than gravity die-casting but not than pressure die-casting.	A new process. Choice of alloys is important. Only a small range yet available. Good metal quality free from inclusions is needed or the property improvement compared with gravity die-casting is lost.
Low pressure die-casting. A process which uses a simple machine to displace metal vertically upwards into the die by means of a low (0.02–0.10 MPa) pressure of air on the liquid metal.	Very good surface finish (aluminium alloys are anodizable) and accuracy. Low internal porosity since little filling turbulence. Exceptionally high yield (90–98%). Best for castings with low surface area/volume ratio. Useful where capital outlay is limited (die cost approx. 20% of that for high pressure). Sandcores can be used. Cast size and weight much higher than high pressure die-casting.	Only for low melting point alloys, especially Zn and Al base. Problems of dross pick-up in metal transfer tube.
High pressure cold chamber die-casting (also known simply as die-casting) A process using a large, complex machine which injects metal into the die at 5–500 MPa pressure by means of a horizontally acting piston. When metal solidifies, die is opened and casting automatically ejected. The pressure chamber may be permanently filled with liquid metal to increase the production rate. Variant is Acurad process in which die is filled slowly by large piston and liquid subsequently subjected to high pressure by second piston.	Extremely good surface finish and accuracy. Rapid production rate. Irons and steels now also possible. Dies can be photoengraved with a relief pattern to simulate various decorative finishes such as leather, woven fabrics or wood grain.	Very high initial costs and delay for making die. Size of part limited. Castings may not be heat treated since air entrainment porosity would cause blistering or distortion on heating. Average die life (number of shots) Zn >100 000 Al or Mg 20–100 000 Cu <10 000 Fe 5000–20 000. After this craze-cracking produces hairline ridges on castings. Al alloys not readily anodized.
Pore-free casting Variant of high pressure die-casting where die cavity is filled with reactive gas which combines with the metal so as to leave no gas pore. The gas–metal compound is finely dispersed throughout the casting.	Improved mechanical properties. Aluminium alloys may be heat treated. Very thin walled castings can be made. Lack of porosity reduces rejects and facilitates electroplating.	Equipment more sophisticated, production ratio lower and costs higher than conventional high pressure die- casting. Too expensive for 80% of die-casting working. Process has been patented.

Table 1.6.1 – *continued*

Table 1.6.1 Production routes for metal parts – *continued*

Process	*Advantages*	*Limitations*
Thixocasting Slugs which have been 'rheocast' (by stirring melt while cooling to about 50/50 solid/liquid state) are reheated to ~ 50/50 solid/liquid and die-cast.	Less solidification shrinkage – better accuracy. Lower die heat input therefore longer die life. Low air entrapment therefore castings may be heat treated.	Close temperature control required. Process has been patented. Not yet extensively applied.
Investment processes Refractory slurry (fine powder in aqueous suspension) is coated over pattern (i.e. pattern is invested) giving optimum resolution of detail, allowed to set and dry. Pattern is then removed and mould is baked prior to casting.	Good surface finish, high dimensional accuracy, almost unlimited intricacy. Can cast into hot mould to fill thin sections, inert mould (no carbon or gas pick-up, no fume) can cast in vacuum or with centifugal assistance.	Some limit to size; requires labour and patterns. Mould making time relatively long.
Plaster mould castings (Antioch process) Slurry of special gypsum plaster, water and other ingredients is poured over pattern and allowed to set to form a block mould. Pattern is removed and mould baked.	As above and materials cheap.	As above plus occasional adherence of air bubbles to pattern causes small spheroids on casting, however easily fettled off; limited to non-ferrous metals and components, up to 1000 kg in weight. Castings cool slowly and may therefore exhibit coarse grain, gas and shrinkage porosity.
Ceramic mould castings (Shaw process) Slurry of refractory powder, binder and gelling agent poured over wood or metal pattern and allowed to set forming a block mould or contoured shell. Mould is stripped whilst still rubbery, ignited to burn off volatiles causing microcrazing and high permeability, and finally fired at approximately 1000°C. Variant is Unicast process in which mould is dipped in a hardening liquid prior to firing.	As in general description above, any metal can be cast.	As in general description above but may have parting line errors; licenses are required to operate both the Shaw and Unicast processes.
Investment castings (lost wax process) Pattern is fabricated from wax, (occasionally plastic or mercury) dipped into refractory slurry and coated with refractory grains successively to build up a shell. When refractory has hardened pattern is melted out and mould fired at approxi-mately 1100°C. Shell sometimes backed up in a flask to produce a block mould.	As in general description above but no parting line (hence no flash and no associated mismatch errors) any metal can be cast; dies to make wax patterns made from aluminium alloy (not expensive die steel). Last approx. 100 000 injections. Almost unlimited freedom of design.	Process intended for quantity production: Minimum economic quantities could be 50, but more usually not less than 200. Labour intensive, although now being automated. Distortion can occur in extensive thin plate or arm sections; can be avoided by special attention to support of waxes.
CLA process Variant of lost wax process. The inverted mould is housed in a vacuum chamber and molten metal is drawn upwards into the mould.	Better metal yield. Better integrity, fine grain size, can produce steel components with walls down to 0.25 mm thick.	Complex equipment required.
Centrifugal processes Centrifugal casting Sand, metal or graphite mould is rotated about a horizontal or vertical axis. Molten metal is held against mould wall by the centrifugal action until solidified.	Good dimensional accuracy; rapid production rate; good soundness and cleanness; ability to produce very large cylindrical parts.	Shape limited to cylinders; spinning equipment expensive; larger castings sometimes show radial segregation and banding; other common defects include laps and hot tears.

Table 1.6.1 – *continued*

Table 1.6.1 Production routes for metal parts – *continued*

Process	***Advantages***	***Limitations***
Semi-centrifugal casting Process for the production of rotationally symmetrical sand or investment castings (such as pulleys and gear blanks, etc.) by rotating mould about its axis arranged vertically using a central feeding head.	Accuracy and finish characteristic of the sand or investment process used; good soundness and structure at the outer rim of the casting. Inner and outer surfaces of casting are formed by the mould. Moulds can be stacked vertically and spun together using common central runner and feeding system to increase production rate and metal yield.	As above, except shape limited to rotationally symmetrical components.
Centrifugally-aided casting (centrifugal pressure casting) Process for the production of small castings (often investment) located off axis. Metal forced into mould under centrifugal pressure.	Accuracy and finish characteristic of sand or investment process used, although often used for jewellery since pressure assistance gives excellent reproduction of detail.	Size limited.
Powder metallurgy Metal powder, produced by reduction of ore, or by comminution of the solid which may be an element, an alloy or a mixture is compacted (by solid phase sintering, or melting of one component). Unsintered powder may form the raw material of a working process such as extrusion or rolling and in this case sintering is contemporaneous with working.	100% material utilization. Alloys may be produced which cannot be obtained by other methods. A controlled amount and form of porosity may be produced for filters or oil impregnated bearings. Low conversion cost (powder to component) for large volume production.	Scrap sometimes difficult to recycle. Powder is more expensive than solid metal.
Cold pressing and sintering Cold die pressing and sintering Powder is blended with a lubricant and fed into dies, pressed to shape, sintered in a controlled atmosphere and sometimes coined or repressed for fine detail and dimensional precision.	Properties may be improved by heat treatment and/or surface treatment, gas carburization, tufftriding, induction hardening (for some compositions) and steam oxidation of pores which may increase wear resistance at less cost than carburizing.	Porous structure may be less ductile than wrought or cast structure. Large volume needed to amortize tool costs (prototype parts machined from blocks may not be representative of fully moulded parts). Height diameter ratio should not exceed 3 : 1. Re-entrant tapers must be machined after sintering. Parts intended for plating should be sealed first. Limited to approximately 14 kg weight.
Cold isostatic pressing and sintering (CIP) Powder is sealed into a flexible die or 'tool' which is then subjected to liquid hydro-static pressure. Tooling may be 'wet bag' (the tool is completely immersed in the fluid) or 'dry bag' with the tool a fixture in the press.	Sizes of over 450 kg weight can be produced with greater complexity, higher density and superior mechanical properties compared with die- pressing. Suitable for 'one-off' or prototype manufacture.	More machining required than for die-pressing.
Hot pressing Hot die pressing Similar to cold pressing except that no lubricant is needed, and sintering takes place during pressing which is at elevated temperature.	Better compaction, strength and ductility than cold pressing and sintering. May be used to improve soundness of castings.	Pressing operation takes longer. Lower throughput than cold pressing and sintering.
Hot isostatic pressing (HIP ping) As cold hydrostatic pressing but (gas) pressure is applied hot.	Advantages of CIP but in addition gives very high density and excellent properties. Hot processes can be carried out at a lower temperature than that required for sintering cold compactions.	More machining required than for die pressing.

Table 1.6.1 – *continued*

Table 1.6.1 Production routes for metal parts – *continued*

Process	***Advantages***	***Limitations***
Powder metal forging Powder of the desired composition is mixed with a lubricant, performed in dies in a press, heated, closed die forged (usually one operation), heat treated and given a minimal amount of machining.	Reduced number of forging operations. Lower pressure, lower cost of equipment. Reduced forging temperature, less fuel. Reduced tool costs, reduced skilled labour requirements, better yield, greater accuracy, less machining. Better material properties. Should become less expensive than conventional forging.	High cost of alloy powder. Difficult to alloy during sintering. More expensive than casting.
Powder Injection moulding Blended powder mixed with equal volume of polymer is injection moulded into shape. Polymer then burnt-out and the powder sintered	Rapid production is possible. Good shape control at the green stage. Robust green compacts produced.	Burn-out kinetics limit application to small components. Process difficult to control.
Deposition from Fluid	Used for production of special shapes difficult or expensive to produce in other ways. May also refine metal to give otherwise unattainable purities and properties.	
Electroforming Production of parts by the electrodeposition of metal on a mandrel or mould which is subsequently removed from the deposited metal leaving the electroformed part. In certain circumstances the mandrel or parts of it may be left as an integral part of the electroform.	Extremely high dimensional accuracy, excellent control over properties; practically no size limitations, considerable intricacy possible; any surface finish attainable.	Relatively slow production rate; scratches and tiny cracks reproduced; skilled techniques required; selection of material limited. Control of process parameters much more critical than for conventional electroplating due to thicker deposits, careful design of mandrel and location of anode necessary to ensure uniform deposits.
Photofabrication Metal selectively removed or deposited by photographic and chemical/electrochemical processes. Part is first prepared as a silhouette several times size for accuracy, then photographically reproduced and reduced for contact on photosensitive resist material.	Low tooling and tool maintenance costs; design versatility; burr-free parts; freedom from metallurgical damage and induced stresses; applicable to most metals and alloys; design changes quickly accommodated.	Part edges have chevron shape. Width/depth ratio for holes, slots, etc., is 1:5.
Deposition from vapour Metal is transmitted by evaporation (aluminium) by breakdown of a halide (zirconium) or by sputtering (gold).	Can produce very thin films, (mostly on substrates) cheaply.	Very specialized application.
Working **Forging** *Hand or open die forging* Bar ingot or billet is heated and shaped under a hammer or press using traditional smithy techniques. Often a part-process resulting in a dummy, this is often the first stage in a metalworking procedure.	Cheaper than closed die forging for small production totals. Used for components too large to be closed die forged or worked by any other process and can also be used for small 'hand-held' components. Grain flow can be arranged to be advantageous to mechanical properties.	Does not produce a very accurate final shape. Requires skilled labour. Only suitable for hot workable materials. Slow production rate. Yield can be as low as 40%
Drop or closed die forging (stamping) A piece of bar or a forged dummy are heated and shaped by blows between mating dies.	Rapid production rates. Less operator skill required. More complex, closer tolerance components than with open die forging.	Expensive for short runs because of die costs. Often requires a predummying operation and final machining. Rounded corners and edges necessary.

Table 1.6.1 – *continued*

Table 1.6.1 Production routes for metal parts – *continued*

Process	***Advantages***	***Limitations***
Close to form forging Most accurate variant of drop or closed die forging	Smaller blanks, higher yields (less flash). Lower machining costs due to general dimensional accuracy. Sharp corners and draftless sections possible. Exploited for production of high grade Al alloy and Ti alloy forgings.	Significant reduction in machining costs necessary to amortise high die costs.
Precise form forging (impact machining) Accurately machined slug is forged in controlled atmosphere furnace with precise die cooled under controlled conditions. Flash is machined off the back end of the forging, location holes are bored and all burrs removed. Used mainly for gear production.	Machining of blank removes and prevents forging of surface defects. Controlled furnace and cooling atmosphere prevent surface decarburization and scale formation. Elimination of expensive machining operation such as gear cutting. Improved strength as grain fibre is not cut by tooth machining.	High die costs and in addition controlled atmosphere costs have to be balanced by elimination of machining.
Draftless forging Variant of drop/or closed die forging with zero draft angle on dies.	Allows production of components with pockets and ribs without the need for draft angles and large fillet radii. Eliminates machining operations.	Size limitations. Higher tooling costs.
Flashless forging Component stock volume is carefully controlled so that when the material is squeezed into the enclosed tools there is just sufficient to fill the die impression.	Elimination of flash which may be source of weakness in components subjected to fatigue, e.g. valves and pump bodies subjected to high pulsating pressures.	High die costs. Generally components made by this method must have barrel configurations with, due to tool design considerations and the avoidance of feather edges, flat sides. Accurately prepared blanks necessary.
Split die forging Variant of drop or closed die forging with negative impression of component on split dies.	Allows production of shapes with bosses, flanges, projections and undercuts on two faces. Shapes formerly restricted to castings or welded fabrications can be forged.	Double die costs.
Reverse flow forging Produces hollow forgings, closed at one end and usually circular in plan form, with parallel internal and external walls, although rectangular or irregular shapes with stepped or fluted walls and bosses, lugs and recesses in the closed end are possible.	Allows production of cup shaped components or bottles with forged product characteristics.	Length to diameter ratio limited to 5 : 1.
Reverse flow split die forging Combination of split die and reverse flow forging	As for split die and reverse flow forging with additional advantage of including features on the outer form which would preclude the possibility of ejection from solid tools.	As for split die and reverse flow forging.
Press forging Heated stock is pressed into shape instead of by hammer blows, otherwise as reverse flow split die forging.	Some intricacy of shape possible. Fast production rate. Precise-to-form components possible which reduces machining cost.	Size of parts limited. High die costs.
Ring rolled forging A special process for forging a ring shaped part from a doughnut shaped blank by hot working between a pair of rollers.	For very large rings this may be the only process.	Maximum ring diameter 1.8 m. Maximum ring height 250 mm. For smaller rings closed die forging, welding or machining may be more economical.

Table 1.6.1 – *continued*

Table 1.6.1 Production routes for metal parts – *continued*

Process	***Advantages***	***Limitations***
High energy rate forging Special machines are used which strike the work at high speed so that the forming is usually completed at a single blow. The velocity at impact can be as much as 8 times that of conventional forging.	Speed of forming. Good accuracy and surface finish. Low draft allowance. Good mechanical properties.	Expensive forming machines required. Special dies required. Not suitable for very large parts. Sharp corners and small radii cause high die wear.
Transverse roll forging (i) Wedge rolling Stock is fed between rolls in which the shape to be produced has been machined. Shape and reduction obtained in single revolution of rolls. (ii) Spur gear rolling A blank is pushed axially between a pair of tapered toothed forming wheels after passing through an induction heating coil.	 Rapid tooling changes. Good tolerances Close to form production.	 Stock diameter $\ngtr$ 2 product diameter Size of parts limited to 250 mm diameter.
Orbital forging Upper curved die reduces billet height by oscillating: the direction of oscillation rotating but not the die. Gives an area of contact that moves around the billet deforming it progressively. Forging is done warm.	Only requires about 1/10 the load of conventional forging. Very good surface finish and detail. Lower machining allowances. Product can be made thinner for a given diameter than a conventional forging. No scale formation.	Limited to symmetrical shape in the horizontal plane. Presently limited to 150 mm diameter.
Isothermal forging Material forged at a slow rate of deformation in superplastic state with dies maintained at forging temperature. The billet may be made by powder metallurgy. Inert atmosphere control required.	Only light capacity presses required. Complicated shapes near to finished dimensions formed with consequent savings in machining costs. Up to 50% less material required as compact with conventional forging.	Only economic for expensive materials where the saving of material offsets the high capital costs and low throughput.
Automatic hot forging Steel bar stock inductively heated to 1250°C in <60 s: descaled, sheared into blanks which are then transferred through three or four closed die stations to produce 70–180 close to form forgings per minute. Forged part ejected at ≃1050°C.	Tolerances better than conventional forging. Small draft and elimination of flash requires less bar stock. Application to all C and alloy steels. Tool contact time of 0.06 s allows for longer tool life.	Large production runs of the order of 25 000–100 000 required – 5 kg size limitation on existing machines. Mainly suited for round, symmetrical shapes. *L/D* ratio <1.
Extrusion (hot) Driving material contained in a chamber through an orifice which determines the shape and cross section of the project. In direct extrusion the die is stationary and a plunger forces the billet to pass through it. In indirect extrusion the die moves and a lower pressure is required: The material temperature is at 0.6–0.8 of melting.	The material is under compression and deformation is continuous. The die can be changed. A wide range of materials can be processed including duplex components with the two materials welded together. Well suited to shapes having multiple re-entrant angles and angled projections. Tubes can be formed. Can be done cold on a wide range of materials. May be operated as a forming process for material intended for finishing. May be used to produce finished parts, tubes, rods, sections, etc.	Power consumption and capital investment high. More expensive than rolling, where both processes are suitable, for wide strip or plate, large diameters or very long lengths.

Table 1.6.1 – *continued*

Table 1.6.1 Production routes for metal parts – *continued*

Process	***Advantages***	***Limitations***
Rolling (hot) Material, which may be a cast billet or a forging is fed through one or a series of rolls which may be plain (to form sheet or plate) or may be grooved (to produce bar). Material temperature up to 0.8 of melting.	Very high production rate for large quantity production. Material can be finished at lower temperature to give enhanced properties. Plates, bar rod, girders, etc., may be produced.	Rolls expensive and roll changing time consuming. Large area required.
Tube forming Tubes are formed by a variety of processes including casting, boring, mannesman piercing, extrusion, hot reducing, push bench drawing, etc., which may produce the final tube or may provide a semi- finished hollow for cold reducing.		
Finishing process **Cold forging (extrusion)** *Open die forging* Matching dies come together to clamp a cut metal rod. A punch then moves forward to bend and shape the metal between the dies.	Ideal for long, thin parts with flats, collars and multiple heads. Parts ejected by dies coming apart therefore no warping by friction as compared with ejection from solid dies.	Sizes limited to approximately 200 mm length × 20 mm diameter. Internal stresses may have to be relieved.
Cold heading Similar to cold forging but deformation restricted to head. Flange type work.	Good strength and surface finish. No scrap loss. Rapid production rate.	Internal stresses may have to be relieved. Applicable to special requirements like forming bolts.
Cold heading – Dynaflow[R] A variant of cold heading is the GKN Dynaflow[R] process where a slug of wire is extruded to give a profiled shank before cold heading.	Very large unit volumes in the upset portion.	Larger volume (nearer to 100 000 parts than 50 000) needed to justify investment. Optimum economies with half a million components.
Cold forging (extrusion) Forming a slug at ambient temperature by force between a punch and a die. Plastic flow produces the required shape. 1. Reverse extrusion – can type work. 2. Forward extrusion – shaft type work. Hydrostatic pressure can be employed to permit brittle materials to be extruded.	Good structure, no scale. No forging draft to shapes. Dimensions close to finished part. High yield of metal in finished forging. High production rate 200–1000 parts/h. Reductions in area up to 75% possible in sizes above 2.5 mm. Allows for higher head to shoulder than cold heading. Aluminium gas bottles up to 60 litres capacity are produced by reverse extrusion.	Powerful presses and expensive tooling essential. Suitable runs of 50 000 parts. Careful material selection – parts must be easily cold deformed. Heat treatment often necessary after forming. Best suited to symmetrical shapes. Some alloys not easily cold forged. Heat treatment may be necessary after forming.
Impact extrusion The slug is extruded either backwards over the punch or forward through an opening in the die, or both on being impacted at high velocity by the punch. Usually carried out cold.	Very rapid production rate. Good finish low scrap loss. Low tool costs. Highly developed for collapsible tubes.	Choice of metal, shape and size of part restricted. *L/D* ratio normally 10:1 or less.
Orbital forging As hot orbital forging but carried out at ambient temperature.	Excellent surface finish. Machine manufacturers claim heading extrusion and gear shaping possible.	Cylindrical shapes performed and size restricted to 100 mm diameter.
Hydroforming High pressure fluid presses tubes outwards into two part metal dies to produce complex shapes.	Complex structural components produced otherwise requiring fabrication from several parts.	Slow rate of production, limited to tube like members.

Table 1.6.1 – *continued*

Table 1.6.1 Production routes for metal parts – *continued*

Process	***Advantages***	***Limitations***
Wire and tube reduction methods *Tube working* *Drawing* A rod (or tube) after pointing is pulled through a die which reduces its diameter at ambient temperature. Tubes may be reduced (sunk) without an inner plug, or drawn over a plug (which may or may not be held in place) or on a mandrel. The stock must be annealed at intervals between passes to permit further reduction.	Excellent process for large-scale manufacture producing good finish and tolerances.	Limited to ductile materials.
Rotary swaging Rod, wire or tube is fed through dies opening and closing 600–1000 times per minute while rotating round the axis of the product.	Slower than wire drawing but more tolerant. Materials too brittle to draw may be reduced by swaging. Large reductions in cross section possible. Easily possible to produce diameter variations along the length of a bar or tube. Ideal for attaching terminals and lugs to cables. Excellent finish. Good production rate.	Circular shapes only may be produced.
Hydrostatic forming T pieces are formed from tubing using oil pressure. Two side rams compress the oil filled tube while the inner of a third ram system is withdrawn to allow for the stem formation. Cap of the stem is then machined off.	No wall thinning occurs. Tolerances of good quality tubing maintained for subsequent welding to same diameter tubing.	Expensive equipment requires large numbers. Presently limited to 200 mm diameter tubing and 12 mm wall thickness.
Cold rolling *Section rolling* Rods and sections may be reduced by cold rolling and annealing.	Higher production rate than swaging or drawing.	Shape and dimensional accuracy inferior to drawing or swaging.
Cold profile rolling Cold finish rolling Gears Pre-hobbed blank is revolved between rotary dies which generate a finished involute form on the gear blank.	Improved strength and metallurgical characteristics compared with conventional methods of cutting and shaving. Higher production rates and better more consistent quality than conventional techniques. (1.5 million parts compared with 30 000 before tool change necessary.)	Dies require careful preparation. Component material should have low sulphur and phosphorus content, be fine grained and have high elongation and low work-hardening properties. Limited at present to spur gears.
Bearing races Annular ring shaped starting blank is located in the bore by split mandrel and inner and outer surfaces formed simultaneously by rotating forming dies.	Improved bearing race strength and metallurgical characteristics. More economic utilization of material – saving up to 40% of the metal required in conventional machining. Faster production rates. Higher more consistent quality. Less distortion after heat treatment.	as above.
Barrel rolling Cylindrical slugs are radially fed between two rotating profiled rollers that shape the slugs into barrel shaped rollers up to about 60 mm diameter.	14% material and 61% manufacturing time saved.	as above.

Table 1.6.1 – *continued*

Table 1.6.1 Production routes for metal parts – *continued*

Process	Advantages	Limitations
Thread rolling Bar is passed between two roller segments each profiled with a thread form and rotating in opposite directions to generate a thread on the bar.	Time and material saved compared with thread cutting. Stronger thread because material grain is worked into thread form.	Dimensionally less accurate than can be obtained by machining or grinding.
Flat product rolling Plate, sheet strip and foil are produced from billet, or from hot rolled plate or sheet by rolling between cylindrical (or very slightly barrelled) rollers. Annealing may be necessary between passes.	Excellent finish and dimensional accuracy.	
Sheet metalworking *Conventional* Sheet metal is cut, bent or drawn to a required shape by suitable presses	Applicable to metals obtainable as sheet. Very rapid production rate, uniformity or product, good surface finish. Versatility in possible sizes and shapes.	High tool costs. Yield may be poor. Thickness limited to sheet sizes. Edges may need dressing.
Rubber press The die contacts the sheet which is supported by a deformed rubber bed.	Lower tool costs. Suitable for shorter runs.	Only possible for some shapes.
Superplastic Plastic deformation at a fixed moderately low temperature of sheet metal by 100–1500% under low stresses.	Metals can be formed to complex shapes with deep draws normally associated with plastic blow moulding.	Metals used limited to specific compositions that exhibit superplasticity. Specialized moulding equipment required.
Spinning A sheet blank is clamped and rotated. The sheet is forced over the form by a spinning tool – a simple piece of rounded wood or small roller.	Deep, hollow objects can be formed readily and cheaply at room or low temperatures. Cheap equipment and tooling. Strength developed by cold work.	High tooling and equipment cost. Only suitable for large production.
Roll forming Strip is successively formed into a shaped section by passage through pairs of shaped rolls.	Very rapid production. Large choice of section shape. High yield. Dimensionally accurate.	High tooling and equipment cost. Only suitable for large production.
Fine blanking Multi-action press plus precise tooling grip flat unblanked material and press out finished component. Punch-die clearance is much smaller than for conventional work.	Production of parallel, cleanly sheared surfaces ready for assembly. (Elimination of finishing processes such as shaving, broaching, milling, reaming, grinding or deburring which may be necessary with conventionally blanked material.) Close tolerances. Multiple blanking operations performed simultaneously with consequent increase in dimensional accuracy. Quiet operation.	Large production runs and/or large savings in finishing costs necessary to justify more expensive machines and dies. Longer time needed to produce dies.
Trimming Fine trimming of rough blanked shape done by very accurate punch and cutting die.	Can achieve perfectly parallel sides.	Slower process than fine blanking. thickness $\leq$ 8 mm.

Table 1.6.1 – *continued*

Table 1.6.1 Production routes for metal parts – *continued*

Process	***Advantages***	***Limitations***
Vibratory trimming 0.05 mm amplitude vibration applied at 700–1500 Hz to tooling.	Cut surfaces improved and allows for increased thickness of workpiece – 30 mm.	Special press required.
Fabrication Making a part by joining together semi-manufactured pieces. Methods include mechanical fastening, (riveting, etc.), brazing, soldering, welding, adhesive joining.	Very great versatility. Low capital costs.	Skilled labour required. Production rate low. Joins may be points of low strength or liable to fatigue and corrosion.
Machining *Conventional machining* Turning, milling, grinding, boring, etc.	Great versatility. Stock can be chosen with isotropic and well established mechanical properties.	May lead to excessive metal waste. Not always possible or economic for very intricate shapes.
Automatic machining Round, hexagonal or shaped bar stock is fed through hollow spindles and is cut by automatically operated tools. Most traditional machining operations are possible.	Very high production rate with excellent dimensional accuracy and surface finish.	High cost of machines. Need for skilled tool setters. Limited to small parts. May be low yield due to loss as swarf.
Broaching Stock is removed by means of a 'broach' which has a number of teeth profiled to cut away (in stages) the precise shape required when forced over the surface to be machined by means of a press.	Great dimensional accuracy of complex shape achieved in one stroke. Rapid production at low cost for large numbers of components.	High cost of press and (consumable) broach.
Spark machining Erosion of a surface caused by the passage of an electric spark immersed in a light oil.	Can produce intricate shapes in very hard materials like hardened steel or carbides.	Relatively slow and expensive compared with conventional methods except for very hard metals. Fire precautions necessary.
Ultrasonic machining Microscopic chipping by repeated tool oscillations at about 20 000 Hz usually associated with an abrasive slurry.	Very complex shapes can be cut into very hard materials – suitable for die-making. Good finish.	Cutting rate slow if materials are not brittle. For anything but very small impressions powerful machines are necessary.
Electrochemical machining Jets of electrolyte plus electrodes are fed over surface of the material. By profiling the paths of the jet electrode, shapes can be generated.	Simple and very fast. No stresses introduced into work piece. Good accuracy (0.0127 mm possible) and surface finish.	Tooling cost may be high.

Table 1.6.1 – *continued*

Table 1.6.2 Comparison of manufacturing processes for metal parts

Process	Size and shape limitations				
	Overall size/weight			Section thickness (mm)	
	Max.	Optimum	Min.	Max.	Min.
Continuous castings	Non-Fe 230 mm Fe 400 mm		12 mm		3
Sand castings	Green no limit Dry 3 ton Al 100 kg Iron 50 ton Steel 200 ton	1–100 kg or higher.	 30 g 30 g 100 g	No limit in floor and pit moulds.	Al 4 Mg 4 Cu 2.5 Iron 2.5 Steel 6–12
	Furane 30 ton			No limit	
	Fluid sand 40 ton castings have been made.		2 kg	No limit	2.5
Shell mould castings (croning process)	Al 15 kg Iron 50 kg Steel 150 kg 600 × 900 mm	0.1–20 kg	30 g 30 g 50 g	50	2–6
Gravity die-castings (permanent mould)	50–200 kg	1–50 kg	100 g	50	Al 2.5–3 Mg 4 Cu 2.5–8 Iron 5
Squeeze-castings	4.5 kg	0.5–12 kg	20 mm dia.	200	6
Low pressure die-castings	200–500 kg	0.5–50 kg	100 g	10	Zn 2 Al 2.5 Mg 3
High pressure die-castings	Pb 7, Sn 5, Zn 35, Mg 20, Al 50 kg (350 × 400 × 1000 mm) Fe 5 kg (75 × 250 × 600 mm)	Al 50 g–10 kg Zn 150 g–3 kg Fe 3 kg	0.1–10 g (Fe 20 g at present)	8 preferable 12 possible	Pb, Sn 0.7–1.5 Zn 0.2–1.2 Mg 1.2–2.3 Al 0.7–2.0 Cu 1.2–2.3 Fe 1.0–2.0
Plaster mould castings	50 kg	0.1–5 kg	25 g		1.0–1.5

Table 1.6.2

Table 1.6.2 Comparison of manufacturing processes for metal parts – *continued*

Size and shape limitations			***Tolerances***			
Undercuts	***Inserts***	***Holes (min. diameter mm)***	***Dimensional tolerances ISO tolerance system IT***[a]	***Draft allowance (degrees)***	***Machine finish allowance (mm)***	***Surface smoothness (μm rms)***
Yes	No	3 mm min. in casting direction	10-12	Straightness tolerance <125 dia. 4 mm per metre length	0–125 0.8–1.5 125–250 2.5	
Yes	Yes	5–7	Al, Mg 13–15 Cu 15–16 Grey iron 14–16 Malleable 13–16 Steel 16–18	1–3	Non-Fe Iron Steel 0–150 mm 1.5 2.5 3 150–300 mm 1.5 3 5 300–500 mm 2.5 4 6 500–1500 mm 3.6 5 6	2.5–25
Yes	Yes	5–7	Intermediate between green sand and shell mould castings.	0–3	Approximately 50% of green sand process	2.5–5
Yes	Yes	6	Steel, often 16 but 18 attainable.	0–0.5	0.8–6	2.5–25
Yes	Yes	3–6	12–14 parting line error 0.25–0.5 mm	0.1 attainable 0.25–3 usual.	Often none required.	1–4
Yes - expensive lower production rate	Yes	Zn 3 Al, Mg 3–10 Cu 5 Iron 3–10	Al 12–14 Iron 12–15	0.1 attainable but high die wear expensive 0.2–3 usual 5 preferred in recesses.	0–100 mm 0.8 over 100 mm 1.5	2.5–6.5
Yes	Yes	Yes	Intermediate between gravity and high pressure die-castings.	As for gravity die-castings.	0.6–1.2	3–4
Yes - expensive lower production rate	Yes	Zn 2 Al 3 Mg 3	Intermediate between gravity and high pressure die-castings.	0.2–3	0.5–1.0	2–4
Yes - expensive lower production rate	Yes- somewhat reduced production rate	Pb, Sn, Zn 0.1–1.2 Al, Mg 2–0 Cu 3.0 also 5–10° draft angle	Zn 11–13 Al 11–14 Fe 11–13 + 0.05 mm parting line error	~20	0.25–0.80	1–2
Yes		12	11–14	0.5–1	0.8	0.7–1.3

[a] See Table 1.6.29

Table 1.6.2 – *continued*

Table 1.6.2 Comparison of manufacturing processes for metal parts – *continued*

Process	Size and shape limitations				
	Overall size/weight			Section thickness (mm)	
	Max.	Optimum	Min.	Max.	Min.
Shaw process castings Ceramic mould castings	Several tons	0.2–5 kg	100 g	No limit	0.6–1.2
Investment castings (lost wax)	Al 10 kg Steel 700 kg max. length 1 m	50 g–5 kg	0.5 g	75	0.25–1.2
Centrifugal castings	Non-ferrous diameter 1.8 m length 8 m Ferrous diameter 1.2 m length 15 m		Non-ferrous diameter 25 mm Ferrous diameter 29 mm	Wall thickness 100	2.5–6.5
Cold die-pressing and sintering	USA UK 258 cm^2 64.5 cm^2 152 mm ht 0.8 mm ht 140 kg 4 kg	0.2 kg	0.8 cm^2 0.8 mm height 0.081 kg		0.5
Isostatic pressing powder	450 kg	10 kg		500	
Powder metal forging	4.5 kg wt			Dia. 200	
Electroforming	Limited by size of bath	Thickness 1.2–0.25 mm		6	0.12
Hand or open die forging	200 t		25 mm	2 m dia. 20 m length	5 mm
Conventional closed die forging	7000 cm^2	3000 cm^2	10 cm^2	No limit	3
Close to form forging	2580 cm^2	1290 cm^2	15 cm^2	No limit	1
Precise form forging (impact machining)	4.9 kg	0.7–3.4 kg	0.22 kg	150 dia.	5
Draftless forging	650 cm^2	250 cm^2	25 cm^2	No limit	2.5
Flashless forging	322 cm^2	250 cm^2	25 cm^2	No limit	10

Table 1.6.2 – *continued*

Table 1.6.2 Comparison of manufacturing processes for metal parts – *continued*

Size and shape limitations			*Tolerances*			
Undercuts	***Inserts***	***Holes (min. diameter mm)***	***Dimensional tolerances ISO tolerance system IT***[a]	***Draft allowance (degrees)***	***Machine finish allowance (mm)***	***Surface smoothness (μm rms)***
Yes	Yes	0.5–1.2	11–13 usual +0.25 mm parting line error	0.1	0.6	2–4
Yes	No	0.5–5 can use pre-formed ceramic cores for deep holes and complex shaped passage ways.	11–13 usual 10 attainable	Usually zero 0.5–1 required for exceptionally long cores	0.8 machining 0.35 grinding	1–3
No	Yes	Yes	11	1	Non-ferrous castings – small 1.5 large 6 Ferrous castings – small 2.5 large 3.0	2.5–5
No – unless isostatically pressed	Yes	Yes – in direction of pressing *L/D* 4:1	9–10 4–8 (sintered and coined)	0.06 Can be zero	Can be zero Can be zero	<1 <1
Yes	Yes	No	2–20 mm depending on size	Zero	N.A.	N.A.
Usually no	Not recommended	Yes	9–10	0.06	Can be zero	<1
Yes with expendable mandrels	Yes	Yes – limited by short thickness – can be as low as 0.01.	3	Zero	Zero	0.025–0.05
No	No	Yes	2–50 mm depending on size	N.A.	N.A.	Rough
No	No	Yes	15–18 (From +0.5 mm –0.25 mm)	5–7 external 7–10 internal	0.8–1.6 to 6.3 depending on size.	<3.2
No	No	Yes	14 (From ±0.25 mm)	5	None on forged faces	<3
No	No	No	11–15	Can be zero	None on forged faces	1–1.5
No	No	Yes	14–15 (+ 0.5 mm –0.25 mm)	From nil to 5	None on forged faces	<2
No	No	Yes	14–15 (+ 0.5 mm –0.25 mm)	Nil to 5	None on forged faces	<2

[a] See Table 1.6.29

Table 1.6.2 – *continued*

Table 1.6.2 Comparison of manufacturing processes for metal parts – *continued*

Process	Size and shape limitations				
	Overall size/weight			Section thickness (mm)	
	Max.	Optimum	Min.	Max.	Min.
Split die forging	322 cm^2	250 cm^2	25 cm^2	No limit	2.5
Reverse flow forging	610 mm dia.	200 mm dia. *L/D* 5:1 max.	20 mm dia.	60	6
Reverse flow split die forging	400 mm dia.	250 mm dia.	50 mm dia.	60	6
Press forging precise to form	800 cm^2	250 cm^2	25 cm^2	No limit	2
Ring rolled forging	250 mm height 1.5 m dia.	100–200 mm height 1 m dia.	100 mm height 350 mm dia.	150	25
Transverse roll forging	40 mm 600 mm length		250 mm length	50	2
Orbital forging	150 mm dia.		50 mm	50	2.5
Automatic hot forging	5 kg wt				
Hot extrusion	Billet size up to 8 x 10^6 kg common max. extruded component cross-section contained within 50 cm circle.			Approx. 100	1
Cold heading	250 mm long 18 mm dia.		1.5 mm long 0.75 mm dia.		
Cold forging (steel)	19 cm^2	12 cm^2	3 cm^2	No limit	0.5
Cold extrusion (other metals)	Billet size up to 7.5 x 10^5 kg. Extruded component up to 4 mm long.				
Impact extrusion	*D* (cm), *L* (m) Al 15, 2 Steel 11, 2 Mg 16.5, 1.3 Pb, Sn 7.6, - Zn 6.4, 0.2		Dia. mm 6 12 9 9 1.3	0.25–12.7 22.0 6.3 0.4 3.5	0.09–1.6 0.9 0.15 0.09 0.25
Hydrostatic forging	200 mm dia.			12	
Cold finish rolling				Gears OD 127 Rings OD 95	Gears OD 38 Rings OD 18
Sheet metal work cutting	2.5 m long x 6.4 mm thick or 3.6 m long x 3.2 mm thick				

Table 1.6.2 – *continued*

Table 1.6.2 Comparison of manufacturing processes for metal parts – *continued*

Size and shape limitations			*Tolerances*			
Undercuts	***Inserts***	***Holes (min. diameter mm)***	***Dimensional tolerances ISO tolerance system IT[a]***	***Draft allowance (degrees)***	***Machine finish allowance (mm)***	***Surface smoothness (μm rms)***
Yes	No	Yes	14–15 (+ 0.5 mm –0.25 mm)	5	None on forged faces	<2
No	No	No	14–15 (Eccentricity 0.5–5 mm)	Nil	None on forged faces Normally 1.5 mm	<2
In one plane	No	No	14–18 (Eccentricity 0.5–5 mm)	5 on some faces	None on forged faces	<2
Yes	No	No	12 (±0.25 mm)	From nil to 1°	None on forged faces	Usually <2 but 1.6 possible
No	No	No	11–16 (3 mm on diameters 3 mm on height)	Nil	3	<3.2
No	No	No	12	~5	~1	<3
No	No	Yes	12	0–3.5	None on forged faces	0.8
No	No	Yes	12	0.5–1	None on forged faces	<2
Yes	Yes	Yes in extrusion direction.	12	Straightness approx. 0.3 mm/m	Usually zero	1–1.25
Yes	Yes	No	Diameter 10–11 Length 9–13	None	Usually zero	<1.0
No	No	Yes	13 (±0.12 mm)	Nil	Nil	<1.5
Yes	Yes	Yes in extrusion direction.	9	Straightness approx. 0.3 mm/m	Usually zero	Fe 0.5–0.075 Non-Fe 0.1–0.5
No	Not recommended		Length 11 Diameter 10	Zero		Fe 0.25–1.7 Non-Fe 0.1–0.5
			Tolerances of cold drawn tubing are maintained.	Zero	Not machined.	
Possible	No	Holes can cause distortion (consult suppliers)	9		Gear – can be zero Rings – subsequent machining can be necessary	<1–12
No	No	Dia. thickness	11–12	Zero	Zero	

[a] See Table 1.6.29

Table 1.6.2 – *continued*

Table 1.6.2 Comparison of manufacturing processes for metal parts – *continued*

Process	Size and shape limitations				
	Overall size/weight			Section thickness (mm)	
	Max.	Optimum	Min.	Max.	Min.
Bending				25	0.075–0.12
Drawing	6 mm diameter 1 m depth		3 mm dia.	25	0.075–0.12
Deep drawing and ironing	600 mm dia.		2 mm dia.	12	0.1
Superplastic forming	2400 x 1200 mm			4	0.8
Spinning	7.5 m dia.		6 mm	75	0.1
Roll forming	Length 12 m Width 1.5 m		Width 10 mm	6.35	0.12
Fine blanking	Biggest to date 12.7 kg Material width 80 cm max. Length of feed 700 mm max.		Pin head size	2.5	0.4

Table 1.6.2 – *continued*

Table 1.6.2 Comparison of manufacturing processes for metal parts – *continued*

Size and shape limitations			*Tolerances*			
Undercuts	***Inserts***	***Holes (min. diameter mm)***	***Dimensional tolerances ISO tolerance system IT***[a]	***Draft allowance (degrees)***	***Machine finish allowance (mm)***	***Surface smoothness (μm rms)***
Yes	Yes	Yes – hole should not be near bend.	15–16 Accurately located holes should be pierced after bending.	Zero	Zero	
Yes	Yes	No	11–12	Zero	Zero	
No	No	No	6–8	Straightness 1 in 14 000	None on ID and OD	<1
Yes	Yes	No	14	2.50		
Yes	No	No – unless in blank	11–12	Zero	Zero	
Yes	No		13–14 (cross-section thickness)	Zero	Zero	
No	No	60% material thickness	6–9	Zero	None	0.3–1.5

[a] See Table 1.6.29

Table 1.6.2 – *continued*

Table 1.6.3 Choice of process for a part depending on quantity required

Process		No. required				
		1–10	*10–10²*	*10²–10³*	*10³–10⁴*	*>10⁴*
Castings	Green sand	*	*	*	*	*
	Dry sand		*	*	*	*
	Sodium silicate		*	*	*	*
	Furane		*	*	*	*
	Full mould	*	*			
	Fluid sand	*	*			
	Shell moulding			*	*	*
	Vacuum moulding			*	*	
	Effset process			*	*	
	Premium quality casting			*	*	*
	Slush-casting				*	*
	Gothias process				*	*
	Gravity die-casting			*	*	*
	Squeeze- casting			*	*	*
	Low pressure die-casting			*	*	*
	High pressure die-casting				*	*
	Pore-free casting				*	*
	Thixocasting				*	*
	Plaster mould casting		*			
	Ceramic mould casting		*	*	*	*
	Investment casting (lost wax process)		*	*	*	*
	CLA process				*	*
	Centrifugal casting			*	*	*
	Semi centrifugal casting			*	*	*
	Centrifugally aided casting			*	*	*
Powder metallurgy	Cold die-pressing and sintering					*
	Cold isostatic pressing and sintering	*	*	*	*	*
	Hot die-pressing					*
	Hot isostatic pressing	*	*	*	*	
	Powder forging			*	*	*
Deposition	Electro forming	*	*	*	*	*
	Photo fabrication	*	*	*	*	*
	Vapour deposition	Special parts one, foil 10^4 m^2				
Hot working	Open die forging	*	*			
	Drop (closed die) forging			*	*	*
	Close to form forging			*	*	
	Impact machining				*	
	Draftless forging			*	*	
	Flashless forging			*	*	
	Split die forging			*	*	

Table 1.6.3

Table 1.6.3 Choice of process for a part depending on quantity required – *continued*

Process		***No. required***				
		1–10	***10–10^2***	***10^2–10^3***	***10^3–10^4***	***$>10^4$***
Hot working *continued*	Reverse flow forging		*	*	*	
	Reverse flow split die forging			*	*	
	Press forging			*	*	
	Ring rolled forging		*	*		
	High energy rate forging			*	*	
	Transverse roll forging				*	*
	Orbital forging			*	*	*
	Isothermal forging			*	*	
	Automatic hot forging					*
	Extrusion	Above 200 metre length				
	Hot rolling	Above 20 metre				
	Tube forming	Above 500 metre				
Cold working	Open die forging				*	*
	Cold heading				*	*
	Cold heading (dynaflow)					*
	Cold forging (extrusion)					*
	Impact extrusion					*
	Orbital forging			*	*	*
	Drawing	Above 20 metre length				
	Rotary swaging	*	*	*	*	
	Hydrostatic forming	*	*	*	*	
	Cold section rolling	Above 200 metre				
	Cold profile rolling				*	*
	Thread rolling				*	*
	Flat product rolling					
Sheet metal working	Conventional sheet metal working				*	*
	Rubber press forming		*	*	*	
	Superplastic	*	*			
	Spinning	*	*	*	*	
	Roll forming	3000 m existing rolls, 10 000 m new rolls				
	Fine blanking					*
	Trimming			*	*	
	Vibratory trimming			*	*	
Machining	Traditional tools	*	*	*	*	
	Numerically controlled					*
	Automatic machining				*	*
	Broaching					*
	Spark machining	*	*	*	*	
	Ultrasonic machining	*	*	*	*	
	Electrochemical machining				*	*

Table 1.6.3 – *continued*

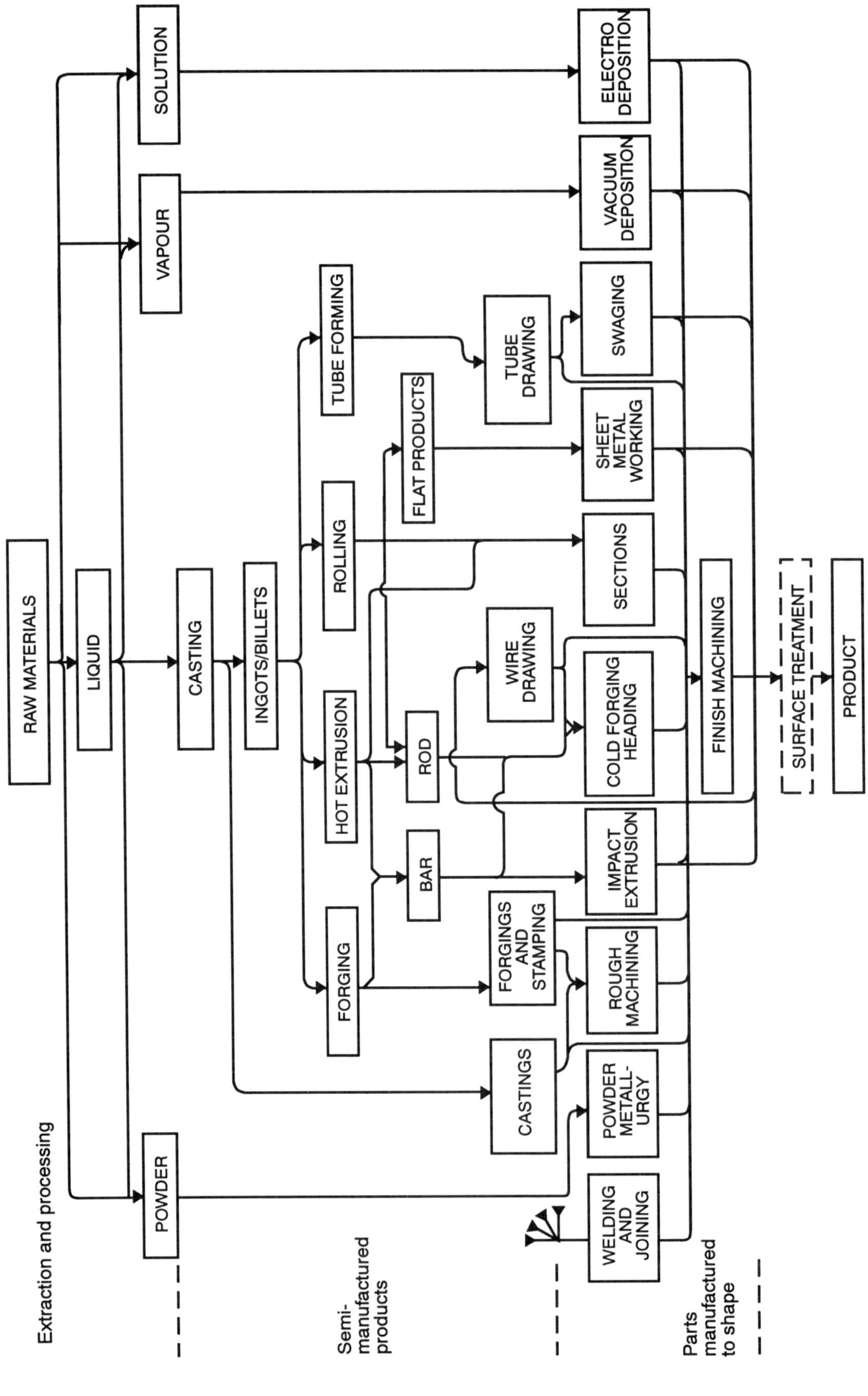

Fig. 1.6.1 Production routes for metal parts

Fig. 1.6.1

1.6.2 Casting

1.6.2.1 Characteristics of casting techniques

Shapes of almost unlimited complexity can be formed by the casting process, which is the only economical route for hollow components having a complex internal shape such as cylinder heads or valve bodies. The internal shapes are relatively easily achieved by the use of cores. Cores are commonly made from sand bonded with a bonding agent which breaks down after casting, allowing the core material to be easily knocked out. Castings can include inserts made from higher melting point metals and thereby produce savings in machining and fitting costs.

Since shapes are formed directly from the liquid phase, casting is especially economical for small quantities of special alloys.

Apart from those components and alloys suited to powder metallurgy, casting is the only economical process for the production of shapes in exotic alloys which are too hard or strong for other forming processes such as rolling, forging or machining, or which may be unweldable.

The mechanical properties of castings are usually more isotropic but somewhat inferior to those of the equivalent wrought product, as shown in:

Table 1.6.4 *Comparison of mechanical properties of cast and wrought alloys*

There are, however, exceptions to this which include pressure die-castings, squeeze castings and so-called 'premium quality aluminium castings' (see Section 1.6.2.2) whose properties approach very closely to those of forgings.

Although wrought components have higher fatigue strengths than castings, fillet welds, which may be used to join two forgings together, have no more than half the fatigue strength of castings.

It is sometimes possible by undirectional solidification to optimize certain desired properties in that direction. Well known instances are the magnetic properties of permanent magnets, and the strength and service life of gas turbine blades.

Certain desirable properties which are not obtainable in wrought materials are readily obtainable in cast materials. Instances are:

a. High temperature creep strength in nickel and cobalt base alloys.
b. Wear resistance of cutting and forming tools, and other components cast in tool steel. This is because of the very hard carbide network which is present in these materials. Similarly, wear resistance can be built in to specific areas of a grey iron casting by providing chills in the mould at that location.
c. The good bearing qualities of lead-containing copper alloys which cannot be forged because of the presence of liquid lead in the structure but are readily cast.
d. Cast iron is an engineering material of the highest importance, combining the desirable properties of cheapness, good castability, good machinability, excellent bearing properties, good damping, i.e. high internal friction, reasonable strength, and, in the case of spheroidal graphite and malleable irons, good ductility. Other specialized properties such as high hardness or corrosion resistance can be obtained in certain grades.

In conclusion, for the vast majority of service conditions with the exception of very highly stressed or shock loaded components good quality castings can fulfil most functional requirements as well as their forged counterparts. They may be superior for certain specific requirements.

The principal casting techniques together with their capabilities are outlined in:

Table 1.6.5 *Summary of the principal metal casting techniques.*

and rough rankings of their performance against a number of criteria are presented in:

Table 1.6.6 *Performance rankings of the principal casting techniques*

The most favourable technique by each criterion is listed at the top of the table, the least favourable at the bottom. The table is of a general nature only and is included to provide an initial guide to process selection. There will be exceptions to this order of listing where specialized techniques are being used. For example, high pressure machine moulded green sand (DISAMATIC Type) provides accuracy in the top 3 or 4 ranks whereas hand rammed green sand moulds invariably result in lowest accuracy. The rankings are also strongly dependent on the type of alloy being cast, e.g. ferrous, copper base, light alloy, etc. Since there is rarely a single process which is best for any given component the design engineer is strongly recommended to consult foundries at an early stage.

Estimates of the quantifiable characteristics of the different processes are presented for further preliminary guidance in:

Table 1.6.7 *Estimates of the quantifiable characteristics of casting processes.*

Three methods of casting aluminium alloy LM25 cylinder heads (weight 9kg) for a production rate of 800 000 per annum are compared in:

Table 1.6.8 *Comparison of techniques for casting aluminium alloy cylinder heads*

The methods considered are sand casting using a continuous jolt and squeeze moulding line, permanent mould gravity die casting using a rotary multi-station system and low pressure die-casting.

The low pressure system may be superior for relatively simple castings with constant section inherently free from cold shuts, oxide inclusions and shrinkage. The carousel system of gravity die-casting may be preferred for large and delicate castings such as cylinder heads, intake manifolds and brake calipers, but careful design of gating and care in pouring are essential.

Factors which will influence the choice between gravity die-casting and high pressure die-casting of aluminium in the weight range 3–15 kg are assessed in:

Table 1.6.9 *A comparison between methods of die-casting aluminium alloys.*

Table 1.6.4 Comparison of mechanical properties of cast and wrought alloys

Material type	*Nominal composition (wt %)*	*Condition C = cast W = wrought ST = solution treated*	*Ultimate tensile strength (MPa)*	*Elongation (%)*	*Impact strength I = Izod C = Charpy (J)*
Aluminium alloys					
Al	Al 99.0	C W	77 85	30/40 45	I 19 I 27
Lo-Ex Alloy	Si 12.0, Cu 1.0 Mg 1.0, Ni 1.0	C ST and aged W ST and aged	200/290 320	1 5	I 1 –
Al–Cu–Mg–Ag–Ti	Cu 5.0, Mg 0.5 Ag 0.75, Ti 0.45	C (Premium quality)	470	4.5	–
Y Alloy	Cu 4.0, Mg 1.5 Ni 2.0	C sand[a] C gravity die[a] W	220 275 400	0.5 1.0 15	I 1 I 11
Zinc alloys					
Zn–Al–Cu	Al 0.8, Cu 0.4	C sand C gravity die C pressure die W annealed	100/137 137/196 196/255 216/293	0.5/1.5 1/3 2.5/4.5 20/60	I 1/2 I 1/3 I 3/9 I 41
Zn–Al–Cu–Mg	Al 4.0, Cu 0.6/1.0 Mg 0.02/0.05	C sand C gravity die C pressure die W extruded	176/235 196/246 330 363/400	0.5/1.5 1.0/2.5 7/12 8/12	I 1.7/4 I 5/8 I 50 I 41
Nickel alloys					
Ni	Ni 99.4 Si 1.5	C sand cast W	348 479	25 40	– –
Hastelloy	Mo 28.0, Fe 5.0 Mn, Si	C W	540 934	8 42	I 15/22 I 92/106
Steels					
Carbon	C 0.2 C 0.3	C W as rolled C annealed W hardened + tempered	371/417 386 556 541/695	16/23 25 27 20/22	C 2/6 – C 8 I 27/54
1% Chrome	C 0.4, Mn 0.7 Cr 1.0	C hardened + tempered W hardened + tempered	834 695/849	16 18/22	I 39 I 95/122
Mn–Mo	C 0.3, Mn 1.3 Mo 0.3	C as cast C hardened + tempered W hardened + tempered	664 927 618	6 18 22	C 22 C 49 I 54
Austenitic stainless	Cr 18.0, Ni 8.0	C softened W softened W normalized	494 541 618	61 30 50	I 115 I 68 I 149
Titanium alloys					
Ti	Commercially pure	C as cast W	448/552 369/708	17.7/10 22/35	– –
Ti–Al–V	Al6, V4, Ti	C as cast W annealed	896/1037 924/1016	6/10 12	– –

[a] ST and aged

Table 1.6.5 Summary of the principal metal casting techniques

Process	***Description***	***Capabilities***
Green sand mould casting (green = moist)	Mould formed from moist clay bonded sand around re-usable patterns. Moisture remains until metal poured. Wooden patterns used for short runs, but metal patterns required for long runs.	< 25 kg, but 1000 kg possible. Mainly cast iron, but also steel and Al alloys. Consistent tolerances, especially with high density moulds on high pressure green sand moulding machines (e.g. Disamatic process capable of 360 moulds/hour). Machining needed for dimensional accuracy.
Cold set mould casting	Mould formed from sand/ binder mixture. Binder activated by agent, added at mixing stage for large or long life moulds; Gaseous agent injected in the rapid production of small moulds (cold box processes)	Castings up to 200 tonnes, and rapid production of small castings. All castable alloy systems. Setting of sand/binder mixture in contact with tooling gives accurate shape, but machining required. Size and finish capability similar to green sand.
Shell casting	Metal patterns at 250°C are covered with sand containing resin which cures to form a rigid shell. Moulds are formed from matching shells which may be backed with steel shot.	Size range up to 50 kg for steel, but most alloys can be shell cast. Surface quality and reproduction of detail good. Steel shot gives high chilling power and good control over structure. Moulds may be stored for long periods before use. Metal patterns required, and 1000 is minimum economic run.
Cosworth process for premium quality aluminium castings	Moulds made from resin bonded zircon sand. Moulds filled slowly from below using an electromagnetic pump which also maintains metal head for feeding during solidification. (See Fig. 1.6.2)	Slow filling avoids entrainment of oxide particles; zircon sand gives high dimensional stability, and high chilling power controls porosity and structure with close dendrite arm spacing. High integrity castings are produced e.g. ± 0.25 mm tolerances on a 800 mm length.
The V- (Vacuum sealed moulding) process	Mould produced from unbonded sand between refractory coated plastic film. Vacuum applied to mould bonds sand and maintains rigidity during casting	Good surface finish and dimensional tolerances. Plastic film forming limitations restrict process to plate like castings, but steel valves have been made. Ferrous castings 50 to 2500 kg in weight, and other alloy systems.

Table 1.6.5

Table 1.6.5 Summary of the principal metal casting techniques – *continued*

Process	***Description***	***Capabilities***
Investment casting	Wax patterns produced from a metal die and mounted on a 'tree' with runners and feeders. Ceramic shell formed around tree by dipping and stuccoing. Wax pattern melted out and shell fired. Metal poured and solidified in shell, often in a vacuum furnace with moving directional temperature gradient. Grain selector spiral can be included in tree to produce single crystal castings	No constraint on shapes that can be cast, and fine detail reproduced both externally and in fine internal channels. Machining not required. Virtually no limit on alloy systems that can be cast. Use of robots in dipping has raised upper weight limit for viable applications. See Fig. 1.6.3 for a cost comparison between investment casting and machining bar stock.
EPC (evaporative pattern casting) process	Mould formed from dry unbounded sand around plastic foam patterns. Vibration causes sand to fill hollow spaces (cores not needed) and form a rigid structure. Patterns decompose on contact with molten metal. CTI steel cast process sucks gaseous decomposition products from mould, avoiding sand fluidization.	Size capability similar to green sand, but greater shape freedom; undercuts and non-tapered sections possible. Used for inlet manifolds aluminium alloy cylinder heads and blocks, and cast iron electric motor cases. Long runs (>5000) required to justify tooling cost for pattern making. 40% cost savings over green sand reported. Coarse bead size in plastic foam limits surface quality.
Replicast CS (ceramic shell) process	Uses fine bead plastic foam to improve surface quality. Investment type ceramic shells protect fragile patterns Carbon pick-up during pouring avoided by decomposition of patterns before metal is poured.	Steel castings 25–100 kg in weight but other alloys also. Precision comparable to investment casting, but thinner shells used with less shell processing required.

Table 1.6.5 – *continued*

Table 1.6.5 Summary of the principal metal casting techniques – *continued*

Process	***Description***	***Capabilities***
Gravity die-casting	Molten metal flows under gravity into a metal mould – often hand poured.	Used for aluminium and other low melting point alloys. Low capital cost, but long cycle times. Economical for 7–15 kg components. but of limited accuracy. See Tables 1.6.8 and 1.6.9 for further characteristics.
Squeeze casting	Metal flows into a two-part metal mould and high pressure is then applied until the metal is solidified.	Similar to gravity casting, but fine surface detail and high internal integrity can be achieved. See Fig. 1.6.4 for improvements in fatigue properties.
Pressure die-casting	Molten metal is forced under pressure into two-part metal die, closed and opened mechanically	Used for aluminium and other low melting point alloys. Large capital cost but short cycle times achievable. Economical for components 3–7 kg, but competes with gravity die casting up to 11 kg due to greater precision and lower machining costs. See Tables 1.6.8. and 1.6.9 for further characteristics.
Semi-solid metal casting or Thixo-forming	Semi-solid alloy is stirred to break-up dendrite structure and produce thixotropic feedstock billets for pressure die casting. Billets reheated into pasty region (60% solid) prior to injection.	60% solidified state of injected alloy reduces die wear and cycle times over normal pressure die-casting. Lower porosity castings – flow entrainment of air avoided – less shrinkage. Long runs (>5000) required for economic viability. Commercial for aluminium alloys and pilot plants cast tool steels.

Table 1.6.5 – *continued*

Table 1.6.6 Performance ranking of the principal casting techniques[a]

Dimensional accuracy		*Surface finish*	*Size limitations*		*Soundness*	*Alloy versatility*	*Cost*	
Small<25 mm	***Large > 300 mm***		***Small***	***Large***			***Per casting***	***Cost of tools***
Investment 1	High pressure die 1=	High pressure die 1=	High pressure die 1=	Replicast CS 1=	Low pressure die 1=	Green sand 1=	High pressure die 1=	Green sand 1=
High pressure die 2=	Semi-solid metal casting 1=	Semi-solid metal casting 1=	Semi-solid metal casting 1=	Green sand 1=	Semi-solid metal casting 1=	Cold set sand 1=	Semi-solid metal casting 1=	Cold set sand 1=
Semi-solid metal casting 2=	Investment 3=	Plaster 3	Investment 3	Cold set sand 1=	Shell 3=	The EPC process 1=	Low pressure die 3	The V-process 1=
Plaster 4=	Plaster 4=	Investment 4	Plaster 4	The V-process 1=	The V-process 3=	The V-process 1=	Gravity die 4=	Plaster (wooden pattern) 4
Replicast CS 4=	Replicast CS 4=	Shell 5=	Green sand 5=	The EPC process 1=	The EPC process 3=	Shell 5	The EPC process 4=	Gravity die 5=
Low pressure die 6=	Shell 6	Replicast CS 5=	Replicast CS 5=	Low pressure die 6	Replicast CS 3=	Investment 6=	Green sand 6=	The EPC process 5=
Green sand 6=	Low pressure die 7	Low pressure die 7	Cold set sand 5=	Shell 7	Green sand 3=	Replicast CS 6=	Cold set sand 6=	Shell 7=
Cold set sand 6=	Gravity die 8=	Gravity die 8=	Shell 8	Gravity die 8	Cold set sand 3=	Gravity die 8	The V-process 6=	Plaster (metal) Replicast CS 7=
The V-process 6=	The V-process 8=	The V-process 8=	Gravity die 9=	Investment 9	Investment 9	High pressure die 9	Shell 9=	Investment 9
The EPC process 6=	The EPC process 8=	The EPC process 8=	The EPC process 9=	High pressure die 10	Gravity die 10	Plaster 10	Replicast CS 9=	Low pressure die 10
Gravity die 6=	Green sand 11=	Green sand 11=	Low pressure die 11	Semi-solid metal casting 11	Plaster 11	Low pressure die 11	Investment 11	High pressure die 11
Shell 7	Cold set sand 11=	Cold set sand 11=	The V-process 12	Plaster 12	High pressure die 12	Semi-solid metal casting 12	Plaster 12	Semi-solid metal casting 12

[a] Order of merit: most favourable top, least favourable bottom.

Table 1.6.6

Table 1.6.7

Table 1.6.7 Estimates of the quantifiable characteristics of casting processes

Casting process	*Alloy range*	*Weight range kg (lb)*	*Economical quantity*	*Thin section capability mm (in)*	*Cast hole dimensions mm (in)*	*Surface finish μm (μin)*	*Dimensional accuracy mm (in)*
Green sand	Most	0.1 to 100 (0.2 to 220)	1 to 100 000	>5 (0.2)	>6 (0.24)	6.3 to 25 (250 to 1000)	—
CO_2/silicate	Most	1 to 100 000 (2.2 to 220 000)	1 to 1000	>5 (0.2)	>6 (0.24)	6.3 to 25 (250 to 1000)	—
Cold set	Most	1 to 100 000 (2.2 to 220 000)	1 to 1000	>5 (0.2)	>6 (0.24)	6.3 to 25 (250 to 1000)	—
Cosworth	Aluminium alloys	0.2 to 50 (0.44 to 110)	1000+	>4 (0.16)	>6 (0.24)	3.2 to 6.3 (125 to 250)	±0.15 up to 100 (±0.006 up to 4)
Shell Mould	Most	0.1 to 50 (0.2 to 110)	1000+	>2 (0.08)	>3 (0.12)	3.2 to 12.5 (125 to 500)	±0.50 up to 100 (±0.02 up to 4)
Evaporative pattern casting	Most	0.5 to 5 000 (1 to 11 000)	5000+	>3 (0.12)	>3.5 (0.14)	6.3 to 12.5 (250 to 500)	—
Vacuum-sealed	Most	100 to 2 500 (220 to 5 500)	50+	>3 (0.12)	—	3.2 to 12.5 (125 to 500)	—
Plaster mould	Aluminium alloys	0.05 to 50 (0.1 to 110)	50+	>1.5 (0.06)	>6 (0.24)	0.8 to 3.2 (32 to 125)	±0.25 up to 100 (±0.01 up to 4)
Ceramic mould	Most	0.1 to 5 000 (0.2 to 11 000)	50+	>1.5 (0.06)	—	0.8 to 3.2 (32 to 125)	±0.38 up to 100 (±0.015 up to 4)
Investment casting	Most	0.05 to 250 (0.1 to 550)	1000+	>1.5 (0.06)	>1 (0.04)	0.8 to 3.2 (32 to 125)	±0.50 up to 100 (±0.02 up to 4)
CLA/CLV	Most	0.05 to 50 (0.1 to 110)	1000+	>0.75 (0.03)	—	1.6 to 3.2 (63 to 125)	±0.125 up to 100 (±0.005 up to 4)
Replicast (CS)	Most	1 to 100 (2 to 220)	50+	—	>9.5 (0.37)	1.6 to 3.2 (63 to 125)	±0.25 up to 100 (±0.01 up to 4)
Gravity die	Principally non-Ferrous + Cast Iron	0.1 to 70 (Al) (0.2 to 155)	1000+	>5 (0.2)	>6.5 (0.26)	3.2 to 6.3 (125 to 250)	±0.25 up to 100 (±0.01 up to 4)
Low pressure die	Aluminium alloys	5 to 25 (11 to 55)	1000+	>4 (0.16)	>6 (0.24)	1.6 to 6.3 (63 to 250)	±0.55 up to 100 (± 0.022 up to 4)
High pressure die	Principally Aluminium and Zinc alloys	0.05 to 25 (0.1 to 55)	5000+	>0.5 (0.02)	>1 (0.04)	0.4 to 3.2 (16 to 125)	±0.15 up to 100 (± 0.006 up to 4)
Squeeze casting	Principally Aluminium alloy	0.1 to 20 (0.2 to 45)	1000+	>6 (0.24)	–	0.4 to 3.2 (16 to 125)	±0.20 up to 100 (±0.008 up to 4)

Reproduced with permission from, *Precision Casting Processes* (1991) edited by D. Enright, Butterworth-Heinemann, Oxford, UK.

Table 1.6.8 Comparison of techniques for casting aluminium alloy cylinder heads

Feature	***Casting process***		
	Sand-casting	***Gravity die***	***Low pressure die***
Productivity	150 moulds/hour produced by single system for entire production requirements.	5 moulds/carousel produce 52 heads/hour. Several carousels required.	11 cylinder heads/hour/ machine. Many machines required.
Machine breakdown	Total production standstill.	Some loss of production.	Little production loss from closure of one machine.
Production flexibility	Most flexible		Least flexible.
Complexity and reproducibility	Core setting not mechanized.	Less complex than low pressure, with high reproducibility.	Most complex and more likely to give production trouble.
Comparative casting quality	6.25 mm Inferior surface finish. Wide dimensional tolerances.	0.75 mm	0.75 mm Less liability to oxide inclusions and cold shuts.
Mechanical properties 0.2% ps, MPa UTS, MPa Ductility, % RA	 25 30 4.5	 27 34 10	 27 34 10
Cost (relative to sand-casting)	100%	75%	75%

Table 1.6.9 A comparison between methods of die-casting aluminium alloys

Feature	***Casting process***	
	Gravity die-casting	***High pressure die-casting***
Casting design	Wider choice of alloy. Greater freedom in core design. Can consider heat treatment. Economy of volume production smaller. Facility to thicken wall section.	Greater definition of small details. Use of retractable steel cores. Thin-walled sections available. Cannot use complex sand cores. Excellent surface finish. Limited wall thickness 10 mm.
Tooling	Dies cheaper and more rapidly changed.	Die changing can take up to a day. Expensive standby equipment may be required to maintain production.
Alloy specifications	Much larger range available. Greatest tonnage cast in LM4, LM27. Higher strengths obtained by thickening walls and heat treatment.	Largest tonnage cast in BS1490, LM2, or LM24 with some LM6 or LM20. Production for heat treatment beyond scope of most foundries.
Production	Manual pouring more common. Production rates and casting quality can be dependent upon operator. More machining may be required to remove the larger flashes and gates. If large volume production allows for machining to final shape of castings on a transfer line, the extra machining might be performed at very low costs.	Metal injection automatically controlled. Process more prone to production loss due to equipment breakdown. None or only nominal machining required to final shape. However fine die details are prone to wear and die replacement may outweigh savings in machining costs.
Size	Very competitive for castings heavier than 12 kg.	Very competitive for castings in the 3–7 kg range.

Table 1.6.9

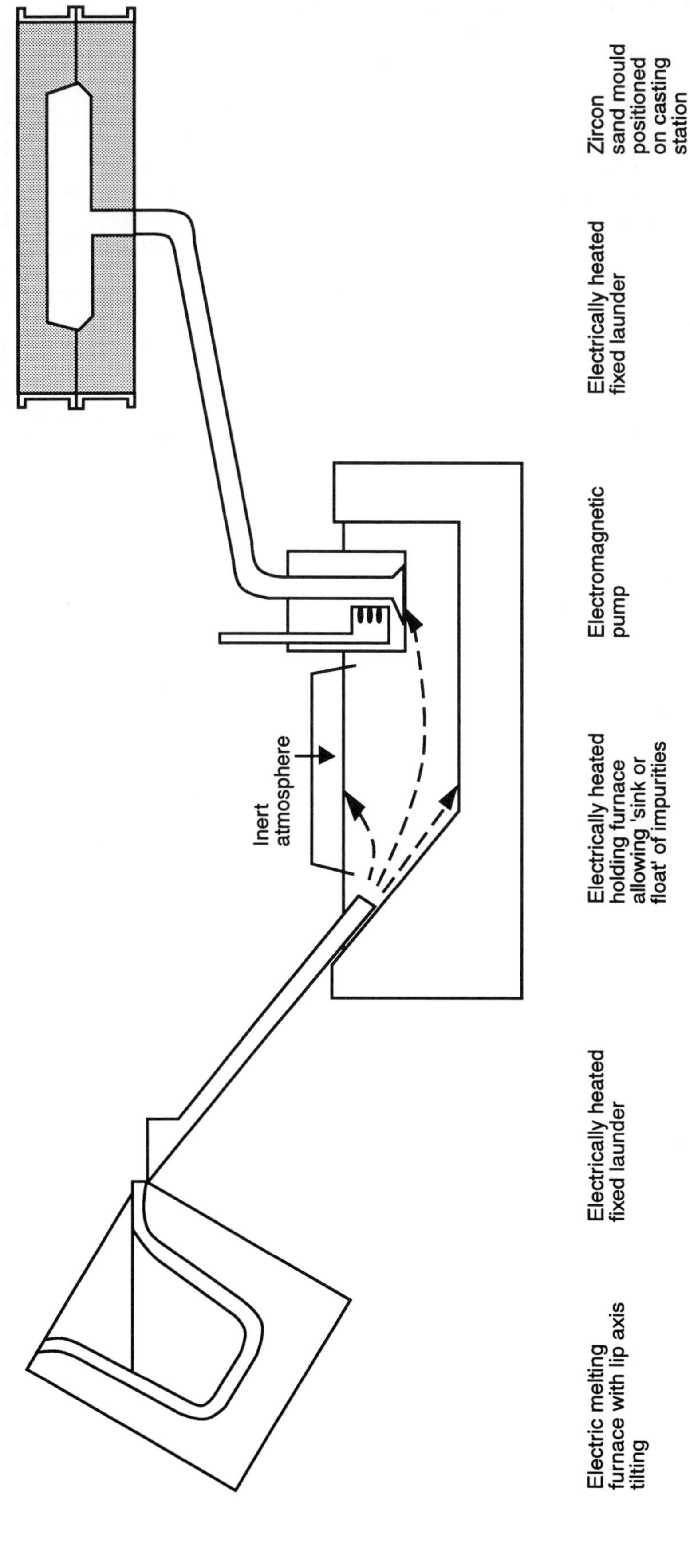

Fig. 1.6.2 Method of pouring employed for 'high precision low pressure sand-casting'

Fig. 1.6.2

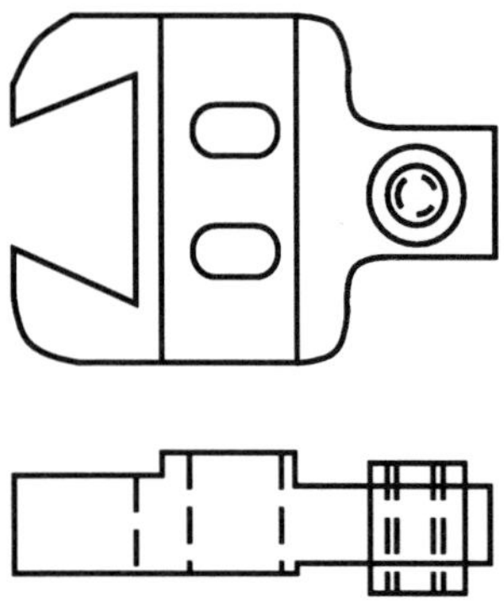

Material – Stainless steel

Tolerances – unless stated | IT 10-11

Scale – 4 : 1

COST COMPARISON

Manufacture by machining from bar	**Relative cost**
Bar stock 45 x 400mm	0.06
Machine to 38 x 25mm	0.02
Mill side relief	0.065
Mill 3 mm radii	0.025
Mill 6 mm radii	0.025
Mill rectangular	0.025
Cut dovetail	0.14
Part off	0.085
Turn 6 mm dia. boss, drill and tap M 3.5 hole	0.16
Turn boss	0.11
Drill four 3 mm holes	0.125
Mill two 3 mm slots	0.30
Total cost/piece	1.495
Tooling cost	40
Total cost/1000	1535

Manufacture by investment casting	**Relative cost**
Investment casting	0.29
Grinding 8 mm thickness	0.06
Tapping M3.5 hole	0.03
Total cost/piece	0.38
Tooling cost	181
Total cost/1000	561

Data by courtesy of B.I.C.T.A.

Fig. 1.6.3 Cost comparison between investment casting and machining bar stock

Fig. 1.6.3

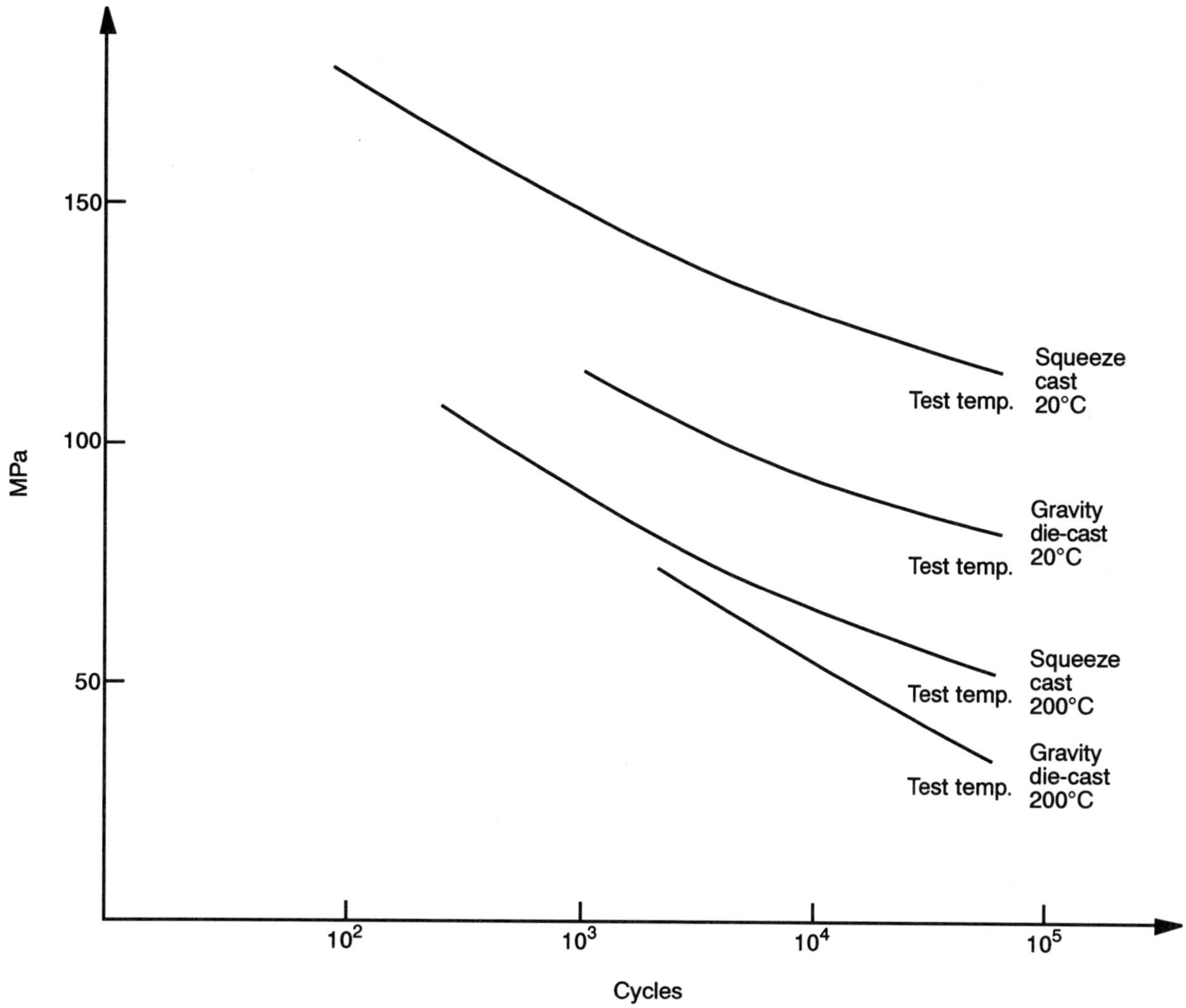

High speed, rotating beam, cantilever bending tests to failure or 50,000 cycles.

Fig. 1.6.4 S–N curves for squeeze cast and gravity die-cast LM26 aluminium diesel engine pistons

Fig. 1.6.4

1.6.3 Powder metallurgy

1.6.3.1 Characteristics of powder processing

The general advantages and disadvantages arising from the characteristics of the powder processing route for manufacture of components are outlined below. Some of the characteristics listed are specific to the more commonly employed cold die pressing and sintering process. They do not necessarily apply to those processes in which compaction and sintering proceed simultaneously, or to isostatic pressing techniques. For example, effectively 100% density, and properties the equivalent of those produced by casting or forging, can be achieved in one operation by powder metal forging or hot pressing. Hot isostatic pressing can be utilized economically for 'one-off' or prototype production.

ADVANTAGES

Control

Powder processing offers precise control of materials and their properties. This permits a wide variation in physical and mechanical properties while assuring performance characteristics of consistent uniformity. Impurities, gas pockets, and similar faults common to other processes, are eliminated.

Versatility

Practically any desired metal, alloy, or mixture of metals – including combinations not available in wrought or cast forms – can be produced. Copper, nickel, brass, bronze, iron, low and medium carbon steels, alloy steels, stainless steels, precious metals, refractory and aerospace metals and carbides and other materials too hard or too brittle to be shaped in any other way are produced by this method. When desired a single part can be made hard and dense in one area, while soft and porous in another. Unique material combinations such as heavy duty metallic friction materials made from copper, tin, iron, lead, graphite and silica, graphite–bronze slip ring segments and copper–carbon brushes can only be produced by powder metallurgy.

Powder metallurgy parts can be produced in a wide range of shapes with irregularly shaped holes, eccentrics, flats, splines, counterbores, and involute gears. Two or more parts can be combined into a single unit, thus eliminating assembly costs and simplifying product design. Keys, keyways and other fastening devices can be made integral with the part, or components can be fabricated in sections and joined by press-fitting, brazing, or resistance welding.

Self-lubrication

The capability of self-lubrication is unique to powder metallurgy parts. The controlled network of small pores can be filled with oil or other lubricants selected for contact with any variety of wearing surfaces. Powder metallurgy bearings hold from 10 to 40% of oil by volume and supply additional lubricant to the bearing surface as heat expands the oil. The oil is reabsorbed on cooling ready again for use when needed. This feature can result in substantial savings by eliminating the need for a costly lubrication system which may not function as satisfactorily.

Economy

Powder metallurgy can be an economical process for large production runs because of the rapid mass-production techniques employed, the reduction or elimination of subsequent machining or finishing, and the reduction or elimination of material and scrap losses. Substantial energy savings (approximately 50%) may also be achieved. An improved product can sometimes be achieved at a lower cost by replacing a mechanically assembled component with a compact which after sintering is physically bonded.

Filtration

Controllable porosity in powder metallurgy parts can be employed for filtration. Filters can separate or pass materials selectively, diffuse the flow of gases or liquids, regulate flow or pressure drop in supply lines, or act as flame arrestors by cooling gases below combustion temperatures. Filters of almost any configuration can be produced.

Controlled density.

The precise control of weight achievable in a component made by powder metallurgy is invaluable for many applications such as, for example, counterbalance weights.

Low porosity.

The mechanical and physical properties, impact strength, tensile strength, yield strength and elongation improve with decreasing porosity. A number of techniques are available for obtaining density within 95% of solid, including double pressing/double sintering; infiltration (see Section 1.6.3.3) promoting high shrinkage rates during sintering by stabilizing the ferrite phase in steel powders at the sintering temperature; warm compaction, in which a special lubricant is used and compaction is carried out at 150°C, followed by conventional sintering in a mesh belt furnace. With subsequent heat treatment, powder metallurgy parts can have tensile strengths of up to 1400 MPa.

The mechanical and physical properties of fully dense powder forged parts are at least equal to those made of comparable wrought alloys. This has been made possible through the development of high purity low alloy powders and improvements in powder forging techniques.

Precise tolerances

Close tolerances and smooth finishes are consistently maintained thus eliminating a considerable amount of the machining required for some other manufacturing methods. For example, a bore can be moulded to a diametric tolerance of IT 6 eliminating machining completely.

Damping

The self-damping nature of powder metallurgy parts minimizes vibration, reduces tool wear, permits quieter operation and smoother action and helps to maintain closer tolerances.

Wear

Less wear can result from the adoption of powder metallurgy because material structures can be produced which have self-lubricating or wear resistant properties. Added wear resistance can be conferred by a steam treatment which oxidizes the surface of the pores.

LIMITATIONS

Overall size

There is a limit to the size to which a component can be produced economically. At present, parts weighing 4 kg and up to 0.02 m^2 in area and 0.1 m in length, are in regular production, by die pressing and sintering, and larger sizes are practicable for specific applications.

Properties

The density produced by a single pressing and sintering operation is usually limited to 90% and the ductility of the material is usually lower than that produced by conventional manufacturing processes. Dimensional tolerances (0.01 mm per mm on a diameter and ±0.04 mm per mm on length) are inferior to those which can be obtained by machining and grinding. These properties can be improved by subsequent processing but at a substantial cost penalty.

Economics

Initial tool costs are high so that production volumes of less than 10 000 are normally not practicable by die processes. However case histories have shown lesser quantities to be economical and part applications must be evaluated on an individual rather than a strict quantity basis.

The raw material powder is usually more costly than the solid ingot used for more conventional manufacturing processes. For example, iron and copper powders retail at at least twice the price of solid metal.

Although process scrap is easy to recycle at the green stage, sintered scrap is more difficult and expensive to recycle than that produced by the casting process.

Purity

Although a controlled composition is normally easy to achieve there are cases where the high surface area of powder, which may lead to unacceptable oxide contents, precludes the use of powder metallurgy.

Corrosion

The presence of porosity, which may be interconnected, may intensify corrosion and require special attention and precautions to overcome it.

Design limitations

There are also shape limitations to die pressings. Weak thin sections, feather edges, narrow and deep splines should be avoided and internal angles must be provided with lands. Chamfers below 45° should be avoided. Undercuts cannot be moulded, but must be machined subsequent to sintering.

Powder processing compared to other forming methods

The characteristics of component manufacture by cold pressing and sintering, and powder forging are compared with other production routes in:

Table 1.6.10 *Comparison of cold pressing and sintering, and powder forging with other forming methods*

1.6.3.2 Principal powder metallurgy processes

The wide variety of powder metallurgy processes available for component manufacture is illustrated in:

Fig. 1.6.5 *Flow sheet of powder metallurgy processes*

The three basic steps of powder metallurgy processing are blending, compacting and sintering.

Blending consists of weighing and mixing the metal powder with lubricants and usually also with alloying elements. Compacting generally comprises filling a die with a metered quantity of blended mixture and compressing it between punches, at pressures varying from 300 to 800 MPa. Alternatively the powder, normally without a lubricant, may be loaded into a shaped flexible polymer container which, after sealing, is isostatically pressed. The pressure used and the production rate achieved vary with size, shape and specified density. The 'briquette' produced by either route can be handled but has low unsintered or 'green' strength, and must be sintered to achieve maximum properties.

Sintering bonds the powder particles together to produce parts with the desired physical and mechanical characteristics. After sintering, most parts are ready for use. However, further treatment may be performed to obtain very close tolerances, increased density, or other physical or surface characteristics. Details of these further treatments are covered in Section 1.6.3.3.

Hot pressing combines compacting and sintering into one operation and may eliminate the requirement for a lubricant. The properties produced may be superior but the throughput is lower than in cold pressing and sintering.

Hot isostatic pressing compresses a powder contained in a preshaped collapsible container. Under isostatic gas pressure sintering may take place at reduced temperatures and a wide range of shapes can be produced of almost theoretical density without the need for a costly die.

Powder forging is an extension of the powder metallurgy processes described. The green briquette may be a simply shaped preform; or its geometry may be similar to the shape of the desired forging. In either case the preform is sintered or heated immediately before being forged in a heated die. To avoid undesirable oxidation, the preform may be coated before heating or may be heated in a protective atmosphere.

More recently developed alternative compacting and sintering procedures include injection moulding, vacuum sintering and spray compaction.

In injection moulding the blended powder, containing an equal volume of a plasticizer/binder mixture, is formed into larger than final dimension parts using a standard injection moulding machine. The plasticizer/binder mixture is then removed in a debinding step – thermal decomposition frequently carried out in the sintering furnace and sometimes assisted by a prior solvent extraction step. The sintering operation is carried out with close control of temperature profile and other processing parameters. Almost isotropic shrinkage of the order or 20–25%, transforms the preform to its net shape with a density of 94–98% of theoretical. Components with intricate configurations such as discontinuous inside threads which currently require secondary operations, can be produced in one operation.

Vacuum sintering is generally applied to alloyed and tool steel powders. Steel powders are pressed into preforms 85% dense and sintered in a vacuum where they shrink about 10% to 98–99% theoretical density. For tool steels more uniform grain and carbide distribution is produced than by conventional forging processes. Tolerances in less complex mechanically pressed parts are ±0.005 mm/mm (IT 10). For highly complex parts, isostatic pressing is used resulting in more liberal tolerances. Parts up to 50 mm thick and having face areas of 100 cm^2 are manufactured by this process.

Whereas injection moulding and vacuum sintering obviate the need for hot working preforms to full density, the Osprey Process of spray compaction eliminates sintering, another costly operation, while facilitating hot working.

Molten metal is atomized and sprayed into a collecting mould to build up a dense (98–99%) preform which can be immediately forged. Accurate weight control appears difficult and it would not, therefore, seem possible to produce flashless high precision forgings by working preforms to full density. However the process is capable of producing very homogeneous large preforms (120 kg) which can be used for conventional forging and extrusion. The process appears to be particularly suitable for producing homogeneous highly workable preforms of high speed steel, nickel alloys, etc. The macrosegregation that can be present in conventionally cast ingots of these alloys is absent in Osprey preforms.

Finally more conventional fabrication processes, for example rolling and extrusion, can be applied to compacted or canned powder to produce finished components or semis which have properties at least the equal of wrought products.

1.6.3.3 Cold die-pressing and sintering

DESIGN

When possible, parts should be designed specifically for fabrication by powder metallurgy, rather than be adaptations of designs originally intended for production by other methods. For advice and assistance the services of an experienced manufacturer of custom P/M parts should be sought.

Adherence to the six basic rules for designing P/M parts listed below will generally result in preferred designs, simplified tooling and production, lower costs, and longer part life.

Rule 1: The shape of the part must permit ejection from the die.

Undercuts and holes at right angles to the direction of pressing, reverse tapers, reentrant angles, threads, and diamond knurls cannot be moulded but are best incorporated by machining after sintering. Zero draft is preferred.

Rule 2: The shape of the part should be such that the powder is not required to flow into thin walls, narrow splines, or sharp corners.

Minor changes in design will usually satisfy the above rule and result in better parts at lower cost. Flanges, bosses and holes are allowed if small compared with overall size and have large draft angles.

Rule 3: The shape of the part should permit the construction of strong tooling.

Considerable economy in both tooling and production, as well as improved parts, can often be effected by simplifying shapes. For example, the design should allow a reasonable clearance between the top and bottom punch faces during the pressing operation. Also, it should be possible to make both the dies and punches without sharp edges. Well rounded grooves are preferable to narrow, deep splines. Generous fillet radii are preferred. Chamfers are preferred to radii on part edges.

Rule 4: The shape of the part should make allowance for the length to which thin-walled parts can be compacted.

The preferred maximum length for thin wall, hollow cylindrical parts is about 2.5 times the diameter of the part. This ratio can be increased to 4 or more times the diameter for parts with thicker walls. Holes should be not less in diameter than 20% of their length with a limit of 1.25 mm. They should not be nearer to the edge of the part than 4 mm. Wall thicknesses should not be less than 0.8 mm for small parts, 1.6 mm for medium length parts.

Rule 5: The part should be designed with as few changes in section thickness as possible.

Uniform density and high strength in a multilevel part can best be maintained when the number of levels does not exceed the number of pressing actions available in the compacting press. When too many levels are attempted, density varies considerably and quality control becomes a problem. Also, unless the diameter differences are large, tool life will be limited. It is better to design for pressing as many levels as is practical to maintain a uniform density, then machine the remaining levels.

Rule 6: Take advantage of the fact that certain forms can be produced by P/M which are impossible, impractical, or uneconomical to obtain by any other method.

For example, true involute gear forms can be readily made by powder metallurgy, but are difficult to make by other methods because of the undercuts at the bases of the teeth. In addition keys and keyways for gears, pulleys, bushings, etc., can be incorporated when pressing the part.

MECHANICAL PROPERTIES OF POWDER METALLURGY COMPONENTS

The mechanical properties of powder metallurgy components increase with increasing density. The relationship between percentage of theoretical density with tensile strength is illustrated in:

Fig. 1.6.6 *Variation of tensile strength with density for carbon steels and iron*

The mechanical properties of the low alloy and stainless steels most commonly used in the pressed and sintered form are given in:

Table 1.6.11 *Mechanical properties of iron and steel P/M parts*

and in:

Table 1.6.12 *Mechanical properties of stainless steel P/M parts*

The mechanical properties of selected heat treatable aluminium and copper alloys are given in:

Table 1.6.13 *Mechanical properties of aluminium and copper P/M alloys*

POST-SINTERING TREATMENTS

Certain end product specifications or applications of cold pressed and sintered components require additional treatment subsequent to sintering.

Infiltration

The process of cold compacting metal powders, followed by sintering of the compacts to produce coherent bodies, leads to materials which contain residual pores which impair mechanical properties not only by the absence of metal but also by the stress raising action.

For some applications the porosity is essential but for structural engineering components, where porosity is not specifically required and where high strength and toughness are important, it is generally desirable to eliminate the residual porosity.

Metal infiltration is one method by which a solid, pore-free metal component may be produced. In this process, a porous metallic body having a substantial degree of interconnected porosity, i.e. the skeleton, is infiltrated with another metal of lower melting point by means of heat treatment above the melting temperature of the infiltrant, but below that of the skeleton. The infiltrant is absorbed as a liquid into the pores of the skeleton by capillary action and produces a component with a composite structure, which may be virtually solid. Such components, accordingly, exhibit physical and mechanical properties which are generally comparable with those of solid metal.

The actual properties will be dependent upon the individual metals which constitute the structure of the infiltrated part, together with the way and the proportions in which they are combined.

Joining

Normal brazing or soldering methods are unsatisfactory because brazing metals, with their low viscosity, have a greater tendency to infiltrate the porous parts than to penetrate and bind the joint. The most successful and practical joining methods are: joining during sintering, infiltration, or projection welding, joining with threaded fasteners, and keys or D-shaped bores and bosses.

Joining during sintering achieves the equivalent of a 'shrink' fit by using two materials with different growth characteristics during sintering. An outer component may be made from a grade of iron powder which has only 0.1% growth during sintering. An inner component may be made from a mixture of iron and copper powders which give 0.6% growth in sintering. The green compacts are produced to give a clearance of 0.05–0.08 mm between the bore and the mandrel. The parts are assembled before sintering and are permanently joined during sintering both by the growth of one part into another and also by the tendency of the copper in one part to diffuse into the other.

By an inversion of the above method, the shrinkage characteristics of a material can be used to close it permanently upon another part having zero shrinkage. In cases where heavy loading will occur or the alignment of teeth between the inner and outer components is important, keys, flats or splines can be formed on the joining faces.

Parts can also be joined during the infiltration process. This is accomplished by assembling the component parts and infiltrating during sintering. Once the porosity within the parts is completely filled with the infiltrant, a bond similar to furnace brazing is achieved.

Sintered parts are reliably joined to sintered parts, or to conventional machined parts by using projection welding. A ring-shaped projection and several small projections, which are formed during moulding, provide the metal necessary for welding. During the welding operation the two parts are kept aligned by an insulated shaft through their respective central holes.

Tumbling

Sharp edges produced at the junction of dies and punches become fine projecting fins or flash as the tools wear. These cannot be removed in the green state and might be forced into the part during sizing. Hand finishing is slow, expensive and variable. Barrel finishing is extensively used for removing flash and radiusing edges and for giving a controlled surface finish to large quantities of small parts.

Whilst chamfering or radiusing may in some cases be incorporated economically into machining operations, the tumbling processes have the additional advantages of increasing surface hardness by up to 12%, and reducing the tendency to corrosion.

Heat treatment

Sintered steel parts respond to heat treatment in the same manner as conventionally produced steel of the same composition. Liquid carburizing or carbonitriding baths are to be avoided because salt absorbed in the pores may cause corrosion. Penetration into the pores may make it difficult to control the depth of the case with gaseous carburizing atmospheres. High hardening and quenching must be done in a protective atmosphere, since even short exposures to air at 850°C will cause excessive oxidation.

Techniques of hardness measurement must make allowances for porosity, particularly with case hardened parts where the indentation may break through the case. For rubbing sliding wear, micro-indentations are to be preferred.

Oil impregnation

Impregnation is best accomplished by means of a vacuum process, but can also be achieved by soaking parts in an oil bath.

Sizing, coining and similar operations

Sizing and coining consist of repressing the part to achieve very close tolerances or increased density or both. Repressing usually improves the surface finish. They both involve a plastic deformation of structural parts or bushings after sintering. The difference between them could be defined as follows:

Sizing is used to obtain high dimensional accuracy, thus compensating for warpage or other dimensional defects occurring in the sintering operation. A slight plastic deformation only is necessary, so the pressures required are normally quite moderate.

Coining has a double purpose. Not only is dimensional accuracy improved as in sizing, but by the use of higher pressures the density of the part is increased. Because the sintering operation has soft annealed the part, considerable plastic flow is possible during coining. The hardness and tensile strength of the part increase so much that a soft, unalloyed, sintered part often gains sufficient strength for use under severe conditions.

In cases where production quantities are small and the shape of the part is simple, sizing can be done in the same press and tools as for compacting. For large quantities it is normally preferred to perform the sizing in special tools. Furthermore it is more economical to use simple sizing presses instead of the much more expensive metal powder compacting presses. Sizing of symmetrical parts in large quantities is often performed in specially designed automatic sizing presses, which operate at very high speeds.

Two processes used to improve the fatigue performance of fully dense materials, shot peening and surface cold rolling, are also applied to sintered compacts. In both cases, they impart local densification and compressive stresses to the surface layers. Surface cold rolling is used together with subsequent case carburizing to improve the fatigue strength and pitting resistance of bearing race and gear parts.

Effects of post-sintering treatments

Whilst post-sintering operations can enhance properties and improve tolerances, as shown in:

Fig. 1.6.7 *Influence of processing on properties and tolerance for P/M components*

they usually incur a cost penalty, as illustrated in:

Fig. 1.6.8 *Trends in cost and quality incurred by the use of common processes for the manufacture of P/M components*

MACHINING

Designing a component for P/M techniques usually reduces machining to such unavoidable operations as undercuts and cross holes. Components are, however, being successfully and economically made where a sintered blank is only 50–90% of the final form and a number of machining operations are necessary to complete the component.

Where it is essential that machined surfaces should retain their porous structure, the cutting tool must be maintained in very sharp condition. A worn cutting tool smears the surface, closing the pores and reducing the capillary action by which oil is brought to the surface of the material.

As the density of sintered materials is increased and the porosity correspondingly reduced, the cutting characteristics change until at about 90% density only isolated pores remain. Above this point the cutting tool passes through virtually solid material which cuts similarly to forged or bar material of the same composition.

Low and medium density P/M components (below 90% density) can be machined without difficulty with sharp tungsten carbide tools using high speeds and fine feeds. High speed steel tools can be used for small quantities or for roughing cuts.

Sintered materials should be machined dry whenever possible, using a low pressure air blast to remove the chips and cool the tool. The use of water or conventional cutting oils should be avoided, or followed by thorough washing and drying operations, because these coolants enter the pores of the material, causing corrosion, and if the part is later impregnated with oil the coolants dilute the oil and spoil its lubricity. Alternatively, a volatile coolant may be used. The components can afterwards be heated to the temperature at which the coolant vaporizes.

All oil-impregnated components should be re-impregnated after machining, to compensate for oil losses due to heat and centrifugal force.

Machining operations which are carried out on P/M components include:

Turning

Shaping and planning

Drilling

Power fed drilling is preferred as it gives uniformity in the rate of feed and so lengthens drill life.

Tapping

Standard taps can be used satisfactorily, but tungsten carbide taps are preferred for large numbers of components. Standard tap-drill sizes give a 75% depth of thread. The tap relief can be increased, if necessary, to nearly double the relief used for tapping conventional ferrous materials.

Milling

Milling cutters should be ground with minimum order radii or, where possible, bevelled corners. Helical teeth provide the best means of chip disposal.

Grinding and lapping

Sintered materials can be ground or lapped, but this operation is not recommended for bearing surfaces as the surface pores tend to become choked. Hard, fine wheels are recommended, running at conventional speeds with light feeds. A light bodied oil should be used to flush away abrasive particles.

CORROSION PROTECTION

Plating

P/M parts are electroplated by standard procedures. However, it is important first to impregnate the pores with a resin or to close them by peening the surface to exclude the plating salts from the pores of the part.

Thermal passivation

This is a technique used for the protection of stainless steel powder metallurgy parts. It consists of heating parts for 30 min in air at 325–500°C during which time a passive film develops. This film improves the resistance of the parts to certain corrosive attacks. Its value lies in passivating the surfaces of the part to the bottom of the pores which improves the resistance of the parts to pitting corrosion.

Steam treatment

This is accomplished by heating parts to 1000°F and subjecting them to superheated steam under pressure.

Cementation

Articles to be coated are heated in a rotating furnace together with powdered coating metal.

Metal spraying

A metal coating is sprayed on the surface to be coated.

Oil impregnation

Powder metallurgy parts, as specified earlier, achieve greater protection against corrosion by being impregnated with oil or other non-metallic materials.

1.6.3.4 Powder metal forging

This process consists of the steps:

1. Preparation of a green preform shape from powder.
2. Heating the pressing in a controlled atmosphere furnace.
3. Forging the hot preform.
4. Quenching or controlled air cooling
5. Machining if necessary.

The process, although having some similarities to powder sintering, is much closer to forging in respect of the properties of the final product and should be regarded as competing with forging rather than sintering. A comparison of flow sheets for powder forging and conventional forging is shown in:

Fig. 1.6.9 *Flow sheets for powder and conventional forging*

Advantages of the process are:

reduced forging steps – often one blow in one die
reduced forging pressure lowers capital cost of forging equipment
reduced forging temperature thus saving fuel
reduced tool costs

reduced skilled layout requirement
better metal yield – improved accuracy thus reducing final machining.

Limitations are the high cost of alloy powder and the difficulty of alloying pure metal powders during sintering. Fully pre-alloyed powders can be used in powder forging, however, since the reduced cold compressability is not a limitation where the preform densities are low.

The powder forging process is usually utilized for carbon, low and medium alloy steels. Because carbon is added to the powder as graphite, any desired content can be produced. Its suitability compared with other forming methods is shown in:

Table 1.6.14 *Comparison of powder forging with other forming methods*

A comparison of the mechanical properties of powder forgings with those for cold pressed and sintered products is illustrated in:

Fig. 1.6.10 *Variation of mechanical properties with porosity for P/M components*

The tensile properties of the steels generally used by the engineering industry are shown in:

Table 1.6.15 *Specification and properties of powder forged steels*

The fatigue behaviour compared with wrought components is shown in:

Fig. 1.6.11 *Fatigue properties of powder forged steel components*

As the fatigue properties of a powder forging lie between the longitudinal and transverse fatigue properties of a conventional forging an actual component operating under complex stresses may well give better performance in the powder forged condition. Examples of this behaviour have been shown for connecting rods and pinion gears.

The fatigue performance of powder forged titanium and its alloys is essentially equivalent to that of wrought products. The fatigue behaviour of a powder forged aluminium alloy is illustrated in:

Fig. 1.6.12 *Fatigue performance (notched specimen) of powder forged aluminium alloy and comparable wrought alloys*

Table 1.6.10 Comparison of cold pressing and sintering, and powder forging with other forming methods

Feature	*Cold pressing and sintering*	*Powder forging*	*Die forging*	*Precision forging*	*Cold forging*	*Precision casting*
Part weight (kg)	0.01–14	0.1–5	0.05–1000	0.05–100	0.01–35	0.1–10 Extreme cases to 35
Height diameter ratio	Height usually lower than width.	Height usually lower than width.	Not limited.	Not limited.	Not limited.	Not limited.
Shape	No large variations in cross-section allowed, openings limited. Thin walls to be avoided.	No large variations in cross-section allowed, openings limited. Thin walls to be avoided.	Any shape, openings limited.	Any shape, openings limited.	Mostly of rotational symmetry.	Any shape, any openings possible.
Material utilization (%)	98	98	50–70 (of bar stock)	85–95 (of bar stock)	95–100 (of bar stock)	65–85
Tolerances[a]	IT 6–8 if sized.	IT 8–10	IT 15–18	IT 10–12	IT 7–11	IT 8–10
Economical production	20 000	100 000	2000	4000	10 000	3000
Main goal	Moderate strength. No machining.	High strength. No machining.	High strength machining to final shape.	High strength minimal machining.	High strength minimal machining.	Intermediate strength minimal machining.
Automation possibilities	Good.	Good	Limited	Limited	Very good.	Limited

[a] See Table 1.6.29

Table 1.6.11 Mechanical properties of iron and steel P/M parts

Nominal composition	*MPIF[a] designation*	*Condition*	*Specific density*	*UTS (MPa)*	*YS (MPa)*	*Elong. % on 25 mm*	*E (GPa)*	*Unnotched impact strength (J)*
Iron 99.7/100% Fe	F–0000–N	As sintered	5.6–6.0	110	76	2	70	4
	F–0000–T	As sintered	7.2–7.6	276	179	15	170	31
Steel 97.4/99.7 Fe– 0.3/0.6 C	F–0005–N	As sintered	5.6–6.0	124	100	1.0	70	3.4
	F–0005–S	As sintered	6.8–7.2	296	193	3.5	130	12
	F–0005–S	Heat treated	6.8–7.2	552	517	<0.5		
Steel 97.0/99.4 Fe– 0.6/1.0 C	F–0008–P	As sintered	6.0–6.4	241	207	1.0	90	4
	F–0008–S	As sintered	6.8–7.2	393	276	2.5	130	9.5
	F–0008–S	Heat treated	6.8–7.2	648	627	<0.5		
Copper steel Fe–2.7 Cu–0.45 C	FC–0205–P	As sintered	6.0–6.4	276	234	1.0	90	4.7
	FC–0205–S	As sintered	6.8–7.2	427	310	3.0	130	13
	FC–0205–S	Heat treated	6.8–7.2	689	655	<0.5		
Iron nickel Fe–4.25 Ni– 1.0 Cu–0.15 C	FN–0400–R	As sintered	6.4–6.8	248	152	5.0	115	22
	FN–0400–T	As sintered	7.2–7.6	400	248	6.5	155	68
Nickel steel Fe–2 Ni–1.25 Cu– 0.75 C	FN–0208–R	As sintered	6.4–6.8	331	207	2.0	11	11
	FN–0208–T	As sintered	7.2–7.6	545	345	3.5	155	30
	FN–0208–T	Heat treated	7.2–7.6	1103	1069	0.5	155	24
Infiltrated steel and iron Fe–11.5 Cu–0.45 C	FX–1005–T	As sintered	7.2–7.6	572	441	4.0		19.0
	FX–1005–T	Heat treated	7.2–7.6	827	738	1.0		9.5
Fe–20 Cu–0.45 C	FX–2005–T	As sintered	7.2–7.6	517	345	1.5		13
	FX–2005–T	Heat treated	7.2–7.6	793	655	<0.5		8
Fe–20 Cu–0.8 C	FX–2008–T	As sintered	7.2–7.6	586	517	1.0		14
	FX–2008–T	Heat treated	7.2–7.6	862	738	<0.5		7
Copper steel Fe, 1–11 Cu, C <0.9			0–2	220–745	190–340	6–7.1		
Phosphorus steel Fe, 0.3–0.9 P, 0.2 C			6.55–7.43	230–584	190–474	4–23		
Nickel steel Fe, 2–7 Ni, C <0.7			6.3–6.9	180–887	105–320	0.5–2		
Low alloy steel Fe, <4.5 Ni, <0.7 C <2 Cr, <0.8 Mo <0.2 Mn			6.4–7.8	260–950	160–920	0–4		
High alloy steel (1) Fe, C 0.3–0.6, Cu 1–3, Ni 3–6, Mo 0.3–0.7			6.4–7	410–680	350–520	0–2		
High alloy steel (2) 8% Ni–1% Mo– 0% Cu			7.35	1350–1400		2		
Fe, C 0.5, Cu 1.5, Ni 4, Mo 0.5			6.8–7.2	850–1130	830–1000	1		
Injection moulded metal Fe–7 Ni–0.2 Si– 0.2 C			7.7	965–1034	758–827	2		

[a] Metal Powder Industries Federation (USA).

Table 1.6.11

Table 1.6.12 Mechanical properties of stainless steel P/M parts

Grade	*MPIF Designation*	*Specific density*	*U T S (MPa)*	*YS (MPa)*	*Elong % on 25 mm*	*Unnotched impact strength (J)*
AISI 303	SS-303-P	6.0–6.4	241	221	1.0	
AISI 316	SS-316-P	6.0–6.4	262	221	2.0	
AISI 410	SS-410-P	6.0–6.4	397	376	0.5	
AISI 316L		6.2–6.6	290		4–10	4–20
AISI 316		6.6–7.1	450		4–21	4–45
	SS-410-90HT	6.5	724	724	0–1	

Table 1.6.13 Mechanical properties of aluminium and copper P/M alloys

Material	Specific density (%th.)	Temperature[a]	UTS (MPa)	YS (MPa)	Notched impact strength (ASTM E-23-66) (J)	Elong. (%)	Fatigue limit (smooth rotating beam 500 × 10⁶ cycles)	
Al–0.25 Cu–0.6 Si–1.0 Mg (Alcoa 601 AB)	2.42 (90)	0	100.0	–	8.8	8.0	–	
		T1	138	88	7.0	5.0	–	
		T4	172	144	7.0	5.0	–	
		T6	232	224	2.7	2.0	–	
		T61	252	247	–	2.0	–	
	2.55 (95)	T1	145	94	– – –	6.0	38 45	c.f. Wrought (Al–0.8/1.2 Mg–0.4/0.8 Si–0.15/0.35 Cr–0.15/0.4 Cu. T6) 97
		T4	177	117		6.0		
		T6	238	230		2.0		
		T61	255	250		2.0		
Al–0.4 Si–0.6 Mg (Alcoa 602 AB)	2.42 (90)	T1	121	59	–	9.0	–	
		T4	121	62	–	7.0	–	
		T6	180	169	–	2.0	–	
		T61	183	169	–	2.5	–	
	2.55 (95)	T1	131	62	–	9.0	–	
		T4	134	65	–	10.0	–	
		T6	186	172	–	3.0	–	
		T61	193	176	–	3.0	–	
Al–6.4 Cu–0.8 Si–0.5 Mg (Alcoa 201 AB)	2.50 (90)	0	120	–	–	7.0	–	
		T1	201	170	5.8	3.0	–	
		T4	245	205	11.4	3.5	–	
		T6	323	–	3.9	–	–	
		T61	349	342	–	0.5	–	
	2.64 (95)	0	128	–	19.4	8.0	– 45 52	c.f. Wrought (Al–3.9/5.0 Cu–0.5/1.2 Si–0.4/1.2 Mn–0.2/0.8 Mg T6) 124
		T1	209	181	7.5	3.0		
		T4	262	214	17.6	5.0		
		T6	332	327	6.0	2.0		
		T61	356	454	–	2.0		
Al–1.6 Cu–0.2 Cr–2.5 Mg–5.6 Zn	2.51		310	275		2.0		
Al–15–40SiC	2.9		689–710	579–689		0.9–5		

Copper base alloys

	ρ		UTS	YS		Elong	
Cu–0.2 C–9–11 Sn	7.2–7.7		150	100		5.0	
Cu–10 Sn	7.9		220	120		6.0	

[a] 0 Annealed 1 h at 412°C, furnace cooled (max, rate 28°C/hr to 260°C).
T1 Cooled from sintering temperature to 426°C (601 AB and 602 AB) or 260°C (201 AB) in N_2, air cooled to RT.
T4 HT 30 min @ 520°C (601 AB and 602 AB) or 504°C (201 AB) in air, cold water quenched and aged minimum 4 days at RT.
T6 HT 30 min, @ 520°C (601 AB and 602 AB) or 504°C (201 AB) in air, cold water quenched and aged 18 h @ 160°C.
T61 Repressed, HT to T6.

Table 1.6.13

Table 1.6.14 Comparison of powder forging with other forming methods

Feature	*Powder forging*	*Sintering*	*Die forging*	*Precision forging*	*Cold forging*	*Precision casting*
Part weight (kg)	0.1–5	0.01–14	0.05–1000	0.05–100	0.01–35	0.1–10 Extreme cases to 35
Height diameter ratio	Height usually lower than width	Height usually lower than width	Not limited	Not limited	Not limited	Not limited
Shape	No large variations in cross-section, allowed openings limited. Thin walls to be avoided.	No large variations in cross-section allowed openings limited. Thin walls to be avoided.	Any shape – openings limited.	Any shape – openings limited.	Mostly of rotational symmetry.	Any shape, any openings possible.
Material utilization %	100	100	50–70 (of bar stock)	85–95 (of bar stock)	95–100 (of bar stock)	65–85
Tolerances[a]	IT 8–10	IT 6–8 if sized	IT 15–18	IT 10–12	IT 7–11	IT 8–10
Economical production	100 000	20 000	2000	4000	10 000	3000
Main goal	High strength. No machining.	Moderate strength. No machining.	High strength. Machining to final shape.	High strength. Minimal machining.	High strength. Minimal machining.	Intermediate strength. Minimal machining.
Automation possibilities	Good	Good	Limited	Limited	Very good	Limited

[a] See Table 1.6.29

Table 1.6.15 Specification and properties of powder forged steels

Material	*Composition*							*Condition*	*UTS (MPa)*	*YS (MPa)*	*Elongation (%)*	*Reduction of area (%)*	*Impact strength (J)*
	C	*Mn*	*Mo*	*Ni*	*Cr*	*S*	*P*						
Plain carbon steel	0.1	0.25 Max.				0.02 Max.	0.02 Max.	As forged	400–800	250–650	20–10	60–15	
Low alloy through hardened steel	0.35/ 0.45	0.3/ 0.4	0.25/ 0.35	0.2/ 0.3	0.1/ 0.25	0.035 Max.	0.02 Max.	14 mm section Quenched and tempered 575°C	700–950	550 min	10 min	22 min	24–10
Low alloy carburizing steels	0.18/ 0.24	0.4/ 0.5	0.25/ 0.35	0.2/ 0.3	0.25/ 0.35	0.03 Max.	0.02 Max.	As forged	450–620		25–17	47–42	
								14 mm section Carburized, quenched and tempered 150°C	650–830	450 min	10 min		15 min
	0.22/ 0.3	0.4/ 0.5	0.25/ 0.35	0.2/ 0.3	0.25/ 0.35	0.03 Max.	0.02 Max.		700–900	475 min	8 min		12 min
Medium alloy carburizing steels	0.17/ 0.23	0.25/ 0.35	0.45/ 0.55	1.8/ 2.2		0.015 Max.	0.02 Max.	As forged	490–615		28–20	59–50	
								14 mm section Carburized, quenched and tempered 150°C	770–1450	500 min	8 min		12 min
	0.20/ 0.26	0.25/ 0.35	0.45/ 0.55			0.015 Max.	0.02 Max.		850–1550	750 min	6 min		9 min

Table 1.6.15

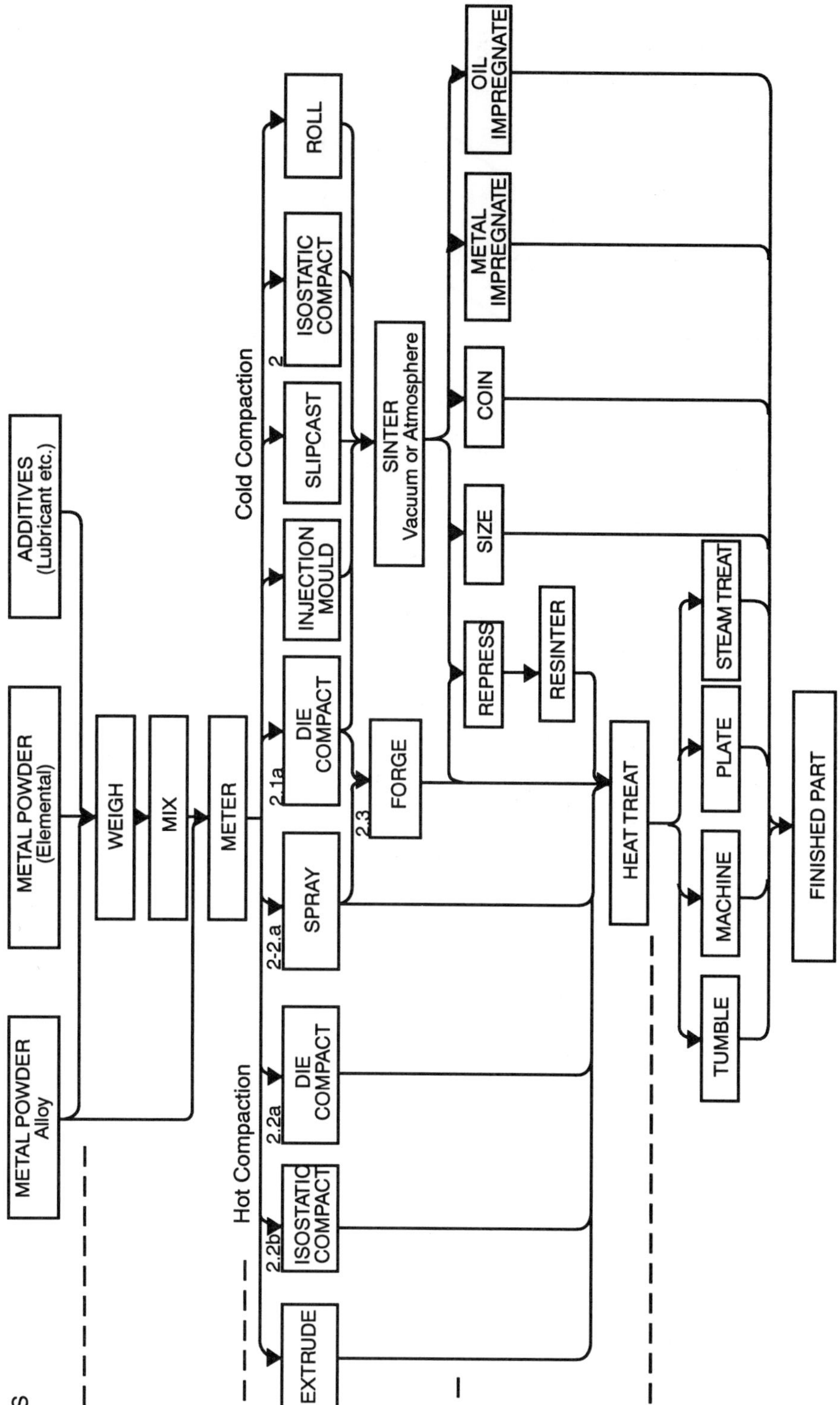

Fig. 1.6.5 Flow sheet of powder metallurgy processes

Fig. 1.6.5

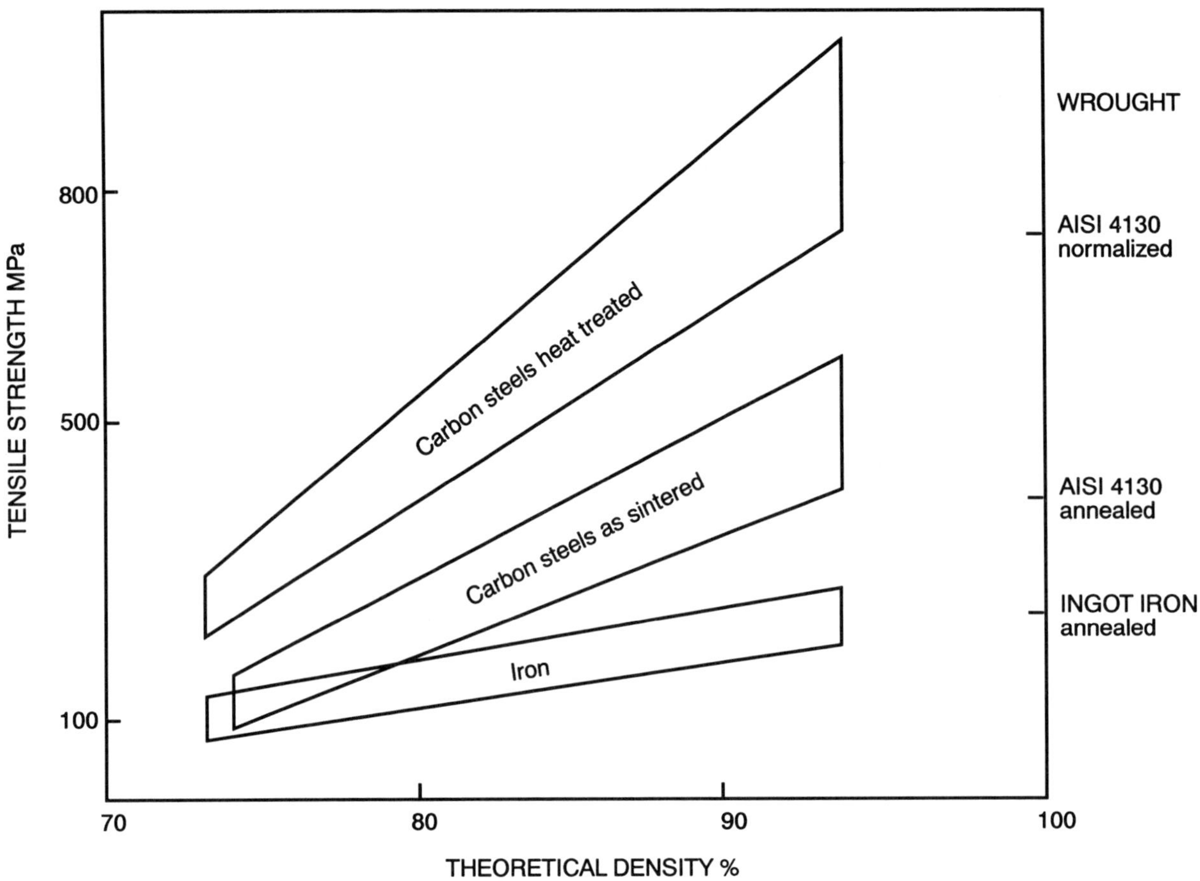

Fig. 1.6.6 Variation of tensile strength with density for carbon steels and iron

Fig. 1.6.6

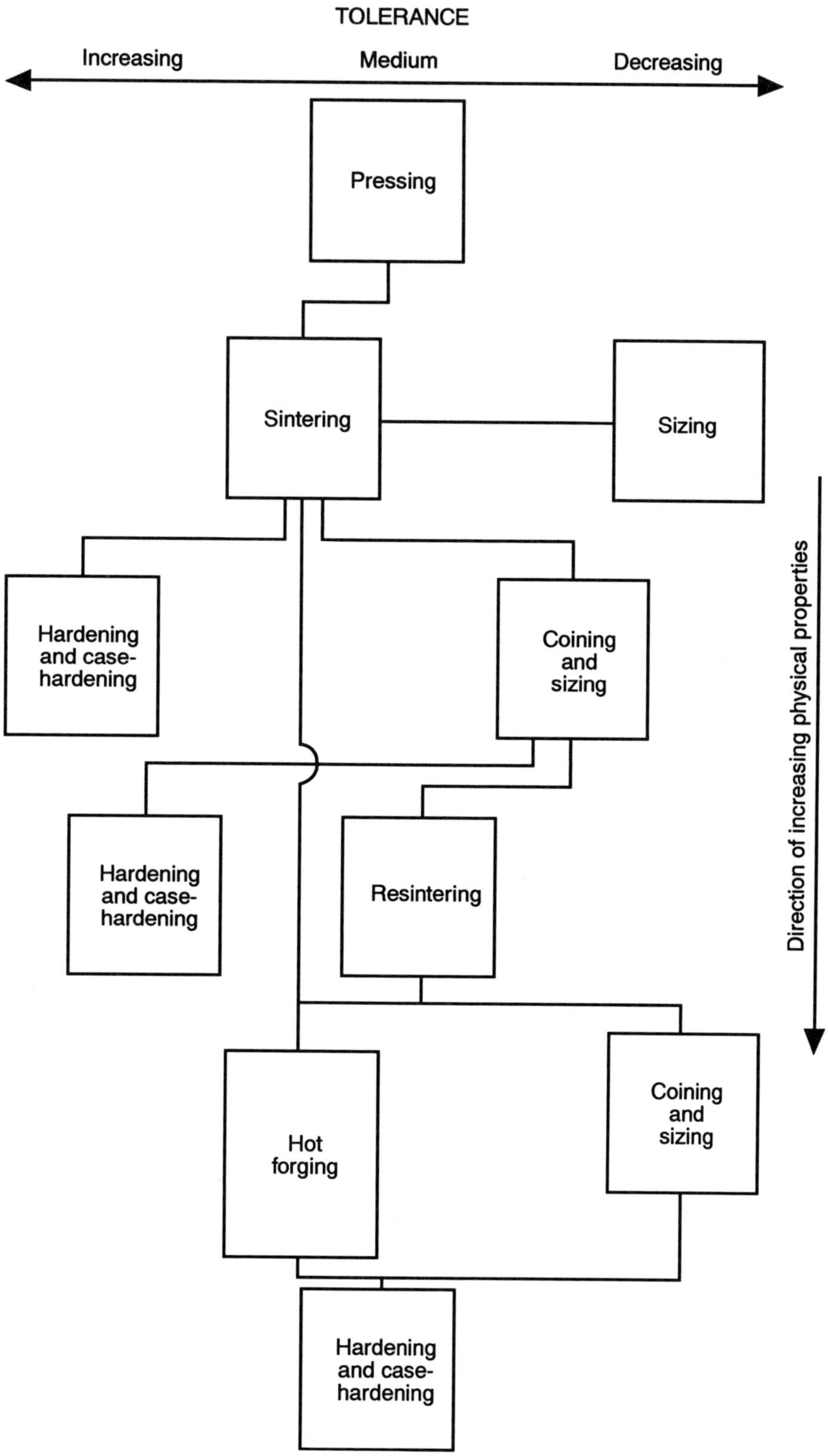

Fig. 1.6.7 Influence of processing on properties and tolerance for P/M components

Fig. 1.6.7

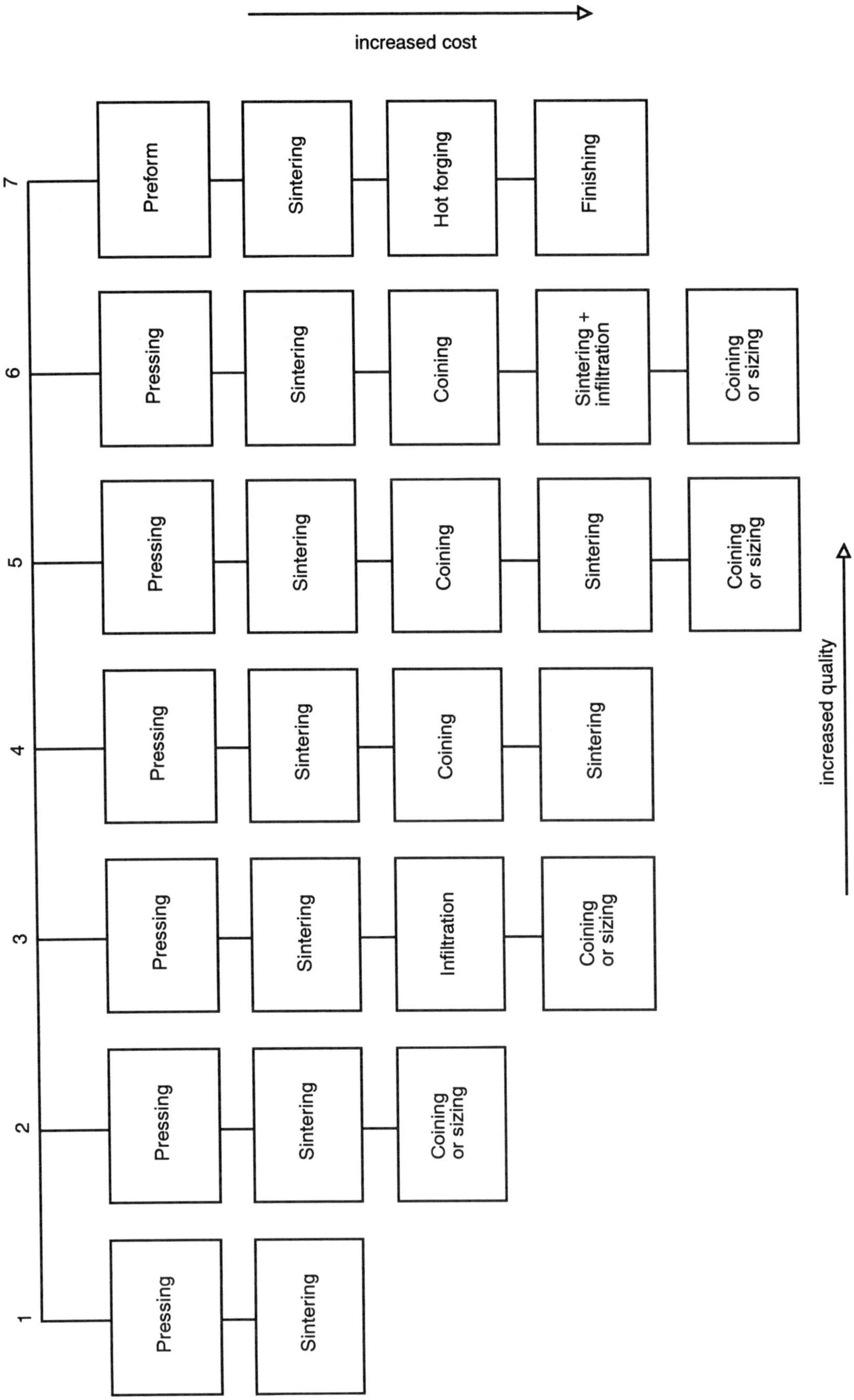

Fig. 1.6.8 Trends in cost and quality incurred by the use of common processes for the manufacture of P/M components

Fig. 1.6.8

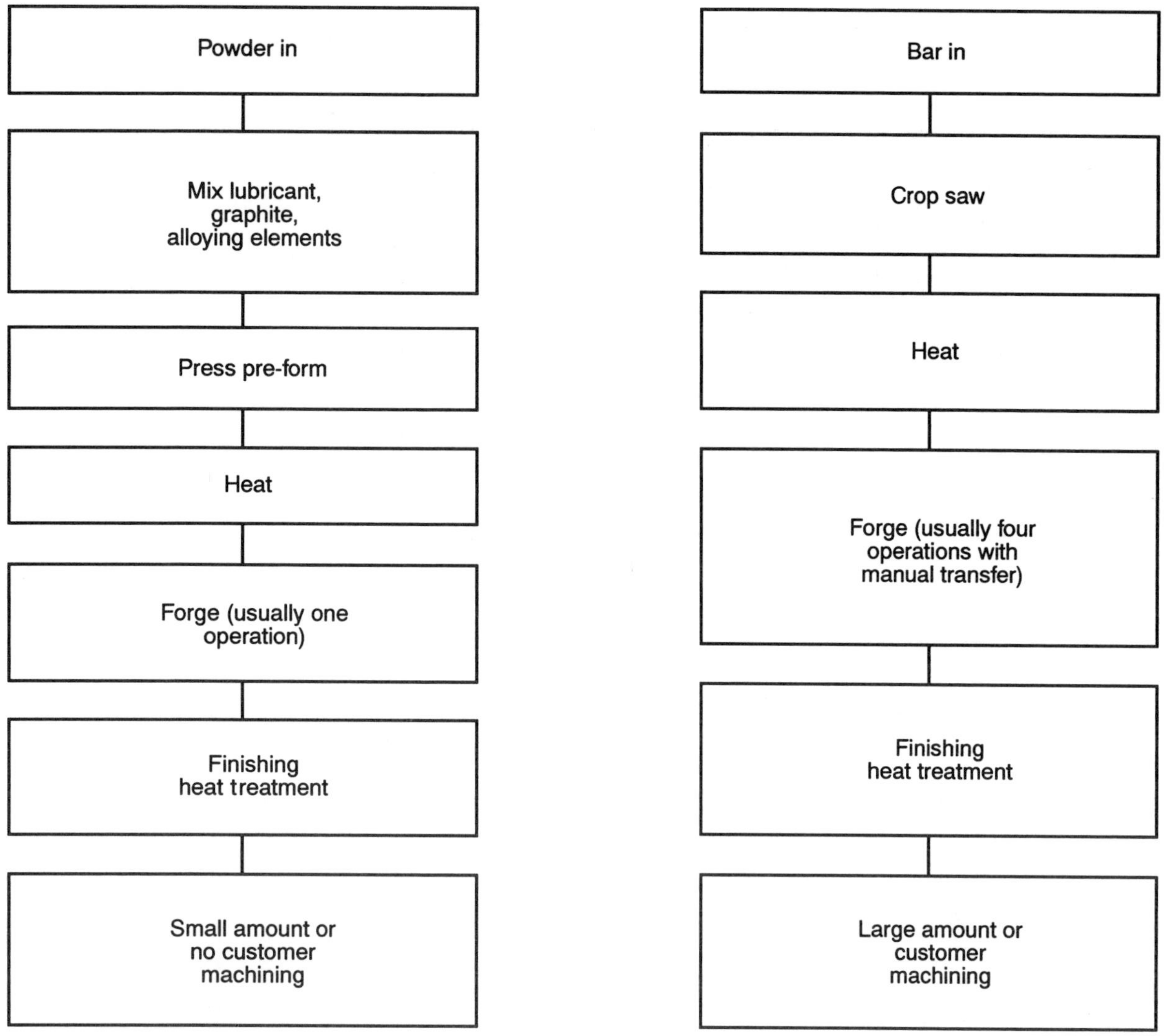

Fig. 1.6.9 Flow sheets for powder and conventional forging

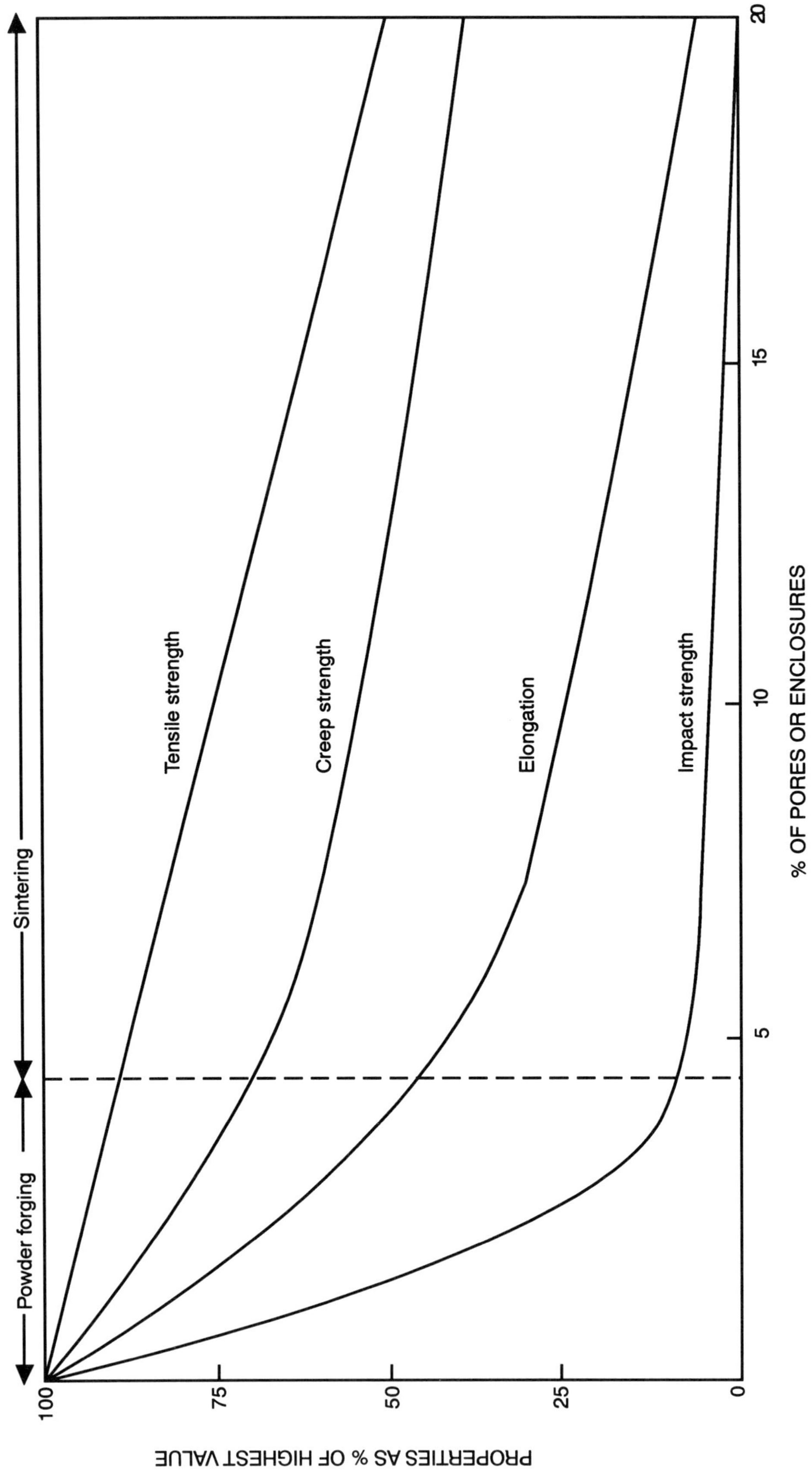

Fig. 1.6.10 Variation of mechanical properties with porosity for P/M components

Fig. 1.6.10

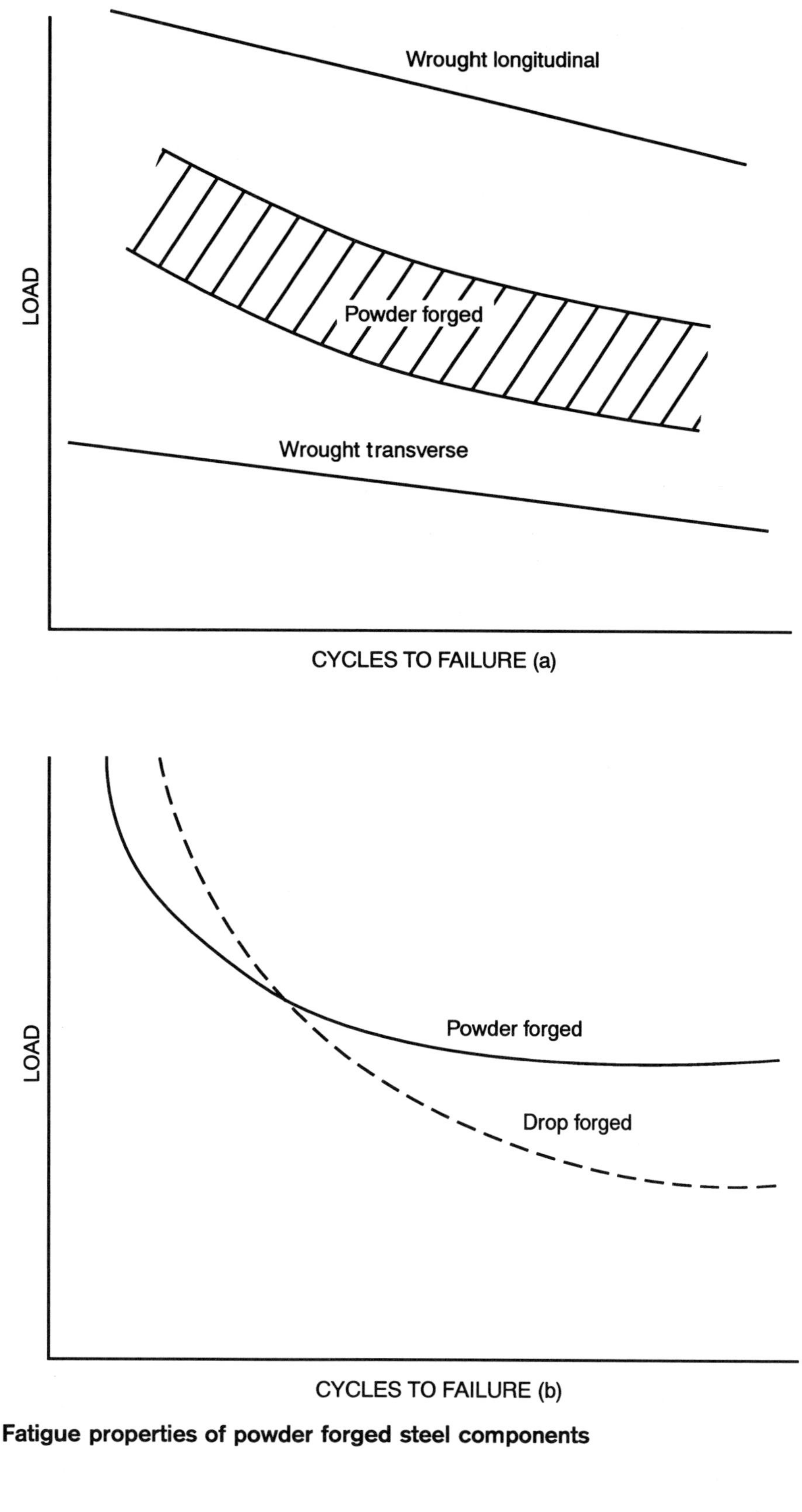

Fig. 1.6.11 Fatigue properties of powder forged steel components

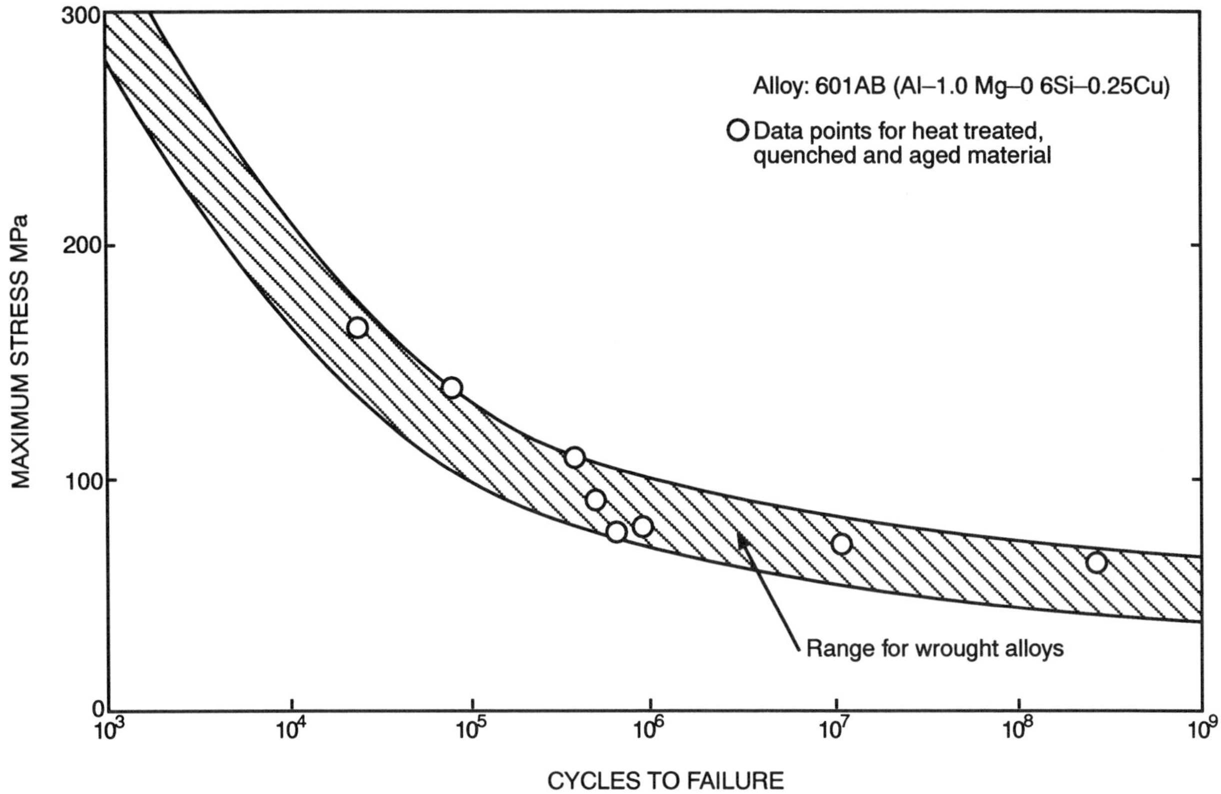

Fig. 1.6.12 Fatigue performance (notched specimen) of powder forged aluminium alloy and comparable wrought alloys

Fig. 1.6.12

1.6.4 Metal working

1.6.4.1 Introduction

ADVANTAGES

The advantages of wrought compared with cast products are described briefly below.

Specific shape requirements

Certain shapes, such as foil and wire which can readily be produced by metal working are difficult if not impossible to cast.

Good mechanical properties

This is the result of many factors, including:

a. grain refinement,
b. homogeneity of internal structure,
c. reduction or elimination of porosity,
d. breakdown and dispersal of inclusions,
e. fine microstructure which gives greater susceptibility to heat treatment,
f. the more regular shape of a forging ingot compared with that of most castings makes it easier to obtain good mechanical properties.

Improvement in reliability

Properties are more predictable as a result either of the welding-up of small defects, or of prior failure arising from large defects during the working operation. The working process therefore acts both to improve and test the material.

Limited surface decarburization in steels

Steel forgings show less surface decarburization than the equivalent castings. Thus, unless the casting is subject to a carburizing treatment, the forging exhibits superior fatigue properties.

Fibrous structures

The mechanical properties of most wrought materials are dependent on their orientation with respect to their fibrous structure. Maximum tensile strength and toughness are obtained along the fibre direction. For this reason a bending beam is strong in this mode of stress if fibres run along the length of the beam; any transverse crack cannot easily propagate across the beam, but is diverted along the fibre direction, thus conferring toughness. However, properties are poor when the load lies across the grain. Properties can therefore be optimized by careful design both of the component and the operations required to form it. Subsequent machining operations must not create notches which might cause the metal to open along the fibre direction under stress.

Cost

Certain wrought products, such as pressings, stamping or coinings from sheet, are amongst the cheapest compared with products from all other routes. This is especially true for large quantity production.

Surface finish

Surface finish can be of an extremely high standard for cold formed components, especially those made from sheet by stamping or coining. Coins are a good example.

MECHANICAL PROPERTIES

In general, many metals have better mechanical properties in the wrought than in the cast state. This is especially true for ductility, as shown by tensile elongation and by impact results. There are other properties, like high temperature creep, etc., where the cast properties are better than those in the wrought condition. Apart from this difference, wrought products and particularly forgings have different properties depending on the direction of the grain flow which has taken place during the working operation. A familiar example is the superiority of the toughness of gear teeth in which the metal flows around the form of the teeth instead of directly across, as might be the case in a gear machined from bar stock.

A guide to the kind of differences in mechanical properties for the longitudinal and transverse grain flow position is given for aluminium forgings in:

Table 1.6.16 *Properties of aluminium alloy forgings in longitudinal and transverse directions of grain flow*

This directionality of mechanical properties is also shown up by the fracture toughness in the three directions (longitudinal, transverse and short transverse). Figures for the fracture toughness of some wrought products in these directions as well as the changes with temperature of these properties can be found in Chapter 2.2.

1.6.4.2 Warm and hot working

CLOSED DIE FORGING

Closed die forging uses carefully machined matching die blocks to produce forgings of close dimensional tolerances. Full filling of the die is ensured by the use of a small excess of metal which forms a flash as the dies come together. Whilst close dimensional tolerances are possible, it is difficult to produce parts with sharp fillets, wide thin webs, and high ribs. In addition, the forging dies must be tapered to facilitate removal of the finished piece.

The tolerances achievable for the thickness of forgings varies with the weight of the forging, as shown in:

Table 1.6.17 *General guide to die closure tolerances*

Guidance in the design of fillets is given in:

Fig. 1.6.13 *Design of fillets in closed die forging*

For fillet heights of greater than 10 mm the required fillet radius is 0.1 of fillet height. For fillets of less than 10 mm a fillet height to radius ratio of 0.2 is necessary. Machining allowances vary from less than 0.75 mm to more than 5 mm dependent upon the size of the casting.

Minimum fillet radius	:	1.5 mm
Minimum corner radius	:	0.76 mm
Angular tolerance	:	$\pm 0°15'$
Contour tolerance	:	± 0.25 mm
Draft, outer faces	:	0°
Draft, inner faces	:	$1° \pm 0.5$
Surface finish	:	3.2 μm (can obtain 1.6 μm)

HOT FORMING

Material is forged above its recrystallization temperature (greater than half its absolute melting point). Dies are either cold or warm, the latter giving less heat loss during the process. Forging machines are of the hammer type, yielding high strain rates between 1 and 10 s^{-1}, or of the press type, yielding low strain rates between 0.1 and 1 s^{-1}.

High strength high temperature materials are often too hard to forge by conventional means but can sometimes be forged by 'isothermal forging'. The dies are maintained at the same temperature as the billet by including the die stack and billet in an electrically heated furnace. For superalloys and titanium alloys this requires molybdenum dies in a hydrogen-rich atmosphere. In order to maintain low forging loads the strain rates are low, often less than 10^{-4} s^{-1}. The process is slow, single disc forgings taking up to an hour, but the greater degree of temperature uniformity can yield more uniform metallurgical structures and parts can be produced nearer to net shape, thus reducing subsequent machining.

CLOSE TO FORM FORGING (ALUMINIUM AND MAGNESIUM)

(Sometimes referred to as 'precise to form' forging or 'near net shape' forging.)

By using composite dies and a very close to final weight billet, a flashless and complex shaped forging which will require only a minimum amount of machining can be produced.

Advantages claimed for such forgings include:

(a) Possibility of zero draft.
(b) No end-grain or flash which may impair transverse ductility and stress corrosion behaviour.
(c) Increased versatility with respect to profile and variation in adjacent section sizes. Curved faces, undercut forms and projections can be placed in external faces which would not be possible with a conventional forging.
(d) Practically no limit to the shape of the profiled form of the forging.

Minimum wall thickness: Rib/wall ratio:

Thickness	Height
1.5 mm	0–38 mm
2.5 mm	38–101 mm
3.5 mm	101–154 mm

Rib/wall ratio over 154 mm height is dependent on adjacent geometry, heat treatment requirement, and area of wall.

Overall length and width tolerances:

0–65 mm	+0.13 mm
65–127 mm	+0.25 mm
127–254 mm	+0.39 mm
254–330 mm	+0.50 mm

Length times width should not exceed 413 cm^2.

Straightness tolerance:

Length or width	Tolerance
0–127 mm	0.13 mm
127–254 mm	0.25 mm
254–330 mm	0.50 mm

Minimum web thickness:

Area of web	Recommended thickness
0–52 cm^2	1.5 mm
52–162 cm^2	2.5 mm
162–413 cm^2	3.5 mm

Die closure or thickness tolerance:

Plan area	
0–700 cm^2	+0.50–0.25 mm

Depending upon forging configuration, thickness tolerance can be reduced to ±0.25 mm.

N.B. The above are a guide only as each compound needs to be assessed individually.

RING ROLLED FORGING

For producing wrought rings a **pancake-shaped** blank is punched out in the centre and rolled to the required shape and size. Since no stock is removed the finished ring is of the same weight as the punched out blank. Inner and outer walls can be contoured.

Mills have been constructed to produce rings ranging from 0.3 to 5 m. Aluminium alloys, steel, titanium alloys and heat resisting alloys can all be treated to produce rings by this method.

Ring rolling often competes favourably with bending and welding because of the ability to produce a shaped section which reduces final machining. The structure of ring rolled parts is both good and homogeneous. Welded rings have a weld with a heat affected zone in the ring structure.

ORBITAL FORGING

The orbital forging process operates at a lower temperature than conventional forging. As a consequence, little or no scale is produced, eliminating the need for shot blasting, and the material is not decarburized. The minimum practicable forging thickness is reduced from 5 mm for conventional forging to 2.5 mm, and the tolerance from ±0.75 to ±0.15 mm. The grain flow within the component is spiral rather than radial and the component can be produced on a much lower rated machine.

An example of production costing, including materials, power, dies, lubricants, labour, etc., is shown for a typical orbital forged component, together with data for

an equivalent conventionally forged component, in:
Fig. 1.6.14 *Comparison of costings for orbital and conventional forging*

1.6.4.3 Finishing processes

COLD FORGING AND COLD HEADING

Cold forming has grown in popularity for wrought steel parts since its introduction in the 1950s. Because of the limitation on cold deformability, steels to be cold formed require careful selection. A balance must be struck between the loss of cold deformation due to alloying with a particular element and the advantages which can be gained from the addition in heat treatment hardenability. Chromium and boron are useful alloying elements for a heat treated cold forging steel while, if no heat treatment is to be given, the normal silicon and manganese contents of steel should be kept as low as possible to minimize working pressure.

Low carbon steels are easy to forge and after processing have a yield strength of approximately 600 MPa. The medium carbon steels, manganese molybdenum steels and 1% Cr steels are more difficult to forge and must be heat treated after forming. The low chromium and nickel chrome case-hardening steels are easier to forge than medium carbon steels, but they must be subsequently case-hardened.

Steels which are suitable for cold forging are detailed in:
Table 1.6.18 *Steels suitable for cold forging*

The steps involved in the production of a pinion shaft with gear teeth by cold forging are illustrated in:
Fig. 1.6.15 *Production of a near net shape pinion shaft with gear teeth*

Ideally the part should be deburred, finish ground and roll formed to bring the gear teeth into tolerance. Less than 2% of the weight of the part is removed in secondary machining.

A comparison of typical features of cylindrical shape production routes for mild steel are shown in:
Table 1.6.19 *Cylindrical shape production routes compared for mild steel*
and for non-ferrous metals in:
Table 1.6.20 *Cylindrical shape production routes compared for non-ferrous metals*

Typical tolerances achieved for a range of cold forged components are given in:
Table 1.6.21 *Typical commercial tolerances for a range of cold forged components*

DEEP DRAWING AND IRONING

During deep drawing and ironing a blank is drawn through a cupping die and then passes through a series of ironing dies of successively smaller diameter. The original blank can be in the form of a flat disc or can be a cup preform, thus a forging can be used which may be 'stretched' in an ironing operation.

Where materials have less than 10% elongation in standard tensile tests, cold drawing and ironing is not recommended.

Cylinder liners are produced by parting off and scrapping the thick base of an open-ended tube formed by the deep drawing and ironing process. In spite of this waste, the accuracy of the component justifies the use of a thick blank and the drawing and ironing process.

1.6.4.4 Superplastic forming

Metals are said to be superplastic when they are capable of considerable plastic deformation of the order of 100–1500% under relatively low forming stresses. They can then be formed in a manner similar to plastic materials and by techniques familiar in the plastics industry. For metals to display superplastic behaviour they must:

1. have the correct microstructure (grain size 1–5 μm, stable at temperature of deformation),
2. be deformed at a temperature of at least half the melting temperature in degrees Kelvin,
3. be deformed slowly, e.g. about 0.01 mm/min.

The main superplastic forming processes are described in:
Table 1.6.22 *Superplastic forming processes*
Mostly vacuum or pressure thermoforming is used starting from sheet material, although compression moulding or forging is also possible. Thermoforming because of lower tool, assembly and finishing costs is competitive with traditional forming methods for low to medium runs. The increase in time for one operation compared with normal presswork reduces its competitiveness for mass production. Forging is especially useful for the shaping of expensive materials difficult to work, for instance, the superalloys which are used for aerospace applications. A saving in material is possible because of reduced machining requirements and powdered starting materials can be used. Simultaneous diffusion bonding may with advantage be combined with superplastic forming.
The superplastic temperature range, the elongation possible and comments on each alloy system which is either commercially important or likely to be used shortly are given in:
Table 1.6.23 *Superplastic alloy systems*
A number of alloys exhibit superplasticity but only those possessing good mechanical properties can be commercially utilized. Of these only a few have so far been employed.

Table 1.6.16 Properties of aluminium alloy forgings in longitudinal and transverse directions of grain flow

Alloy	*Condition: W = solution treated P = precipitation-hardened*	*Longitudinal or transverse*	*0.1% PS (MPa)*	*UTS (MPa)*	*Elongation % on $4\sqrt{A}$*
Al–1Mg–1Si–0.75Mn		L	270	293	8
(HF30: BS 1472)		T	255	278	4
Al–2Cu–1Fe–1Ni–1.5Mg	WP	L	332	417	6
(DTD 731 A)		T	317	386	4
Al–4.4Cu–1.5Mg–0.6Mn	W	L	293	432	14
(DTD 5090)	Plate	T	263	417	10
(AA No. 2024)		Short T	247	386	4
Al–5.7Zn–2.5Mg–1.6Cu– 0.3Mn–0.25Cr	WP	L	445	515	8
(AA No. 7075)		T	425	490	4

Table 1.6.17 General guide to die closure tolerances

Forging weight (kg)	*+ Tolerance (mm)*
0.23	0.50
1.36	0.75
2.72	1.00
5.45	1.25
9.07	1.50
18.15	1.75
45.40	2.00
90.75	2.25
181.50	2.50
363.00	3.75

Table 1.6.18

Table 1.6.18 Steels suitable for cold forging

Steel type	*UK BS 970 specifications*		*USA SAE or AISI specifications*	*Germany DIN 17210 or SEL specifications*	*France AFNOR specifications*
(a) *For most shapes*					
		Soft iron	Armco	Soft iron	Soft iron
		–	–	Mbk 6	–
Carbon–manganese carburizing	045M10	En 32 A	1008/1010	Ck 10	XC 8 and 10
Carbon–manganese carburizing	080M15	En 32 B and 32 C	1015	Ck 15	XC 12
Carbon–manganese	070M20	En 3 A and B	1017/1020/1025	Ck 22	XC 18
Carbon–manganese	080A25	En 4 and 4 A			
(b) *For some shapes*					
Carbon–manganese	080 M30	En 5 and 5 D	1030/1033/1035	Ck 35	XC 32 and 35
Low alloy	523A14 and 527A19	En 206 and 207	5115/5120	(EF 60) 15 Cr 3	12 C 3 and 18 C 3
		–	–	(EC 80) 16 Mn Cr 5	16 MC 3
		–	–	(EC 100) 20 Mn Cr 5	20 MC 5
		–	3115	15 Cr Ni 6	16 NC 6
(c) *Only for screws, nuts, bolts etc.*					
				DIN 1654 or SEL	
Carbon–manganese carburizing	045M10	En 32 A	1010	Muk 7	XC 8 and 10
Carbon–manganese	070M20	En 3 A and B	1025	C 25 and CK 25	XC 18
Carbon–manganese	080M30	En 5 and 5 D	1035	C 35 and CK 35	XC 32 and 35
Carbon–manganese	080M40 and 080M50	En 8 and 43 A	1045	C 45 and CK 45	XC 38, 42 and 48
Low alloy direct hardening	709M40 and 708A37	En 19 and 19 B	4120/4132/4135	(VC Mo. 135) 34 Cr Mo 4	25 CD 4 and 35 CD 4
Low alloy direct hardening	640M40 and 640A35	En 111 and 111 A	3120/3130/3135	(V EN 15) 36 Ni Cr 6	20 NC 6 and 25 NC 6
					35 NC 6
Carbon–manganese	150M36	En 15 and 15 A	1041	40 Mn 4	45 M5
		–	–	(VMS 135) 37 Mn Si 5	38 MS 5
Low alloy direct hardening	530A30 and 530A32	En 18 A and 18B	5130/5130	34 Cr 4	32 C 4 and 38 C 4
Low alloy direct hardening	530A36	En 18C			
Alloy spring steel	735A50	En 47	6150	50 Cr V 4	50 CV 4

Table 1.6.19 Cylindrical shape production routes compared for mild steel

Feature	*Deep draw and iron*	*Back extrusion (cold)*	*Flow forming*	*Hot drawing*	*Cold drawn seamless*
Size	Up to 600 mm dia.	Up to 270 mm dia.		100–450 mm dia.	150–650 mm dia.
Wall thickness	0.1–12 mm	5–7 mm	Up to 37 mm	16–40 mm	150–650 mm dia
Lengths	Length/dia. ratio 2 : 1–10 : 1	1000 mm		25 mm	8.5 mm
Tolerances	50 mm dia. ±0.06 mm 3 mm w.t. ±0.06 mm	100 mm dia. ±0.07 mm 1.65 mm w.t. ±0.8 mm		All sizes +3.2 mm −1.6 mm 1.0 mm w.t. +0.25 mm −0 mm	200 mm dia. +0.25 mm −0.75 mm 6 mm w.t. +0.6 mm −0 mm
Surface finish	0.1–0.3 μm	0.3–0.5 μm		1 μm	0.1–0.3 μm
Straightness	1 : 14 000	1 : 2000			1 : 2500
Number required	5000	20 000	Very low tooling costs and flexibility in design can justify single component		

Products: gas bottles, fire extinguisher bodies, shock absorber bodies, cylinder liners, etc.

Table 1.6.20 Cylindrical shape production routes compared for non-ferrous metals

Feature	*Back extrusion*		*Forward extrusion*	*Open die forging*
	Impact	*Low strain*		
Shape	*Hollow closed end*		*(a) Tubular* *(b) Solid*	*Long and thin*
Sizes	100 mm dia. 300 mm length 0.1–2.00 mm wall thickness max. length dia. ratio 10 : 1	70–270 mm dia. 1300 mm length 5–17 mm wall thickness	(a) 2.5–130 mm dia. 4 mm length 0.1–9.5 mm wall thickness longitudinal fins, keyways and slots (b) +0–18 mm dia. +0–180 mm length max. extrusion ratio Al 100 : 1 Cu 10 : 1	0.5–18 mm dia. 9–220 mm length

Table 1.6.21 Typical commercial tolerances for a range of cold forged components

Components	*Alloy type (%)*	*Dimensions and tolerances (mm)*			
		Outside dia.	*Length*	*Wall thickness*	*Base thickness*
Alumium alloys					
Cylindrical can	Al 99.99	25 ± 0.15	125 ± 0.8	0.1 ± 0.01	0.63 ± 0.08
Filter bowl	Al–Mn 1	75 ± 0.15		1.0 ± 0.05	
Open tube	Al–Si 1–Mg–Mn	63 ± 0.1		1.6 ± 0.08	
High pressure gas cylinders	Al–Si 1–Mg–Mn	150 + 0.5		4 + 0.4	
Ordnance part	Al–Mn1	100 + 0.25 ID	780 ± 0.5	1.3 ± 0.1	40.6
Copper alloys					
Diesel engine injector sleeve	Cu–0.5/1.2 Cd	25.4 ± 0.1	105 ± 2.0		2.5 ± 0.15
Semi-conductor base stud	Cu–0.5/1.2 Cd	31.5 ± 0.15 Across hexagon head flats	8 ± 0.15 Head thickness	1.0 ± 0.15 Shank diameter	

Table 1.6.22 Superplastic forming processes

Process	*Method*	*Purpose*	*Advantages*	*Applications*
Sheet thermoforming	Clamped heated sheet material forced into contact with male or female mould by pneumatic pressure.	Manufacture of complex shaped parts. Re-entrant angles possible.	Lower tool costs than pressing and stamping. Diffusion bonding can be combined with sheet thermo-forming.	Deep covers, trays, architectural panel-ling, aircraft compo-nents.
Blow moulding	Heated hollow cup or tube is internally pressurized to force material into contact with mould.	Hollow shapes with externally applied form.	Complex shapes with small wall thickness possible.	Vacuum chambers, pressure vessels.
Forging (Pratt and Whitney Gatorizing)	Heated material formed by mechan-ical pressure between dies. Tech-nique and equipment similar to that for conventional isother-mal forging or com-pression moulding of polymers but lower ram speeds neces-sary.	Manufacture of functional and structural components of complex shape.	Use of powdered metals permits reduced machining costs in production of complex shapes. Materials difficult to work can be forged.	Integral blade turbine discs.
Die-less drawing	Bar, tube or section locally heated by in-duction whilst under tension until necks. Coil moved progres-sively along workpiece.	To produce reduced sections.	Materials difficult to deform can be reduced by this technique.	Not yet in commercial use.

Table 1.6.21 and Table 1.6.22

Particulars of fillet radii,* etc.

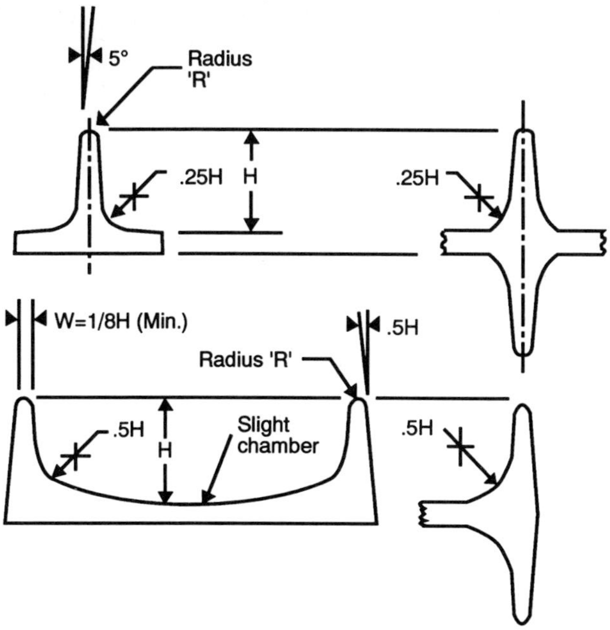

HEIGHT 'H'		RADIUS 'R'	
INCHES	mm	INCHES	mm
0–1	0–25.2	1/16	1.5
1–1½	25.5–38.0	3/32	2.25
1½–2	38.0–51.0	1/3	3.0
2–3	51.0–76.5	3/16	4.5

For small forgings where 'H' is 0.5 in. (12.75 mm.) or less, above proportions of radii are to be increased by one-half.
Draft angle and fillet radii can be reduced if the forgings are manufactured on such plant as hydraulic presses.
*Where the web area is large it is, if possible, preferable to allow for a pierced hole.

Machining allowances & corner radii*

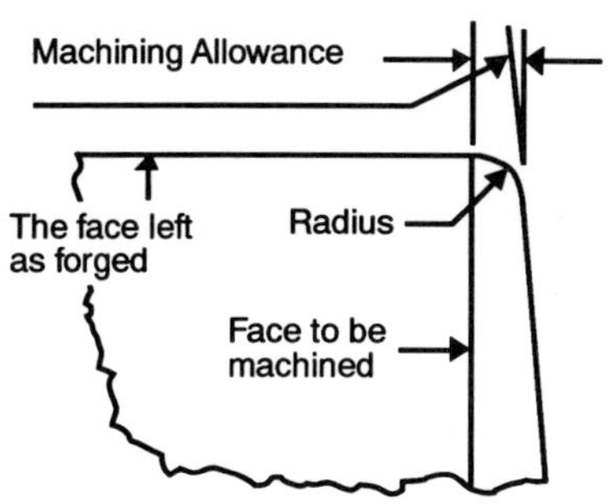

When adjacent faces fo a forging are to be machined, the corner radius is the same as the machining allowance.

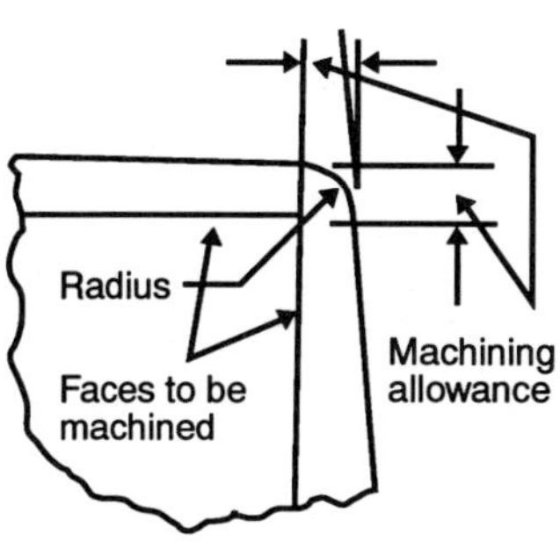

* These values are for light alloys.

When one face is machined, the radius is smaller than the machining allowance.
This ensures that, after machining, a sharp corner is obtained, although the forging may not have been fully made up or may have been damaged.

MACHINING ALLOWANCE		CORNER RADIUS	
INCH	mm	INCH	mm
0.03	0.75	0.02	0.5
0.05	1.25	0.03	0.75
0.06	1.5	0.04	1.0
0.10	2.5	0.06	1.5
0.15	3.75	0.10	2.5
0.20	5.0	0.15	3.75

Machining allowances may be less than 0.03 in. (0.75 mm.) or more than 0.20 in. (5 mm.) dependent upon the size of the forging.

Fig. 1.6.13 Design of fillets in closed die forging

Table 1.6.23 Superplastic alloy systems

Alloy system	*Superplastic temperature range (°C)*	*Elongation (%)*	*Comments*
Zinc			Available as sheet. Cheaper than aluminium alloy sheet. Excellent forming characteristics. Modest forming temperature. Suitable only for low temperature and low stress applications. Properties similar to zinc die- casting alloys. Very low creep strength at room temperature.
Zn–Al eutectoid (SPZ Grade 22, (20–22% Al) Super Z, Zilon C Korloy)	200–300	1500	
Magnesium			Not yet commercially available but promising for applications (forgings) requiring low forming temperatures.
Mg–Zr	500	150	
Mg–Zn–Zr	270/310	1000	
Mg–Al	350/400	2100	
Aluminium			Available both unclad or clad both sides with pure aluminium.
Al–6 Cu–0.5 Zr (Ti Superform Supral 150)	480	1000	Components made with superplastic alloys are now commercially available.
Al–5 Ca–5 Zn (Alcan 05050)	450		Developed, but unlikely to be used due to recycling problems.
Al–7.6 Ca (Euratom)			
Al–4.5% Mg (AA 5083)			Available commercially from several aluminium companies.
Al–5.5% Zn–2.5% Mg–1.6% Cu (AA 7075)	≃ 500°C	500%	
Al–2.5% Li–1.5% Cu–0.6% Mg (AA 8090)			
Copper			
Cu–7P	500–600	1000	Phenomenon of temporary superplasticity used to produce brazing wire.
Cu–2.8 Al–1.8 Si 0.4 Co	600	200	Replaces more expensive alloys for contact springs, connectors and terminals, not yet used for its superplastic properties.
Nickel			These superalloys can be isothermically forged in the superplastic range.
Ni–10 Cr–15 Co–3 Mo–5 Ti–5.5 Al–1 Fe (IN100)			
Ni–15 Cr–15 Co–5.2 Mo–3.5 Ti–4.4 Al	950–1100	700	Expensive starting materials but powders can be utilized. Used for gas turbine discs and other aerospace applications.
Ni–19 Cr–14 Co–4.3 Mo–3 Ti–1.3 Al–1 Fe			
Ni–16 Cr–8 Co–1.8 Mo–3.5 Ti–3.4 Al–2.6 W–1.8 Ta–1 Nb (IN738)			In commercial use.
Ni–35 Fe–1.3 Cr–5.6 Mo–2.5 Ti	900–100		
Iron			No commercial relevance in sheet forming applications.
Carbon and alloy steels Range 0.1 to 1.9 C	540–825	500	
White cast iron	650		
Fe–2.25 C–1.5 Mn		500	
Stainless steel			
18/8 Type	870–980	600	Special alloy containing dispersals developed, but not yet exploited.
'Super duplex' stainless steel. Fe–25% Cr–7% Ni–3% Mo	≃1000	500%	May be formed in air.
Titanium[a]			
Ti–6 Al–4 V	800–1000	2000	Sheet forming, blow moulding and isothermal forging commercially available. Application mainly for aerospace.
Ti–6 Al–2 Sn–4 Zn–6 Mo	900–1100	450	Isothermal forging of these two alloys in commercial practice.
Ti–8 Al–1 Mo–1 V	1000	450	

[a] Many other commercial titanium alloys exhibit superplasticity.

Table 1.6.23

ORBITAL FORGED

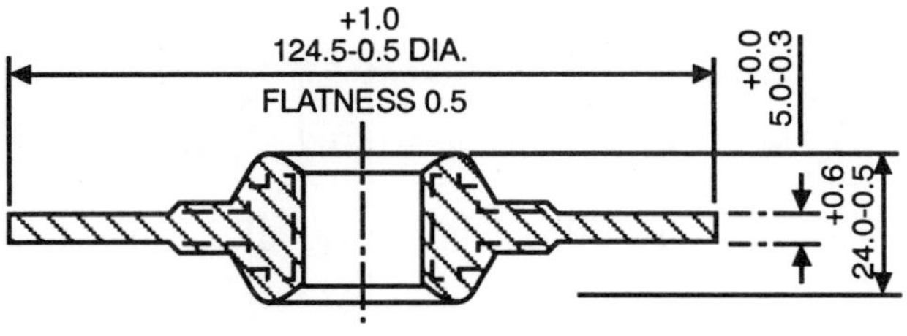

CONVENTIONAL

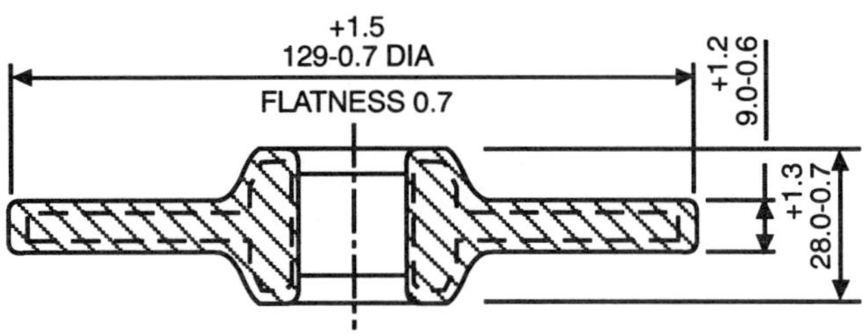

FORGING TYPE: CLUTCH DISC MATERIAL: EN 180 SAE 5140 DIN 41Cr4

		Conventional	Rotaform
Forging machine		1500 tonne	150mm
Billet weight	kg	1.10	0.7
Square or round		square	round
Forging weight	kg	1.05	0.66
Production rate	pces/h	500	550
Forging temperature	°C	1250	850
Batch size		10000	10000

CONTRIBUTION CALCULATION

		Conventional	Rotaform
h	SELLING PRICE APPROX	100	100
i	billet material	47	36
j	dies and tools	8	7
k	electricity	3	2
l	net scrap	4	2
m	DIRECT COST	62	47
n	DIRECT ADDED VALUE	38	53
o	ADDED VALUE/ MACHINE HOUR*	191	289
p	INCREASE AV/ MACHINE HOUR		51%

*Note 'o' does not include 'n'

The contribution calculations shown here concentrate on direct added value i.e. selling price minus materials and other volume-variable costs. They are based operating experience.

The contribution table shows relative costs as a proportion of the selling price of the conventional forging.

The lines in the contribution table are defined as follows
h selling price – ex works, as forged, unboxed
i billet material i.e. billet weight multiplied by materials cost
j dies and tools – materials content assuming own tool making facilities
k electricity – assumes electric billet heating and forging machine power
l net scrap – bar ends, cutting loss, forgings scrapped, flash etc. less recoveries by sale of scrap
m direct cost – the sum of *i* – *l*
n direct added value – selling price *h*-direct cost *m*
o added value per machine hour – unit direct added value *n* multiplied by production rate *e*
p increase in added value per machine hour – the difference between the orbital forged value and conventional value at line *o*, as a percentage of the conventional value.
The direct added value must be sufficient to cover all labour, overheads and profit.

Fig.1.6.14 Comparison of costings for orbital and conventional forging

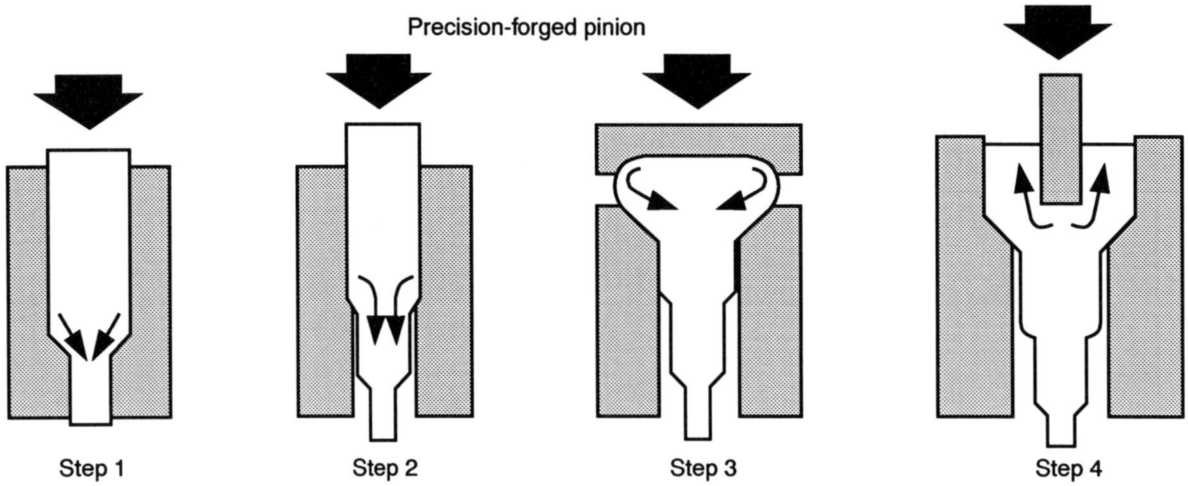

Step 1. Blank forward extruded to form first diameter.
Step 2. Second diameter is extruded.
Step 3. Heading and upsetting to rough head and rough form teeth on diagonal head face.
Step 4. Back extrusion to form blind hole and size O.D.

Fig.1.6.15 Production of a near net shape pinion shaft with gear teeth

Fig. 1.6.15

1.6.5 Sheet metal forming – press work

1.6.5.1 Conventional sheet forming

A very large number of parts can be made rapidly and economically using sheet metal as the starting material. The processes can be divided into those in which:

(a) the shape is blanked out and/or pierced;
(b) the sheet is bent into the required shape and
(c) the sheet is deformed so that combinations of stretching, drawing and bending occur which lead to changes in thickness as the required shape is produced.

In addition, markings can be applied in a fully closed engraved die as in coining.

BLANKING

Fairly simple presses and tools make this an inexpensive mass production method of producing flat, thin section parts. If holes are required without stringent tolerances these can be pierced by the press. It is often good practice to use several presses, each performing one operation to produce the finished part.

A wide choice of sheet thicknesses and sheet material is possible.

Typical parts which can be made in this way are laminations, washers, shaped plates, etc. The addition of a coining operation can apply decoration or lettering to the blanked out piece. Very accurate reproduction and finish is possible economically by coining. Metals suitable for coining include:

Mild steel	(carbon 0.3)	
Stainless steels	Cr 17/18, Ni 7/8, C 0.1	Cr 19, Ni 9, C 0.08
	Cr 18, Ni 12, C 0.12	Cr 12, C 0.15
	Cr 16, C 0.12	
Sterling silver	Ag 92.5, Cu 7.5	
Copper, gold and their alloys		

As with many processes the cost per part falls off rapidly with the number required. Some indication of how this cost falls with the size of the run is shown in:
Fig. 1.6.16 *Variation of cost per part with number for coining or blanking*

PRESS BENDING

The shaping of the blanked piece by bending enormously increases the scope of the press forming process.

Since some metal deformation is being introduced the sheet used may need to be in the annealed condition. Some work-hardening will be introduced and there is often a change in the surface appearance where severe bending has taken place.

For complex parts, dies require specialized engineering and are nearly always provided with ejector mechanisms. For some parts it is possible to dispense with a lower die and substitute a rubber pad. The sheet forms over the punch as it and the work is forced into the rubber. Not only can a wide range of materials be formed into parts by press work, but there is also a large latitude on size. The part formed can vary from a very small bracket to a complete motor car body section. Some idea of how cost per piece will fall for very large numbers of parts is given in:
Fig. 1.6.17 *Variation of cost per part with number for press bending using multiple separate dies*

DEEP DRAWING AND STRETCH FORMING

The sheet is forced into a die so that a cup or box is produced, changes of thickness occurring during the process. In some cases it is necessary to restrain the outside of the blank by a surrounding blank holder to prevent wrinkles from forming during the draw. Ironing, in which the space between the punch and die is less than the blank thickness, leads to uniformity of wall thickness in the finished part. Good lubrication is required in ironing operations in order to avoid galling and damage to die and workpiece.

The forming of cup-like shapes often involves both stretch forming, which is a purely tensile deformation, and the drawing of the metal between die and punch.

The structure of the sheet metal is often extremely important in deep drawing processes. Grain size, hardness and crystal orientation (texture) are all important considerations as well as the particular metal chosen.

The drawing reduction is defined by the blank diameter and the diameter of the shell being produced. This is illustrated in:
Fig. 1.6.18 *Relationship of draw ratio to diameter (D/d) and height (H/d) for a single drawn, shallow, flat-bottomed circular shell*

An approximate relationship between the depth of a deep drawn shell and the maximum recommended draw reduction, assuming anneals between draws and the use of a die holder, is given in:
Table 1.6.24 *Variation of shell depth with draw reduction for ductile metals*
The maximum load L, for drawing a round cup is given approximately by the relation:

$$L = dts\left(\frac{D}{k} - k\right)$$

where d = cup diameter, D = blank diameter, t = sheet thickness, s = tensile strength and $k = 0.6/0.7$ depending on the lubricant used.

RUBBER PRESS FORMING

In rubber press forming a rubber block or pad forms one platen while the form block is attached to the other platen. The rubber acts like a fluid and squeezes the sheet around the form nearly equally in all directions. Advantages of the method are economy of tooling and that thinning of the sheet is avoided. Disadvantages are high wear of rubber pads, lack of precision in forming and the possibility of wrinkling. The process can be used for thin gauges of aluminium alloys and austenitic stainless steels.

For forming an edge around a part it must exceed a certain width so that a sufficient forming force is developed. An indication of this requirement is the minimum flange width in millimetres as follows:

Annealed stainless steel	$5 + 4.5\,t$;
1/4 Hard stainless steel	15;
4% Cu 1.5% Mg aluminium alloys – soft	$2 + 2.5\,t$.

where t is sheet thickness

Although this process can be used for quantity production, it has attractions for prototypes because of the simplicity of the tooling.

1.6.5.2 Spinning, including flow forming and pipe spinning

Spinning is the progressive shaping of a metal component over a former of the required axi-symmetric geometry starting with a simpler or thicker shape of adequately ductile metal. Sections cylindrical, conical, spherical, elliptical or parabolic in form, or combinations of these can be produced either with open ends or as dome-shaped pieces. For some shapes a succession of formers are necessary and it is possible to produce bowl-shaped parts with rolled over or re-entrant edges.

The process(es) have been used for shapes up to 2 m diameter and pipes up to 3 m long.

The equipment ranges from a light lathe used for hand spinning to very large machines with power or hydraulic feed.

Typical section thicknesses are shown in:

Table 1.6.25 *Section thicknesses for differing materials and spinning processes*

The minimum radius that can be produced is about 50% greater than the thickness of the material.

Notwithstanding the large size of some of the machines, tooling costs are low and there is production versatility over a wide range of sizes.

One-off components can be produced for design and experimental purposes.

Manual spinning is suitable where limited numbers are required. The former can be of wood, the spinning equipment a lathe and the tools a hand held steel bar with rounded end. Manual spinning is suitable for production runs of up to 1000 parts. For larger production runs or where greater thicknesses of blanks are involved, power operated spinning machines would be used. Above 1000 pieces in a run, deep drawing may be more economical than spinning.

Parts previously made by casting, forging and machining are now produced by welding together spun halves. The risk of porosity is eliminated and a smaller wall section can be used because of the higher material strength produced by cold working. All ductile metals obtainable as sheet can be spun.

The three distinct processes of spinning, flow forming and pipe spinning may be employed separately or may be combined. The principle of the three processes is illustrated in:

Fig. 1.6.19 *Spinning, flow forming and pipe spinning*

SPINNING (OR MANUAL SPINNING)

In spinning the operator uses a tool of wood or metal to form a blank, usually a flat disc, progressively over a mandrel of wood or metal. The technique of spinning a spherical shell of uniform thickness from a flow formed blank is shown in:

Fig. 1.6.20 *Spinning combined with flow forming*

A hard roller at the end of the tool reduces the force required and is used when forming hard materials, thick blanks and uneven shapes. The thickness of the blank that can be formed manually depends on the type of material used and is generally less than 4 mm. During spinning the radius of each element of the blank is reduced significantly compared with its original radius, but the material thickness changes only to a small extent. Manual spinning is used in small-scale operations where high dimensional accuracy is not a primary requirement.

FLOW FORMING

Flow or shear forming is often used in the manufacture of conical, spherical or oval shells. It differs from spinning in that the radial position of an element is unchanged but the thickness is reduced and the material strain hardens considerably. Reductions of up to 75% are used in flow forming and pipe spinning.

Where a material is not amenable cold to one of the spinning processes it can sometimes be spun hot. A list of materials which have been fabricated by spinning, the conditions under which they can be spun and a guide to the percentage of reduction that may be used is given in:

Table 1.6.26 *Percentage of reduction for flow forming or pipe spinning without intermediate anneals*

Flow forming may be used to produce articles with uniform or varying wall thickness provided that the preform is machined to the necessary shape.

The procedure for the manufacture of a spherical configuration shell of uniform wall thickness is illustrated in:

Fig. 1.6.21 *Flow forming of a shell of uniform thickness and spherical configuration*

Shells of spherical configuration but non-uniform wall thickness can be flow formed from flat discs as shown in:

Fig. 1.6.22 *Flow forming of a spherical configuration shell of non-uniform thickness*

A typical procedure used to manufacture a 1.52 m aluminium hemisphere is illustrated in:

Fig. 1.6.23 *Two-stage flow forming of aluminium hemisphere*

After blank machining the flange is shaped to the required diameter and reduced 40% in wall thickness. The material is then solution treated, annealed or stress relieved as appropriate, flow formed in one pass and finally heat treated or aged.

PIPE SPINNING

Pipe spinning is a process used in the manufacture of precision stainless cylindical shapes with or without closed ends for military aerospace and commercial use usually in steel, stainless steel or aluminium alloys.

The process resembles flow forming in that the products manufactured are axially symmetric and intentional thinning of the preform is carried out during the process. The preform is in the shape of a thick-walled cylinder or cup made by forging or machining. Forming is carried out by means of one, two or three rollers.

There are forward and backward variants of the process as shown in:

Fig. 1.6.24 *Pipe spinning*

In forward spinning, the forming tool moves in a direction parallel to the mandrel towards the free end of the pipe. Both the tool and the undeformed part of the pipe move in the same direction. In backward spinning the tool moves towards the fixed end of the mandrel and the deformed pipe moves in the opposite direction. In forward spinning, the deformed tube is under tension, while in backward spinning, the deformed tube is stress-free except for residual stresses. The undeformed tube in backward spinning is under compression.

Pipe spinning may be used for the manufacture of dimensionally accurate thin-walled pipes which cannot be produced satisfactorily by normal pipe forming processes but the spiral feed of the tool induces a corresponding waviness in the outer surface and a corresponding variation in the plastic deformation. Swaging is a possible alternative process for small diameter pipes.

The wall thickness of the preform may be reduced by over 75% with a concomitant increase in length. The maximum thickness to which a preform can be spun depends on the ductility and the strain-hardening characteristics of the preform material.

1.6.5.3 Roll forming

Strip is fed continuously into shaped pairs of rolls, often successively, to form continuous lengths of complex shaped channel. A very wide range of shapes is possible and the process can also produce shaped tubes. These can be sealed by a continuous welding operation. Although usually operated for thin strips, heavy duty roll stands can be used for thick material. The speed of forming is high, usually in the range 25–30 m/min. Some examples of the type of section that can be produced are shown in:

Fig. 1.6.25 *Some typical roll formed sections*

In many cases a part could be made either from a roll formed length or by press work. A comparison of cost per piece for press and roll form manufacture of a steel shape like that marked (a) in Fig. 1.6.25 is given in:

Fig. 1.6.26 *Comparison of cost per piece for part made by roll forming and by press forming*

The piece length for the comparison was about 1 m.

The process is applicable to nearly all metals which can be obtained as rolled strip and have reasonably good formability. Steel, aluminium alloys and copper alloys are very suitable for roll forming. Titanium alloys, nickel base alloys and heat resistant alloys present difficulties and would need to be specially investigated for mass production roll formability.

1.6.5.4 Fine blanking

Fine blanking is a relatively new process in which extremely close clearances (typically 0.005 mm), and very rigid presses and die sets (to prevent sheet bending) produce a blank with no tearing at the sheared edge. These parts can be used as gears, cams, etc., and do not require any machining of the edges. Details of the cost savings which can be achieved for production of a typical component by fine blanking as compared with conventional pressing are given in:

Fig. 1.6.27 *Cost advantages for production of a plunger by fine blanking*

Manufacturing the 'plunger' parts in a progressive fine-blanking tool eliminates the time consuming shaving operation required to achieve cleanly sheared surfaces. As the chamfers are coined in a station before shearing the costly milling operation could also be excluded. From the table cited above it can be seen that approximately 7.6 h manufacturing time can be saved in the production of 1000 parts. The extra tooling costs (780 h for building the progressive fine-blanking tool) is amortized after the production of 100 000 to a maximum of 200 000 parts – dependent on the hourly rates involved. As at least 500 000 plungers have to be made each year the change of working method paid for itself in a short period.

1.6.5.5 Trimming and vibratory trimming

The characteristics of trimming and vibratory trimming are compared with those of fine blanking in:

Table 1.6.27 *The characteristics of fine blanking, trimming and vibratory trimming compared*

Table 1.6.24 Variation of shell depth with draw reduction for ductile metals

Depth of shell in diameters	*No. of draws*	*Percentage reduction*			
		1st	*2nd*	*3rd*	*4th*
0.5	1	40	–	–	–
1.0	2	40	25	–	–
1.5	3	40	25	15	–
2.0	4	40	25	15	11

Table 1.6.24

Table 1.6.25 Section thicknesses for differing materials and spinning processes

Material	*Section size*		
	Manual	*Hydraulic*	*Numerical control*
Aluminum	5 mm	75 mm	5 mm
Mild steel	2.5 mm	37 mm	3 mm
Stainless steel	–	18 mm	1.5 mm

Table 1.6.26 Percentage of reduction for flow forming or pipe spinning[a] without intermediate anneals

Material	*Flow forming*		*Pipe spinning*
	Cone	*Hemisphere*	
4130	75	50	75
6434	70	50	75
6340	65	50	75
D6AC	70	50	75
Rene 41	40	35	60
A286	70	55	70
Waspaloy	40	35	60
18% Ni Steel	65	50	75
321 Stainless	75	50	75
17–7 PH Stainless	65	45	65
347 Stainless	75	50	75
410 Stainless	60	50	65
H11 Tool steel	50	35	60
6–4 Ti[b]	55	–	75
B120 VCA Ti[b]	30	–	30
6–6–4 Ti[b]	50	–	70
Commercially pure Ti[b]	45	–	65
Molybdenum[b]	60	45	60
Pure beryllium[b]	35	–	–
Tungsten[b]	45	–	–
2014 Al	50	40	70
2024 Al	50	–	70
5256 Al	50	35	75
5086 Al	65	50	60
6061 Al	75	50	75
7075 Al	65	50	75

[a] These percentages of reduction should be refined to suit specific conditions. They are listed as a guide which, if followed, will reduce development time.
[b] Spun hot.

Table 1.6.26

Table 1.6.27 The characteristics of fine blanking, trimming and vibratory trimming compared

Feature	*Fine blanking*	*Trimming*	*Vibratory trimming*
Max size (mm)	800 × 700	250 dia.	250 dia.
Max thickness (mm)	16–25	8	30
Min thickness (mm)	0.4	0.1	0.1
Min hole dia. (S = sheet thickness)	(0.6–0.8) S ≥ 0.8 mm	(0.6–0.8) S ≥ 2 mm	(0.6–0.8) S ≥ 2 mm
Tolerances IT Inner and outer surfaces	6–9	7–9	7–9
Cut surface finish (μm)	0.3–1.5	1	0.5
Sharp corner radii	(0–0.2) S	0	0
Materials	Unalloyed lead-free C steels with C < 0.7% (0.7–1.0% spheroidized cementite).	Steels with a maximum 1.0% C, malleable iron, non-ferrous metals with poor deformability but good ductility.	
	Alloy steels with good deformation properties.		
	Brass (max. 40% Zn), copper, bronze, monel, aluminium and some aluminium alloys		
Least number of pieces	2000–10 000	5000–10 000	10 000–50 000
Parts per minute	15–60	6–14	6–14

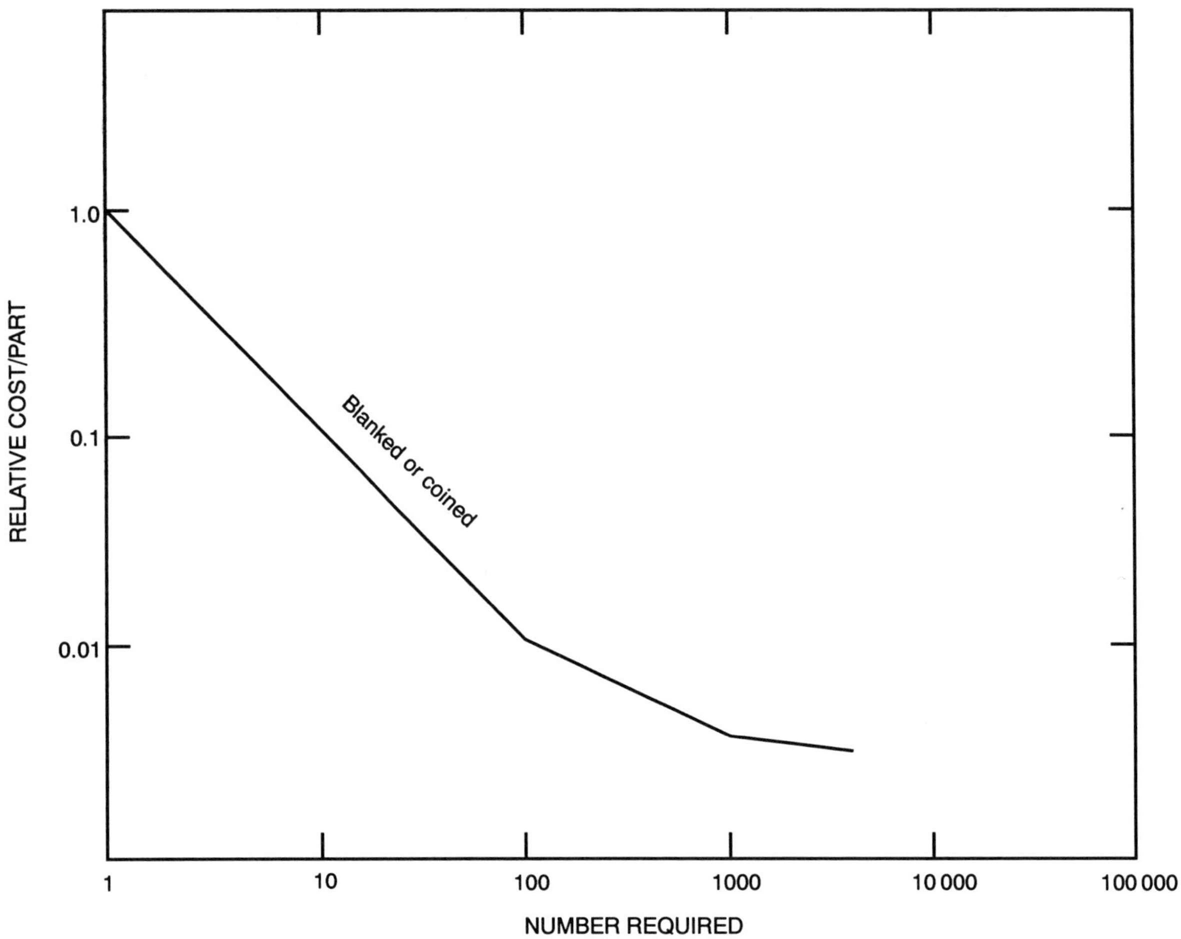

Fig. 1.6.16 Variation of cost per part with number for coining or blanking

Fig. 1.6.16

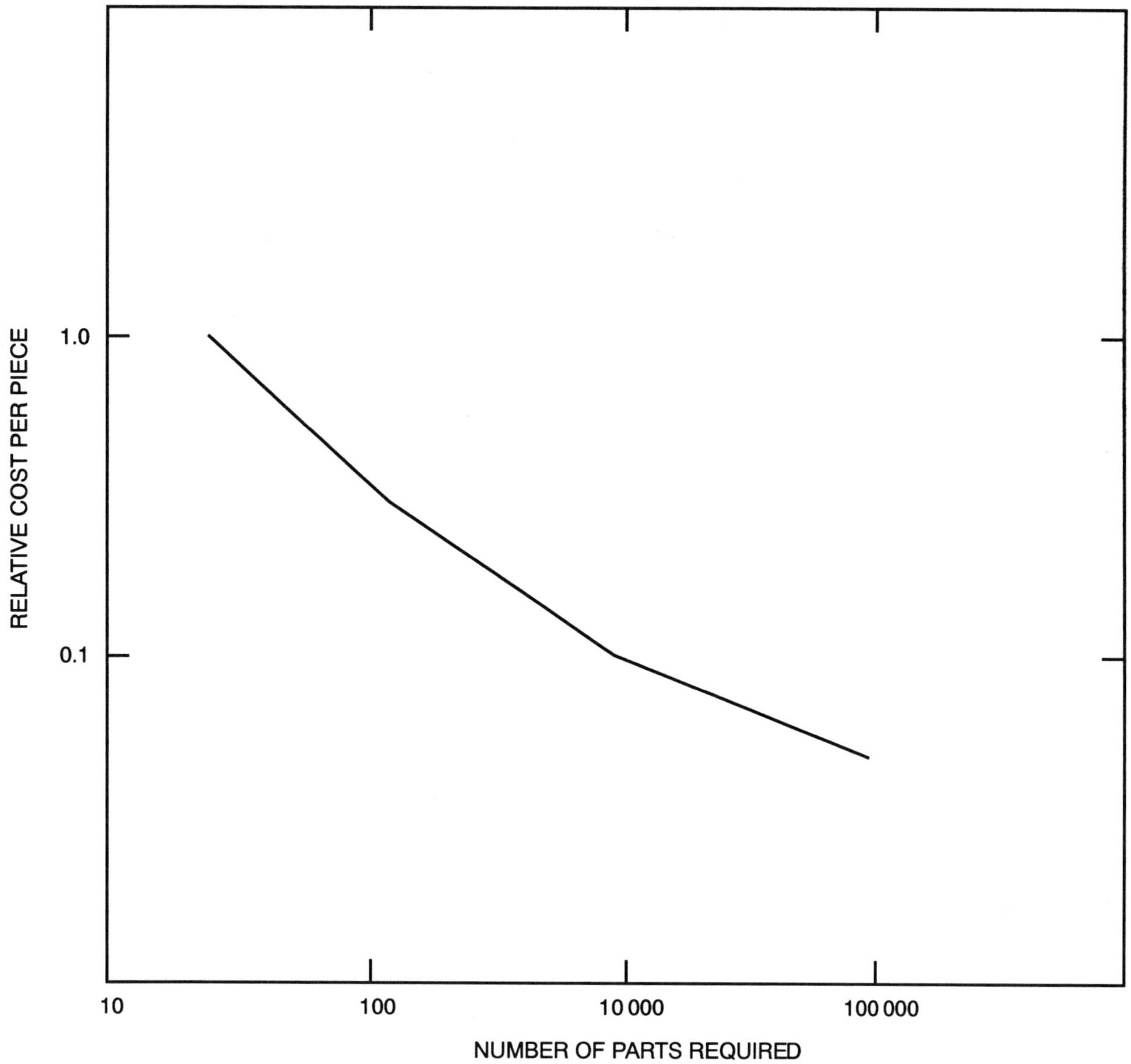

Fig. 1.6.17 Variation of cost per part with number for press bending using multiple separate dies

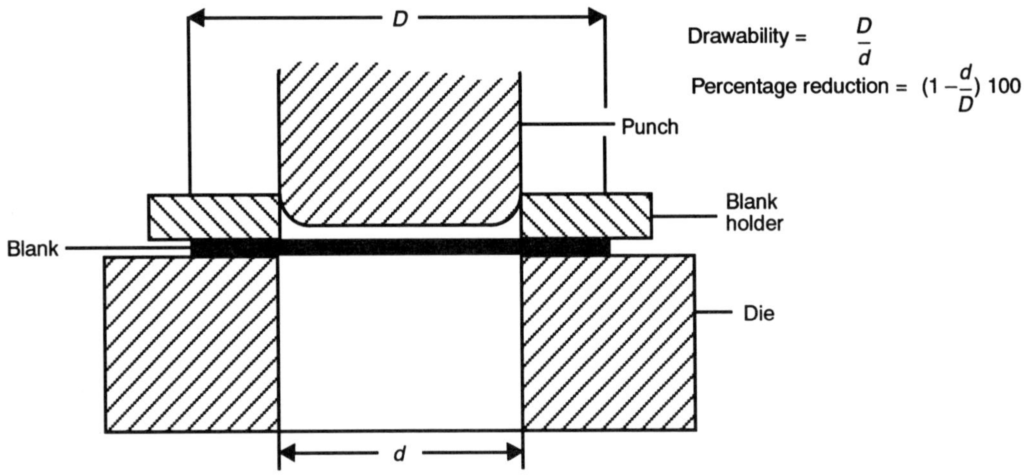

Material Grade	Drawing quality, rimmed steel	Deep drawing quality aluminium stabilised steel	Low carbon, niobium stabilised cold rolled steel	Hot rolled drawing quality steel	Austentic stainless steel Type 302	Ferritic stainless steel Type 444 (Ti stab.)	Brass and copper	Aluminium (1100 series)	Zinc
Maximum D/d	2.15	2.2	2.25	2.1	2.2	2.3	2.2	2.1	1.8
Maximum H/d	0.9	0.95	1.0	0.85	0.95	1.0	0.95	0.85	0.6

Fig. 1.6.18 Relationship of draw ratio to diameter (*D/d*) and height (*H/d*) for a single drawn, shallow, flat-bottomed circular shell

Fig. 1.6.18

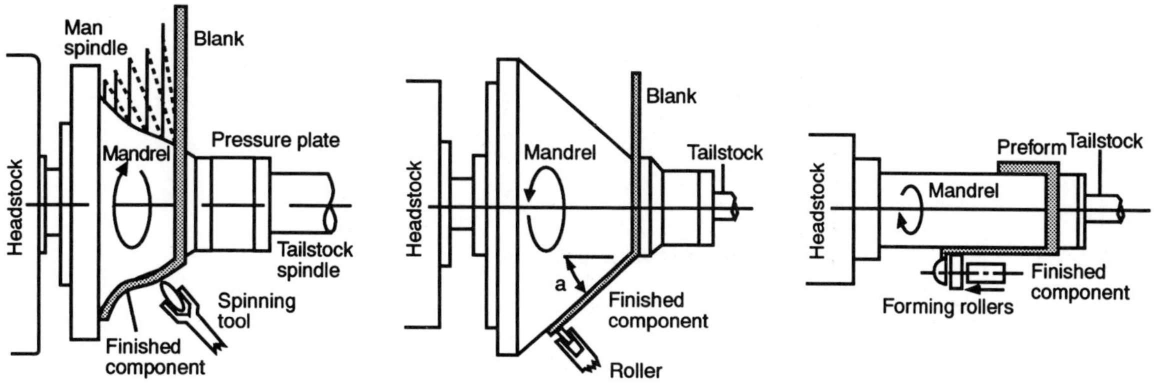

Spinning
In metal spinning, the thickness of the finished component and blank are the same. Tool can be manually or hydraulically controlled.

Flow Forming
Using hydraulic loading wall thinning occurs to increase strength/weight ratio. $\alpha \nless 15°$

Pipe Spinning
Preformed blank used to form cold worked thin walled cylinders. Wall thickness reduction may reach 90%.

Fig. 1.6.19 Spinning, flow forming and pipe spinning

Fig. 1.6.19

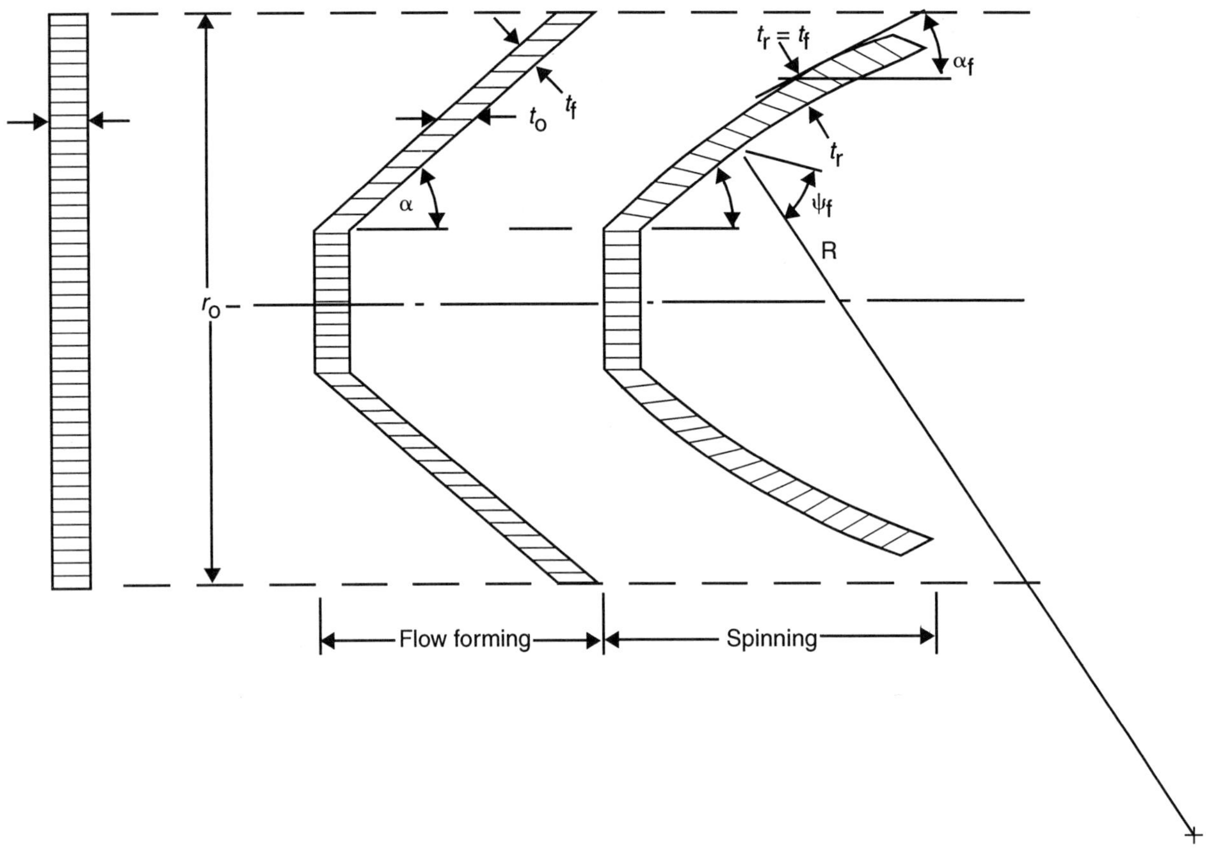

Fig. 1.6.20 Spinning combined with flow forming

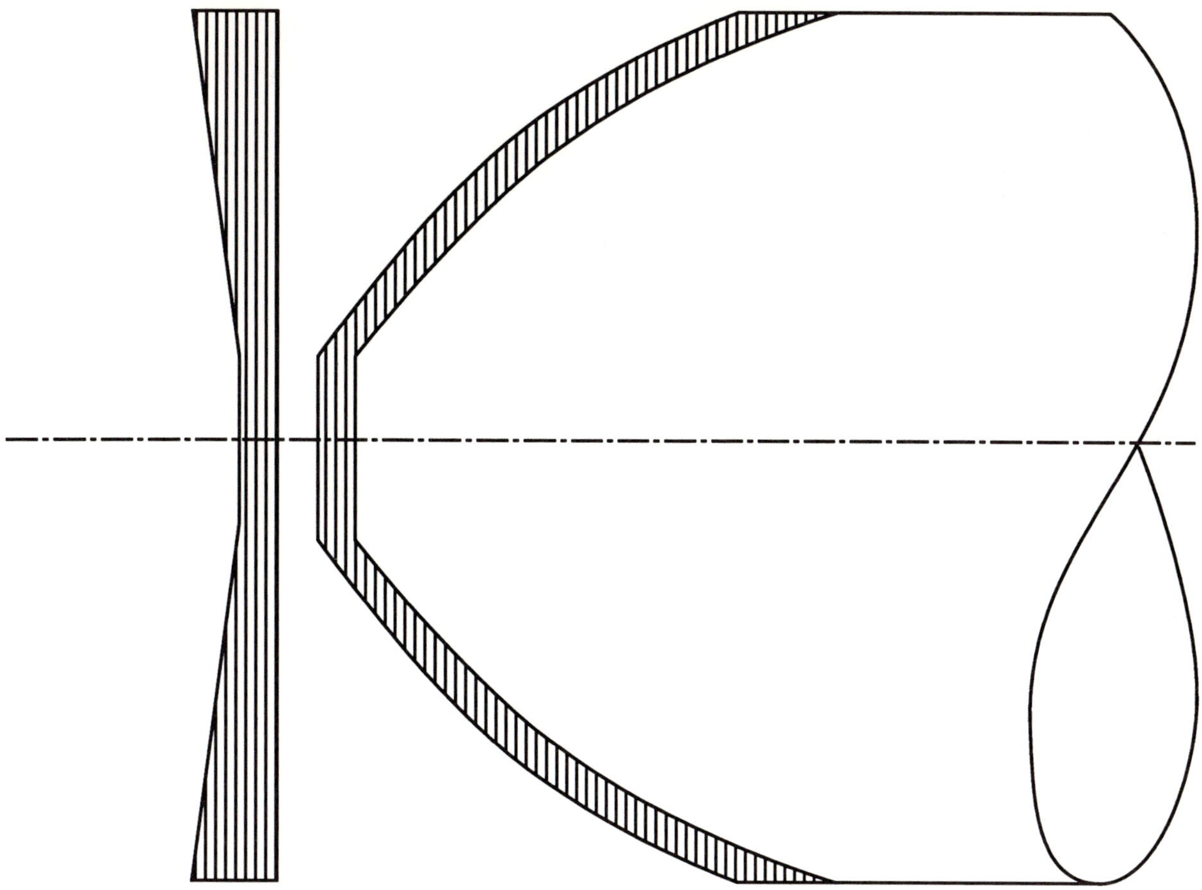

Fig. 1.6.21 Flow forming of a shell of uniform thickness and spherical configuration

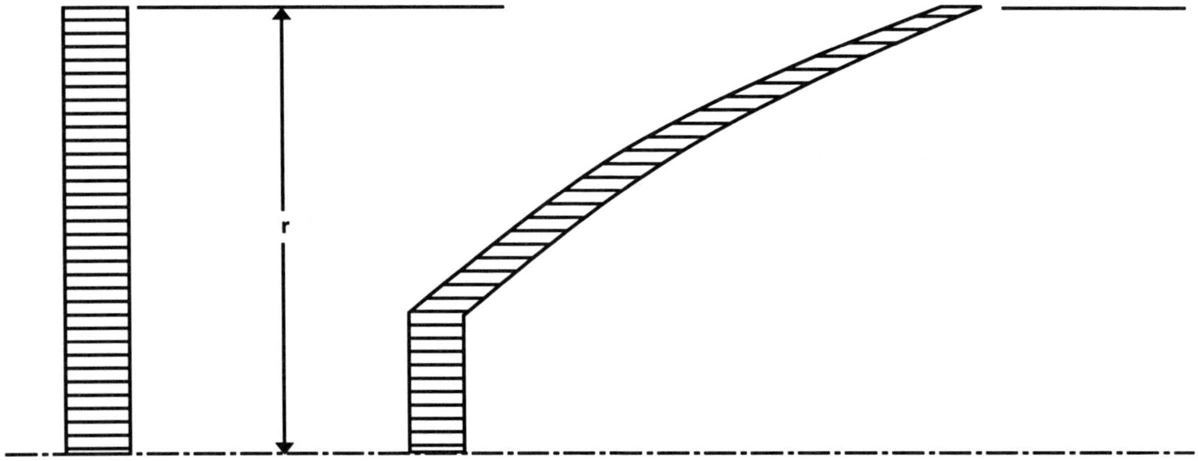

Fig. 1.6.22 Flow forming of a spherical configuration shell of non-uniform thickness

Fig. 1.6.22

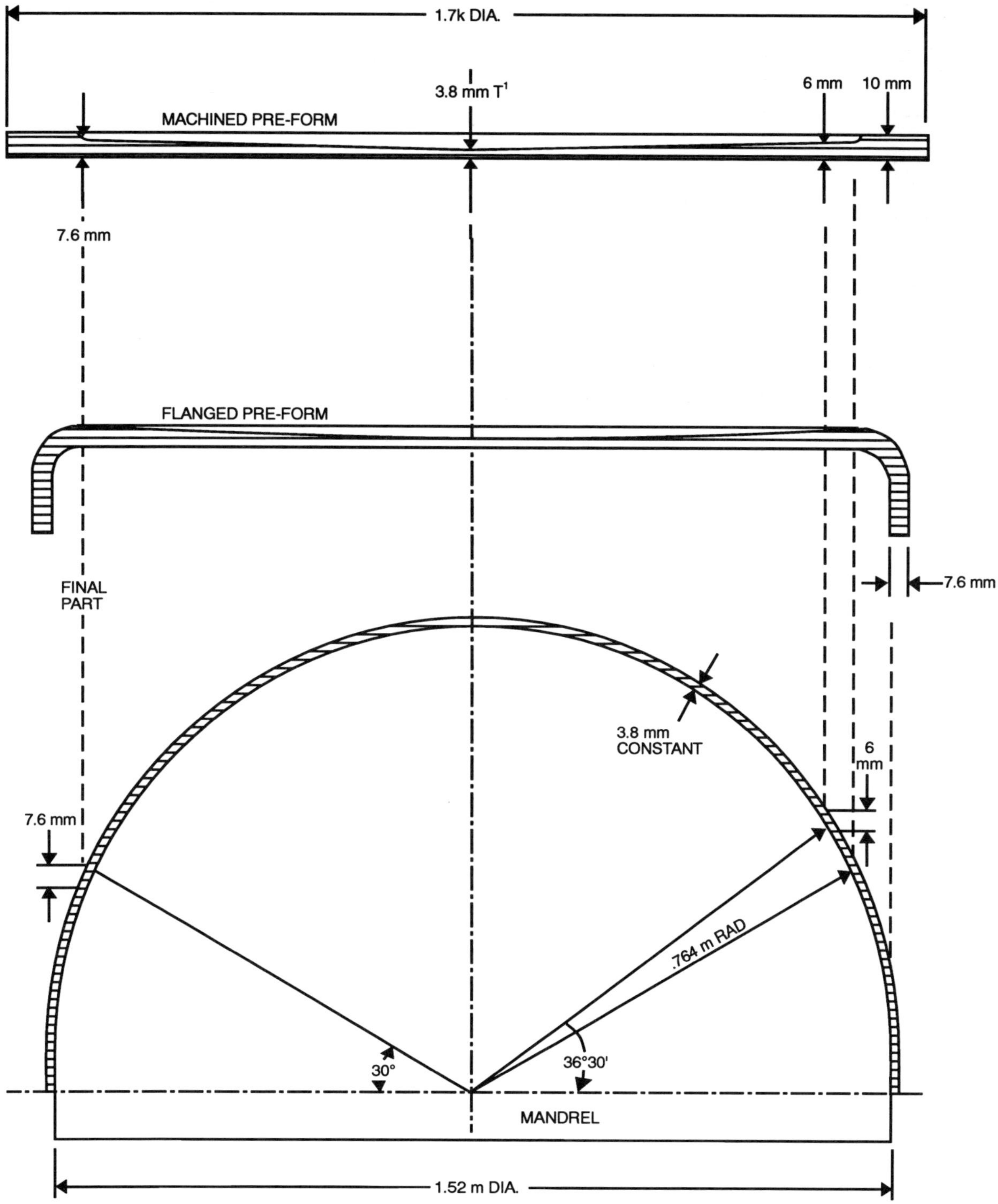

Fig. 1.6.23 Two-stage flow forming of aluminium hemisphere

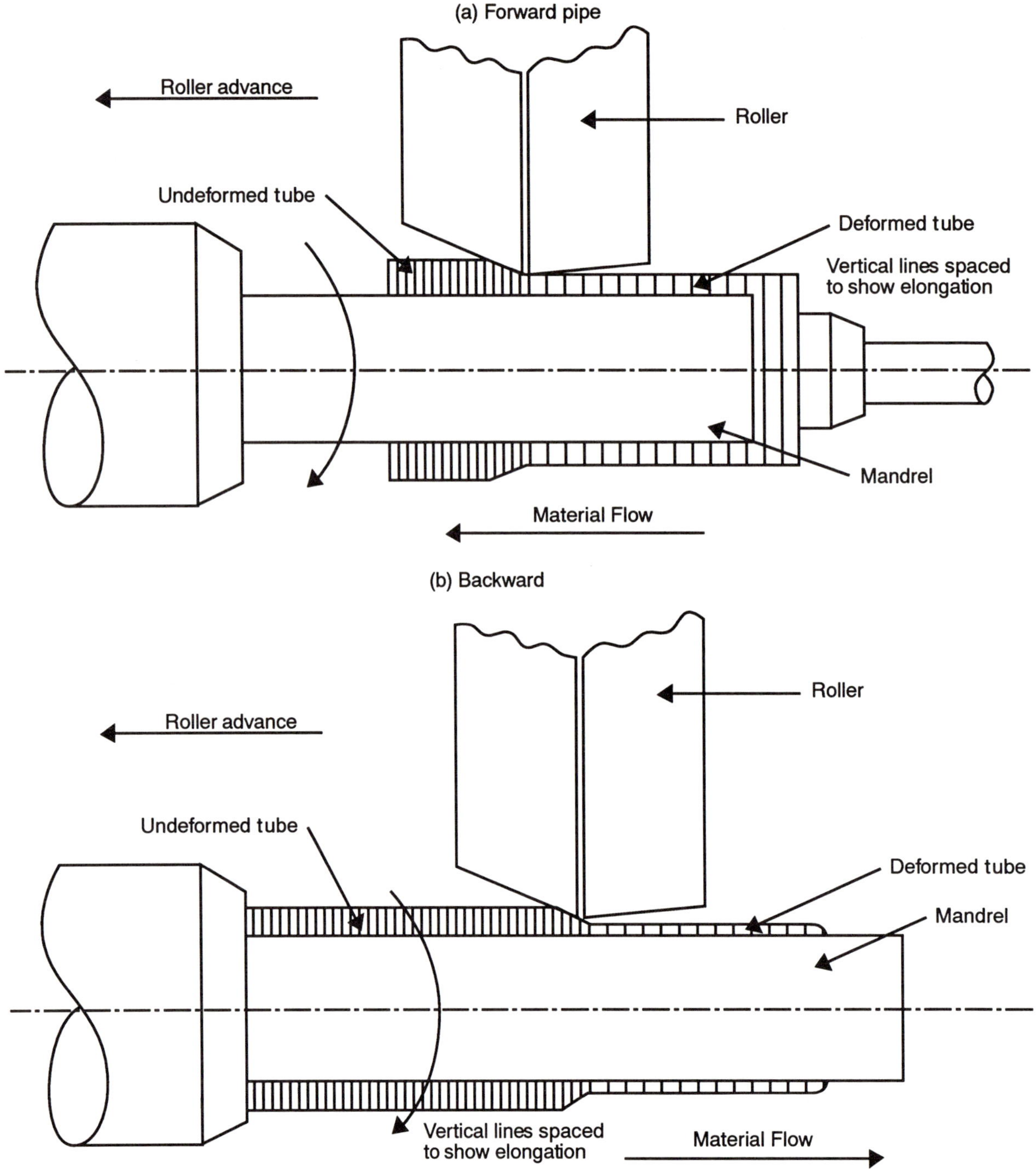

Fig. 1.6.24 Pipe spinning

Fig. 1.6.24

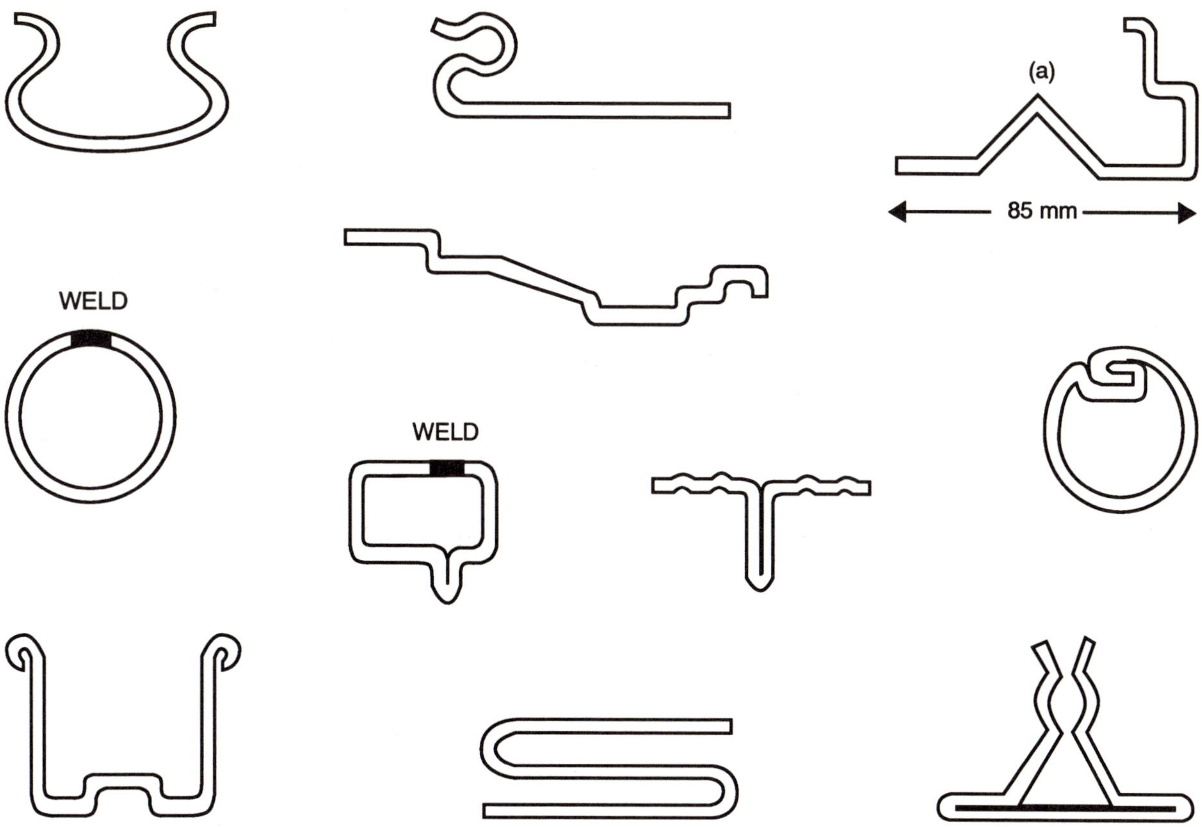

Fig. 1.6.25 Some typical roll formed sections

Fig. 1.6.25

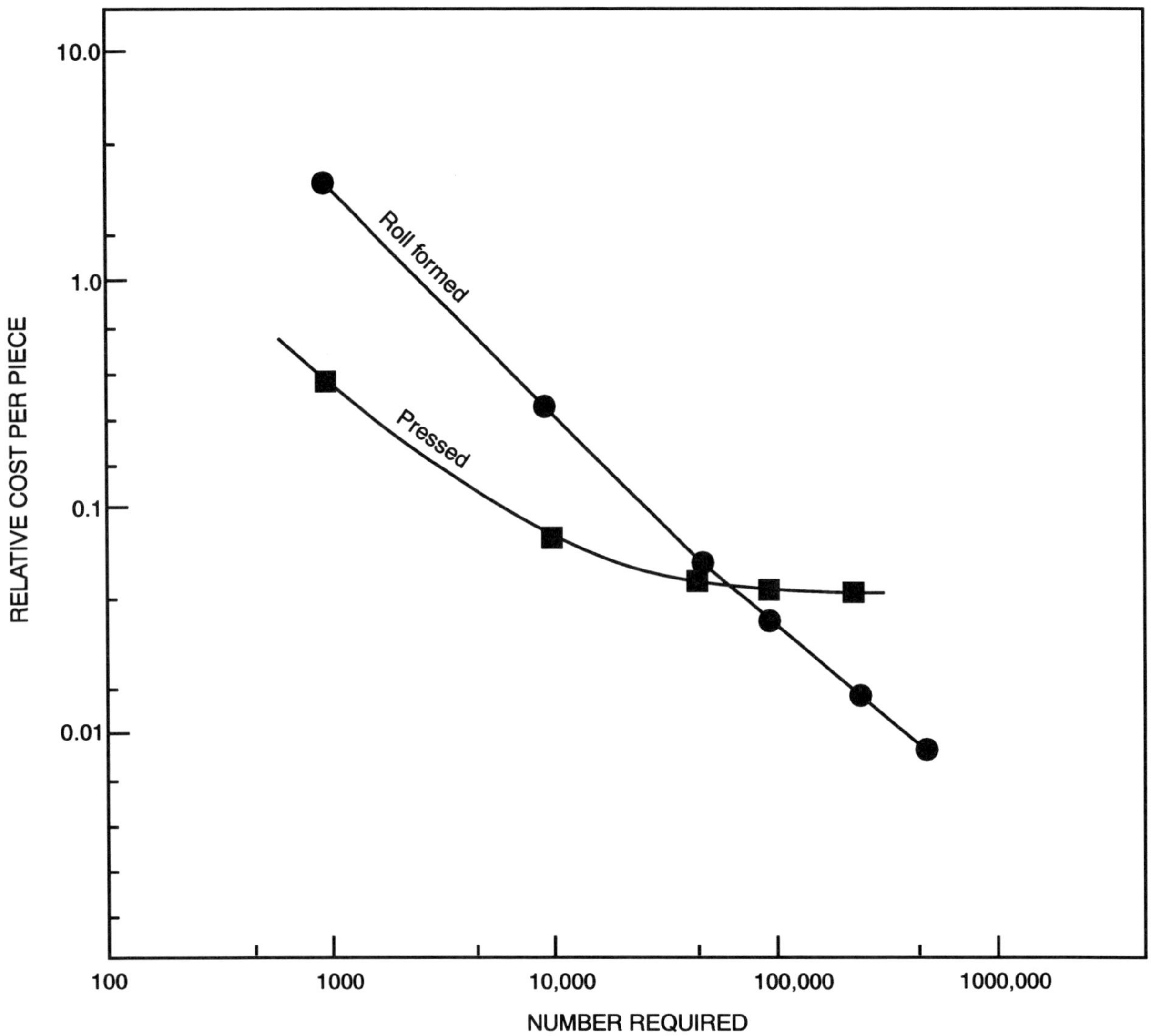

Fig. 1.6.26 Comparison of cost per piece for part made by roll forming and by press forming

Fig. 1.6.26

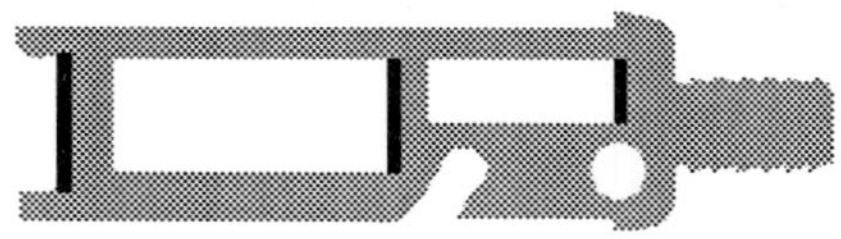

Plunger, scale 1:1

PLUNGER	CONVENTIONAL	FINE BLANKING	
– Blank (single punch tool)	20	– Fine-Blank (twin punch tool)	14.3
– Shave	150		
– Mill	300		
– De-burr	30	– De-burr	30
	500 min./1000 parts		44.3 min./1000 parts

Production time for tools

– Normal blanking tool (Single punch)	150	– Progressive fine blanking tool (twin punch)	
– Shaving tool	120		
– Milling jig	50		1100
	320 hours		1100 hours

Fig. 1.6.27 Cost advantages for production of a plunger by fine blanking

1.6.6 Machining

1.6.6.1 Introduction

More than 80% of all manufactured parts must be machined before they are completed. The ease and accuracy with which a material can be machined depends on how well the surface of the material can be matched to the movements of the machine and the cutting edge of the tool used. The machined surface can be generated by a form tool which matches the shape to be produced in the surface or it can be generated by feeding the cutting tool in and out along the length of the workpiece. Although form tools lack flexibility and are relatively expensive their use is economically favourable in high volume production.

The cost of machining a component will depend on the machinability of the material, a parameter which is very difficult to define, on the tolerances required and on the desired surface finish.

1.6.6.2 Machinability

The machinability of a material is usually assessed in terms of four factors: tool life, cutting forces and power required, surface finish and swarf removal. Small changes in a material (usually in the microstructure) which do not affect the in-service performance can markedly alter the machining characteristics. This is particularly serious for mass production where unpredictable machinability can cause chaos.

A rough comparison of the machinability of different metals (in terms of metal removal rates) compared with free cutting mild steel (=100) is given in:

Table 1.6.28 *Machinability of common metals*

Due to the wide variation of machine efficiency and utilization, the determination of the cost per unit weight of metal removed for a given machining operation is best done on a local level. As a general guide, the cost of copy milling mild steel may be 15 times greater per unit weight of material removed than repetition turning.

1.6.6.3 Tolerances

Tolerances for machined components are quoted in terms of ISO (International Standards Organization) grades, as shown in:

Table 1.6.29 *Standard tolerances for machining*

Increasing the tolerance grade will increase the cost of the part. The relation between tolerances and production costs is shown in:

Fig. 1.6.28 *Relationship between tolerances and production costs*

Since production costs increase rapidly as tolerances are tightened, a systematic approach to tolerance specification will attempt to keep production costs down by maximizing allowed production tolerances.

1.6.6.4 Surface finish

The cost of a machined part will depend on the surface finish specified. The roughness obtained on a component by a range of metal removal operations, together with the approximate cost of obtaining that surface finish relative to a very rough machined (very coarse feed) surface, is given in:

Table 1.6.30 *Surface finishing costs*

This table illustrates why care must be taken to avoid the high cost penalties involved in over-specifying surface finish quality.

1.6.6.5 Tool materials and cutting speeds

Guidance on the selection of machine tool tip material is given in:
Table 1.6.31 *Guide to cutting tool materials and speeds for general purpose machining*

Cutting speed ranges are given for each material, the lower speed being for machining with a high speed tool steel at a large depth of cut (typically 16 mm). The upper speed limit is for machining under good conditions with a coated carbide tool at a small depth of cut (typically 1 mm). Other things being equal, slower speed machining is more economic since it saves on tool life.

Table 1.6.28 Machinability of common metals

Material	***Machinability index***
Ball-bearing steel: BS970–534A99	30
Inconel: Ni 77.5–Cr 16.0–Fe 6.5	35
Phosphor bronze: Cu 9.5–Sn 5–P 0.2	40
Stainless steel: Cr 18–Ni 8	45
Nickel steel: Ni 3.5–C 0.3	50
Nickel chrome steel: C 0.45	50
Pearlite cast iron	50
Wrought iron	50
Monel: Ni 70–Cu 30	55
Structural steel	60
Copper – 1/4 hard rolled	60
Aluminium bronze: 5% Al	60
Chrome–molybdenum steel: BS970–830M31	65
Cast steel: 0.35 C	70
Chrome steel – free-cutting	70
Cast copper	70
α Brass	80
Cast iron – soft	80
Free-cutting mild steel	100
Malleable cast iron – ferritic	120
Free cutting α/β brass	200–400
Aluminium – half hard	300–1500
Magnesium: Al 6.5	500–2000

Table 1.6.29 Standard tolerances for machining

Nominal sizes mm		*Tolerance grades*																	
Over	*To*	*IT 01*	*IT 0*	*IT 1*	*IT 2*	*IT 3*	*IT 4*	*IT 5*	*IT 6*[a]	*IT 7*	*IT 8*	*IT 9*	*IT 10*	*IT 11*	*IT 12*	*IT 13*	*IT 14*	*IT 15*[b]	*IT 16*[b]
–	3	0.3	0.5	0.8	1.2	2	3	4	6	10	14	25	40	60	100	140	250	400	600
3	6	0.4	0.6	1	1.5	2.5	4	5	8	12	18	30	48	75	120	180	300	480	750
6	10	0.4	0.6	1	1.5	2.5	4	6	9	15	22	36	58	90	150	220	360	580	900
10	18	0.5	0.8	1.2	2	3	5	8	11	18	27	43	70	110	180	270	430	700	1100
18	30	0.6	1	1.5	2.5	4	6	9	13	21	33	52	84	130	210	330	520	840	1300
30	50	0.6	1	1.5	2.5	4	7	11	16	25	39	62	100	160	250	390	620	1000	1600
50	80	0.8	1.2	2	3	5	8	13	19	30	46	74	120	190	300	460	740	1200	1900
80	120	1	1.5	2.5	4	6	10	15	22	35	54	87	140	220	350	540	870	1400	2200
120	180	1.2	2	3.5	5	8	12	18	25	40	63	100	160	250	400	630	1000	1600	2500
180	250	2	3	4.5	7	10	14	20	29	46	72	115	185	290	460	720	1150	1850	2900
250	315	2.5	4	6	8	12	16	23	32	52	81	130	210	320	520	810	1300	2100	3200
315	400	3	5	7	9	13	18	25	36	57	89	140	230	360	570	890	1400	2300	3600
400	500	4	6	8	10	15	20	27	40	63	97	155	250	400	630	970	1550	2500	4000
500	630	–	–	–	–	–	–	–	44	70	110	175	280	440	700	1100	1750	2800	4400
630	800	–	–	–	–	–	–	–	50	80	125	200	320	500	800	1250	2000	3200	5000
800	1000	–	–	–	–	–	–	–	56	90	140	230	360	560	900	1400	2300	3600	5600
1000	1250	–	–	–	–	–	–	–	66	105	165	260	420	660	1050	1650	2600	4200	6600
1250	1600	–	–	–	–	–	–	–	78	125	195	310	500	780	1250	1950	3100	5000	7800
1600	2000	–	–	–	–	–	–	–	92	150	230	370	600	920	1500	2300	3700	6000	9200
2000	2500	–	–	–	–	–	–	–	110	175	280	440	700	1100	1750	2800	4400	7000	11 000
2500	3150	–	–	–	–	–	–	–	135	210	330	540	860	1350	2100	3300	5400	8600	13 500

Tolerance unit 0.001 mm.
[a] Not recommended for fits in sizes above 500 mm.
[b] Not applicable to sizes below 1 mm.

Table 1.6.30 Surface finishing costs

Surface finish	*RMS (μm)*	*Relative cost*
Very rough machined (very coarse feed)	50	1
Rough machined	25.6	3
Semi-rough machined	12.8	6
Medium machined	6.4	9
Semi-fine machined	3.2	13
Fine machined	1.6	18
Coarse ground	0.8	20
Medium ground	0.4	30
Fine ground	0.2	35
Superfine lapped	0.1	40

Table 1.6.30

Table 1.6.31 Guide to cutting tool materials and speeds for general purpose machining

Machined material	*Tool material*					*Surface cutting speed (m/s)*	*Remarks*
	Carbon steel	*High speed steel*	*Cast cobalt alloys*	*Sintered carbide* *WC/Co*	*Sintered carbide* *WC + TIC/TaC*		
Steels Free-machining low carbon steels	√	√	√		√	0.5–7.0	
Low carbon steel	√	√	√		√	0.5–5.0	
Medium carbon steel		√	√		√	0.2–4.5	Machinable to HRc 38/42
High carbon steel		√	√		√	0.13–4.0	As above
Cast steel (medium carbon)	√	√	√		√	0.25–4.0	As above (125 → 300 BHN)
High manganese steel		√	√		√	0.25–0.5	
High speed steel			√		√	0.12–2.5	Annealed
Hot work die steel			√		√	0.2–3.0	Annealed; sometimes as hardened
Tool steel			√		√	0.22–1.75	Annealed
Ferritic stainless steel[a]	√	√	√		√	0.4–4.0	
Martensitic stainless steel[a]		√	√		√	0.1–4.0	Annealed; sometimes as hardened
Austenitic stainless steel	√	√	√	√		0.15–3.6	
PH stainless steel		√	√	√		0.13–2.7	In range HRc 32 to 48 (150 → 440 BHN)
Cast irons Grey cast iron	√	√	√	√		0.5–5.0	Free carbides cause tool chipping
Nodular cast iron	√	√	√	√		0.5–5.0	
Chilled cast iron			√	√		0.75–3.0	
Non-ferrous materials Aluminium alloys – wrought	√	√		√		2.5–10	
– cast	√	√		√		2.0–10	
Beryllium				√		0.5–3.0	Dust and chips highly toxic
Brass and leaded copper	√	√	√	√		2–8	
Bronzes	√	√	√	√		1–3	
Cobalt base alloys		√		√		0.25–0.7	
Copper alloys – cast	√	√	√	√		0.4–5.5	
Magnesium alloys		√		√		2.5–13	Fire risk with fine chips
Manganese alloys		√	√	√		0.25–0.3	
Molybdenum alloys				√		1.5–1.75	
Nickel base alloys		√	√	√		0.1–2.25	Often difficult to machine
Tantalum		√	√	√		1–4	Chlorinated or subzero coolant preferred
Thallium		√		√		0.75–1	Oxide debris poisonous. Mineral oil coolant
Thorium		√		√		2.5–3	Pyrophoric – flood water coolant required
Titanium alloys		√		√		0.13–1.3	Coolant essential
Tungsten alloys				√		0.3–0.8	Machining improved at > 300°C
Zinc alloys		√	√	√		1–8	
Zirconium				√		0.4–1.6	Pyrophoric – water-soluble coolant required

[a] includes free-machining grades.

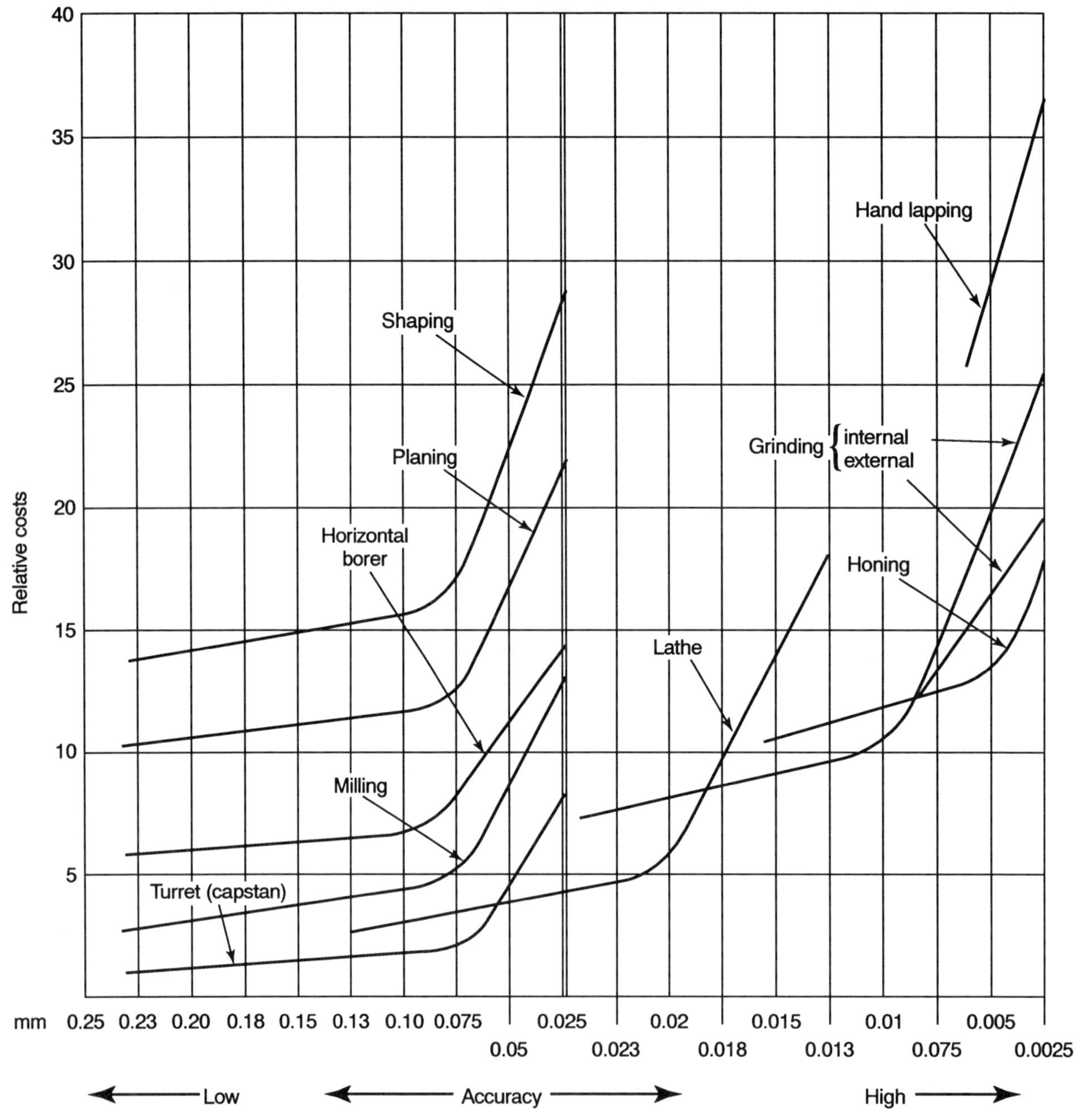

Fig. 1.6.28 Relationship between tolerances and production costs

Reproduced with permission from *The Management of Design for Economic Production*, BSI PD6470; published by BSI, 1996.

Fig. 1.6.28

1.6.7 Surface treatments for metal components

1.6.7.1 Introduction

Surface treatments are applied to metallic substrates in order to enhance some aspect of performance.

Most commonly, the performance criteria to be enhanced are corrosion and/or wear resistance.

However, surface treatments can also be applied to enhance a wide range of other criteria e.g. frictional characteristics, fatigue strength, thermal conductivity, electrical conductivity, magnetic properties, surface finish or merely aesthetic appearance.

Sometimes, surface treatment is needed because the product requires a combination of properties (e.g. high surface hardness and good toughness) which cannot readily be attained in a single material.

In other applications, the selection of an alternative (and more expensive) bulk material would be an option. In these cases, surface treatment would be used only when it is the more cost effective option.

The relative costs of the two options must be evaluated in terms of:

(i) raw material cost
(ii) manufacturing cost
(iii) maintenance requirements (e.g. in some cases, it may be cheaper to replace a low cost component on a frequent basis; in others, the use of surface treatment may be desirable to reduce expensive down-time)
(iv) desirability of an *in situ* deposited coating

1.6.7.2 Surface treatment processes

The wide range of available surface treatment processes can be divided into five different categories, differentiated by their influence on the original substrate material:

Coatings

Deposition of a layer of a separate material onto the surface of the substrate, without any chemical reaction between coating and substrate. Specific examples of coating processes are:

Paints, powder coatings and other organic coatings.
Electrodeposition and electroless (autocatalytic) plating.
Dip coating methods (including hot dip galvanizing).
Other zinc coating methods.
Enamelling.
Thermal spraying.
Weld deposition.
Roll cladding.
Chemical vapour deposition (CVD), with and without plasma assistance.
Physical vapour deposition (PVD), with and without plasma assistance.

These coating process types are described in:
Table 1.6.32 *Coating processes*

Chemical conversion treatments

Chemical reaction between the surface layers of the original substrate and an external reactant. Examples of treatments of this type are given in:
Table 1.6.33 *Chemical conversion coatings*

Microstructural modification of the surface layers of the substrate

These processes can involve microstructural changes created by the diffusion of a different chemical species from the substrate surface (thermochemical treatments) or by phase changes created by thermal cycling only (thermal treatments).

The leading examples of each of these categories are given in:

Table 1.6.34 *Microstructural modifications*

Material removal methods

Removal of material from the original substrate surface, to influence surface finish. Examples of material removal methods are:

Chemical etching.
Electropolishing.
Shot blasting.
Vibratory polishing.
Grinding and texturing.

Mechanical deformation methods

Mechanical deformation of the surface layers of the original substrate, mainly to induce compressive residual stresses or to influence surface finish. Examples are given in:

Table 1.6.35 *Mechanical deformation methods*

The benefits, limitations and main areas of application of the specific treatment processes are described in Tables 1.6.32 to 1.6.35.

A comparison of their important features is given in:

Table 1.6.36 *Comparison of surface treatment processes*

The most industrially significant treatment processes (in descending order) are:

Engineering paints.
Electroplating.
Hot dip galvanizing and other zinc coating methods.
Surface treatments (i.e. thermochemical and thermal treatments).
Powder coatings.

Table 1.6.32 Coating processes

Process	***Brief description***	***Benefits/ limitations/ applications***
Organic coatings		
Engineering paints	See Vol. 1, Chapter 1-4	
Powder coatings	See Vol. 1, Chapter 1-4	
Other organic coatings	See Vol. 1, Chapter 1-4	
Electrodeposition		
	In all electrodeposition processes, which take place in a bath of suitable electrolyte, the substrate is the cathode and the material to be deposited is the anode. The application of an electrical current between the electrodes results in deposition on the cathode.	Many variations of metal coatings can be applied to confer a range of performance benefits (see below). Most electrodeposits are relatively cheap. No alloy formation at the interface, so adhesion is not as strong as for diffusion processes.
Hard chromium	Chromium deposited direct onto substrate surface. Plated thicknesses can be up to 500 μm, but most applications call for a thickness in the range 10 to 50 μm.	Depending on deposition conditions, coating hardnesses in the range 450 to 1100 H_v achievable. Therefore, wear resistance conferred. Chromium deposits generally contain microcracks, therefore, with no underlayer, corrosion resistance is modest. Throwing power and control of coating thickness for thicker layers (above 100 μm) are limited.
Nickel	Nickel deposited direct onto substrate at rates of up to 400 μm/h. For most applications, coating thickness will be within the range 10 to 50 μm.	Relatively soft (150 to 400 H_v) but ductile deposit. Used mainly for build-up of worn surfaces.
Nickel underlay + hard chromium	Typical coating thicknesses are 50 to 100 μm of nickel under 25 to 50 μm of chromium.	Considerably enhances the corrosion resistance of hard chromium e.g. a factor of 3 improvement (to around 400 hours) in acid salt spray chamber test.

Table 1.6.32 Coating processes – *continued*

Process	***Brief description***	***Benefits/ limitations/ applications***
Copper		Ductile coating for build-up of worn machine parts. Also, selective stop-off for parts to be case-hardened. Useful on steel to assist cold extrusion operations.
Silver	Can be applied up to 500 μm thick but usually at 50 μm or below.	Gives high thermal and electrical conductivity – used in special applications e.g. faying surfaces of copper bus-bar joints. Also used as a high temperature lubricant on stainless steel threads.
Tin	Can be applied to components for various reasons (see next column). As a separate (and very significant application), can be deposited in-line onto steel strip intended for food canning ('tinplate').	Soft, ductile, corrosion resistant, easily solderable and non-toxic coating. Apart from protection of steel cans, also useful as a thread lubricant and as a pre-treatment to assist soldering. Extensively used to prevent copper corrosion on electrical switchgear and bus-bars.
Tin–lead alloy		Often applied over nickel coatings to aid soldering e.g. radio equipment chassis.
Lead		Extensively used for corrosion prevention of battery connectors and terminals.
Zinc (electrogalvanizing)	Applied to many household and DIY items, wire and machinery components. Also, extensively applied in-line to steel sheets for forming for automotive and domestic appliance applications.	Zinc is more electronegative than iron so provides sacrificial cathodic protection to steel substrates.
Indium		Used on bearings and dry Al connectors in telecommunications.

Table 1.6.32 – *continued*

Table 1.6.32 Coating processes – *continued*

Process	***Brief description***	***Benefits/ limitations/ applications***
Palladium		Ductile coating used on light duty contacts in flexible electronic connectors.
Platinum		Inert anodes in cathodic protection systems and in industrial electrolytic processes.
Rhodium		Used in electronics for extreme wear resistance e.g. in high duty relay contacts.
Cadmium		Gives good corrosion resistance, low coefficient of friction and low electrical contact resistance. Cadmium plating has been extensively applied to threaded fasteners. But, its use is under severe pressure for environmental reasons (it is rated as a deadly poison), so alternatives are now being introduced.
Electroless (autocatalytic) plating		
	Autocatalytic chemical reaction of metal ions at a catalytic surface from an aqueous solution. The deposited metal is also catalytic as the process becomes self-perpetuating.	Excellent throwing power, uniform coatings on complex shapes. Low energy costs. But, deposition rates lower than for electroplating (up to around 25 μm/h) and is commercially viable only for a few metals.
Electroless nickel	The deposit is actually a nickel–phosphorus alloy with 8–12% P. Normally applied up to a maximum coating thickness of 100 μm.	Deposition uniform on all surfaces with no edge or corner build-up. Hardness is 500 H_v as-deposited, can be increased to 900 H_v with a brief treatment at 400°C. Offers a good combination of wear and corrosion resistance.

Table 1.6.32 – *continued*

Table 1.6.32 Coating processes – *continued*

Process	***Brief description***	***Benefits/ limitations/ applications***
'Composite platings' based on electroless nickel	Co-deposition of electroless nickel with physically suspended, finely divided particles.	Wear resistant carbides (e.g. SiC) can be integrated in this way. Also, fine (1 μm) PTFE particles can be introduced to give hard coatings with very low coefficients of friction.
Nickel–boron		Gives higher hardness than Ni–P i.e. 650–700 H_v as-deposited, around 1200 H_v after heat treatment. Ni–B coatings also exhibit good lubricity.
Copper		Thick electroless copper has been used for many years in making printed circuit boards. Thin coatings are used to control electromagnetic interference.
Cobalt	Co–P and Co–B alloys are deposited.	Used in electronic and magnetic recording applications.
Dip coating methods		
Hot dip methods	Immersion of workpiece in molten coating metal. Interfacial alloying occurs and the coating is solidified in a cold air stream after removal from the bath.	Gives a good metallurgical bond and is a rapid process. But, energy costs are high, uniformity of coatings and control over thickness are inferior to plated finishes and there is a limitation to low melting point metals.
Zinc (hot dip galvanizing)		Applied to steel in sheet, rod, pipe or fabricated form for corrosion protection.
Aluminium, Al–Zn		For corrosion protection of steel substrates.
Tin		For corrosion protection – now being superceded by electrodeposition method for tinplate applications.

Table 1.6.32 – *continued*

Table 1.6.32 Coating processes – *continued*

Process	***Brief description***	***Benefits/ limitations/ applications***
Lead–tin (Babbit's)		Gives low friction coefficient, prevents galling or scoring.
Sol–gel coatings	Range of ceramic or glass materials applied to metallic substrate surfaces in the form of a sol by dipping or spin coating. Coating then dried and fired at temperatures up to 850°C (higher firing temperatures can be used for increased coating hardness).	Used for corrosion and wear protection.
Other zinc coating methods		
Organometallic zinc coatings e.g. Dacromet, Deltatone	Bichromated zinc– aluminium flakes, admixed with epoxy resin and applied by a dip/spin/cure method.	These coatings offer excellent salt spray corrosion resistance. As a dry process, there is no possibility of hydrogen cracking problems. These coatings are finding application as alternatives to cadmium plating e.g. in automotive fasteners.
Sherardizing	Components are tumbled with zinc powder and fine sand in a sealed drum for several hours at around 375°C. The zinc diffuses into the substrate material giving a fine grained zinc alloy coating.	The uniform and strongly adherent coating gives corrosion protection to the substrate. Typical applications include scaffold clips, hinges, nuts. bolts, chains, washers, screws and nails.
Mechanical plating	Tumbling of small parts in a drum containing zinc dust, tiny glass balls and an activating solution. The balls hammer the zinc onto the substrate surfaces, to which it becomes cold welded. In this case, no heating is required.	As with other zinc coating processes, corrosion protection to steel substrates is conferred.

Table 1.6.32 – *continued*

Table 1.6.32 Coating processes – ***continued***

Process	***Brief description***	***Benefits/ limitations/ applications***
Vitreous enamelling	Fine powdered glass is mixed with clay, water and pigment to form a slip. This slip is applied to the surface by dipping or spraying, then dried and fired at around 800°C to fuse the glass to the metal surface.	Gives decorative finishes with good heat and wear resistance mainly on mild steels for applications such as architectural panels, signs, cookware, heat exchangers, hot water tanks, mechanical processing equipment, engine exhaust silencers, baths, stoves, catering equipment, kitchen sinks and shower receptors.
Thermal spraying		
Flame spraying (gas)	Wire or powder fed into an oxy-acetylene flame and the molten particles are projected by compressed air onto the substrate. Sometimes preceded by bonding coat of Mo or Ni–Al and followed by sealing coat or polymer or paint to improve corrosion resistance. Equipment used is low cost and readily portable, high operator skill not required. Substrate is heated to around 200°C.	Both metals and ceramics can be flame sprayed, but the process is limited to materials which will melt at the flame temperature (around 3000°C). Widely used for corrosion resistance and reclamation of worn parts. Stainless steel and nichrome flame sprayed coatings are used for their oxidation resistance. Ceramic flame sprayed coatings are also used for thermal insulation and protection against erosion.
Flame spraying (ARC)	Arc is struck between two wires (which can be of different materials) and the metals are projected by compressed air onto the substrate. Lower equipment cost than for plasma spraying.	Similar range of performance benefits to gas flame spraying, but higher temperature of arc gives better bond strength between coating and substrate. Lower substrate temperatures can be maintained.
Spray and fuse	Powder alloys are deposited usually through an oxy-acetylene flame to form a coating which is then subsequently fused.	Gives better bond strength than gas flame spraying alone, but fusing operation can cause distortion and limited to self-fluxing alloys.

Table 1.6.32 – ***continued***

Table 1.6.32 Coating processes – *continued*

Process	***Brief description***	***Benefits/ limitations/ applications***
Plasma spraying	An arc is struck in an inert gas and a plasma (gas raised to such a temperature, 1500– 20 000°C that ionization takes place) and powdered metal is projected through the plasma onto the substrate. Substrate temperatures generated are low at around 120–200°C. Equipment is expensive and required operator skill is high.	Ideal for depositing high melting point materials such as tungsten.
HVOF (high velocity oxy-fuel) spraying	This process uses an internal combustion (rocket) jet or detonation with spark plug to generate hyper-sonic gas velocities to project powder at the substrate.	Very smooth pore free coatings with excellent adhesion are produced. Low substrate temperatures are maintained.
Weld deposition processes		
Gas welding	Deposition of coatings from rod or wire using oxy-acetylene heat sources. Equipment costs are low and the deposits are easy to apply. Suited to thick deposits and thin sections, small components and on-site work. The high heat inputs required can cause distortion.	Excellent metallurgical bond is produced. Typical applications include seating surfaces for valves in engines and compressors, rolls for precision rolling, power plant valves and nozzles and bearing sleeves operating in unusual environments of temperature and corrosion.
Powder welding	Deposition of coatings from powder using tool with oxy-acetylene heat source and powder container. Low equipment costs and lower operator skill level than for rod or wire source. Suited for small areas, thin deposits, thin sections, small parts and on-site work.	Excellent metallurgical bond produced. Less suited to thick deposits than gas welding.

Table 1.6.32 – *continued*

Table 1.6.32 Coating processes – ***continued***

Process	***Brief description***	***Benefits/ limitations/ applications***
Arc welding	Deposition of coatings from rod, wire or strip. Electric arc heat source. Rapid deposition of large amounts of metal. Suited to thick deposits, thick sections and massive work.	Very high heat inputs with consequent distortion possibilities. Excellent metallurgical bond produced.
Friction surfacing	A solid phase deposition process in which a rotating consumable bar usually in the range 10–40 mm diameter is pressed onto a laterally moving substrate. The process allows materials to be deposited which would otherwise be metallurgically incompatible with the substrate yet still provides a deposit with a high degree of adherence.	The process is used to deposit corrosion and/or wear resistant materials. It has been used for reconditioning of worn shafts, for anti-corrosion overlay of slide plate valves, surfacing of the annular contact face of composite pipe flanges, disc brakes etc., inlay cladding of strategic materials to suit bearing or seal contact areas on shafts and cladding of exposed regions of shafts.
Vapour deposited coatings		
Chemical vapour deposition (CVD)	Gas or gas mixture deposits coating onto heated substrate as a result of vapour phase chemical reaction. Temperature of deposition in range 200–1200°C depending on the particular reaction. Produces high quality coatings of refractory metals and inorganic compounds (borides, carbides, nitrides, oxides etc). Good throwing power.	Deposit thicknesses can be in the range 1 μm to 2 mm. The coating materials are aimed at enhancing resistance to wear, corrosion or oxidation. Substrates must be compatible with processing temperatures (which can raise problems, for instance, with the heat treatment condition of steel substrates). The process requires a containing vessel for gases and careful control and is therefore relatively expensive.

Table 1.6.32 – ***continued***

Table 1.6.32 Coating processes – ***continued***

Process	***Brief description***	***Benefits/ limitations/ applications***
Plasma assisted chemical vapour deposition (PACVD)	The process involves reactions of precursors gases as for conventional CVD but within a plasma created by electromagnetic field around the substrate. The presence of the plasma allows reaction temperatures to be reduced considerably to 100 to 600°C.	Essentially any material can be deposited onto any non-gassing substrate which can withstand the reaction temperature. In practice, most PACVD coatings are of hard, wear resistant materials. PACVD is a relatively expensive batch process.
Physical vapour deposition (PVD)	PVD is a collective name for a range of techniques with the fundamental characteristic that one or more of the reaction species undergoes a change in state from solid to vapour phase within the vacuum chamber. The processes can be either evaporative (with the energy source being resistive heating, induction, electron beam gun or arc) or sputtering (involving diode, triode, ion beam or magnetron).	High purity films of almost any metal and some non-metals onto any non-gassing substrate possible, due to the vacuum techniques. Vapour condenses on all cool surfaces in the line of sight of the source. Properties of the coating are highly susceptible to variations in process parameters. Difficult with low vapour pressure metals such as tungsten. Without plasma assistance, PVD is typically in applications such as metallization of plastics.
Plasma assisted physical vapour deposition (PAPVD)	The generation of a plasma between the coating material source and the substrate enhances deposition rates, coating adhesion and control over coating quality. Energy sources can be DC diode, triode, RF or ion beam.	Virtually any coating material can be deposited onto any non-gassing substrate. Typical coatings are for wear resistance, thermal barriers, high temperature corrosion resistance or electronic/ electro-magnetic properties. Deposition rates are readily controllable. PAPVD is an expensive, batch process.

Table 1.6.32 – ***continued***

Table 1.6.33 Chemical conversion coatings

Process	***Brief description***	***Benefits/ limitations/ applications***
Phosphating	Treatment of cast iron, steel, galvanized steel or aluminium surfaces with a dilute solution of phosphoric acid and accelerators such that chemical reaction occurs to produce a dense, continuous layer of a metal phosphate.	Phosphate coatings provide mild corrosion resistance, a base for subsequent painting or coating, a surface that facilitates cold forming or lubricity and wear resistance.
Chromating	Chemical interaction between the substrate and a chromate-containing aqueous solution. This causes dissolution of some surface metal and the formation of a protective film containing complex chromium compounds. The reaction is generally between the metal and hexavalent Cr, which oxidizes the metal and is itself reduced to trivalent state. (This is becoming a problem as environmental legislation is now rating hexavalent chromium as a carcinogen).	Chromate coatings can be complete in their own right e.g. on Al substrates for low electrical resistivity, or as a surface 'seal' for other coatings (e.g. Zn or Cd) to improve atmospheric corrosion resistance, particularly in marine environments.
Anodizing	Anodic oxidation of the substrate in aqueous solution under selected voltage current density conditions to form a spongy oxide layer which is often subsequently coloured by a pigment.	Limited to strong oxide formers – mainly commercially applied to Al but sometimes also to Mg or Zn. Anodized surfaces give protection against corrosion, are wear resistant and durable and, if impregnated with a lubricant, can give some protection against fretting under dry running conditions.

Table 1.6.33

Table 1.6.33 Chemical conversion coatings – *continued*

Process	***Brief description***	***Benefits/ limitations/ applications***
Steam tempering	Heating a ferrous substrate in a steam atmosphere at 530–550°C to form a surface layer of magnetite (Fe_3O_4)	Applied to some steel tools (e.g. twist drills) and to ferrous powder metallurgy parts. The treatment improves wear and frictional properties and provides some degree of rust protection.

Table 1.6.33 – *continued*

Table 1.6.34 Microstructural modifications

Process	***Brief description***	***Benefits/ limitations/ applications***
Thermochemical treatments		
Carburizing (sometimes referred to as case hardening)	Austenitic thermochemical treatment for steels in which carbon is diffused into the surface layers at a temperature in the range 825–950°C to produce a case which can be hardened by quenching from the treatment temperature. The source of the carbon can be solid, molten salt, gas or plasma.	Carburized case depths vary depending on the application but are often in the range 0.5 to 2 mm. Carburizing gives a combination of high surface hardness, up to about 850 H_V, for wear resistance and good core toughness. Compressive residual stresses are created in the case, which, together with the high case hardness, improve fatigue strength. Typical applications are cams and camshafts, rockers and rocker shafts, bushes, bearings and transmission gears.
Carbonitriding	A modified form of carburizing in which both carbon and nitrogen are diffused from the surface at temperatures between 800 and 870°C. The carbon source can be a liquid, gas or plasma. Again, the component is quenched from the treatment temperature.	Carbonitrided case depths are generally limited to around 750 μm. Surface hardness is similar to carburizing. The nitrogen gives better surface hardenability than carburizing so that oil rather water quenching can be used, therefore reducing quench cracking problems. Carbonitriding is usually used to impart wear resistance; benefits in fatigue performance are more modest than for carburizing. Typical applications are bearing cages gears, air-compressor cylinders and camshafts.

Table 1.6.34 Microstructural modifications – *continued*

Process	***Brief description***	***Benefits/ limitations/ applications***
Nitriding	Ferritic treatment of low alloy and tool steels, involving diffusion of atomic nitrogen from the surface at temperatures in the range 500 to 590°C, so there is no phase transformation on cooling. Chemical reaction of the nitrogen with the steel produces hard iron alloy nitrogen compounds on the surface and a diffusion layer in which nitrogen imparts solid solution hardening. The source of the nitrogen can be solid, fused salts, gas or plasma.	Case depths are lower than for carburizing (up to around 500 μm) but surface hardnesses can be higher (up to 1300 H_v). Because of the lack of phase transformation, quenching distortion is eliminated and nitriding can be used on finished parts. Fatigue strength is improved by the dissolved nitrogen in the diffusion layer and by establishment of compressive residual stresses.
Nitrocarburizing	A modified form of nitriding in which both nitrogen and carbon are diffused from the surface, occasionally with small amounts of sulphur. Treatment temperatures are in the range 540 to 600°C. Nitrocarburizing can be applied to mild steel as well as alloy steels.	The surface compound layer is thinner than for nitriding (at about 10 to 25 μm) and has a hardness of 600–900 H_v. The diffusion zone can extend to 250 μm or more. The main advantages of nitrocarburizing over carburizing or carbonitriding are better wear resistance and anti-scuffing characteristics resulting from the formation of the compound layer.
Boriding (or Boronizing)	Diffusion of boron into the surface layers of plain carbon and low alloy steel substrates at temperatures between 900 and 1100°C for 1 to 6 h. The source of boron can be solid, liquid or gas, but is most commonly solid (pack boriding). Compound and diffusion layers are created.	Borided steel surfaces exhibit high hardness (1500 to 2300 H_v), good wear resistance and high hot hardness. Ti, Ta and Ni alloys can also be borided. Ti alloys can exhibit even higher surface hardnesses (up to 3000 H_v).

Table 1.6.34 – *continued*

Table 1.6.34 Microstructural modifications – *continued*

Process	***Brief description***	***Benefits/ limitations/ applications***
Chromizing	Diffusion of chromium into a steel substrate by heating at 950–1300°C in a chromium-rich atmosphere. The chromium goes into solid solution or, in the presence of carbon, forms carbides.	Chromizing provides oxidation and wear resistance. In high carbon steels, the surface layer is a complex chromium carbide with hardness of 1200–1300 H_v. Chromized surfaces on low carbon steels are ductile and can be rolled, pressed or drawn. Typical applications include turbine buckets, steel components to resist nitric acid attack and containers for copper brazing.
Aluminizing (also known as calorizing)	Aluminium is diffused into a low or medium carbon steel surface at 750 to 1000°C for 3 to 6 h, followed by an annealing treatment at 800 to 980°C for 12 to 48 hours.	The aluminized layer is a solid solution of aluminium and iron. The process imparts resistance to corrosion and oxidation at high temperature. Typical applications are gas turbine blades, burner pipes, pyrometer protection tubes, flue stacks, boiler tubing, high temperature fasteners and chemical reactor tubing.
Thermal treatments		
Induction hardening	Heating of the surface layers of a steel or cast iron substrate to a temperature in the range 850 to 1000°C (to create transformation to austenite) by electromagnetic induction. The part is then quenched to transform the austenitic surface to martensite.	Case depths from 250 μm to 5 mm with hardness around 500–750 H_v are created. An advantage over thermochemical treatments is that parts of a component can be selectively treated. Compressive residual stresses give fatigue strength benefits. Typical applications include cams, rotating shafts, machine ways, swivel pins, steering joints, gears, valves and bearing surfaces.

Table 1.6.34 – *continued*

Table 1.6.34 Microstructural modifications – *continued*

Process	***Brief description***	***Benefits/ limitations/ application***
Flame hardening	Similar process to induction hardening except that the means of heating the surface is by flame impingement.	A very simple process but less controllable than induction hardening i.e. possible problems are overheating, spotty and uneven hardening, shallow depth of hardening, cracking or distortion of large components and excessive scaling.
Laser hardening	Similar principle to induction hardening except that surface heating is by scanning of a laser beam (1 to 15 kW CO_2). The cold interior produces a self-quenching of the heated surface layers.	Case depths are lower than for induction hardening at 100 μm or below. Because of this and the self-quenching effect, distortion levels are very low. The process is most effectively used for very localized treatments.
Electron beam hardening	Similar to laser hardening except that the heating source is an electron beam.	Electron beam hardening is more efficient than laser (90% vs. 10%) and the equipment is lower cost. The disadvantage is that depth of penetration of an electron beam is much higher so that case depths are 1 to 3 mm.
Laser glazing	Creation of a very thin melted layer on the surface of the substrate, using a laser. The rapid cooling rates in subsequent self-quenching can solidify this material to an amorphous state.	

Table 1.6.34 – *continued*

Table 1.6.35 Mechanical deformation methods

Process	***Brief description***	***Benefits/ limitations/ applications***
Shot peening	A controlled blasting of the substrate surface with a high velocity stream of shot. This creates plastic deformation of a surface layer that can be about 0.12 to 0.25 mm deep.	Shot peening provides a surface finish which is highly receptive to oil retention for lubrication and rust prevention. The plastic deformation of the surface layers induces compressive residual stresses which are beneficial for fatigue performance. Also, tensile stresses which could cause stress corrosion cracking can be relieved. Typical applications are improving of fatigue strength of structural components of aircraft, salvaging (restoring perfect roundness) of large ring gears and restoring straightness of shafts.
Surface cold rolling	Rolling the surface of components (perhaps in selected areas) to create plastic deformation (and induce compressive residual stresses) and influence surface finish.	The compressive residual stresses and improved surface finish both contribute to enhancing fatigue strength. Typical application areas are in finish rolling of transmission gears and fillet rolling of crankshafts.
Ballizing	Passing an oversize cemented carbide ball through the bore of a component.	Improves tolerances, surface finish and roundness. By work-hardening of the surface layers and imparting compressive residual stresses, fatigue strength can be improved.

Table 1.6.35

Table 1.6.36 Comparison of surface treatment processes (extension of 1.6.31)

Process	*Principal imparted characteristic*	*Additional beneficial properties*	*Dimension restoration*	*Surface preparation*	*Cost*	*Accessibility*	*Distortion*	*Limiting thickness (μm)*	*Temperature attained by substrate (°C)*
Painting	Corrosion resistance		No	Brushing	Low	Good	Nil	–	RT
Powder (organic) coatings	Corrosion resistance	Better abrasion resistance than paint in some cases	No	See Vol. 1, Chapter 1.4	Low to moderate	Restricted	Nil	See Vol. 1, Chapter 1.4	Variable, mainly low
Electroplating	Corrosion and/or wear resistance	Lubricity, thermal and electrical conductivity	Sometimes	Degreasing	Low to moderate	Good	Nil	250 (around 100 for most practical applications)	Room temperature
Electroless plating	Corrosion and/or wear resistance	Lubricity, electrical and magnetic properties	No	Degreasing	Low to moderate	Excellent	Nil	25	<100
Dip coating	Corrosion resistance	Aluminium gives oxidation resistance	No	Plating (Al) Degreasing (Zn)	Moderate	Good	Moderate	100	740 (Al) 430 to 480 (Zn) 850 (sol–gel)
Other Zn coating methods	Corrosion resistance		No	Degreasing	Moderate	Good	Low to moderate		450
Vitreous enamelling	Corrosion and wear resistance	Heat resistance, chemical inertness	No	Blasting	Moderate	Good	Moderate	200	800 to 900
Thermal spraying	Wear resistance	Thermal insulation, erosion	Yes	Blasting	Low to high	Not inside bores	Low	500	200

Table 1.6.36 – *continued*

Table 1.6.36 Comparison of surface treatment processes – *continued*

Process	***Principal imparted characteristic***	***Additional beneficial properties***	***Dimension restoration***	***Surface preparation***	***Cost***	***Accessibility***	***Distortion***	***Limiting thickness (μm)***	***Temperature attained by substrate (°C)***
Weld deposition	Wear resistance	Corrosion resistance	Yes	Blasting	Low to high	Not inside bores	Low to moderate	None	
Chemical vapour deposition (CVD)	Wear and corrosion resistance		Yes	Special	High	Excellent	Variable	2000	Up to 1200
Plasma assisted CVD (PACVD)	Wear and corrosion resistance		No	Special	High	Excellent	Low	10	Up to 600
Physical vapour deposition (PVD)	Wear and corrosion resistance		No	Special	High	Line of sight	Low	10	Up to 500
Plasma assisted PVD (PAPVD)	Wear and corrosion resistance	Thermal barriers, oxidation and erosion resistance, electronic/ magnetic properties	No	Special	High	Line of sight	Low	25	Up to 500
Roll bonding	Corrosion resistance		No	Brushing	Low to moderate	Sheet only	Nil	None	20
Chemical conversion coatings	Corrosion resistance, wear resistance	Lubricity, base for painting (phosphating), electrical resistivity (chromating)	No		Low	Excellent	Nil	700	80
Carburizing	Wear resistance	Fatigue strength	Yes	Machining	Low	Good	Moderate	2000	825 to 950

Table 1.6.36 Comparison of surface treatment processes – *continued*

Process	***Principal imparted characteristic***	***Additional beneficial properties***	***Dimension restoration***	***Surface preparation***	***Cost***	***Accessibility***	***Distortion***	***Limiting thickness (μm)***	***Temperature attained by substrate (°C)***
Carbonitriding	Wear resistance		Yes	Machining	Low	Good	Low to moderate	750	800 to 870
Nitriding	Wear resistance	Fatigue strength	No	Machining	Low	Good	Low	500	500 to 590
Nitro-carburizing	Wear resistance	Anti-scuffing	No	Machining	Low	Good	Low	250	540 to 600
Boriding	Wear resistance	Hot hardness	No	Machining	Moderate	Good	Moderate	25	1100
Chromizing	Oxidation and wear resistance		No	Machining	Moderate	Good	Moderate		1300
Aluminizing	Oxidation and corrosion resistance		No	Machining	Moderate	Good	Moderate	1000	1000
Thermal hardening	Wear resistance	Fatigue strength	No	Machining	Low	Moderate	Low to moderate	5000	1000
Surface cold deformation	Fatigue strength		No	Machining	Low	Moderate	Low	250	Room temperature

Table 1.6.36 – *continued*

1.6.8 Forming of plastics

1.6.8.1 Introduction

There are many processes available for the production of plastics components. In broad terms the choice depends on: whether the material is thermosetting or thermoplastic, the design requirements of the component, and the quantities required. The factors influencing process selection and the importance of the process itself on the results obtainable are reviewed below.

When describing processing techniques, it is convenient to use the conventional classification of polymers into thermoplastics, thermosets, and elastomers, although some processes (with minor modifications) are feasible with all categories.

1.6.8.2 Processing techniques

HOT MELT PROCESSES

In the hot melt processes the polymeric material in the form of granules, powder or pellets is heated above its softening point to a fluid state. It is then formed by a die or a mould or a combination of both. Hot melt processes can be highly automated and complex-shaped components are easily produced from suitable moulds. Although tooling is usually expensive it may not be prohibitive if large numbers of components are to be produced. A brief description of the various hot melting processes together with their advantages and limitations is given in:
Table 1.6.37 *Hot melt processes*

POWDER SINTERING TECHNIQUES

Powder sintering techniques basically are limited to using thermoplastics in the form of powders. They are interesting for development and production because several materials are fairly readily available in powder form. In fact, some come initially from the chemical reactors as fine powders and for this reason powder technology is of particular interest and should have good prospects.

A brief description of the principal powder sintering techniques together with their advantages and limitations is given in:
Table 1.6.38 *Powder sintering techniques*

THERMOFORMING

In various related forms thermoforming is a very old transformation technique. The commonest starting preform is extruded film or sheet. Variations of the basic techniques can be employed on rods and tubes.

A brief description of the principal thermoforming processes together with their advantages and limitations is given in:
Table 1.6.39 *Thermoforming processes*

SOLID-PHASE FORMING

If a thermoplastics material is held at a temperature not too far below its 'softening point' it is feasible to apply some of the characteristic cold forming processes for metals to bring about deformation. As applied to plastics this is called broadly 'solid-phase' forming. The term 'cold forming' has been used, but results at ambient temperature are disappointing. Care has to be taken to control snap-back which arises because of the extent to which plastics materials are elastic at temperatures below the softening region. It is convenient to distinguish three different forms of the technique, each of

which is described together with their respective advantages and limitations in:
Table 1.6.40 *Solid-phase forming techniques*

FABRICATION

A number of fabrication techniques are available whereby articles or structures can be produced. In some instances they are useful for prototype development (e.g. machining) or in other situations they may be applicable to large-scale work. The most common use is for small numbers, since a high degree of adaptability is possible. The principal fabricating processes, together with their advantages and limitations, are examined in:
Table 1.6.41 *Fabrication processes*

PROCESSES INVOLVING PASTES, EMULSIONS AND LIQUIDS

A wide variety of processes involving use of pastes, emulsions and liquids are feasible for development work. Some have been established on a large scale and these, together with their advantages and limitations, are given in:
Table 1.6.42 *Processes involving pastes, emulsions and liquids*

COATING BY SOLUTION PROCESSES

Whilst solution processes exhibit some similarity to processes involving pastes, emulsions and liquids, true solutions are employed, and solvent removal, and perhaps recovery, forms an essential part of the overall solution process. The solution may be of reactive or unreactive materials giving thermoset or thermoplastic coatings, respectively. The principal coating processes, together with their advantages and limitations, are given in:
Table 1.6.43 *Solution processes for coatings*

COMPOSITE PRODUCTION

The principal production techniques for composite materials involve placement of the fibres and resin impregnation, followed by a consolidation/curing operation which proceeds simultaneously. Brief details of each process, together with their respective advantages and limitations, are given in:
Table 1.6.44 *Composite production techniques*

RESIN TECHNIQUES

The title 'resin techniques' is used to group together a number of casting processes involving conversion of a monomer or a prepolymer to a polymerized formed state. It also includes polymer casting into a mould. The principal techniques, together with their advantages and limitations, are described in:
Table 1.6.45 *Resin techniques*

FOAM MOULDING

The two major techniques of production of structural foams, with their advantages and limitations, are described in:
Table 1.6.46 *Forming of structural foams*

1.6.8.3 Selection of processing technique

A guide to the suitability of the various classes of polymers (thermoplastics, thermo-

sets and elastomers) for processing by a standard production route is given in:
Table 1.6.47 *Polymer types and suitable processing techniques*

The general characteristics of the various techniques for processing plastics are detailed in:
Table 1.6.48 *Comparative characteristics of plastic processing techniques*

A more detailed analysis of the suitability of the processing techniques for specific polymer materials is given in:
Table 1.6.49 *Suitable processing techniques for specific polymer materials*

1.6.8.4 Dimensional tolerances

Typical tolerances obtainable by plastics processing methods are as follows:

Machining	0.1–0.3%
Injection and compression moulding	0.2–1.0%
Extrusion	2.5%
Extrusion blowing operation	2.5%
Thermoforming	5–10%
Hand or spray lay-up	5–10%

The above figures are given for comparison purposes and should not be taken as true for a particular material or component. Factors affecting dimensional tolerances include the thermal expansion of the material and the temperature of the process, the occurrence of phase changes due to crystallization or polymerization taking place during the process, mould pressure and mould accuracy. In addition, post-moulding dimensional changes must also be considered.

The first three factors affect dimensional tolerances via shrinkage. As an approximate guide:

1. Crystalline materials which undergo a phase change during processing will show high shrinkage and dimensional tolerances will be difficult to predict.
2. The higher the process temperature the poorer the dimensional accuracy.
3. The higher the pressure the better the dimensional accuracy, but this will limit die life.
4. Most fillers reduce thermal expansion and moulding shrinkage.
5. Amorphous filled materials will have the lowest shrinkage and highest obtainable dimensional accuracy.

Information on moulding tolerances for particular plastic materials are given in the relevant section in Vol. 3.

1.6.8.5 Plastic containers

The manufacture of shaped, formed plastic containers and packages includes those made from plastic sheet, which differs from plastic film only in thickness. There is some overlap in nomenclature, but the dividing line is usually drawn at 250 μm, anything thinner than this being called 'film'.

A very wide range of materials and forms of container are available and it is not possible to draw up a general algorithm for selection which could cover all eventualities within this category of packaging. However, it may be possible to narrow the field by using the following procedure:

1. Decide on the general shape and form of container needed – tub, bottle, tube, tray, and the likely quantity of production.
2. Choose the process(es) suitable for making this form of container in the necessary quantities and size.

3. Decide which materials it would be possible to use for the forming process(es) chosen. Note that it may be possible to use other material/process combinations if it is likely to be difficult, inconvenient or expensive.
4. Select which of the chosen materials have the necessary combination of properties for the application under consideration.
5. Establish which material meets the requirements at minimum cost.

This should be regarded as a preliminary selection process.

Many of the materials may be stiffened and often strengthened by inclusion of fillers such as wood meal, chopped fibre, talc, etc., and others may be toughened by adding plasticizers. Although sometimes more expensive as materials, such modified materials often allow the final article to be produced at lower overall cost.

Finally, careful consideration should be given to the dangers of effects of interactions of environment, packaging and contents. Thus surfactants can cause stress-cracking of some plastics, absorption of oxygen from a dead-space in a bottle by the packaged fluid can cause the bottle to collapse under outside atmospheric pressure, contents may absorb plasticizers from the package causing it to be excessively brittle if the temperature falls.

The forming processes suitable for the production of each container type are identified in:

Table 1.6.50 *Container type and applicability of forming processes*

and their characteristics, advantages and disadvantages in:

Table 1.6.51 *Characteristics of container-forming processes for plastics*

A guide to the suitability of various polymers for use in the container-forming processes is shown in:

Table 1.6.52 *Suitability of various polymers for use in container-forming processes*

Table 1.6.37 Hot melt processes

Description	*Advantages*	*Limitations*
Compression moulding		
Material, as powder or compressed pellets, is loaded by hand or automatically into female part of moulds. Material may be pre-heated to improve cure, and cut time. Mould is continuously hot Moulding extracted hot.	For thermosets and elastomers. Useful for highly and coarsely filled materials. Very wide size range. Relatively low tooling costs for good accuracy in formed parts.	Not generally applicable to thermoplastics, heating and cooling cycles being required as mouldings must be extracted cold. As thickness is increased cycle times extend severely. For most moulds flash line must be cleaned off. Scrap not reusable.
Transfer moulding		
Variation of injection moulding for thermosetting resins. Material as compressed pellet is dropped into a pot in mould above the moulding cavity and transferred under pressure through a sprue. Mould is continuously hot. Pellets may be pre-heated. Final hardening is completed out of the mould.	Allows use of thermosetting resins at lower cost than injection moulding at adequate throughput. Improves heat distribution through mass as compared with compression moulding and provides better flow into mould. Better for flash line cleaning.	Tooling more complicated than compression moulding leading to higher mould and machine costs. Size range more restricted than compression moulding. Waste material not useable.
Injection moulding		
Granules, powder or elastomer strip or dough are fed to heating cylinder of machine, usually by means of rotating extrusion screw or plunger which subsequently moves axially to force hot material into thermally controlled mould. For thermosets and elastomers the cylinder is just hot enough to melt the material.	Low machine cost/piece in large-volume production. Product requires very little finishing and can be intricate by reason of mould features. Processes can be highly automated and can be accurately controlled. Wide range of sizes possible – 1 g–50 kg. Wall thickness no limitation; except on cycle and on design (uniformity of thickness at junctions). Scrap reusable.	Moulds should be of special mould steel unless only few mouldings required. Machine expensive, especially if precision required. Very small machines with cheap moulds used for low output, small mouldings, up to 15 g.
Extrusion blow moulding		
A hot parison, usually a circular tube is produced from an extruder with a suitable die. In a variety of ways the parison is fed to a blowing mould where it is inflated and cooled against the mould walls. In advanced equipment parison wall thickness can be controlled to suit the design of the required article; cooling rates can be increased by water spray or CO_2 spray internally for rapid production of large	Can produce hollow articles with small openings (e.g. bottles with small necks) or large. Capable of being highly automated for large quantity production. Relatively simple tooling for limited numbers. One extruder can feed a number of parison-producing die heads. Can be used for 'industrial' mouldings of suitable shapes (e.g. petrol tanks, large toys).	Applicable to materials which have reasonable melt strength; as parison sag must not be too great while moulds close. There is a bottom weld. Unless parison control is used, the wall thickness is not wholly controllable and wide part of a moulding would be thinned on blowing. Mouldings from 1 g to 60 kg.

Table 1.6.37

Table 1.6.37 Hot melt processes – *continued*

Description	*Advantages*	*Limitations*
Extrusion blow moulding – cont'd quantities. Parisons may be of other shapes than circular cross-section (even two side-by-side sheets).		
Injection blow moulding The parisons required for blowing are formed by injection moulding into moulds of related shape and then transferred to blowing cavities within a multiple mould. Uses either adapted injection machines, or specially designed injection blow machines. Parison may be moulded separately and stored or transported before use. A re-heating stage is then required before blowing.	The chief advantages are increased extent of biaxial orientation, so as to toughen products, and that scrap generation is very small. Better distribution of thickness calibration of orifices such as necks – shaping of parison before blowing allows compensation for thinning in stretching.	Moulds are more complicated than for extrusion blowing. Sizes limited by machine availability and economics.
Other blowing processes for hollow articles (a) Tube stretch-blowing The parison is an accurate extruded tube or moulded test-tube and the blow process can be very simple (like early celluloid processes) or highly specialized for high speed production (the latter so far little used).	In special machines high outputs possible. Good orientation by axial and longitudinal stretching if required.	Reheating is necessary, there are several stages in process, making production machines complicated and special. Wall thickness control minimal.
(b) Dip-blowing The parison is formed on a mandrel designed to allow subsequent blowing within a mould cavity. A new process with special equipment, the idea might be adaptable in certain circumstances.	By adjusting the rate of movement of mandrel in the dip pot wall thickness can be controlled. No scrap, no welds.	Limited to fairly stable materials and shapes which depart only a limited amount from circular in cross-section.
Extrusion Using a screw extruder (sometimes, though seldom, a ram extruder) material is heated for forcing through a die of required design. Used for pipes and other profiles, for sheet and film. For thin films die is often annular ('blow film'). (More plastics pass through extruders than any other machine.)	Products of constant cross-section can be made in continuous lengths. Accuracy is high with well controlled equipment. Most thermoplastics, elastomers and some thermosets can be handled under proper conditions. Can be adapted to make 'foamed' profiles. A highly adaptable process.	Requires good equipment and careful operation to give best results. Slow cooling allows polymers to crystallize, spoiling optical properties in some cases. Cannot make closed containers.
(a) Covering Wire, woodsections, glass rovings, etc., can be covered.	For insulation and protection of sensitive materials.	Cross-head dies are required to allow passage of material to be covered into extrusion stream.

Table 1.6.37 – *continued*

Table 1.6.37 Hot melt processes – *continued*

Description	*Advantages*	*Limitations*
Extrusion – cont'd (b) Fibre-forming processes Filamets, fibres, fibrillated sheet or tapes can be made.	For bristle, packaging band, ropes, textiles, etc.	Draw down and winding equipment required. Need to obtain constant output from extruder dies. May need additional gear pumps.
Extrusion combination		
(a) Extrusion coating Using an extruder and flat die a thin layer of second material is fed to a web of first material to give adherent, continuous coat. First material can be organic (e.g. cellophane or polymeric) or metallic (e.g. Al) or textile.	Use of the first material can be extended as second coat can be for sealing barrier or print protection layer, etc. Small to very large operations are possible.	Special machinery required. Process conditions are required to be well controlled, usually limited to thinner gauges.
(b) Post-extrusion laminating Can be considered as (a) with materials reversed. A warm extruded sheet is given a thin second material layer by roll-pressing integral with extrusion plant.	As 'extrusion coating'	Easier to adapt a sheet extrusion line to this process than to co- extrusion. Layer on surface only.
(c) Co-extrusion Using specially designed dies or blending units the extrusion of two or more materials can be achieved to control product properties.	'Sandwich' cross-section possible. Economic use of expensive special materials.	Requires special equipment – more than one extruder. Skilled operation.
Calendering A series of heavy, temperature controlled rolls is used to convert a plastics mass into sheet form. These rolls are mounted in various configurations; production machines are usually large. The calender is fed from mixing rolls, extruders or other blenders.	A good calender is accurate and robust and runs continuously at high output rates. Ancillary equipment can be fitted for cutting, embossing, printing, etc. Used for very large quantities of film, foil, sheet from about 0.05 to 1.33 mm thick.	Capital outlay is high and process is not very versatile. Limitations regarding materials exist (mainly ABS and PVC).
Sandwich moulding As for injection moulding but two injection units each with a different thermoplastic material inject their melt sequentially into a standard mould via a switched tap.	Material injected second remains totally encapsulated within the first. The core material may be foamed or cheaper/lower quality (eg regrind) or recycled with contaminants (eg paper, print). Surface quality is of first material.	Volume of core material generally 25 to 40% of total. Added cost of second injection unit. Design geometry for good core distribution.
Gas-melt injection Using a standard injection moulding machine, melt is injected which may only partly fill the mould. At a second stage an inert gas (nitrogen) is introduced under pressure into the melt via a nozzle or nozzles. The gas completes mould filling and/or compensates for shrinkage during cooling.	The gas flows preferentially in thicker sections which are hollowed. Weight saving, shorter cooling times and avoidance of sink marks for thick sections, eg reinforcing ribs.	Process controlled by patents. High degree of expertise in mould design and optimum air injection effects. Added cost of air injection unit.

Table 1.6.37 – *continued*

Table 1.6.38 Powder sintering techniques

Process	*Advantages*	*Limitations*
Rotational moulding Used for larger items, containers and industrial mouldings. Powder is melted and fused against inside wall of light weight mould, then cooled. During these stages the mould is rotated on two axes for containers and hollow objects, on one axis only for centrifugally cast pipe.	Very low strain in mouldings. Relatively easy to introduce as capital outlay is low. Very large sizes possible. Also suitable for small runs.	Cycle time long, limits on choice of materials due to (a) availability of suitable powders, (b) temperature of processing. Sinterability. Powders can be expensive.
Pressure sintering Not a large-scale process. Removable second components can be incorporated to control pore size and number.	By controlling the process conditions, the extent of sintering can be controlled to adjust density and porosity. Adaptable for development work, as equipment can be simple.	Slow procedure even when automated for large-scale production. Limitations on choice of materials exist.
Fluidized-bed coating More strictly a 'powder' process this technique is dip coating using fine powder fluidized in a container with a porous floor through which air is forced. Object to be coated is heated prior to dipping.		Ability of objects to withstand heating. Minimum sheet thickness limitations. Clogging of small holes. Out-gassing of entrapped air causes blisters.
Powder spraying This can be classed as a surface coating process and is then not usually considered a polymer process in the sense used here. The process can be used in various ways to produce articles e.g. by spraying into a hot cavity.	It may provide a relatively simple way of producing some types of component requiring inexpensive equipment or small numbers.	Could be difficult to control. Not fully developed.
Impact forming As yet little used, it is feasible to produce certain shapes (eg. billets for forging) by impact forming, where repeated severe blows cause the powder to warm up and sinter. (A process of this type has actually been employed as a basic means of melting material in an injection machine).	Useful when powders are readily available or opportunity of incorporation of other components into powder mix. Probably useful with heat-sensitive materials.	Limitations: powders only, choice of material size.

Table 1.6.39 Thermoforming processes

Process	*Advantages*	*Limitations*
Vacuum forming While basically this process is simple, it is capable of being extended in a number of ways either to improve depth of draw, control of thickness or complexity of pattern. Major variations in technique are: plug assist, drape forming, pressure bubble, twin-sheet. Technique generally improves distribution of wall thickness.	A wide range of sheet thickness and area and of materials can be handled on hand or auto, small or large machines for few or millions of articles. Machinery and dies not expensive.	Sheet has to be made. Know-how required to get good control of thickness distribution. Sheet should have low orientation or shrink recovery.
Pressure forming With some materials or forms, vacuum differential is not adequate. Then pressure form (with or without vacuum on other side) can be introduced.	Stiffer materials can be handled.	Sheet having little orientation is needed.
Orientation of sheet Reference has already been made to orientation in connection with bottle blowing, but orientation can be applied as a specific improvement to sheet, usually called biaxial orientation, by the use of stretching apparatus, called 'stenters'.	This is done to increase toughness, clarity and barrier properties.	Special equipment required that is expensive, needing good control. Process not commonly used except as large-scale operation.
Combination processes (a) Injection plus sheet forming This is the principle involved in the patented 'top former' process whereby a disc of material is moulded, transferred and formed, transferred and ejected on compact rotary-style machine.	The disc preform leads to minimum waste and favourable distribution of wall thickness.	There may be some limitation on materials.
(b) Injection plus 'slug' moulding This is the basis of another process, useful for symmetrical shapes such as containers. Patented as 'jetforming'.	The technique could be adapted for experimental or development purposes and in small scale production without too much complication.	Would require special equipment. Better to use other process usually.

Table 1.6.40 Solid-phase forming techniques

Process	*Advantages*	*Limitations*
Sheet forming		
The process is applied to thinner material to form dish and tube shapes in a forming die set.	Little heating and no process cooling is required. Operation can be very fast, though spring-back presents difficulty.	Limitations exist in respect of materials. Sheet of good quality is required.
'Cold' rolling		
The processes of rolling can be undertaken with some materials at similar temperatures to those for billet forging and sheet forming.	Properties can be improved by orientation for special needs.	Process is, as yet, of limited usefulness.

Table 1.6.41 Fabrication processes

Process	***Advantages***	***Limitations***
Machining		
A number of techniques commonly used with conventional materials and applicable to plastics. Several workshop techniques can be used. Machining such as turning, milling drilling, shaping, is feasible with due regard to tools, cutting speed, Lubricants and coolants.	With correct conditions all plastics can be machined. Flexible plastics and elastomers may require special cooling.	Suitable stock material (sheets, blocks, rods, etc.), need to be available. This is not so for all materials. This may mean prototypes are made up in alternative materials. Inapplicable in normal conditions to large quantities. For a larger number a moulding process is almost certain to be more economical.
Welding (see Vol. 1, Chapter 1-7)		
There are several applicable techniques: (a) heat i hot air or gas, ii welding rod, iii hot bar, (b) high frequency, (c) ultrasonic, (d) spin or friction.	These processes can be applied to sheet material to fabricate components or structures, or to extend the application or injection of forming processes. Consult material supplier for favoured technique.	Special units required and appropriate holding and application jigs. Technique dependent on material. HF and ultrasonic restricted by frequency response of material.
Joining (see Vol. 1 Chapter 1-7)		
Additional material component is required to bring about the combination of different pieces; this can be either:		
(a) A solvent Edges or surfaces are soaked or coated with suitable solvent, brought together and held, often in a jig.	Surfaces to be joined have greatly increased surface energy from applied solvent which subsequently evaporates. Solvent cement uses solvent containing percentage of parent material, which provides gap filling or thixotropy.	Some solvents cause surface deterioration ('crazing'). Limited choice of materials. Usual hazards of volatile fluids. Fairly slow process. Difficult to remove all solvent.
(b) An adhesive Surfaces are coated and brought together, held for sufficient time.	Many available: water-based, anaerobic epoxy mastic, hot melts, pressure sensitive. Consult adhesive specialist.	Limited pot life, shelf life, special equipment for rapid application.
(c) Mechanical fasteners	Self-tapping screws, moulded inserts, staples, rivets.	Limited by stresses imposed.
(d) Integral features Careful design of components enables this kind of joining.	Keys, snap fitting tabs.	Limited by stress or repetition of stress.
Hot forming		
Material (sheet, rod or tube) is heat softened locally to permit deformation, and cooled in new shapes.	Similar to sheet-forming techniques, but located with strip heaters – hot air along bend, etc. Versatile for small numbers and specials. Large range of sizes possible.	Some limitations on materials by virtue of availability. Sharp melting materials can prove difficult.

Table 1.6.42 Processes involving pastes, emulsions, liquids

Process	*Advantages*	*Limitations*
Slush moulding		
The 'slush' is a dispersion (paste) of polymer particles in a liquid plasticizer. Heat is employed to cause solution/gelation. Two-part shell mould is loaded with measured amount of paste, closed, heated, rotated about two axes, cooled.	Flexible, hollow articles (such as dolls, balls, other toys) can be made. Equipment relatively simple, moulds cheap.	Limited generally to PVC and a few other elastomeric materials.
Dip coating		
Objects, large and small, discrete or continuous may be coated by dipping in a dispersion or solution, the thickness of deposit depending on the rate of evaporation of solvent and on the temperature of the object. Hot objects promote gelation and increase deposit thickness.	Reasonable coated-over coverage with flexible layer. Intricate shapes (wirework, etc.), can be protected.	Limited to PVC and some rubbers usually.
Spreading or knife coating		
Machine has conveyor carrying web to be coated under 'doctor blade'. Polymer paste in bank behind blade covers moving web to thickness governed by gap under blade.	Substantial applications in rubber and plastics for fabric coating. Also used for making polyester sheet moulding compound (SMC).	Applicability depends on materials. Usually set up as large-scale process.

Table 1.6.43 Solution processes for coatings

Process	Advantages	Limitations
Film casting		
Filtered, gel-free solutions of polymers are cast, usually continuously on to a polished surface (metal drum/band or non-attacked polymer film) and dried to low solvent content.	A considerable range of thicknesses possible (2–250 μm), good thickness control and excellent surface. Multiple layers are possible from similar or dissimilar materials. Sintered emulsion casting is a possible varient.	Polymers must be soluble and form good solutions. Solvent recovery required on economic grounds. Thicker films slow because of diffusion/evaporation drying.
Dip coating		
This is similar to the dip coating process described in Table 1.6.42 but uses a solution instead of other forms of dispersion, and therefore requires solvent drying and recovery.	A wide variety of materials can be employed to give 'tailored' coatings. Wide range of sizes and quantities feasible.	Solvent recovery. Solvent handling hazards.
Impregnation		
Various materials are impregnated by passing a web, tape or tow through a bath of solution.	Properties of web, etc., improved, material protected or decorated. Wide variety of possibilities and sizes.	Solvent drying and recovery. Usual solvent hazards.
Coating		
Similar in pattern the dip coating process described in Table 1.6.42 but coats applied by rolls, knives or bars according to solution properties.	Advantages are similar to impregnation.	As for impregnation

Table 1.6.44 Composite production techniques

Process	*Advantages*	*Limitations*
Manual: hand lay-up Pattern, model, or reverse coated with release agent is given a 'gel coat' of resin using brush or roller, then a number of layers of glass and resin to required thickness. Pattern usually wood. Resins usually polyester or woven, depending on strength requirements. Considerable finishing, trimming, needed.	Very versatile, low-cost method of making complex objects small or large.	Highly labour-intensive with low output and inherently variable quality.
Manual: spray lay-up As 'manual: hand lay-up' except that deposition of glass is achieved using gun which projects mixture of chopped glass, resin, catalyst, over gel coat, thus reducing handling ops. on glass.	Has advantages when higher output needed on repetitive jobs.	For small quantities somewhat less versatile.
Preform lay-up Strictly a compression moulding process. Chopped glass and a bonding agent are projected at a porous model mounted on a vacuum chamber. The object formed is oven-cured, removed as dense basket-like preform. This is placed in a matched hot metal mould in a press. Moulding compound is poured over the preform and the press is closed.	Important production process, especially where distribution of reinforcement is required to be highly predictable and where moderately high output is required from fairly complex tools.	High capital investment on presses and tools, limited to high quantities.
SMC sheet moulding compound and DMC dough moulding compound Compression moulding at high temperature in matched metal mould. Essential feature is nature of charge. Sheet, often made in-house, is tailored and set in mould to form high-volume products. DMC in charge form of loaves, similar formulation. Both have low shrinkage, good surface finish.	High volume process.	High capital cost. Expensive moulds using positive shut-off. High moulding pressure (up to 30 MPa)
Cold press moulding Essentially room temperature cure process using procedure of 'manual: hand lay-up' but with mould surface top and bottom in low pressure press.	Spans gap between 'manual: spray lay-up' and 'preform lay-up' in cost. Moderate capital.	

Table 1.6.44

Table 1.6.44 Composite production techniques – ***continued***

Process	***Advantages***	***Limitations***
Resin injection Room temperature or warm cure of contents of mould which can be manual-lock or press mounted. Preform is placed in open mould. Catalysed resin injected into closed mould through preform.	Low capital outlay. Moderately costed moulds. High precision possible.	Wet-out problems with high levels of fibre reinforcement unless special techniques are employed.
Pultrusion Continuous drawing of glass rovings through catalyzed resin bath and die to heated cure zone.	Continuous lengths of high axial strength and stiffness, products with desired cross-section.	High capital costs. High die costs.
Filament winding Continuous drawing of glass rovings through catalyzed resin bath to guided distributor which winds it over a mandrel, sometimes collapsible, somewhat in the manner of a coil winding machine, the mandrel rotating.	Essentially for products of rotational symmetry with very high hoop strength such as pressure containers, rocket noses and pipes. Can be highly automated and very reproducible.	High cost of mandrels and production equipment.

Table 1.6.44 – ***continued***

Table 1.6.45 Resin techniques

Process	***Advantages***	***Limitations***
Melt casting The prepared melt, usually from an extruder, is fed to a mould which is operated in a compression mode, preferably with following pressure as melt cools and shrinks.	A technique for making thick sections, rods, tubes and bars. Moulds are basically simple.	Slow, requiring know- how and good operators. Extruder (not necessarily of high quality) is required.
Encapsulation Liquid form pre-polymer is cast around article to completely encapsulate and protect.	Electrical and other protection provided. Can be employed for small or large numbers depending on sophistication of equipment.	Details of process depend on materials employed and quantities required.
Monomer casting Almost entirely associated with caprolactam to form massive pieces and articles of nylon 6.	Chief advantages are in simplicity of equipment and thick sections possible.	Limited to nylon 6. High conversion of caprolactam important.
Pre-polymer casting Monomers or pre-polymers are cast into moulds (e.g. glass-plate moulds for PMMA).	A fairly wide range of materials possibilities exist, (both TP and TS) with care a wide range of thicknesses possible.	Care required to limit exothermic and shrinkage effects. 'Solvent' hazards.

Table 1.6.46 Forming of structural foams

Process	***Advantages***	***Limitations***
Melt foam Thermoplastics with effervescent additive injected into closed mould in injection machine of conventional or special design.	Low clamp pressure, high injection velocity results in mouldings of substantial wall thickness, varying density from max. on skin to 1/2 max. at centre, as material expands against mould walls. Important new method of making structures with many but not all properties of wood.	Some limitations on finish. Output limited by longish cooling time.
Chemical foam Two components, polyol and isocyanate, injected simultaneously into closed mould. Exothermic reaction reacts effervescent additive to expand charge to fill mould.	Low pressures (0.4 MPa), no press needed. Very simple method for prototype making. Important method of making both rigid and elastomeric structures. Very wide use in motor industry.	Critical stoichiometry for successful production using sophisticated machinery.

Table 1.6.47 Polymer types and suitable processing techniques

Process		***Polymer type***		
		TS	***E***	***TP***[a]
Melt processes	Compression moulding	X	X	X
	Transfer moulding	X	X	
	Injection moulding	X	X	X
	Extrusion blow moulding			X
	Injection blow moulding			X
	Tube stretch-blowing			X
	Dip-blowing			X
	Extrusion	X	X	X
	(a) covering			
	(b) filaments, fibres, tape fibres	X	X	X
	Extrusion combining	X	X	X
	(a) coating			X
	(b) laminating			X
	(c) co-extrusion			X
	Calendering		X	X
Powder sintering processes	Rotational moulding			X
	Pressure sintering			X
	Fluidized-bed coating	X		X
	Powder spraying	X		X
	Impact forming			X
Thermoforming	Vacuum forming			X
	Pressure forming			X
	Fabricating			X
	Orientation			X
	Combination processes			X
'Solid-phase' forming	Billet forging			X
	Sheet forming			X
	'Cold' calendering			X

Table 1.6.47

Table 1.6.47 Polymer types and suitable processing techniques – *continued*

Process			***Polymer type***		
			TS	***E***	***TP***[a]
Fabricating	Machining		X	X	X
	Welding	(a) heat			X
		(b) H.F.			X
		(c) ultrasonic			X
		(d) spin			X
	Joining	(a) solvent			X
		(b) adhesive	X	X	X
		(c) mech. fastening			X
		(d) integral features	X		X
	Hot forming				X
Liquid paste and emulsion processes	Slush moulding				X
	Dip coating			X	X
	Spreading			X	X
Solution processes	Film casting			X	X
	Dip coating		X	X	X
	Impregnation		X	X	X
	Coating		X	X	X
Composite-producing techniques	Hand lay-up		X		
	Spray lay-up		X		
	Preform moulding		X		
	SMC and DMC		X		
	Cold press moulding		X		
	Resin injection		X		
	Pultrusion		X		
	Filament winding		X		
Resin techniques	Melt casting		X	X	
	Encapsulation		X	X	
	Monomer casting		X		X
	Polymerization casting		X		X
Foam moulding 'structural' foam	Melt foams				X
	Chemical foams		X	X	

[a] TS — thermoset E — elastomer TP — thermoplastics
N.B. 'X' signifies common processing method for class of material. It does not mean that all types of material in that class can be processed by that method.

Table 1.6.47 – *continued*

Table 1.6.48 Comparative characteristics of plastic processing techniques

Process	Suitable material type	Form of raw material	Size		General tolerances	Surface finish
			Max.	Min.		
Injection moulding	Thermoplastic	Granules Pellets Powder	12 kg (solid) 25 kg (structural foam)	0.1 g	±0.05 mm up to 25 mm dia. ±0.2 mm up to 150 mm dia. >150 mm dia: 0.25 + 0.1 mm per 25 mm over 150 mm dia.	Good to excellent
Injection moulding	Thermoset	Granules Pellets Powder	6 kg	<15 g	±0.05 mm to 25 mm dia. ±0.15 mm to 150 mm dia. >150 mm dia: 0.25 + 0.075 per 25 mm over 150 mm dia.	Good to excellent
Compression moulding	Almost entirely thermosetting	Pellets Powders Preforms	16 kg or 450 mm depth	3.2 mm square	±0.05 mm up to 25 mm dia. ±0.1 mm up to 150 mm dia. ±0.15 mm + 0.05 mm per 25 mm over 150 mm dia.	Good
Transfer moulding	Thermosetting	Pellets Powders Preforms	As above	As above	±0.05 mm up to 25 mm dia. ±0.15 mm up to 150 mm dia. >150 mm dia: as injection moulding (TS)	Good
Casting	Thermosetting and thermoplastic	Liquids	900 kg 600 mm thick	6.4 mm cubed	±0.8 mm ±0.05 mm (special handling)	Fair
Cold moulding	Non-refractory bitumen or refractory phenolic based materials with inorganic binders.	Powders Fillers		6 mm	±0.08 mm	Poor
Coating	Thermosetting and thermoplastic	Liquids Powders			±0.05 mm	Good
Cellular plastics (foam expanded)	Thermosetting and thermoplastic	Granules Pellets Liquids Powders	6 mm dia.	None	±0.1 mm	Varies greatly
Extrusion (a) Sheet and profile (b) Tubing	Thermoplastic and thermosetting Thermoplastic	Granules Powders Pellets	1800 mm wide 150 mm wide	4 mm square 1.12 mm dia.	±0.05 mm or better	Fair
Laminating (high pressure)	Thermosetting and thermoplastic	Liquid resin and fillers	Sheet: 900× 1800× 200 mm Tube: 900× 150 mm dia. Rod: 900× 100 mm dia.	 900× 1800 0.25 900 3 900 6	±0.1 mm	Good
Sheet forming	Thermoplastic	Sheet	2.5×7.5 m	25×25 mm	±0.8 to ±1.6 mm ±0.25 mm thickness	Good
Hand lay-up and related processes	Thermosetting resin with glass reinforcement.	Liquids, glass fibres or glass cloth.	24×24×3 m	Small 100× 100 mm	±0.8 mm to ±6.45 mm (Can be ±0.05 mm)	Varies. Can be good.
Filament winding	Thermosetting resin and glass strand.	Liquids and glass strands.	4.5× 12 m	300×20 mm	±1.6 mm	Interior good. Exterior poor.
Dip and slush moulding	Thermoplastic	Powders			±0.8 mm	Poor
Blow moulding	Thermoplastic	Granules Pellets Powders	0.86 mm ×0.09 m	12.7 mm long	±0.08 mm O.D.	Good
Rotational moulding	Thermoplastic	Powders	3.6 m cubed		±0.08 to ±6.45 mm	Fair
Calendering	Thermoplastic with/without fillers.	Dough preheated	2 m wide	Unlim. length	±1.0 mm width ±0.1 mm	Good

Table 1.6.48

Table 1.6.48 Comparative characteristics of plastic processing techniques – *continued*

Production rate	*Tooling cost*	*Labour cost*	*Material waste*	*Floor area (m^2)*	*Installed power (kVA)*	*Shape classifications possible*[a]	*Economic production range*
Cycles 1–50 per/min up to 5 kg 1–5 per/min over 5 kg	Up to 10% of m/c	Low (in relative terms)	Very low	~50	up to 100 kVA	1–8	5000–20 000
1–4 cycles/min up to 1 kg 1–3 cycle/min over 1 kg	High as above	Low as above	Moderate	As above	As above	1–8	>20 000
20–140 cycles/h	High (up to 25% of m/c)	Fairly low	Fairly low	60	170	1–8	>2000
20–160 cycles/h	High	Fair	Moderate			1–8	>50 000 (more economic than IM at 10 000 off)
Low	Low	High	Moderate			1–8	<1000
3500–4000 cycles/h	Fairly high	Fairly high	Moderate			1–8	<200
Short	Low	Fairly high	Moderate			1–3	N/A
High (spray-on) Medium (moulded)	Variable	Moderate to fairly high	Moderate	50 (moulding)	100 (moulding)	1–5, 7, 8	>100 but never cheap.
1–65 m/min (profile and tube) 2–5 m/min (sheet and rod)	Variable	Moderate	Moderate	50–100	35–100	2–8	5000 kg (sheet) 1000 kg (profile)
Low	Low	High	High			2, 3, 5, 7–9	<1000 (ideally)
Up to 10 000/h (thin-wall cups and containers).	Low	Low	High		3–5	2–5, 7	Economic over all ranges, but the higher the better.
Usually low, depending on method.	Low	Very high	High	Depends on mould size.	Negligible	2–4, 6–9	1–2000 for small items. 1 off for very large items.
Up to 60/h (multispindle). Low (depends on size)	Low	High	Moderate	Depends on mould size.	Low	2, 3, 9	As above.
Low	Moderate	Moderate	Low			2–4, 9	>100
100–2500/h	High	Low	Low	50	35–100	2–4, 9	>1000
12–50 cycles/h	Low	Moderate	Low	40	200	2–4, 7–9	Generally expensive
10–100 m/min	N/A	Moderate	Low	100	200	Flat sheet	>10 000 m

[a] Shape classification
1 – Solid concentric
2 – Hollow concentric
3 – Cone or cup concentric
4 – Cup, dish or cone non-concentric
5 – Hollow or solid non-concentric
6 – Spirals, repetitive irregular concentric
7 – Flanged and flat
8 – Complex miscellaneous
9 – Tanks or closed tubes concentric or non-concentric

Table 1.6.48 – *continued*

Table 1.6.49 Suitable processing techniques for specific polymer materials

Material	Process																									
								Extrusion											Sheet thermo-forming							
	Blow moulding	Calendering	Casting (liquid resin)	Compression moulding (high P)	Compression moulding (low P)	Dip moulding	Dip coating	Cable and wire covering	Film	Paper coating	Monofilament	Pipe	Profile section	Sheet	Tubular film	Injection moulding	Rotational casting	Sinter moulding	Matched mould forming	Vacuum forming	Pressure forming	Transfer moulding	Laminating	Welding and sealing	Wet hand lay-up	Machining
Acrylics	2		Sheet 1 Film									2	2	1		1			2	2	2			2		2
Cellulosics – Cellulose acetate	2		Sheet 1 Film						1					1		1	1		1	1	2			1		3
– CAB							2		2				2	2		1			2	2	2			2		
– Cellulose propionate	2								2				2	2		1			2	2	2					
– Ethyl cellulose									1				1	1		1										
Polyacetals – Homopolymer	2											2	2	2		1								2		2
– Copolymer	2											2	2	2		1								2		2
– Terpolymer	1											2	2			2										
Fluoroplastics – PTFE				2			2	2				2	2					1								2
– FEP								2	2			2	3	3		1										3
– PFA																										
– PVDF	2			2								2	2			1				2	2					
– PTFCE							2	3																		
– PETFE								2																		
– ECTFE									2																	
Chlorinated polyether							2					1	2	1		1								2		2
Polyamides – Nylon 6	2		Spec. 2 Type						1		2	2	2	3		1	2			2	2			3		3
Nylon 6.6	2							2	1		2	2	2	3		1								2		2
Nylon 6.10	2							2			2	2	2	3		1								2		2
Nylon 6.12																1										
Nylon 11	2					2	1	2	1		2	2	2	3	2	1	2							2		2
Nylon 12	2					2	2	2	1			2	2	3	2	1	2							2		2
Transparent	2												2			1										
Polyamide-imide						2							2			1										
Polycarbonate	2								2			3	3	3		1								3		3
Thermoplastic polyesters PBT													2			1										
PETP									1	1						2			2	2	2			3		3
Polyolefin copolymers EVA	2							2	1			2	2	1	2	1	2							3		
Ionomer	1											2	2	2		1				2	2		2	3		3
Polyallomer								2	2			2	1	1	2	1				1	1					
Polyphenylene oxide (modified)												2	2	2		1								3		3
Polyphenylene sulphide				2								2	2	2		1										
Polyethylene LDPE	1						2	1	2	1		2	2	2	1	1	2			3	3			2		
HDPE	1						2	2	2		2	2	2	2	2	1	2			2	2			2		3
Polypropylene	2								2		2	2	2	2		1				2	2			3		

Table 1.6.49

Table 1.6.49 Suitable processing techniques for specific polymer materials – ***continued***

Material		***Process***																									
									Extrusion											***Sheet thermo-forming***							
		Blow moulding	***Calendering***	***Casting (liquid resin)***	***Compression moulding (high P)***	***Compression moulding (low P)***	***Dip moulding***	***Dip coating***	***Cable and wire covering***	***Film***	***Paper coating***	***Monofilament***	***Pipe***	***Profile section***	***Sheet***	***Tubular film***	***Injection moulding***	***Rotational casting***	***Sinter moulding***	***Matched mould forming***	***Vacuum forming***	***Pressure forming***	***Transfer moulding***	***Laminating***	***Welding and sealing***	***Wet hand lay-up***	***Machining***
Polysulphones	Polysulphone	3							2	2		2	2	2	2	2	1										
	PES									2							1				2						
	Polyarylsulphone				2									2			1										
	Polyaryl ether													2			1				2	2					
Thermoplastic polyurethanes	Polyester		2		2					2					2		1				2		2				
	Polyether		2		2					2					2		1				2		2				
Polyvinyl chloride	Flexible		1	Film grade						1 3 Blown				1		2	2	2							1		
	Rigid	2	2	Film grade	2			2		2			1	2	1	2	2				2	2			2		3
	CPVC												1														
	PVDC							2	2	1														1			
Styrene based polymers	Polystyrene	2								1				2	1		1				1	2					3
	HIPS	2/3												2	1		1				1	2					3
	SAN	2												2			1				3	3					
	ABS	2	2							2			1	2	1		1				1	1			2		3
	BDS	Special 2 Grade													1		1				1	1					
TPX		2												2	1		1				1	1					
Alkyds					1												2						1				
Aminos – Urea formaldehyde					1												2						1				
	Melamine formaldehyde				1												2						1				
Diallyl phthalate					1												2						1				
Epoxides				2	1			2									2						1	2			3
Furanes																											
Phenolics					1									2			1								3		3
Polybutadiene				2	1												1										
Unsaturated polyesters	DMC				1												3						2				
	SMC				3	2																		3		1	3
Polyimide						Special 1 grade			2	2							2		1					2			1
Polyurethanes				1				2											2				2				
Silicones				2	1																						

KEY: 1 – Process or processes most widely adopted.
2 – Other suitable processes; may not be adopted as widely due to either technical, economic or marketing reasons. Thus it should not be assumed that these processes are less feasible than those designated 1 without reference to the relevant section, or the manufacturer.
3 – Restricted processes, but not impractical.

Table 1.6.49 – ***continued***

Table 1.6.50 Container type and applicability of forming processes

Type of container	*Forming process*										
	Injection moulding	*Transfer moulding*	*Compression moulding*	*Blow moulding: Injection*	*Blow moulding: Extrusion*	*Extrusion*	*Casting*	*Calendering*	*Rotational moulding*	*Slush castings*	*Thermoforming*
Flat sheet			√			√	√	√			
Tray	√	√	√						√	√	√
Dish	√	√	√						√	√	√
Box, jar, tub	√								√	√	?
Squeeze tube	√					√					
Small bottle				√	√						
Box (lidded)	√								√		?
Sleeve	√					√					
Insert (block)	√	√	√								
Drum					√				√	√	
Large liquid containers					√				√	√	

Table 1.6.51 Characteristics of container-forming processes for plastics

Process	Method	Advantages	Disadvantages
Injection moulding	Thermoplastic material is softened by heat, and often by mechanical working in a screwfeed, and forced by a ram into shaped, often multiple moulds where it solidifies by cooling. Formed object removed by opening the mould.	Low machine cost/piece in large-volume production. High output. Good finish to all surfaces. Scrap can be re-used.	Only open necked shapes with no large undercuts can be made. Moulds are expensive.
Transfer moulding	A variation of injection moulding for thermosetting resins. Mould is kept hot to promote the setting reaction and moulded article ejected while still hot. Final hardening is completed out of the mould.	Allows use of thermosetting resins at lower cost than thermoforming for adequate throughput.	Higher mould and machine costs than thermoforming. Waste material cannot be re-used.
Compression moulding	Slight excess of thermosetting powder is put into a heated female part of mould. Closing of mould by pressing in male component melts and flows thermoset to fill the mould. Resin sets while hot and is removed after opening mould.	Relatively low tooling costs for good accuracy in formed parts. Filled thermosets can be processed easily.	Slow throughput. Surplus material forms flash which has to be removed in extra operation and is not reusable.
Injection blow moulding	Thermoplastic is injection moulded into a cavity surrounding a hollow tube. While still hot it is transferred to inside of a split mould and plastic blown to shape of mould by air pressure through the tube. Removed by opening the mould.	Allows bottles to be made with good accuracy of neck form. Shaping of parison before blowing allows compensation for thinning in stretching giving more uniform wall thickness.	High mould costs and limitation on size – usually not more than 1 litre containers.
Extrusion blow moulding	Simpler process of extruding a section of tube which is then held in a split mould for blowing. The coming together of the mould seals off the bottom of the bottle. More complex process extrudes varying wall thickness to compensate for subsequent blowing.	Lower tooling costs than injection blow moulding. Can have multiple heads to extruder, or several extruders to feed one die to produced striped effects.	Lower accuracy than above, and variable wall thickness unless compensated. 'Pinch off' at bottle bottom may need removing.
Extrusion	Thermoplastic softened by heat and forced out through shaped die continuously by screw action in feed. Hot extrudate is cooled to solidify into shape or die.	Allows parallel sided shapes such as sleeves, tubes etc. of practically any cross-material shape to be made accurately in continuous lengths.	Slow cooling allows polymers crystallize, spoiling optical properties in some cases. Cannot make closed containers.
Casting	Molten thermoplastic is extruded through a slot onto a chilled roll or moving steel belt. After cooling the sheet is peeled off and wound into rolls.	Rapid cooling allows clear film to be made from crystallizable polymers. Surface contacting belt takes on a high gloss.	Slow output. No modification of properties by stretching possible. Only flat sheet produced.
Calendering	Sheet is formed by forcing thermoplastic through slot between temperature-controlled rolls, cooling and setting in air on emergence.	High output and suitable for thicker films and sheet materials.	High capital cost. Only flat sheet produced and suitable for only a few materials (mainly ABS and PVC).
Rotational moulding	A powdered thermoplastic is placed in a heated mould mounted so that it can be rotated in two directions at right angles. The plastic melts and spreads over the inside of the mould. Part is removed after mould is cooled by water spraying and cold air.	Suitable for large parts or for small runs. Mould and machine costs relatively small. Technique avoids internal stresses and gives uniform wall thickness.	Production rate is low and material costs (powders) higher than beads for extrusion. Some materials hard to get in powder form.

Table 1.6.51

Table 1.6.51 Characteristics of container-forming processes for plastics – *continued*

Process	*Method*	*Advantages*	*Disadvantages*
Slush moulding	Resins mixed with plasticizers, fillers etc. to form viscous fluid. Poured into heated mould, gel forms on walls. Excess is poured out and coating on wall is allowed to set, by heating further.	Similar to rotational casting for polymers for which suitable solvating plasticizers are available.	Low production rate and limited range of suitable materials.
Thermo-forming	Sheet thermoplastic is softened by heating and then forced over a shaped mould either by vacuum through the mould, or by air pressure behind the sheet, or by a matched mould pressing the sheet down.	Versatile process. Tooling costs low but big production possible. Stiff materials can be handled by pressure or matched-die processes.	Limited complexity of shapes possible. Good know-how needed to avoid excessive thinning. Post-forming recovery may be troublesome.

Table 1.6.51 – *continued*

Table 1.6.52 Suitability of various polymers for use in container-forming processes

Type of container	*Forming process*										
				Blow moulding							
	Injection moulding	*Transfer moulding*	*Compression moulding*	*Injection*	*Extrusion*	*Extrusion*	*Casting*	*Calendering*	*Rotational moulding*	*Slush castings*	*Thermoforming*
ABS copolymer	√			√	√	√		√		√	√
Acetal	√			√	√	√					
Acrylics	√					√	√		√		√
Cellulose acetate	√		√	√	√		√		√		√
Cellulose propionate	√		√	√	√		√		√		√
Phenol formaldehyde		√	√								
Urea formaldehyde		√	√								
Melamine formaldehyde		√	√								
Polyamides (nylon)	√			√	√	√	√				√
Polycarbonate	√			√	√	√	√		√		√
Polythene (L.D.)	√			√	√	√	√		√		
Polythene (H.D.)	√			√	√	√	√		√		√
Polypropylene	√			√	√	√	√		√		√
Polystyrene	√			√	√	√					√
Polystyrene (toughened)	√			√	√	√					√
Polystyrene (expanded)						√					√
Polyurethane (expanded)	√[a]										
PVC (rigid)				√	√	√	√	√		√	√
PVC (plasticized)				√	√	√	√	√		√	√

Two-part cold liquid mix expands into mould by reaction products.
[a]Absence of a tick (√) does not necessarily mean that the plastic is impossible to process by the method sought, but that it is not normally done for some reason such as liability to thermal degradation, or unsuitability of the material for containers made by the process (e.g. L.D. polythene is too floppy for shapes made by thermoforming sheet).

1.6.9 Forming of ceramics

1.6.9.1 Ceramics processing techniques

Engineering ceramic materials, with the exception of glass, are manufactured from powder of controlled particle size. The general procedure involved in manufacture of a solid article is outlined in:

Fig. 1.6.29 *Salient features of manufacture of ceramic components*

The type of shaping operation used in the forming process determines the subsequent process operations necessary to complete manufacture of an article, as shown in:

Fig. 1.6.30 *Some ceramic manufacturing routes (machining operations are omitted)*

Descriptions of the various forming processes, together with the type of shape for which the method is suitable and its applications and limitations are given in:

Table 1.6.53 *Forming methods for ceramics*

When a cold shaping method is used the 'green' (unfired) article may be machined before firing, to modify its shape. This capability can be a useful adjunct to cold shaping techniques.

FIRING

After forming by one of the methods previously described ceramic components, after drying if necessary, are fired at temperatures between 700 and 2000°C depending on composition. Typically, during this sintering process, porosity in the compact is reduced and the density of the ceramic increases. This is accompanied by a decrease in volume (firing shrinkage) of up to about 45%, and an increase in strength and hardness so that after firing engineering ceramics can only be further shaped by diamond tools.

Silicon carbide and silicon nitride ceramics can also be densified by a process known as 'reaction sintering' or 'reaction bonding' in which densification occurs by a chemical reaction rather than a solid-state diffusion process.

In the case of silicon nitride, compacts of silicon powder are subjected to a lengthy firing schedule at temperatures near 1400°C in a nitrogen atmosphere. A chemical reaction ensues in which the silicon compact is converted to silicon nitride. Densification and increased strength occurs without large firing shrinkage.

Reaction sintered silicon carbide is made by a similar process in which a compact of silicon carbide and graphite is impregnated with molten silicon which converts the graphite to silicon carbide. There is little or no dimensional change during the sintering process.

HOT PRESSING AND HOT ISOSTATIC PRESSING

An alternative route for production of a dense ceramic material from the initial powder is by hot pressing. Effectively, forming and firing are carried out in a single operation by the simultaneous application of heat and pressure. The process is commonly carried out in graphite dies heated by induction.

Almost fully dense, strong ceramic materials are produced but the process is limited by the complexity of the shapes that can be produced. Subsequent shaping can only be carried out by diamond tools. This fact, and the fact that graphite dies wear rapidly, means that hot pressed ceramics are currently more expensive than ceramics processed by other methods.

Hot isostatic pressing is a variant of the hot pressing process in which the compaction pressure is applied by a gas, the powders to be compacted being encapsulated in metal or glass. In addition, fired ceramic articles with no open porosity can be further densified without encapsulation.

1.6.9.2 Forming methods for glass

Glass components can be formed by five general processes, namely blowing, pressing, drawing, rolling and casting.

Pressing and blowing operations are used for the manufacture of containers. Glass tubing is made by drawing a supply of molten glass over a hollow metal mandrel. Air flowing through the mandrel fills the glass tube as it is drawn off, forming a tube.

All glass products are annealed (i.e. reheated to temperatures of about 500°C and allowed to cool slowly) directly after forming so that stresses set up by differential cooling after manufacture are relieved.

Processes for the production of sheet glass, together with their advantages and limitations are described in:
Table 1.6.54 *Processes for production of sheet glass*

1.6.9.3 Joining of ceramics

Generally the range of processes available for joining ceramics is very limited, but particular ceramics may be joined by one of the following techniques.

ADHESIVE BONDING

For low temperature applications, adhesives can be used to make ceramic–ceramic joints or ceramic–metal assemblies. Epoxy adhesives have shear strengths up to 35 MPa but are limited to temperatures of about 150°C.

GLASS BRAZING

Glass brazes can be used to make relatively low pressure/low temperature metal-ceramic seals. Glass powders of matched expansion coefficient are used. In a typical application, that of sealed terminals for passing electrical conductors through the walls of sealed assemblies, an alumina–glass–metal seal withstands pressures of 2 MPa and will withstand temperatures higher than the melting point of the soft solder used to make electrical connection to the conductor.

Aluminas and porcelains may be joined to metals by a metallizing-and-brazing process.

The commonest metallizing technique is the molybdenum–manganese process. A slurry of molybdenum powder, manganese powder and a suitable binder is applied to the ceramic and fired under oxidizing conditions to form a metallized layer, which after plating, usually with nickel, will take a braze or solder. Selection of a metal with a matched thermal expansion coefficient (e.g. Kovar, nickel monel, molybdenum and some stainless steels) allows joints to be made to withstand operating pressures up to 70 MPa.

REACTIVE METAL BRAZING

Reactive metal brazes containing titanium, titanium hydride or zirconium can also be used to make joints in oxide ceramics such as alumina.

Because, in general, metals have higher expansion coefficients than ceramics and the latter have high compressive strengths, shrink-fitting of metal components to ceramics can sometimes be employed, though this technique is not recommended for assemblies experiencing high in-service temperatures.

WELDING

Assemblies of reaction bonded silicon nitride can be fabricated from smaller components by a joining process in which the unreacted compacts are joined by flame-spraying silicon powder to the joins. The complete assembly is then reaction sintered in the usual way to produce the final composite component of silicon nitride.

Because of their softening at elevated temperatures, glass components can be joined to one another by a process analogous to the fusion welding of metallic materials.

GLASS–METAL SEALING

Glass–metal seals can readily be made because of the ease with which molten glass wets oxidized metal surfaces, but the inherent brittleness of glass means that tensile stresses due to thermal mismatch lead to fracture of the glass. Either metals with matched thermal expansion coefficients must be used or the seal designed in such a way that relatively soft metallic components like foils or filaments of platinum or copper are used. Plastic flow in the metal component then occurs before large stresses can develop in the glass.

1.6.9.4 Guide to design of sintered ceramic articles (see also Vol. 2, Chapter 2-15)

This guide has been prepared for the use of engineers and draughtsmen in designing ceramic parts. Ceramic materials exhibit a number of common property features which necessitate particular consideration when designing ceramic components. These are listed in:

Table 1.6.55 *The effect of ceramic properties on design and construction methods*

The data included represent that which is considered good design practice, and do not necessarily imply that features contrary to these recommendations cannot be provided by special handling in some form or another. These recommendations will secure not only cheaper articles, but increased production by reducing the possibility of waste, assembly difficulties and delay. In order to ensure the most desirable design and to facilitate production at minimum cost, it is strongly recommended that designers consult the ceramic manufacturer in the initial stages concerning specific design details applicable to the part under consideration.

TOLERANCES

In the course of manufacture the ceramic material is shaped in a raw state. The articles are then fired, converting the material into its final state. This conversion is accompanied by a relatively large contraction which cannot be completely controlled. Every effort should therefore be made to adopt the normal tolerances, detailed in this recommendation, which are considered to be the minimum practicable for normal bulk production. Where closer tolerances are needed, these should be specified only for those dimensions where they are essential, as this generally means selection or the introduction of additional processes. The ceramic manufacturer is in a much better position to interpret the purchaser's requirements and to provide him with trouble-free assembly, if he has full information regarding the size and character of the parts to be used in conjunction with the ceramic. In addition to drawings of such associated parts, a sample assembly is a great help, and should be sent to the ceramic manufacturer whenever possible. Frequently, as a result of an examination of the assembly, the ceramic manufacturer is able to suggest detail design alterations of the ceramic parts which will improve production and assembly.

METHODS OF SHAPING

The methods by which articles are made may be grouped according to whether they are by nature a 'pressing' or a 'plastic' method. The description 'plastic' covers processes such as extrusion, turning, jolleying and casting. The characteristics and desirable features of the most commonly used methods are described as follows:

Extruded shapes

The cross-section of extruded rods should be as symmetrical as possible.

Tubular shapes which are produced by extrusion should have a wall thickness as uniform as possible. All outside shapes should be made as symmetrical as possible, and the inside preferably cylindrical and concentric with the outside.

It is desirable to maintain a reasonable wall thickness whenever possible, and under no condition should walls be excessively thin. If the outside diameter exceeds 10 times the wall thickness, wider tolerances will be required.

All transverse holes are by the nature of the process somewhat oval. This fact should be taken into consideration by the designer.

Die-pressed parts

Uniaxial pressing is particularly suitable for articles of simple shape. Articles with many non-planar surfaces with at the same time varying thicknesses, cause such complexity of the dies that they become non-economic and an alternative technique such as injection moulding has to be employed. The die costs are relatively high and therefore the process is only economic where large quantities are involved. Automatic pressing is often possible.

It is recommended that die-pressed parts be so designed as to maintain as nearly as possible a uniform thickness over the full area of the piece. The minimum thickness of simple flat shapes, discs, plates, and the like, should be computed as follows:

Where the ratio of the maximum superficial dimension to the largest dimension at right angles to it is less than 5,

$$\text{minimum thickness} = \frac{\sqrt{A}}{10} \text{ mm}$$

Where the same ratio is 5 or greater,

$$\text{minimum thickness} = \frac{\sqrt{A}}{8} \text{ mm}$$

In both cases A is the outline area.

Exceedingly small holes and blind holes should be avoided where possible. For economical production, holes should not be less than 1.5 mm diameter. Reasonable tapers and radii (draft and fillets) should be specified for all depressions or bosses and thin walls, i.e. less than 1.0 mm, should be avoided where possible.

Where parts are to be mounted on some flat surface they should be so designed that they can be ground flat after firing to prevent breakage during the mounting operation. The area to be ground should be reduced as much as possible, for example, by providing small bosses round the fixing holes. The height specified for these bosses should take into consideration the camber tolerance. Their diameter should be about twice that of the hole which they surround, with a minimum of 2.0 mm plus the diameter of the hole.

To minimize warping, grooves or channels which run right across a part should be kept as shallow as possible and should preferably be not deeper than one-third of the thickness of the piece. Whenever possible, a recess which does not run right across the part should be substituted for a channel.

Tool damage may occur if the faces of the top and bottom punches come into contact in the course of the pressing operation. The component should be designed to avoid this possibility.

HOLES, THREADS AND FASTENINGS

Holes

The distance between hole centres may change due to variations in shrinkage. This can generally be accommodated by expanding one or both holes by an amount equal to the total tolerance on the hole centres. If selective assembly is to be avoided additional expansions may have to be provided to accommodate the variations in the stud or screw spacing of the part to which the ceramic article is assembled. If pin gauges are used for checking the hole centres, the design of gauge should be such as to meet the assembly requirements, but such pin gauges should also take into consideration the tolerance on the hole centres as well as the tolerance on the diameter of the holes.

Threads

It is most economical for the threads of coil formers to be continuous for the entire length. Any portion which must be left unthreaded should preferably not be of greater diameter than the diameter at the bottom of the thread unless the wall thereby becomes too thin.

It is essential that a slight radius be provided at the bottom of 'V' threads and, where possible, the included angle should not be less than 55°.

For the purpose of these recommendations the thickness of a threaded former is considered to be that between the root of the thread and the bore.

A pitch finer than 0.8 mm introduces manufacturing difficulties.

Fastenings

Tapped holes in ceramic parts and external screw threads for assembly purposes should, wherever possible, be avoided. If, however, these are essential the ceramic manufacturer should be consulted.

In securing ceramic parts to other materials, consideration should be given to the differences in the coefficient of expansion which may exist between the two materials.

Mounting at the minimum number of points is desirable, preferably not more than three.

Radial serrations around holes where eyelets and rivets are to be used, help in eliminating any tendency of the metal parts to turn, reduce breakage during assembly and serve to take up variations in thickness of the ceramic. Pinning is less likely to damage the ceramic than riveting.

When securing ceramic parts by clamping, care should be taken to avoid excessive stress concentration. The interposition of a resilient material between the clamping surfaces and the ceramic is desirable.

SURFACES

Although it is deemed impracticable to specify even general levels of quality for surface defects, since both size and application of ceramic parts are subject to wide variations, the surfaces should not be chipped, blistered, pimpled or speckled to an extent that would impair the life and usefulness of the articles. It is recommended that questions of surface defects should be considered on a common sense basis rather than by any empirical formula.

LIMITS FOR UNGROUND ARTICLES

Dimensional tolerances

Ceramic articles are supplied to a normal tolerance of ±2%, but ±1% can be achieved. The percentage tolerance is subject to a minimum value which varies with the manufacturing process, the material and the type of article. Details of the tolerances and the minimum values are given in:
Table 1.6.56 *Dimensional tolerances for unground articles*

Geometrical tolerances

(i) Roundness. Major and minor diameter at any cross-section should both be within the specified limits for diametral tolerance.
(ii) Parallel limits. Plane surfaces shall be considered to be parallel if the thickness tolerances are satisfied and the angle between the two surfaces does not exceed 2°.
(iii) Camber. For the purpose of this recommendation, camber is defined as the ratio e/L

where
e = the maximum perpendicular distance from the reference plane to the corresponding curved surface,
L = a reference length measured between the sector extremities.

The maximum camber for various ratios of L/D, where D is the minimum dimension or diameter in the same direction as e, is given in:
Table 1.6.57 *Maximum cambers for ceramic articles*
The measurements of L, D and e to be used in conjunction with Table 1.6.57 for a number of basic shapes is shown in:
Fig. 1.6.31 *L, D and e for basic shapes*
For flat shapes, discs and plates the above limits should apply only where the thickness is equal to or greater than the thickness determined by the formula. Where the thickness is less, the camber should be specified in the relevant drawing.

LIMITS FOR GROUND ARTICLES

The grinding processes and the tolerances in general use are given below, but closer tolerances may be agreed between purchaser and manufacturer.

Dimensional tolerances

(i) Centreless grinding.
The tolerances on diameter for centreless grinding are given in:
Table 1.6.58 *Dimensional tolerances for centreless grinding*
(ii) Cylindrical grinding of external and internal diameter.
±0.1% but not less than ±0.07 mm
(iii) Flat grinding.
Tolerance on thickness ± 0.05 mm
(iv) Honing of bores.
Diametral tolerances ± 0.025 mm

Geometrical tolerances

(i) Roundness.
The difference between major and minor diameters at any cross-section of a centreless or cylindrical ground article should not exceed the dimensions in:

Table 1.6.59 *Roundness tolerances for ground articles*

(ii) Parallel limits.
The difference between the maximum and minimum thickness should not exceed 0.05 mm.

(iii) Camber.
Grinding processes do not necessarily correct camber, particularly with slender articles; it should also be noted that centreless grinding does not necessarily straighten a rod or tube which is bowed before grinding.

Camber tolerances should be agreed between purchaser and manufacturer.

The need to adopt different design principles when designing for the use of ceramic materials, as compared with those principles which are used when designing for the use of metals, will now have become apparent.

DOS AND DON'TS FOR DESIGNERS IN ENGINEERING CERAMICS

The following simple guidelines for design will be helpful. Although most of them will aid good design in metal components, they are even more important for engineering ceramics because of the brittleness problem.

DON'T use sharp corners, notches, slots which can act as stress raisers.

DO use generous radii.

DO pay attention to surface finish. A good surface finish is essential.

DO take advantage of the higher compressive strengths of ceramics by preferring compression joints and connections to threads and flanges.

DO disperse stresses at assembly faces by using soft gaskets where possible to avoid high spot loadings.

DO allow for thermal mismatch, particularly at ceramic/metal joints. Metals have expansion coefficients up to 10 times those of ceramics.

DON'T incorporate abrupt changes of section which may lead to high thermal stresses in temperature gradients.

DO make use of proof-testing procedures if possible.

DO remember the batch-to-batch variability of ceramics and make trials on the grade and surface finish to be used in production.

DO specify dimensions carefully in view of the greater difficulties in final machining than with metals.

Table 1.6.53 Forming methods for ceramics

Process	*Description*	*Applications and limitations*	*Type of shape for which method is suitable*
Soft plastic forming ('jiggering', 'jollying' and their roller head versions)	Used for clay-based ceramics (white-wares and other pottery). The plastic body placed on or in a plaster of Paris mould which is rotated and forms the surface of the object. The other surface is formed by either a rotating or a non-rotating tool.	Plates, bowls and cups etc., high voltage insulators, crucibles, refractory sleeves and nozzles. Broadly limited to circular or, with jiggering, oval components.	Large diameter, hollow cylindrical.
Slip casting and pressure casting	In slip casting, a slurry (i.e. a very fluid suspension of powder in water or organic liquid) is poured into a plaster of Paris mould. The porous mould absorbs water so that a solid product builds up on the mould walls. Solid products, or hollow ware, can be produced by pouring off excess slurry after an appropriate time. In pressure casting, pressure is applied to a slurry contained in a porous polymer mould until a solid product is formed between the opposing mould faces.	Suitable for any ceramic powder provided sedimentation can be prevented and adequate permeability can be established in the cast layer. Used for white-wares, particularly sanitary ware, and specialist laboratory ware. Effective at producing thin-walled shapes and economical for small production runs due to low mould costs.	Wide range of thin-wall articles but solid casting of 'blocky' shapes is possible.
Extrusion	A stiff plastic mix is forced through a die orifice. Non-clay-containing ceramics require the addition of an organic plasticizer for successful extrusion. Extrusion is at higher pressure for such ceramics and low-porosity compacts with little shrinkage on firing are produced.	Production of rod and tube in aluminas, clay based ceramics, silicon nitride and silicon carbide ceramics. Production of brick and sewer pipe.	Bar, rod or tube (column of uniform cross-section).
Injection moulding	Injection moulding techniques similar to those used in the plastics industry can be used to produce components of complex shape at high output rates. Organic additives are added to non-clay-containing ceramics to ensure easy flow properties.	Metal moulds used are expensive so that injection moulding is only economic if large numbers of similar components are to be fabricated.	Wide range of small shapes in large quantities.
Uniaxial pressing	Dry powder containing binder and plasticizer is pressed in a metal die at sufficiently high pressures to form a dense and strong compact. Dies are usually of hardened steel or tungsten carbide. Engineering ceramics are usually pressed at 50–150 MPa to tolerances of $\pm 1\%$ or 100 μm, whichever is greater.	Probably the most widely used process for the mass production of engineering components. Complex shapes can be produced accurately and compacts can be machined by conventional techniques.	
Isostatic pressing	The press body is encased in a rubber sheath and pressure applied through a fluid. This gives more uniform pressure distribution throughout the compact so that uniform density ceramics with low firing shrinkage are produced. Pressures of 30–35 MPa are used and automatic production rates of 1000–1500 pieces per hour are feasible.	Less complex shapes can be formed in isostatic pressing. Compacts can be machined by conventional techniques, including grinding of green compacts.	Simple shapes of very uniform high density.
Flame spraying	Compacts of silicon for manufacture of reaction bonded silicon nitride can be made by flame spraying silicon onto formers which are later removed.	Products have similar strength and density to isostatically pressed material.	
Other forming methods	Techniques involving ceramic–plastic technology have been developed particularly for the manufacture of reaction bonded silicon nitride. Mixing powders with thermoplastic or thermosetting resin allows the production of silicon nitride foams. Organic material is removed before firing.	Complex and intricate shapes can be produced. Produces final product of somewhat lower density and strength than isostatic pressing.	

Table 1.6.53

Table 1.6.54 Processes for production of sheet glass

Process	*Description*	*Advantages and limitations*
Float glass process	Molten glass pours on to a surface of liquid tin enclosed in a protective atmosphere where it gradually cools to form flat, parallel sheet with good surface finish.	Sheet is exceptionally free from distortion and with better surface finish than polished sheet.
Casting	A ribbon of molten glass pours on to metal rollers and travels on more rollers to annealing furnace.	Glass of constant thickness is produced but one surface has poor finish because of contact with metal rollers. Used to produce patterned or textured glass and wired glass.
Flat-drawn sheet	A continuous sheet of glass is drawn from a tank furnace by electrically driven rollers. Air cooled rollers cool the edges of the sheet to prevent waisting.	Sheet glass up to 20 mm thickness is produced. Surface finish is good but thickness varies. Used in windows where some optical distortion is permissible. Largely replaced by the float glass process.

Table 1.6.55 The effect of ceramic properties on design and construction methods

General property of ceramics	*Consequence*
Wear resistance and lack of ductility, malleability and toughness.	Form grinding of ceramic articles must be done, in most cases, with diamond wheels and is time consuming and costly. Avoid as far as possible. Finish grinding should be cut to a minimum. Diamond wheel cut-off operations are, however, nearly always economically acceptable.
Low tolerance for deformation.	Susceptibility of ceramics to the stress-concentrating effect of fixing holes, threads, sharp changes in section, etc. Designs should spread fixing or clamping stresses and avoid sharp corners by using suitable radii.
Generally low thermal conductivity; often (but not always) poor resistance to thermal stress or thermal shock.	Choose a low expansion ceramic (e.g. lithium aluminosilicate glass ceramic or magnesium aluminosilicate glass ceramic) in cases where good resistance to thermal stress or shock is required. Alternatively, choose one of the higher thermal conductivity types (e.g. silicon carbide, cemented tungsten carbide or graphite). If neither of these courses is possible and only resistance to thermal *stress* is necessary, maintain uniform minimum cross-section.

Table 1.6.56 Dimensional tolerances for unground articles

Process	*Normal tolerance*	*Fine tolerance*
Plastic methods	±2% with a minimum of ±0.8 mm	
Dry pressing	±2% on thickness with a minimum of ±0.25 mm unglazed and ±0.35 mm glazed	±1% with a minimum of ±0.15 mm unglazed and ±0.25 mm glazed on all dimensions
	±2% on other dimensions with a minimum of ±0.2 mm unglazed and ±0.3 mm glazed	
Extrusion	±2% with a minimum of ±0.2 mm unglazed and ±0.3 mm glazed	±1% with a minimum of ±0.15 mm unglazed

Table 1.6.56

Table 1.6.57 Maximum cambers for ceramic articles

$\frac{L}{D}$	Maximum camber as a percentage is: $\frac{100e}{L}$	
	Unglazed articles (%)	Glazed articles (%)
Less than 2	0.25	0.35
From 2 to less than 3	0.30	0.40
From 3 to less than 5	0.35	0.45
From 5 to less than 8	0.40	0.50
From 8 to less than 13	0.45	0.55
From 13 to less than 20	0.50	0.60
From 20 to less than 30	0.55	0.65
From 30 to less than 40	0.60	0.70
From 40 to less than 50	0.65	0.75

Note: If the value of *e* calculated from the above is less than 0.15 mm then the value of *e* permitted should be 0.15 mm unless otherwise agreed between the purchaser and manufacturer.

Table 1.6.58 Dimensional tolerances for centreless grinding

Diameter (mm)	Tolerance (mm)
0–6.3	±0.025
6.3–12.7	±0.05
Above 12.7	To be agreed between the purchaser and manufacturer.

Table 1.6.59 Roundness tolerances for ground articles

Method of grinding	Diameter (mm)	Maximum difference between major and minor diameters (mm)
Centreless	0–6.3	0.025
Centreless	6.3–12.7	0.05
Centreless	Above 12.7	Should be agreed between the purchaser and manufacturer.
Cylindrical	0–76.2	0.07
Cylindrical	Above 76.2	0.1%

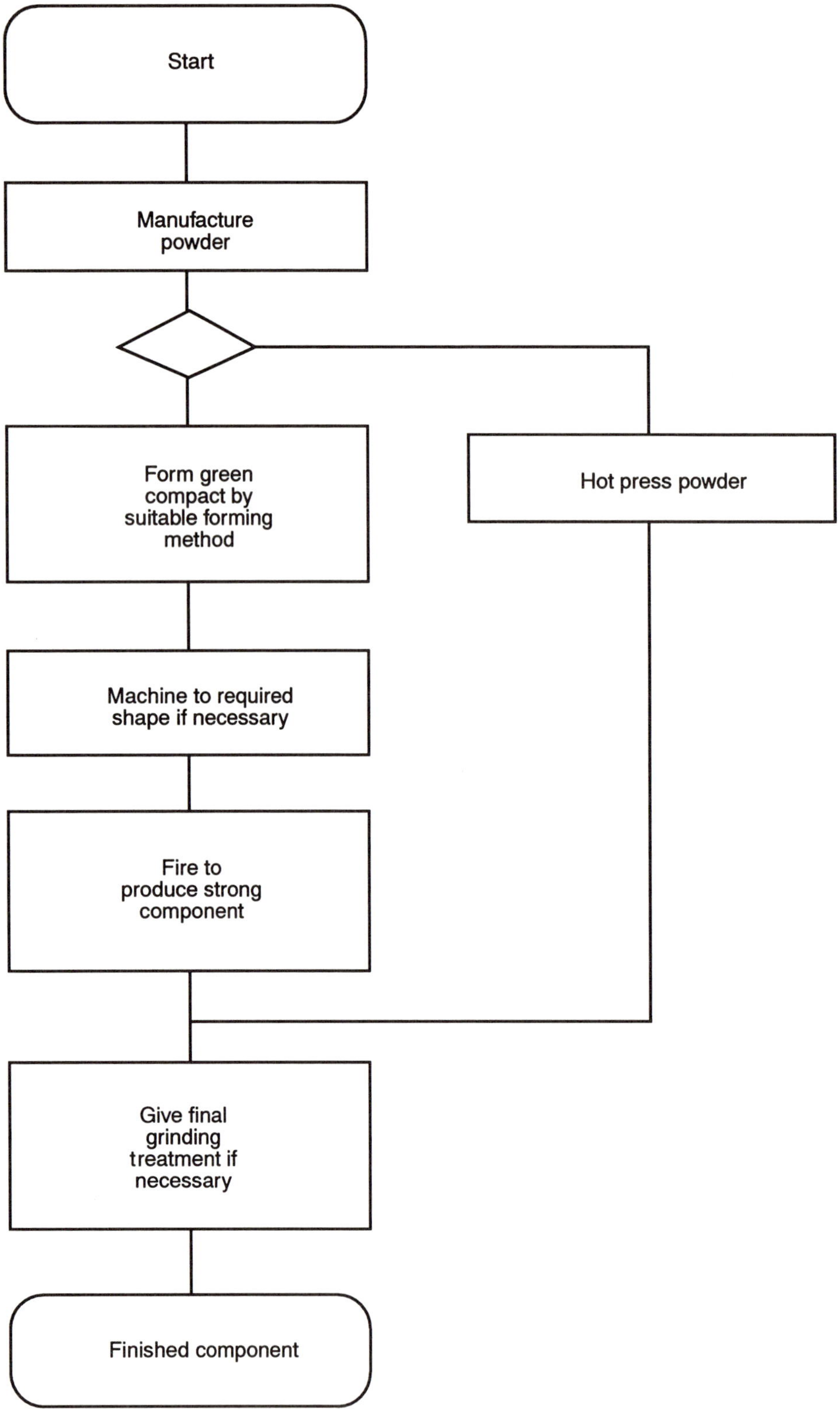

Fig. 1.6.29 Salient features of manufacture of ceramic components

Fig. 1.6.29

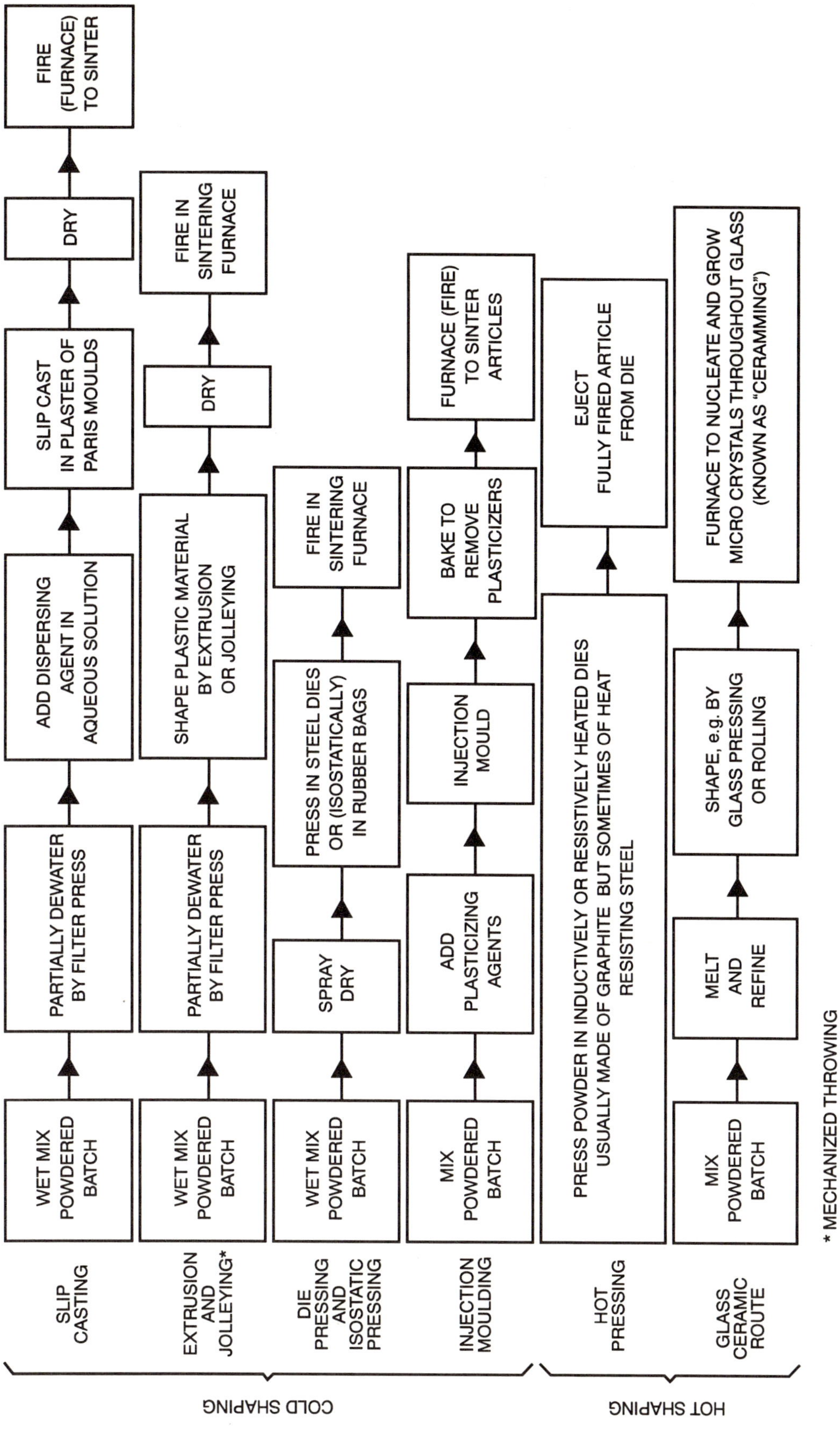

Fig. 1.6.30 Some ceramic manufacturing routes (machining operations are omitted)

Fig. 1.6.30

(a)

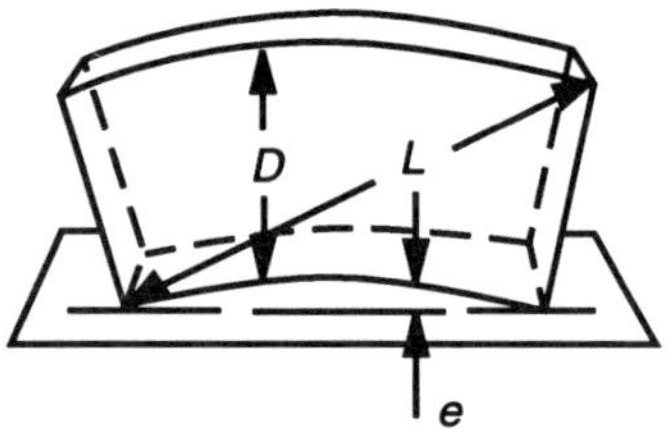

L = Maximum diagonal length on face
D = Minimum dimension in same direction as *e*

(c)

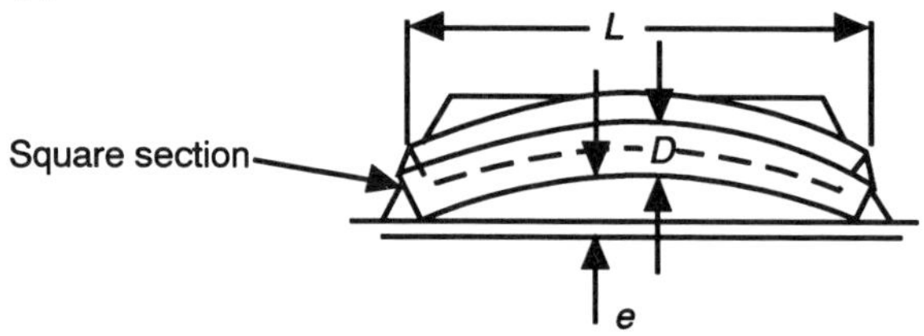

L = Maximum length of piece
D = Minimum dimension in same direction as *e*

(b)

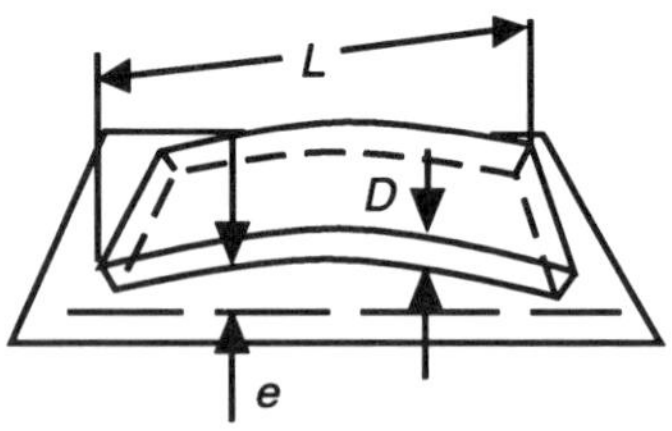

L = Maximum diagonal length on face
D = Minimum dimension in same direction as *e*

(d)

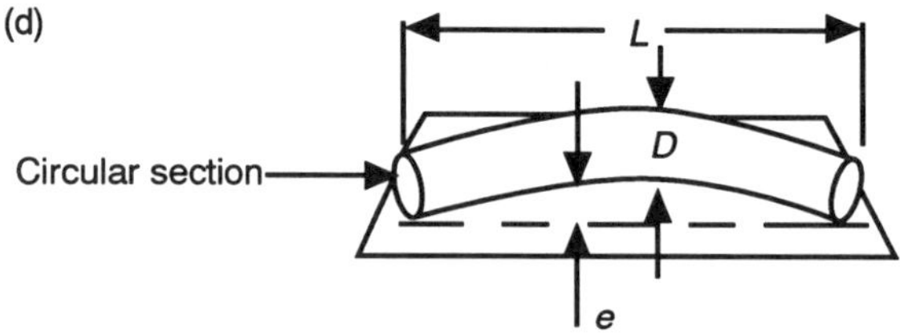

L = Maximum length of piece
D = Minimum diameter in same direction as *e*

Fig. 1.6.31 *L, D* and *e* for basic shapes

Fig. 1.6.31

1–7

Joining

Contents

List of tables

List of figures

1.7.1 Choice of joining method

The performance of a joint in service will depend on the stresses and environment to which it is subjected. These, to a greater or lesser extent, depend upon the mechanical design of the joint, which must be fully compatible with the proposed joining process and the materials which will be used to make it.

Hence there are three mutually dependent factors upon which a successful joint depends:

materials,
design,
joining process.

In any joining processes the materials to be joined may either be the same, or different. Hence either one, two or three materials might be involved in the process.

The first step is the choice of joining method, see:

Table 1.7.1 *The suitability of joining methods for various pairs of materials*

In the table, the number found at the intersections of the columns and rows indicate the appropriate joining methods. No consideration as to the suitability of design is given in this table.

The common joining processes are compared in:

Table 1.7.2 *Characteristics of common joining processes*

The suitability of these processes for metallic materials is given in:

Table 1.7.3 *Joining processes for metallic materials*

The joining of dissimilar metals is complicated by the different physical characteristics of the metals (melting point, thermal conductivity, etc.), and the danger of preferential attack (galvanic corrosion). Where galvanic corrosion is likely to be a serious problem, a joining method should be chosen which does not provide an electrically conducting path across the joint, or which provides an intermediate metal to slow down the corrosion rate. Processes such as brazing and welding can often be complicated if working near to the melting point of one base metal since excessive dissolution will take place. A guide is given in:

Table 1.7.4 *Suitability of joining processes for dissimilar metals*

Table 1.7.1 The suitability of joining methods for various pairs of materials

Materials to be joined	Metals	Plastic elastomers	Plastic thermosets	Plastic thermoplastics	Ceramics	Cement and concrete	Glass	Wood	Paper	Leather	Textiles
Metals	1 2 3 5 7 8	5	5	5	1 (2) 4 5 6	1 4 5	1 5 6	1 5	5	1 5	1 5
Plastic elastomers		4 5	5	5 9	5	5	5	5	5	1 5	1 5
Plastic thermosets			5	5	5	5	5	5	5	5	5
Plastic thermoplastics				3 5 9	5	5	5	5	5	5	5
Ceramics					1 3 4 5 6	1 4 5	1 4 5 6	5	5	5	5
Cement and concrete						1 4	1 4	1	5	5	5
Glass							4 5 6	5	5	5	5
Wood								1 5	5	5	5
Paper									1 5	1 5	1 5
Leather										1 5	1 5
Textiles											1 5

Key
1. Mechanical fixing – including sewing.
2. Soldering and brazing.
3. Welding – metals and plastics.
4. Cementing – hydraulic and heat setting.
5. Adhesives.
6. Glass/ceramic–metal sealing–glass working.
7. Diffusion bonding.
8. Roll bonding.
9. Solvent bonding.

Table 1.7.1

Table 1.7.2 Characteristics of common joining processes

Joint features	*Welding*	*Brazing, soldering*	*Mechanical fastening*	*Adhesive bonding*
Permanence	Permanent joints.	Usually permanent (soldering may be non-permanent)	Threaded fasteners can be disassembled.	Mainly permanent, some available with specific strength levels.
Stress distribution	Local stress points in structure.	Fairly good stress distribution.	Points of high stress at fasteners – can be centres for fatigue cracks.	Good uniform load distribution over joint area (except in peel).
Appearance	Joint appearance usually acceptable. Some dressing necessary for smooth surfaces.	Good appearance joints. Neat and unobtrusive joints obtained.	Surface discontinuities sometimes unacceptable. Machining sometimes required.	No surface markings. Joint sometimes invisible.
Joint strength	Highest strength joints (in structural applications there is frequently no alternative).	Soldering recommended for joining where the joint strength is not critical. In brazing, joints only as strong as parent metal.	Careful calculation of stresses at design stage.	Moderate to high strengths obtainable from structural adhesives depending on temperature and environment.
Materials	Generally, although not always, limited to similar material groups. Mainly metals, some plastics, ceramics.	Some capability for joining dissimilar metals. Generally applicable to metals only, some ceramics.	Most forms and combinations of materials can be fastened. Well established methods and techniques for joining most metals.	Ideal for joining most dissimilar metals. Used for joining metal, plastics, ceramics, textiles, etc. Heat sensitive materials can be joined.
Environmental resistance	Very high temperature resistance. Danger of corrosion, particularly with dissimilar metals, or flux residue.	Temperature resistance limited by filler metal. Danger of corrosion, particularly with dissimilar metals or flux residue.	High temperature resistance. Danger of corrosion fatigue, stress corrosion, galvanic attack.	Poor resistance to elevated temperatures. Variable resistance to corrosive environments. Electrical resistance reduces corrosion at joints.
Fatigue	Special provision often necessary to enhance fatigue resistance.	Fairly good resistance to vibration.	Special provision for fatigue and resistance to loosening.	Excellent fatigue properties.
Joint preparation	Little or none on thin material. Edge preparation on thick plates.	Prefluxing often required (except for special brazing processes)	Hole preparations and tapping for threaded fasteners.	Cleaning often necessary. Degreasing and/or abrasion pretreatments often prescribed.
Post-processing	Heat treatment sometimes necessary.	Corrosive fluxes must be cleaned off.	Usually no post-processing – occasionally retightening in service.	Not often required.
Equipment	Relatively expensive. Bulky and often requires heavy power supply.	Manual equipment cheap. Special furnaces and automatic equipment expensive.	Relatively cheap, portable.	Multi-feature dispensers. Dedicated equipment used for mass production.
Consumables	Wire, rods, etc., fairly cheap.	Some special brazing fillers expensive. Soft solders cheap.	Quite expensive.	Structural adhesives quite expensive.
Production rate	Can be very fast.	Automatic processes quite fast.	Joint preparation and manual tightening slow. Mechanical tightening fairly rapid.	Seconds to hours, according to type.
Quality assurance	NDT methods applicable to most processes.	Inspection difficult, particularly on soldered electrical joints.	Reasonable confidence in torque control tightening.	NDT methods limited.
Other remarks	Versatile, widely used. Design flexibility. Extreme care required in selecting materials and processes.	Some limitation on joint design.		Suitable for materials of different geometries. Weight reduction. Variable joint durability. Special joint design.

Table 1.7.3 Joining processes for metallic materials

	Mechanical fastening		Low temperature joining		Welding				Adhesives	
	Threaded fastening	Riveting and metal stitching	Soldering	Brazing	Arc welding	Gas welding	Electron beam welding	Resistance welding	Adhesive bonding	Adhesive bonding and metal stitching
Cast iron	R	R	N	P/N[a]	R	R	N	P/N	R	R
Low carbon and low alloy steels	R	R	N/P	R[a]	R	R	R	R	R	R
Med–high carbon steels	R	R	P/N	P/N	P	P	R	N	R	R
Stainless steel	R	R	P	R	R/P	P	P	R	R	R
Aluminium	R	R	P	R	R	R/P	R	R	R	R
Magnesium	R	R	P	P[a]	P	P[a]	R	P	R	R
Copper and copper alloys	R	R	R	R	R	R[a]	P	R	R	R
Nickel and nickel alloys	R	R	R	R	R[a]	R[a]	R	R[a]	R	R
Titanium	R	N	N	N	R	N	R	R	R	R
Lead	R	R	R	N	R	R	N	N	P/N	R
Zinc	R	R	R	N	N	R	N	N/P	P/N	R

R = Recommended. P = Possible. N = Not recommended. [a] = Only suitable for certain materials or alloys.
In all cases certain exceptions or special conditions may prevail.

Table 1.7.3

Table 1.7.4 Suitability of joining processes for dissimilar metals

Process		Suitability for dissimilar metals
Mechanical fastening		Common – danger of galvanic action.
Soldering, diffusion bonding		Recommended – danger of galvanic action (lower with diffusion bonding).
Brazing		Can be common for systems where the degree of dissimilarity is not great. Furnace brazing of small fabrications is practised.
Welding of metals	All arc, gas methods	Difficult, only used where melting points are within about 35°C. Also danger of galvanic action.
	Electron beam, laser	Possible, but care if melting point spread is great.
	Solid phase and flash welding, resistance welding	Recommended method if no galvanic action danger. Difficult due to differing mechanical and thermal properties.
Adhesive bonding		Recommended with suitable compatible adhesive and after satisfactory pretreatment. Joint properties may be limited for certain applications.

1.7.2 Mechanical fasteners

1.7.2.1 Selection of fastener type

Mechanical fasteners are classified into two groups, permanent (rivets etc.) and semi-permanent (threaded fasteners).

The material for a mechanical fixing may be determined by:

mechanical properties,
environment,
cost,
appearance,
availability.

Fastener selection is generally based on achieving the lowest installed cost while providing the required performance.

The various types of available mechanical fastener are described, with typical applications, in:

Table 1.7.5 *Basic types of mechanical fastener*

The suitability of fastener types depends on the materials to be joined. A guide is given in:

Table 1.7.6 *Suitability of fasteners for joining various materials*

Assembly costs can sometimes be reduced by use of particular fastener types, such as:

Thread rolling screw	– several contact points around the starting thread. Minimize the need for lock washers and other locking devices.
Self-drilling screws	– drill their own holes.
Flanged bolts and nuts	– permit tightening to higher preloads and eliminate the need for washers by providing a greater load-bearing area. Particularly useful for bolting soft materials.

Maintenance costs may also be important. Semi-permanent fasteners which require special equipment for removal prevent unauthorized removal and vandalism. Fasteners can be coated or treated to minimize corrosion during service.

Locking screws should be selected when it is important to maintain preload, minimize fatigue problems and prevent loosening, rattles and squeaks. Locking action is provided by plastic pellets or strips fixed to the thread, deformed threads, a resilient bulge on one side of the thread and adhesive bonding in conjunction with a mechanical fastener.

In some applications, such as aircraft, weight saving is important. Titanium and aluminium fasteners have lower weight than steel, although steel gives a higher strength on a small diameter.

To prevent fatigue failure of a bolt or bolted material, the service load should be less than the preload, the bolts should be properly preloaded, the fasteners should be as flexible as possible and the joint members as rigid as possible. Vibration failure may occur at low loads at frequencies up to 2 kHz, particularly if there is resonant vibration. Resonance should be accounted for at the design stage.

To avoid hydrogen embrittlement of fasteners after plating, susceptible materials should be baked after plating or alternative coating methods should be used.

If thread stripping is a problem, the fasteners should be checked for excessive decarburization and the value of the applied torque should be rechecked.

1.7.2.2 Fastener materials

The characteristics, availability and applications of common fastener materials are given in:

Table 1.7.7 *Characteristics of fastener materials*

Typical mechanical properties are given in:

Table 1.7.8 *Typical mechanical properties of metallic fastener materials*

Designations for carbon steel screws and bolts, with their associated mechanical properties, are given in:

Table 1.7.9 *ISO strength grades for screws and bolts (carbon steels)*

SELECTION OF MATERIALS FOR HIGH-TEMPERATURE USE

The two main problems with fasteners at high temperatures are rapid fall in strength and rapid oxidation. The following provides a guide to material selection:

Up to 200°C	Medium alloy steel. Cadmium plate for corrosion protection.
200–500°C	Medium alloy steel (Cr–Mo–V). Nickel and nickel alloy plate for corrosion protection. Heat resistant stainless steels for low-strength applications.
500–650°C	Iron–chromium–nickel alloys. Superalloys. Silver plating used to prevent thread galling upon removal.
650–900°C	Nickel base superalloys, with high alloy contents. Coatings with high oxidation resistance necessary.
900–1100°C	Dispersion strengthened nickel and cobalt base alloys
Above 1100°C	Molybdenum, niobium, tantalum with coatings for short-time applications. Tungsten used above 2000°C despite oxidation. No suitable coatings available for many applications.

SELECTION OF MATERIALS FOR LOW TEMPERATURE USE

Threaded fasteners made from nickel and austenitic steels and titanium alloys have excellent mechanical properties at cryogenic temperatures. Bolts made from nickel alloys thus have a very wide temperature range applicability.

Many standard fasteners are unsuitable for use at cryogenic temperature due to brittle failure. Austenitic stainless steels and titanium alloys are suitable for certain applications down to −200°C.

SELECTION OF MATERIALS TO REDUCE CORROSION

Mechanical fasteners may be subject to:

uniform slow oxidation,
galvanic corrosion,
pitting,
stress corrosion,
stray current,
demetallification (particularly brass fasteners).

Methods of reducing corrosion include:

1. Use materials to resist the environment to which the fastener is exposed.
2. Use metals close together in the galvanic series if possible.
3. Where galvanic corrosion is likely, the area of the less noble metal should be as large as possible.

4. Irregular stresses in the design are hazardous.
5. Paint, coat or insulate dissimilar metals.
6. Inhibitors should be added to corrosive electrolytes
7. Dissimilar metals should not be connected by fasteners, otherwise a corrosion cell will be activated.
8. Sacrifically protect by attaching non-functional less noble pieces of metal to the design.

Guidance to finishing and coating fasteners for corrosion resistance is given in:
Table 1.7.10 *Fastener finishes and coatings*

1.7.2.3 Adhesive fastening

Combinations of adhesive bonding and mechanical fastening can overcome some of the limitations of purely bonded joints. In these cases, the adhesive bond is used to relieve the fastener of part of the load. Solid rivets or spot welds are not recommended in situations where the fastener is intended to contribute significantly to joint strength.

Bolts or hollow rivets are preferred.

Anaerobic adhesives provide the best means of securing threaded components, as they:

can be permanent or dismountable,
provide an effective seal against most corrosive environments,
are suitable for common clearances.

Two-part epoxy adhesives are also used with mechanical fasteners. Each ingredient may be pre-coated separately, or microencapsulation of the active ingredient may be employed.

Precautions to be applied:

surfaces must be free of excessive contamination,
a minimum of three engaged threads must be treated.

1.7.2.4 Snap-fits

Thermoplastic components can be assembled by incorporating snap-fit features on the mating parts. Snap-fitting relies on the relatively large elastic deformations (compared with metals or ceramics) which most thermoplastics exhibit. Tough, partially crystalline plastics such as the acetals and polyamides can be used for snap-fits.

The design of snap-fit joints is a specialized business. A balance must be struck between:

adequate elastic recovery to anchor the two parts securely,
avoidance of exceeding the elastic limit of the male component.

Snap-fits are normally a one-way process, i.e. once assembled the parts cannot be separated without destroying the joint.

Table 1.7.5 Basic types of mechanical fastener

Type	*Definition*	*Applications*
Bolts	An externally threaded headed fastener used in a through hole with a mating nut.	General applications. Heavy duty machinery. Electrical contacts. Wood. Masonry.
Nuts and locknuts	A piece with an internal thread and at least one pair of parallel faces. Locknuts are able to grip a bolt to prevent loosening.	As above.
Screws	An externally threaded headed fastener used in a threaded hole.	As above.
Set screws	A screw used as a semi-permanent fastener to hold collars, gears, etc., into a shaft.	As above, particularly in cases where frequent assembly/disassembly is required.
Thread-producing screws self-tapping self-drilling thread-forming	Screws provided with cutting edges so that they produce a mating thread. Other forms require no hole at all.	As above.
Wood screws	Screw with pointed end. Some threaded only 60% giving plain dowel section below head. Others have two start thread.	Used for joining wood or metal to wood.
Rivets solid tubular split	A headed fastener secured by hot or cold working of the end. Cheaper than bolts.	General applications. Very strong fabrications. Can get tight tolerances and good surface finish. Permanent.
Nails	Straight shanked friction fasteners.	Generally for wood, masonry, etc.
Blind rivets	Installed in a joint from one side only. Set by self-contained mechanical or chemical means.	Applications where joint is accessible from one side only. Also used to improve appearance or simplify assembly.
Pin fasteners	Semi-permanent. Require pressure for removal. Ring or T handles.	Split pins, safety devices, machinery.
Self-sealing fasteners	Fasteners incorporating pre-assembled mastic or plastic or soft metal seals.	Steel–rubber, steel–PTFE, steel–lead.
Retaining rings	Used to lock components to shafts or into bores by providing a resilient removable shoulder. Manufactured from spring steel, heat treated to increase strength.	
Quick acting fasteners	Draw-pull fasteners. Slide action fasteners.	Geometric locks, etc.
Washers	Discs with centre holes applied under the head of the part to be tightened.	Used for locking bolts during application of nut. Can be used to compensate for loss of tension between parts.
Plastic fasteners	Fixing devices manufactured from thermoplastics. Generally made from nylon.	Strong, relatively hard surface, corrosion resistant, tough.
Surface thread fasteners	Forged by stamping a thread-engaging impression on a flat piece of metal.	High C steel. Any metal.
Threaded inserts	Nut which seals a tapped hole in blind or through-hole locations.	

Table 1.7.6 Suitability of fasteners for joining various materials

Materials to be joined	***Permanent***	***Semi-permanent***
Cast iron	Recommended in structural uses – large solid rivets.	Common – self-tapping screws, studs, hooks, etc.
Carbon and low alloy steels	Recommended with solid rivets for high shear or medium tensile stress uses.	Recommended with all types. Most made of same materials.
Stainless steel	Common with solid stainless steel rivets for high shear stress uses.	Recommended with bolts and nuts. Common with others.
Aluminium and magnesium	Common with small solid aluminium rivets, or semi-tubular at low stresses.	Common with self-tapping screws, inserts, others.
Copper and its alloys	Common with small solid copper alloy rivets, seldom with semi-tubular.	Common with fasteners available in similar materials.
Nickel and its alloys	Common with small solid nickel alloy rivets, seldom with semi-tubular.	Recommended with fasteners available in similar materials.
Titanium	Not used	Common, some Ti fasteners available.
Lead and zinc	Recommended with semi-tubular or split rivets, seldom with solid.	Recommended with self-tapping screws inserts, special fasteners.
Thermoplastics	Recommended with tubular rivets stitching common. Limited thickness.	Seldom – inserts are used, also special fasteners.
Thermosets	Common with all types.	Common with self-tapping screws, elevator bolts, inserts.
Elastomers	Recommended with tubular or split rivets, stitching common.	Seldom – inserts are used, also special fasteners.
Ceramics	Seldom – semi-tubular rivets used.	Common with nuts, washers and bolts, especially load indicating types.
Glass	Seldom – thermoplastic semi- tubular rivets used.	Common with plastic fasteners, seldom with elevator bolts.
Wood	Common, semi-tubular rivets for hard woods, tubular rivets for soft.	Recommended with wood and lag screws, self-tapping screws, inserts, dowels, etc.
Dissimilar metals	Common, but galvanic action may be a danger.	Common, but galvanic action may be a danger.
Metals to non-metallics	Recommended with self-tapping screws, inserts, stitching.	Common using self-tapping screws, inserts, special fasteners.
Dissimilar non- metallics	Recommended with lag screws, stitching, special fasteners.	Seldom.

Table 1.7.7 Characteristics of fastener materials

Material	Advantages	Limitations	Availability															Remarks, applications
			Bolts	Nuts	Screws	Set screws	Tapping screws	Wood screws	Rivets	Quick acting fasteners	Blind rivets	Pin fasteners	Self-sealing fasteners	Plastic fasteners	Nails	Retaining rings	Washers	
Carbon steels	Strength range.	Low corrosion resistance. Low acid resistance.	X	X	X	X	X	X	X	X	X	X			X	X		Which steel dictated by heat treatability, manufacturing process, availability and cost; also ability to be staked, riveted, spin riveted, and ability to take special finishes. Mild steel for general applications, and bright finished screws and fasteners. Low carbon martensitic grades (boron-treated) used for high strength bolts/cap screws. Free-cutting grades used for fasteners.
Stainless steels	High temperature properties. Corrosion resistant. Large strength range.	Expensive, not always available.	X	X	X	X	X	X	X		X				X	X		Made from martensitic stainless steel – hardened and tempered by heat treatment. Ferritic grades can be used hot worked and cold forged. Austenitic are non-magnetic, very corrosion resistant and have good high temperature properties.
Copper alloys	Electrical/thermal conductivity. Non-magnetic. Corrosion resistant (marine applications).	Expensive. Lower mechanical properties in some alloys.	X	X	X			X	X		X					X	X	High silicon bronze for hot forged bolts, nuts and set screws. Low silicon bronze for cold formed bolts, nuts, rivets and screws. Silicon–aluminium bronzes for hot forged products requiring special properties. Copper–nickel alloys are recommended for fasteners used in seawater. Brass commonly used for bolts, screws, wood screws, tapping screws and rivets. Brass fasteners often require stress relieving to eliminate embrittlement and improve corrosion resistance.
Nickel and nickel alloys	Immune to discoloration. Very corrosion resistant. Suitable for high temperature. Non-magnetic.	Expensive.	X	X	X				X		X							Use pure nickel, Monel, Inconel. Monel and Inconel possess strength, hardness and corrosion resistance. Used in food and chemical processing industries. Monel is cheap and easy to cold head and roll thread – bolts, nuts, screws, and rivets. Inconel excellent for high strength and oxidation resistance up to 900°C.
Aluminium alloys	High strength/weight ratio. Good corrosion resistance. Non-magnetic.	Not suitable for alkaline environments. Overtightening of threads leads to stripping.	X	X	X	X					X							Rivets are produced from various alloys (ASTM B316). Non-heat treatable – driven cold. High strength, heat treatable driven hot or freshly quenched (threading after heat treatment).
Titanium and titanium base alloys	Very high strength/weight ratio. Suitable at temperatures up to 350°C. Corrosion resistant. Non-magnetic.	Very expensive. Threads subject to galling. Notch-sensitive.	X	X	X			X		X								Used in aircraft, and GRP racing boats. Mainly from Ti–6Al–4V. Bolts are also made from Ti–6A–6V–2Sn. Higher grade alloys have better fabricating control, part reliability. Titanium locknuts exist for use with steel bolts. Buckable rivets have been made from titanium beta alloy. Titanium when in contact with aluminium should always be coated.
Nylon	Strong, corrosion resistant, insulator. Low friction. Self-extinguishing. Vibration resistant.	Not fire resistant. Not suitable for contact with mineral acids. Embrittlement in sunlight.	X	X	X								X	X		X		Nylon 6/6 most common plastic used for mechanical fasteners. Nylon 6/10 is used where resistance to ultraviolet light or high temperature are required.

Table 1.7.7

Table 1.7.7 Characteristics of fastener materials – *continued*

PVC	Flexible or rigid. Corrosion resistant.		X	X	X								X	X				Flexible: Usually used in conjunction with metals such as for a sealant in a cap-fastener. Rigid: Industrial applications.
Acetal	Dimensional stability. Strong. Resistant to moderate heating.	Attacked by strong acids. Inflammable. Affected by UV.	X	X	X								X	X		X		
Polycarbonate	Heat rigidity and toughness. Dimensional stability. Insulating. Self-extinguishing.	Expensive.	X	X	X				X				X	X		X		High Impact applications as rivets or nails.
Polyethylene	Low cost. Rubberlike. Excellent insulator.	Only suitable for non-critical applications.	X	X	X													Used in electrical industry.
Polypropylene	Low cost. Rigid. High impact strength. Good electrical properties.				X					X			X	X		X		Electrical low strength applications.
Polystyrene	Good dielectric.	Brittle. Only suitable for non-critical applications.	X	X	X													
Polyester (thermoplas-tic)	High temperature and dimensional stability.	Use in water above 70°C not recommended. Attacked by strong acids and alkalis. Notch-sensitive.	X	X	X				X									Special glass-filled fasteners.
ABS	Versatile Plateable (some grades). Low cost.	Limited weather resis-tance. Poor solvent resistance.	X	X	X									X				
Acrylic	Transparent	Brittle												X				Model work and toys. Otherwise limited.
Fluorocarbons	Extremely corrosion resistant. Highest temperature resistance of non-metallic fasteners. Good dielectric.	Low strength: about one-tenth of nylon.	X	X	X								X	X		X		

Table 1.7.8 Typical mechanical properties of metallic fastener materials

Material	Grade	Composition	Condition	0.2% Proof strength (MPa)	UTS (MPa)	Temperature limit to longterm use
Steels	Carbon	0.1% C	–	200	400	300°C
		0.15% to 0.30% C	OQ	560	820	
		0.28% C	–	360	520	
		0.28% to 0.55% C	OQ	700–820	900–1000	
		0.28% to 0.55% C	QT	510–550	700–820	
	Low alloy	0.15 to 0.25% C–0.77% Mn	QT (450°C)	560	820	400°C
		1% Cr–Mo	–	200	430	
	Martensitic stainless	410	HT	400–1020	630–1350	500°C
		416	HT	400–1020	630–1350	
		431	–	650–1100	875–1400	
	Ferritic stainless	430	Annealed	300	520	450°C
		430	Cold worked (CW)	550	650	
	Austenitic stainless	300 series	Sol. annealed (SA)	200	500	600°C
			SA and CW	340	680	
			Strain hardened	680	850	
Aluminium alloys	5251	Al–2% Mg	H3	150	200	100°C
	5154A	Al–3.5% Mg	H2	185	270	
	5083	Al–5% Mg	Rolled	90	200	
	6060	Al–Mg–Si	T6	180	210	
	2014A	Al–4% Cu–Mg	T6	420	470	
	6061	Al–Mg–1% Si–Cu	T6	265	305	
	DTD5074A		T6	470	540	
Copper alloys	Silicon bronze	3 Si–1 Mn	–	92–620	350–770	200°C
	Cupro-nickel	up to 30% Ni	–	up to 540	up to 650	
	Brass	60% Cu–40% Zn	H	–	310	
	High tensile brass	+Al, Fe, Mn, Sn, Ni	H	195	460	
	Naval brass		H	100	340	
	Bronze	5–8% Sn	–	220	400	
	Aluminium bronze	5% Al	–	260–340	520–700	
Nickel alloys	Nickel	–	–	100–210	380–550	300°C
	Monel	–	–	170–350	480–620	400°C
	Inconel	–	–	210–340	550–670	600°C
	Nimonic 80A	–	–	740	1200–1220	700°C
	Nickel alloys	+ C, Mn, Fe, Mn	–	110–210	340–550	300°C
Titanium alloy	Titanium	–	Annealed	323	497	400–500°C
	Ti Al 6 V	–	TF	924–1028	1000–1155	

OQ = Oil quenched
HT = Hardened and tempered.
H3 = Strain hardened.
H = Hardened.
SA = Solution annealed.
CW = Cold worked.
QT = Quenched and tempered.
TF6 = Solution heat treated and precipitation treated.
H2 = Strain hardened, partially annealed.

Table 1.7.9 ISO strength grades for screws and bolts (carbon steels)

Strength grade (marked on bolt)[a]	*4–6*	*4–8*	*5–6*	*5–8*	*6–6*	*6–8*	*8–8*	*10–9*	*12–9*	*14–9*
Tensile strength (MPa)	392–539		490–686		588–785		785–981	981–1177	1177–1373	1372–1569
Hardness HB	110–170		140–215		170–245		225–300	280–365	330–425	–
Hardness HRB	62–88		77–97		88–102		–	–	–	–
Hardness HRC	–		–		–		18–31	27–38	34–44	40–47
Hardness HV30	110–170		140–215		190–245		225–300	280–370	330–440	400–510
Elongation (%)	25	14	20	10	16	8	12	9	8	7

(1) First number $= \frac{\text{UTS}}{10}$ (kgf/mm^2) minimum

(2) Second number $= 0.2\% \frac{\text{Proof Stress}}{10(\text{min.UTS})}(\%)$

Hence (1) × (2) = 0.2% P.S. (kgf/mm^2)
100 (1) ≈ min UTS (MPa)
(since 1 kgf/mm^2= 9.81 MPa)

Nuts have grade designations as the first number of the screws and bolt grade. It is recommended that when a nut is used with a bolt the nut number should be the same as the first number of the bolt or screw grade.

[a]These grades also apply for stainless steels where the UTS will normally be grades 5, 6 or 8.

Table 1.7.10 Fastener finishes and coatings

Finish	Used on	Corrosion resistance	Applications, characteristics, remarks	
Anodizing	Aluminium	Excellent	Hard oxide surface, gives excellent protection.	
Black oxide	Steel	Good for indoor application. Not recommended outdoor. Wax or oil coatings normally applied for subsidiary protection.	Waxed or oiled. Produces a rust inhibited surface.	
Black chromate	Zinc plated steel Cadmium plated steel	Additional corrosion protection.	Applied by chemical dip. Decorative outdoor purposes.	
Blueing	Steel	Indoor application. Not used outdoors. Wax or oil coatings normally applied.	Decorative use, black to blue. Waxed or oiled.	
Brass plate (lacquered)	Mainly steel	Fair. Indoor applications only.	Electroplated, then lacquered.	
Bronze plate (lacquered)	Mainly steel	Fair. Indoor applications only.	Similar colour to 80% copper, 20% zinc brass. Electroplated and lacquered.	
Cadmium plate	Most metals	Excellent. Preferred to zinc in marine applications. Also used indoors.	Electroplated. Bright silver grey, dull grey or black. Used decoratively. Requires an environmentally isolated plant.	
Clear chromate	Cadmium and zinc plated parts	Very good. Coloured chromate better than clear.	Clear or iridescent.	
Coloured chromate	Cadmium and zinc plated parts	Very good. Superior to clear chromate.	Green, gold, bronze.	
Chromate	Cadmium and zinc plated parts	Very good	Yellow, brown, green or iridescent.	
Chromate plate	Most metals	Good, improved by copper and nickel preplate	Electroplated. Bright blue-white. Hard surface. Used for decorative purposes.	
Copper plate	Most metals	Fair	Electroplated. Undercoated for nickel and/or chrome. Can be blackened.	
Copper brass, bronze finishes	Most metals	Very good for indoor applications. Require lacquering to maintain colour.	Decorative, applied for matching colours. Colour tones vary from black to original colour.	
Lacquering	All metals	Improves corrosion resistance. Can be designed for severe applications.	Decorative. Can be clear or coloured for matching purposes.	
Lead–tin	Mainly steel	Fair to good	Hot dip. Silver grey, dull. Also lubricate fasteners.	
Bright nickel	Most metals	Excellent indoors. Good outdoors if sufficient thickness (not less than 0.012 mm).	Electroplated. Silver finish. Used for appliances, hardware, etc.	
Dull nickel	Most metals	As above	Mechanical surface finishing or special plating bath.	
Passivating	Stainless steel	Excellent	Chemical treatment.	
Phosphating	Mainly steel	Good, particularly after paint or lacquer applications.	Less expensive than zinc or cadmium plating. Good lubricity.	
Coloured phosphates	Steel	Superior to regular phosphates and oiled surfaces.	Available in variety of colours.	
Rust preventatives	All metals	Varies with function of oil.	Variable colour. Film thickness depends on type. Usually applied for added protection with black oxide or phosphate coatings.	
Silver plate	All metals	Excellent	Decorative, expensive, excellent electrical conductor.	
Electroplated tin	All metals	Excellent, particularly in contact with food.	Silver-grey colour.	
Hot dip tin	All metals	Excellent, particularly in contact with food.	Silver-grey colour. The thickness is more difficult to control than with electroplated tin, especially with fine thread fasteners.	
Electroplated zinc	All metals, particularly steel	Very good	Blue-white-grey coating.	Preferred to cadmium for outdoor applications, other than marine.
Electrogalvanized zinc	All metals	Very good	Dull grey.	
Hot dip zinc	All metals	Very good, proportional to coating thickness.	Maximum corrosion protection. Dull grey. Coating thickness cannot be controlled.	
Hot dip aluminium	Steel	Very good	Maximum corrosion protection. Dull grey. More expensive than hot dip zinc, but superior.	

Table 1.7.10

1.7.3 Soldering

1.7.3.1 Introduction

Soft soldering is joining by applying a molten low melting point metal between workpieces. Slight alloying occurs between workpiece and soldering alloy.
Soldering requires:

(1) Preparation and fitting together of metal parts.
(2) Application of flux.
(3) Pre-heating of assembly – good heat transfer is required for adequate flow of solder.
(4) Application of molten solder. Solder fills gaps by wetting and capillary action.
(5) Cooling of joint.
(6) Removal of flux/residue.

Fluxing aids the cleaning of surfaces and wetting. There are broadly three types of flux: corrosive, most of which are based on zinc chloride, non-corrosive, such as those based on resins, and intermediate/low residue giving residues which do not cause damaging corrosion under many conditions of service.

The ability to satisfy the requirements of a good soldered joint depends upon the combination of substrate, solder, flux and heating conditions employed.

Soldered joints are used to:

(a) Provide mechanical strength between components.
(b) Act as a gas/liquid seal
(c) Provide a conducting path for electric current/heat

Soft solders are low strength, low melting point alloys and should not be employed at temperatures exceeding 40°C below the solidus temperature.

Soldering of metallic materials is considered in this section; soldering of glazed surfaces is considered in Section 1.7.8.

1.7.3.2 Selection of solder and soldering method

The suitability of solders for joining common metals is given in:
Table 1.7.11 *Solderability of common metallic materials*

A guide to the costs and process parameters for the various soldering methods is given in:
Table 1.7.12 *Characteristics of soldering methods*

1.7.3.3 Solder materials and their properties

General purpose solders are usually a lead–tin base. High service temperature solders are available. These include Sn–5%Sb, Sn–3.5% Ag and Pb–5% Sn–1.5% Ag.

Lead-based solders containing less than 10% Sn, including Pb–5% Sn–1.5% Ag, are suitable for use at very low (cryogenic) temperatures, whilst Sn–5% Sb is used in normal refrigeration equipment.

Special solders are available for soldering aluminium, gold, glazed surfaces, and for specialized applications such as the food industry and nuclear energy.

The effect of solder composition on properties of solders is shown in:
Table 1.7.13 *Effect of alloying elements on solder properties*

The properties of bulk solders are given in:
Table 1.7.14 *Tensile strength of bulk solders*

It must be remembered that the properties of the joint will depend as much on joint design, process and substrate material as on the solder itself.

1.7.3.4 Surface coatings for soldered components

Components to be soldered may be coated before soldering to protect the components in storage, to aid the soldering process or to prevent soldering of specific areas. Typical coatings are given in:

Table 1.7.15 *Surface coatings for soldered components*

Fusible, non-soluble or soluble coatings are applied by electroplating or hot dip coating. Hot dip coating has the advantage that the metal is required to wet and bond to the substrate in the same way during a soldering operation, so that this is a built-in solderability test.

The use of pure tin coatings is not recommended for electronics applications because of a tendency to whisker formation and circuit bridging.

Freshly plated cadmium presents no soldering problem, but passivated cadmium will not wet unless corrosive fluids are employed.

Tin–nickel is a 65% tin–35% nickel alloy only obtainable by electroplating and intermediate flux is required for successful soldering.

1.7.3.5 Joint design, joint defects and testing methods

The soundness and properties of a soldered joint depend on good joint design. Examples are shown in:

Fig. 1.7.1 *Typical soldered joint forms*

Defects in soldered joints may arise because of poor solderability of the substrate or inappropriate process conditions. Common defects and their causes are given in:

Table 1.7.16 *Defects in soldered joints*

1.7.3.6 Properties of soldered joints

Mechanical properties of soldered joints are given in:

Table 1.7.17 *Mechanical properties of soldered joints*

Solders are low strength materials. The strength of a joint may be enhanced by:

1. Additional mechanical fixing methods.
2. Avoiding service temperatures in excess of that recommended for the solder.
3. Design of joint
   - (i) Generous overlap of lap joints. Lap joints are preferable to butt joints.
   - (ii) Shallow angle between solder and metal for good wetting.
   - (iii) Joining clearances – small enough to retain solder by capillarity – large enough to allow flow of solder into joint.
   - (iv) Easy provision and removal of solder and flux.
4. Avoiding excess solder.
5. When a corrosive flux is used, washing joints free of flux after soldering to prevent subsequent corrosion.

Failure of soldered joints may occur by creep, fatigue or tearing. Simulative tests may enable failures to be predicted.

Table 1.7.11 Solderability of common metallic materials

Material		*Typical solder*	*Remarks*	
Carbon and low alloy steels (>0.5% C seldom soldered)		Pb–Sn (<5% Sb)	Not common. Readily solderable. Often tin plated or galvanized.	
Stainless steels sometimes pre-tinned		Pb–Sn (<5% Sb)	Highly corrosive flux. Not suitable for austenitic or high tensile steels.	
Cast iron		High tin or Sn–Pb	Usually hot tinned or coated with Fe, Ni, Cu. Uneven heating must be avoided. Immersion or oven soldering preferred.	
Aluminium and some aluminium alloys.	General	Sn 8–40% Zn; Cd–Zn	Special fluxes necessary.	Poor solderability on alloys containing significant amounts of Mg and/or Si. Sn-containing solder joints on aluminium corrode in moist environments.
	Abrasion solder	Sn–30% Zn	No flux	
	Joining to dissimilar metals.	Zn–5% Ag		
	High strength aluminium joints	Zn–5% Al		
Copper and copper alloys	Copper	Pb–Sn (Sb)	Use corrosive flux or preplate and non-corrosive flux.	
	Brass	Pb–Sn (<1% Sb)		
	Gunmetal, tin bronze	Sn–Pb (Sb)		
	Aluminium and silicon bronze	Sn–Zn	Special preparation and fluxing	
	Beryllium copper	Pb–Sn (Sb)	Corrosive fluxes	
	Manganese bronze	Not recommended	Pre-plating necessary	
	Nickel silver	Pb–Sn	Tendency to de-wet	
Nickel and nickel alloys	Nickel	Pb–Sn (Sb)	Highly solderable.	Solder must be compatible with service temperature and environment
	High temperature alloys	Pb–Ag–Sb		
	Age hardened alloys	Pb–Sn–Sb	Solder after heat treatment.	

Table 1.7.11 Solderability of common metallic materials – *continued*

Material		***Typical solder***	***Remarks***
Lead and lead alloys		Pb–Sn	Highly solderable. Normally by soldering iron or wiping using flames and blowtorch.
Zinc and zinc alloys	Zinc	Pb–Sn	Easy to solder. Do not use antimonial solders.
	Zinc based die-castings	Pb–Sn	Contain small proportion of aluminium. Special fluxes or pre-coating necessary. No antimonial solders.
Tin and high tin alloys	Tin	Sn–Pb, Sn–Bi or same material as parent metal.	Very easy to solder. Need to match colour of parent and filler.
	Pewter		
Silver		Pb–Sn–Ag.	
Gold		Sn–Pb–Cd Sn–30 Pb–17 In–0.5 Zn	Gold plate $>2\ \mu$m thick. Quick soldering avoids excessive gold solution. Brittle joints unless precautions taken.
		Pb–Sn	Satisfactory joints on gold plate $\leq 1.5\ \mu$m thick.
Beryllium, chromium, titanium, magnesium, manganese bronze, refractory metals		Soldering not recommended	

Table 1.7.11 – *continued*

Table 1.7.12 Characteristics of soldering methods

Method	*Basic capital cost*	*Output*	*Versatility*	*Skill required*	*Solder form*
Flame	L	L	H	H	Hand-fed wire or strip
Electrical resistance	L/M	M/H	L	L	Preform
RF induction	H	M/H	L	L	Preform
Furnace/oven	M/H	H	M	L	Preform or paste
Dipping bath	L/M	L	L	H	Melted ingots or bars
Soldering iron	L	L	H	H	Hand-fed wire or strip.

L = low; M = medium; H = high.

Table 1.7.13 Effect of alloying elements on solder properties

Alloying elements		*Advantages*	*Limitations*
Major alloying elements	Pb	Low melting point.	Low strength.
	Sn	Low melting point. Good wetting agent.	Becomes brittle below −70°C.
Minor alloying elements	Sb	Prevents recrystallization under very cold conditions increases creep strength.	0.2% causes brittleness in substrates containing zinc.
	Ag	Increases creep strength. Slows down alloying of silver plated parent metal with solder.	–
	Cu	Slows down alloying of fine copper wires and copper soldering iron bits with solder.	
	Cd, Bi, In	Lower melting point of solders. Reduces dissolution of low melting point substrates and gold.	Reduces strength of common solder alloys.

Table 1.7.14 Tensile strength of bulk solders

Solder composition (%)	*Tensile strength at test speed of 1 mm min^{-1} (MPa)*
100 Sn	15
96.5 Sn–3.5 Ag	44.1
95 Sn–5 Sb	36.3
60 Sn–40 Pb	38.2
50 Sn–50 Pb	35.0
40 Sn–60 Pb	28.5
10 Sn–90 Pb	27.5
62 Sn–36 Pb–2 Ag	65.4
50 Sn–32 Pb–18 Cd	43.0
40 Sn–58 Pb–2 Sb	35.7
93.5 Pb–5 Sn–1.5 Ag	35.0

Data by permission of ITRI Ltd.

Table 1.7.15 Surface coatings for soldered components

Coating	*Description*	*Examples*
Protective	Temporary surface protection during storage. Early removal by melting or abrasion.	Resin lacquers. Polyurethane enamels.
Fusible	Coating which retains solderability during storage and fuses with solder.	Tin rich solder alloys.
Soluble	Coatings which do not melt at soldering temperatures but are soluble in molten solder and alloy with it. Often specified for use in the electronics industry.	Gold (1 μm applied over 5–8 μm of tin–nickel). Silver, tin, cadmium (copper).
Low solubility	Intermediate barrier on difficult metals.[a]	Copper, nickel, iron, tin–nickel.
Stop–off	To protect certain areas from wetting by solder.	Graphite, magnesium oxide, epoxy paints.

[a] Less easy to solder rapidly.

Table 1.7.16 Defects in soldered joints

	Defect	*Cause*
Defects due to poor solderability	Non-wetting (base metal visible, discontinuous solder film).	Poor surface preparation. Inefficient flux.
	De-wetting (globules of solder adhering to solder-coloured surface).	Surface contamination. Contamination of solder bath with Al, Zn, Cd. Inefficient flux.
Defects due to wrong soldering conditions	Non-wetting	Low soldering temperature (see also above). Soldering time too short.
	De-wetting	Low soldering temperature (see also above). Soldering time too short. Solder impurities e.g. Al, Zn, Cd.
	Excessive solder pick- up.	Soldering temperature. Impurity elements in solder.
	Rough and dull solder film.	High soldering temperature.
	Blackened viscous flux.	High soldering temperature, overheating.
	Porosity	Excessive flux. Poor design for escape of air and vapours from flux

Table 1.7.17 Mechanical properties of soldered joints

Solder composition (%)	*Shear strength of copper ring and plug joints at test speed of 1 mm min^{-1} (MPa)*	*Stress to rupture stress for 1000 h life failure of copper ring and plug joint. (MPa)*	*Mean tearing strength (Chadwick test) for joints. (N mm^{-1} width)*	
			0.28 mm tin plate	*0.45 mm copper*
100 Sn	22.1	7.5	15.4	8.2
99 Sn–1 Cu	28.0	8	–	–
96.5 Sn–3.5 Ag	37.7	14	–	–
95 Sn–5 Sb	37.2	12	–	5.9
60 Sn–40 Pb	35.0	4	6.2	7.7
50 Sn–50 Pb	30.0	3	7.3	8.7
40 Sn–60 Pb	34.3	3	7.1	7.7
10 Sn–90 Pb	28.9	–	8.0	6.2
42 Sn–58 Bi	50.0	4	–	–
62 Sn–36 Pb–2 Ag	43.0	5	6.1	6.9
50 Sn–32 Pb–18 Cd	32.0	3	–	–
40 Sn–58 Pb–2 Sb	30.4	3	7.3	6.5
93.5 Pb–5 Sn–1.5 Ag	23.8	5	–	13.7

Data by permission of ITRI Ltd.

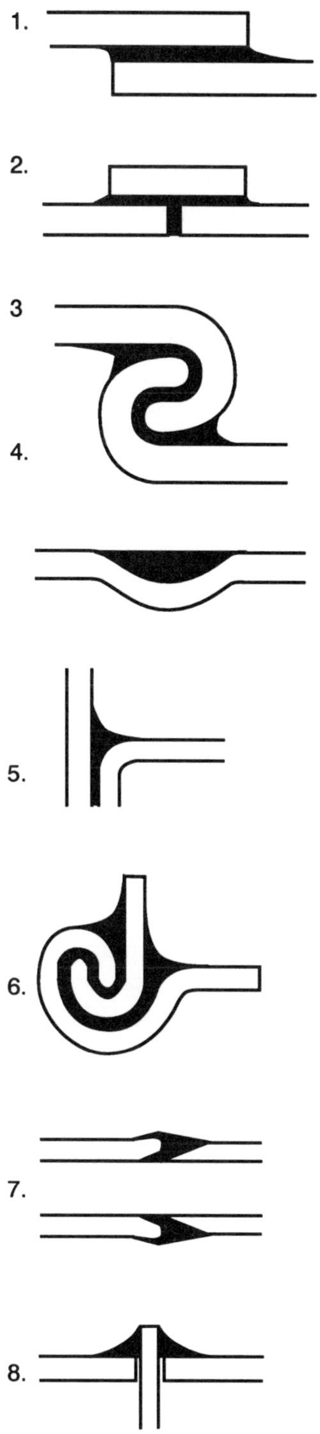

Lap	Give generous overlap to ensure strength.
Strap	One side smooth. Solder angles must be shallow – good wetting.
Lockseam	Strength from joint rather than from solder.
Filling	Sheet metal work. Use high lead solders.
Corner	Easily cracked. Food containers use high tin, lead-free solders.
Lock corner	Strength from joint form.
Tube join	Scarf joint gives greater strength than end butt.
Electronics	Plated through hole in printed circuit board

Fig. 1.7.1 Typical soldered joint forms

Fig. 1.7.1

1.7.4 Brazing

1.7.4.1 Introduction

Brazing is a process carried out at temperatures above about 450°C, but below the melting point of any of the materials being joined. The joint relies upon the capillary penetration of the filler material into fine parallel gaps between the components. Hence the brazing temperature must exceed the liquids of the braze metal. Brazing is used for most metals encountered in engineering practice.

Brazing is particularly useful for:

1. Joining of dissimilar metals.
2. Joining in a stress-free condition.
3. Complex assemblies joined by sequential brazing with filler materials of progressively lower melting points.
4. Joining materials of unequal thickness
5. Little or no finishing operations.

1.7.4.2 Selection of braze and brazing method

The suitability of brazing for joining common metals is given in:
Table 1.7.18 *Brazeability of metallic materials*

Brazing methods are described, and their advantages and limitations listed, in:
Table 1.7.19 *Characteristics of brazing methods*

1.7.4.3 Brazing materials and their properties

Braze filler metal selection is influenced by the base metal, the type of joint and service conditions. The filler metal must have an appropriate melting point and corrosion resistance for the service conditions.

Typical applications and properties of common brazing alloy systems are given in:
Table 1.7.20 *Applications and tensile strengths of brazing alloys*

Brazing fluxes are necessary to remove oxides from the surface of the workpieces. For minimum oxidation the flux should have a low melting point and a high thermal stability. In practice, fluxes are used which are molten about 50°C below, and stable about 50°C above, the brazing temperature.

The main non-gaseous fluxes are:

1. Borax- and fluoroborate-containing fluxes for uses at temperatures greater than 750°C (except refractory oxides).
2. Fluoride-type fluxes for temperatures below 750°C and for refractory oxides.
3. Alkali halide fluxes for all processes involving aluminium and aluminium base alloys.

1.7.4.4 Design of joints for brazing

The strength of brazed joints depends largely on efficient penetration of molten filler metal into a capillary gap between the parts. Lap joints are preferred to butt joints, and for best results the overlap distance should be between three and four times the thickness of the material being joined.

Joint designs suitable for brazing are illustrated in:
Fig. 1.7.2 *Joint designs for brazing*

As efficient capillary penetration into the joint gap is very important, clearances must be carefully selected and controlled. Recommended clearances are given in:
Table 1.7.21 *Recommended joint clearances*

1.7.4.5 Failure of brazed joints

Failure in joints is caused by deformation or fatigue in the joint or by thermal stresses resulting from the joining process. Common failures and their remedies are given in:
Table 1.7.22 *Failures in brazed joints*

The risk of failure can be minimized by the following factors:

Design	Design to recognized configuration.
Strength	Design to strength of brazing alloy. The tensile strength of a completely sound joint is about equal to the cast strength of the brazing material. DO NOT design to empirical joint strength – although thin joints are very strong the extra strength cannot always be reliably reproduced. With copper and copper alloy brazing processes joints often show higher strength than the mild steel components. In these cases it is considered safe to depend on a higher strength than could be obtained from the brazing material itself. Even in this case it is imperative to ensure that the fit of the parts is maintained within the narrow limits where this effect is regularly obtained (between 0.025 mm interference and 0.05 mm clearance).
Temperature	Do not overheat during brazing – excessive alloying can lead to weakness.
Joints	Use recommended joint clearances.
Fillets	In most cases the presence of external fillets does not affect the joint soundness. However, in the case of a pure T-butt joint a fillet can act as a stress distributor, and it is wise to assume the filler material is ductile and to design to allow for a fillet. In most cases, a T-joint would be preferred.

1.7.4.6 Brazing of ceramics

Brazing is a widely accepted method of joining ceramics to other materials.

CERAMIC – METAL JOINT

(a) Metallize ceramic surface	–	use molybdenum and manganese metallizing mixture. Braze conventionally.
(b) Active metals or hydrides	–	unmetallized BeO is joined to metals using a nickel-clad titanium filler. Zr, V are also used as fillers.
(c) Tungsten carbide	–	joining of tungsten carbide is included in Table 1.7.18

BRAZING OF DIAMOND

See Table 1.7.18, brazeability of metals.

GRAPHITE–METAL JOINTS

Graphite is not wetted by conventional fillers. Carbide-forming fillers (Ti, Zr, Cr, Si, Mo, Ta, Nb) are used.

Table 1.7.18 Brazeability of metallic materials

<table>
<tr><th colspan="2">Material</th><th>Methods</th><th colspan="2">Remarks</th></tr>
<tr><td colspan="2">Carbon and low alloy steels</td><td>All methods. Use silver or copper base filler. Nickel used for high strength or corrosion resistance.</td><td colspan="2">Not recommended for high carbon or heat treated alloy steels. Can combine brazing and heat treating. Avoid overheating – grain growth decarburization.</td></tr>
<tr><td rowspan="4">Stainless steel</td><td>Austenitic</td><td>Torch or furnace. Ag, Cu or Ni base fillers.</td><td colspan="2">Danger of stress corrosion and carbide precipitation unless stabilized.</td></tr>
<tr><td>Ferritic</td><td>Special Ag-base fillers with Ni additions.</td><td colspan="2">Danger of interfacial corrosion without special fillers.</td></tr>
<tr><td>Martensitic</td><td>Filler must be compatible with heat treatment.</td><td colspan="2">Heat treat during or after brazing.</td></tr>
<tr><td>Precipitation hardened</td><td>As above, Ni or Pd fillers for high temperature. Dry H_2 or vacuum brazing.</td><td colspan="2">Heat treat after brazing or solution treat during brazing.</td></tr>
<tr><td rowspan="4">Cast iron</td><td>Malleable</td><td rowspan="3">Silver brazing alloys. Furnace, torch, induction or dip methods.</td><td>Readily brazed after shot blasting or grinding.</td><td rowspan="3">High carbon and silicon reduce weldability. Keep brazing temperature low and uniform to avoid cracks. Special pre-cleaning.</td></tr>
<tr><td>Spheroidal</td><td>Easier to braze than grey.</td></tr>
<tr><td>Grey</td><td>Not wetted by silver alloys without oxidation and reduction of surface.</td></tr>
<tr><td>White</td><td>NOT RECOMMENDED</td><td colspan="2"></td></tr>
<tr><td colspan="2">Aluminium and aluminium alloys, magnesium</td><td>Aluminium base, zinc base or cadmium base fillers.
Active flux. Some aluminium alloys can be brazed with aluminium silicon alloys. All methods used, high skill and low melting point fillers with torch.
DO NOT BRAZE – most Al–Cu–X; Al–Zn–X; many cast alloys.</td><td colspan="2">Only suitable for aluminium, aluminium alloys and Mg–Mn.
Brazing temperature at least 30° below solidus temperature. Corrosion risk from flux residues. Wider clearances are employed than with most other materials. Special brazing sheet is available with high silicon braze roll bonded to aluminium alloy sheet. High Mn inhibits wetting.</td></tr>
<tr><td colspan="2" rowspan="2">Carbon (diamond)</td><td>Orthodox silver brazing alloys.</td><td colspan="2">To secure diamond tips and dies to supports.</td></tr>
<tr><td>Silver alloys containing titanium or zirconium; gold–nickel–chromium alloys. Heat in dry hydrogen or argon.</td><td colspan="2">True brazing of diamond.</td></tr>
<tr><td colspan="2">Chromium</td><td>Heat in dry hydrogen or air. Special flux.</td><td colspan="2">Not commonly used. Brazing of chromium plated surfaces destroys or discolours plating.</td></tr>
<tr><td colspan="2">Cobalt</td><td></td><td colspan="2">Straightforward.</td></tr>
<tr><td rowspan="5">Copper alloys</td><td>Copper</td><td>Orthodox methods with flux using Cu–Zn–Ag alloys.
No flux required with phosphorus bearing brazing alloys.</td><td colspan="2">Oxygen-bearing coppers prone to embrittlement. These should be rapidly brazed with low melting point fillers (Cu–P or Ag base).</td></tr>
<tr><td>Cu–Al
Cu–Si</td><td>These can be readily brazed with silver Cu–P, Cu–Zn alloys.
Special flux necessary, not more than 10% Al.</td><td colspan="2">Brittle joints arise when brazing to steels.
Stress relieve Cu–Si to avoid integranular and hot short cracks. Readily brazeable.</td></tr>
<tr><td>Cu–Be</td><td>Up to 2% beryllium can be silver brazed during heat treatment.</td><td colspan="2">Effects on age hardening characteristics.
Brittle joints when brazed to steels.</td></tr>
<tr><td>Cu–Cr</td><td>Readily silver brazeable.</td><td colspan="2">Effects on age hardening characteristics.
Avoid corrosion by flux residues.</td></tr>
<tr><td>Cu–Ni</td><td>Silver brazed (Cu–P fillers not recommended).</td><td colspan="2">Intergranular penetration by solder. Avoid stressing joints. Avoid corrosion by flux residues.</td></tr>
</table>

Table 1.7.18 Brazeability of metallic materials – *continued*

Material		***Methods***	***Remarks***	
	Cu–Pd, Cu–Ag	Silver alloys used.	Readily brazeable.	
	Cu–Sn (bronze)	Silver brazing. Not recommended for Al, Mg, Zr. Provide excess flux for 10Si, Cr. Pb-containing bronze not recommended for brazing.	Heat slowly and evenly to avoid cracks. Joints containing Si, Cr are less ductile.	
	Cu–W	Special flux. Pre-cleaning. Silver alloys.	Powder metallurgy compacts.	
	Cu–Zn (brasses) Cu–Zn–Ni	Readily brazeable with good fluxing. Gradual heating necessary.		
Rhodium, platinum, palladium, gold and gold alloys.		Silver alloys. All methods.	Straightforward.	
Nickel and nickel alloys	Nickel Ni–Be Ni–Cu Ni–Fe Ni–Mn Nickel-silver	Strong fluxes and special filters. Vacuum or induction brazing employed. Fillers are: Ni base alloys – avoid excessive erosion. Pd base alloys. Au base alloys.	Intergranular penetration by brazing alloys. Avoid stressing joints. Stress relieving before brazing is advantageous. With Ni–Be brazing can affect age hardening characteristics.	
	Ni–Cr Ni–Cr–Al Ni–Cr–Fe	Ag base (+ flux) or Ni base (small clearances) can be used.	Nickel alloys containing chromium and/or titanium are used for high temperature strength. Special fluxes used to achieve maximum joint strength.	
Lead		Use low melting point. Pb base solders.		
Zinc	NOT RECOMMENDED			
Refractory metals	Beryllium	Pure silver or silver–copper eutectic brazing alloy. Other Be base alloys available. Vacuum brazing.	Brazing of beryllium to itself and nickel–iron alloys. Not common.	Inert vacuum brazing under all conditions to avoid brittleness. Can be brazed for low temperature service with conventional fillers. At high temperatures operation is more difficult.
	Molybdenum	Not normally recommended. Some silver brazing or special alloys and fluxes, vacuum brazing.	Braze below recrystallization temperature.	
	Tantalum Tungsten	NOT RECOMMENDED. Can be brazed with Ni base fillers. Special fillers for high temperatures.	Tungsten – braze below recrystallization temperature.	
	Titanium Zirconium	Vacuum brazing with active unpleasant flux and silver, silver–copper, silver–manganese fillers, Zr base fillers for Zr, match brazing and heat treating procedures with Ti.	Not normally recommended. Tendency to form brittle intermetallics with orthodox brazing alloys. Surface may be pre-tinned. Long brazing times give highest strength joints.	
Tungsten carbide		Special silver base fillers and fluxes. HF induction normally used.	Brazing to a steel backing. Large difference in thermal expansion coefficient, requires skillful joint design. Recommended method for joining carbide tips to steels. Overheating must be avoided.	

Table 1.7.18 – *continued*

Table 1.7.19 Characteristics of brazing methods

Brazing type		*Description*	*Advantages*	*Limitations*
Furnace	Without protective atmosphere	Parts must be assembled with filler metal pre-placed. Filler melts and fills joint space.	Mass produced parts (self-jigged). No flux. Temperature uniform. Several joints simultaneously. Little operator skill. Recommended for joints involving dissimilar metals.	High capital outlay – uneconomical if only small portion of workpiece has to be heated for brazing. Development skill to establish optimum working conditions. If heating involves loss of parent metal mechanical properties then this occurs in entire assembly. Size limited. Furnace muffles have limited life.
	With protective atmosphere	As above with protective atmosphere.	As above. A number of atmospheres are available suitable for particular operations.	As above.
	Vacuum	As above with vacuum for no oxidation, or hydrogen or nitrogen atmosphere.	As above. Suitable for refractory metals. Only solution for superalloys, nimonics, electronic components, etc.	As above. Expensive furnace/pumping equipment. No volatiles in braze.
Hand operated torches	Oxyacetylene, coal gas and compressed air.	Heat workpieces by torch. Filler either foreplaced or faceted. Joint clearance 50–125 μm for capillary flow. Flux for wetting and oxidation.	Low capital and upkeep costs. Moderate running cost. Flexibility in operation. High heating rates and temperatures (oxyacetylene).	Skilled/semi-skilled operators. Variable results due to human element. Non-uniform heating. Flux required. Oxidation hazards. High temperatures can endanger work (oxyacetylene). Some developmental skill necessary.
Fixed gas burners		Heat workpieces by fixed burners, otherwise similar to torch. The number of burners may be varied.	Moderate capital, upkeep and running costs. Flexible. Variable heating rates and temperatures. Process timer can be employed. Little operator skill.	Some developmental skill necessary.
HF induction brazing		Parts heated to brazing temperature by induction using frequency of the order of 10 000 Hz.	Maximum rate of output attainable with suitable work. No temperature limitation. Heating can be localized. No human element. Little operator skill. Change in bulk properties can be minimized. Particularly adaptable to vacuum heating.	High capital cost. Specialized knowledge for servicing. Coils to heat difficult shapes without overheating difficult to design. Fixtures near inductor must be non-magnetic. Not suitable for dissimilar metals. Slow for low resistance metals.
Electric resistance brazing		Heating by passage of electric current through joint. Similar to resistance welding machine. Joint and pre-placed metal held together by pressure of machine electrodes.	High speed of heating attainable. Localized heating. No human element. Temperatures lower than for resistance welding. Automated.	Good results only for high conductivity metals and equality of section from electrode to joint face. Flux fouls electrode which causes frequent cleaning stoppages. Only suitable for small area joints. Not suitable for dissimilar metals. Parts must be capable of being pressed together by working electrodes.

Table 1.7.19

Table 1.7.20 Applications and tensile strengths of brazing alloys

Brazing alloy system	***Applications***	***Tensile strength (MPa)***		
		Copper and copper alloys	***Ferrous materials***	***Aluminium and aluminium alloys***
Copper	General purpose brazing.	–	175–325	–
Copper–zinc Bronze welding type	General purpose brazing.	Similar to that of parent metals.	205–345	–
Silver copper Copper–silver–zinc Copper–silver–phosphorus Copper–phosphorus Copper–silver–tin Copper–silver–zinc–cadmium–nickel Copper–silver–zinc–cadmium Silver	General low-temperature brazing. Fluxless brazing of copper alloys.	Exceeds strength of parent metal. Exceeds strength of parent metal.	235–380	–
Cobalt–chromium–boron Nickel–chromium–boron Nickel–chromium–silicon–boron Nickel–silicon–boron Nickel–phosphorus	For high temperature oxidation resistant alloys and stainless steels.	–	Value is largely determined by the brazing conditions. Refer to alloy manufacturer for further details.	–
Silver–palladium–manganese Nickel–palladium–manganese Copper–palladium–nickel–manganese Palladium–nickel Copper–nickel Copper–palladium Silver–palladium	Nickel, molybdenum, tungsten alloys, stainless steels. Vacuum brazing.	–	Value is dependent upon type of parent materials and brazing conditions. Typical values in the range 400–1000.	–
Silver–copper–palladium Gold–nickel Gold–copper	High temperature brazing of iron, nickel, cobalt, gold alloys. Vacuum brazing. Au–Ni has high temperature strength.	Exceeds strength of parent metal.	500–1100; refer to manufacturer for more exact information.	–
Aluminium–silver–nickel (or manganese) Aluminium–silicon Aluminium–silicon–copper	For aluminium alloys.	–	–	Usually exceeds that of metals joined.

Table 1.7.21 Recommended joint clearances

Alloy system	*Recommended joint clearance (mm)*			
	Copper	*Copper base alloys*	*Ferrous metals*	*Aluminium and alloys*
Copper			Interference to 0.075	
Silver	0.025–0.10	0.025–0.10	rarely used	
Copper–zinc Bronze–welding type	0.075–0.40	0.075–0.40	0.05–0.25	
Copper–phosphorus	0.075–0.40	0.075–0.40		
Silver–copper	0.025–0.20	0.025–0.20	0.05–0.20	
Copper–silver–zinc Copper–silver–phosphorus Copper–silver–tin Copper–silver–zinc–cadmium–nickel Copper–silver–zinc–cadmium	0.025–0.20		0.05–0.20	
Cobalt–chromium–boron Nickel–chromium–boron Nickel–chromium–silicon–boron Nickel–silicon–boron Nickel–phosphorous			0.025–0.15	
Silver–palladium–manganese Nickel–palladium–manganese Copper–palladium–nickel–manganese			0.025–0.15	
Palladium–nickel Copper–nickel Copper–palladium			0.025–0.15	
Silver–palladium			0.025–0.10	
Silver–copper–palladium	0.025–0.10	0.025–0.10	0.025–0.10	
Gold–nickel Gold–copper	0.05–0.20	0.05–0.20	0.05–0.20	
Aluminium–silver–nickel (or manganese) Aluminium–silicon Aluminium–silicon–copper				0.10–0.60[a]

[a] Joints whenever possible are designed half fillet and half lap, and wider clearances are normally employed with aluminium brazing. This is due to similar melting points of filler and substrate which result in the danger of high parent metal losses during brazing.

Reproduced by kind permission from P. M. Roberts, *Brazing*; published for the Design Council by Oxford University press.

Table 1.7.22 Failures in brazed joints

Type of failure	*Cause*	*Remedy*
Fatigue/deformation in joint area	Stress concentration at areas where there is a marked section change. 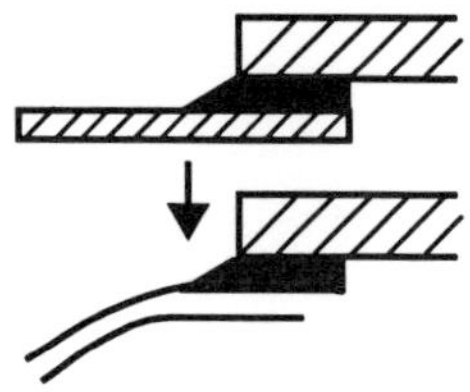	Equalize section thickness (i.e. weaken stronger member) to redistribute stress. 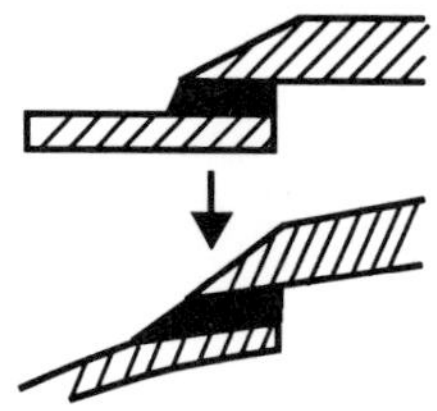
Solidification shrinkage	In butt joints contraction leads to stress-raiser formation.	Redesign joint – it is impossible to avoid this effect if design specifies an edge to edge plain butt joint.
Differential coefficient of expansion of parent metals	Relative contraction of one parent metal leads to a larger joint gap than specified (only occurs with dissimilar substrates).	Female member of joint should have greater expansion coefficient. Contraction during cooling leads to shrinkage of joint gap. Refer to recommended joint clearance tables. Can use low temperature brazing alloy (silver solder).

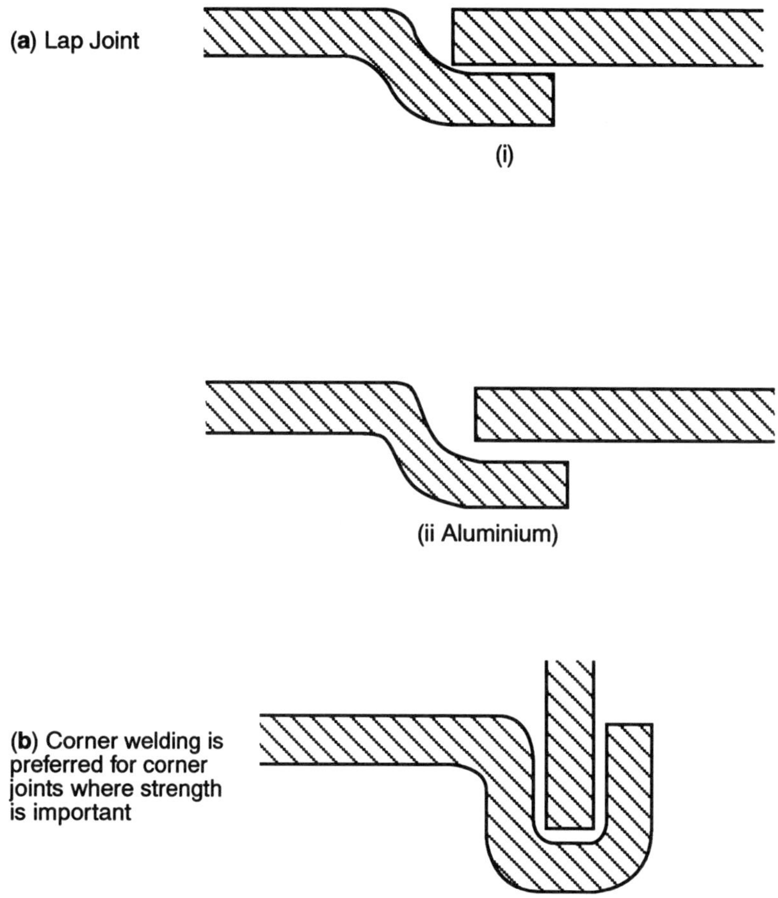

Fig. 1.7.2 Joint designs for brazing

Fig. 1.7.2

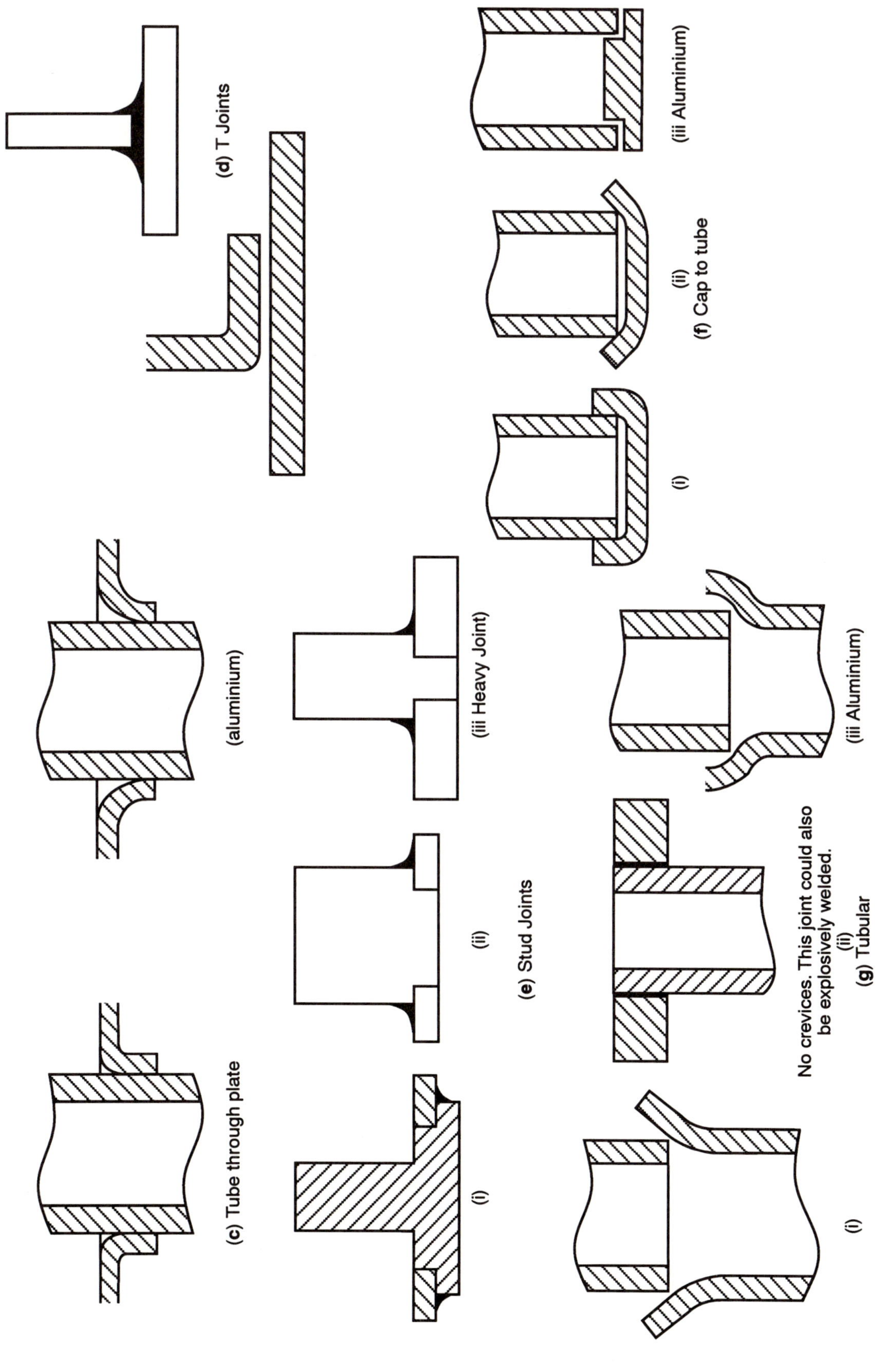

Fig. 1.7.2 Joint designs for brazing – ***continued***

Fig. 1.7.2 – ***continued***

1.7.5 Welding

1.7.5.1 Introduction

Welding is the most versatile and widely used of all joining methods and, although its uses are mainly in joining most of the metals in engineering practice, it is used extensively in the joining of plastics and to some extent in the joining of ceramics. Welding of metals is considered in Sections 1.7.5.2–1.7.5.9, plastics in Section 1.7.5.10 and ceramics in Section 1.7.5.11.

For any welding operation, design, material and fabricating processes cannot be selected independently. Broad decisions on design and material are taken first and these are then modified as the decision on process selection is developed.

A welding specification must fulfil the following functions:

(a) Provide a weld with satisfactory mechanical properties under service conditions, affected by materials selection, geometrical specification.
(b) Specify a welding process compatible with the materials and geometrical factors selected.
(c) Have accessibility for the welding operation.
(d) Be such that the properties of the parent material are not adversely affected during or after the welding process.
(e) Ensure that the weld is accessible for post-weld inspection – usually essential for critical applications.
(f) Be designed to avoid premature fatigue failure – usually requiring consideration of joint design, welding procedure and weld quality.

1.7.5.2 Selection of welding process

There are many welding processes available, each suitable for a variety of materials and applications. The major welding processes for joining metals are described, and their advantages, limitations and typical applications given in:

Table 1.7.23 *Characteristics of welding processes*

The major process types are pressure, electrical resistance, electric arc, gas, power beam and chemical reaction processes. Operating principles of the process types are show schematically in:

Fig. 1.7.3 *Operating principles of the major welding processes*

Welding processes may be manual or automatic. The advantages and limitations of automatic processes are given in:

Table 1.7.24 *Characteristics of automatic welding processes*

1.7.5.3 Weldability of metals

From the viewpoint of weldability most metals can be classified broadly as:

solution treated and aged alloys,
cold-worked alloys,
alloys in which a thermal cycle does not greatly affect mechanical properties,
carbon and low alloy steel – most commonly welded.

SOLUTION TREATED AND AGED ALLOYS

Examples: Aluminium–copper alloys.
Aluminium–zinc alloys.
Beryllium–copper alloys.
α–β Titanium alloys.

Some brasses and bronzes.
Marageing steels.
Precipitation hardened stainless steels.

If the alloy is welded in the solution treated condition, the effect is to re-solution treat the weld metal and HAZ. At distances from the weld, overageing and ageing occurs. The over-aged zone has similar mechanical properties to the weld zone, but may have inferior corrosion resistance.

If the alloy is welded and then solution treated and aged, the weld metal will be resolution treated and at distances from the weld, overageing and ageing progressively occur. The weld metal is soft and ductile. Re-solution treatment, quenching and ageing should be effected. Or use filler metal with higher mechanical properties than base metal or which ages at room temperature. The overaged zone should be minimized by minimizing the weld heat input.

COLD WORKED ALLOYS

Examples: 300 series stainless steels
brasses
aluminium alloys

Welding removes work hardening characteristics. Either deposit high strength as-cast alloy, in which case, steps should be taken to minimize the heat-affected (annealed) zone, or rework after welding (not always practical).

NON-HEAT TREATABLE ALLOYS

Examples All pure metals
α–Titanium alloys
Unworked cold workable alloys

The only effects are grain-size increase and segregation of impurities to the grain boundaries. These alloys are more susceptible to flaws resulting from residual cooling stresses.

STEELS

There are many steels which can be welded and the welding processes and their effects on the material vary. Details are given in Section 1.7.5.5.

1.7.5.4 Welding of non-ferrous metals

The welding processes recommended for non-ferrous metals, with comments on the weldability of groups of metals and specific alloys, are given in:
Table 1.7.25 *Weldability of non-ferrous metals*

1.7.5.5 Welding of ferrous metals

WELDING OF CARBON AND LOW ALLOY STEELS

Welding of carbon steels or high strength, low alloy steels depends upon hardenability, which may be defined as ease of martensite transformation. Martensites and bainites can be brittle and hence an increased hardenability results in an increased likelihood of brittle microstructures forming in the area affected by the welding process.

It is important to consider whether a brittle microstructure, such as martensite, will form in the heat-affected zone (HAZ) during cooling. The formation of the microstructure, and its mechanical properties will depend upon the alloy content and the thermal cycle during the welding operation.

All common alloying elements increase the hardenability and hence a carbon equivalent scale has been devised as an approximate guide to weldability.

$$\text{C.E} = \text{C\%} + \frac{\text{Mn\%}}{6} + \frac{\text{Cr\%} + \text{Mo\%} + \text{V\%}}{5} + \frac{\text{Ni\%} + \text{Cu\%}}{15}$$

If C.E < 14	Excellent weldability, no special precautions (e.g. low carbon steel).
0.14 < C.E < 0.45	Medium carbon or low-alloy steels. Martensite is more likely to form. Modest pre-heats and low hydrogen electrodes become necessary.
C.E > 0.45	Extreme complications. Pre-heat, use of low hydrogen electrodes and post-heat with slow cooling may be required.

Pre-heating slows the cooling rate and lessens the tendency of martensite formation. In addition the martensite which forms can temper to a less brittle structure as it cools. Low hydrogen content minimizes the likelihood of martensite cracking.

In low alloy steels martensite may be tempered to much higher strengths. Because of the danger of untempered martensite forming in the HAZ pre-heat and post-weld stress-relieving are often specified.

Preheating temperatures to avoid hydrogen cracking will depend on:

base steel composition (carbon equivalent),
hydrogen content of weld metal (afterweld),
section thickness,
arc energy input.

Recommended welding processes for steels are given in:

Table 1.7.26 *Weldability of steels*

Alloying elements can affect weldability markedly, as shown in:

Table 1.7.27 *Effect of microalloying additions on weldability of steels*

An example of the relationship between the necessary pre- heating temperature, the thickness of the base material and the composition, is for a given arc energy input and hydrogen content is shown in:

Fig. 1.7.4 *Pre-heating temperatures for carbon steel welded by the manual arc process*

Post-weld treatment may be required, depending on thickness, composition, and the application, as shown in:

Fig. 1.7.5 *Post-weld heat treatment of carbon steels. Temperature of 600–625°C; 1 h per 25 mm thickness*

WELDING OF STAINLESS STEELS

A useful method of assessing the general welding characteristics of stainless steels is by means of the Schaeffler diagram, shown in:

Fig. 1.7.6 *Schaeffler diagram for welding of stainless steels*

The alloying elements are expressed in terms of nickel and chromium equivalents and hence a point of the diagram will indicate the main phases to be found in a stainless steel of a particular composition. The diagram here has been used to indicate the main problems likely to be incurred as a result of welding these alloys. The effects of dilution and the amount of ferrite may be estimated from the diagram.

In fully austenitic weld metal, cracking (solidification or liquation) can occur. The presence of retained δ-ferrite (5–10%) is extremely beneficial. The presence of δ-ferrite can be detrimental in severely corrosive environments or for high temperature service conditions. Intergranular corrosion in the HAZ, through carbide precipitation, can be avoided by using stainless steels containing small amounts of Ti or Nb which form carbides which are stable at high temperatures (550 to 850°C)

WELDING OF OTHER FERROUS METALS

The welding processes recommended for ferrous metals other than steels are given in:
Table 1.7.28 *Weldability of other ferrous metals*

1.7.5.6 Design of welded joints

An extremely wide range of joints are possible with welding. The type of joint is in part determined by the process and materials employed, and it is important that the design, material and process are compatible. There are five basic types of joint: butt, tee, corner, lap, edge. Features of the different types of joint and their applicability are given in:
Table 1.7.29 *Design of butt joints*
Table 1.7.30 *Design of lap and corner joints*
Table 1.7.31 *Design of tee joints*
Table 1.7.32 *Design of insert and stud joints*
Table 1.7.33 *Design of structural and heavy joints*
Table 1.7.34 *Design of tube and sheet joints*

Factors which affect weld design include dissimilar thicknesses, dissimilar metals, inaccessible joints and multiple welds.

DISSIMILAR THICKNESSES

With joints involving dissimilar thicknesses (e.g. rings, diaphragms and flanges) the main problem is distortion of the thinner member. Possible solutions are to:

1. Machine thicker member to match thinner.
2. Reheat thicker member.
3. If size allows, furnace braze whole joint.
4. Use rapid heating process, e.g. electron beam or pulsed TIG to minimize heat affected region to width as that of thin member.

Mechanized welding is recommended.

DISSIMILAR METALS

Dissimilar metals are best welded using solid phase processes such as flash welding, friction welding, diffusion welding, explosive welding and cold welding.

Power beam processes are the only recommended fusion welding processes. Certain material combinations, e.g. stainless steel to C–Mn steel in lined vessels can be fusion welded using electric arc processes. In cladding, an electrode with higher alloy content is used whilst, in welding, the C–Mn steel is 'buttered' with stainless steel before completing the weld with a stainless steel electrode.

INACCESSIBLE JOINTS

Good weld design is especially important in welding inaccessible joints. Possible techniques include:

1. sequential assembly,
2. special purpose mechanized equipment.
3. electron beam welding, which can be operated over long distances due to the fine focus of the electron beam. Keyhole welding,
4. furnace brazing of preassembled parts,
5. projection welding,
6. some variations of metal-arc welding, 'firecracker welding'.

MULTIPLE WELDS

Methods of producing multiple welds include:

1. projection welding (not more than six),
2. spot welding,
3. multiple electron beam (keyhole) welding.

Multiple welds may be required in cases where the workpiece is too large or the workpiece is inaccessible from one or both sides.

1.7.5.7 Defects in welds

A weld is essentially a small casting and many of the defects inherent in welds are similar to those found in castings. There are two main regions which are affected by the welding operation:

fusion zone	– the weld area itself,
heat affected zone	– area in which the microstructure of the parent metal is affected by heat from the welding process. In large multi-pass welds, the early welds become part of the heat affected zone (HAZ) for subsequent welds.

Descriptions of common weld defects and their causes and remedies are given in:
Table 1.7.35 *Weld defects*

1.7.5.8 Inspection of welded joints

For critical joints requiring high quality the joint must be designed to allow the desired inspection method to be applied.

With resistance spot welds, friction welds, flash welds, diffusion bonded and brazed joints inspection is extremely difficult. Destructive testing and regular checks are employed, or various 'in-process' monitoring techniques.

Weld inspection methods include visual inspection, dye penetration and eddy-current techniques, radiography and ultrasonics. These are described in:
Table 1.7.36 *Weld inspection methods*

1.7.5.9 Hybrid joining processes based on welding

WELD BONDING

Resistance welding through high strength structural adhesives to give exceptionally high mechanical strength properties.

Requirements of adhesive

- Must be capable of being moved under pressure of the welding electrodes in order for metal to metal contact to occur at the joint interface.
- The heat resulting from the spot weld must only cause limited detrimental effects on the bond strength.

Modified epoxy resins are commonly employed.

Process

Adhesive in paste form is applied to the parts. Parts brought together and temporarily clamped. Parts placed between the electrodes of a conventional resistance spot welder and welded together. Then placed in a low temperature oven and the adhesive cured for about 1 h (the exact time and temperature depends on the adhesive and alloy being weld bonded).

Applications

Aircraft joints for aluminium and titanium alloys 0.5–6 mm thickness.

The process gives higher strength bonds than adhesive bonding only and eliminates sealing problems encountered with spot welds and rivets. It has highlighted a need for high strength adhesives to enable higher strength joints to be produced by this process.

WELD BRAZING

Spot welding to position parts followed by final fabrication by capillary flow of brazing alloy.

Advantages

Superior in tensile, shear, stress rupture, fatigue to fabrication by spot welding or brazing alone.
Ease of fabrication lowers cost.
Brazing eliminates sealing problems encountered with spot welds or rivets.

Applications

Can be used with most metals, but the more notable applications are for joining nickel base superalloys, refractories and titanium.

1.7.5.10 Welding of plastics

Plastics can be welded by a number of techniques, including hot plate, ultrasonic, hot gas, high frequency and friction welding. These processes are described in:
Table 1.7.37 *Processes for welding thermoplastics*

The suitability of the welding processes depends on the materials to be joined. Guidance for hotplate and ultrasonic welding is given in:
Table 1.7.38 *Weldability of thermoplastics by the hotplate welding process*
Table 1.7.39 *Weldability of thermoplastics by the ultrasonic welding process*

1.7.5.11 Welding of ceramics

While the welding of ceramics is not widely practised it is possible for some systems and can confer the following advantages:

Sealing	–	impermeability at elevated temperature not possible by means of mechanical seals.
Mechanical continuity	–	levelling of stress distribution in a ceramic structure which would not occur with a mechanical joint.
Retention of properties	–	refractory properties of the ceramic are retained across the weld.

Types of ceramic welding are:

Solid state	–	heating the parts together under pressure so that a sinter/weld is produced. Heating can be by means of a furnace, friction or microwaves. Matching thermal expansions and chemical compat-

ibility required. The method is not suitable for boron nitride, silicon nitride or silica.

Fusion – heating can be by electron beam, image furnace, laser, or, under some conditions, electric arc.
Prevention of cracking is the major problem which can only be overcome by pre-heating over a wide area around the weld and very slow cooling to prevent the build-up of large thermal stresses. Ceramics must be free from phase transitions. For alumina, heating to about 1000°C before welding and cooling at 28°C/min has been found satisfactory.

Details of welding processes for ceramics are shown in:
Table 1.7.40 *Characteristics of solid state welding processes for ceramics*

Table 1.7.23 Characteristics of welding processes

Process type	*General characteristics*	*Process name*	*Description*	*Advantages*	*Limitations*	*Applications/remarks*	*Unit cost of plant*[a]
Pressure welding	Processes where the two sides of a joint are brought into intimate contact by mechanical deformation or atomic diffusion. Fluxes are not used. Also known as 'solid-phase welding processes'. Absence of cast metal at joints, no intermetallics formed. Minimal disturbance of parent metal properties. Suitable for joining dissimilar metals.	Cold joining	Mechanical deformation at room temperature.	Low cost.	Only few ductile metals and those without surface layers. Deleterious effects on mechanical properties.	Al, Cu, Ag, Au	NK
		Press servicing. Self–piercing rivets.	Mechanical interlocking using punch or self piercing steel rivet.	Low cost, simple. Mechanical performance of rivets similar to resistance weld.	Button on reverse side. Thin sheet material only.	All sheet materials. Dissimilar metal combinations.	NK
		Hot forging	Mechanical deformation at higher temperatures.	More useful than cold joining. Less severe effect on mechanical properties.	Only applicable with forging operations. Not generally suitable.	General forging processes.	NK
		Friction welding	Heat developed between a stationary and a rotating workpiece pushed together. After rotation stage, high pressure applied to forge the weld.	Excellent joint properties. Low power requirements. High production rates. Good reproducibility. No surface preparation.	High equipment costs. Need for finish machining. Small area butt joints only. Difficult to inspect welded joint.	Areas up to 20 000 mm^2. Refractory metals. Oxidizable metals. Microfriction version for miniature sizes.	4–100 A
		Explosive	Controlled explosion. Deformation and welding.	Low cost for specialist uses. Excellent joint properties. No need for power supplies.	Limited to lap joints. Blast damage noise. Explosives used.	Cladding of tubes and plate, and welding tubes to plates. On-site welding.	10+ A
		Diffusion bonding (See section 1.7.9)	Workpieces heated and pressed together with interlayer. Welding by atomic interdiffusion.	Large area joints. Low deformations. Differing geometry joining. Vacuum-tight, heat and oxidation resistant.	Need vacuum chambers. Long bonding time. Post-weld inspection difficult.		NK
		Ultrasonic vibration	Pieces pass between welding heads, one of which imparts ultrasonic vibrations to cause welding.	No arc, spark, atmosphere. No surface preparation.	Specialized, expensive.	High integrity welds. Mainly thin sheet, foil. Continuous seam welds.	1.5–15 A
Electrical resistance (pressure) welding	Heat is produced by the passage of an electric current across the joint interface. Pre-set pressure is applied to ensure intimate contact and weld consolidated. General advantages: (a) Easily automated. (b) High production rate. (c) Low capital cost except for full automation. (d) Very little pre-or post-weld Heat treatment. (e) Accurate joint location (f) No bulk fusion – can use for dissimilar metals (g) Low metal wastage. (h) Lower mechanical stresses than in fusion welding.	Projection	One workpiece has projections pressed against the other. Pieces forced together.	Simultaneous positioning and welding. Cleaning not critical.	Machining necessary for projections.	Stamped low carbon or low alloy steel parts 0.5–3.5 mm thickness. Not >0.2%C.	1.5–15 A
		Spot	Copper alloy electrodes pressed against pair of faces opposite to those to be welded. Pre-heat.	Fast. Not expensive. Automated	Post weld inspection difficult. Exterior welds not tight. Susceptible to corrosion. With steel, risk of hard brittle welds above 0.10%C.	Sheet only, 0.3–4 mm thickness. Cap, stitch and seam welds. Most metals. Can weld Zn coated steel (v. important).	1.5–15 A
		Seam	Work pressed together. Rotating wheel electrodes. Pre-heated.	Tight seal joints. Smaller seam than spot or induction.	Straight lines, slow curves only. Pre-cleaning. With steel, risk of hard brittle welds above 0.10%C.	1.1–2.0 m/min. 0.25–3.2 thick mild steel. Most steels, nickel, aluminium, magnesium. Some dissimilar metals.	NK
		Induction	Heating to fusion by HF induction.	Fast. Mass production. No electrodes.	Expensive equipment.	Making tubes from sheet.	NK

Table 1.7.23 Characteristics of welding processes –*continued*

		Flash Butt	Workpiece held lightly together and melted. Then pressed together, expelling molten metal.	Very strong joints. Variety of applications. Comparable with pressure gas.	Hazardous (hot molten particles). Flash must be removed after welding. Postweld inspection difficult.	Joining and forging heavy sections. Dissimilar metals, especially aluminium, copper, special steels. Range of sections.	4–500 A 0.5–70 Resistance Butt A
Electric arc	Arc between electrode and workpiece. Electrode may be consumable (metal) or non-consumable (tungsten) to form arc plasma and shield the weld pool from oxidation. Flux or gas used. Argon commonly used. For TIG and plasma. Argon + CO_2 or O_2 and CO_2 commonly used for MIG but organ used for pulse metals and argon–helium + CO_2 or O_2 mixture gives hotter arc. Adequate joint properties obtainable in mild and low alloy steels and many non-ferrous alloys. Most welds amenable to non-destructive test examination. Presence of cast metal and heat-affected zone can require pre- and post-weld treatment.	Manual arc. (MMA)	Rod electrode flux-covered, protective gases generated by flux.	Simple, cheap, versatile. Excellent, joint properties. Welding inaccessible joints. Low accuracy setting-up required.	Manual – high operator skill. Electrodes require frequent changing. Multirun welds + slag chipping necessary on thick plate.	Very versatile. With high carbon steels low hydrogen electrodes must be used. Not suitable for most aluminium alloys.	1–3 M
		Flux core arc. (FCA)	Wire electrode has flux-core. Protective gases may be used.	Can be automated. Fast. Self-shielded wires suitable for site work.	Weld wires more expensive than solid wires.	Production applications. Repair of castings. General heavy/medium structures. Alloy steels. Flux composition depends on application.	1–3 M
		Submerged arc. (SA)	Arc between consumable wire and work. Weld pool protected by flux from travelling hopper.	High deposition welding. Additives with flux. Welding wire inexpensive. Carried out in exposed places.	Automatic equipment required. Flux must be clean/dry. Flat position only. Careful joint alignment. Danger of slag entrapment.	Heavy gauge >5mm, usually multi-welds. Mild steel, medium, stainless steel. Unsuitable for some cast irons, tool steel, Hadfield steel, Al, Mg, Cu alloys. Can add grain size refiners, alloy elements. Also for building up abrasion resistant coatings on steel.	10 A
		Metal inert gas. (MIG)	Bare wire consumable electrode. Stream of inert gas used to shield the weld pool from air.	Fast, continuous. No slag. Low H_2. Deep penetration, any position. High quality welds.	High cost of gas. High operator skill. Equipment non-portable. Weld metal cracking with hardenable steels. Care in joint alignment. Not suited to high winds.	Used with 0.5 mm sheet (lower limit). Most metals, although some require special procedures.	1.5–5 M
		Metal active gas. (MAG) spray metal transfer	Argon-CO_2 used as shielding gas. High welding currents.	Deep penetration and high deposition rates. Low cost of shielding gas. High production rates.	Limited to flat and horizontal–vertical positions (in steels). Careful joint alignment.	Mild/low alloy steels above 6 mm thickness. Second quality welds.	>MIG M/A
		MAG transfer dip metal	CO_2 or argon – CO_2 as shielding gas. Low welding currents.	Versatile (manual or automatic). All positions. Low gas (CO_2) costs. Joint gaps and misalignments can be accommodated.	Danger of lack of fusion with semi-skilled operators. Spatter.	Fabrication of sheet, multi-runs in thicker steels. Thin steel sheet 1.4 mm.	>MIG M/A
		MIG or MAG pulse welding	Spray metal transfer at low currents	Versatile, manual or automatic. All positions. Good weld quality and appearance.	Power sources more expensive than non-pulsed MIG or MAG Some equipment complexity.	Particularly useful for stainless steels and non-ferrous alloys.	>MIG M/A

Table 1.7.23 – *continued*

Table 1.7.23 – *continued*

Table 1.7.23 Characteristics of welding processes – *continued*

Process type	General characteristics	Process name	Description	Advantages	Limitations	Applications/remarks	Unit cost of plant[a]
		Tungsten inert gas. (TIG)	W electrode, stream of inert gas, usually argon with or without filler metal. Argon–H_2 and He used to improve performance.	Versatile. Manual/automatic. Very high quality weld deposits.	Limited speed and expense. High operator skill. Multiruns necessary in butt welds > 3 mm thickness.	High quality welds, most metals. Especially suitable for thin stainless steel tubes using automatic welding equipment.	1.5–10 M/A
		Plasma arc.	Arc between W electrode and work in gas stream, constricted to form plasma. Second outer gas for shielding.	Concentrated very hot arc. No W contamination. Deep penetration up to 10 mm using keyhole, (automatic only).	High equipment cost, High Gas consumption. Careful torch maintenance required. Flat position only. Good joint alignment.	Very thin foil (50 μ m). Keyhole welds to 10 mm stainless steel, 12 mm titanium. Not used on light alloys. Also materials too thick for TIG.	>2 M/A
		Stud and percussion welding.	Arc drawn between stud workpiece and base electrode. Stud then forced into welding contact. Capacitor discharge can replace arc.	Very fast. Inaccessible positions. Good appearance.	Stud size and shape restricted to permit chucking and limited by power source capacity.	Common for steel fixing. Percussion for electrical connections and contents. Dissimilar metals.	4 with discharge 0.2–15
		Spot arc.	Weld made in lap joint through one piece of metal into other piece. Flux, MIG, TIG, sub-merged arc versions are used.	Fast. Relatively inexpensive Automated. No preparation required in thin sheets, but may need to drill hole in top sheet in thicker plate.	Weld quality can be poor if sheet surfaces insufficiently clean. Post-weld inspection difficult Susceptible to corro-sion. Brittle welds >0.1%C.	Suitable for sheets of different thickness. Flux-cored are give higher strength and smoother surfaces than other forms.	NK M
		Electro-gas.	Arc between consumable wire with or without flux core. Weld pool shielded by inert gas and pool cavity formed by sliding water-cooled dams. Usually single pass weld.	Thick plates. Join length unlimited. Vertical joints. Better starting and visibility than with electroslag.	Heavy work only. Vertical position only. Special equipment required.	Low medium carbon steels (tanks, shipbuilding). Sometimes on austenitic steels Plates (12–100 mm) of differing thicknesses. better toughness than electroslag in 15–40 mm thickness range.	NK A
Gas welding	Heat to fuse joint provided by gas flame (mixture of oxygen and acetylene) Adequate joint properties. Welds amenable to non-destructive examination. Nozzle size and gas flow determines amount of heat produced. Cast metal and heat-affected zone effects (HAZ).	Manual gas.	Flame tip melts edges to be joined to form continuous weld. Filler material may be added to pool.	Versatile in joint design. Adequate joint properties. Welds amenable to non-destructive examination. Low cost. Portable. (oxidizing-carburizing). Adjustable flame. Can be used for difficult material e.g. cast iron.	Slow, mainly manual. Low products rates. Poor surface appearance and wide HAZ.	Short runs, field work. Mainly sheet materials up to 5 mm stainless steels, most non-ferrous metals, except lead, zinc, some precious metals, refractory metals. Inert gas processes usually preferred for non-ferrous metals.	0.2 M/A
		Pressure gas.	Parts heated by gas to form molten ends, then forced together and upset. Pressures c.7–28 MPa	No filler metal. Weld has worked structure.	Limited section thickness and to suitable shaped work. Post weld heat treatment.	Mild steel, low alloy and stainless steel. Non-ferrous alloys. Suitable for joining rods, sec-tions, tubes, rails, rings.	NK M/A

Table 1.7.23 Characteristics of welding processes – *continued*

Power beam	Focused beam of energy to give deep narrow welds. Energy intensive. The beam can pass through the material to form keyhole.	Electron beam	Narrow focused electron beam. Workpieces held in contact, *in vacuo*. Local melting.	Excellent quality and purity welds – small HAZ. High welding speed. Full protection for reactive metals. No filler.	Small pieces (vac. chamber). Expensive, time-consuming. Large equipment. Very expensive. Very close fitting parts. Prealignment necessary.	Recommended for high quality work of precision components. Very large components (100 mm) can be welded also. Keyhole welding. Dissimilar metals.	10–150 A
		Laser	Focused coherent light beam, (0.1 mm spot size). No vacuum. Two types of laser, CO_2 and Nd: YAG. CO_2 laser beam is focused and manipulated with mirrors, Nd: YAG is transmitted via a fibreoptic cable.	Narrow weld beads. Operates in air. Very high welding speeds, small HAZ.	Small joins only. More difficult on reflective surfaces. (coatings may be employed). Joint gap <0.1 mm. Special jigging and beam manipulation equipment.	Keyhole welding 3–10 mm thickness. Almost no distortion – fine welds, with small HAZ enables welding near to heat-sensitive components (gaskets, electronic).	NK A
Chemical reaction processes (casting-on)	Welding by chemical reaction between workpieces. Very high temperatures generated.	Electroslag	Welding heat generated resistively between consumable electrode and molten bath of slag on molten weld pool.	Heavy-duty work *in situ*. Vertical joints. Very high quality. Minimum distortion. Avoids the need for multi-run techniques.	Only heavy sections, long runs, expensive. Vertical position only. Prolonged heating leads to adverse effects around weld.	Very thick plates (3–30 cm). Low or medium carbon steels, copper alloys. Heavy fabrications. Toughness of welds inferior to manual arc unless heat treated.	1.5–9 A
		Thermit	Casting techniques in which heat and weld metal are supplied from the reduction of an oxide (usually Fe_3O_4 by aluminium).	Needs inexpensive equipment. Simple to operate, no skill required.	Slow, only suitable for filling cavities and repair work. Not very high quality.	Repair of welds, heavy sections (non-critical structural joints). Rails, housings, frames, etc.	0.2+ A

[a] Figures relative to cost of manual arc M = Manual (semi-automatic).
NK = Not Known.
A = Automatic

Table 1.7.24

Table 1.7.24 Characteristics of automatic welding processes

Type	***Processes***	***Features***	***Advantages***	***Disadvantages***
Tractors	MIG/MAG, SCA, SA, oxy-arc plasma electrogas	Tractor mounted on track to move welding head. Systems normally designed to be portable. Tractor may have automatic features, e.g. oxilation, sensors.	Can be applied to both straight ream and circular paths. Less operator skill required than manual welding and less tiring for operators.	Although portable may be time–consuming to set up. Not cost effective for short runs. Joint fit-up tolerances smaller than for manual welding.
Positioners	MIG/MAG. SCA, TIG, SA, plasma	Simple device for holding and rotating components. Also, heavy equipment (rollers) for welding large structures.	Components, such as pipes, can be welded in the most convenient and productive position, usually in the flat.	Requires welding head to be mounted separately, e.g. on a column and boom.
Orbital heads	TIG/MIG/MAG	Welding torch can be rotated around the pipe. Clamping arrangements vary from horseshoe clamps for small diameter pipes to ring clamps for large diameter pipes.	The pipe can be in a fixed position and nests of pipes can be welded. All positional welding, but programming of welding parameters is normally required.	Normally requires a number of orbital heads to accomodate range of pipes sizes. More sophisticated power source/ controller. Spacing between pipes must have sufficient clearance for head rotation.
Robots	Resistance, MIG/MAG, TIG, Laser	Basic system comprises – mechanical arm – component holding fixture or component manipulator – intelligent controller for programming. Various system types available including cartesian or gantry, and articulated arm.	General purpose machine can be re-programmed for change in work. Articulated arm can reach areas of restricted access. Very high production rates.	Multi-axis machines (up to five axes), with sophisticated controllers can be expensive.
Dedicated systems	MIG/MAG, resistance, TIG, laser, plasma, oxy-acetylene	Usually multi-head welding station for manufacturing high volume components. (Automotive)	Sensors can be included for remote operation.	Difficult to adapt for different components. Expensive and usually single task.
		Load-unload and finishing operations often included.		
		Also specialized equipment, for example for welding in hostile (nuclear) environments.		

Table 1.7.25 Weldability of non-ferrous metals

Material	***Typical compositions***	***Welding processes***		***Pre-heat (°C)***
		Recommended	***Others***	
Aluminium alloys	Series 1000, pure Al	MIG, TIG, Gas, Resistance	Laser	Only for thick section
	Series 2000, Al–Cu	Resistance		
	Series 3000, Al–Mn	MIG, TIG, Gas, Resistance	Laser	
	Series 4000, Al–Si			
	Series 5000, Al–Mg	MIG, TIG, Resistance	Gas, Laser	
	Series 6000, Al–Mg–Si	Resistance	MIG, TIG	
	Series 7000, Al–Zn–Mg	Resistance, MIG, TIG		
	Series 8000, Al–Li	MIG, TIG, Laser	EB	
Magnesium alloys	Mg 8 Al 0.3 Mn 0.5 Zn	TIG	Resistance welding	Heat thin sections to prevent cracking. 260–380°C depending on alloy recommendations.
	Mg 6 Al 3 Zn			
	Mg 6 Zn 0.8 Zr (2 Th)			
	Mg 2.5 Ag 0.7 Zr (2.2 Ce)			
	Mg 2.2 Zn 0.6 Zr (2.7 Ce)			
	Mg 2.2 Zn 0.7 Zr (3 Th)			
	Mg 0.1 Al 2 Mn 0.1 Zn			
Copper and copper alloys	Tough pitch copper Deoxidized copper	MIG, TIG, electron beam.	Resistance, MIG, MMA	For arc welding preheat thick sections up to 700°C and conserve heat with insulating blanket.
	High conductivity copper			
	Brasses: <20% Zn	MIG Resistance	TIG	250–400 °C
	Brasses: 20% <Zn < 40%			
	Brasses: High tensile brass Additions Pb–Fe + Al			
	Tin bronzes: Phosphor bronze Cu 5–8% Sn	TIG, Resistance MIG, Electron Beam		Heavy sections 175–290 °C
	Tin bronzes: Gunmetal 5–10% Sn, 2–5% Zn			
	Tin bronzes: 9% Sn	Resistance good Arc welding fair		–
	Silicon bronzes: Cu 3% Si 1% Mn	TIG MIG Resistance		Pre-heat 20–65 °C

Table 1.7.25

Table 1.7.25 Weldability of non-ferrous metals - *continued*

Post-weld (°C)	***Remarks***	
Most welds left in as-welded condition. Heat treatable alloys may be re-heat treated.	Readily weldable, sheets and pressings	
	Not normally fusion welded	
	Readily weldable, building and transport applications	
	Not normally welded	
	Readily weldable, high strength welded applications	
	Widely welded but welds have low strength and care must be taken to avoid cracks.	
	Can be welded, structures with high mechanical strength through age-hardening	
	Can be welded, careful choice of filler wire to avoid cracking	
Stress relieve anneal wherever possible, particularly when likely to be exposed to wet or corrosive environments.	Weldable, but prone to stress corrosion.	Stress-corrosion in alloys containing aluminium. Ce, Th are used in Mg Zn Zr alloy to improve weldability.
	Prone to weld cracking	
	Welding of alloys containing zinc only satisfactory with thorium or rare earth additions (shown in brackets).	
	Weldable	Use helium-rich gas (MIG, TIG).
	Difficult to weld – high temperature weakness, porosity, low weld ductility.	If metal of identical composition is not required, use fillers of silicon or phosphor bronze or brazing J.b.2–2. Beware of galvanic attack in hostile environments.
	More suitable for welding, especially with copper fillers.	
15% Zn 260–280°C stress relieve to avoid contraction cracking.	Good weldability.	Zinc causes porosity. Use oxidizing conditions. Filler metal { tin bronze, phosphor-bronze / aluminium bronze } Toxic fumes evolved.
	Excessive zinc or 2% Pb gives hot shortness (weld-metal cracking)	
	Susceptible to lead and sulphur embrittlement. Aluminium impedes coalescence of liquid metal.	
Peen at 480–500 °C then rapid cool for maximum ductility.	Relatively good weldability.	
	Hot cracks and porosity (avoid overheating or joint stressing). Greater than 2% Pb gives hot shortness.	
	WELDING NOT RECOMMENDED AT OR ABOVE 9% Sn	
	Can be welded by arc, resistance methods. Avoid stressing joints. Filler as parent metal.	

Table 1.7.25 – *continued*

Table 1.7.25 Weldability of non-ferrous metals – *continued*

Material	***Typical compositions***		***Welding processes***		***Pre-heat (°C)***
			Recommended	***Others***	
Copper and copper alloys continued	Aluminium bronzes	Cu 5% Al Cu 10% Al 5% Fe 5% Ni	TIG, Resistance MIG, Electron Beam		Pre-heat 20–150°C
	Nickel silver	10–18% Ni 25% Zn	TIG, Resistance		20–110 °C
	Cupro-nickel	Cu–10 Ni Fe Cu–30 Ni Fe	TIG, MIG	MMA	Not normally required
	Copper beryllium	1.0–2.75% Be 2.5–0.2% Ni	Electron beam		
Nickel alloys	Commercial purity nickel Nickel–copper (Monels), e.g. 66 Ni–31.5 Cu–0.9 Mn–1.35 Fe 55 Ni–44 Cu Nickel–molybdenum–chromium (Hastelloys) Nickel chromium (Inconel 600) 60.5 Ni–23 Cr–14 Fe–1.35 Al Ni–Cr–Mo–Nb (Inconel 625)		TIG, MIG Manual arc Electron beam		Not normally required.
	Invar Fe–36 Ni Permalloy 79 Ni–4 Mo–1 Fe Mumetal Ni–14 Fe–5 Cu–4 Mo				
	Nickel–chromium–cobalt–aluminium–titanium (Nimonics). Ni–20 Cr 18 Co max. 10 Mo max. 5 Al max. 4 Ti max.		TIG, MIG		Anneal in inert atmosphere before welding.
Titanium alloys	α Alloys	CP grades	TIG, MIG, electron beam.	Resistance pressure, flash butt.	Weld in annealed condition.
	α–β Alloys	Ti–6 Al–4 V	TIG. Electron beam	Resistance pressure, flash butt.	
		Ti–4 Al–4 Mo–0.5 Si–2–4 Sn Ti–2.25 Al–11 Sn–Si Ti–2.25 Al–11 Sn–5–Zr–Si Ti–6 Al – 5 Zn–Mo–Si Ti–6 Al – 6 V – 2–Sn	WELDING NOT RECOMMENDED		
	β Alloys	Ti–13 V–11 Cr–3 Al		Electron beam, resistance pressure, flash butt.	

Table 1.725 – *continued*

Table 1.7.25 Weldability of non-ferrous metals - *continued*

<table>
<tr><th>Post-weld (°C)</th><th colspan="2">Remarks</th></tr>
<tr><td></td><td colspan="2">Good weldability</td></tr>
<tr><td>Not required</td><td colspan="2">Good weldability</td></tr>
<tr><td>Not required</td><td colspan="2">Readily weldable, argon gas</td></tr>
<tr><td>–</td><td colspan="2">Increasing beryllium decreases weldability. Poor heat conductors. Toxic fumes.</td></tr>
<tr><td rowspan="2">Necessary to avoid intergranular corrosion.</td><td>Solution hardened nickel alloys</td><td rowspan="3">Low ductility temperature range; avoid stressing welds.
Particularly susceptible to hot shortness with Pb, S.
Danger of oxide formation.
Generally good weldability – generally more viscous than steel except Hastelloys. This should be allowed for in joint designs.</td></tr>
<tr><td>Special purpose alloys.</td></tr>
<tr><td>Resolution treat and age after welding (some alloys can be solution treated before and aged after welding).</td><td>Precipitation-hardened alloys. Susceptible to heat-affected zone cracking.</td></tr>
<tr><td>Stress relieve 400–650°C</td><td>Good weldability. Slight reduction in weldability with increasing alloying additions. Good ductility.</td><td rowspan="4">Highly reactive, requires good shielding.
No filler is used below 2.5 mm thickness.
Otherwise parent metal or commercial purity titanium is used.</td></tr>
<tr><td>2 h at 538°C
Complete ageing during stress relief.</td><td></td></tr>
<tr><td></td><td>Air hardenable alloys.</td></tr>
<tr><td>Cold work and fully reheat treat.</td><td>Low strength welds, ductile as welded. Not widely used.</td></tr>
</table>

Table 1.7.25 – *continued*

Table 1.7.25 Weldability of non-ferrous metals – ***continued***

Material	*Typical compositions*	*Welding processes*		*Pre-heat (°C)*
		Recommended	*Others*	
Lead and zinc–lead alloys	Lead, zinc Lead–antimony Lead traces Cu, Ag, Te	Gas resistance	TIG	
Refractory metals	Molybdenum and tungsten alloys	Electron beam.	TIG	150 for TIG welded tungsten alloys.
	Niobium Tantalum Tantalum alloys With up to 15%, Hf, Mo, W, Re.	TIG, electron beam.		
	Zirconium	TIG		Weld in annealed condition.
	Beryllium	Electron beam pressure welding, TIG		

Table 1.7.25 – ***continued***

Table 1.7.25 Weldability of non-ferrous metals - *continued*

Post-weld (°C)	***Remarks***		
	Equal to base metal in corrosion resistance and strength. Lead easy, Pb–Sb more difficult. Zinc difficult. Fillers of parent metal.		
	Brittle – avoid restraint on joints. 0.05% oxygen in argon arc. Grain growth occurs.	Strong reactivity demands very thorough protection. Weld pool prone to recrystallization and has low ductility.	
Stress relieve 1 h at 816°C	Broadly similar to titanium. Can weld tantalum alloys to give very high room temperature bend strength.		
Stress relieve 400–650°C	See titanium. Zirconium requires even better protection (chamber).		
	Brittle – avoid restraint on joints. 0.05% oxygen TIG. Grain growth occurs.	Satisfactory joints obtainable.	Toxic fumes

Table 1.7.25 – *continued*

Table 1.7.26 Weldability of steels

Steel type	*% Typical composition*						*Recommended welding processes*
	C	*Mn*	*Cr*	*Ni*	*Mo*	*Others*	
Mild steel	<0.15	<0.5	–	–	–	–	Arc welding, (MIG, TIG, MMA, SA), oxy-acetylene, resistance, power beam. Electroslag.
Vanadium steel	<0.25	6.9–1.4	–	–	–	Si 0.25–0.9 V 0.03–0.08	
Carbon–manganese steel	0.15–0.025	<1.4	–	–	–	–	
	<0.2		–	–	–	–	
Medium carbon steel	0.25–0.6	0.6–1.3	–	–	–	Si<0.35	
High strength low alloy low carbon (HSLA)	0.2	0.4	1.5	3.0	0.7	–	
Carbon–molybdenum	0.13	0.5	–	–	0.5	Si 0.25	
Medium carbon and low alloy	0.35	<1.8	<4	<3	<0.5	–	
Chromium–molybdenum steels	0.15	–	1.9	–	0.5–1.5	–	
'Cryogenic' steels	<0.1	–	–	9	–	–	
	<0.2	–	–	5	–	–	
Martensitic stainless steels AISI 410 AISI 420	<0.15 0.15–0.2	<1.0 <1.0	11.5–13.5 12.0–14.0	– –	– –	Si<1.0 Si<1.0	Arc welding, (MIG, TIG, MMA) Resistance, power beam.
Martensitic stainless steels AISI 440	0.6–1.2	<1.0	16.0–18.0	–	–	Si<1.0	
Ferritic stainless steels	0.08–0.2 max.	1.0–1.5 max.	12–27	–	–	Si 1.0 max.	

Table 1.7.25 Weldability of steels - *continued*

<table>
<tr><th>Pre-heat (°C)</th><th>Interpass (°C)</th><th>Hold cool (°C)</th><th>Post-weld (°C)</th><th colspan="2">Remarks</th></tr>
<tr><td>–</td><td>–</td><td>–</td><td>–</td><td colspan="2">Excellent weldability – no precautions necessary with rutile and basic electrodes.</td></tr>
<tr><td>–</td><td>–</td><td>–</td><td>–</td><td colspan="2">Low carbon grade electrodes.
Minimum heat input 30 kJ/mm for 30 mm thickness, unless preheating or low hydrogen electrodes are employed</td></tr>
<tr><td>100–200</td><td>–</td><td>–</td><td>–</td><td colspan="2" rowspan="2">For thin sections, no pre-heat required.
For thick sections, low hydrogen electrode, high heat input, pre-heat.</td></tr>
<tr><td>100–250</td><td>–</td><td>–</td><td>–</td></tr>
<tr><td>250–350</td><td>–</td><td>–</td><td>–</td><td colspan="2">High heat input. Low hydrogen electrodes.</td></tr>
<tr><td>250–300</td><td>–</td><td>–</td><td>–</td><td colspan="2">High heat input. Low hydrogen electrodes.</td></tr>
<tr><td>150–300</td><td>–</td><td>–</td><td>650–700</td><td colspan="2">Filler as parent metal.</td></tr>
<tr><td>200–350 thick
150 thin</td><td>200–350</td><td>–</td><td>550–600</td><td colspan="2">Filler rod lower carbon than parent metal.
Make test welds with highly hardenable low alloy steels.</td></tr>
<tr><td>250–300</td><td>–</td><td>–</td><td>Normalize 870–980
Temper 680–770
Anneal 840–900
} Stress relieve 700-730</td><td colspan="2">Filler as parent metal, but lower Cr.
Post-weld treatment depends on properties.</td></tr>
<tr><td>–</td><td>–</td><td>–</td><td>–</td><td>Electrodes 50 Ni–16 Cr/Fe</td><td rowspan="2">Pre-heat/post-weld not required up to 50 mm thickness.</td></tr>
<tr><td>–</td><td>–</td><td>–</td><td>–</td><td>Electrodes 25 Cr–20 Ni Fe–25Cr–12 Ni Fe</td></tr>
<tr><td>250
250</td><td>250–350
250–350</td><td>80–100
150–180</td><td>700–750
700–750</td><td colspan="2" rowspan="2">No special precautions if C<0.1%.
Austenitic fillers or fillers for matching strength.
Require pre-heat to avoid H_2 cracking.
Tendency to harden when cooled from high temperatures.</td></tr>
<tr><td colspan="4">WELDING NOT RECOMMENDED</td></tr>
<tr><td>200°C</td><td>–</td><td>–</td><td>Anneal after welding 700–850°C.</td><td colspan="2">Limited transformation to martensite via austenite can occur.
Grain growth. Two factors reduce ductility.
Most weldable grades contain Al, Nb to retain ferrite.</td></tr>
</table>

Table 1.7.26 – *continued*

Table 1.7.26 Weldability of steels – *continued*

Steel type		*% Typical composition* C	*Mn*	*Cr*	*Ni*	*Mo*	*Others*	*Recommended welding processes*
Precipitation-hardened stainless steels	W	0.07	–	17	7	–	0.2 Al 0.7 Ti	Arc welding (MIG, TIG, MMA) resistance, power beam.
	17–4 PH	0.04	–	17	4	–	4 Cu–0.8 Nb	
	15–5 PH	0.04	–	15	5	–	4 Cu–0.3 Nb	
	17–7 PH	0.07	–	17	7	–	AR 1.0	
	15–7 Mo	0.07	–	15	7	2	AR 1.0	
	14–8 Mo	0.04	–	14	8	2	AR 1.0	
	AM 350	0.08	–	17	4	3	–	
	AM 355	0.12	–	16	5	3	–	
	17–10 P	0.12	–	17	10	–	0.25 P	
	HN.M	0.3	3.5	19	9	–	0.3 P	
Austenitic stainless steels	AISI 302	0.15	–	18	8	–		Arc welding (MIG, TIG, MMA, SA,) resistance, power beam.
	301	0.15	–	17	7	1		
	303	0.15	–	18	8	–	S or Se 0.15 min.	
	304 L	0.08	–	19	10	–		
	304	0.03	–	19	10	–		
	305	0.12	–	18	11.5	–		
	309	0.20	–	23	13.5	–		
	310	0.25	–	25	20.5	–		
	314	0.25	–	24.5	20.5	–	Si 2.75	
	316	0.08	–	17	12	2.5	321	
	321	0.08	–	18	10.5	–	Ti 5 × C min.	
	347	0.08	–	18	11	–	Nb + Ta = 10 × C max.	
	–	0.15	3.5–10	18	3.5–6	–	N 0.25 max.	
Maraging steels		0.03	–	–	18	5	8 Co	Arc, (MIG, TIG).

Table 1.7.26 – *continued*

Table 1.7.26 Weldability of steels - *continued*

Pre-heat (°C)	***Interpass (°C)***	***Hold cool (°C)***	***Post-weld (°C)***	***Remarks***
	–	–	+ +	Good weldability. If joint strength not required equal to parent metal, use 18 Cr 8 Ni type filler. For $^{+}_{+}$ precipitation-hardened joint use similar composition and precipitation harden after welding.
–	–	–	650–870 then refrigerate or cold work.	Good weldability. Weld in annealed condition. Filler depends on mechanical properties required. Can be precipitation hardened stainless steel. Post-weld treatment precipitates carbides and martensite.
ARC WELDING NOT RECOMMENDED				Flash welding possible.
Not recommended			950–1150°C If undertaken – or stress relieve below 650°C (to avoid weld decay).	Excellent weldability. Filler materials of similar composition but higher alloy content. **Weld cracking** is avoided by ensuring that the weld contains 5–10% ferrite. **Weld decay** (chromium carbide precipitation) leads to loss of corrosion resistance. Minimized by C <0.03% or add Nb, Ti (stabilized steel) and pay regard to post-weld heat treatment. **HAZ liquation cracking**: low energy input fine grain size ferrite $\geq$ 5%
–	–	–	Age after welding at 500°C.	Soft martensite matrix. No risk of decarburization, distortion or cracking. Very poor weldability.

Table 1.7.26 – *continued*

Table 1.7.27 Effects of microalloying additions on weldability of steels

Microalloying addition	*Hydrogen cracking*	*Solidification cracking*	*Heat-affected zone toughness*	*Weld metal toughness*
Vanadium	Reductions in carbon content can offset the increased risk of cracking.	Possibility of advantage	Can be beneficial in as–welded condition, but in presence of Nb, may be detrimental, especially when heat treated.	Complex. Grain refinement. Precipitation-hardening
Niobium	Not known to be detrimental.	Detrimental above 0.13%.	Generally, can be tolerated up to 0.04 %	As above
Titanium	NK	NK	NK	As above.
Zirconium	NK	NK	NK	Thought to be detrimental.
Boron	Risk of cracking at higher carbon contents.	Detrimental with increasing carbon content.	NK	Variable.

NK = Not known.

Table 1.7.28 Weldability of other ferrous metals

	Composition	*Weld processes*	*Procedure*	*Remarks*
Grey cast iron	3.66 C 1–2.1 Si 0.50 Mn 0.30 S 0.024P High carbon, high silicon, grey irons are ferritic. Low carbon, low silicon, are pearlitic.	MMA, MIG, TIG, oxy-acetylene. Filler rods 3.5C–3Si–0.6P	Ferritic grey irons are readily welded but complex irons are difficult to weld. Increasing silicon above 2% also decreases weldability. Pre-heat to 400–600°C or to 1000°C with 60/40 brass filler (bronze welding). Pre-heat to avoid hard brittle deposit. Difficult to machine unless pre-heated. Prone to cracking – cool slowly and/or stress relieve at 600°C when hot. Can also peen when red hot.	Less weldable than carbon steel owing to higher carbon and silicon content leading to lower ductility. More metallurgical complications in weld metal and heat-affected zones. Massive free carbide and martensite in weld deposit. Welding used in repair of castings and production of welded assemblies. Filler materials are cast iron, low carbon steel, nickel base alloys, copper base alloys. With MIG, base wire electrodes of low carbon steel and nickel base alloys are also used.
Ductile irons nodular or spheroidal	C 3.3–4.0 Si 2–3.0 Mn 0.2–0.6 P 0.08 max. Ni 0–2.5 Mg 0.02–0.07	Most fusion processes	Welding should be done on fully annealed material. Composition affects weld quality. For MIG a 60Ni–40Fe filler is best. Pre-heat to 250°C. Interpass 80–100°C.	
Austenitic nodular irons	C 3.0 Si 3.2 max. Mn 0.8–1.5 Ni 18–22 Cr 0–2.5			
Malleable cast iron	Ferritic C 2.0–2.7 Si 1.4–0.8 Mn 0.55 max. P 0.18 S 0.15	Use low hydrogen rods. Can be successfully brazed or soldered.	Pre-heat or other procedures to minimize heat buildup. Heat concentration should be held to a minimum. Very important to clean before welding to avoid porosity and low joint strength.	
	Pearlitic C 2.35–2.50 Si 0–1.5 Mn 2.28–0.48 S 0.16–0.2 P 0.06–0.10			
White and alloy cast irons	Chilled and white cast irons	WELDING NOT RECOMMENDED		
	Corrosion resistant cast irons high Si, high Cr, high Ni		Welding confined to minor repair of castings subject to availability of filler materials. For more extensive welding properties of material should be determined.	
	Heat resistant cast irons		As above.	
Iron nickel chromium heat resisting alloys	Precipitation-hardened (e.g. Incoloy) 800: 0.4C–32Ni–20Cr–46Fe Al Ti 802: 0.35C–32.5Ni–21Cr–46Fe Al Ti 901: 0.05C–43Ni–13.5Cr–34Fe–6Mb Al Ti	TIG most widely used especially in thin sections. Also manual arc (<12 mm thickness) is used. MIG is employed where TIG is not practical. Submerged arc is employed for thick sections provided a suitable flux is available. Fillers are employed to produce weld metal of similar composition to the base metal.	Heat input kept low. Rapidly cool to maintain weld ductility. Precipitation-hardenable alloys containing Al, Ti, Ni are very difficult to weld. Can obtain high joint efficiency up to 2.5 mm thickness. Sensitive to cracking and require pre-weld and post-weld heat treatment and selection of most suitable filler material. Solid solution strengthened alloys are more easily welded.	Usual range of temperature 1200–1400°C. Limits the selection of filler materials and possible pre- and post-weld treatment.

Table 1.7.29 Design of butt joints

Joint type	***Configuration***	***Illustration***	***Remarks***
Butt joints	Planar – may be truly planar or meet in a curve or angle. Smooth transmission of stress across joint. Bad weld shape or poorly fused welds can reduce fatigue life.	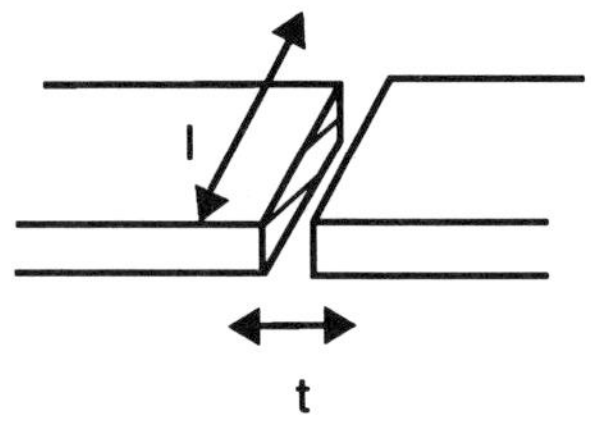 In general l > 10t	(a) Thin sheet. Up to 3 mm thickness with TIG. Up to 8 mm thickness with plasma. With electron beam and laser thickness depends upon power available. With these processes careful alignment of parts is required (jigging), especially in mechanized welding. Metal arc, MIG, MAG can also be used. The fit-up is not so important in manual welding and gaps up to sheet thickness can be tolerated (maximum 3 mm).
		 V groove weld 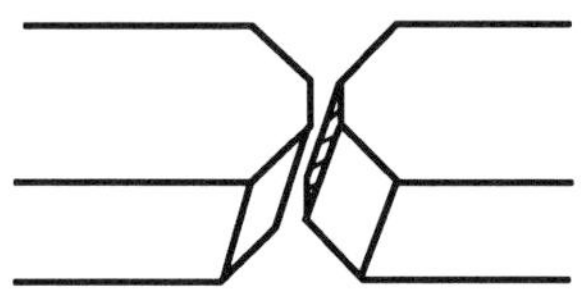 Double V groove weld	(b) Thicker welds. Weld from one or both sides TIG up to 5 mm thick. MIG, submerged arc, high current CO_2 from one or both sides up to 12 mm thick. Use grooving-metal arc up to 5 mm, mechanized welding to 25 mm. Multi-pass processes with arc methods using single or double grooves (i.e. weld from one or both sides). There is less distortion welding from both sides. Electroslag welding – single pass welds up to 200 mm thickness but restricted range of materials. Electrogas – higher toughness welds then with electroslag but restricted to range 15–40 mm thickness.
			(c) Short butt joints Normally made by resistance butt, or flash welding because it is difficult to rapidly start and stop a fusion welding process.
	Tubular – large tubes are effectively planar, provided they may be rotated. Differences to the above occur as the diameter is reduced.	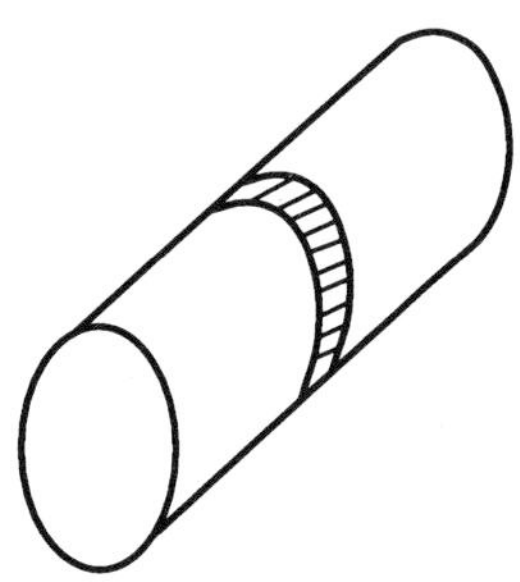 Tubular butt	(a) Large diameter pipes can be treated as planar, if they can be turned. Use MIG, TIG, metal arc, submerged arc, electron and laser beam.
			(b) Small diameter pipes cause problems; positioning the molten pool and heat build up. Can use EB laser pulsed TIG, friction or flash welding. Diffusion bonding is also a possibility if flash from friction or flash welding is not required, although these two processes give best results.

Table 1.7.29 Design of butt joints – *continued*

Joint type	*Configuration*	*Illustration*	*Remarks*
Butt joints (*continued*)	Wires, rods and sections.	 Section butt	(a) Wire – resistance butt, spark discharge, or cold welding where suitable. Friction welding also possible. (b) Bar – resistance butt, spark discharge or more usually, flash and friction. Arc welding is also possible with precautions. (c) Section – Solid shapes – flash, thermal diffusion bonding. Extruded shapes – flash or fusion.

Table 1.7.29 – *continued*

Table 1.7.30 Design of lap and corner joints

Joint type	*Configuration*	*Illustration*	*Remarks*
Lap joints	Planar – normally in sheet metal. Joint is inferior to butt joint when stressed tranversely. Useful when there is a requirement for a close fitting tolerance. Stiffening attachments in sheet are often made by lap joints.	 Fillet weld	Normally recommended with brazed joints. Fillet welds may be made by TIG, MIG, CO_2, submerged arc, metal arc. Resistance spot and seam welding only suitable for lap joints. Mash seam welding is a process where the overlap is subsequently crushed down to bring the sheets into the same plane.
	Non-planar – joints between rod, wire or tube and a flat surface.	 Fusion weld	Cross wire welding – projection welding Wires to flat shapes – projection welding or fusion welding, e.g. metal arc, TIG, MIG, CO_2.
Corner joints	Planar corner joints	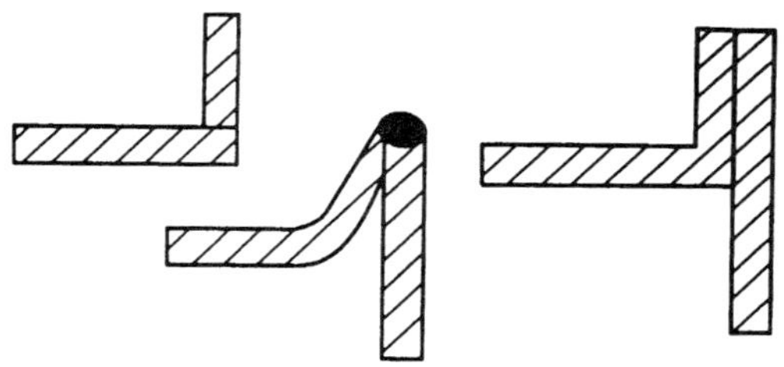 Thin sheet corner weld	Thin sheet or plate – gas, TIG or power beam welding. Metal arc or MIG only suitable for thickness over 2 mm. With lap corner joints it is possible to employ brazing or resistance spot/seam welding. Thick plate – any arc process or electron beam welding are possible. With arc processes a multipass weld is made, whereas a single weld is possible with EB, but this gives poorer through thickness weld properties.
	Frame joints – corners made with sections of rod bar or tube.	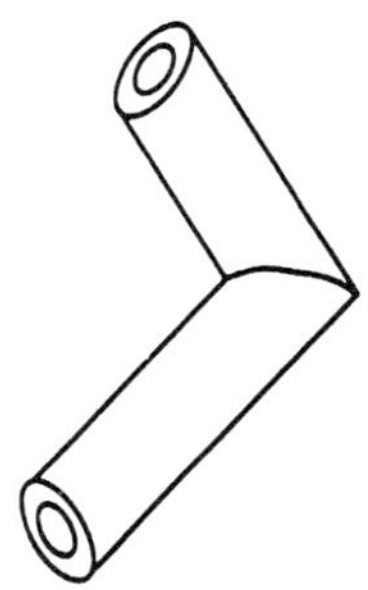 Frame joint made from tube	Flash welding gives strongest and quickest joints. Brazing gives good appearance but poor strength. Fusion weld processes are suitable in many cases, but edge preparation to accommodate deposited metal may be required. High current MIG, CO_2 are used except on small sections. Cored wire may be used on heavy sections. Very thick joints are made by metal arc, electroslag or thermit.

Table 1.7.31 Design of tee joints

Joint type	***Configuration***	***Illustration***	***Remarks***
Tee joints	**Planar** A number of variations on the basic form exist which in part determines the type of welding operation used.	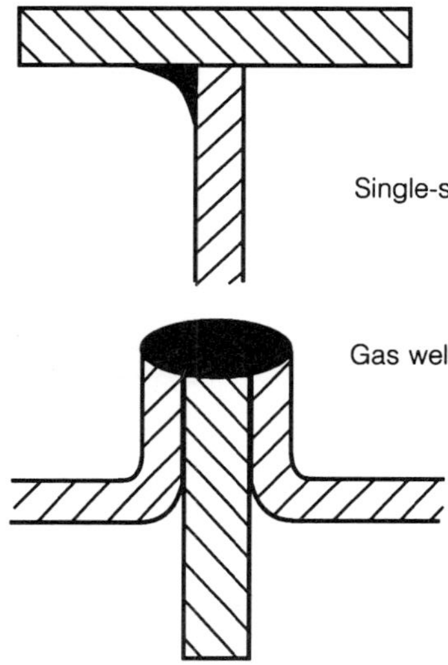 	Double- and single-sided fillets. – Fusion welding processes with filler material employed. For forms other than sheet metal arc, MIG, CO_2, and submerged arc are more suitable than gas and TIG. Spot welds. – Arc spot or projection welding. Short-length thick welds. – Electroslag is used. T-joints in heavy plate. – Metal arc, CO_2, cored wire or submerged arc may be used. See also brazed joints (size of fillet large compared with sheet thickness).
	Structural Two structural members at right angles.	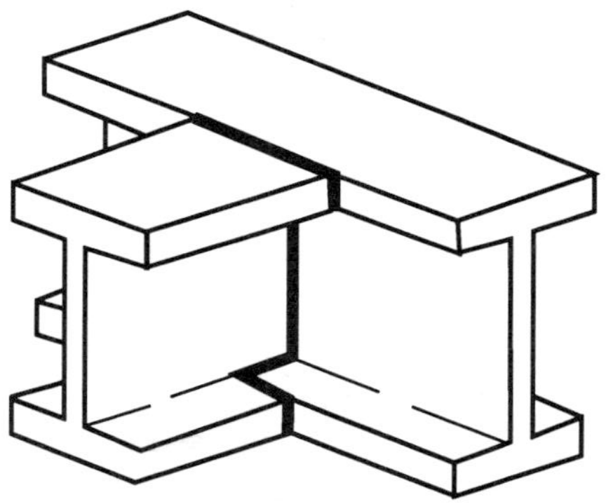	Generally the preferred processes are those which give good access and are operable *in situ*. These are metal arc, MIG, CO_2, and cored wire. Structural members usually give a mixture of simple butt and fillet welds.

Table 1.7.32 Design of insert and stud joints

Wires and tubes attached to sheet		***Inserts in sheet***	
 (a)	Friction weld	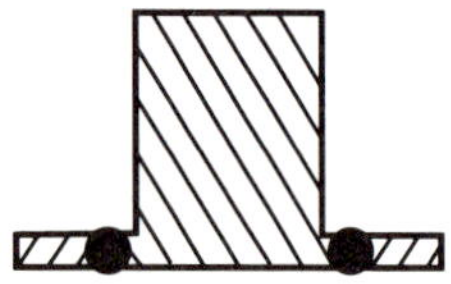 (a)	Mechanical fusion weld. Very good fatigue properties but difficult to make.
 (b)	Not penetrating sheet – microfriction, brazing, capacitor discharge.	 (b)	Where weld inaccesible from one side an integral backing is used. As (a), but more prone to root defects.
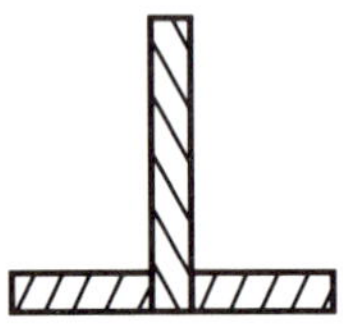 (c)	Penetrating sheet – TIG, plasma, brazing (arc spot, power beam).	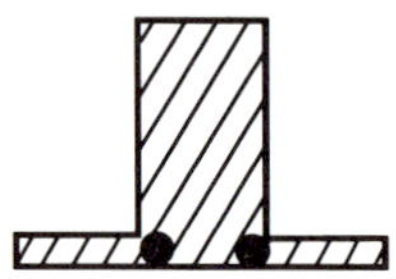 (c)	It would also be possible to braze both these designs by slight modifications.
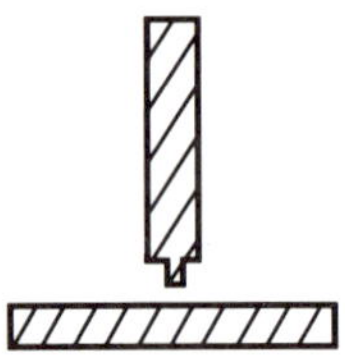 (d)	Projection weld	 (d)	TIG, plasma, power beam weld. Heat source may be magnetically deflected to avoid rotation of work-piece.
 (e)	Fillet weld – metal arc, TIG, CO_2. Tendency to distort flat member.	(e)	See Table 1.7.31, Planar
 (f)	Arc stud – weld. Circular cross-section only.		
 (g)	Fusion weld with no colour – recommended in preference to (e) due to improved heat distribution around joint.		

Table 1.7.32

Table 1.7.33 Design of structural and heavy joints

Joint type	***Illustration***	***Remarks***
Structural and heavy joints	(a) 	Metal arc, MIG, CO_2, TIG is used with non-ferrous metals but with thin sheet there is a risk of distortion. A mechanically weaker version of the joint excludes the upper weld. Submerged arc or cored wire arc used for larger joints.
	(b) 	Flash weld (any cross-section but requires stub-raised on plate).
	(c) 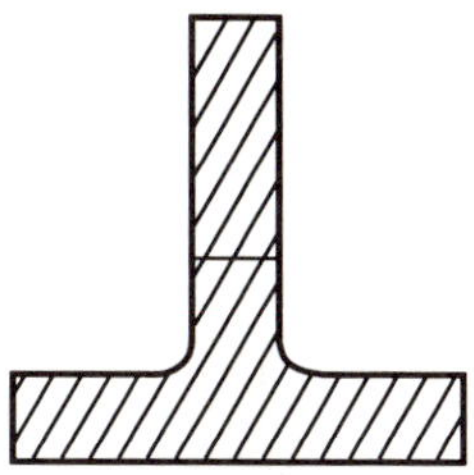	Friction weld – (circular cross-section)
	(d) 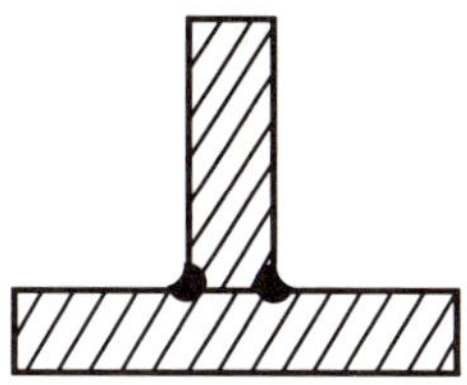	Set-on connection. One fillet weld. Gas, TIG or MIG processes.
	(e) 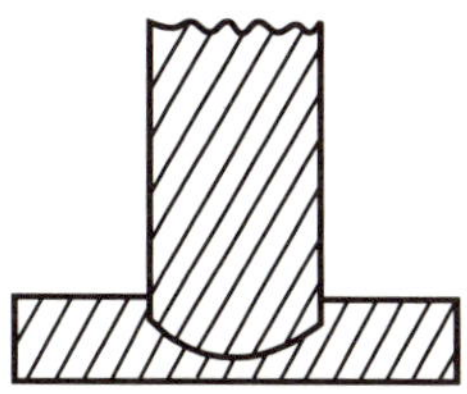	Projection weld; in making the weld the corner collapses, giving a good annular joint.
	(f) 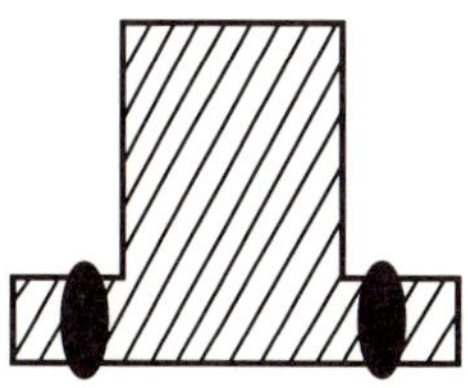	Butt welded nozzle. Gas, TIG, MIG processes and electron beam (for high penetration in thick sections).

Table 1.7.34 Design of tube and sheet joints

Tube through plate	*Tube, closures, cap on tube*
(a)  Thick tube sheet. Metal arc or TIG (crevice left)	(a)  TIG, plasma or gas (no filler). For larger sizes metal arc, MIG and CO_2 are used, but the disc is reduced in diameter to allow for filler deposit.
(b) 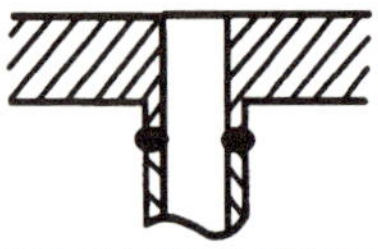 Thick tube sheet. TIG usually, using bore welding torch (no crevice).	(b)  As for (a); electron beam and laser welding are additional possibilities.
(c)  Thick tube sheet. Brazed or explosively welded – no crevices.	(c)  Resistance seam weld if accessible to seam welding wheel. Can use resistance stitching method.
(d) 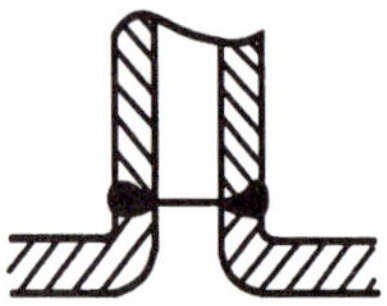 Sheet or light plate. Sheet is pressed out to allow a butt weld, metal arc or TIG process.	(d)  Metal arc, MIG or CO_2 are used. TIG or gas melt welds are unsuitable because of thermal expansion problems.
(e)  Fillet weld with metal arc, MIG, CO_2, cored-wire. Inferior to (c).	(e)  If there is a gap metal arc, MIG or CO_2 may be employed. TIG or EB may be used for close joints
	(f) 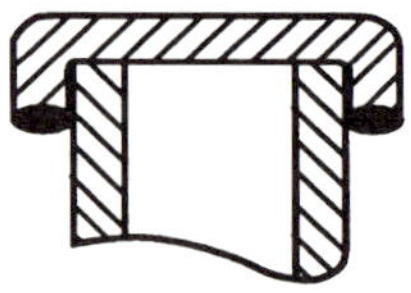 Suitable for TIG, MIG CO_2 metal arc, high speed submerged arc (large diameters).
	(g)  Projection weld with the insert providing natural current concentration for the weld.
Various modifications on these designs are available for improving the heat distribution obtainable with this process.	Tube closure joints are often mechanically welded as this merely involves reducing the joint under the welding heads. Structures (c), (d), (e) and (g) could also be brazed. It is also possible to design tube closures for friction/flash welding.

Table 1.7.34

Table 1.7.35 Weld defects

Defect	*Description*	*Cause*	*Remedy*
Porosity	Cavities from freezing-in of gas in the weld.	Causes are material or process. 1. Moisture or surface coatings, e.g. paints 2. Inadequately dried electrodes or fluxes (MMA, SA) 3. Poor gas shielding (TIG, MIG)	1. Clean joint edges immediately before welding 2. Dry electrode/fluxes according to the manufacturer's recommendations 3. Check gas flow rate; clean nozzle of spatter.
Hydrogen-induced cracking	Cracking develops by the diffusion of hydrogen from the weld metal to the highly stressed, hardened HAZ	Factors increasing the risk: 1. Hydrogen generated from electrode coating and flux, moisture and grease. 2. Hardness of the weld metal or HAZ, (CE value and arc energy input). 3. High level of residual stress after welding. 4. Thick material increases cooling rate.	1. Choice of less susceptible material or weld metal. 2. Increase energy input of welding process. 3. Use thinner sections. 4. Use low hydrogen electrodes and/or gas shielded process. 5. Pre-heat and maintain interpass temperature, if all else fails.
Solidification cracking	Lowering of ductility of weld metal by the presence of residual liquid films, combined with solidification strains leads to cracking. Hot cracking susceptibility of steels= $\frac{\%C[\%S + \%P + \%Si/25 + \%Ni/40]}{3\%Mn + \%Cr + 2(\%Mo + \%V)}$	1. High restraint 2. Contamination with sulphur and phosphorous. 3. Poor weld size or weld bead shape. 4. Incorrect weld metal for accommodating plate impurity.	1. Reduce strain by improving fit-up or step by step welding. 2. Weld bead should have adequate depth for the root thickness. 3. Use basic-coated electrodes. 4. Use more resistant weld metals and adjust welding procedures to minimize dilution from the parent metal.
Heat-affected zone cracking (liquation, burning).	Local melting in region adjacent to fusion boundary. Microcracks formed which do not heal in cooling.	Sulphur and phosphorus.	Keep Mn%>20S% in low C(<0.2%) steels. In other steels may need to be higher. Avoid submerged arc, electroslag processes if possible.
Lamellar tearing	Cracking in systems where the through thickness ductility is very low. Stepped appearance parallel to the plate surface generally associated with planar inclusion.	Joint design is such as to cause high strains through the thickness. Planar inclusions. For a given inclusion population lamellar tearing is more likely as the strength of the steel increases.	Improve design. Use forged products (reduces lamellar distribution of inclusions). Plate selection – vacuum degassed plate. 'Grooving and buttering'
Reheat cracking	During heating a temperature range exists where a marked strength decrease develops.	During stress-relieving or service. Time necessary for cracking depends upon temperature. Alloying additions made to provide good low and high temperature strength often lower high temperature ductility. Geometry. Presence of long-range stresses.	Profile the toes of fillet welds. Control sulphur content. Use low strength weld metal. Butter weld surfaces with weld metal of improved creep ductility. Optimize alloy compositions.

Table 1.7.35 Weld defects – *continued*

Defect	*Description*	*Cause*	*Remedy*
Slag inclusions	Slag or foreign matter entrapped in weld bead usually between adjacent runs.	Poor welding technique. Inaccessible joints. Inadequate cleaning of weld between runs. Flux type. Dirty surface condition.	Correct electrode size, shape and overlap of weld beads to prevent pockets for slag. Accessibility for correct electrode manipulation is essential. Change flux. Clean surfaces between runs.
Lack of root fusion.	Lack of root fusion in butt welds and in the corner of fillet welds, more likely in MMA, MIG, SA, where weld metal automatically deposited.	Incorrect joint design too large root face. Incorrect electrode size. Current too low to penetrate thickness.	Joint must be accessible for manipulation required to make the joint. Thickness of parts must not differ such that the thinnest section melts away before fusion on the thicker sections can occur. Too large or small an electrode can cause lack of fusion or inadequate penetration. Correct current to penetrate but not too great to cause flooding.
Undercut	Wastage of parent metal alongside a weld. May lead to slag entrapment and the initiation of cracks.	Excessive welding current. Joint design. Damp electrodes. Heavy mill-scale. Heavy buildup.	Reduce current. Joints must be accessible. Use dry electrodes. No mill-scale. No heat buildup.

Table 1.7.35 – *continued*

Table 1.7.36 Weld inspection methods

Method	*Principles*	*Advantages*	*Limitations*
Visual	Examination for obvious flaws. Usually carried out before detailed inspection.	Good assessment of quality of workmanship from its general appearance.	Only large, surface flaws can be detected. Time consuming. Messy.
Penetrant	Penetrant applied to surface which penetrates into cracks. After cleaning, an absorbant coating is used to locate flaws.	Very sensitive technique for fine, surface flaws. Can be applied to all metallic and non–metallic materials.	Interpretation can be difficult. Surface preparation needs to be adequate, otherwise flaws will be masked.
Magnetic particle	Component is magnetized to produce magnetic lines of force. Flaws will produce secondary magnetic poles which create flux leakage across the flaw. Magnetic particles applied to the surface will be attracted to flux leakage showing location of flaw.	Only limited training required to be proficient. Simple and inexpensive technique. Immediate indication of flaws. Surface and sub–surface defects revealed. Only limited pre–cleaning. Requires limited training.	Limited to ferromagnetic materials, i.e. not stainless steel. Only surface and sub-surface flaws revealed. Requires electricity. Messy.
Eddy current	Eddy currents induced in metals by alternating magnetic fields. The fields which are parallel to the surface are disturbed by the presence of flaws. Changes in eddy current field recorded on meter.	Simple technique for detecting surface and near-surface flaws. Clean technique.	Requires operator skill in interpretation. Requires electricity.

Table 1.7.36 Weld inspection methods – *continued*

Method	***Principles***	***Advantages***	***Limitations***
Radiography	X–ray or gamma radiation is passed through the component and the emerging radiation is recorded on plastic film coated with silver halide. Differences in thickness and presence of flaws in the body of the component are recorded.	Internal flaws can be detected. Skilled radiographer can identify type of flaw, e.g. cracks, porosity, slag inclusions and root defects. Permanent record.	Training and experience required in interpretation of radiographs. Stringent safety precautions for radiation. Time consuming, disruptive to production. Expensive. Access required from both sides.
Ultrasonic	Ultrasonic mechanical vibrations (2–5 MHz) transmitted into the material. Reflections or echoes, from flaws and surfaces are picked up by a probe. Signal is analysed by means of cathode ray tubes (CRT).	Internal flaws can be detected and sized. No limitation on section thickness. No radiation hazards. Planar defects can be detected irrespective of their orientation.	Surface defects difficult to detect. Multiple reflections cause difficulty in interpretation. Requires significant level of operator skill in interpretation of signals. Spurious indications may lead to false interpretation and unnecessary repairs.

Table 1.7.37 Processes for welding thermoplastics

Process type	***Description***	***Advantages***	***Limitations***
Heated tool welding	Heated tool, e.g. plate, clamped between surfaces to be joined until they melt. Surfaces then brought together under controlled pressure during cooling.	Joint has at least 90% of strength of substrate. Machines may be semi- or fully automatic; pneumatic or hydraulic. Close tolerance moulding not required. Batch consistency not critical.	Relatively slow process, 30–50 s, (can be offset by multi-impression tooling).
Ultrasonic welding	High frequency vibrations (20–40 Hz) are applied to two surfaces causing them to rub together. Generated heat produces firm weld by melting. Also used for staking metal or other material to plastics; and insertion of metal into plastics.	Very short cycle times (<1 s). Very clean process Very good results with polystyrene, ABS, polycarbonate, PPO, acetal.	Special design considerations necessary, for both parts and joint design. Plastics require high Elastic Modulus to transmit energy. Less suitable materials include nylon, PVC, polypropylene, acrylics.
Hot gas welding	A stream of heated gas is used to heat the parent material (joint) and filler rod. Components are cleaned and a rod of parent material is used. Parent material and rod are heated simultaneously and softening surfaces are forced together to form homogenous bond.	For many purposes compressed air is satisfactory; nitrogen can be used for higher joint strength. Suitable substrates include PVC, polyethylene, acrylic, polycarbonate,	Cleaning required. Oxygen sensitive plastics require nitrogen gas to be used.
Dielectric (CRF) Welding	Surfaces to be bonded are placed between electrodes. AC supply (1–100 MHz) causes polar molecules to spin and the material to heat up. Radio transmitter (27 MHz) connected to electrodes which clamp sheets as in resistance welding and melting occurs to form weld.	High speed process (seconds). Suitable for thin sheets. Most thermoplastics readily welded including PVC, EVA, polyurethane, nylon, ABS.	Not suitable for thick sections. Intermittent welding operation, not continuous process. Not suitable for many thermoplastics, e.g. polyolefins.
Friction	Heat is generated, by rotating or linear oscillating, one of the parts whilst holding the other stationary. As soon as interface is molten, movement stopped and weld formed under pressure. Motion can also be linear, i.e. by oscillating one of the parts.	Fast cycle times (10 to 15 s) No joint preparation. Suitable for large items.	One part must be rotated or moved linearly.

Table 1.7.37

Table 1.7.38 Weldability of thermoplastics by the hotplate welding process

	Polypropylene	Acetal	Nylon	Polycarbonate	Thermoplastic polyester	PPO (Noryl)	ABS	SAN	Polyethylene	Polystyrene	Polysulphone	Polyethersulphone	Polyurethane (TP)	EVA	PVC	Acrylic
Polypropylene	X													X	X	
Acetal		X														
Nylon																
Polycarbonate				X			X									
Thermoplastic polyester					X											X
PPO (Noryl)						X										
ABS				X			X									X
SAN								X		X						
Polyethylene									X					X		
Polystyrene								X		X						
Polysulphone											X					
Polyethersulphone												X				
Polyurethane (TP)													X			
EVA	X								X					X		
PVC	X														X	
Acrylic					X		X									X

X = suitable.

Table 1.7.39 Weldability of thermoplastics by the ultrasonic welding process

	ABS	ABS/PVC	Acetal	Acrylics	Acrylic/PVC	ASA	Celleulosics	PPO (Noryl)	Polycarbonate	Polyethylene	Polyimide	Polypropylene	Polystyrene	Polysulphone	PVC	SAN
ABS	X	(X)		X	(X)	(X)							(X)			(X)
ABS/PVC	(X)	(X)		(X)	(X)	(X)			X						(X)	(X)
Acetal			X													
Acrylics	X	(X)		X	(X)	(X)			(X)							(X)
Acrylic/PVC	(X)	(X)		(X)	X	(X)									(X)	
ASA	(X)	(X)		(X)	(X)	X							(X)			(X)
Cellulosics							X									
PPO (Noryl)								X					X			(X)
Polycarbonate		X		(X)					X							
Polyethylene										X						
Polyimide											X					
Polypropylene												X				
Polystyrene	(X)					(X)		X					X			(X)
Polysulphone														X		
PVC		(X)			(X)										X	
SAN	(X)	(X)		(X)		(X)		(X)					(X)			X

X = suitable. (X) = under special conditions.

Table 1.7.39

Table 1.7.40 Characteristics of solid state welding processes for ceramics

Process	*General characteristics*	*Limitations*	*Applications*
Diffusion	Solid phase bond achieved via atomic migration. No macro-deformation although limited plastic deformation may occur at the interface. For materials of differing thermal expansion, thin interlayer material, of intermediate thermal expansion coefficient, is required.	High temperatures (0.6 to 0.8 T_m) and high pressures required. Components to have similar thermal expansions otherwise interlayer required.	Silicon nitride and silicon carbide and also, to metals such as aluminium- and nickel-based alloys (aerospace). Aluminium and aluminium nitride bonded to copper (electronics) using gold interlayer.
Friction	One component held stationary (usually the ceramic) while the other rotated or moved linearly. Slight pressure to give friction heating, but after rotation stage force increased to forge the weld. Final weld contains flash or collar from material extruded during rotation and forging stages.	Ceramic to metal components. Restricted component shape, i.e. cross-section suitable for rotating or linear motion.	Various ceramic to metal joints, especially aluminium, including Al_2O_3–Al, SiC–Al, ZrO_2–Al, AlN–Al
Microwaves	Microwaves will selectively heat the interface between ceramics (ceramics have dielectric properties) to form glassy bond. Alternatively, glass can be added to interface to enhance bonding. Bond strength usually in excess of 95% of parent material strength.	Process depends on interaction between microwaves and ceramics which is complex.	Various ceramic-to-ceramic bonds have been produced. Alumina using an oxide glass. Silicon nitride using an oxynitride glass.

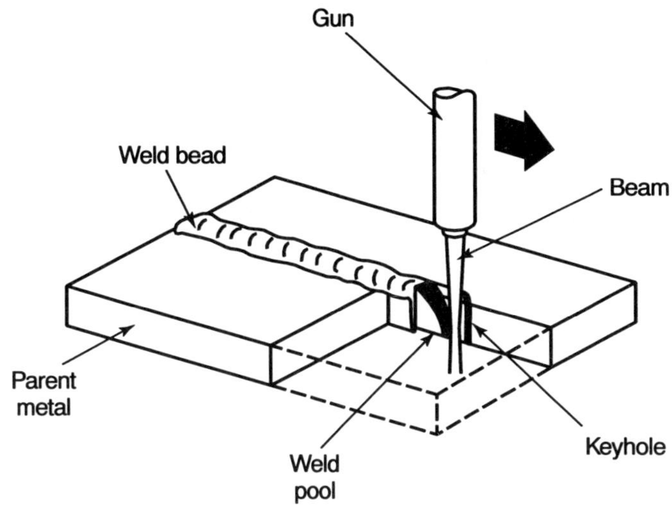

(a) Power beam

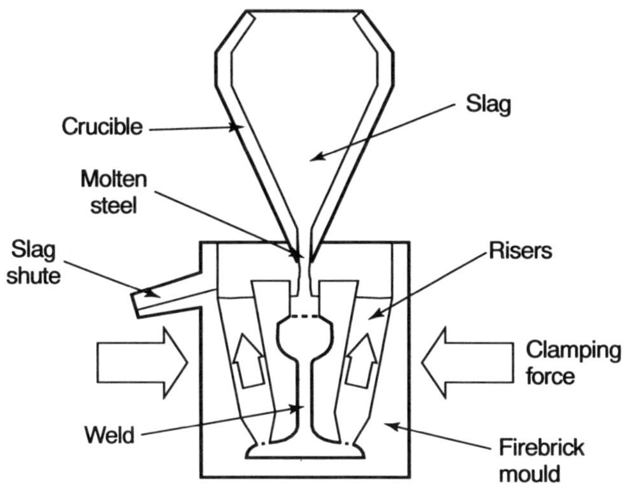

(b) Chemical reaction (Thermit welding)

Fig. 1.7.3 Operating principles of the major welding processes

Fig. 1.7.3

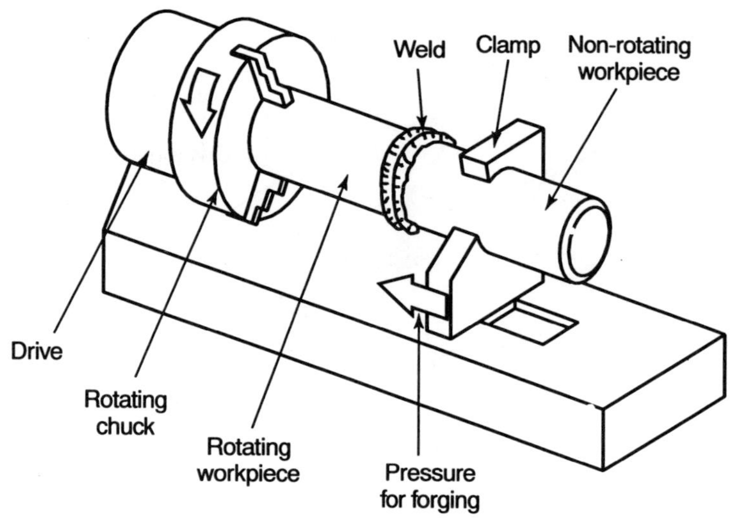

(c) Pressure (friction welding)

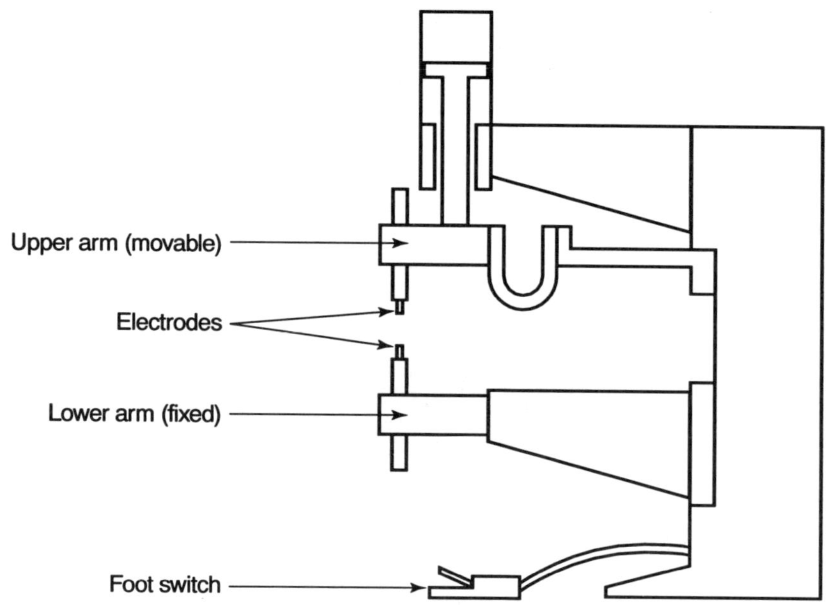

(d) Electrical resistance (spot welding)

Fig. 1.7.3 Operating principles of the major welding processes – *continued*

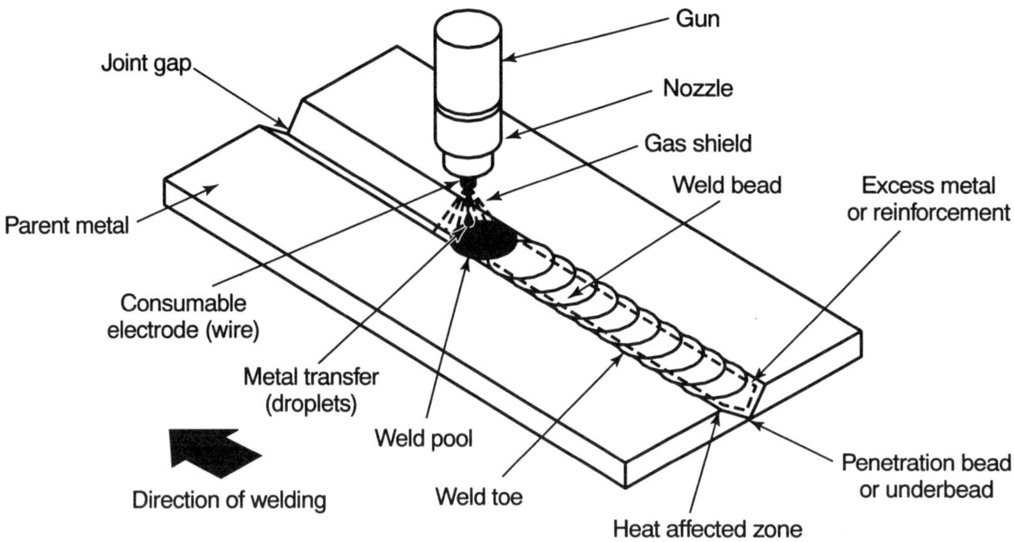

(e) Electric arc (MIG welding)

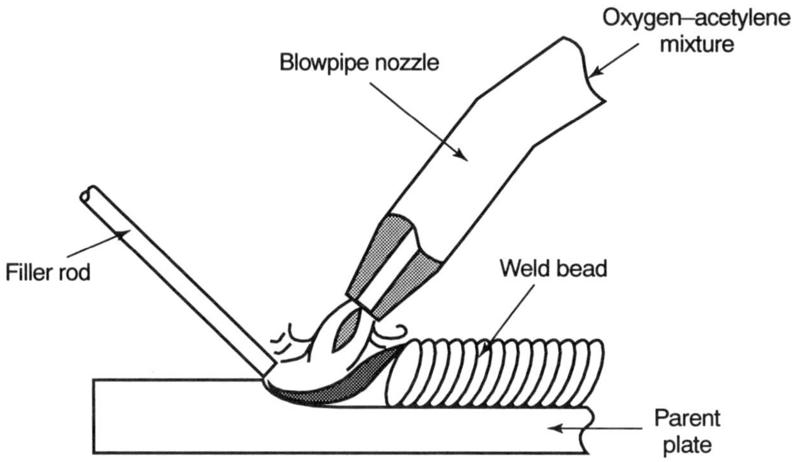

(f) Gas (oxy-acetylene welding)

Fig. 1.7.3 Operating principles of the major welding processes – *continued*

Fig. 1.7.3 – *continued*

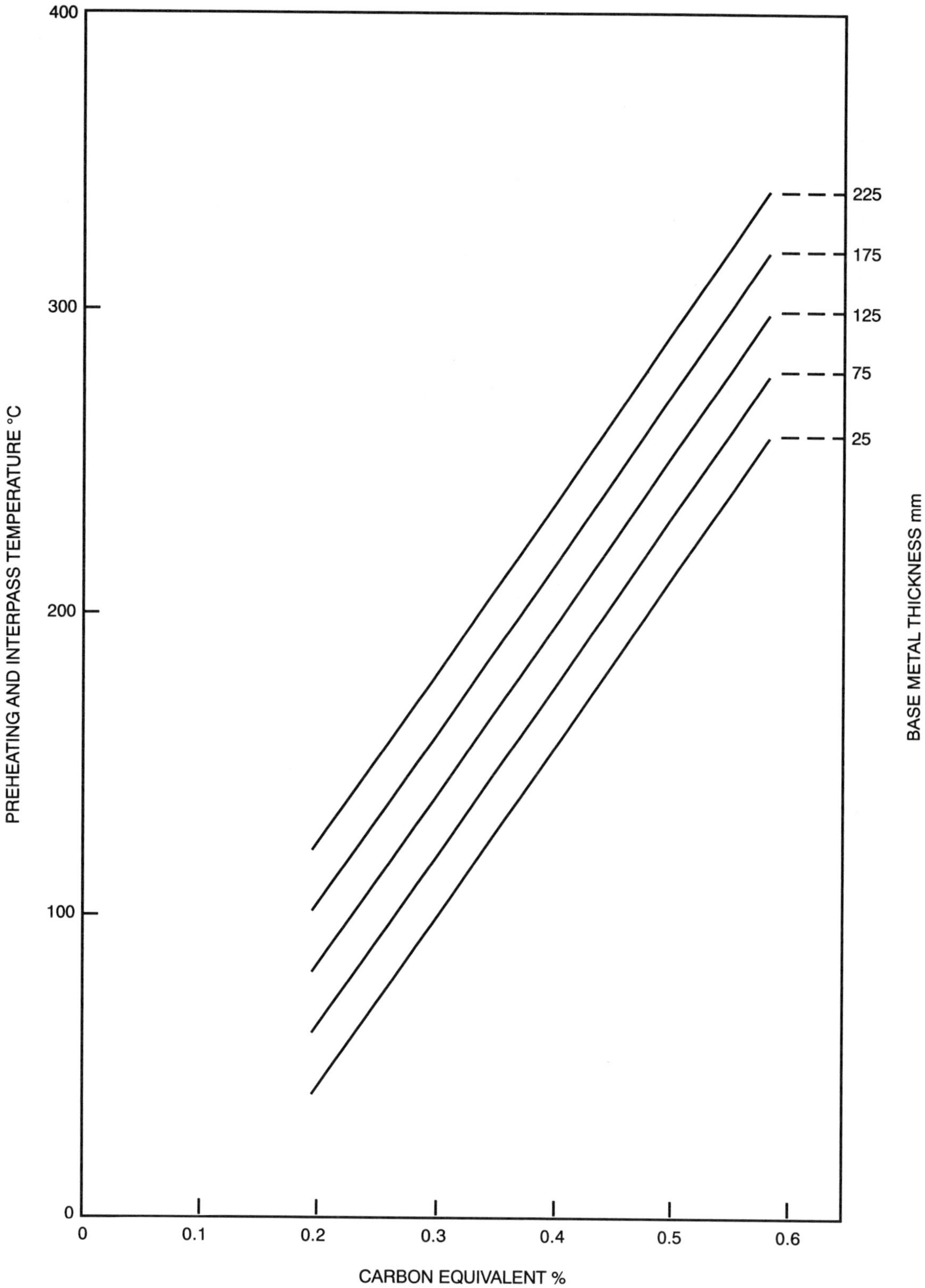

Fig. 1.7.4 Pre-heating temperatures for carbon steel welded by the manual arc process

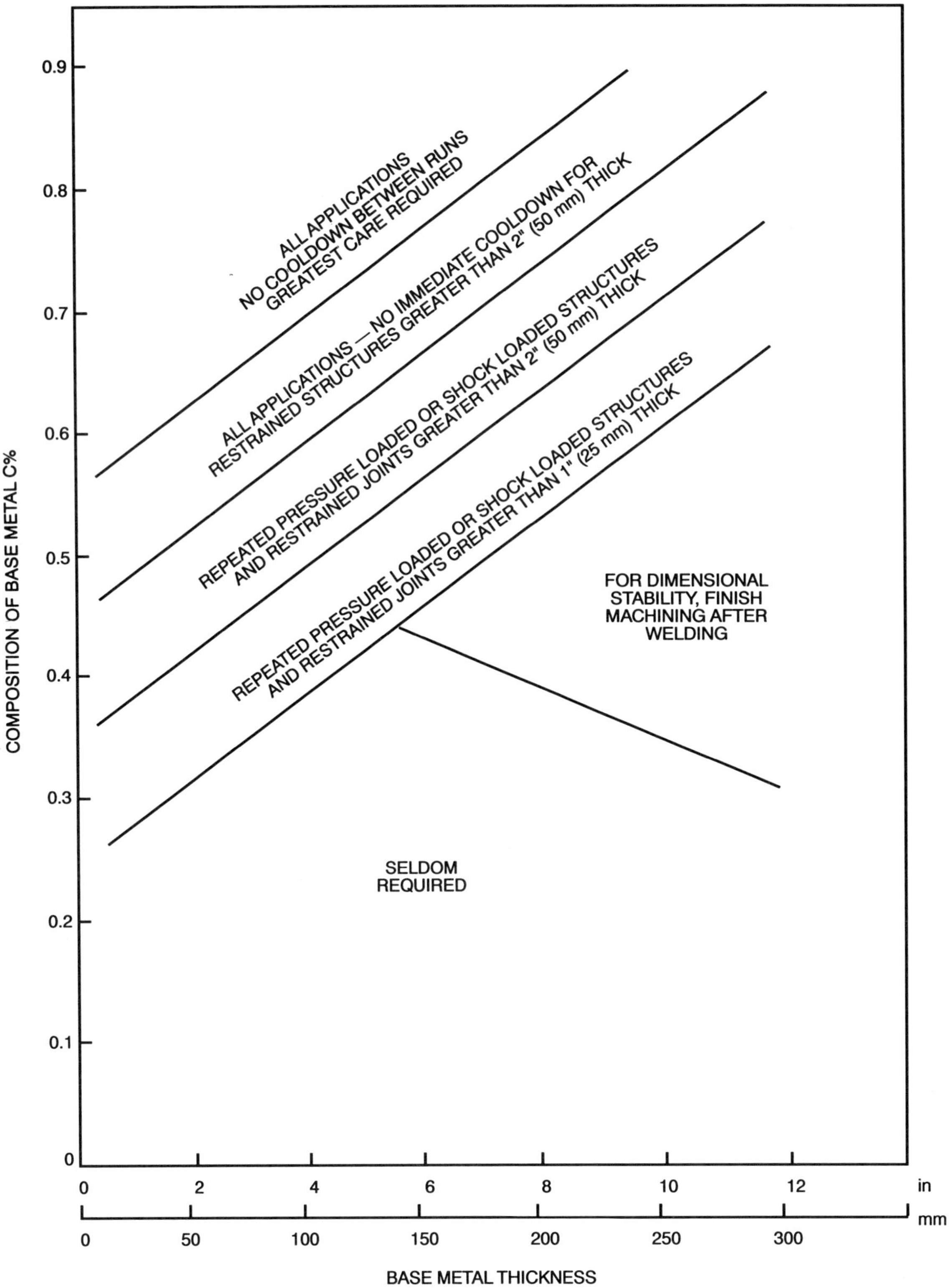

Fig. 1.7.5 Post-weld heat treatment of carbon steels. Temperature of 600–650°C; 1 h per 25 mm thickness

Fig. 1.7.5

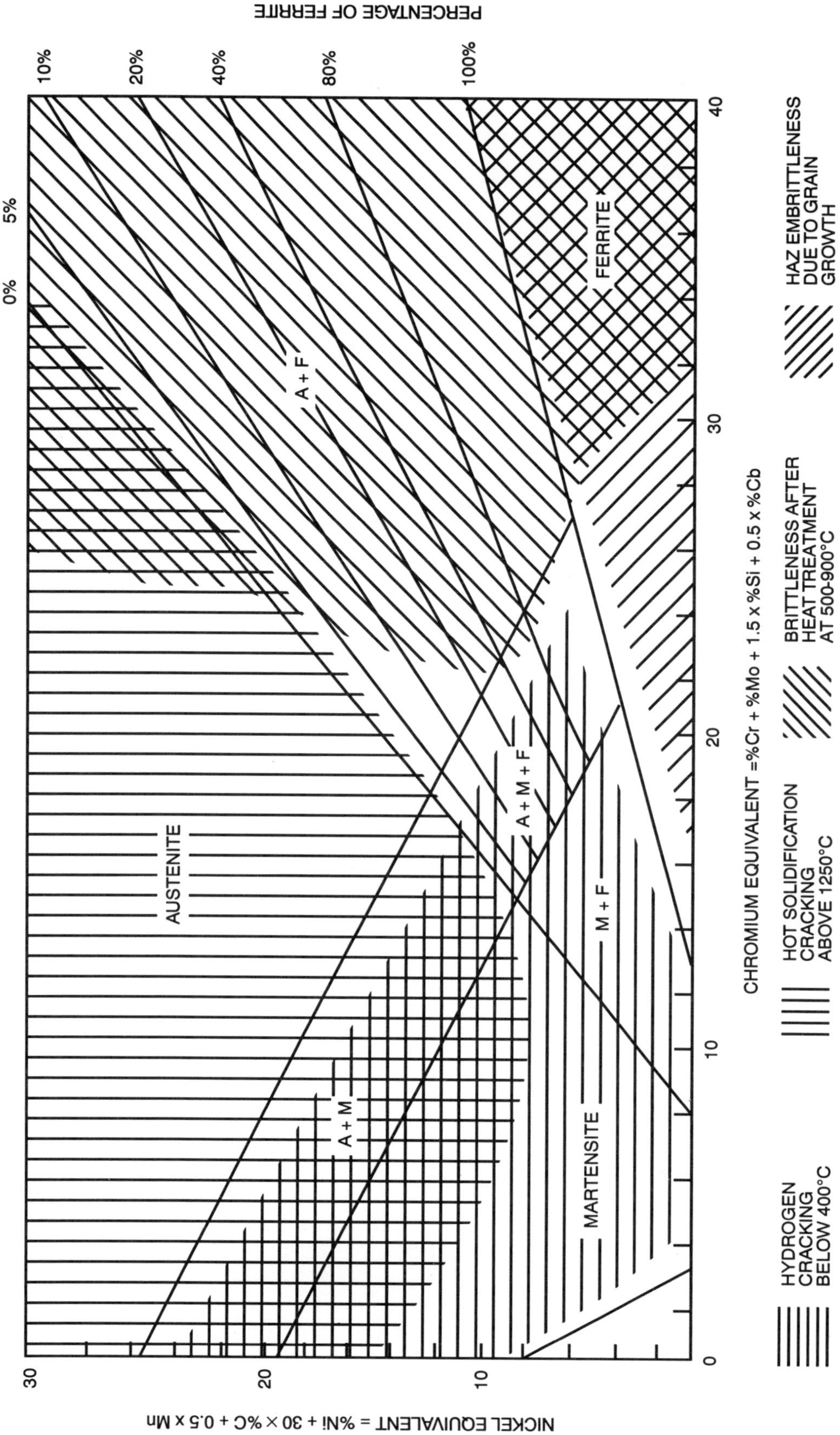

Fig. 1.7.6 Schaeffler diagram for welding of stainless steels

Fig. 1.7.6

1.7.6 Cementing

Cements are mixtures of inorganic materials which harden by chemical reaction to form coherent solids with good adhesion properties.

Hydraulic cements set by a hydration reaction on addition of water (e.g. Portland cement).

Heat or air setting cements set by chemical reactions usually involving loss of water. Once the water has been removed by drying, the reaction is not reversible.

The cohesive and adhesive properties of cements derive from the formation of complex atomic networks of certain oxides or oxygen-containing materials, e.g. silicates, aluminates, phosphates and oxychlorides. Cements may be classified on the basis of the chemical bond on which their strength is based.

Although in principle cements can be used to join almost any combination of materials, in practice their applications as adhesives exploit the ceramic-like properties of cements, namely:

- ability to withstand very high temperatures,
- excellent resistance to chemical attack, especially by organic materials and strong acids (except hydrofluoric acid),
- good electrical insulating and dielectric properties.

The characteristics of cements are given in:

Table 1.7.41 *Characteristics of cements*

Table 1.7.41 Characteristics of cements

Classification	***Type of cement***	***Method of applying***	***Curing procedure***	***Maximum service temperature (°C)***	***Density (10^3 kg/m^3)***
High temperature adhesive cements	Silicate-bonded cements	Brush Spray	Air dry and heat	1000–1400	–
	Alumina based cements	Brush Spray Dip	Air dry and heat	1700–1800	2.5
	Zirconia based cements	Brush Spray	Air dry and heat	2400	2.8
	SiC filled cements	Brush Spray Dip	Air dry and heat	3000 for short times	–
High temperature conducting cements	Metal-filled cements	Brush Spray	Air dry and heat	1000	3.2
Chemically resistant cements	Silicate-bonded cements	Trowel	Air dry	1000–1400	–

Table 1.7.41 Characteristics of cements – *continued*

Thermal expansion coefficient (10^{-6}/K)	***Thermal conductivity (W/m per K)***	***Volume resistivity (10^2 Ωm)***	***Corrosion resistance***	***Suggested uses***
7–11	2	10^9	Resistant to organic materials. Good resistance to all acids except hydrofluoric. Fair resistance to alkalis.	Bonding of ceramics, metals, glasses, graphite, quartz, refractory coatings.
8	5	10^9		Refractory coatings, end-seals for thermocouple wires and electrical lead-throughs. Insulating coatings for electrical and electronic components.
7.4	0.7	10^{13}–10^{15}	Resistant to acids except HF and boiling H_2SO_4. Fair alkali resistance. Will react with carbon above 1500°C	Very high temperature applications.
6	–	10^{10}	Good resistance to mild acids and organics. Not recommended for strong acid or alkali.	For use at very high temperature. Used as a bonding agent for manufacture of high-temperature carbon composites.
–	150	3	–	Produces high-strength bonds of good thermal and electrical conductivity to ceramics and metals.
–	–	–	Resistant to all acids (oxidizing and non-oxidizing) except those containing fluorides. Not resistant to alkaline conditions.	Used as bedding and jointing compound for vitrified bricks and tiles in floors, benches, pipes, channels, acid plants, towers and vats. Should not be used other than for joining vitrified tiles except in special circumstances.

Table 1.7.41 – *continued*

1.7.7 Adhesive bonding

1.7.7.1 Introduction

In certain situations, adhesive bonding is the most economic and convenient form of assembly, and may be used alone or in conjunction with other fasteners.

Many adhesive types are available, and significant important recent developments in adhesive technology have given rise to new possibilities for using adhesives in assembly. Synthetic adhesives are used for many structural and mechanical applications, although care must be taken in the choice of adhesive for particular materials, stress patterns and operating environments. Natural adhesives are used for traditional applications in less critical areas. Surface preparation prior to joining and sensible joint design are also very important, and in many cases markedly affect the performance of bonded joints.

Examples of areas in which adhesives are particularly useful are:

Dissimilar materials – combinations of metals, plastics, glass, rubbers, wood, etc. Joinability does not necessarily depend on similarity of physical or mechanical properties. The use of adhesive bonding is not necessarily restricted to adherends with similar physical or mechanical properties.

Laminates – based on plastics, wood, and metals; sandwich and honeycomb structures, printed circuit boards and core laminates in electric motors and transformers.

Structural assemblies – load-bearing structures in aircraft, automotive, marine and civil engineering industries.

Sealed joints – pipes, tubes, casing lids, pressure seals and seams for cans.

Delicate components – thin films and foils (aluminium alloy foils down to 0.35 mm, metal films down to 0.05 mm).

Securing of inserts – studs, shafts, concentric rotors joints, metal inserts, frame constructions, tube fittings and mortise joints.

Temporary fasteners – pressure-sensitive tapes for labels, masking and fixation of parts.

Other advantages and limitations of adhesives are:

Advantages – no irregular surface contours as with fasteners,
improvement in surface finish,
maintenance of integrity of structural members, i.e. no holes,
complex shapes are readily assembled,
minimal stress concentrations in bonded area,
no effects on parent metal properties as with welding, brazing, etc,
high damping capacity (improved fatigue resistance, vibration damping),
reduced finishing requirements caused by freedom from surface blemishes,
good sealing and insulation from moisture, chemicals, electricity, heat and sound.

Disadvantages – dependence of joint durability on processing conditions,
changes in design may be required to avoid cleavage failure,
limited resistance to extreme service conditions, especially heat,
optimum bond strength is not instantaneous,
poor joint designs may be susceptible to failure by peel or cleavage,
thermoplastic adhesives may creep,
degradation in severe environments (temperature, radiation, chemical),

may be physiologically active and flammable,
apart from a few special adhesives (anaerobics), not easily dismantled for repair or salvage,
rigorous process control and sample testing may be needed,
special stages may be required for preparation of components,
older types of adhesive may require heat and pressure to cure them.

THE FUNCTION OF ADHESIVES

Individual atoms and molecules are surrounded by fields of force. These forces are attractive at close range, and responsible for holding solids and liquids together as cohesive masses. Separate masses do not cohere and adhere unless they can approach each other closely enough for the attractive forces to come into play; this occurs readily for separate liquid bodies but not for solid masses. The action of an adhesive, therefore, is to make close contact with the bonding surfaces while it is liquid, and then to solidify, forming a cohesive joint holding the solids together. Chemical bonds are formed in a few special cases.

Adhesives come in many forms, e.g. liquids of high or low viscosity (including aqueous dispersions), pastes or mastics, films, tapes, powders (which require heating or mixing with liquids), and solids such as granules and rods which are heated to liquefaction. Curing of adhesives (i.e. solidification) may occur in a variety of ways, depending on the formulation of the adhesive. Solvent- and dispersion-based adhesives set by the evaporation of solvent or water, while temperature-setting adhesives cure during a heating and cooling cycle. The action of heating may melt the adhesive (hot melts) or allow dissolution of particles in solvent (plastisols); in both cases a solid remains after cooling. Many adhesives are cured by a chemical reaction of which there are several possible types, some requiring heat and pressure. Once reaction is condensation polymerization, in which water is driven off during curing (e.g. phenolics). Another possibility is a reaction which occurs when two components of the adhesive are brought together; either the two components react with each other or one component catalyses the setting action of the other. Finally some adhesives set in particular environmental conditions, e.g. absence of air (anaerobics) or contact with absorbed moisture (cyanoacrylates).

For most adhesives the surfaces of the adherends must be clean, in order that a good bond may be produced, since the adhesive cannot make close contact with the underlying surface unless it is free from oil, grease and solid debris. Some modern adhesives (for example some of the toughened types) are able to dissolve surface contamination and to 'wet-out' and absorb solid debris.

Adhesive joints tend to be the weakest link in bonded structures, and care must be taken to design the joint so that it is not likely to be overloaded in normal service (including accidental transient stresses). Adhesives for use in structural applications must be capable of supporting high loads. Traditional structural adhesives (e.g. epoxies) are hard and prone to brittle failure under shock loading. Modern toughened adhesives incorporate a 'rubber' phase in the load-bearing hard matrix and are better able to withstand shock. Adhesives for mechanical uses such as locking and sealing threaded components, sealing plane interfaces (gasketting) and press-fit assembly do not normally need such good adhesive properties though they can be incorporated.

1.7.7.2 The adhesives

Adhesives may be classified by a number of criteria, chemical composition, adherends, physical state, setting action, method of application or by other properties. For purposes of adhesive selection, it is most convenient to classify as follows:

Acrylic group	anaerobic
	cyanoacrylate
	radiation cured
	two-component
	toughened variants
Amino	
Emulsion	
Epoxide group	single-component
	two-component
	toughened variants
Hot melt group	conventional (non-reacting)
	reactive
Natural	
Phenolic	
Plastisol	
Polyimide	
Polyurethane group	single-component
	two-component
Rubber solution	
Tape	

The properties and applications of these groups of adhesives are given below and a general summary of the types available in each class is given in Table 1.7.44. A selection procedure which can be used to select the best candidate adhesive for a particular application is given in Section 1.7.7.5 based on the adhesive classes listed above.

ACRYLIC GROUP

This family is most conveniently reviewed by distinguishing between the major types and then examining the toughened variants.

ACRYLIC – ANAEROBIC

Anaerobic adhesives, based on the acrylic polyester resins, are stabilized by atmospheric oxygen when in the liquid state. They are produced in viscosities ranging from thin liquids to viscous, thixotropic pastes and harden to give materials with various predetermined levels of performance. (See also 'toughened variants').

Cure mechanism

For the whole family, 100 % liquid-to-solid conversion, catalytically initiated by iron and copper and their various alloys.

Normal applications

Joint sealing and retention of coaxial components, often both in the same application, are prime examples. Components can be threaded (screws or pipes) or splined or smooth (e.g. bearings). Special variants act as gaskets, others can seal leaking gas pipes etc.

Design criteria

These adhesives come in a wide range of viscosities and strengths to facilitate assembly and to make dismantling possible. The majority of them function in a jamming mode (compression) though some are true adhesives. Even the weakest of them will normally be capable of retaining a screwed joint in the presence of severe vibration. Although high axial shear strengths (some 35 MPa averaged over the bonded area) are often quoted, like the true structural adhesives the anaerobic types should not be constantly stressed beyond 10 % of their quoted performance. Their primary function (and strength) is to be seen in the carrying of very high compression loads. The stronger grades should be used in bad environments.

Adherend compatibility

Anaerobic adhesives are compatible with all metals; some such as copper, accelerate curing while others, such as zinc, may reduce the cure rate of some types and may necessitate the use of a secondary catalyst or primer. Performance falls off with the softer, more ductile metals and alloys and particularly on plastic parts. The latter are usually only used after treatment with a primer. Plastics may stress-crack.

Surface preparation

Although oil contamination reduces performance, thin oil films are usually tolerated so well that strict surface preparation for most coaxial assemblies is unnecessary. Wiping to remove excess cutting, lubricating or preserving oil is often sufficient, with de-greasing only necessary to maximize coaxial performance. Load-bearing lap joints demand the special adhesive types and good surface preparation. See 'toughened variants' below.

Major benefits

No other adhesive type has the versatility of this family for assembly (with dismantling possible) of coaxially fitting parts. This is not so for lap joints. The ultimate performance of even the best of the truly adhesive anaerobics is readily surpassed by other groups, particularly the toughened, non-anaerobic acrylics and toughened epoxies. However, both of these may be less convenient to use.

Major restraints

Essentially limited as a class to coaxial mechanical assembly, retention and sealing, they also make an excellent general purpose gasketing medium. The cure rate depends on surface activity and may require a supplementary catalyst. The family copes with the gaps of normal engineering practice. As clearances increase beyond 0.5 mm their capacity to cope falls rapidly.

Fault finding

In established production lines, characteristic problems are caused by cure rate variation induced by changes in temperature or the chemical activity of the surface. A change in gap may well reduce or increase the handling time. The liquid adhesives deteriorate under hot storage conditions which may lead to premature polymerization or a reduction in the curing rate.

Equipment

Anaerobics are usually dispensed from purpose-made bottles, cartridges etc. Metering and dispensing equipment must be specially designed to prevent polymerization caused by air exclusion and metal catalysis.

Health and safety

No major toxicity problems have been associated with this family of materials and their general physiological activity can be considered to be very low. However some faster curing systems appear to be more aggressive. Products of the major manufacturers meet the requirements of the Water Research Council and may be used for sealing pipework carrying potable water.

Use in accordance with the manufacturer's instructions, Health and Safety data and ensure that all COSHH requirements are in place.

ACRYLIC – CYANOACRYLATE

Cyanoacrylate adhesives are usually relatively low viscosity fluids based on acrylic monomers and characterized by extremely fast rates of cure. When placed between closely fitting surfaces, some will cure to give a strong joint in two to three seconds. (See also 'toughened variants').

Cure mechanism

The whole family relies on the catalytic anionic polymerization of the vinyl group of the acrylic ester, giving 100 % liquid-to-solid conversion. The catalyst normally takes the form of traces of surface moisture, which are, in fact, usually slightly alkaline in nature. Acids strongly inhibit curing.

Normal applications

The assembly of all kinds of small rubber, plastic and metal parts are typical and they are often used to bond lengthy rubber strips and mouldings. Occasionally they are used in the assembly of all metal mechanical components such as small bearings for use in benign conditions.

Design criteria

Although nominally very strong, the thermoplastic nature of most cyanoacrylates, coupled with their low softening point and sensitivity to humidity on unpainted metallic surfaces, generally restricts their use to very lightly loaded assemblies. The majority are brittle and joints should not be flexed or subject to shock loads unless at least one of the adherends is sufficiently compliant to absorb the deformation without transferring it to the bond line.

Adherend compatibility

The adhesives will bond almost all materials though some plastics (low surface energy) will cause difficulty unless specially treated. Alkaline glass may cause premature bond failure. All glasses should be silane primed if at all possible, as this considerably improves the joint's humidity resistance. They bond metal well, but joints based on unpainted metal are at risk under hot, humid conditions.

Surface preparation

With the exception of polyolefins and other low surface energy plastics which should be corona treated, most plastic parts need no surface preparation. Metal should be degreased though, again, small components may often be bonded as delivered. Some rubbers, containing various compounding oils (e.g. EPDM) will only bond well on freshly cut or abraded surfaces. Chlorinated cleaning solvents should be avoided as these sometimes leave acid traces which prevent cure.

Major benefits

Speed of cure and almost unrivalled ability to bond many plastics and rubbers. Very clean and easy to use.

Major restraints

The rapid cure rate hinders bonding of large areas (though the bonding of long strips presents no problems). They cannot cope with large gaps. These adhesives are liable to brittle failure on stiff adherends and are moisture sensitive when used directly on metal exposed to warm humid conditions.

Fault finding

On established production lines, problems usually stem from changes in dimensions or humidity. Very humid or very dry conditions may cause weak bonds and can prevent curing altogether. Similarly, an inadvertent increase in joint gap may either hinder or prevent the surface initiated curing mechanism. Poor storage (warm, humid) may cause curing problems.

Equipment

Usually purpose-made bottles. Metering and dispensing equipment must be designed to prevent premature polymerization.

Health and safety

All the adhesives in the family bond skin very readily and rapidly, presenting a real hazard which must be anticipated but not exaggerated. Industrial experience shows that, with correct training and, where necessary, appropriate equipment, the difficulties and danger are minimal. However, several of the more popular materials have irritant vapours. These must be extracted if the adhesives are in constant use.

Use in accordance with the manufacturer's instructions, Health and Safety data and ensure that all COSHH requirements are in place.

ACRYLIC – RADIATION CURED

These variants are always single-component adhesives which are primarily, though not always, dependent upon some form of radiation to initiate polymerization. Between them they cover a wide viscosity range and in some there is very precise rheological control. They are usually limited in use to the assembly of small components one of which is normally transparent to the incident radiation. Electron beam-induced polymerization is known though this is not common practice. (See also 'toughened variants'.)

Cure mechanism

Acrylic monomers can polymerize spontaneously in normal daylight. However, for most practical purposes the rate is not good enough and so light-sensitive catalysts are provided in the various types and these respond very positively to radiation at specific wave lengths in the UV and visible (blue) regions. Irridation splits the catalyst which then initiates an extremely rapid radical polymerization. In many systems polymerization stops when radiation ceases and with it the otherwise 100% liquid to solid conversion. However, new products, based upon special epoxy and vinyl chemistry, are becoming increasingly available whose polymerization, once initiated, will continue reasonably quickly even though irridation has ceased. This is the so cold 'dark cure' reaction.

Direct irradiation of the exposed surface of all types can cause extremely rapid polymerization which can be extremely useful when forming a hardened 'blob' on the surface of a single adherend. However, the surface of many formulations is left in a tacky state due to oxygen inhibition of the polymerization process.

Some formulations will also behave anaerobically and this means that not only will they polymerize spontaneously within a joint (in the dark and in the absence of atmospheric oxygen) but this process will become very rapid if the assembly is heated.

Of late, the natural tendency of the acrylic molecule to polymerize spontaneously when either heated or in an anaerobic condition, has been exploited extensively. Products have been created which extend the simple concept of tacking circuit board wires in place into the initially 'tacked' construction of large multi-component assemblies. In the construction step the various components are held in place by the almost instantaneous curing of exposed fillets of pre-placed adhesive. The fillets are cured as a consequence of their exposure to intense UV or blue light. While the hardened fillets hold everything together the adhesive within the joint remains uncured at this stage. Polymerization within these shadowed interior zones takes place by either or both anaerobic and thermal cure. Such processes are not related to the special 'dark cure' mechanisms referred to earlier.

Design criteria

Initially these adhesives were limited to the use of unfilled monomers, thin bond lines and the necessity of having at least one adherend composed of an efficient radiation transmitter. While this is still largely true the increasing availability of more powerful lamps and 'blue light'-sensitive catalysts means that thicker bond lines can be accommodated. The use of blue light is particularly relevant because it is not absorbed, as is UV, by some of the fillers which need to be used to obtain rheological control and other important functional characteristics. Additionally the secondary 'dark', anaerobic and thermal cure processes extend the range of potential applications enormously. While relatively expensive these adhesives should always be given consideration when there is a need to mass produce components particularly when they are comprised of a multiplicity of relatively small repeating units.

Adherend compatibility

Good with glass, most metals and plastics, other than those possessing low surface energy. The latter must be treated.

Surface preparation

Surfaces must be clean, dry and oil free. Normally silane priming of the glass is not required for these materials which often contain their own inbuilt coupling agents.

Major benefits

They bond glass well and are very useful for the extremely rapid assembly of small parts, especially plastic components. Because their cure is very rapid they are less prone to stress crack plastic materials than the other acrylic types.

Major restraints

Generally limited to the smaller assemblies and where one component at least is a radiation transmitter. Some varieties do not cure to give a tack-free surface.

Fault finding

Unexpected levels of radiation absorption by the adherends can cause problems. Shadows too can cause difficulties.

Equipment

Normally applied from either the bottle or an anaerobic or cyanoacrylate type of dispenser.

Health and safety

Because many of these adhesives need to incorporate the basic acrylic monomers they are more prone to cause physiological problems (dermatitis and associated sensitization) than the other types of adhesive within this family group. It is particularly important therefore that they are used in accordance with the manufacturer's instructions and the requirements of their Health and Safety data. Particular regard must be given to the COSHH requirements.

In addition, it needs to be noted that UV radiation is dangerous and that proper precautions must be taken. In this context, the possibility of reflection should not be overlooked. By contrast, there are no known problems associated with the blue wavelengths other than the sheer intensity of the light which may, on occasions, need to be used.

ACRYLIC – TWO COMPONENT

The two component versions use a variety of chemical mechanisms to initiate their 100% conversion from liquid to solid.

They are available in a variety of rheological forms and will react to give handling strength in anything from five seconds to thirty or so minutes according to type and grade. Most can be considered to be toughened or crack resistant due to either flexibilization or true toughening and for this reason they are coded as both 'w' and 'x' in Table 1.7.44 (See also 'Acrylic – toughened variant').

Cure mechanism

When these adhesives were first developed in their toughened form one of the components could be considered to be the adhesive proper while the other comprised the catalytic initiator of cure. They were used in a ratio of approximately 50:1 respectively to give a 100% conversion of the liquid adhesive to solid. Because of this high ratio, and the way in which they were used without mixing, they were often called 'pseudo one-component' adhesives. Nonetheless, the ratio required was not convenient and as a consequence there has been a progressive move towards 1:1 systems. Here, the distinction between adhesive and curing agent (the catalyst) has been lost. These adhesives do not require the precise stoichiometric proportioning of the two component epoxy-based adhesives since the influence of one component upon the other is essentially catalytic. Hence, the ability to create 'dot-by-dot' and 'bead-on-bead' systems where the adhesive functions reasonably well even without proper mixing of its two parts.

As a consequence of the sophisticated manipulation of these types of formulations curing may be initiated in any one of a number of ways: true mixing, partial mixing (dot-by-dot and bead-on-bead) and by the spreading of a catalytic initiator over one of the adherend surfaces prior to the closing of the joint. In this latter instance the adhesive proper, on the other surface, spreads over the initiator, absorbs it and polymerizes as a consequence. This film absorption has proved to be very effective in thin bond lines but only complete two component mixing is effective when the bond lines are thick. Bead-on-bead and dot-by-dot techniques are not suitable for large gaps as there is insufficient mixing induced by the physical placement of the parts.

Normal applications

As the individual grades vary extensively in cure rate, viscosity and strength they can, as a product group, cover a vast range of applications. At one extreme they can rival the cyanoacrylate for speed yet offer superior resistance to harsh conditions. Thus, they may be used for the rapid assembly of small and medium sized components which need to be used in hot and possibly humid conditions. Some, particularly the toughened variants, can be used to bond load-bearing structures where gap-fill and extended cure times are essential. The designer should really give the entire product group detailed consideration before making a final selection.

Design criteria

The product group can be divided into those types which are truly toughened and the rest. The toughened group lend themselves to the assembly of substantial structures, such as bus and coach panel work, where the rather stringy, viscous nature of some versions is not too much of a disadvantage. Their shear modulus lends itself to the bonding of GRP whose surface, especially the gel coat, is prone to be brittle. Individual cure rates and gap-filling capabilities will always tend to dictate which particular grade is the most suitable for any particular application.

Adherend compatibility

Apart from rubber, and the lower surface energy plastics, most materials and particularly the common engineering alloys bond well. Unfortunately, zinc often presents a problem unless it is either passivated or special initiators and hardening agents are used. Some types are more likely to cause stress cracking of susceptible thermoplastics than others, especially when the latter contain residual moulding stresses. In such situations prolonged contact with uncured adhesive/hardener/initiator must be avoided.

Surface preparation

Of all the adhesive types these acrylics appear to cope better with badly prepared and contaminated surfaces than any other adhesive family. Although oil contamination does reduce their performance they can still perform well. However, this ability should be seen as a bonus which will counter poor workmanship rather than a reason for tolerating poor preparation with its attendant uncertainties. Zinc surfaces must be passivated or special initiators used and the low surface energy plastics must also be properly prepared.

Major benefits

Speed of cure and the capacity to bond a variety of poorly prepared surfaces set these materials apart. Furthermore, some versions are particularly resistant to demanding outdoor environmental conditions. As a consequence, they have been successfully used to both build and repair substantial structures under less than ideal circumstances.

Major restraints

Ultimately their strength/modulus characteristics limit their load bearing capacity to a level below that of the best epoxides. Yet, in the correct design where loads are spread widely they will serve very well indeed. Unlike epoxides, their chemical nature and polymerization processes prevent the incorporation of property modifying fillers. Hence, adjustments for density, expansion, thermal transfer and electrical properties are very difficult to make. Even rheological manipulation is restricted. Thus, the group as a whole is limited in this particular sense and cannot be compared with the epoxy-based materials. The latter can accommodate almost infinite subtleties of formulation manipulation.

Fault finding

Problems are usually associated with poor mixing or chemical inhibition of the polymerization processes. Where rapid cure is required some sensitivity to temperature and gap size may be noted. The liquid adhesives must be stored under cool conditions otherwise premature polymerization or other forms of deterioration may occur.

Equipment

This ranges in nature from the direct use of the supplied tubes, bottles and brushes through static mixer equipped cartridges to complex fully automated systems.

Health and safety

No major toxicity problems have been associated with this family of materials and their general level of physiological activity appears to be low.

Use in accordance with the manufacturer's instructions, Health and Safety data and ensure that all COSHH requirements are in place.

ACRYLIC – TOUGHENED VARIANTS

It is convenient to separate out the toughened variants of the four foregoing acrylic types because it allows a more direct comparison of their relative capabilities, which vary widely. Despite the latter comment all employ, to some degree, the use of 'rubbery' flexibilizers or the distinct concept of a secondary 'rubbery' phase within the stiff, load bearing polymer of the major phase of the hardened adhesive. The flexible rubbery phase distorts under load and thus reduces the stress intensity at the tip of any invasive crack. This stress reduction mechanism stops the crack from developing and thus prevents further damage to the joint.

Toughening can be used to improve the shock resistance of cyanoacrylates but, because of their inherent sensitivity to modification, the level of improvement gained can probably best be described as modest. However, major benefits can be obtained in the anaerobic types which can be converted from being essentially jamming agents into true adhesives by means of this mechanism. If necessary, very effective and robust gasketing materials can be formulated from the toughened anaerobics. Similar manipulation can be performed in the case of the radiation cured adhesives and with some of the two part systems the transformation is dramatic. Extremely tough, robust and versatile materials have been develop which challenge the technical supremacy of the toughened epoxies because of the latter's slower cure rate and generally, though not always, poorer performance on badly prepared surfaces.

AMINO

These urea and melamine formaldehyde adhesives come as water-based syrups or powdered solids which need mixing with water. Catalysts may be required. Both cold and heat cured versions are available.

Cure mechanism

Under the influence of heat and/or a catalyst polymerization takes place. Water is produced as a by-product of the curing reaction.

Normal applications

These adhesives have few conventional engineering applications, being commonly used in the woodworking industry for the bonding and lamination of wood and wood-associated materials. The melamine versions have superior water resistance and better temperature stability.

Design criteria

They are available in different viscosities, reaction speeds, etc. to suit a multiplicity of uses. However, they may not be used with metal without special primers, they almost invariably require at least one surface to be porous and will not fill large gaps successfully. Because of their high water content, pressure must be applied during curing. They form good joints with wood but the poor environmental resistance of the urea versions renders them unsuitable for exterior use.

Adherend compatibility

These adhesives are compatible with a wide range of wood and wood-associated materials such as the common laminated plastics sheets. They are effective on absorbent protein-based and cellulose-based materials, but unsuitable for thermoplastics. Almost invariably one surface must be absorbent.

Surface preparation

None is normally required, except on metal surfaces. The best results are obtained from cleanly cut, well fitting cellulose-based components.

Major benefits

Although some formulations may require complex handling, these adhesives are normally simple and reliable.

Major restriants

Little application outside the woodworking-associated industries. The need for the surfaces to be clamped tightly and sometimes heated is also limiting. They may not be used on alkaline surfaces. Poor gap-filling and the requirement that one surface should be absorbent further limit their use.

Fault finding

The resins on which these adhesives are based are unstable in the liquid state and care must be taken to ensure that the material is in good condition before use. Otherwise, little difficulty is expected unless the quantity of the catalyst is incorrect.

Equipment

Ranges from extremely simple, to complex and expensive, according to the nature and quantity of bonded components being produced.

Health and safety

Based on formaldehyde, these resins and the liquid solutions made from powder should be regarded as being physiologically active and likely to cause irritation and possibly dermatitis.

Use in accordance with the manufacturer's instructions, Health and Safety data and ensure that all COSHH requirements are in place.

EMULSION

Emulsion adhesives are generally modest viscosity, milky white dispersions. They can be based on synthetic rubbers or polyvinyl acetate or related polymers. Their solids contents varies according to type from approximately 20 to 65%. They can be made up as pastes and mastics. They are becoming of increased interest and are the subject of intense technical development due to the environmental pressures which are now suppressing the use of the more convenient solvent-based rubber adhesives, with their short evaporation times.

Cure mechanism

The water carrier is usually absorbed by at least one of the bonded surfaces though evaporation can also play some part. As water is removed, the dispersed plastic phase is forced into contact and congeals; the resulting film binds the surfaces together. Cure rate depends on the speed of water extraction. Ultimately, the absorbed water is lost by indirect evaporation.

Normal applications

Extensively used in the woodworking and packaging industries, the adhesives have applications on fabrics and the fabric-based components used in engineered products (e.g. loudspeaker cones). The paste-like variant bonds both plastic and ceramic tiles well. They are normally used for relatively lightly loaded materials, preferably in good environments.

Design criteria

The water carrier of the emulsion must be lost either by evaporation or absorption and subsequent evaporation. Generally, therefore, at least one surface must be absorbent.

The adhesive may corrode metal, which should always be painted. Clamps must normally be used to maintain pressure during absorption. The family as a whole is suitable only for lightly loaded components and, except for the cross-linked (two-part) variants they should not be exposed to warm, wet conditions. Heat accelerates their rate of degradation.

Adherend compatibility

Emulsions bond most materials which absorb water. Non-absorbent adherends may be bonded if the other surface provides an escape route. Thus, some adhesion can be obtained on most metal and plastics surfaces, though care should be taken to ensure that metals are adequately painted to prevent corrosion. Polyolefin plastics and other low surface-energy types are not generally bonded with these adhesives.

Surface preparation

Surfaces must be free of oil and grease and the low surface-energy plastics must be properly prepared.

Major benefits

These adhesives are particularly suited to large scale, low cost assembly of wood, fabrics and other absorbent materials. They are clean and simple to use, effective and present no major problems or hazards.

Major restraints

One surface at least must normally be absorbent, most types are readily damaged by warm, moist conditions and they may not be heavily loaded without careful evaluation. Except for the tacky paste versions, handling times may be prolonged. May corrode metal.

Fault finding

There are no characteristic faults.

Equipment

There is a variety of metering and dispensing equipment available. Complexity depends upon the scale of the operation. Small quantities are very conveniently dispensed from purpose-made containers.

Health and safety

There are no particular problems with the single-part members of this family though the acid catalyst of the two-part version can be dangerous and must be assumed to be physiologically active, corrosive and toxic.

Use in accordance with the manufacturer's instructions, Health and Safety data and ensure that all COSHH requirements are in place.

EPOXIDE GROUP

This family is most conveniently reviewed by examining the single- and two-component versions and their toughened derivatives prior to considering general features.

EPOXIDE – SINGLE COMPONENT: HEAT CURED

These catalyzed adhesives must be heated to effect cure when they form robust, thermoset and usually stiff polymers. The untoughened variants can be brittle and the more flexible versions are more prone to damage in hot, wet conditions than the toughened types. Nonetheless, the older, untoughened versions, have proved popular in the past though they are being progressively replaced whenever the load-bearing, crack-resisting properties of the toughened types offer advantages.

EPOXIDE – SINGLE COMPONENT: RADIATION CURED

Recently a new type of single component adhesive has been introduced which depends upon UV-induced ionic polymerization.

The reaction is particularly useful since it will continue after illumination has ceased. As a consequence, it is not necessary to have transparent adherends. Because of their cost and, as yet, limited development these adhesives have seen only modest use to date. Further progress is anticipated. See also 'Acrylic – radiation cured.'

EPOXIDE – TWO COMPONENT

This group exploits the capability of the epoxide resin to react with a wide variety of 'hardeners' to give the very precise characteristics needed for some applications. Thus, they may be very soft and pliable when cured or very hard, brittle, highly cross-linked and thus resistant to severe environmental attack from aggressive chemicals. It is possible to match expansion, electrical and fire-resisting requirements very accurately. While the more pliable forms are shock-resisting they are prone to environmental damage.

Most members of the two component group will polymerize at normal ambient temperature. Almost all can be warmed to increase their rate of cure though some must be heated to effect cure at all. These latter are usually intended for use in harsh environments.

EPOXIDE – TOUGHENED

These adhesives owe their crack-resisting properties to the inclusion of low molecular weight, rubbery monomers in the compositions of the foregoing types. Materials such as these are more robust than the conventional single and two part epoxides and will resist very harsh environments. They are, as a class, one of the most robust and versatile of all the synthetic adhesives. The heat cured versions are exceptional within the epoxy group as a whole in that they are capable of dissolving thin oil films. They have proved to be very reliable in this regard and proper industrial processes have been based upon this fact.

EPOXIDE – IN GENERAL

Epoxy adhesives are thermosetting resins which solidify by polymerization and, once set, will soften but not melt on heating. Two part resin/hardener system solidify on mixing (sometimes accelerated by heat), while one part materials require heat to initiate the reaction of a latent catalyst. Their properties vary with the type of curing agent (amine, amide, acid anhyride) and the resin type used. Epoxies generally have high cohesive strength, are resistant to oils and solvents and exhibit little shrinkage during curing.

Cure mechanism

Polymerization is by either of two processes. In two-component formulations, an addition reaction takes place between the active oxygen of the epoxide group and a labile hydrogen, usually from an hydroxy, amine, amide or acid group in the hardener. These reactions usually proceed at room temperature and may be speeded by warming. Some two part versions (anhydride cured) need heating. The single part heat cured versions are cured catalytically and many are toughened, a process which beneficially suppresses crack propagation. All mechanisms give rise to complete liquid to solid conversion.

Normal applications

Strength, durability and great versatility give epoxies a wide diversity of applications.

Design criteria

Because of this family's versatility it is probably more useful to summarize its limitations. A few formulations are brittle, cure rate varies enormously with formulation and their viscosity can make use difficult on very small assemblies. In general, their naturally very high strength may not be modified (reduced) readily.

Film-based forms are available and these can be used whenever strict weight control and accurate placement is necessary.

Adherend compatibility

This is excellent, except with thermoplastics and rubber where performance is substantially reduced. Low surface energy materials (polyolefins and fluoropolymers) may not be bonded without special surface preparation such as corona discharge.

Surface preparation

This family generally needs good surface preparation, except for a few special formulations and the heat cured, toughened variants, which have proved very tolerant of mineral oil-based contamination. Some special versions can bond unprepared polyester composite.

Major benefits

Strength, durability, robustness and general versatility. Unlike the surface-activated acrylic adhesives, cure rate is generally quite independent of surface chemistry.

Major restraints

Cure rates are generally slow unless the adhesives are heated, and the special formulations which cure quickly at ambient temperatures tend to be rather brittle. The untoughened formulations may have low peel strengths and may exhibit brittle failure on impact. Weighing and mixing of the two component systems must be accurate and thorough and they can be messy and difficult to clear up after use. However, special packing and dispensing systems do alleviate such difficulties.

Fault finding

Problems can arise from poor proportioning, mixing or temperature variations. Adhesive failure is almost invariably associated with contamination.

Equipment

Small quantities may be readily hand mixed and the single part, heat cured versions may be applied directly from a tube or cartridge. Similarly, special packing and dispensing systems may be used to dispense the two component adhesives on the cartridge scale, but large volume metering and mixing equipment is often complex and expensive.

Health and safety

Epoxy resins and hardeners are known to cause dermatitis but, used properly, have proved to be relatively hazard free. The dermatitic and related problems which occasionally arise with these products, are often linked with either, or both, nitrogen-based hardeners or the powerful solvents used for cleaning. Some amine-based hardeners fume strongly and must be used with effective ventilation.

Use in accordance with the manufacturer's instructions, Health and Safety data and ensure that all COSHH requirements are in place.

HOT MELT GROUP

HOT MELT – CONVENTIONAL

Many synthetic thermoplastic polymers can be used as hot melt adhesives. Such adhesives, which often comprise blends of several materials, are to be found in a variety of forms: tape, film, powder, rod pellet, block and also liquid, solvent-based versions in which the solvent evaporates after placement but before bonding. Some can be foamed on application.

Cure mechanism

The thermoplastic mix of materials in the adhesive melts on heating (usually in the range 65–180°C) and solidifies on cooling to form a strong bond.

Normal applications

They lend themselves particularly well to the high speed and repetitive assembly of all types of lightly stressed components unlikely to be exposed to elevated temperatures or extreme environments.

Design criteria

Component design must accommodate the extremely rapid hardening of the adhesives as they cool. Jigs may be required for accuracy and good access is vital. Gap-filling is excellent (especially with the foaming versions). Substrate materials must be chosen to match the adhesives application temperature. Loads may only be light, especially if exposure to elevated temperatures is anticipated.

Adherend compatibility

These adhesives are very versatile and appropriate formulations will even bond the difficult polyolefins. However, they are unsuitable for most rubbers.

Surface preparation

Surfaces must be dry and oil free.

Major benefits

Speed and versatility.

Major restraints

They may harden too quickly for accuracy, cannot sustain high loads and are temperature sensitive. Their generally high melt-viscosities may make close work on small objects difficult.

Fault finding

The adhesive may chill too quickly or simply fail to wet a contaminated surface.

Equipment

May not always be required where manual placing of sheet film is sufficient. Otherwise, equipment varies in size and complexity according to application, ranging from modest hand guns to extremely large, complex metering and dispensing equipment coupled to fume and vapour extraction facilities.

Health and safety

The major problem is the possibility of severe burns. According to formulation, the vapour and fumes may vary in physiological activity but, in principle, these must always be extracted.

Use in accordance with the manufacturer's instructions, Health and Safety data and ensure that all COSHH requirements are in place.

HOT MELT – REACTIVE

Adhesives in this group comprise a special sub-set of the 'hot melt family'. They are distinguished by the fact that they are based on a variety of isocyanate pre-polymers. These, while forming tacky solids at normal ambient temperatures, are capable of reacting further on exposure to atmospheric moisture. Thus, they will bond components quickly as they cool and then proceed to increase their strength through further reaction.

According to the circumstances governing atmospheric exposure, full cure will take up to two days, but generally this is not a disadvantage.

Within this group a wide variety of properties have been made available. Differential cooling rates, ability to reflow under IR heating, levels of (or even absence of) tack, melt temperature variations and their associated strength levels. All serve to aid specialized types of mass production.

Unlike the conventional hot melt they will function well at high atmospheric temperatures and it is for this reason they are seeing an ever extending use in the motor industry for attaching trim components and even lightly loaded parts.

Like the polyurethane adhesives they are sensitive to moisture both in the pre-use storage stage and after application. They should not be used on unpainted metal if they are likely to be exposed to exterior conditions.

Because they really combine the properties of hot melt and single-part polyurethanes care needs to be exercised in their use. Special attention must be paid to ventilation. Reference should be made to the Health and Safety requirements of each of these types.

NATURAL

Adhesives made from natural substances have been in use for a very long time. Many of them are being progressively replaced by synthetic types, but some are still being used and examples are given in:
Table 1.7.42 *Natural adhesives*

PHENOLIC

Adhesives of this type are usually medium or high viscosity liquids, but powdered and film forms are available, the latter being particularly important for the aircraft industry. The metal-compatible forms are usually heat cured and, along with the acid-catalyzed, cold-curing variants, are often used on wood. Water release during cure demands high pressure to maintain contact between the surfaces being bonded and to prevent the formation of bubble-based voids.

Cure mechanism

The adhesives of this type are phenol and/or resorcinol-based, part-reacted with formaldehyde and usually dissolved in water. Dry powder and film forms are available. The reaction is completed during bonding and this stage may be initiated either by elevated temperature or a catalyst. More water is released during this process as a by-product of the reaction.

Normal applications

Bonding and laminating wood and metal especially for use in warm wet conditions. Speciality use with friction materials and ceramics where high temperatures may be encountered.

Design criteria

The group is limited by its cure mechanism and the consequent need to press surfaces together to accommodate the loss of water. Thus, the **process** lends itself to bonding large flat or gently shaped areas which may be readily compressed and, if required, heated. While these adhesives are quite unsuitable for bonding coaxial assemblies they lend themselves particularly well in the film form to the fabrication of large structures such as those seen in the aircraft industry.

Adherend compatibility

Between them the various sub-groups cope with the common industrial materials, particularly steel and aluminium alloys and with many absorbent materials, especially wood and its derivatives. With a few exceptions, such as polyamides, the family as a whole cannot cope with most thermoplastics though some of the thermoset plastics can be bonded extremely well.

Surface preparation

The surface of cleanly cut wooden components and absorbent materials should need no special preparation. However, roughly abraded wood may not absorb the adhesive adequately and performance should be carefully checked if the bonded components are to be used in a demanding situation.

Major benefits

High shear strength and excellent durability, especially in harsh environments.

Major restraints

The family is prone to brittle failure and even the much improve, slightly flexibilized and toughened versions should be used cautiously if mechanical abuse is likely. The usual requirements of heat-induced polymerization, high clamping pressures and the need for very careful metal preparation greatly restrict the use of these adhesives in the mass production, metal-based industries.

Fault finding

Since the adhesives are supplied in a part-reacted state some types may continue to react during storage. This may cause poor adhesion because the substrate will not wet properly. Otherwise, most problems are likely to be associated with improper mixing (where required), poor surface preparation and failures during the heating and pressing stages.

Equipment

Varies in size and complexity according to the formulation and application. Large complex installations are required for many woodworking applications, but despite this, the use of these materials remains a cost effective and attractive proposition in this industry.

Health and safety

The various resins, hardeners and fumes encountered must be considered to be potentially irritant, dermatitic and toxic.

Use in accordance with the manufacturer's instructions, Health and Safety data and ensure that all COSHH requirements are in place.

PLASTISOL

Usually prepared as viscous, immobile pastes based on suspensions of PVC particles in a combination of liquids, these adhesives are normally composed of plasticizers and reactive monomers.

Cure mechanism

Heating to around 180°C swells the PVC which absorbs the liquid phase of the suspension. Any monomers present react and cross-link to give a semi-thermoset polymeric mass.

Normal applications

Assembly of large metal components, typically the lightly stressed panels of motor vehicles, bonnets, boot lids etc.

Design criteria

Compared with the elegant use of some adhesives, plastisols are relatively crude and do not lend themselves readily to the assembly of smaller components.

Adherend compatibility

Adhesives in this group are traditionally applied to sheet steel but other adherends can also be bonded successfully. The main limitation is the high curing temperature.

Surface preparation

Not normally necessary.

Major benefits

The whole family is simple and easy to use and their tolerance of poorly prepared surfaces ensures that they offer a robust and reliable performance.

Major restraints

Their thick paste-like nature prevents fine work and their aggressive tendency to dissolve adherends during the heated curing cycle tends to limit their use to fairly large metal components. Although widely used in the vehicle industry, they are becoming less popular because they induce acid based corrosion in the immediate area of spot welds.

Fault finding

They are not prone to any particular problem, though prolonged storage may cause settlement.

Equipment

Very simple dispensing apparatus is required but fume extraction must be used during curing.

Health and safety

There are no particular hazards associated with the cold, unreacted adhesive though some of the monomers used may be irritant.

Use in accordance with the manufacturer's instructions, Health and Safety data and ensure that all COSHH requirements are in place.

POLYIMIDES

These adhesives are usually single component, heat cured materials which are supplied in a film or paste form. Their primary application is bonding at high temperatures.

Cure mechanism

This can be either a condensation or an addition reaction. Processing is difficult compared to acrylic and epoxy adhesives.

Normal applications

Wherever high service temperatures demand that these expensive, intractable and brittle materials are used. They see little service outside the aerospace and electronic industries. Depending on type, they can function at 350°C for short periods or as long as 3000 h at 220°C. This latter temperature is the maximum that conventional epoxides can normally withstand, for a short period, before they rapidly oxidize and degenerate.

Adherend compatibility

Normally only used for metal and ceramic bonding.

Surface preparation

Surfaces must be thoroughly prepared.

Major benefits

Their performance at elevated temperature.

Major restraints

They are difficult to use, most are brittle and all are relatively expensive.

Fault finding

Process requirements, which can be complex, not being met.

Equipment

Can be complex.

Health and safety

Some versions involve materials which are physiologically active and these need to be handled cautiously.

Use in accordance with the manufacturer's instructions, Health and Safety data and ensure that all COSHH requirements are in place.

POLYURETHANE (BOTH SINGLE- AND TWO- COMPONENT)

The family group is named after the polymer type formed on completion of the reaction. The adhesives are always isocyanate base. The hardeners of the two-component versions are usually based on amines or glycols. Some single component versions can be used as hot melts. (See Hot melt – reactive). Some solvent borne, non-reacting versions are also available. (See 'Rubber solution').

Cure mechanism

On mixing, the isocyanate and hardener undergo a powerful, rapid addition reaction at room temperature, normally resulting in a 100% conversion to the polymeric adhesive form. However, if required, gas-producing reactions can be incorporated to foam the adhesive. This is often done to both bond and insulate. Single-component variants react with atmospheric moisture in order to effect cure. However, this reaction can take a long time to complete; 8 hours or sometimes even more, to handling strength is not uncommon.

Normal applications

The bonding of a wide range of components fabricated from metal, wood, various plastics (both thermoset and thermoplastic) and rubber. The adhesives have general utility but their viscosity and rate of cure can limit use on the smaller components.

Design criteria

These strong resilient adhesives are capable of structural use, though the paste-like constitution of many versions prevents their ready use on small objects. The family has excellent gap-filling characteristics particularly enhanced in foaming forms. The single part versions are often used for both sealing and bonding when flexibility is required and loads are either very light or nominal.

Adherend compatibility

The family as a whole is very versatile. While performance on some plastics is only modest, the group may be used with advantage on plastics materials sensitive to stress-cracking. The polyolefins and other low surface-energy plastics present major difficulties unless specially prepared.

Surface preparation

May be critical and depends to a large degree on the nature of the adherend and the individual adhesive. Like cold cured epoxies the polyurethane adhesives do not cope well with oily surfaces. Generally, it is necessary to prime otherwise bare metal surfaces to enhance performance in hot and humid conditions. Some special versions are said to be capable of bonding unprepared polyester composite.

Major benefits

The family as a whole will bond a wide range of adherends. The various formulations may be readily manipulated to give excellent gap-filling and high reaction speed coupled to a robust structural performance.

Major restraints

Both the raw materials and the final polymeric form of the adhesives are sensitive to moisture-induced damage, particularly if used on unpainted metal. Atmospheric moisture reacts readily with the isocyanate resin rapidly rendering it unusable. The cured adhesive should not be used for high loads in hot, humid environments unless surfaces have been especially prepared and the assembly has been thoroughly evaluated. The single-component versions cure very slowly and the rate depends upon the availability of atmospheric moisture.

Fault finding

The family as a whole can suffer from inadequate surface preparation and poor wetting. Variations in reactivity can be induced by moisture, and temperature changes will affect critically the timing and efficiency of some fast setting machine-mixed and dispensed versions.

Equiment

Modern formulations have substantially reduced the cost and complexity of the equipment required to handle these adhesives.

Health and safety

Recent developments have reduced the physiological hazards which have long been associated with these types of materials. Nonetheless, their use should be supported by good industrial hygiene and the very reactive versions should be treated with caution.

Use in accordance with the manufacturer's instructions, Health and Safety data and ensure that all COSHH requirements are in place.

RUBBER SOLUTION

The family is based on natural and synthetic rubber solutions whose viscosity ranges from the relatively low to viscous pastes and semi-solids. They are becoming less popular because of the environmental implications of the loss of organic solvents by evaporation.

Cure mechanism

These thermoplastic and flexible adhesives function when the carrier either evaporates or is absorbed. The residual polymer, formed as the solvent is lost, gives a slightly mobile and tacky film. This is capable of migrating into a second such film to provide a coherent structure.

Normal applications

Laminating plastics sheet and the bonding of an extremely wide variety of materials, particularly where flexibility is required and the operating environment is not too demanding.

Design criteria

The family should be considered only for very light or nominal loading. The formulations within the family vary widely in tack, durability, strength and resistance to fluids and elevated temperatures. Components must be designed to allow precise placement because instantaneous grab is displayed by many formulations. Adequate means of pressing the parts together after initial contact must often be available.

Adherend compatibility

The family as a whole may be used on almost any article not damaged by the solvent used to dissolve the rubber. A distinction must be made here between an article itself and the material from which it is made. Used in thick section, some plastics will resist solvent damage readily whereas otherwise identical thin sheets may be badly affected.

Surface preparation

Load-bearing, metal bonding normally involves those special formulations which are cured by the vulcanization process, and are not really part of this group of adhesives. In such cases surface preparation is critical and the manufacturer's advice should always be sought.

Rubber solutions will bond well to most surfaces, other than metal and polyolefin plastics. Normally preparation is not so critical, but oil and grease must be removed and fibrous materials, such as leather, are best bonded after coarse roughening with a rasp.

Major benefits

The basic forms are effective and very simple to use.

Major restraints

In the main, the adhesives are restricted to light or nominal loads in benign environments. Solvent loss may present both fire and health hazards. Although the bonding process is almost instantaneous, long periods may be needed for the necessary solvent evaporation to take place prior to joint closure; this can be inconvenient. They are becoming less acceptable because of their potential for environmental damage.

Fault finding

Assembling either too soon or too late, is the major hazard. With premature assembly, the surfaces will not be retained in position because residual solvent reduces the adhesive film strength and natural stresses in the adherend may cause joint cleavage. Delayed assembly allows excessive solvent loss when the dry films are not tacky enough to be effective.

Equipment

This ranges from very simple to extremely complex depending upon the application and its scale. However, the more complex equipment is usually only used by specialist organizations making components for which these materials were traditionally developed. If used on any appreciable scale, ventilation must be provided.

Health and safety

The physiological activity of the solvents used and the vapours released during the evaporation must be recognized. Solvent vapour will irritate the eyes and induce narcosis if inhaled in quantity. However, used sensibly, these adhesives cause few problems, hence the earlier popularity of the simpler versions that were employed for domestic repairs.

Use in accordance with the manufacturer's instructions, Health and Safety data and ensure that all COSHH requirements are in place.

TAPE

Single and double sided self-adhesive tapes come in many forms and sizes. A variety of backing materials may be used and common examples are cellulose, polyester, foamed polyurethane and PVC. Those of modest performance tend to be based on moisture activated gummed paper or fabric. Better performances are obtained from tapes based on a tacky, semi-solid acrylic or rubber based polymer which may be used with a release strip. Some special higher performance tapes are heat activated.

Cure mechanism

Apart from the cross-linking variants, the permanently tacky surface of the tape adheres firmly to most surfaces. However, the common materials can only carry nominal loads, though the cross-linked structure of the specialized forms may be lightly loaded.

Normal applications

Between them the single, double sided and foam-supported versions will bond almost any objects whose surface geometry is not too demanding and where the loads involved are either nominal or very light. Specialized types will cope with greater loads and the more demanding external environments.

Design criteria

The family as a whole presents an effective and fast method of attaching light parts and components. Creep can be a problem at temperatures much above normal atmospheric and solvents must be avoided. Surfaces to be bonded must be clean and if not flat, preferably curved only in one plane, rough surfaces do not bond well since the tape can only attach itself to the high spots.

Adherend compatibility

The family as a whole will bond almost any surface to some degree.

Surface preparation

The family as a whole will bond almost any surface to some degree.

Major benefits

The main advantages are speed, ease of use and freedom from hazards.

Major restraints

Not even the cross-linked versions can carry substantial sustained loads and the family is particularly susceptible to creep. It is also sensitive to heat and solvents. Tapes are quite unsuitable for use on 'fitted' parts (where one piece fits into another) and are difficult to use on three-dimensional shapes. They are not normally effective on contaminated surfaces.

Fault finding

Failure to adhere is usually associated with dust or oily surface contamination. A common failing in use is caused by plasticizing oils migrating from the bonded substrate (often flexible PVC sheet) into the adhesive layer, softening it and thereby leading to premature failure.

Equipment

Very simple mechanical dispensers are available if required.

Health and safety

The family as a whole is generally considered to be free of user hazard.

1.7.7.3 The adherends

Some materials will bond far better than others and some will not bond at all without special treatment. The suitability for bonding depends on surface preparation, joint design and the function to be performed (joining, sealing, etc.).

Most common metals can be bonded satisfactorily. Really careful surface preparation (e.g. cleaning, abrasion, and chemical treatment) is normally only needed to maximize the performance of critical structures used in demanding environments. Were it not for contamination, the thermoset plastics (polyepoxy and polyester composites) would bond well without light abrasion and solvent washing. Thermoplastic materials are another matter altogether because of their generally low surface-energy. Such materials must normally be treated to obtain an adequate bond.

The primary adhesive used in structural bonding, where demands are at their highest are the acrylic, epoxy and PU types. The first two usually being used in their toughened versions for the more highly stressed applications.

A broad impression of the compatibility of these adhesives with a number of commonly encountered adherends is provided in:

Table 1.7.43 *Compatibility of the principal structural adhesives with a variety of composite and associated materials*

The information given should be interpreted carefully in the light of the following observations:

Ceramic — Will usually bond well unless conditions are warm and moist when a coupling agent should be used (see glass).

Glass — Will only bond well if a coupling agent is used. This may be incorporated into the adhesive or used separately.

Metal — Both the heat cured epoxides and some of the toughened acrylic types will bond oil-contaminated metal surfaces very successfully. However, safety-critical, loaded joints must be properly prepared to guarantee a long durable performance. PU based adhesives must not be used directly on bare metal, or oil-contaminated, surfaces which will be exposed to demanding external environments.

Aluminium — Bonds well with the epoxides and particularly well with the pseudo-single-part acrylics. Maximum durability requires either a silane, phosphate or chrome passivation treatment unless the adhesive is formulated so as to contain a coupling agent of the silane type.

Steel — Bonds well with all three types. Maximum durability requires silane or chrome passivation treatment although a very good performance can be obtained from grit blasting. Silane fortified adhesives can be effective without direct chemical modification.

Zinc — This is a weak, readily oxidized metal that behaves badly unless properly prepared and loaded correctly. Chrome passivation give bonded joints of excellent durability thought if these are to be stressed low modulus adhesives should be used to spread the load. (Note: PUs will normally need a separate primer coating).

Thermoplastics

ABS — Tendency to stress-crack with pseudo-single-part acrylics. Two-part mixed acrylics and PUs are better. Epoxides are generally too stiff for this type of plastic.

Polyamide (nylon) — Bonds poorly unless bonded hot or pre-treated by shot blasting or corona discharge. (Whenever corona discharge is mentioned plasma treatment may also be considered. Although the latter is said to be more effective it requires the use of a vacuum chamber.) When hot bonding consider release of absorbed moisture.

Polycarbonate	Strong tendency for unalloyed material to stress crack. Evaluate acrylic adhesives carefully. Surface preparation may be needed; shot blast, corona discharge could be better.
Polyphenylene oxide	Bonds poorly unless bonded hot or pre-treated by shot blasting or perhaps better by corona discharge. When hot bonding consider release of absorbed moisture. May stress-crack; evaluate acrylic adhesive carefully.
Polypropylene	Surface preparation needed; shot blast, better being corona discharge, or chlorinated primer.
Polyurethane	Some of these composites may bond reasonably well with the acrylic adhesives. Epoxides are unsuitable, polyurethane best.
PVC (including plasticized version)	Bonds very well without surface preparation with most of the acrylic adhesives. Take care to evaluate longterm performance if the PVC contains an unreacted plasticizer.

Thermosets

Epoxy	Usually used for only the most demanding circumstances and consequently bonded with either the two-part or single-part epoxides. When the latter are used care should be taken to ensure that any absorbed water is released prior to bonding. While the acrylics and PUs can bond them well the usually lower moduli of these adhesives generally makes them unsuitable for the type of work normally involved. Check carefully for the presence of mould release agent.
Polyester	Will bond well with all types of adhesive but single-part epoxides not normally used because of the higher curing temperatures required. If single-part epoxides are used then care should be taken over the release of absorbed water prior to bonding. Surface preparation needs very careful consideration because of mould release agent.

Paint	Cataphoretic paints are probably the only paint materials with sufficient integrity to be used as a substrate for structural bonding. Even then, the underlying surface will need proper preparation if the paint itself is to adhere adequately. Paint, or a specific primer, needs to be used to ensure the chemical stability of PU adhesives on metal.

Set against this background it will be readily appreciated that the cheapest adhesives may not actually provide the most economical solution which is technically acceptable when everything is taken into account. Each application requires careful consideration of the costs incurred when extra painting processes may have to be invoked or heating equipment is required to cut production cycle times to an acceptable figure.

The subject of adherend–adhesive compatibility is included in:

Section 1.7.7.5 *The Adhesive Selector* in a broader, though less specific framework than has been considered here.

1.7.7.4 Joint design for adhesive bonding

A joint must be designed to withstand the type of load encountered in its specific situation. Different stress configurations require the use of different types of joints.

STRESS CONFIGURATIONS

Compressive

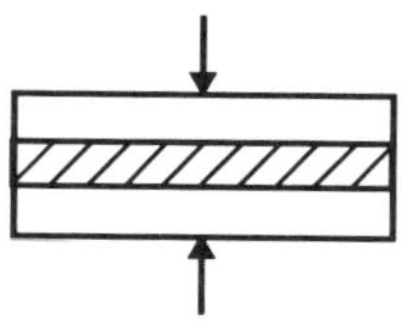

Adhesive joints perform best under compressive stress but pure compression may not be achievable in practice.

Shear

(i) Compressive shear

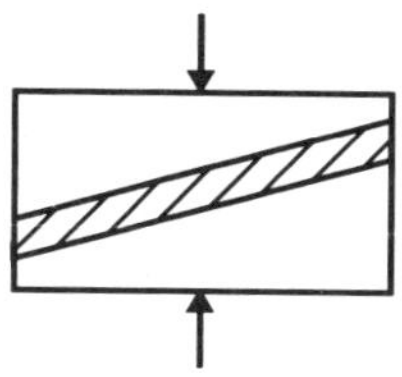

A very high performance is obtained from this type of joint and joints in shear should be designed whenever possible to introduce compression.

(ii) Uniform shears

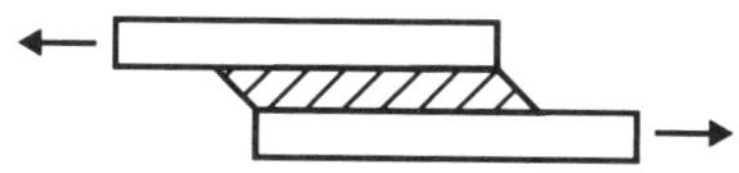

Joint area is used to advantage to give an economical bond that is resistant to failure. Most frequently used joint type. (Note: True shear is rarely achieved and there is usually an element of peel.)

(iii) Torsional

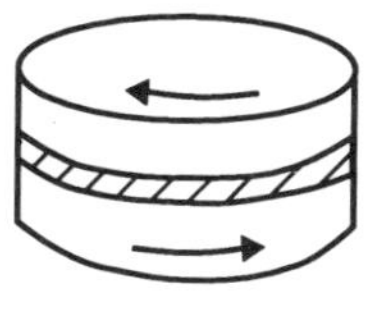

Not often met with in practice.

Tensile

Must be designed to ensure that no point of joint carries more than its share of the load. To prevent non-uniform stressing, adherends should be thick and not liable to deflection under load.

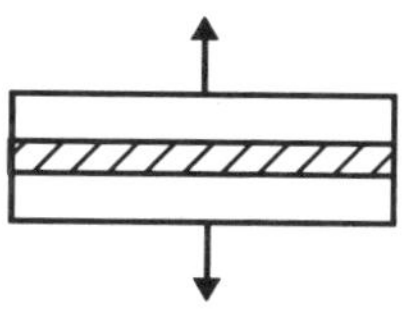

It is very easy with joints loaded in tension to introduce cleavage stresses, inadvertently. These stresses give rise to premature and unnecessary failure.

Cleavage

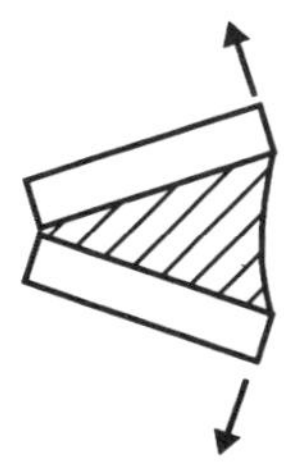

High stress concentration.

NOT RECOMMENDED.

Peel

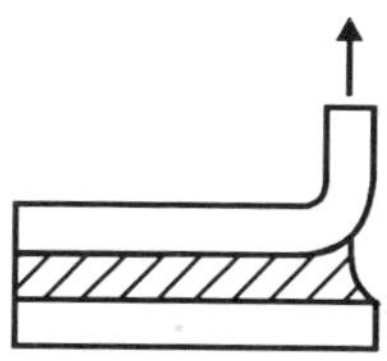

High stress concentration.

NOT RECOMMENDED.

JOINT TYPES

Lap joints

The mean rupture strength is determined by the maximum stress values at the joint ends.

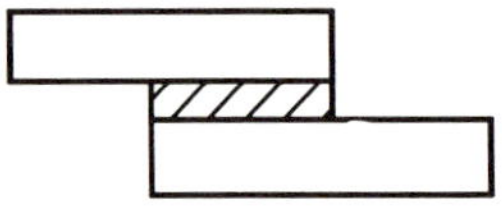

(i) Simple lap joint

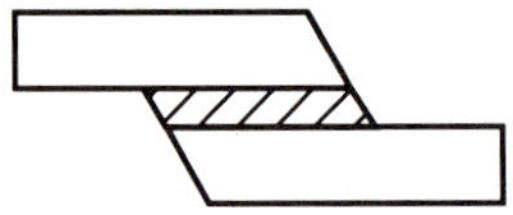

(ii) Bevelled lap
The introduction of toughened adhesives has rendered this type of joint obsolete for many applications. Angles depend on adhesives, stresses and materials.

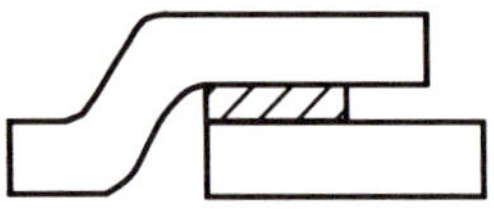

(iii) Rebated lap
The rebated joint has the best strength properties in this class, because it is one type of compressive shear design.

(iv) Scraft joint
This design of joint should only be used to join very thick sections.

Butt joints

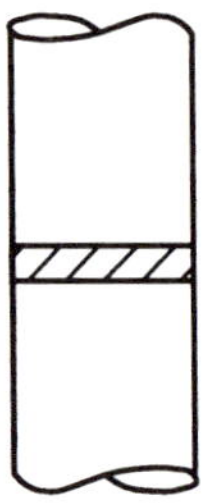

(i) Simple butt joint. Satisfactory in compression but tends to fail by cleavage in tension.

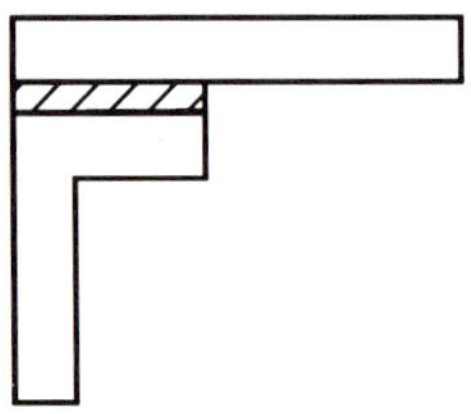

(ii) Right angle butt joint suitable for bearing low loads only.

(iii) The performance of a butt joint may be improved by distributing the load.

Strap joints

Joints comprising the advantages of butt and lap joints. These have better strength properties but incur higher machining costs.

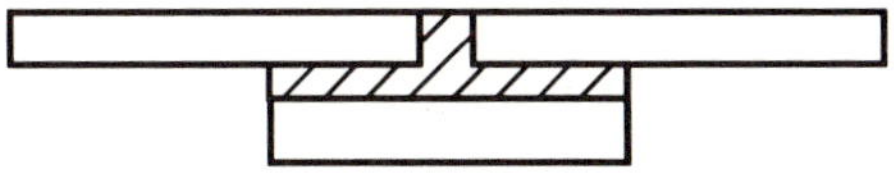

(i)
Single strap

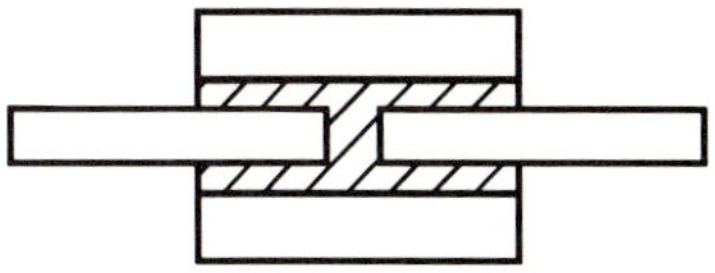

(ii)
Double strap

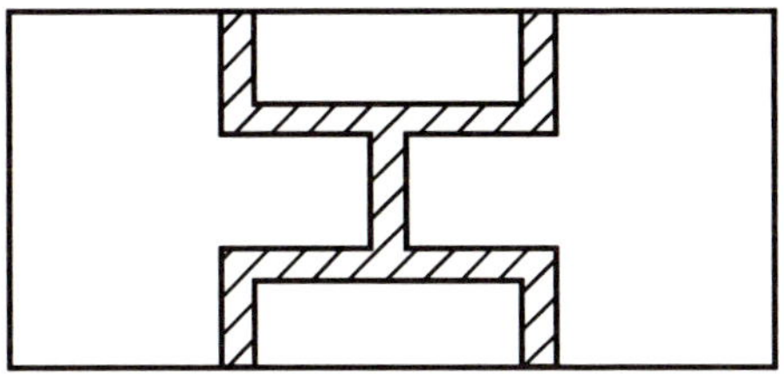

(iii)
Stepped double strap

Hybrid joints

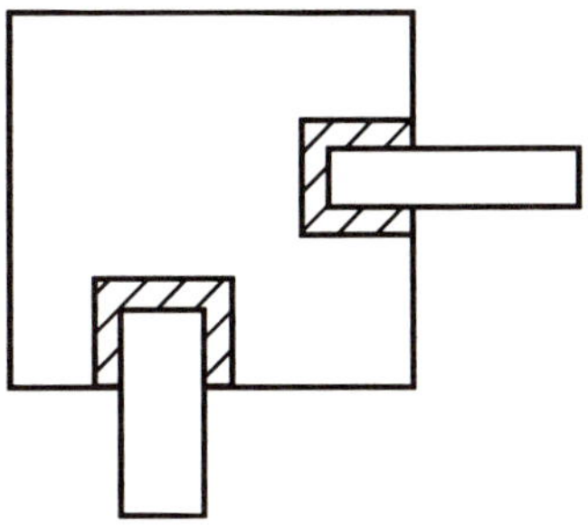

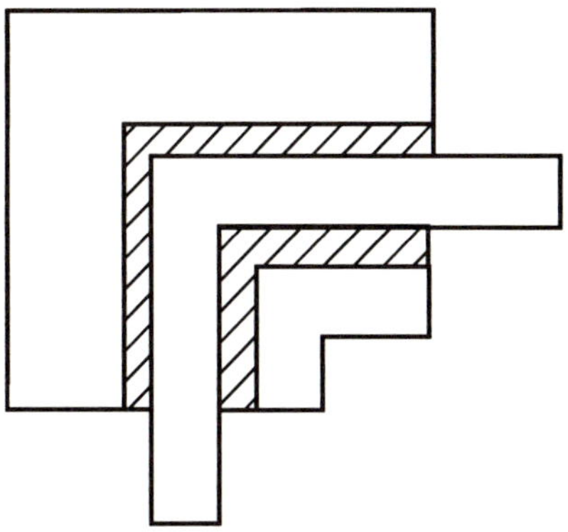

Stepped double strap and hybrid joints are economic only for use with extruded sections.

Coaxial joints

Alternative to interference or press fits. Usually made by anaerobic adhesives because their cure time facilitates assembly and adjustment and they are suitable with the joint clearances commonly met in conventional engineering practice. Best results are obtained around 0.05 mm clearance.

(i) Ideal $L = 2D$
In any case $L > D/2$
If $L < 2$ mm satisfactory bonds may not form owing to oxygen inhibition.

(ii) Systems which might induce peel forces in the joint should be avoided.

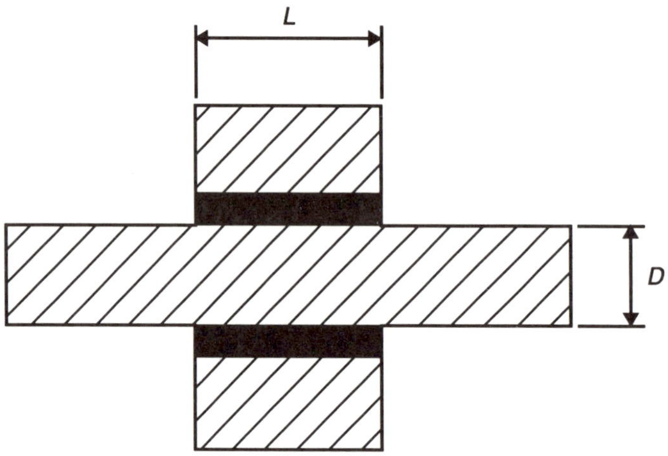

Joints in peel and shear situations

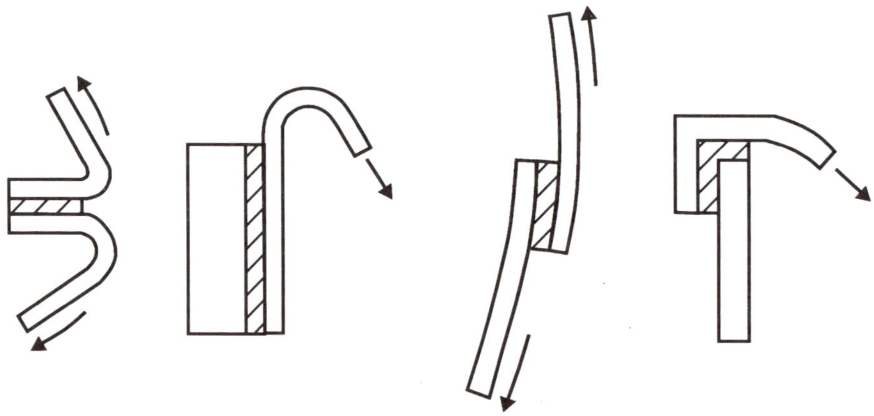

(a) (b) (c) (d)

(a), (b) NOT RECOMMENDED – peeling.

(c), (d) Preferred types for resisting shear stresses because they introduce compressive stresses.

1.7.7.5 The Adhesive Selector

Many factors influence the selection of an adhesive for a particular application. In order to select the best type the 'Adhesive Selector' procedure given here is in the form of an elimination questionnaire. An extended and detailed version of the 'Adhesive Selector' is also available in the form of a computer program†. This procedure will indicate which adhesive types are the most suitable, or the least objectionable, candidates for any particular application.

Its use will enable the enquirer to have a much more informed discussion with a potential supplier, a practice which is always recommended because a selection procedure cannot cover all eventualities.

The 'Adhesive Selector' comprizes the following tables:

Table 1.7.44 *List of adhesives, codes and related special notes*
Table 1.7.45 *Adhesive/adherend compatibility table and related special notes*
Table 1.7.46 *Adhesive Selector application questionnaire*
Table 1.7.47 *Adhesive Selector assessment table*
Table 1.7.48 *Special notes to Adhesive selector*

OPERATING INSTRUCTIONS

1. Consult Table 1.7.45 (Adhesive/Adherend Compatibility Table) and enter on a copy of Table 1.7.47 (Assessment Table) the status of each of the adhesives listed for each of the two adherend surfaces. Note, if there is only one surface material involved only complete one line. Also, enter in the column provided the reference number(s) of the Special Notes given for each material (adherend) involved.
2. Consider the questions presented in Table 1.7.46 (Application Questionnaire) and, in view of the decision taken, mark the relevant line and column with an × for each individual adhesive rejected. Note: often, there will be a nil entry – no adhesive having been rejected by the decision taken. Enter the reference number of any Special Notes raised by a question and in particular take action on Notes 31, 34 and 35 during the process of entering the rejects.
3. Having completed the questionnaire and having noted all of the rejected adhesives in Table 1.7.47 count the number of times each adhesive has been rejected and enter the number in the last row. Take particular care to log the zero entries – adhesives which have not been rejected at any stage during the progression through Table 1.7.46.
4. Prepare a separate list of all the adhesives which have suffered no more than two rejections.
5. Consult Table 1.7.44 (List of adhesives, Codes and Related Special Notes) and annotate each of the adhesives listed (Table 1.7.48) with the relevant Special Note.
6. The most likely adhesive to suit the application in question will be zero rated (in terms of number of rejections) but the final choice must be based on an assessment of the significance of the Special Notes associated with the adhesive in question and those related to the adherends involved.

   Sometimes, though not often, the Notes will indicate that there is some form of incompatibility. In which case, consideration should be given to those adhesives which suffered no more than two rejections – provided that the rejections were relatively trivial in nature. Alternatively design modifications or production changes could avoid the issue which caused the rejection.

   If consideration of the less desirable adhesives fails to produce an acceptable solution, either consideration must be given to changing the materials to be

† Information on the application and availability of this computer program version of the selector may be obtained from Permabond Division, NSC Ltd., Woodside Road, Eastleigh, Hampshire, UK.

bonded, or a major design change is required.

As a generalization it can be said that the greater the number of adhesives allowed by the selection procedure the greater the likelihood that a suitable individual adhesive will be found for the application in question. The more restrictive the selection becomes the less likely there is to be a satisfactory solution to the problem.

7. Finally, discuss the problem with an adhesive manufacturer.

1.7.7.6 Using adhesives

PRODUCTION CONSIDERATIONS

The specific characteristics and requirements of individual adhesive types need to be considered against the realities of cost, space and the possible conflict of subsequent production processes. For example, some adhesives must be used with relatively expensive metering equipment, and extractors must often be fitted to remove fumes. Such costs, coupled with the provision of jigs, can be enough to force a design change or, at least, a reappraisal of other adhesives and their associated construction techniques.

Some processes employ conditions subsequent to adhesive assembly which are detrimental to the adhesive itself, e.g. welding next to a bonded joint will clearly ruin the latter.

A list of some of the more important production parameters affecting choice of adhesives is given in:

Table 1.7.49 *Production factors affecting choice of adhesives*

A summary of the more common factors responsible for production problems is given in:

Table 1.7.50 *Identification of production problems*

In this latter connection it is worth pointing out that the most likely time for problems to occur is when a new bonding process is introduced. Active liaison between user and supplier is strongly advocated for this can eliminate or at least minimize the teething problems which are otherwise likely.

With a fundamentally sound application and a correctly installed manufacturing process, trouble free production can be anticipated once all concerned have become familiar with what is often, to them, a completely new production technology. Consequently, it is strongly recommended that written procedures and specifications are established and that these are supported by regular audit. A check list for safety critical issues is given in:

Table 1.7.51 *Safety critical check list.*

ADHESIVE STATUS AND SURFACE CONDITION

The successful use of this Selector, and ultimately the proper use of the adhesives themselves, depend upon an appreciation of how 'adhesive status' (primary, secondary and reject) has been considered and defined (albeit in a subjective manner by the Selector's authors) in relation to a surface and its condition.

When an adhesive is classified as 'reject' (R) in relation to a surface then this is usually because there is some fundamental problem which is difficult to overcome i.e. it dissolves readily or will not stick at all, as is the case with silicone surfaces.

The classification for both 'primary and secondary' (P/S) assumes that the surface has been prepared for bonding to the minimum standard of 'Method 1' (See 'Surface preparation'), or as required if more is considered necessary. The distinction between a 'P' and 'S' classification is that in relation to the surface under consideration a primary adhesive is considered to have greater utility within that relationship than an adhesive classed as secondary. A secondary adhesive is not, and should not be considered to be, 'second rate'. The adhesive in question may be more difficult to use or its performance may be repressed in some way, but given that there is good practice, an adequate performance can be anticipated. It is essential that this point is appreciated for sometimes other considerations may prevent a 'primary' from actually being used (for example, heat curing may be impractical in the production area concerned).

THE OPERATING ENVIRONMENT

Hot, humid conditions can severely degrade some adhesives in contact with metal, glass, ferrites, rubber and plastics. Glass is a particularly difficult surface and invariably needs either a special adhesive or special treatment if a long bond life is to be expected. On the other hand, some adhesives and many substrates are barely affected by bad environments, although the possibility should always be borne in mind and discussed with the supplier.

Similar, many adhesives are only suitable for short duration service at temperatures above 80°C. Again, caution is required and detailed advice should be sought. The performance of some of the toughened types is noteworthy, for they can cope successfully with temperatures in the 180–200°C range.

Shock, peel and impact forces are often met by chance rather than by design and their effect can be disastrous unless the basic concept is adequate. The resistance of the toughened adhesives to shock, cleavage and peel forces is outstanding.

SURFACE PREPARATION

Simple surface preparation such as cleaning and degreasing is required for most adhesives. Materials which are prone to weak or loose surface layers, stress-cracking, solvent attack or water migration require special treatment.

Certain adhesives work well on unprepared surfaces. Cyanoacrylates are capable of penetrating or dissolving surface debris, especially on plastics. Surface cleanliness is not too important when coaxial joints are fabricated with anaerobic adhesives, as the adhesives are supported by a locking or jamming action, and in the majority of cases, this is sufficient to hold the components securely together even though the level of adhesion gained may be low. Some of the toughened adhesives can cope with oil films. The toughened acrylic and toughened heat cured, single-component epoxides are particularly good in this latter regard.

The preparative techniques given below have been drawn from the best of current practice and some of the latest methods which have been developed. In this latter regard the requirements of current, and forthcoming, environmental legislation are particularly pertinent. Users must ensure that the methods they choose will meet both their technical needs and their legal obligations.

The following notes are used extensively and are frequently referred to.

General notes

G1 Remove oil and grease based residues using aqueous materials if possible. Non-ionic detergents give good results. Proprietary alkaline cleaners may be used. Always rinse and dry thoroughly in a stream of warm air for about 10 minutes at 60°C. If solvents must be used then wherever possible use isopropyl alcohol. Otherwise, a ketone (MEK or MIBK) or an approved halogenated solvent. Alcohols and particularly ketones are flammable. All are narcotic in concentration. Ventilate properly, take account of vapour density and draw fumes away from the operator.

G2 Do not allow any cleaning materials to contact the skin. Abuse may lead to dermatitis.

G3 Wear eye protection and protective clothing.

Special notes

Some of the methods given below employ dangerous techniques, materials and proprietary chemicals. It is essential therefore that the suppliers' instructions are followed, Health and Safety Data Sheets are studied and appropriate COSHH procedures established.

Waste and spent materials, chemicals and solutions must be disposed of in accordance with legal requirements and preferably through the services of an authorized disposal organization, whose advice should always be sought. Whenever solutions need to be made up, or utilized, then equipment made from only polyethylene, polypropylene or tetrafluoroethylene should be used.

METHOD 1

The following method (hereafter referred to as Method 1) is suitable for a number of different materials.

1 Remove oil or grease contamination.
2 Abrade or shot blast. Use medium grit.
3 Remove coarse debris with vacuum.
4 Remove fine debris. Use clean tissue and solvent.
5 Ensure surface is dry.
6 Bond without delay.

ASBESTOS BOARD AND ASBESTOS CEMENT.

Use Method 1.

Notes

G 1-3 and
4
Handle asbestos with great care – avoid breathing dust.

CELLULOSE including wood and chip/particle boards (paper does not normally need preparation)

1 Plane surface to remove contamination.
2 Remove debris using a vacuum.
3 Bond without delay.

Notes

1 Wood with a moisture content of more than 20% should be dried before bonding.
2 Abrasion, rather than planing, has been found to reduce joint strength.
3 Surfaces may be slightly acid. If a cyanoacrylate is to be used it may be necessary to employ a special activator which will neutralize the surface. Consult the adhesive manufacturer.

CEMENTITIOUS – concrete, mortar etc.

1 Remove oil or grease contamination by scrubbing with a 2% solution of a non-ionic detergent.
2 Wash thoroughly with water
3 Sand blast and remove 1–2 mm from the bonding surface.
4 Remove debris using a vacuum.
5 Allow the surface to dry thoroughly.
6 Bond without further delay.

Notes

1 If limestone, dolomite or other carbonate aggregates are present then the surface must be acid etched, after sand blasting, and then thoroughly washed.
2 Use dilute sulphuric acid. Take great care in handling and use full protective clothing. See SPECIAL NOTES above.

BITUMINOUS CONCRETE

The surface must always be scrubbed with a detergent solution. When all contaminants and loose debris are removed wash off detergent with a high pressure hose. Continue hosing until the surface no longer feels slippery.

CERAMICS – ferrite, masonry, pottery

1 Remove oil or grease contamination.
2 Abrade – see Note 4.
3 Remove coarse debris with vacuum.
4 Remove fine debris. Use clean tissue and solvent.
5 Ensure surface is dry.
6 Bond without delay.

Notes

G 1-3 and
4 Abrade bricks or other similar non-glazed ceramics with a wire brush. With glazed surfaces use emery paper or sand blast (medium grit) to remove glaze. A possible alternative is the use of a silance coupling agent.
5 Ferrites usually bond well without abrasion and cleaning is often effected well enough without items 2–4 above.

FABRICS

1 Remove oil or grease contamination.
2 Ensure material is dry.
3 Bond without delay.

Notes

G 1–3 and
4 Check that solvent chosen is compatible with fabric in the event that this is synthetic.

FRICTION MATERIALS (for example brake parts)

Use Method 1

Notes

G 1–3 and

4 Handle any asbestos based materials with great care – always avoid breathing dust from any source.

GLASS – glass and quartz

1 Remove oil or grease contamination.
2 Immerse for 10–15 minutes, at room temperature, in the etch solution.
3 Rinse thoroughly with distilled or deionized water.
4 Dry for 30 minutes at 98 ± 1°C.
5 Apply the selected adhesive preferably while the surface is hot.

Etch solution (See SPECIAL NOTES above.)

Material	Parts by weight
Chromium trioxide	1.0
Distilled/deionized water	4.0

Preparation:
Add chromium trioxide to water and stir to dissolve.

Notes

G 1–3 and

4 Cyanoacrylate adhesives may be inhibited by the etched surface.
5 An alternative to the etch treatment is the use of a proprietary silane coupling agent.

LEATHER

1 Remove oil or grease contamination.
2 Abrade using coarse sand paper or a rasp.
3 Remove coarse debris with vacuum.
4 Remove fine debris. Use clean tissue and solvent.
5 Ensure surface is dry.
6 Bond without delay.

Notes

G 1–3

PAINT – Cataphoretic

Use Method 1

Notes

G 1–3

PAINT – Epoxy powder

Use Method 1

Notes

G 1–3

PAINT – Other than cataphoretic paint

Use Method 1

Notes

G 1–3

METAL

ALUMINIUM

A Military

Use DEF STAN 03–2/3 or alternatively Boeing PAA5555

B Recent proprietary developments:

The aircraft procedures tend to be complex and expensive. Commercial alternatives are proprietary silane coupling agents and Accomet C (ex Albright and Wilson).

C Traditional:

1 Remove oil or grease contamination.
2 Immerse the item to be etched for 10 minutes at 68 ± 3°C in the etch solution.
3 Remove item and rinse thoroughly in cold distilled or deionized water.
4 Dry with hot air for about 10 minutes at a temperature not exceeding 65°C.
5 Bond as soon as possible after drying.

Etch solution (See SPECIAL NOTES above.)

Material	Parts by weight
Potassium dichromate or sodium dichromate	2.0
Concentrated sulphuric acid (S.G. 1.84)	10.0
Distilled/deionized water	30.0

Preparation:

With constant stirring, cautiously add the acid to 60% of the water and then add the dichromate.

ADD ACID TO WATER. NEVER WATER TO ACID

Stir until dissolution complete and then add the remaining water.

Notes

1–3 and
4 Treated aluminium should never be exposed to the atmosphere of a plating shop.
5 Although some special variations on the foregoing procedure will produce surfaces that may be successfully bonded after both oiling and component forming it should be assumed that, in general, treated surfaces are readily damaged and should not be touched, even when gloves are worn, if very high performance, safety critical structures are being created.
6 Cyanoacrylate adhesives may be inhibited by the etched surface.

ALUMINIUM – anodized

For general purpose work no preparation is normally required other than the removal of any contamination. However, surfaces should be bonded as quickly as possible after anodizing and preferably within four hours. For critical structural applications or where the surface is either 'hard' anodized or friable then the oxide should be removed by abrasion when one of the preparation procedures given for aluminium should then be employed. If the surface oxide needs to be removed proceed as follows.

1 Remove oil or grease contamination.
2 Abrade or shot blast. Use medium grit.
3 Remove debris with vacuum.

Notes

G 1–3

ALUMINIUM – passivated

Prepare as anodized aluminium.

CADMIUM – irrespective of base material

(A) Critical applications

This metal has often proved difficult to bond and for the more critical applications it has often been bonded after being electroplated with nickel. Where parts can be bonded immediately after electroplating they should be washed with deionized water, dried in a hot air stream for about 10 minutes at a temperature not exceeding 65°C and then immediately bonded. Otherwise, if a delay is incurred, proceed as required under the heading nickel, taking care not to remove an excessive quantity of the plate.

Other procedures which should be given consideration are DEF STAN 03–2/8, the use of Accomet C (Albright and Wilson) and conventional chrome passification procedures.

Notes

G 1–3 and
4 Conventional passification procedures are normally dedicated to reducing corrosion or providing a base for painting. It is important for bonding that the passivate is not loose or friable and that all plating residues have been thoroughly removed by adequate washing. See also Cadmium – passivated.

(B) General

Use Method 1

Notes

G 1–3

CADMIUM – passivated

It should be remembered that this surface is normally produced to reduce corrosion or to provide a base for paint. It normally bonds well unless the surface is loose or friable. Care should be taken to ensure that the surface is free of plating residues. If the surface is unsatisfactory it should be removed by abrasion and re-prepared. Either re-plate and passivate or use the procedures given under the heading Cadmium. If passivate needs to be removed proceed as follows:

1 Remove oil or grease contamination.
2 Abrade very lightly (fine/medium grit). Take care not to penetrate the plated cadmium layer.
3 Remove debris with vacuum.

Notes

G 1–3

CHROME PLATE irrespective of base material

1 Remove oil or grease contamination.
2 Immerse the item for 1–5 minutes at 93 ± 2°C in the etching solution.
3 Remove item and rinse thoroughly in cold distilled or deionized water.
4 Dry in a hot air stream for about 10 minutes at a temperature not exceeding 65°C.
5 Bond without delay.

Etch solution (See SPECIAL NOTES above.)

Material	Parts by weight
Concentrated hydrochloric acid (S.G. 1.18)	1.0
Distilled/deionized water	1.0

Preparation:
With constant stirring cautiously add the acid to the water.

ADD ACID TO WATER. NEVER WATER TO ACID

Notes

G 1–3

COPPER AND COPPER alloys (includes brass and bronze)

(A) Military

Use the procedure defined in DEF STAN 03–2/4.

(B) General

1 Remove oil or grease contamination.
2 Immerse the item at room temperature for the appropriate, specified time in the preferred etching solution.
3 Remove item and rinse thoroughly in cold distilled or deionized water
4 Dry using clean, cold pressurized air. Hot air will discolour the surface.
5 Bond without delay

Etch solutions (See SPECIAL NOTES above.)

1. Ammonium persulphate etch.

Material	Parts by weight
Ammonium persulphate	1.0
Water	3.0

Preparation:

With constant stirring dissolve the persulphate in the water. Immerse the parts for one minute.

2. Ferric chloride etch.

Material	Parts by weight
Aqueous ferric chloride (42% by weight)	15.0
Concentrated nitric acid (S.G. 1.42)	30.0
Distilled/de-ionized water	195.0

Preparation:

Add ferric chloride solution to water and then cautiously add the acid stirring continuously. Immerse the parts for 1–2 minutes.

3. Nitric acid etch.

Material	Parts by weight
Concentrated nitric acid (S.G. 1.42)	7.0
Distilled/deionized water	15.0

Preparation:

While stirring continuously add the acid cautiously to the water. Immerse the parts for 30 seconds.

ADD ACID TO WATER. NEVER WATER TO ACID

LEAD including pewter

Use Method 1

Notes

G 1–3

MAGNESIUM AND ITS ALLOYS

Preparation

(A) Military

Use the procedure defined in DEF STAN 03–2/7

(B) Traditional

1 Remove oil or grease contamination.
2 Immerse the item in a sodium hydroxide solution for 10 minutes at 70 ± 5°C. Wash in cold tap water.
3 Immerse the item in the etching solution for 10 minutes at room temperature.
4 Wash with cold tap water. Wash again with distilled or deionized water. Dry in hot air at a temperature not exceeding 65°C.
5 Bond immediately to reduce surface oxidation.

Sodium hydroxide solution (See SPECIAL NOTES above.)

Material	Parts by weight
Sodium hydroxide	1.0
Distilled/deionized water	12.0

Preparation:
Add sodium hydroxide to water and stir until dissolution is complete.

Etch solution (See SPECIAL NOTES above.)

Material	Parts by weight
Sodium sulphate (anhydrous)	1.8
Calcium nitrate	2.2
Chromium trioxide	24.0
Distilled/deionized water	122.0

Preparation:
Add the materials to water in the order given above and stir until dissolution is complete.

Notes

G 1–3 and
4 FIRE HAZARD – never place magnesium components in a vapour bath and never abrade.

NICKEL AND ITS ALLOYS (including plate)

1 Remove oil or grease contamination.
2 If necessary abrade or shot blast. Use medium grit.
3 Immerse the item for 5 seconds at room temperature in concentrated nitric acid (S.G. 1.42). See SPECIAL NOTES above.
4 Remove item and rinse thoroughly in cold distilled or deionized water.
5 Dry in a hot air stream for about 10 minutes at a temperature not exceeding 65°C.
6 Bond as soon as possible after drying.

Notes

G 1–3

STEEL – mild and iron but excluding stainless steel

(A) Military

Use the procedure defined in DEF STAN 03-2/2A-B.

(B) Structural

1 Remove oil or grease contamination.
2 Abrade or shot blast. Use medium grit.
3 Remove coarse debris with vacuum.
4 Immerse for 10 minutes at 60±2°C in the etching solution.
5 Brush off the black deposit with a stiff nylon brush under clean cold running water.
6 Blow dry with clean air.
7 Wipe over with isopropyl alcohol and allow surface to dry.
8 Then heat for 1 hour at 120°C.
9 Bond without further delay.

Etch solution (See SPECIAL NOTES above.)

Material	Parts by weight
Industrial methylated spirit	1.0
Orthophosphoric acid (S.G. 1.7)	1.0

Preparation:
With constant stirring dissolve the acid in the methylated spirit.

Notes

G 1–3

(C) General

Use Method 1.

Notes

G 1–3 and
4 For exterior use and other demanding environments the Method 1 should be supplemented by following stage 4 with a treatment based upon a proprietary silane coupling agent or Accomet C (Albright & Wilson). These supplementary procedures will enhance the performance considerably.
5 Oily (not lanolin based) steel can be bonded very well with certain types of adhesives. However, for safety critical structures preparation should be thorough.

STEEL – stainless

(A) Recent development

It has recently been shown that grit blasting (180/220 alumina grit) followed by treatment with Accomet C (Albright & Wilson) may be advantageously employed.

(B) Traditional

1. Remove oil or grease contamination.
2. Abrade using medium grit emery or shot blast using alumina (Note: metallic agents must not be used to roughen the surface)
3. Remove coarse debris with vacuum.
4. Remove fine debris. Use clean tissue and solvent.
5. Wash in detergent solution for 10 minutes at 75 ± 5°C.
6. Rinse in cold water.
7. Rinse in distilled or deionized water.
8. Immerse the item in the etching solution for 5–10 minutes at a temperature of 60 ±3°C.
9. Rinse in cold water.
10. Rinse in cold distilled or deionized water.
11. Dry. Use hot air stream, oven or infra- red lamps for about 10 minutes at a temperature not exceeding 95°C.
12. Bond without delay.

Detergent solution

Material	Parts by weight
Nansa S40/S (Albright & Wilson)	0.5
Tetrasodium pyrophosphate	1.5
Sodium hydroxide	1.5
Sodium metasilicate	3.0
Distilled/deionized water	133.5

Preparation:
Add materials in order given to water and stir to dissolve.

Etch solution (see SPECIAL NOTES above.)

Material	Parts by weight
Oxalic acid	1.0
Concentrated sulphuric acid (S.G. 1.84)	6.0
Distilled/deionized water	7.0

Preparation:
Add the sulphuric acid very cautiously to the water with continuous stirring. Then, with further stirring add and dissolve the oxalic acid at the etching temperature of 60°C.

ADD ACID TO WATER. NEVER WATER TO ACID

Notes

G 1–3

TIN

Use Method 1

Notes

G 1–3

TITANIUM AND ITS ALLOYS

1 Vapour degrease the surface.
2 Abrade using medium grit emery or shot blast using alumina (Note: metallic agents must not be used to roughen the surface).
3 Vapour degrease again.
4 Wash the item in a detergent solution for 10 minutes at 75± 5°C.
5 Rinse in cold distilled or deionized water.
6 Dry. Use hot air stream oven or infra-red at a temperature not exceeding 95°C.
7 Immerse the item in an etching solution for 5–10 minutes at room temperature.
8 Wash in cold distilled or deionized water.
9 Dry (oven) at 75±5°C for 10–15 minutes.
10 Bond without delay.

Detergent solution (See SPECIAL NOTES above.)

Material	Parts by weight
Nansa S40/S (Albright & Wilson)	0.5
Tetrasodium pyrophosphate	1.5
Sodium hydroxide	1.5
Sodium metasilicate	3.0
Distilled/deionized water	133.5

Preparation:
Add materials to water in the order given and stir until solution is complete.

Etch Solution (See SPECIAL NOTES above.)

Material	Parts by weight
Chromium trioxide	5.0
Sodium fluoride	10.0
Concentrated sulphuric acid (S.G. 1.84)	50.0
Distilled/deionized water	250.0

Preparation:
With constant stirring cautiously add the acid to 60% of the water and then add the other materials in the order given and finally make up the balance with the residual water. Acid and chromium trioxide are both toxic and corrosive; follow approved procedure for disposal.

ADD ACID TO WATER. NEVER WATER TO ACID

Notes

G 1–3 and
4 Recent work has shown that the steps 4–8 above may be replaced advantageously by the SHA process which requires anodization at 10 V in a 5 M solution of sodium hydroxide at room temperature.

ZINC AND ITS alloys including hot dipped galvanized steel, electro galvanized steel and iron–zinc alloy coated steel

(A) Military

Use the procedure defined in DEF STAN 03–2/8.

(B) Structural bonding

1 Remove oil or grease contamination.
2 Immerse item for 5 minutes in the alkaline cleaner Bonderite 71 in accordance with the manufacturers instructions (Pyrene Chemical Services Limited). See SPECIAL NOTES above.
3 Wash in distilled or deionized water.
4 Treat item with a 5% volume solution of Accomet C (Albright & Wilson) according to the manufacturers instructions to yield a dry coating weight of oxide of 0.5 g/m^2. See SPECIAL NOTES above.
5 Dry in oven or air stream at 115±5°C.
6 Bond immediately.

Notes

G 1–3

(C) Non-structural

The standard anti-corrosion and pre-painting passivation processes will suffice well for most non-structural bonding applications. However, care needs to be taken to ensure that the passivate layer is not too thick (a dark yellow or greenish appearance is an indication of this) and therefore weak and friable. Similarly, ensure it is not covered in a loose residue from a poor post-passivating washing process. Excellent results have been reported from the use of the various passivation materials supplied by W. Cannings Materials Limited. See SPECIAL NOTES above.

(D) Minimum treatment

Use Method 1

Notes

G 1–3

Zinc passivated – irrespective of base material

This surface normally bonds well unless the passivate film is loose or friable. If it is proceed as follows:

1 Remove oil or grease contamination.
2 Abrade or shot blast. Use medium grit.
3 Remove coarse debris with vacuum.
4 Select one of the preparative methods given under the heading Zinc and its alloys and proceed as appropriate.

Notes

G 1–3

METALS IN GENERAL

Use Method 1

Notes

G 1–3 and
4 Consider the use of proprietary silane coupling agents.

PLASTICS AND RUBBER

ABS

Cyanoacrylate, toughened acrylic and conventional solvent based adhesives will normally bond this surface without preparation. Nonetheless, care should be taken to ensure stress cracking is avoided. If cyanoacrylates are to be used and preparation is required ensure that only alcohol is used as a solvent.

For maximum performance with other adhesives either use the following method or appropriate sections in conjunction with proprietary corona discharge techniques.

1 Remove oil or grease contamination, take care to avoid stress-cracking.
2 Immerse the item in the etching solution for 15 minutes at room temperature.
3 Remove the item and rinse thoroughly in cold water.
4 Rinse again in hot water.
5 Dry in a hot air stream at a maximum of 60°C.
6 Bond without delay.

Etch solution (See SPECIAL NOTES above.)

Material	Parts by weight
Potassium dichromate or sodium dichromate	1.0
Concentrated sulphuric acid (S.G. 1.84)	10.0
Distilled/deionized water	30.0

Preparation:
With constant stirring, cautiously add the acid to 60% of the water and then add the dichromate. Stir until dissolution complete and then add the remaining water.

ADD ACID TO WATER. NEVER WATER TO ACID

Notes

G1–3 and
4 When using toughened two part acrylics ensure that the minimum amount of initiator is used and that after the bonding process no wet initiator and/or uncured adhesive remains on the surface. Otherwise, stress cracking may occur.

CELLULOID

Use Method 1

Notes

G1–3 and

4 Check that any solvent selected will not damage the plastic paying particular attention to the problem of stress-cracking.

5 If preparing for use with epoxides heat the plastic parts for 1 hour at 93°C and apply adhesive while the parts are still warm. Care, elevated temperatures will increase the cure rate of the adhesive used.

POLYACETAL

First, relieve any moulding stresses then follow the method given for ABS. Note that solvent based adhesives are not likely to be as effective here and that cyanoacrylate adhesives may be inhibited by the etched surface. Alternatively, a proprietary corona discharge technique may be employed following either step 1 or 5 of Method 1 as is appropriate.

POLYACRYLATE

Cyanoacrylate, toughened acrylics and conventional solvent based adhesives will normally bond this surface without preparation. However, care should be taken to ensure that stress cracking is avoided.

Use Method 1

Notes

G1–3 and

4 Check that any solvent selected will not damage the plastic paying particular attention to the problem of stress-cracking

POLYALKYD including paint

Use Method 1 or proprietary corona discharge techniques following either step 1 or 5 of Method 1 as is appropriate.

Notes

G1–3 and

4 Check that any solvent selected will not damage the plastic paying particular attention to the problem of stress-cracking.

5 If preparation for cyanoacrylate adhesive is necessary then only use alcohol.

POLYALLYL PHALATE

Prepare as polyalkyd.

POLYAMIDE

A Prepare as polyalkyd taking special care to ensure that absorbed water is not a problem by heating if necessary.

B Etch

1 Remove oil or grease contamination, take care to avoid stress cracking.
2 Immerse the item in the etch solution for 8 ± 2 seconds at room temperature.
3 Dry in a well ventilated atmosphere at 23 ±2°C for note more than 30 minutes.
4 Bond as soon as the surface is dry.

Etch Solution (See SPECIAL NOTES above.)

Material	Parts by weight
Ethyl acetate	91.0
Resorcinol	9.0

Preparation:
Stir the resorcinol into the ethyl acetate until dissolution is complete.

Notes

G1–3 and
4 Cyanoacrylates may be inhibited by the etched surface.

POLYAMINO

Prepare as polyalkyd.

POLYCARBONATE

Prepare as polyalkyd.

POLYEPOXY - including composites

Prepare as polyalkyd but take additional precautions to check for the presence of mould release agent. Solvent-induced stress-cracking is not likely to occur but the possibility of damage should not be ignored.

POLYESTER (THERMOSET) including composites

Prepare as polyepoxy. It needs be noted that some types of polyester plastic contain mould release agent within their structure. Specific toughened acrylic and epoxy adhesive may compensate for this to an acceptable degree.

POLYESTER (THERMOPLASTIC)

Use proprietary corona discharge techniques to prepare this surface.

POLYETHER ETHER KETONE

Use proprietary corona discharge techniques to prepare this surface.

POLYETHYLENE including halogenated derivatives

Use proprietary corona discharge techniques to prepare this plastic or a proprietary etch solution of sodium–naphthalene for the halo derivatives. Some of the recently developed primers may also be suitable.

CAUTION: WHEN USING SODIUM–NAPHTHALENE FOLLOW THE MANUFACTURER'S INSTRUCTIONS MOST CAREFULLY.

POLYIMIDE PLASTIC

Use proprietary corona discharge techniques to prepare this surface.

POLYMETHYL METHACRYLATE

Prepare as polyacrylate.

POLYPHENOLIC

Prepare as polyalkyd. Solvent-induced stress-cracking is not likely to occur but the possibility of damage should not be ignored.

POLYPHENOLIC FOAM

This surface does not normally require preparation beyond the removal of any loose material.

POLYPHENYLENE OXIDE

Prepare as polyalkyd.

POLYPROPYLENE

Prepare as polyethylene.

POLYSTYRENE

Prepare as polyacrylate.

POLYSTYRENE FOAM

This surface does not normally require preparation beyond the removal of any loose material. The material is particularly susceptible to damage by all forms of non-aqueous solvent.

POLYSULPHONE

Prepare as polyalkyd.

POLYETRAFLUOROETHYLENE

Prepare as polyethylene.

POLYVINYL CHLORIDE (PVC)

Prepare as polyacrylate.

PVC FOAM

This surface does not normally require preparation beyond the removal of any loose material. The material is particularly susceptible to damage by all forms of non-aqueous solvent.

POLYURETHANE (PU)

Prepare as polyalkyd.

PU FOAM

This surface does not normally require preparation beyond the removal of any loose material. The material is particularly susceptible to damage by all forms of non-aqueous solvent.

PLASTIC GENERAL INFORMATION

Either use the following process or appropriate sections in conjunction with proprietary corona discharge techniques.

1 Remove oil or grease contamination.
2 Abrade or shot blast. Use medium grit.
3 Remove coarse debris with vacuum.
4 Remove fine debris. Use clean tissue and solvent.
5 Ensure surface is dry.
6 Bond without delay.

Notes

G1–3 and
4 If preparation for cyanoacrylate is necessary use only alcohol.

RUBBER

Most types of rubber may be prepared as follows:

1 Remove oil or grease contamination.
2 If possible, abrade with a suitable material or cut cleanly to obtain a fresh surface.
3 If appropriate, remove coarse debris with vacuum.
4 Wipe surface using clean tissue and solvent.
5 Ensure surface is dry.
6 Bond without delay.

Notes

G1–3 and
4 Check that any solvent selected will not damage the rubber.
5 If preparation for cyanoacrylate adhesive is necessary then only use alcohol.

SILICONE RUBBER

1 Remove oil or grease contamination.
2 If possible, abrade with a suitable material or cut cleanly to obtain a fresh surface.
3 If appropriate, remove coarse debris with vacuum.
4 Wipe surface using clean tissue and solvent.
5 Ensure surface is dry.
6 Bond without delay.

Notes

G1–3 and
4 Silicone rubber does not bond readily and usually only does so with silicone based compositions. Consult the supplier for specific advice.

1.7.7.7 Test methods and standards

TESTING JOINTS

Destructive testing	(1) Standard specimen	– not representative but inexpensive.
	(2) Mechanical specimens resembling joints in the actual assembly	– not entirely reproducible.
	(3) Final assemblies	– representative results but high cost.
Non-destructive testing	(1) Proof loading	– subjected to in-service loads but time dependent effects not detected.
	(2) Sonic testing	Detection of glueline
	(3) Ultrasonic	imperfections and voids.
	(4) Thermal infrared	Techniques have yet to be perfected.
	(5) Visual inspection	– gluelines in transparent joints. Spot thickness checks.

TEST METHODS AND SPECIFICATIONS

A number of European Standards have been recently introduced and others are in the course of preparation. The more important, which can be presently identified, are listed below together with other significant ISO and ASTM standards. They are presented in terms of property determination, tensile and shear stress evaluation, peel and cleavage assessment, general durability and surface preparation.

I. *Tests relating to properties of the adhesive*

a. ASTM D1084 Viscosity of adhesives.
b. ASTM D2183 Flow properties of adhesives.
c. ASTM D1338 Working life of liquid or paste adhesives by consistency and bond strength.
d. ASTM D2556 Adhesives having shear rate dependent flow properties.
e. EN 6041 DSC Analysis: T_g glass transition temperature.
f. EN 542 Specific gravity.
g. EN 924 Flashpoint.
h. ISO 10364 Work life/pot life determination.

II. *Tests emphasizing tensile stresses*

a. ASTM D897 Tensile properties of adhesive bonds.
b. ASTM D209 Tensile strength of adhesives by means of bar and rod specimens.
c. ASTM D1344 Testing cross-lap specimens for tensile properties.
d. EN 26922 Determination of the tensile strength of butt joints.
e. ISO 527 Tensile characteristics.

III. *Tests emphasizing shear stresses*

a. ASTM D1002 Strength properties of adhesives in shear by tension loading (metal-to-metal).
b. ASTM D3165 Strength properties of adhesives in shear by tension loading of laminated assemblies.
c. ASTM D2182 Strength properties of metal- to-metal adhesives by compression loading (disk shear).
d. ASTM D2295 Strength properties of adhesives in shear by tension loading at elevated temperature (metal-to-metal).
e. ASTM D2557 Strength properties of adhesives in shear by tension loading in the temperature range from −267.8 to −55°C (−450 to −67°F).
f. ASTM D3164 Determining the strength of adhesively bonded plastic lap-shear sandwich joints in shear by tension loading.
g. ASTM D3163 Determining the strength of adhesively bonded rigid plastic lap-shear joints in shear by tension loading.
h. ASTM D3166 Fatigue properties of adhesives in shear by tension loading (metal/metal).
i. ASTM D2293 Creep properties of adhesives in shear by compression loading (metal-to-metal).
j. ASTM D1780 Conducting creep tests of metal-to-metal adhesives.
k. ASTM D Shear properties of structural adhesives between rigid adherends using the modified rail test.
l. ASTM D4501 Shear strength of adhesive bonds between rigid substrates using the block–shear method.
m. ASTM D4562 Shear strength of adhesives using pin-and-collar method.
n. ASTM D905 Comparative strength properties of adhesive bonds in shear using the compression loading test.
o. ASTM D3983 Shear strength/modulus of non-rigid adhesives by the thick adherend tensile lap specimen method.
p. ASTM E229 Shear strength/modulus of structural adhesives.
q. EN 2243–1 Overlap shear strength (single lap).
r. EN 2243–6 Thick adherend test (stress/strain curve determination) using aluminium test pieces.
s. EN 1465 Overlap shear strength (single lap).
t. EN 29653 Shear impact resistance.
u. EN 30123 Shear strength of anaerobic adhesives using the pin and collar specimens.
v. ISO 10964 Determination of torque strength of anaerobic adhesives on threaded fasteners.
w. EN 31003 Thick adherend tensile test (steel).

IV. *Tests emphasizing peel stresses*

a. ASTM D903 Peel or stripping strength of adhesive bonds.
b. ASTM D1876 Peel resistance of adhesives.
c. ASTM D1781 Climbing drum peel test for adhesives.
d. ASTM D3167 Floating roller peel resistance of adhesives.
e. ASTM D3807 Strength properties of adhesives in cleavage peel, by tension loading engineering plastics-to-engineering plastics.
f. EN 28510 Peel test for a flexible to rigid bonded assembly (Part 1 90 peel: Part 2 180 peel).
g. EN 1464 Peel test using the floating roller method.
h. ISO 11339 180 peel test for metals (T. Peel).
i. EN 2243–2 Peel strength

V. *Tests employing cleavage stresses*

a. ASTM D1062 Cleavage strength of metal-to-metal adhesive bonds.
b. ASTM D950 Impact strength of adhesive bonds.
c. ASTM D5041 Fracture strength in cleavage of adhesives in bonded joints.
d. ASTM D3433 Fracture strength in cleavage of adhesives in bonded metal joints using the contoured double-cantilever beam.
e. ISO 11343 Dynamic resistance to cleavage using the impact wedge test method.

VI. *Tests relating to bond durability*

a. ASTM D1151 Effect of moisture and temperature on adhesive bonds.
b. ASTM D1828 Atmospheric exposure of adhesive-bonded joints and structures.
c. ASTM D3310 Determining corrosivity of adhesive materials.
d. ASTM D1879 Exposure of adhesive specimens to high-energy radiation.
e. ASTM D2918 Determining durability of adhesive joints stressed in peel.
f. ASTM D2919 Determining durability of adhesive joints stressed in shear by tension loading.
g. ASTM D3929 Stress-cracking by adhesives-plastics, using the bent-beam method.
h. ASTM D3762 Adhesive-bonded surface durability of aluminium at metal–adhesive/primer interface using the wedge test.
i. ASTM D896 Adhesive bond resistance to chemical reagents.
j. ASTM D1151 Effect of moisture/temperature on adhesive bonds.
k. ASTM D1183 Resistance of adhesives using cyclic laboratory ageing conditions.
l. EN 29142 Guide to the selection of standard laboratory ageing conditions for testing bonded joints.
m. EN 2243–5 Structural adhesive test methods: ageing test.

VII. *Tests relating to surface preparation and characterization*

a. ASTM D2651 Preparation of metal surfaces for adhesive bonding.
b. ASTM D2093 Preparation (pre-treating) of plastic surfaces prior to adhesive bonding.
c. ASTM D3933 Structural adhesive bonding – preparation of aluminium surfaces using phosphoric acid anodizing.
d. ISO 10354 Wedge rupture test – characterization of durability.
e. EN 828 Wettability – determination by measurement of contact angle (critical surface tension of solids).
f. EN 2334 Acid chromate pickling of aluminium alloys.
g. EN 2497 Dry abrasive blasting of titanium and titanium alloys.
h. EN 3002 Chromic acid anodize.

Table 1.7.42 Natural adhesives

Types	***Examples***	***Applications***
Animal glues (Higher strength than vegetable glues. Solidify on loss of solvent.)	Albumen Animal glue	Furniture woodworking, bookbinding, leather, textiles. Abrasive binder for abrasive papers and wheels.
	Casein	Woodwork, furniture, joinery. Interior grade plywood fabrication.
	Shellac Beeswax	
Vegetable glues	Gum arabic Tragacanth Colophony Canada balsam Oils and waxes, (carnauba wax, linseed oils) Proteins (soyabean)	Interior grade plywood and softwood panelling.
	Carbohydrates (starch, dextrines)	Paper carton and bottle labelling, stationery, interior grade plywood fabrication.
Mineral	Mineral resins and waxes Bitumes Ceramics, refractory inorganics, glasses and enamels.	Mainly used for electrical applications.
Inorganic adhesives and cements	Sodium silicate	Wood bonding, metal/metal bonding, glass, thermal insulation materials, refractory cements and foundry moulds.
	Zinc phosphate	Dental cement
	Litharge cements	Repair of sinks, pipe valves glass, stoneware.
	Sulphur cements	Acid tank construction
	Hydraulic (e.g. Portland) cement	Building, roads, bridges.

Table 1.7.43 Compatibility of the principal structural adhesives with a variety of composite and associated materials.

Material to be bonded	***Acrylic***		***Epoxy***		***PU***
	Pseudo-1-part	***2-part***	***1-part (heat cured)***	***2-part***	***2-part***
Glass/ceramic	1	1	4	2	1
Metal (also see Paint)					
– Aluminium	1	1	1	1	4(1)
– Steel	1	1	1	1	4(1)
– Zinc	2	2	2	2	4(2)
Thermoplastic					
– ABS	3	1	4	3	3
– Polyamide (nylon)	2	2	3	2	2
– Polycarbonate	3	3	4	3	2
– Polyphenylene oxide	3	3	3	2	1
– Polypropylene	2	2	4	3	1
– Polyurethane (elastomer)	3	3	4	4	2
– PVC	1	1	4	3	2
– PVC–plasticized	1	1	4	4	2
Thermoset					
– Epoxy	2	2	1	1	2
– Polyester – hand lay	1	2	3	1	2
– VARI	1	2	3	1	2
– SMC	1	2	3	1	2
– cold press	1	2	3	1	2
– Polyurethane – RIM	3	3	4	2	2
Paint – Cataphroetic	1	1	1	1	1
Scale: 1=Excellent 2=Good 3=Good but possible problems 4=Unsuitable ()=If pre–primed or painted	Note: This categorization is given in relation to the types of application usually seen in association with the materials nominated. Therefore, each line should be considered to be unique and be read in conjunction with the preceding notes.				

Table 1.7.44

Table 1.7.44 List of adhesives, codes and related special notes

Adhesive type	*Sub-groups of main types*	*C code*	*Special notes*
Amino	Cold cured, two-part	a	2 8 10 26 32
	Hot cured (or warmed), two-part	b	1 2 10 26 32
Anaerobic (non- toughened)	Cold cured	c	2 6 9 15 27
	Cold cured plus accelerator	d	2 6 10 15
Cyanoacrylate	Cold cured	e	3 9 30 31 34
Emulsion	Cold cured	f	2 9 26
Epoxide	Single-part, tape or film form (always heat cured)	g	4 5 9 16 32
	Single- part, liquid/paste form (always heat cured)	h	2 4 9 32
	Two-part, cold cured	i	8 10 28
	Two-part, heated to some degree	j	4 8 10
Hot melt	Conventional, always heat activated	k	6 32
	Reactive, always heat activated, but with later moisture cure	l	6 7 10 32 34
Phenolic	Cold cured, two-part	m	26 32
	Hot cured, two-part	n	1 26 32
Polyurethane	Cold cured, one-part	o	10 16 32 34
	Cold cured, two-part	p	6 8 10 16 32 34
Rubber solution	Cold cure	q	11 29
Tape (pressure sensitive)	Cold cure	r	2 5 6 9
Toughened adhesives	Acrylic-based		
	anaerobic, thermoplastic, cold cured	s	2 9 27
	anaerobic, thermoplastic, cold cured plus accelerator	t	2 10
	anaerobic, thermoset, cold cured	u	2 9 27
	anaerobic, thermoset, cold cured plus accelerator	v	2 10
	non-anaerobic, two-part, cold cured (could be mixed)	w	2 10 11 35
	non-anaerobic, two-part, cold cured (must mix)	x	2 10 11
	Epoxide-based		
	one-part liquid/paste form (always heat cure)	y	2 4 9 32
	two-part, cold cured	z	8 10
	two-part, heated to some degree	$	4 8 10
UV & visible rad. cured		#	33

Because of their limited or specialized use, plastisols and the polyimides have not been incorporated in this table which has been arranged to present the more commonly used engineering adhesives by the names in common use.

Sometimes known as pseudo-single-part acrylic (toughened) adhesives.

Table 1.7.45 Adhesive/adherend compatibility table and related special notes

Reference number	*Material (adherend)/*	*Amino: cold cured, two part*	*Amino: hot cured, two part*	*Anaerobic: cold cured*	*Anaerobic: cold cured plus accelerator*	*Cyanoacrylate*	*Emulsion*	*Epoxide: single-part, tape or film form, heat cured*	*Epoxide: single-part, liquid/paste form, heat cured*	*Epoxide: two-part, cold cured*	*Epoxide: two-part, heated to some degree*	*Hot melt: conventional, always heat activated*	*Hot melt: reactive, always heat activated*	*Phenolic: cold cured, two-part*	*Phenolic: hot cured, two-part*	*Polyurethane: cold cured, one-part*	*Polyurethane: cold cured, two-part*	*Rubber solution*	*Tape (pressure sensitive)*	*Toughened ads.: acryl. anaer. thermopl., cold cure*	*T'd ads.: acr. anaer. thmopl. cold cure plus accel*	*Toughened ads.: acryl. anaer. thermoset / cold cure*	*T'd ads.: acr. anaer. thmoset / cold cure plus accel*	*Tough'd ads.: acryl. non-anaer., cold cure, could mix*	*Tough'd ads.: acryl. non-anaer., cold cure, must mix*	*Tough'd ads.: epoxy one-part, liquid/paste (heat cure)*	*Tough'd ads.: epoxy two-part, cold cured*	*Tough'd ads.: epoxy two-part, heated to some degree*	*UV and visible radiation cured*	*Reference number*	
	/Adhesive Type Code	a	b	c	d	e	f	g	h	i	j	k	l	m	n	o	p	q	r	s	t	u	v	w	x	y	z	$	#		Sp'l Notes
1	Cellulose (wood/paper)	P	P	R	R	P	P	S	S	P	S	P	P	P	P	P	P	S	S	S	R	R	P	P	P	S	P	S	S	1	23
2	Cementitious	R	R	R	R	R	P	R	S	P	S	P	P	R	R	P	P	R	R	R	R	R	P	P	P	S	P	R	R	2	12 23
3	Ceramic	R	R	R	S	S	P	P	P	P	P	P	P	S	S	P	P	S	S	S	S	P	S	S	P	P	P	P	P	3	23
4	Fabric	S	S	R	R	S	P	S	S	P	P	P	P	P	P	P	P	R	R	R	R	R	S	S	P	S	P	S	S	4	23
5	Friction material	R	R	R	S	S	R	S	P	P	P	S	S	P	P	S	S	S	S	S	S	S	S	S	S	P	P	S	S	5	12 23
6	Glass	R	R	R	S	S	S	P	P	P	P	P	P	S	S	P	P	S	S	S	S	S	S	S	P	P	P	P	P	6	13
7	Leather	S	S	R	R	S	P	R	R	P	S	P	P	S	S	P	P	S	S	S	R	R	S	S	P	R	P	S	S	7	23
8	Metals[a]	R	R	P	P	P	R	P	P	P	P	P	S	S	S	S	S	P	P	P	P	P	P	P	P	P	P	P	P	8	14
	Plastics[b]																														
9	ABS	R	R	R	S	S	S	R	R	S	S	S	P	R	R	P	P	S	S	S	S	S	P	S	P	R	P	P	P	9	15
10	Polyacetal	R	R	R	S	P	S	R	R	P	P	S	S	R	R	P	P	S	S	S	S	S	P	S	P	R	P	S	S	10	16
11	Polyacrylate	R	R	R	S	S	S	R	R	S	S	S	P	R	R	P	P	S	S	S	S	S	P	S	P	R	P	P	P	11	15
12	Polyalkyd	R	R	R	S	P	S	S	P	P	P	S	P	R	R	S	S	R	R	R	R	R	S	S	S	P	P	P	P	12	16
13	Polyallyl phthalate	R	R	R	S	P	S	R	S	P	P	S	S	R	R	S	S	S	S	S	S	S	S	S	S	S	P	S	S	13	16
14	Polyamide	R	R	R	S	S	S	R	S	S	S	P	P	P	P	S	S	S	S	S	S	S	S	S	S	S	S	P	P	14	15
15	Polyamino	R	R	R	S	P	S	R	S	P	P	S	S	R	R	S	S	S	S	S	S	S	S	S	S	S	P	S	S	15	16
16	Polycarbonate	R	R	R	S	S	S	R	R	S	S	S	P	R	R	P	P	S	S	S	S	S	P	S	P	R	P	P	P	16	17
17	Polyepoxy (inc Comp'te)	R	R	R	S	P	S	P	P	P	P	S	S	R	R	P	P	S	S	S	S	S	S	S	S	P	P	P	P	17	16
18	Polyester (inc Comp'te)	R	R	R	S	P	S	P	P	S	S	S	P	R	R	P	P	S	S	S	S	S	P	P	P	P	P	S	S	18	16
19	Polyethylene	R	R	R	S	S	S	S	S	S	S	S	P	R	R	S	S	S	S	S	S	S	S	S	S	S	S	S	S	19	18
20	Polyimide	R	R	R	S	P	S	P	P	P	P	P	S	R	R	P	P	S	S	S	S	S	S	S	S	P	P	S	S	20	16
21	Polymethylmethacrylate	R	R	R	S	S	S	R	R	S	S	S	P	R	R	P	P	S	S	S	S	S	P	S	P	R	P	S	S	21	15
22	Polyphenolic (inc Comp'te)	S	S	R	S	P	S	P	P	P	P	P	S	S	S	P	P	P	P	P	S	S	S	S	S	P	P	P	P	22	16
23	Polyphenylene Oxide	R	R	R	S	P	S	S	S	P	P	S	S	R	R	P	P	S	S	S	S	S	P	P	P	S	P	S	S	23	20
24	Polypropylene	R	R	R	S	S	S	S	S	S	S	S	P	R	R	S	P	S	S	S	S	S	S	S	S	S	S	S	S	24	18

Table 1.7.45

Table 1.7.45 Adhesive/adherend compatibility table and related special notes – *continued*

Reference number	Material (adherend)/	Amino: cold cured, two part	Amino: hot cured, two part	Anaerobic: cold cured	Anaerobic: cold cured plus accelerator	Cyanoacrylate	Emulsion	Epoxide: single-part, tape or film form, heat cured	Epoxide: single-part, liquid/paste form, heat cured	Epoxide: two-part, cold cured	Epoxide: two-part, heated to some degree	Hot melt: conventional, always heat activated	Hot melt: reactive, always heat activated	Phenolic: cold cured, two-part	Phenolic: hot cured, two-part	Polyurethane: cold cured, one-part	Polyurethane: cold cured, two-part	Rubber solution	Tape (pressure sensitive)	Toughened ads.: acryl. anaer. thermopl., cold cure	T'd ads.: acr. anaer. thmopl. cold cure plus accel	Toughened ads.: acryl. anaer. thermoset / cold cure	T'd ads.: acr. anaer. thmoset / cold cure plus accel	Tough'd ads.: acryl. non-anaer., cold cure, could mix	Tough'd ads.: acryl. non-anaer., cold cure, must mix	Tough'd ads.: epoxy one-part, liquid/paste (heat cure)	Tough'd ads.: epoxy two-part, cold cured	Tough'd ads.: epoxy two-part, heated to some degree	UV and visible radiation cured	Reference number	
	/Adhesive Type Code	a	b	c	d	e	f	g	h	i	j	k	l	m	n	o	p	q	r	s	t	u	v	w	x	y	z	$	#		Sp'l Notes
25	Polystyrene (inc Foam)	R	R	R	S	S	S	R	R	S	S	S	P	R	R	P	P	S	S	S	S	S	P	S	P	R	P	P	P	25	15 19
26	Polysulphone	R	R	R	S	S	S	R	R	S	S	R	R	R	R	S	S	P	P	P	S	S	S	S	S	R	S	S	S	26	20
27	Polytetrafluoroethylene	R	R	R	S	S	S	S	S	S	S	S	S	R	R	S	S	S	S	S	S	S	S	S	S	S	S	S	S	27	18
28	Polyvinyl chloride	R	R	R	S	P	S	R	R	S	S	S	P	R	R	S	S	S	S	S	S	S	P	P	P	R	S	P	P	28	15
29	Polyurethane	R	R	R	R	S	S	R	R	S	S	S	P	R	R	P	P	S	S	S	S	S	S	S	S	R	S	S	S	29	18 19 23
	Rubber																														
30	Butyl	R	R	R	R	P	R	R	R	R	R	R	R	R	R	R	R	R	R	R	R	R	R	R	R	R	R	R	R	30	21
31	Chlosulpolyethylene	R	R	R	R	P	R	R	R	R	R	R	R	R	R	R	R	R	R	R	R	R	R	R	R	R	R	R	R	31	21
32	EPM	R	R	R	R	P	R	R	R	R	R	R	R	R	R	R	R	R	R	R	R	R	R	R	R	R	R	R	R	32	21
33	EPDM	R	R	R	R	P	R	R	R	R	R	P	S	R	R	S	P	R	R	R	R	R	R	R	R	R	R	R	R	33	21
34	Fluoroinated	R	R	R	R	S	R	R	R	R	R	R	R	R	R	R	R	R	R	R	R	R	R	R	R	R	R	R	R	34	25
35	Chloroprene	R	R	R	R	P	R	R	R	S	S	R	S	R	S	S	P	R	R	R	R	R	R	R	R	R	R	R	R	35	21
36	Cyclized	R	R	R	R	P	R	R	R	S	S	R	R	R	S	R	R	R	R	R	R	R	R	R	R	R	R	R	R	36	21
37	Hard structural	R	R	R	R	P	R	R	R	S	S	R	S	R	S	S	P	R	R	R	R	R	R	R	R	R	R	R	R	37	21
38	Natural	R	R	R	R	P	R	R	R	R	R	R	S	R	R	S	P	R	R	R	R	R	R	R	R	R	R	R	R	38	21
39	NBR	R	R	R	R	P	R	R	R	S	S	R	S	R	S	S	P	R	R	R	R	R	R	R	R	R	R	R	R	39	21
40	Neoprene	R	R	R	R	P	R	R	R	S	S	P	S	R	S	S	P	R	R	R	R	R	R	R	R	R	R	R	R	40	21
41	Nitrile	R	R	R	R	P	R	R	R	R	R	P	S	R	R	S	R	R	R	R	R	R	R	R	R	R	R	R	R	41	21
42	SBR	R	R	R	R	P	R	R	R	R	R	R	S	R	R	S	P	R	R	R	R	R	R	R	R	R	R	R	R	42	21
43	Soft non-structural	R	R	R	R	P	R	R	R	S	S	R	S	R	S	S	P	R	R	R	R	R	R	R	R	R	R	R	R	43	21
44	Silicone	R	R	R	R	R	R	R	R	R	R	R	R	R	R	R	R	R	R	R	R	R	R	R	R	R	R	R	R	44	20

[a]The common engineering metals all tend to behave in a similar manner. See also Table 1.7.43.
[b]Some plastics, especially those with a low surface energy are particularly difficult to bond unless properly treated.

Table 1.7.46 Adhesive Selector application questionnaire

Question number	Questions	Yes/No	Reject	Go to
1.	Do you intend to operate at temperatures above 220°C?	Yes No	See Special Note 24 –	 2
2.	Is the joint gap greater than 0.125 mm (0.005 in)?	Yes No	e –	3 4
3.	Is the joint gap greater than 0.50 mm (0.020 in)?	Yes No	bcdemnstuv See special Note 35 –	4 4
4.	Is the joint coaxial, i.e. composed of round, turned, threaded or fitted parts? (note: **not** axial butt joined parts)	Yes No	bgnqr cd	6 5
5.	Is the width of the bond greater than 50 mm (2 in)?	Yes No	e –	7 7
6[a]	Is the joint intended to be permanent with no possibility of dismantling for maintenance?	Yes No	r Reject all but cdstuv	8 24
7[a]	Is the joint intended to be permanent with no possibility of dismantling for maintenance?	Yes No	kr Reject all but cdkopqrstuv	8 8
8.	If relevant will the joint be subject to peel, cleavage or impact forces? See Special Note 22.	Yes No	abcdeghlmnop See Special Notes 31(e) & 34(lop) –	9 10
9.	Is the bonded assembly intended to flex readily and repeatedly in a manner that will distort the adhesive in the bond line?	Yes No	Reject all but fklopqrstwx –	10 10
10.	Do you require the adhesive to provide a positive gas or fluid seal?	Yes No	abefmnqr –	11 11
11.	Will the joint be exposed to the weather?	Yes No	elop See Special Notes 31 (e) & 34 (lop) –	12 12
12.	Are aggressive chemicals involved or will there be exposure to liquids above 90°C?	Yes No	abefklopqrst –	13 13
13.	Is the cure time of the adhesive important?	Yes No	– –	14 19

Table 1.7.46 Adhesive Selector application questionnaire – *continued*

14.	Is instantaneous adhesion on contact required?	Yes No	Reject all but fklqr –	19 15
15.	Is it absolutely essential that handling strength be developed within 10 s of joint assembly?	Yes No	Reject all but befklnqrwx# See Special Note 35 –	19 16
16.	Must handling strength be developed within 60 s of joint assembly?	Yes No	acijmosuz$ –	19 17
17.	Must handling strength be developed between 1 and 15 min after joint assembly?	Yes No	eklmoqrz –	19 18
18.	Must handling strength be developed 15 min or more after joint assembly.	Yes No	bdefklqrtv# –	19 19
19.	Must the joint withstand a permanent load greater than 3.5 MPa (500 psi)?	Yes No	abefkloqrst –	20 20
20.	Must occasional extremely high shear or rotational loads greater than 28 MPa (4000 psi) be borne during use?	Yes No	Reject all but cdghijnuvyz$ –	21 21
21.	Are thin materials being bonded which may be distorted?	Yes No	abcdghmn –	22 22
22.	Is warming or heat curing possible even if not desirable?	Yes No	– bghjny$	23 23
23.	Is it possible to use a two–part mixed system?	Yes No	– abijmnpxz$ See Special Note 35	24 24
24.	Is the joint continuously exposed to water?	Yes No	abeflop See Special Notes 31(e) & 34(lop) –	

[a] A possible alternative version of this question could be: 'Do you wish to be able to dismantle the joint readily, e.g. for maintenance or for some other reason?'

Table 1.7.46 – *continued*

Table 1.7.47 Adhesive Selector assessment table

	a	b	c	d	e	f	g	h	i	j	k	l	m	n	o	p	q	r	s	t	u	v	w	x	y	z	$	#	Special Note Ref.
Adherend 1																													
Adherend 2																													
Question 1																													
2																													
3																													
4																													
5																													
6																													
7																													
8																													
9																													
10																													
11																													
12																													
13																													
14																													
15																													
16																													
17																													
18																													
19																													
20																													
21																													
22																													
23																													
24																													
Total																													

Table 1.7.48 Special notes to Adhesive Selector

Number	*Notes*
1.	Hot presses are expensive.
2.	Often used without rigorous surface preparation.
3.	Plastics and rubber parts not normally cleaned and never with chlorinated solvents.
4.	Inexpensive heating techniques available.
5.	May need pre-shaping, curved surfaces difficult.
6.	Equipment costs vary with scale.
7.	Opened containers likely to react with atmospheric moisture.
8.	Small volumes may be hand mixed.
9.	Physiological activity usually low.
10.	Possible physiological hazards.
11.	Vapour extraction often needed. Some materials have hazardous flashpoint.
12.	Resistance to cleavage and tensile forces low due to nature of adherends surface.
13.	Glass joints only reliable in dry environments. Otherwise, surface preparation or specialized adhesives necessary.
14.	Most metal components will bond well in coaxial (slip-fitted or threaded) joint configurations – usually without rigorous surface preparation. However, this is not so for variants on the lap and butt joint. Here performance depends on: basic metal or alloy, modulus, joint geometry, mode of loading, surface preparation. Steel alloys generally bond well and often, when toughened adhesives are used, do not require surface preparation. However, other common engineering metals and alloys can be difficult and often surface preparation is required to maximize performance.
15.	CARE – stressed plastic may crack.
16.	Surface preparation may be needed.
17.	CARE – some forms particularly prone to stress-cracking.
18.	Requires special preparation for use in lap/butt joints. Rough or knurled surfaces usually acceptable for coaxial (slip-fitted) joints without chemical pre- treatment.
19.	Some adhesives dissolve this plastic – take special care with the foam form.
20.	Special problems – consult supplier.
21.	Freshly cut, or lightly abraded, close fitting surfaces usually give the best results.
22.	Butt joints are susceptible to damage and fail readily if abused. Extreme forms – such as bonded edges – must be avoided except in special circumstances, e.g. honeycomb sandwich panels.
23.	CARE – absorption likely.

Table 1.7.48

Table 1.7.48 Special notes to Adhesive Selector – *continued*

Number	***Notes***
24.	Apart from a few specialized adhesives (such as the polyimides – consult specialist supplier) all conventional types decompose above this temperature.
25.	Success depends on degree of modification – consult manufacturer.
26.	A porous surface may be necessary.
27.	Narrow bond lines may not cure properly.
28.	Quick cure versions are brittle.
29.	High performance levels only from vulcanized systems.
30.	Cyanoacrylates are brittle and readily degrade in warm wet conditions, particularly when unpainted metal surfaces involved.
31.	Do not reject 'e' (cyanoacrylate) if: either (a) it is a primary for both surfaces *and* (b) neither surface is metal, or (c) one surface is a flexible rubber-based (30–43 inc.) material.
32.	Vapour extraction may be required.
33.	With the exception of 'dark cure' variants, or where secondary curing is to take place (anaerobic/heat), one adherend must always be transparent to incident radiation (UV/blue).
34.	Only reject l, o and p (reactive hot melt, single-and two-component polyurethane) if unpainted or unprimed material is being bonded and there is a working load.
35.	Performance of some type 'w' (acrylic, toughened, two-part, could mix) adhesives is impaired when gaps are greater than 0.5 mm unless they actually are mixed.

Table 1.7.48 – *continued*

Table 1.7.49 Production factors affecting choice of adhesives

Factor	***Questions***
Components	What is the nature of the surfaces? Is surface preparation necessary and if so, is it compatible with the nature of the components? Are the facing surfaces accessible, vertical or horizontal? Are the components liable to suffer major temperature fluctuations? Are there liable to be major dimensional changes which could materially affect the separation of the facing surfaces?
Adhesive	Does it have to be mixed and if so, is equipment necessary? If a mixed system, what is the 'pot life'? What is the handling time of the bonded parts? Is the adhesive hazardous and if so, is special equipment necessary? Will subsequent manufacturing processes conflict with the use of this particular adhesive?
Assembly process	How much space is required by the components, the assembly process and the curing cycle? Consider storage racks, ovens, fixtures, jigs and all equipment. What is the true cost of employing any one particular adhesive compared with other adhesive types or other assembly techniques?

Table 1.7.50 Identification of production problems

Issues to be addressed
Is it the correct adhesive?
Was there enough adhesive?
If mixing is necessary has it been done correctly?
Has the adhesive been contaminated?
Has the viscosity/thixotropy altered?
If a primer is needed has it been used and did it work?
Has the joint been made and then re-made, allowing the formation of voids?
If heated, has the joint been moved prior to cooling?
Is the joint securely jigged and clamped?
Have the dimensions changed significantly, even though they may still be within mechanical design limits?
Has the surface changed?
Are there differences in the plating and its passivation?
Are the components oxidized, rusted or corroded in some way?
Have the components become contaminated?
Are the components still composed of the same basic material(s)?
Does the adhesive wet the surface?
Is the assembly at its normal temperature?
Is the process as it should be in all respects?
Is the process at the right temperature and humidity?
Has the cure schedule been maintained?

Table 1.7.51 Safety critical check list

Audit: factors for consideration
Assess newly delivered adhesives using batch tests
Random stock checks (using batch acceptance tests)
Periodic sequential rotation through type tests.
Constant monitoring of ALL aspects of surface preparation.
Wherever possible use application sensor on dispensing equipment.
Monitor function of dispensing equipment
Monitor assembly environment for major temperature and, or humidity fluctuations
Monitor bonded surfaces and ensure freedom from voids
Prepare test coupons in parallel with actual components
Constant monitoring of test coupon performance
Carry out periodic destructive tests using actual components wherever possible.

1.7.8 Glass–ceramic–metal sealing techniques

1.7.8.1 Introduction

This section deals with the processes for making vacuum-tight permanent seals between:

glass–glass
glass–metal
ceramic–metal
ceramic–glass
glass–ceramic–metal

All these techniques involve the fusion of one component of the seal (the glass in conventional glass–metal seals, or a metallic solder in the case of ceramic–metal seals) and subsequent cooling to form an adherent join.

(For glass and ceramic forming techniques which do not involve fusion, see the Sections 1.7.6 and 1.7.7 on adhesives and cements respectively)

Because of the high temperatures used in seal manufacture and brittle nature of glass and ceramic materials, successful design is concerned to minimize the tensile stresses developed in the brittle component of the seal by the different thermal expansions of the component parts.

1.7.8.2 Glass–glass seals

Glass components for which the thermal expansion coefficients differ by less than 10% (for low expansion glasses) or 15% (for high expansion glasses) can be sealed directly to each other by glass-blowing techniques. Low expansion glasses have a thermal expansion coefficient around $3–5 \times 10^{-6}$ per K and high expansion glasses have a thermal expansion coefficient around $8\text{-}12 \times 10^{-6}$ per K.

GLASS-BLOWING

Glass-blowing produces strong, permanent joints between glass components. The process essentially consists of:

1 Heating both components locally (usually by means of a flame), until the glass in the joint region has sufficiently low viscosity.
2 Connecting the parts in this state so that they 'weld' together.
3 Cooling and annealing the seal in such a way as to minimize or avoid stresses in the cooled joint.

Advantages include:

Joints are all-glass and are:

- as strong,
- as corrosion resistant and
- as temperature resistant,

as the individual components.

Joints between glasses of identical compositions will be strain-free and capable of thermal cycling to the same degree as the individual glasses.

Disadvantages include:

Graded seals (i.e. using glasses of intermediate expansion, see below) will be required to join glasses of very different thermal expansion coefficients.
Relatively high temperature (600–900°C) are required to make the joint and glass-blowing generally requires a high degree of operator skill.
'Annealing' of the finished joint is necessary to minimize thermal strains in the cooled joint.

'Free cooling' of a glass joint will give rise to additional thermal stresses in the cold joint due to temperature gradients during cooling. Annealing consists of holding the completed joint at a temperature at which the glass is sufficiently ductile to allow the stresses to be relieved by plastic flow in the glass (the 'annealing point' corresponds to a glass viscosity of about 10^{12} Nsm^{-2}.

High expansion glasses tend to have annealing points between 350 and 550°C, while low expansion glasses have annealing points in the range 500–1100°C.

Annealing may be carried out by flame-heating or by furnace-heating, the latter being move usual. A simple annealing schedule for furnace annealing comprises:

1 Heat the work to a temperature some 5–20°K above the annealing point.
2 Hold at this temperature for 5–10 min.
3 Cool at least the first 100 K from this temperature at a rate less than:
(a) $20/d^2$ K/min for high expansion glasses;
(b) $100/d^2$ K/min for low expansion glasses;

where d is the wall thickness in mm.

GRADED SEALS

Joints between glasses of widely different expansion coefficients can be formed by graded seals consisting of a number of segments of glass each with progressively slightly different expansion coefficients. Each tube section of a graded seal should have a length/diameter ratio of at least 1/5 to avoid thermal cracking.

Graded seals manufactured by the sintering of powdered glasses pressed in layers are also commercially available.

1.7.8.3 Glass–metal seals

Glass–metal seals are widely used in the electrical and electronics industry for electrical lead-throughs in lamps, valves and vacuum devices.

They can be classified as follows:

(a) matched seals – in which the thermal expansion coefficients of metal and glass components are similar so that thermal stresses on cooling from the sealing temperature are avoided.
(b) unmatched seals – in which thermal stresses are avoided by other design features of the joint, e.g. the use of thin, ductile metal components able to relieve thermal strain by plastic deformation, or the use of compression joints.

For glass–metal seals made by softening of the glass component, the metal selected must have a higher melting point than the softening point of the glass. As this is at least 600°C, glass seals to metals such as tin and lead, etc., are impossible and to silver and aluminium, very difficult.

Matching pairs for glass–metal seals are given in:
Table 1.7.52 *Matched glass–metal seals*
Designs of seals for unmatched glass–metal pairs are given in:
Table 1.7.53 *Unmatched glass–metal seals*
Glass–metal seals can also be made by use of metal solders. Characteristics of the process are given in:
Table 1.7.54 *Unmatched glass–metal seals using solder metals*

1.7.8.4 Ceramic–glass and ceramic–metal seals

Successful design will be based on choosing metal-ceramic pairs with close thermal expansion coefficients. In addition the following guidelines will be helpful:

1. Use relatively thick-walled ceramic components, but thin-walled metal parts to minimize the development of tensile stresses in the ceramic.
2. Choose relatively soft metals with good ductility if possible.
3. Choose a brazing material with an expansion coefficient near to that of the ceramic parts, especially if the metal component is sealed inside the ceramic.

Joining techniques are given in:
Table 1.7.55 *Ceramic–glass and ceramic–metal seals*
Solder glasses are low melting point glasses which can be used to join glass and ceramic components in applications where the high temperatures involved in other sealing techniques cannot be tolerated. They can also be used to seal mica components.
The solder glass is usually applied in powder form in a suitable organic solvent although pre-formed shapes such as washers and sleeves are also available.
Suitable combinations of ceramic, glass and metal are given in:
Table 1.7.56 *Possible ceramic-glass-metal combinations*

1.7.8.5 Glass–ceramic seals

A specialized sealing technology is available for certain applications, particularly useful when the component needs to be used at a temperature higher than the sealing temperature. Making the seal with a glass, but then controllably crystallizing the glass to a glass–ceramic, the joint can be made more refractory than the parent glass. An example of this technique is found in the manufacture of TV CRTs where a low temperature joining process is used to seal the face plate to the funnel to protect the phosphor screen. By subsequently crystallizing the glass, a higher bake-out temperature can be used during evacuation.
Joints can also be made directly between glass–ceramics and metals by casting the glass into a mould containing the metal parts, cooling the moulding to a little below the annealing temperature and then re-heating it to crystallize the glass–ceramic. This is especially useful for joining to high expansion alloys.

Table 1.7.52 Matched glass–metal seals

Metal	*Room temperature expansion coefficient*	*Suggested matching glass*	*Remarks*
Metals for sealing to low expansion glasses, i.e. quartz, borosilicate, aluminosilicate glasses. Tungsten	(4.5×10^{-6}/K)	Corning 3320, 7720, 7780	Metal should be surface-ground, pickled and oxidized to blue-green colour to promote good adhesion. Nickel or nickel and chromium plating may be used for protection of the seal against high humidity or alkaline metal vapours.
Molybdenum	(4.9×10^{-6}/K)	Corning 7082, 7040, 1826	Metal should be cleaned and lightly oxidized to a light yellowish colour. Excessive oxidation during sealing must be avoided.
Fe–Ni–Co alloys (typically 54 28 18%) e.g. Nilo K, Sealvac A, Rodar, Therlo, Fernico I, Kovar A, etc. Kaser.	($5.6–6.0 \times 10^{-6}$/K)	Corning 7052, 7040, 8800 Chance Pilkington Type ME, 1.	Pre-treatment of metal in wet hydrogen for 4 h at 900°C or in vacuum to ensure freedom from gas bubbles in the seal.
Fe–Ni alloys e.g. Nilo 40 (40 Ni–60 Fe)	($4.0–6.0 \times 10^{-6}$/K)	Corning 1826	Cleaned by heating to 950°C in hydrogen.
Zirconium	(6.3×10^{-6}/K)	Corning 7052	May be useful if non-magnetic seals are desired or extreme resistance is necessary.
Tantalum	(6.5×10^{-6}/K)	Corning 7720, 7052	
Metals for sealing to high expansion glasses (soda-lime-silica and lead glasses) Platinum	(9×10^{-6}/K)[a]	Corning 0280, 0041, 7570, 7560, 0050	Used mainly in scientific instruments. Clean wire in *aqua regia* before use. No oxide formation.
'Dumet alloys'	(7.1 longitudinal) (9.0 radial) $\times 10^{-6}$/K	Corning 7570, 0120, 0010	A copper-clad Fe-Ni wire widely used in lead throughs as a cheap substitute for platinum (cf. electric lamps and valves). Wire size limited to about 0.5 mm diameter.
Fe–Ni (50/50 or 54–46)	($7.0–9.0 \times 10^{-6}$/K)	Corning 0010, 0120, 0080	Wires limited to 0.5 mm diameter
Fe–Cr alloys 72–28 to 84–16	($8.4–9.3 \times 10^{-6}$/K)	Corning 0050, 0060, 0080	Used in TV tubes. Pre-oxidize by heating in wet hydrogen (950–1100°C, 15–30 mm).
Fe–Ni–Cr alloys	(-9×10^{-6}/K)	Corning 8870, 0080, 0014, 0050	Relatively large seals can be made. Reoxidize in wet H_2 at 1050–1250°C.
Nickel	(13×10^{-6}/K)	Corning 1990	Metal surface pickled and oxidized in a hydrogen flame. Seal is best made in an oxidizing gas flame at about 650°C.
Copper	(16×10^{-6}/K)	Corning 9776	Difficult to form matched seals.
Iron	(12×10^{-6}/K)	Corning 1990, 0110	Cu, Ni or Cr plating may be used to prevent excessive oxidation. Surface decarburization in wet H_2 prevents bubble formation in the seal.

[a] Expansion coefficients below the asterisk refer to 0–100°C

Table 1.7.53 Unmatched glass–metal seals

Type of seal	Design type	Design considerations and limitations
Compression seals (glass component is subject only to comprehensive stresses since glass is much stronger in compression than tension.	Window seals Window seals with multiple rod or pipe seals 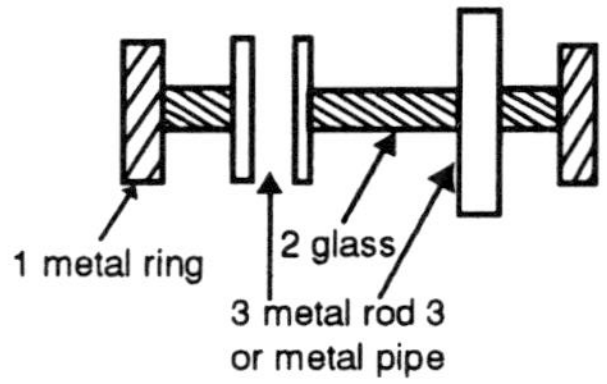 	Expansion coefficient of ring must be greater than that of glass (a) Metal ring matches expansion of glass but is greater than rod or pipe. i.e. $\alpha = \alpha 2 > \alpha 3$ or (b) $\alpha > \alpha 2 = \alpha 3$ (c) If ring and rods are of the same material then ring wall thickness > 1/4 ring outer radius if tensile stresses are to be avoided.
Housekeeper seals (tensile stresses are avoided by plastic deformation in thin sections of ductile metals).	Seals are restricted to copper, platinum, some steels, and molybdenum materials. (1) Wire seal 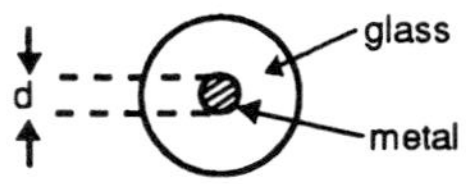 	Housekeeper seals can be made with any type of glass (1) The expansion of the metal part is greater than that of the glass so that stresses normal to the interface are always compressive. (2) The metal used is soft and thin enough to deform plastically to follow the dimensional changes in the glass as it cools. (3) The metal must form a strong bond with the glass. (4) The shape and dimensions of the seal should provide a large contact area between metal and glass. Critical dimensions for seal designs (1)–(4).
	(2) Ribbon seal 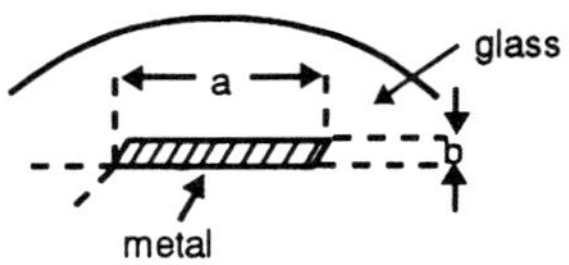 	(see critical dimensions below)
	(3) Tube seal 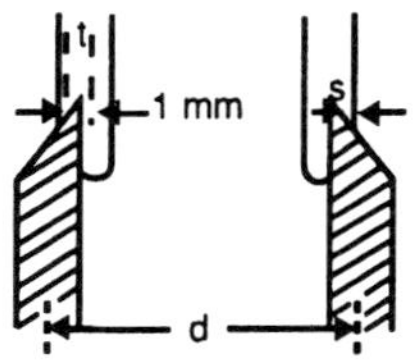 	(see critical dimensions below)
	(4) Disc seal 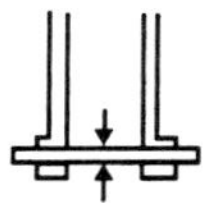	(5) Possible shapes to tubular housekeeper seals  Designs in which sharp changes of section should be avoided.

Critical dimensions for seal designs (1)–(4):

	Copper				Platinum	Iron and steel	Molybdenum
(1) Wire	*d* max. = 0.05 mm				*d* max. = 0.2 mm flattened to 0.1 mm or *d*-2 mm or 0.075m mm tubing.	stainless steel *d* = 0.5 mm	–
(2) Ribbon	*a* = 25 mm *b* = 0.4 mm				*a* = 0.1 mm *b* = 0.008 mm	–	*a* = 1.3 mm *b* = 0.01–0.05 mm sp. for sealing into quartz.
(3) Feather edge tubular	*d*	*t*	Θ	s	1°	1°	–
	10–50 mm 50–100 100–175	0.07–0.09 mm 0.11–0.13 0.13–0.75	2°–3° 2°–3° 3°–4°	2.5–3.5 3.5–4.0 4.5–5.0			
(4) Disc	*t* = 0.3–0.4 mm				–	–	–

Table 1.7.53

Table 1.7.54 Unmatched glass–metal seals using solder metals

Type of seal	***Design considerations and limitations***
Seals using solder metals (tensile stresses are avoided by deformation of the solder)	Solder may be used to join glass or ceramic components to glass, ceramic or metal components. A wide range of joint designs is possible, but applications are restricted to low temperatures by the low melting point metallic solders used.
Soldering with indium and indium alloys	No flux is necessary, but both components must be thoroughly clean and metal components tinned using a zinc chloride flux and washed. Molten indium does not wet glass but can be applied by wiping to the preheated component. Adherence is impaired by too high a temperature but a smooth coating should be obtained as the glass cools. Protection of the soldered joint with lacquer is recommended.
Soft soldering of metallized components	Glass can be metallized by: (a) chemical silvering, (b) using metallizing pastes of silver, platinum, gold or palladium, (c) by vacuum evaporation or sputtering. Electroplating the deposited film with copper allows soldered joints to metal components to be made with conventional lead–tin solder alloys, although careful heating is required to avoid cracking of the glass component during soldering.
Mechanical joints using solder metals	Simple glass–metal joints, in which solder fills the annulus between concentric glass and metal components, rely on the contraction of the solder to form a mechanical seal to the glass component.

Table 1.7.55 Ceramic–glass and ceramic–metal seals

Type of seal	*Sealing technique*	*Design considerations and limitations*
Ceramic–glass	Matched seals	Expansion differential not more than 10% but exceptionally 30% differences can be tolerated if the compressive stress is in the glass part and the tensile stress in the ceramic part. Seals can be made by conventional glass-fusion techniques or by pre-coating the ceramic part with a suspension of glass powder in water. After drying this is fired to form a glaze on the ceramic surface to which the glass components is joined.
Ceramic–glass–metal	Glass used as an intermediate between ceramic and metal	Metals rods or plates (usually Fe–N–Co) are first embedded in an appropriate glass (see glass–metal seals) and the inside surface of the ceramic coated with a powdered glass suspensions. The assembled components are then furnance-heated until glass softens and seals. Cylindrical ceramic–glass–metal seals can also be made. 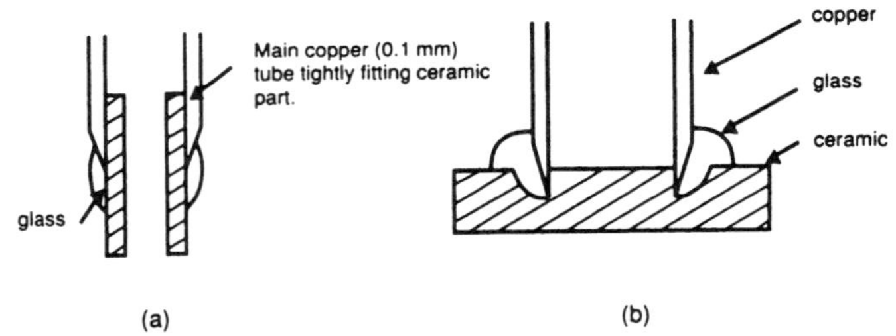 Ceramic tube (a) and disc (b) modified housekeeper seals.
Ceramic–glaze–metal	A very thin coating of glaze or enamel (or solder glass), is used as an intermediary. The glaze is incorporated into the ceramic/metal structure because of its thinness so that after fabrication the seal can be heated to above the melting point of the glaze.	A layer of solder glass or enamel is applied in an organic binder to the ceramic part and heated to form a glaze layer which may be ground to dimensions. Controlled oxidation of the metal may be necessary to form a strong bond.
Ceramic–metal by soft-soldering	(a) Direct soldering with indium (b) Soft-solder used to make a seal between metal and the ceramic which has been coated with metal.	As for glass–metal sealing using soft solder.
Brazed ceramic–metal seal	Brazing to a sintered metal coating in the ceramic parts.	A very widely used process for joining a variety of ceramic components to metal parts. Ceramic component is first coated with a suspension containing molybdenum (and/or manganese) and fired to produce a sintered coating. This may then be plated with nickel to ease brazeability. Brazing using Au–Ag eutectic alloy.
Ceramic–metal seal	'Active metal' processes	A brazing alloy is used with a 'reactive metal' (Ti or Zr) which when molten wets the ceramic and alloys with the braze so that a strong bond is formed. The chief advantage of this process over the sintered metal brazing technique is that only one heating operation is required to make the seal. The active metal can be applied by: (a) applying a metal hydride layer to the ceramic components, (b) applying a titanium–nickel powder mixture, (c) using a brazing rod with a reactive metal core, (d) using washers of active metal or active metal alloys.
Ceramic–metal seals	Pressure seals ('ram seals')	Seals are made by diffusion of the metal into the ceramic under pressure. High alumina ceramics in diameters up to 5 mm have been bonded to copper-clad steel pipes by heating at 1000°C for 2 h under a pressure of 15–20 MPa

Table 1.7.56 Possible ceramic–glass–metal combinations

Ceramic	***Glass***	***Metal***
Forsterites	High expansion glasses	Fe–Ni (46–51% Ni) Fe–Cr (16% Cr) Fe Fe–Ni–Cr
Steatites		Fe–Ni (42–46% Ni) Fe–Ni–Cr
Aluminas		Ni Fe–Ni
Zircon ceramics	Low expansion glasses	Mo Kovar
Aluminosilicate ceramics		

1.7.9 Diffusion bonding

1.7.9.1 Introduction

Diffusion bonding is a method of joining metallic materials without forming the large volumes of cast metal typical of fusion welding or causing the bulk deformation and recrystallization which occurs with pressure welding.

In diffusion bonding, the time allowed for diffusion of liquid metal at the interface is sufficiently long to allow the liquid phase to disappear into the bulk, whereas in soldering and brazing the joint is formed by solification of the filler metal as a predominantly separate interfacial layer. It is thus a sophisticated solid-phase welding process. The methods of diffusion bonding are:

(a) Surfaces brought together under high pressure for a short time.
(b) Surfaces brought together under low pressure for a long time.
(c) Use of a liquid phase interlayer which allows the use of low pressures and short times.

Titanium alloys, niobium alloys, copper, steels, uranium, dispersion-strengthened nickel alloys and aluminium alloys have been joined by diffusion bonding.

1.7.9.2 High pressure diffusion bonding

In diffusion bonding, the two surfaces to be joined are pressed into intimate contact at a temperature high enough for interdiffusion to occur at an acceptable rate.

High pressures are used when it is necessary to disrupt a surface oxide film or when the surfaces are so rough that contact is made only at asperities. The pressure has to be sufficient to cause creep at the tips to ensure good contact over the whole surface to be bonded, otherwise cavities will be left at the interface. The pressures needed will be up to 20 MPa. Normally the process will be carried out in a high vacuum or a protective atmosphere. The times and temperatures should be judged so as not to form brittle intermetallic layers.

1.7.9.3 Low pressure diffusion bonding

This is preferred when surfaces are very smooth and when any oxide film can dissolve in the matrix at the bonding temperature. This is especially applicable to titanium, copper and zirconium alloys.

Pressures are usually 5–15 MPa, depending on the material, temperature and the time available to complete the bond.

Diffusion bonding may be combined with superplastic forming to manufacture complex structures, especially in titanium alloys. A non-bonding layer is applied to the surfaces at which bonding is not required. Where superplastically formed structures come into contact, diffusion bonding will occur as the temperatures and pressures needed for superplastic forming are the same as those for diffusion bonding.

1.7.9.4 The interlayer method

An interlayer of other metal(s) is used between the two surfaces. It is frequently arranged that the interlayer melts, either by raising the bonding temperature to the metal's melting point, or by choosing an interlayer material which will form a eutectic alloy with the substrate (e.g. copper interlayer for aluminium). The interlayer is applied either as an electroplate or a foil.

Upon liquefaction, the interface is wetted, cavities are removed and continued diffusion into the base ultimately absorbs all the liquid metal and forms a strong bond. The time taken can vary from minutes to hours, depending upon conditions adopted. The advantages and limitations of diffusion are given in:

Table 1.7.57 *Characteristics of diffusion bonding*

The requirements of the interlayer are as follows:

1 The interlayer should melt below the temperature at which bonding is to be carried out or, preferably, should readily form a eutectic liquid with the base alloy which melts below this temperature.
2 Liquid should wet interfaces and be capable of diffusion into the base metal after bond formation.
3 Liquid should reduce or dissolve oxide films (alternatively surface oxides must be removed as with aluminium).
4 Components of liquid phase should diffuse into base metal at compatible rates.
5 Local alloying should not markedly affect the structure of the workpiece. No intermetallic phases should be formed.
6 The melting points of the component metals of the interlayer should be as near as possible to that of the parent metal to avoid differential vacancy migration and void formation at interface (compromise with 1).

1.7.9.5 Surface preparation

The surfaces are required to conform closely with one another. There is some evidence, however, that ductile metals bond more readily following coarse preparation (emery 120 grit) than with finer preparation. This is attributed to the deeper layer of worked material promoting recrystallization, which contributes to bonding.

With hard metals, a soft interlayer allows greater contact due to deformation under pressure.

Surfaces are normally prepared by grinding and degreasing and immediate transfer into an inert atmosphere.

Table 1.7.57 Characteristics of diffusion bonding

Advantages	*Limitations*
Alloys which are not normally considered to be weldable can be joined without a change of structure at the interface.	Need for controlled atmosphere to heat and press components (limits size).
Dissimilar metals (e.g. those liable to galvanic attack) may be joined by the interlayer method.	Accurate fit-up and satisfactory surface preparation necessary.
Pressures in the interlayer process can be as low as 70 KPa, but are more usually 500–1000 kPa.	Considerable experience exists with adhesive bonding, which is used to satisfy similar requirements.
Interlayers can be chosen which do not change the structural balance of phases in the base material.	Desirable to test bond quality.
Creep resistant alloys like dispersion-strengthened nickel can be joined at temperatures at which the creep resistant structure is not destroyed.	Geometry of joints can cause difficulty.
	Cavities or oxide globules along interface can impair toughness and ductility.
	Lack of fatigue data on diffusion bonded joints.

1.7.10 Solvent bonding of plastics

1.7.10.1 Introduction

There are three types of solvent bonding as detailed below.

ORGANIC SOLVENT

Solvent welding employs the property of many thermoplastics of being softened by certain organic solvents. The selected solvent is applied to the surface being joined and after assembly evaporates from the joining surfaces, leaving a true weld.

Suitable materials	– PVC, polystyrene, ABS, acrylics.
Not normally suitable	– polyethylene, polypropylene, nylons (polyamides), acetals (good resistance to solvents).

The major disadvantages are long setting times and no capability for gap-filling.

POLYMER/SOLVENT CEMENT ('BODIED CEMENTS' OR 'DOPES')

PVC, polystyrenes and acrylics can also be bonded by using elements which are solutions of a polymer in a solvent. As the solvent evaporates, polymer is deposited in the joint. Long setting times are still involved, but a degree of gap-filling is possible.

POLYMER/MONOMER CEMENT

Acrylics can be joined by using a solution of polymer dissolved in its own monomer as a cement. The monomer dissolves the joint surfaces and is then polymerized in the joint. The 'curing' is achieved by chemical activation, UV radiation or heating.

The system has good gap-filling capability and completely homogeneous joints may be obtained with marked reductions in hardening time. Properties are close to parent material and joint efficiencies approaching 100% can be achieved.

THE PROCESS

1. Application of solvent to the surface to be bonded (applied by dropper hypodermic needle, brush or immersion).
2. Joint left open for a short time (about 10 s)
3. Parts brought together under pressure (usually 0.3–1.0 MPa, with polycarbonates up to 4 MPa).
4. Post-treatment required to consolidate joint. This can take several hours and may be done either during or after application of pressure.

The advantages and limitations of solvent bonding are given in:
Table 1.7.58 *Characteristics of solvent bonding (welding)*

1.7.10.2 The solvents

In solvent bonding, the function of the bonding agent is a transient one. Even in bodied cements the dissolved polymer does not ultimately function as an adhesive. The characteristics of the solvent which have the most effect on the joint are its solvent power for the polymer, its diffusion behaviour in the polymer and its volatility.

Examples of typical solvents are given in:
Table 1.7.59 *Typical solvents for bonding plastics*

1.7.10.3 Joinability of materials

Solvent bonding is most commonly used with non-crystalline polymers:

polystyrene and styrene copolymers,
acrylic polymers and copolymers,
polycarbonate and copolymers,
polycarbonate and polysulphone,
polyphenylene oxide,
polyvinyl chloride,
some vinyl chloride copolymers.

Some partly crystalline polymers can also sometimes be bonded:

cellulosics,
acetals,
nylons.

Solvent bonding is normally used to join components of the same material, but bonding of dissimilar materials may be possible if the solvent used affects both materials in approximately the same way and to the same extent, or if a compatible mixture of solvents can be found to give this result.

1.7.10.4 Joint design

Wide variety of joint shapes and geometries. Works best with joints in which the contact (interface) areas are moderate in relation to the total size and local thickness of the components.

The mating surfaces should give the closest possible match-voidage as imperfections resulting from local mismatching can impair the strength and stability of a joint. A bodied cement has some ability to fill the structural voids, but the filling is never complete because of the comparatively low polymer content of the cement.

1.7.10.5 Joint inspection

Normally a visual inspection is carried out for general finish and presence of voids. Dimensional checks are sometimes carried out for local or overall distortion.

Joint strength may be monitored (sample checks) by mechanical testing.

Table 1.7.58 Characteristics of solvent bonding (welding)

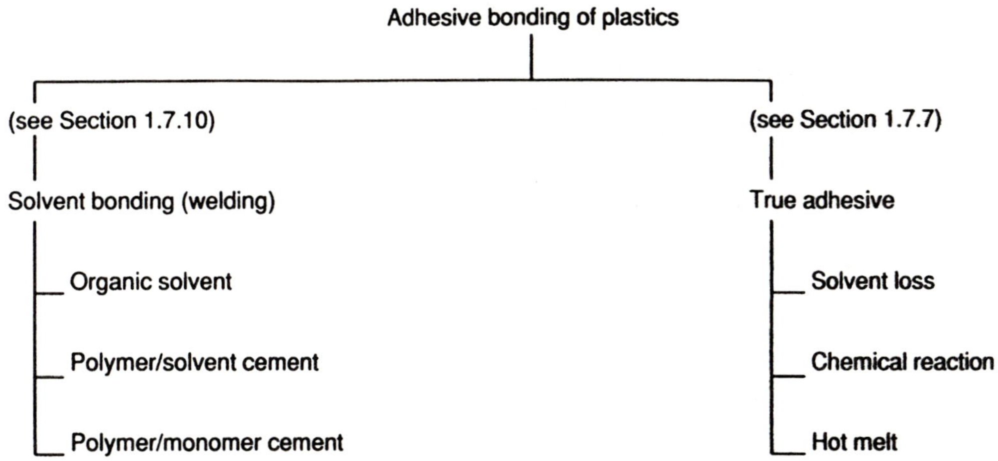

Advantages of solvent bonding
1. High bond strengths (almost equal to bulk material strength). 2. Simple and inexpensive equipment. 3. Complicated joint interface shapes and small areas can be joined.
Limitations of solvent bonding
1. Post-treatment (hours) required to consolidate joint. 2. Clamping of joint to apply pressure. 3. Danger of crazing (cracking when bonding internally stressed components anneal before joining). 4. Health and fire hazard.

Table 1.7.59 Typical solvents for bonding plastics

Polyvinyl alcohol	*Styrene butadiene (BOS)*	*Styrene acrylonitrile (SAN)*	*PVC and copolymers (acetate)*	*Polysulphone*	*Polystyrene*	*Polyphenylene oxide*	*Polycarbonate*	*Polyamide (nylon)*	*Ethyl cellulose*	*Cellulose nitrate*	*Cellulose propionate*	*Cellulose acetate butyrate*	*Cellulose acetate film*	*ABS*	*Acrylic (polymethyl methacrylate)*	*Substrate* / *Solvent*
								X								Acetic acid (glacial)
	X									X		X	X			Acetone
												X				Acetone/ethyl acetate/cellulose acetate butyrate (40/40/20)
												X	X			Acetone/ethyl lactate (90/10)
												X	X			Acetone/methoxyl acetate (80/20)
											X	X	X			Acetone/methyl acetate (70/30)
												X	X			Butyl acetate/acetone/methyl acetate (50/30/20)
		X														Butyl acetate/methyl methacrylate monomer (40/60)
								X								Calcium chloride solution (alcoholic)
					X											Chloromethane
							X								X	Chloromethane/methyl methacrylate mon. (60/40)
															X	Chloromethane/methyl methacrylate mon. (50/50)
				X	X		X								X	Dichloroethane (ethylene dichloride)
							X									Dichloroethane/chloromethane (50/50)
				X			X			X	X	X			X	Dichloromethane (methylene dichloride)
		X			X					X		X	X			Ethyl acetate
									X							Ethyl acetate/ethanol (80/20)
										X	X	X				Ethylene glycol
X																Glycerine/water (15/85)
										X		X				Methyl acetate
										X	X	X				Methyl ethyl ether (monomethyl ether)
	X	X	X		X					X	X	X		X		Methyl ethyl ketone (butanone)
	X		X											X		Methyl isobutyl ketone
															X	Methyl methacrylate monomer
								X								Phenol (aq. or alc. soln)
								X								Resorcinol (aq. or alc. soln)
					X											Tetrachloroethene (-ylene)
							X									Tetrachloroethane (carbon tetrachloride)
			X													Tetrahydrofuran/cyclohexanone (80/20)
					X											Toluene
									X							Toluene/ethanol (90/10)
														X		Toluene/methyl ethyl ketone
							X									1,1,2-Trichloroethane
					X											Trichloroethylene
						X										Trichloromethane (chloroform)
			X													Xylene
						X								X		Xylene/methyl isobutyl ketone (25/75)

Table 1.7.59

1–8

The significance of materials properties and testing

Contents

List of tables

List of figures

The following materials properties and testing procedures are covered in other sections:

Hardness – Vol. 1, Chapter 1-11 Units and conversion factors.
Fracture toughness – Vol. 1, Chapter 1-2.
Corrosion and degradation – Vol. 1, Chapter 1-4 and Vol. 3, Chapter 3-1.
Abrasion – Vol. 1, Chapter 1-5 and Vol. 3, Chapter 3-1.

1.8.1 Introduction

This section of The Materials Selector shows how a knowledge of material properties is essential to the production of efficient designs. The significance of each property is discussed, the terms used to describe it defined, and the appropriate test methods indicated. Mechanical properties are introduced in terms of the behaviour of metals, which is then compared with that of plastics and ceramics which have different test methods and characteristics.

1.8.2 Short-term mechanical testing

1.8.2.1 The tensile test

Tensile tests are usually conducted by measuring the force required to extend a standard test-piece at a constant rate. Full details of the test-pieces and the methods of conducting the tests on metals are given in National Standards, e.g. BS18 and ASTM E8. Figure 1.8.1(a) shows the way in which the load varies with extension for a mild steel testpiece.

Fig. 1.8.1(a) *Schematic load–extension curve*

From O to A the load is proportional to the extension, and thus Hooke's law is obeyed. At the point A, the limit of proportionality, the extension increases more rapidly until a maximum load is reached at B, the upper yield point. The force then drops to a value at C known as the lower yield point, and remains constant for some further extension before increasing as the extension is continued. The load–extension curve passes through the point D which is the load required to give a specific permanent offset (that is, extension after unloading); D is obtained by drawing XD parallel to OA where OX is the specific offset. As the extension is continued beyond D, the load reaches a maximum at E before falling to a value at F, when the test-piece breaks.

In order to normalize the behaviour of different sized test-pieces, it is usual to divide the force, P by the original cross-section area, S_o, to obtain the nominal stress σ_n

$$\sigma_n = P/S_o \tag{1}$$

The extension is divided by the gauge length, that is the original length, L_o, over which the extension is measured, to obtain the engineering or nominal strain, e

$$e = \frac{(L - L_o)}{L_o} \quad \text{or} \quad e\,(\%) = \frac{(L - L_o)}{L_o} \times 100 \tag{2}$$

The standard value of $L_o = 5.65\ \sqrt{S_o}$.

A load–extension curve such as Fig. 1.8.1(a) is therefore a stress–strain curve with the y-axis scaled by $1/S_o$ and the x-axis by $1/L_o$. Direct plots of stress against strain, as in Fig. 1.8.1 (b), represent the tensile properties of unit volume.

Fig. 1.8.1(b) *Schematic stress–strain curve*

INTERPRETATION OF NOMINAL STRESS-STRAIN CURVES

From the stress–strain curves the following properties may be determined:

(a) Limit of proportionality and Young's modulus

This is the stress at which purely elastic behaviour is replaced by a combination of elastic and plastic behaviour (point A, Fig. 1.8.1). The plastic contribution is permanent and irreversible.

The slope of the line OA is the tensile elastic, or Young's modulus, E

$$E = \sigma/e \tag{3}$$

Young's modulus is used in designs based upon stiffness considerations in tension or bending where a limited deflection is required.

(b) Upper and lower yield points

Points B and C, Fig. 1.8.1 (a), designated σ_{yu} and σ_{ye}, respectively.

(c) Proof stress

Point D, Fig. 1.8.1, designated $\sigma_{p0.2}$, where the permanent plastic offset is 0.2%. Other values of offset (e.g. 0.1%, 0.5%) are sometimes quoted. The proof stress is a limiting stress above which substantial deformation will occur.

(d) Tensile strength

Point E, Fig. 1.8.1 (a), designated σ_u. This is the maximum stress which a component can sustain without fracture (at the expense of considerable deformation). The difference between proof stress and tensile strength is a measure of the safety margin against failure by accidental overload when the working stress approximates to the proof stress.

(e) Elongation at fracture and reduction of area

These properties are obtained by fitting together the broken test-piece, measuring the final gauge length L_f, and final gauge area S_f.

$$\text{Elongation at fracture}, A\% = \frac{(L_f - L_o)}{L_o} \times 100 \qquad (4)$$

$$\text{Reduction of area}, Z\% = \frac{(S_o - S_f)}{S_o} \times 100 \qquad (5)$$

These parameters are measures of material ductility. Ductility has two important consequences:

(i) the ability to absorb plastic strain and accommodate local stress concentrations in load-bearing service,
(ii) the ability to accommodate strain in cold forming operations.

(f) Poisson's ratio

Young's modulus is determined from the longitudinal strain of the test-piece along the tensile axis. If the radial strain in a circular cross-section test-piece is also determined, Poisson's ratio, ν can be evaluated as the ratio (radial strain)/(longitudinal strain) at a given stress (ν = 0.3 for most metals).

(g) Secant modulus

The limit of proportionality occurs at a very low stress and Hooke's law is only valid for small strains. Consequently, the use of Young's modulus will lead to an underestimate of the strain at stresses between the limit of proportionality and the proof stress. A convenient method of estimating the strain at points within this range is by means of the secant modulus OG – one of a number of moduli shown in Fig. 1.8.1(b). It should be noted that unloading a test-piece from a stress greater than that represented by point A will leave a small permanent strain so that on reloading the stress–strain curve will be linear up to the original applied stress. The original flow stress as defined by the limit of proportionality has thus increased owing to the strain-hardening or work-hardening of this material.

THE SIGNIFICANCE OF TRUE STRESS AND TRUE STRAIN

(a) True stress

During a tensile test the load P required to continue the extension of the test-piece is given by:
$P = S\sigma_t$ (6) where S is the current cross-section area of the test-piece and σ_t is the true stress.
Thus

$$dP = S d\sigma_t + \sigma_t dS \tag{7}$$

(b) True strain

The true strain at any point during the test is defined as dL/L and consequently the total true strain ϵ_t at gauge length L' (where L' is the extended length) is given by

$$\epsilon_t = \int_{L_o}^{L'} dL/L = \ln (L'/L_o) \tag{8}$$

(c) Relationship between true stress and true strain

Most metals are deformed under constant volume so that

$$S dL + L dS = 0 \tag{9}$$

Initially the decrease in cross-sectional area, dS, is low and the increase in stress $d\sigma$ due to work hardening is high and dP is positive, so that the force to continue extending the specimen increases. The hardening increment $d\sigma$, however, decreases with increasing strain and eventually dP becomes negative and the load passes through a maximum when $dP = 0$

Hence from eqn (7)

$$\sigma_t dS = -S d\sigma_t \tag{10}$$

Combining (9) with (10)

$$\frac{d\sigma_t}{\sigma_t} = \frac{dL}{L} = d\epsilon_t \tag{11}$$

Or

$$\frac{d\sigma_t}{\sigma_t} = \frac{dL}{L} = \frac{dLL_o}{L_o L} \frac{de}{1+e} \tag{12}$$

where e is the nominal (engineering) strain.

(d) Formability index

The stress–strain curve of most metals and alloys can be represented by an equation of the form

$$\sigma_t = K\epsilon_t^n \tag{13}$$

Hence

$$d\sigma_t = Kn\epsilon_t^{n-1} d\epsilon_t \tag{14}$$

From (13) and (14)

$$\frac{d\sigma_t}{\sigma_t} = \frac{n d\epsilon_t}{\epsilon_t} = d\epsilon_t \text{ (when } dP = 0 \text{ from (11))} \tag{15}$$

Hence instability occurs when $\epsilon_t = n$

Thus the uniform true strain (sometimes called logarithmic or natural strain) is numerically equal to the exponent n, referred to as an n-value, which is obtained experimentally from a log–log plot of true stress and true strain. The n value is a direct measure of uniform elongation (the true strain to maximum load) and is a formability index especially relevant to stretch-forming operations.

A second indicator of formability, which can be obtained from the tensile test, is the R value. R is defined as the ratio (width strain)/(thickness strain) in sheet specimens and is a measure of resistance to thinning. It is an indicator of deep drawability; high values of R are required for good performance in deep-drawing operations. By assuming a constant volume of material, if a volume V of length L width W breadth B then:

$$dV = d(LBW) = 0 \tag{16}$$

Hence
$$\frac{dL}{L} + \frac{dB}{B} + \frac{dW}{W} = \epsilon_L + \epsilon_B + \epsilon_W = 0 \tag{17}$$

i.e. $|\epsilon_L| = \epsilon_B + \epsilon_W$

Since
$$R = \frac{\epsilon_W}{\epsilon_B} \text{ then } |\epsilon_L| = \epsilon_B\,(1 + R) \tag{18}$$

If $R = 1, \epsilon_B = \epsilon_W = \dfrac{\epsilon_L}{2}$

and this is equivalent to a Poisson's ratio of 0.5 for plastic flow. R values of 1 (when the material is said to be isotropic) are seldom encountered in practice.

TENSILE BEHAVIOUR OF METALLIC MATERIALS

The four values frequently quoted in specifications for metallic materials are proof stress, tensile strength, elongation and reduction in area. When the material is a wrought product, a sample having tensile properties that are representative of the product as a whole can usually be obtained. A sample for tensile testing cannot, however, always be obtained directly from a finished cast product, in which case a separate sample bar may be cast either from the melt or on to the product for subsequent tensile tests. Such test-pieces, which are unlikely to have the same tensile properties as the castings they represent on account of their different section sizes and cooling rates, are frequently used for quality control.

The stress–strain curve shown in Fig. 1.8.1 (a) is typical of well annealed mild steel. Cold-worked mild steels, high strength steels, austenitic stainless steel, and most non-ferrous alloys, however, have a different form of stress–strain curve in which the upper and lower yield points do not appear. In the absence of a yield drop, such as the point B, the onset of plastic flow is not clearly defined, and it is customary to express the yield or flow strength as a proof stress. Fig. 1.8.1 (b) shows the initial part of a stress-strain curve in which no yield points occur. It is a curve typical of cast iron.

TENSILE TESTING OF PLASTICS

Tensile tests for plastics materials are generally similar to those for metals, but follow the methods and use the specimens described in National Standards e.g. BS 2782: Part 3 or ASTM D638 or in some other standard more appropriate for a specific material. Some of the stress–strain curves appear the same as those shown as examples for metals, but as the term 'plastics materials' embraces a very wide range of substances it is not, perhaps, surprising to find that their response to tensile stress is equally wide.

In general, all the tensile properties listed for metals can be determined for plastics, but in practice it is usual to ascertain only the following:

(a) Tensile strength	–	(breaking load)/(initial cross-sectional area.)
(b) Elongation at break	–	(final gauge length – initial gauge length)/(original length)
(c) Yield strength	–	(load at deviation from linearity)/(initial cross-sectional area). Deviation is often very loosely defined; for exacting circumstances an offset yield (proof stress) may be used.
(d) Young's modulus	–	defined as for metals (eqn (3)). It is common to quote 100 (or 200)% modulus; this is a secant modulus at 100 (or 200)% strain.

Young's modulus is the most useful tensile property because parts should be designed to accommodate stresses to a degree well below this. For some applications where rubbery elasticity is desirable, a high ultimate elongation may be an asset; for rigid parts it has little value. There is great value in moderate elongation, however, since this quality is a measure of the ability to absorb rapid impact and shock. Thus, the total area under the stress–strain curve is indicative of toughness. A material of very high tensile strength and a little elongation tends to be brittle in service.

TENSILE BEHAVIOUR OF PLASTIC MATERIALS

Thermosets such as phenolics and melamines behave generally in a completely brittle fashion. Their very limited extension is proportional to the stress applied, and fracture occurs in the Hooke's law region of the stress–strain curve. There is no yield as would be expected with even a moderately ductile material. At the other end of the spectrum a polythene will first deform in a Hookean mode and then yield, subsequently, it will neck at a fairly constant load, and with really extensible varieties it may neck again when the initial neck has reached the widened shoulders of the specimen. By this means elongations of several hundred per cent may occur. The vast majority of thermoplastics fall between these two extremes, extending in a linear mode to the limit of proportionality and then yielding up to perhaps 100%.

There are a few materials (or strictly speaking composites) that may disobey (or appear to disobey) these rules. Most plastics compounds used in practice, as distinct from pure polymers, are mixtures of resin and other constituents, some of which (for example plasticizers) can be completely miscible with the polymer so that the whole behaves as one homogeneous mass, whose behaviour is none the less influenced by the additive. Other components remain as discrete elements within the polymer matrix, and thus polymer and additive will behave as two dissimilar components. In many cases this is undetectable or unimportant, but for the special group of fibre-reinforced plastics it is necessary to consider the contribution of each element to the behaviour of the whole, as it is the interaction between their contributions which accounts for the high strengths achieved by such composites. Consider, for example, the extension of a glass-reinforced polyester resin. Here it may well be found that up to 1% extension the material will behave as if it were homogeneous. At this stage, however, the resin may no longer continue to extend and it will therefore progressively crack, usually without any very obvious visual sign, although detectable noise may be emitted. A decrease in the slope of the stress–strain curve caused by the effective loss of resistance of one element accompanies further elongation, but (depending on the glass:resin ratio and the configuration of the glass fibres in the test-piece) this may not be particularly noticeable as the glass is generally the stronger component. Finally, at 2 or 3% elongation, the testpiece will yield slightly and fracture, leaving a specimen consisting of loose glass fibre and small pieces of resin throughout its gauge length. In these circumstances it is quite usual to report the final value of tensile strength (that is, load at fracture/initial cross-sectional area) and completely ignore the intermediate stage at which the material for all practical purposes became useless.

Considerable effort has been devoted to producing data to enable the stress–strain characteristics of resin and glass to be matched, but this technique is by no means universally applied. There are also many different types of fibrous reinforcement, ranging from random fibre mat, through square weave cloth, to cloths that are not equally strong in warp and weft, and finally, to unidirectional reinforcement as used in filament-wound structures and pultrusions.

Glass-reinforced plastics with more or less continuous fibres are usually tested in the form of machined specimens, so that it is necessary to recognize that different specimen orientations may give different results. Most other materials (including thermoplastics, such as nylon, that are reinforced by very short glass fibres) are moulded for test. Usually this process does not cause difficulties in testing, but this is not always so. For example, if a specimen is injection moulded with the molten material entering a relatively long, thin mould from one end, it is probable that it will be oriented with the polymer chains tending to run the length of the mould; thus the strength test results will probably be higher than those obtained from a sample produced by a technique that does not orient the polymer.

The physical properties of most plastics materials, unlike those of most metals, are acutely dependent on temperature (and sometimes humidity) and thus most available data are for measurements made at 'ambient' temperature, usually defined as 20°C or 23°C. Changes from this temperature as small as 20°C result in significant changes in tensile properties for some thermoplastics and all show a deterioration in properties as the temperature of test increases. Thermosets, although initially much less temperature-dependent, start to degrade chemically at a certain temperature and ultimately fail.

It must be remembered that most plastics behave as viscoelastic materials. After an initial instantaneous response to stress they will continue to deform, very slowly indeed, until the stress is either deliberately removed or reduced to a negligible level by the continued deformation. When the stress is taken off a material it again shows an instantaneous response followed by a gradual recovery towards its original unstressed state. See Section 1.8.3.

1.8.2.2 Strain rate sensitivity

The tensile properties of materials are temperature- and rate-sensitive to a greater or lesser degree since the mechanisms by which deformation takes place involve a localized rearrangement of atoms. The strain rate defines the time available for the atoms to move in order to produce the required strain, whilst the temperature controls the mobility (the rate at which the atoms move). Thus, high strain rates and low temperatures which, respectively, decrease the time available for the rearrangement of the atoms and the rate at which they move, require higher applied stresses to activate the deformation mechanism; conversely, lower strain rates and higher temperatures will reduce the flow stress. This, the most general type of behaviour, is exhibited by pure metals. In some alloys a change in the strengthening mechanism (which depends on solute diffusion, that is the atomic mobility of alloying elements) can lead to an inversion of the usual dependence of flow stress on temperature and strain rate over a limited temperature range. An example of this phenomenon is the so-called 'blue brittleness' exhibited by plain carbon steels when the yield and tensile stresses rise to a maximum as the ductility falls to a minimum in the temperature range 100–300°C, as shown in Fig. 1.8.2. The more usual type of behaviour is exemplified by the curves (also in Fig. 1.8.2) for proof stress, $\sigma_{p0.2}$ and tensile strength, σ_u of an austenitic stainless steel.

Fig. 1.8.2 *Proof stress, tensile strength and elongation for plain carbon steel, and proof stress and tensile strength for austenitic stainless steel*

The sensitivity of flow stress to temperature and strain rate has several important practical consequences.

The flow stress is a function of strain rate and temperature which may be written in the general form:

$$\sigma = f_1(\dot{\epsilon}_t) f_2(T) \quad (1)$$

where $\dot{\epsilon}_t = d\epsilon_t/dt$ = strain rate. $f_1(\epsilon)$ increases with increasing strain rate and $f_2(T)$ decreases with increasing temperature.

By differentiating eqn (1) it can be seen that both the temperature and strain rate sensitivities are greater for higher strain rates and lower temperatures. This means that very high strain rates, particularly at low temperatures, require high stress for flow to occur, and this can have the effect of leading to brittle fractures with low ductilities for certain materials having high strain rate sensitivities. This aspect of strain rate sensitivity is discussed in more detail in Section 1.8.5.

EFFECT OF STRAIN RATE SENSITIVITY ON TENSILE TESTING

Since tensile tests are usually conducted by imposing a fixed rate of extension on the specimen, the precise level of the stress–strain curve will depend on the rate of testing. Tensile tests are usually conducted at constant velocity of specimen grip separation so that the true strain rate decreases throughout the test as the specimen gauge length increases. In order to minimize the effect of strain rate on the stresses determined during a tensile test, standard strain rates are specified for tensile testing. For example, for room temperature tensile tests BS 18 specifies a plastic strain rate ϵ_p less than 0.5/min, whereas for hot tensile tests, BS 3688 specifies a strain rate at 0.2% proof stress, $\epsilon_{p0.2}$ of $1 - 3 \times 10^{-3}$/min.

Equation (1) may also be rearranged to predict a strain rate for a given stress and temperature and may thus be written as

$$\dot{\epsilon} = f_3(\sigma) f_4(T) \quad (2)$$

Function $f_3(\sigma)$ increases with stress and $f_4(T)$ increases with temperature. Equation (2) is significant in the following ways:

1. An applied stress lower than that measured for the flow stress in a conventional tensile test will cause plastic flow at a strain rate less than that employed in the test. Whether this strain rate is of practical importance or not will depend on its magnitude. There will, therefore, always be some plastic flow under an applied load and this time-dependent strain is usually called *creep*. The creep rate will be greater for higher stresses and temperatures.
2. It can be shown that the stress and temperature sensitivity of the strain rate are also greatest for high stress and temperature. Thus the practical significance of creep deformation will increase with stress and temperature. A more extensive discussion of creep and the associated phenomenon of *stress relaxation* is given in Section 1.8.3.

The effect of time on the flow stress is greatest at high temperature, although it varies from one material to another. For example, a temperature of 200°C has a large effect on the tensile properties of aluminium, but little effect on those of nickel. A useful way of assessing the effect of a given temperature on flow stress is to compare the temperature of interest with the melting point of the material, that is to express the temperature as T/T_M where T is the temperature and T_M the melting temperature in kelvins (K). Thus for aluminium at 200°C:

$$\frac{T}{T_M} = \frac{(200 + 273)\,°C}{(660 + 273)\,°C} = \frac{473\ K}{933\ K} = 0.51$$

and for nickel,

$$\frac{T}{T_M} = \frac{(200 + 273)\,°C}{(1453 + 273)\,°C} = \frac{473\ K}{1726\ K} = 0.27$$

With pure metals it is usually found that the effect of temperature on the flow stress is not particularly marked for values of T/T_M less than 0.5. In those alloys where additional strengthening mechanisms operate, the limiting value of T/T_M can be increased above 0.5. The temperature or strain rate sensitivity also depends on the crystal structure of the metal or alloy: face-centred cubic structures such as copper, aluminium, nickel, and austenitic stainless steels are less rate sensitive than body-centred cubic structures, such as ferritic steels.

1.8.2.3 Superplasticity (see also Section 1.6.4.4)

Another aspect of strain rate sensitivity that is directly relevant to the tensile test is the possibility of strain rate hardening, which is responsible for the phenomenon known as superplasticity. The distinguishing feature of superplasticity, which is implicit in its name, is the ability of a material to undergo large elongations of the order of 1000% during a tensile test. Most metallic materials undergo less than a twentieth of this elongation before fracture, and a neck forms when the rate of reduction of cross-section exceeds the rate of work-hardening, that is the increase in flow stress with strain. In materials with a high strain rate sensitivity, localized deformation in an incipient neck would cause a higher than average strain rate and would require a higher than average stress level, so that the incipient neck would be unable to develop. One such alloy is Al–22%Zn (Prestal). In hot polymers, depending upon temperature and strain rate substantial amounts of neck-free flow are possible.

1.8.2.4 Tear resistance (elastomers) – ASTM D624

SIGNIFICANCE

This method determines the tear resistance of the usual grades of vulcanized rubber, but not of hard rubber. Since tear resistance may be affected to a large degree by a mechanical fibering of the rubber under stress as well as by stretch distribution, by strain rate, and by the size of the specimen, the results obtained in the test can be regarded only as a measure of the resistance under the conditions of the test rather than necessarily as having any direct relation to service value.

SPECIMEN

The test describes the sizes and shapes of three specimens, each of them with curve and contour. Two of them have a slit cut in the edge.

PROCEDURE

The specimen is clamped in the jaws of a testing machine and the jaws then separated at a speed of 500 mm (20 inch) per min. After rupture of the specimen, the breaking force in newtons (pounds force) is noted from the scale in the test machine. The resistance to tear is calculated from the force and the median thickness of the specimen. Values are given in N/m, or in lbf/in for tearing the specimen of 1 cm (or 1 inch) in thickness.

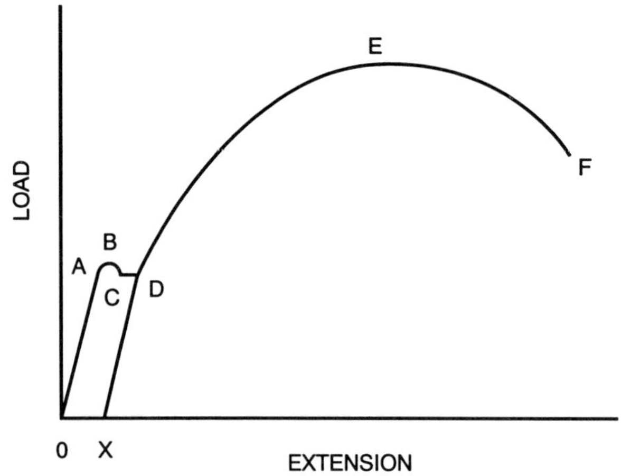

(a)

A = limit of proportionality;
B = upper yield point;
C = lower yield point;
D = 0.2% proof stress (OX = $\frac{0.2\ L_o}{100}$); $P_{0.2} = \frac{\text{Load at D}}{S_o}$; ($L_o$ = initial length; S_o = initial stress).
E = tensile strength; $\sigma_u = \frac{\text{Load at E}}{S_o}$;
F = fracture

Fig. 1.8.1(a) Schematic load–extension curve

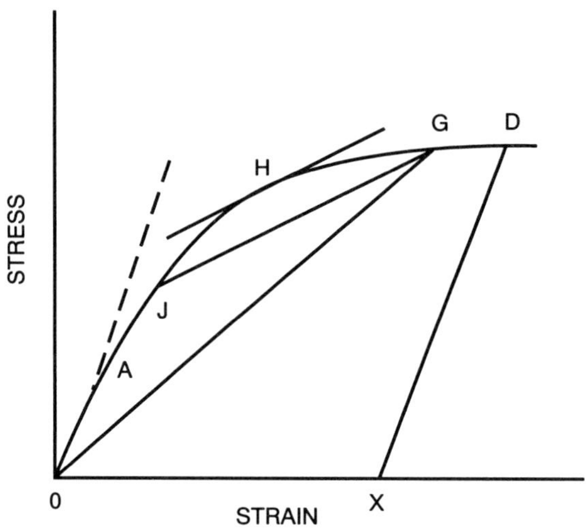

(b)

A = limit of proportionality;
D = proof stress (OX = 0.2%);
G, H and J are arbitrary points between A and D;
OG = secant modulus;
JG = chord modulus;
slope of line through H = tangent modulus.

Fig. 1.8.1(b) Schematic stress–strain curve

Fig. 1.8.1

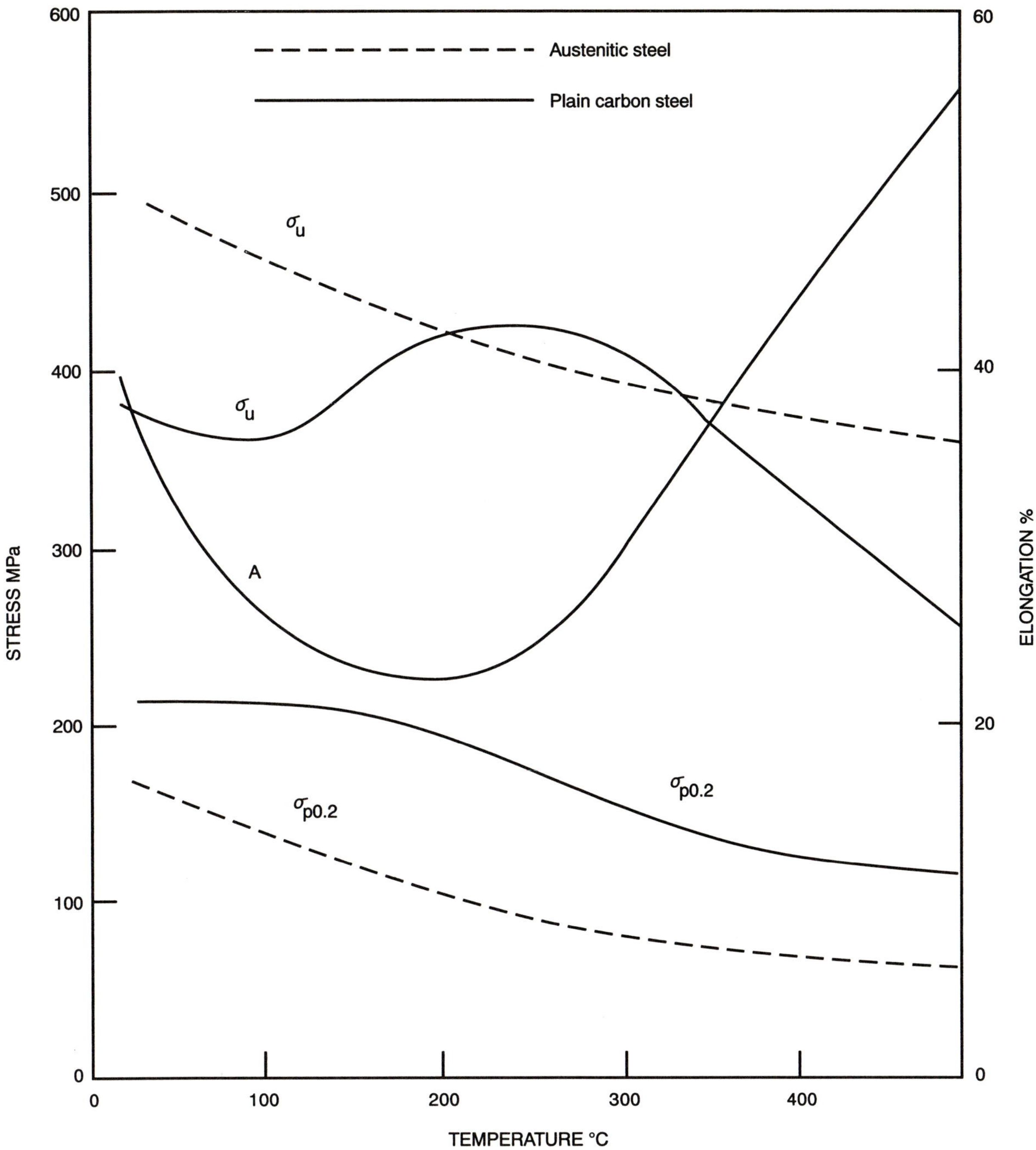

Fig. 1.8.2 Proof stress, tensile strength, and elongation for plain carbon steel, and proof stress and tensile strength for austenitic stainless steel

1.8.3 Creep

The time-dependent strain which occurs when a material is subjected to a sustained stress is known as creep. For a component under a constantly applied external load, creep becomes apparent as the component changes shape or fractures. When a device is loaded by the application of a fixed strain as, for example, in bolts and springs, creep becomes apparent as a stress relaxation. The term 'static fatigue' is sometimes used to describe the behaviour of plastics under sustained loading (Section 1.8.4.5).

Design stresses for metallic components intended for use at ambient temperatures are based either on the 0.2% proof stress or on the tensile strength; at higher service temperatures the likelihood of creep deformation has to be considered. Short-term tensile test data can only be used for design purposes when creep deformation is negligible. Otherwise the appropriate creep property (rupture stress or creep strength) must be used.

Useful creep lives may be of the order of 100 000 h and the experimental determination of data spanning this time is both costly and time consuming. Numerous methods have been proposed to permit short-term test data to be extrapolated in order to predict long-term creep properties. Most of these methods, however, take no account of any structural changes which might take place during long service lives and thus change the creep properties of the material. Their use is, therefore, better suited to interpolation between existing data than to extrapolation beyond their range of time or temperature.

The viscoelastic creep behaviour of some plastics also makes creep a possibility even at ambient temperatures.

1.8.3.1 Creep testing of metallic materials

The standard methods of creep and relaxation testing of metallic materials in tension are described in National Standards e.g. BS 3500 and ASTM E139.

Creep tests are usually performed with test-pieces similar to those used in the tensile test, but instead of imposing a fixed rate of extension on the test-piece and measuring the force required to maintain that extension rate as in the tensile test, the creep test consists of imposing a fixed force on the specimen and measuring the consequent extension. There is also a difference in time scale between the two tests; a tensile test takes a few minutes to complete, whereas a creep test can continue for any time between a few hours and hundreds of thousands of hours. Creep tests are usually performed at elevated temperatures and the standard requires the temperature to be held steady within a few degrees throughout the duration of the test. Usually the test is performed under a constant applied load, so that there is an increase in true stress throughout the test as the cross-sectional area decreases in tensile creep; conversely, in compression testing, the true stress will decrease. This is typical of most service conditions, however, and is generally acceptable. For fundamental studies of creep, it is desirable to maintain constant true stress by arranging for the effective applied load to be decreased in proportion to the reduction in area of the test-piece. This can be done by reducing either the applied load or the length of the lever arm through which the load is applied to the test-piece. In relaxation testing the test-piece is generally preloaded to a given strain level and the force required to maintain the strain is measured throughout the test. Creep testing can take one of the following forms:

(1) Rupture testing, in which the total elapsed time to failure at constant nominal stress and temperature is recorded and the elongation, A, is determined after completion of the test in the same fashion as in the tensile test.

(2) Creep testing proper, in which the total plastic strain, A_p, is measured as a function of time at intervals throughout the test.

As an example the procedure recommended in BS 3500 for determining the plastic strain is shown schematically in Fig. 1.8.3.

Fig. 1.8.3 *Schematic stress–strain diagram of loading a creep test specimen – BS 3500: Part 3*

The total plastic strain–time relationship is shown schematically in Fig. 1.8.4 for four stress levels, at the same constant temperature.

Fig. 1.8.4 *Schematic creep curves for four stress levels* $\sigma_1 > \sigma_2 > \sigma_3 > \sigma_4$, *showing the three stages of creep*

It is interesting to note that the rate of change of total plastic strain is greatest at the highest stress. A similar family of creep curves would be obtained for constant stress and increasing temperature levels.

Creep data for metallic materials are sometimes plotted in the form of a master curve of time and temperature such as the Larson–Miller Parameter.

1.8.3.2 The three stages of creep in metallic materials

The creep curves shown in Fig. 1.8.4 are typical of high-temperature creep at temperatures above $T_M/2$, where three well defined stages of creep occur: primary creep, from O to A, in which the rate decreases with time, followed by secondary creep, from A to B, in which the rate is constant, and finally tertiary creep, from B to C, in which the rate increases until the material finally ruptures. At temperatures lower than $T_M/2$, (1) the creep rate of metallic materials decreases with time, (2) the total strain tends to a limiting value, and (3) failure of the material by creep rupture is unlikely to occur.

The measurements which can be made from a single creep test, at a given stress and temperature, are (1) the time to rupture and (2) the time to achieve any given strain less than the strain to rupture. Creep properties of metallic materials are usually expressed as rupture stress, that is the stress to cause rupture in a given time at a given temperature (which may be thought of as the tensile strength for a given life), and creep strength, that is the stress required to produce a given strain in a given time (which may be regarded as a proof stress for a given life). These properties are derived from a series of creep tests by plotting the results of the individual tests in the form of log stress vs log time, as shown in Fig. 1.8.5. From the resulting curves for stress against rupture life and a given creep strain, rupture stresses and creep strengths for a given life can be obtained.

Fig. 1.8.5 *Schematic log stress–log time plot of creep test results illustrating the derivation of rupture stress and creep strength for lives of 10 000 h*

1.8.3.3 Creep testing of plastics

The stress–strain time relationship, which characterizes the creep of plastics, is strongly affected by temperature, by other environmental conditions such as humidity, and by processing. A knowledge of the stress–strain time relationship is useful to designers in predicting the time-dependent deformation of a component which is subjected to a known system of stresses. Unlike most metallic materials, the creep of plastics is viscoelastic so that recovery (that is, reversal of strain) to varying degrees takes place even after the longest period of creep once the load is removed. In many instances recovery is complete, but rarely where visible defects such as necking or crazing have occurred. It may well be necessary to allow for the phenomenon of recovery in the design of components that will be subject to conditions of intermittent loading.

The primary results of the creep test are usually presented as creep curves showing total strain vs. logarithmic time at different stress levels as shown in Fig. 1.8.6, although the data can be presented in other ways in order to facilitate the selection of information for a particular requirement.

Fig. 1.8.6 *Creep data – three methods of presentation for plastics: (a) isometric stress against* *time; (b) creep curves; (c) isochronous stress against strain – BS 4618: Part 1: Section 1.1*

The so-called isochronous stress–strain curves (also shown in Fig. 1.8.6) can be obtained by constructing sections of constant strain on the creep curves. The same data can also be presented as creep moduli, given by the ratios of applied stresses to creep strains. Creep modulus–time curves are obtained from the creep curves by calculating (stress): (creep strain) ratios for constant strain.

Although BS 4618 deals most comprehensively with the design of creep tests for plastics materials, there are as yet few data for plastics compared with those available for metals. Because of the scarcity of plastics creep data it is usually found necessary to circumvent the effects of creep by conservative design.

1.8.3.4 Recovery

A study of recovery usually follows a creep test. The stress is removed and the diminution of strain is measured for a period of at least 0.1 of the time under stress (usually longer). During the period of creep, the strain increases with time until the end of the test when the load is removed. The strain then decreases with time for the remainder of the test throughout the recovery period. The results of this test are usually presented as fractional recovery (or fraction of recovered strain) reduced time curves where the fractional recovery is defined as the strain recovered by the creep strain when the load is removed.

1.8.3.5 Stress relaxation

Laboratory creep tests are conducted at constant nominal applied stress. However, under practical service conditions, the load acting on a component can vary; a component which creeps is generally associated with other components that may deform only elastically and thus, even with a constant external load, the forces between the components of the system are redistributed with time.

A common example of relaxation is that of the bolted flange joint in pipework. The bolts are initially tightened to a fixed stress level which will decrease with time as a result of relaxation. Relaxation is typical in structure elements subjected to constant external load for long periods, especially at elevated temperatures.

Pure stress relaxation of metals can occur in welded structures or other components which have a residual stress distribution. When such components are stress relieved by heating to an elevated temperature for a given period, the residual stresses are relieved by relaxation. Relaxation tests are thus of interest for indicating the stress reduction that may occur in a given time at a given temperature.

The technical difficulties of relaxation testing are considerably greater than those of creep testing and for this reason, relaxation testing is not often carried out. Whereas the objective of a creep test is clearly defined, that of the relaxation test is not so definite, since the results of relaxation testing are used more often in providing a comparative rating of materials than directly in design calculations.

1.8.3.6 Compression set (elastomers) – ASTM D395

SIGNIFICANCE

The compression set (i.e. residual deformation) measures the ability of compounds to retain elastic properties after prolonged action of compressive stresses. Compression set tests should be limited to those involving static loading, i.e. hysteresis effects confuse the results in dynamic-stress testing.

SPECIMENS

These are 0.51 inch cylindrical discs cut from a laboratory prepared slab of between 0.49 and 0.51 inch (12.5 and 13.0 mm).

PROCEDURE

The test is designed to measure the residual deformation of a test specimen after it has been stressed under either a constant load or a constant deflection. A dial micrometer measures the deformation remaining 30 min after the removal of the loads. The constant load method specifies a force of 1.8 kN (400 lbf); the constant deflection procedure calls for a compression of approximately 25%.

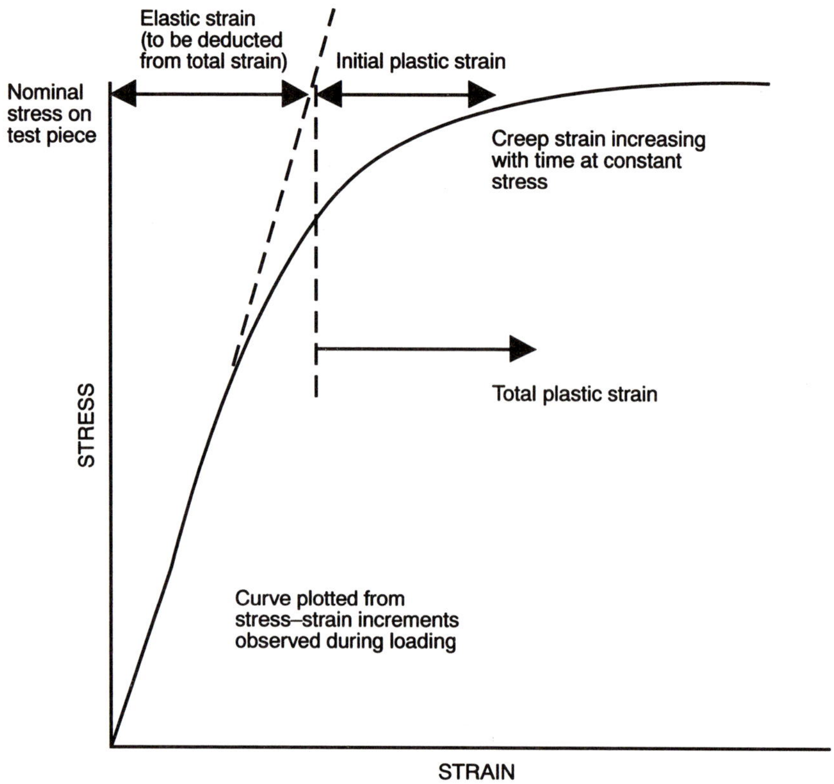

Fig.1.8.3 Schematic stress–strain diagram of loading a creep test specimen – BS 3500: Part 3

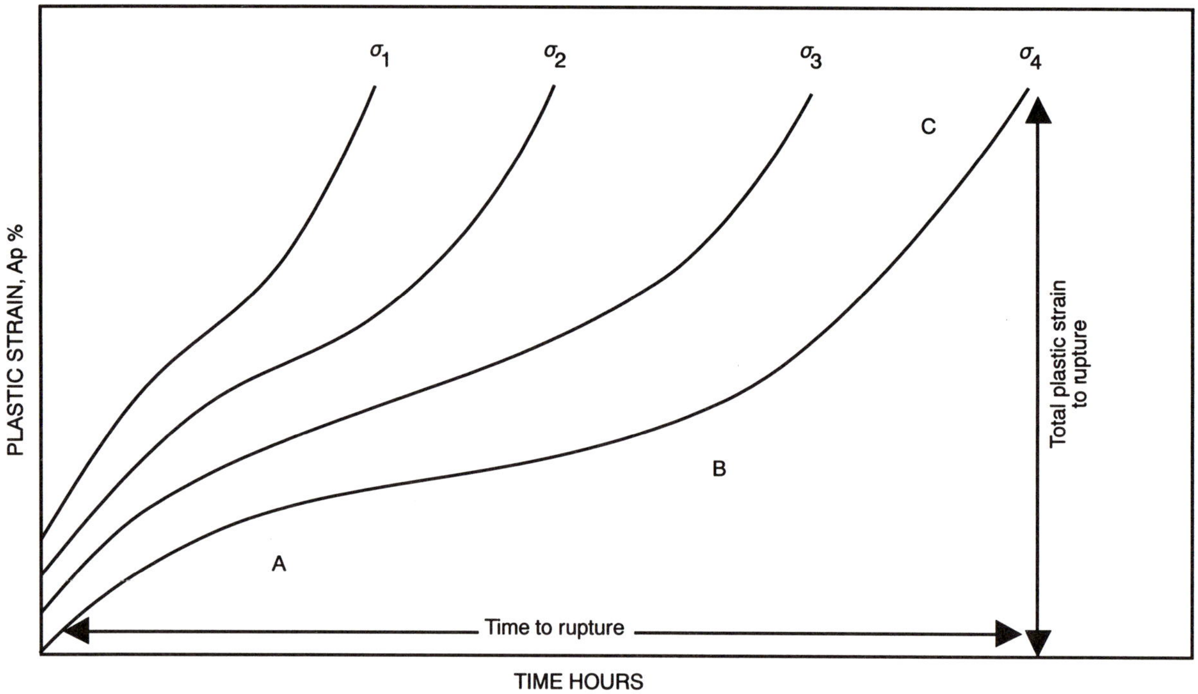

Fig.1.8.4 Schematic creep curves for four stress levels $\sigma_1 > \sigma_2 > \sigma_3 > \sigma_4$, showing the three stages of creep

Fig. 1.8.3 and Fig. 1.8.4

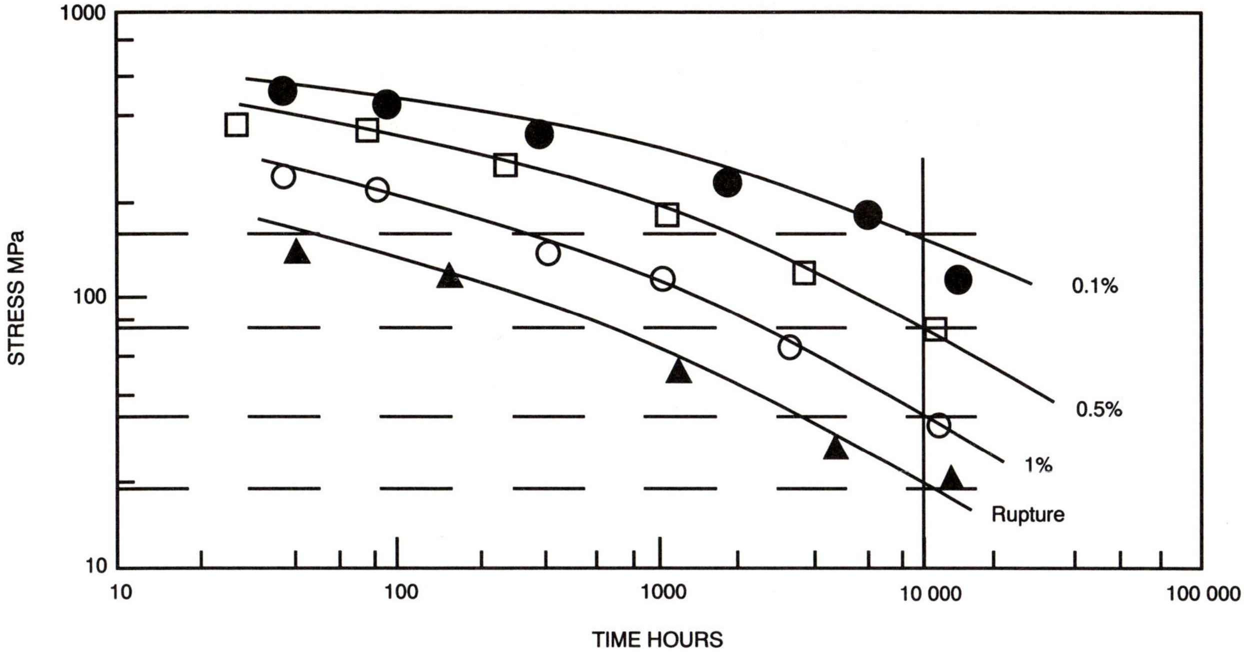

Fig.1.8.5 Schematic log stress–log time plot of creep test results illustrating the derivation of rupture stress and creep strength for lives of 10 000 h

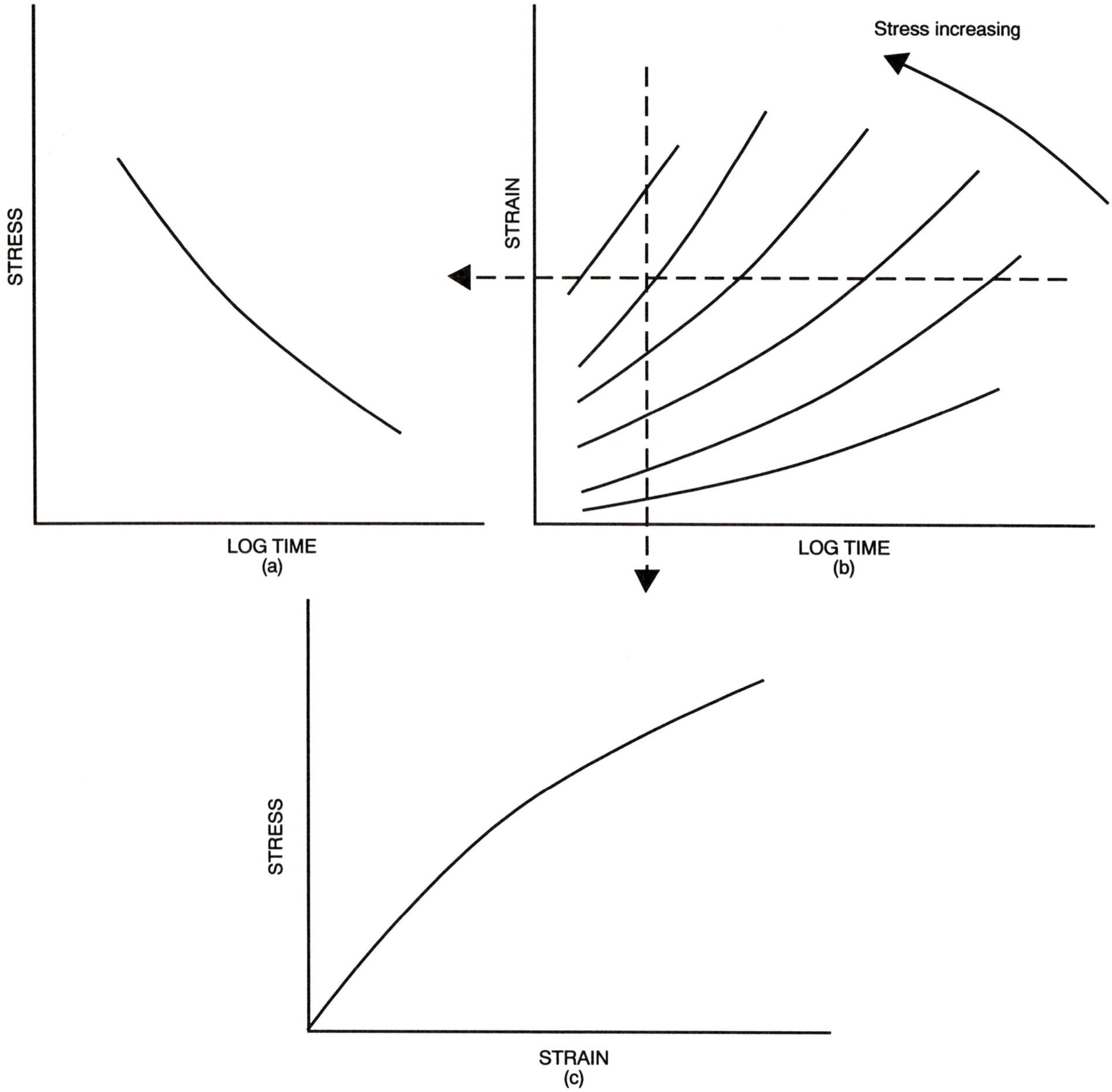

Fig.1.8.6 Creep data – three methods of presentation for plastics: (a) isometric stress against time; (b) creep curves; (c) isochronous stress against strain – BS 4618: Part 1: Section 1.1

Fig. 1.8.6

1.8.4 Fatigue

The term 'fatigue' may be defined as the process of progressive localized permanent structural change that occurs when a material is subjected to conditions that produce fluctuating stresses and strains, possibly culminating in cracks or complete fracture after a sufficient number of fluctuations have taken place. As a consequence of fatigue, fluctuating stresses less than the tensile strength can cause failure if sufficiently repeated. Thus, under fluctuating load conditions design calculations based on the tensile strength will be insufficient to predict a safe working stress.

Examples of fatigue loading fall into one or more of the following simple classifications:

(1) direct stress, under which the material is repeatedly loaded along its major axis (for example, the loading experienced by a piston rod),

(2) plane bending, in which the material is repeatedly bent about a particular neutral plane (for example, the type of loading applied to leaf springs),

(3) rotating bending, in which the material is bent about a neutral plane which rotates with respect to the test-piece (for example, the loading applied to a rotating axle of a railway waggon),

(4) torsion, in which the material is repeatedly twisted about a fixed axis (for example, a motor vehicle suspension torsion bar).

These different types of loading can occur in combination.

1.8.4.1 Fatigue–stress cycle

In the context of fatigue, stress is taken to be the nominal stress calculated from the net cross-section under consideration. The applied stress cycle is the smallest complete range of values in the time–stress pattern that is repeated periodically. Reference to Fig. 1.8.7 shows that any stress varying periodically over a given range can be regarded as a combination of (1) a component alternating between two values opposite in sign, or direction, but equal in magnitude (the stress amplitude), and (2) a static stress (the mean stress). The stress cycle may be wholly tensile, or wholly compressive, or any state between these, depending on the mean stress level.

Fig. 1.8.7 *Fatigue–stress cycle – BS 3518: Part 1: 1962*

INTERPRETATION OF FATIGUE–STRESS CYCLE

Fatigue testing is described in detail in National Standards e.g. BS 3518 and ASTM E466/468. In addition to those described in Fig. 1.8.7, the following terms are employed:

Theoretical stress concentration factor, K_t

The theoretical elastic or geometrical stress concentration factor based on net section of area.

Stress ratio, R

The algebraic ratio between the minimum and maximum stress in 1 cycle, that is $R = \sigma_{min}/\sigma_{max}$.

Fatigue life or endurance, N

The number of stress cycles to cause failure.

Endurance limit, σ_N

The value of the stress condition under which the test specimen has a life of N cycles.

Cycle ratio, n/N

The ratio (cycles applied)/(fatigue life).

Fatigue strength reduction factor, K_f

The ratio of the stress amplitude at the fatigue limit for plain polished specimens to that for test-pieces with a stress concentration.

$$K_f = \sigma_a / \sigma'_a$$

where σ'_a is the stress amplitude for the test-piece containing the stress concentration.

$K_f \neq K_t$ because the response of materials to stress concentrations under fatigue loading varies and this response is described by the

$$\text{Notch sensitivity factor, } q = (K_f - 1)/(K_t - 1)(0 \leq q \leq 1)$$

If $K_t = K_f, q = 1$ and the material is fully notch-sensitive (q is not a material constant, and is also a function of experimental parameters).

1.8.4.2 Fatigue testing

Fatigue tests may be performed on samples of a material, components, or complete structures. When machined samples are tested, special care has to be taken in preparing the test-pieces to ensure a stress-free specimen. Recommended methods of machining and standard test-pieces are described in National Standards e.g. BS 3518: Parts 2 and 3 and ASTM E466.

Apart from tests on plain specimens, tests are frequently carried out to study the effect of notches, fillets, holes, and other forms of discontinuity that can occur in components. In order to obtain information on the sensitivity in fatigue of a material or structure due to notches, it is usual to determine the stress-endurance curve for test-pieces with and without notches. When these tests are run at zero mean stress, the fatigue strength reduction factor, K_f, is obtained by dividing the fatigue strength of the unnotched test-piece by the fatigue strength of the notched test-piece. Welded joints are a special type of discontinuity and frequently their fatigue properties are compared with those of unwelded material in order to determine the effect of the stress concentrations associated with this form of joint. In a similar fashion riveted and screwed joints can be compared with unjoined material.

The number of test-pieces used in a fatigue investigation varies considerably, according to its objective. Where the experiment is of a statistical nature the number of specimens can be very large; whereas, on the other hand, if each test-piece is a costly machined component, the number of test-pieces may of necessity be small.

The failure criterion in fatigue testing is usually taken to be the occurrence of a crack or complete failure, although it may sometimes be a given rate of crack propagation or the development of a given plastic strain. The usual limiting values of endurance tests are 10×10^6 cycles for structural steels and up to 100×10^6 cycles for other steels and non-ferrous metals and alloys.

Most of the available fatigue data have been determined from alternating stress tests in which the mean stress is zero. This is because of the greater convenience of obtaining fatigue data from rotating bending fatigue machines which do not permit the application of a non-zero mean stress. In practice, however, many components and structures are subjected to stresses that fluctuate between unequal values of tension and compression and a number of empirical relationships have been proposed in order to permit an estimate to be made of the fluctuating fatigue strength from the alternating fatigue strength and the tensile strength of the material.

The fatigue strength of a material under fluctuating stresses, that is with a static stress superimposed on the alternating stress, can be represented graphically by plotting the alternating stress σ_a against the mean stress σ_m as shown in Fig. 1.8.8:

Fig. 1.8.8 *Fatigue strength–static strength (or R–M) diagram*

The curve joining the alternating fatigue strength (that is, fatigue strength for a given endurance), σ_N and the static tensile strength, σ_u, or yield strength σ_y, represents the combinations of static and alternating stresses giving the same endurance. In order to determine this curve experimentally a number of stress-endurance curves are determined, each for a constant value of σ_m, σ_a or R.

The two straight lines and the curve shown in Fig. 1.8.8 represent the three most widely used empirical relationships. The straight line joining the alternating fatigue strength to the tensile strength represents Goodman's law. The curve, joining the alternating fatigue strength to the tensile strength, is known as the Gerber parabola. The third relationship, known as Soderberg's law is represented by a straight line joining the alternating fatigue strength and the static proof stress of the material; this relationship is intended to fulfil the condition that neither fatigue failure nor yielding will occur. The three relationships are given by the following equations:

Goodman's law, $$\sigma_a = \sigma_N\,(1 - \sigma_m/\sigma_u), \tag{1}$$

Gerber's law, $$\sigma_a = \sigma_N\,[1 - (\sigma_m/\sigma_u)^2], \tag{2}$$

Soderberg's law, $$\sigma_a = \sigma_N\,(1 - \sigma_m/\sigma_y), \tag{3}$$

where σ_a is the amplitude of the alternating stress associated with the mean stress σ_m, σ_N is the alternating fatigue strength, σ_u is the static tensile strength, and σ_y the static yield strength.

An alternative method of presenting fluctuating stress fatigue data is shown in Fig. 1.8.9. In this diagram the limits of the fluctuating stress for endurances of 10^4, 10^5, and 10^7 cycles are plotted against the mean stress:

Fig. 1.8.9 *Maximum and minimum stresses, σ_{max} and σ_{min} plotted against mean stress σ_m – BS 3518: Part 1: 1962*

1.8.4.3 Hysteresis and damping

Fatigue cracks appear to be brittle, but in fact are produced by plastic deformation. Cyclic stressing at a stress amplitude comparable with the yield strength of a material continuously produces plastic deformation, alternately positive and negative in each cycle. The elastic limit in compression after pre-loading in tension is reduced (Bauschinger effect), so that plastic yielding resumes in the reverse direction at a numerically smaller stress. This continual to and fro plastic deformation in localized regions is eventually responsible for generating and spreading the fatigue crack.

If the stress alternates between equal tension and compression, that is with zero mean load, the tensile strain will be mostly nullified by the compression strain and after a few cycles of stress a closed loop will be formed, as shown in Fig. 1.8.10. The closed loop in which the strain lags behind the stress is known as a hysteresis loop. The range of strain, during the stress cycles, is equal to the sum of the elastic and plastic strains.

Fig. 1.8.10 *Stress–strain hysteresis loop*

When a fatigue test is conducted over a constant stress range the plastic strain associated with each stress cycle may vary during the test. The amount of plastic strain depends strongly on the stress level. Some materials, such as plain carbon and austenitic stainless steels, exhibit considerable plastic strain in each cycle, even at stress ranges below their fatigue strengths. On the other hand, high strength alloy steels and aluminium alloys, for example, show no detectable strain unless the stress is greater than the range for failure in $10^5 - 10^6$ cycles.

The area of the hysteresis loop is a measure of the work done on unit volume of the material during the stress cycle. Usually, only a small proportion of this energy is stored in the material as a result of a permanent distortion of the structure, and the greater part is dissipated as heat. When fatigue tests are carried out on materials which have a large hysteresis loop the accompanying dissipation of heat may cause an appreciable rise in the temperature of the test-piece, which can affect the fatigue strength. Some means of cooling the test-pieces is necessary under these circumstances and it is desirable to record their temperatures. This problem is particularly acute with some plastics materials, especially when higher frequency stress cycles reduce the time available for the heat to be dissipated. This heating effect can also be of practical significance when cooling of the part is insufficient to prevent a temperature rise sufficient to impair the material's resistance to fatigue.

The area of the hysteresis loop is also a measure of the damping capacity of the material, and this can be an important property in applications where fatigue failure may result from resonant vibrations in a component or structure.

1.8.4.4 Low cycle fatigue – ASTM D606

In applications such as pressure vessels, aircraft landing gear and guns the total life required may be of the order of some hundreds or thousands of stress cycles, that is a factor of 10^3 or 10^4 less than that required for high cycle fatigue. In these circumstances a material can often withstand stresses and strains that are appreciably greater than the high endurance fatigue strengths, so that a knowledge of the behaviour of materials at low endurance is required if economic designs are to be achieved for components with a low cycle life.

At the high stresses that materials can withstand for low endurances there is often a large plastic strain associated with each cycle. The term 'progressive fracture' has been used to describe failures resulting from a small number of stress reversals, and it may be that the failure mechanism is different in low and high cycle fatigue, which hypothesis is certainly supported by the appearance of the fracture surfaces. The fracture resulting from up to several thousands of stress cycles often resembles a static tensile fracture and is quite distinct from the typical high endurance fatigue fracture, consisting of a smooth area of crack propagation and a rough fibrous area of final fracture.

The high plastic strains during fatigue at low endurances mean that stress is not directly proportional to strain and it is, therefore, necessary to draw a distinction between the resistance to alternating stress and the resistance to alternating strain; in practice it is often the latter which is more important. In an engineering part, fatigue failure usually propagates from regions of stress concentration, and, if a part is subjected to a fluctuating load, the material in the region of the stress concentration will be constrained to a given strain range by the surrounding elastic material.

ALTERNATING STRESS

If the properties of a material are considered from the viewpoint of its resistance to alternating stress, the static tensile strength represents one point on the stress–endurance curve at 1/4 cycle. It is desirable to connect this point to the conventional stress–endurance curve, which is usually determined for endurances greater than 10^4–10^5 cycles. Conventional fatigue machines which run at high frequencies ($\approx$ 50 Hz), are unsuitable for tests lasting less than 10^4 cycles. Fortunately the fatigue strength of most metallic materials exhibit little sensitivity to frequency at ambient temperatures, so that low-cycle tests run at low frequencies give results which correlate satisfactorily with those for larger endurances at higher frequencies.

The available information on the fatigue strength of metals at low endurances can be summarized by plotting the ratio (fatigue strength for a given endurance, σ_N)/(static tensile strength, σ_u) against the endurance, as shown in Fig. 1.8.11.

Fig. 1.8.11 *Alternating fatigue strength/tensile strength (metals)*

Different limits are shown for axial stress, rotating bend, and reversed bend and these include data for several steels, aluminium alloys and magnesium alloys. The axial stress fatigue strength for a low number of cycles can be estimated approximately by extrapolating the σ_N -log N curve backwards from 10^4 to 10^5 cycles by means of a straight line to the tensile strength at 1/4 cycle; this estimate is usually conservative.

The values of fatigue stress for the bend specimens shown in Fig. 1.8.11 were calculated from elastic theory making no allowance for the effect of plasticity on the stress distribution and thus overestimating the maximum surface stresses. At low endurances this is the most important reason for the apparent difference in the results between bend and direct stress; it also accounts for the values of fatigue strength in bending which appear to be greater than the tensile strength.

In view of the wide variety of test materials represented in Fig. 1.8.11, the spread of results is surprisingly small, and it appears that the fatigue strengths of metals at low endurances are quite closely related to their tensile strengths. In contrast, the resistance of a material to alternating strain at low endurances cannot be related to its tensile strength.

ALTERNATING STRAIN

The fatigue life in bending within the range 10–10^4 cycles depends upon the range of strain. The resistance of a metallic material to alternating plastic strain depends upon ductility and that to elastic strain on the tensile strength. The resistance to total alternating strain at high stresses, where plastic strain predominates, therefore depends upon ductility, and at low stresses, where elastic strain predominates, upon strength.

1.8.4.5 Static fatigue

The term 'static fatigue' is frequently used to describe the failure of non-metallic materials after a sustained period of loading. On metallic materials this type of failure is usually referred to as 'stress rupture' or 'stress corrosion'. Although not widely investigated, the phenomenon of static fatigue (stress rupture) is of greater importance with plastics materials than fatigue resulting from dynamic loading. Closely related to creep, it is the phenomenon of failure (by virtually any mechanism) after a period under a constant stress. The conditions which should be applied to investigations of static failure are: (1) test temperatures should be kept constant to within ±1°C, (2) relative humidity for certain materials (such as nylons) should be maintained constant to within ±2%, and (3) applied stresses should be known to an accuracy of 1%. When designing the test, it is of course necessary to consider any anisotropy of the material, the effects of such factors as production variables, and, if relevant, notch sensitivity.

Whilst static fatigue is not widely investigated, it has been studied fairly extensively with fibre-reinforced plastics, where the frequent use of these materials in sophisticated, highly stressed engineering applications renders it imperative for assessments of their stress life to be made both at ambient and elevated temperatures. However, by far the largest collection of static fatigue data is directly associated with the use of plastics water pipes. Water supply installations are usually designed in the expectation that they will be in use continuously for 50 years, and although, in practice they may be subject to small variations in pressure, their use is justified by long-term constant-hydraulic pressure tests at constant temperature. Early work on PVC involved tests on specimens that were continued for 10 or more years, but once reliable relationships between actual performance and relatively short-term tests (up to 10 000 h) had been established, first for PVC and subsequently for polythene, short-term tests (sometimes only up to 1000 h) were finally accepted as giving sufficient background for confident use of new varieties of these materials.

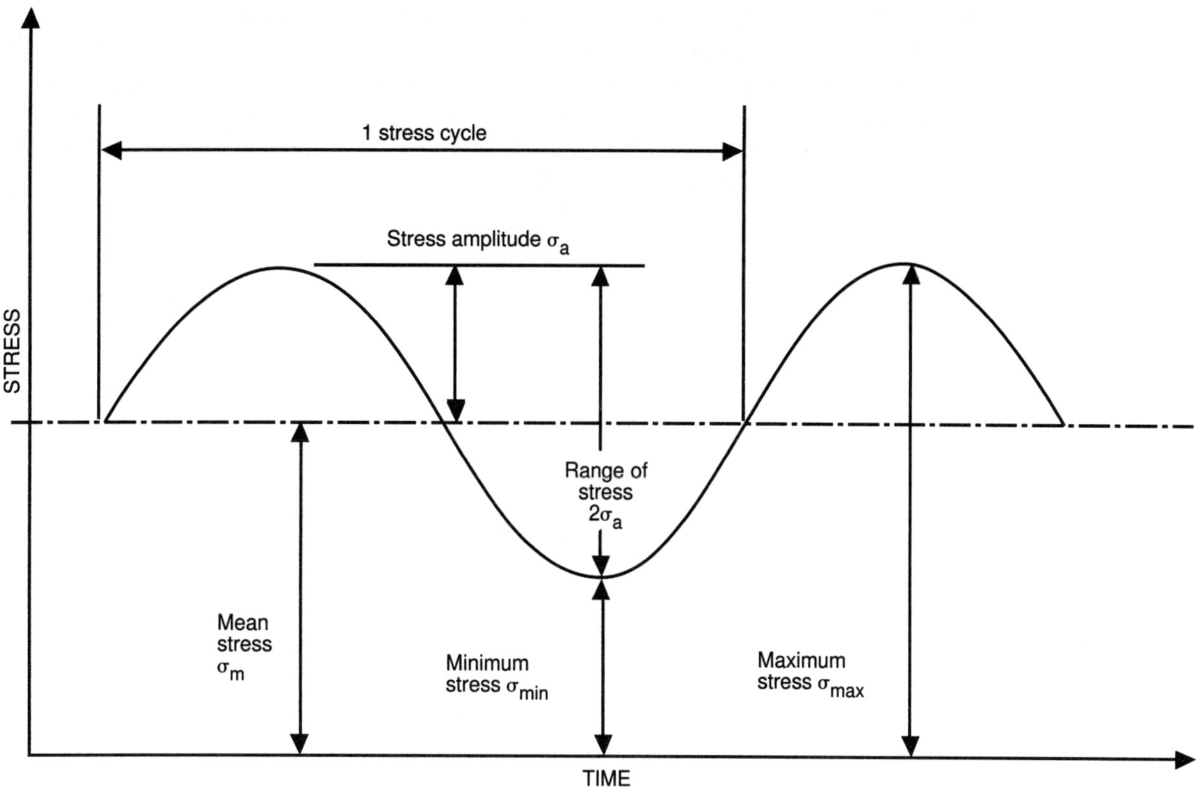

Fig.1.8.7 Fatigue–stress cycle – BS 3518: Part 1: 1962

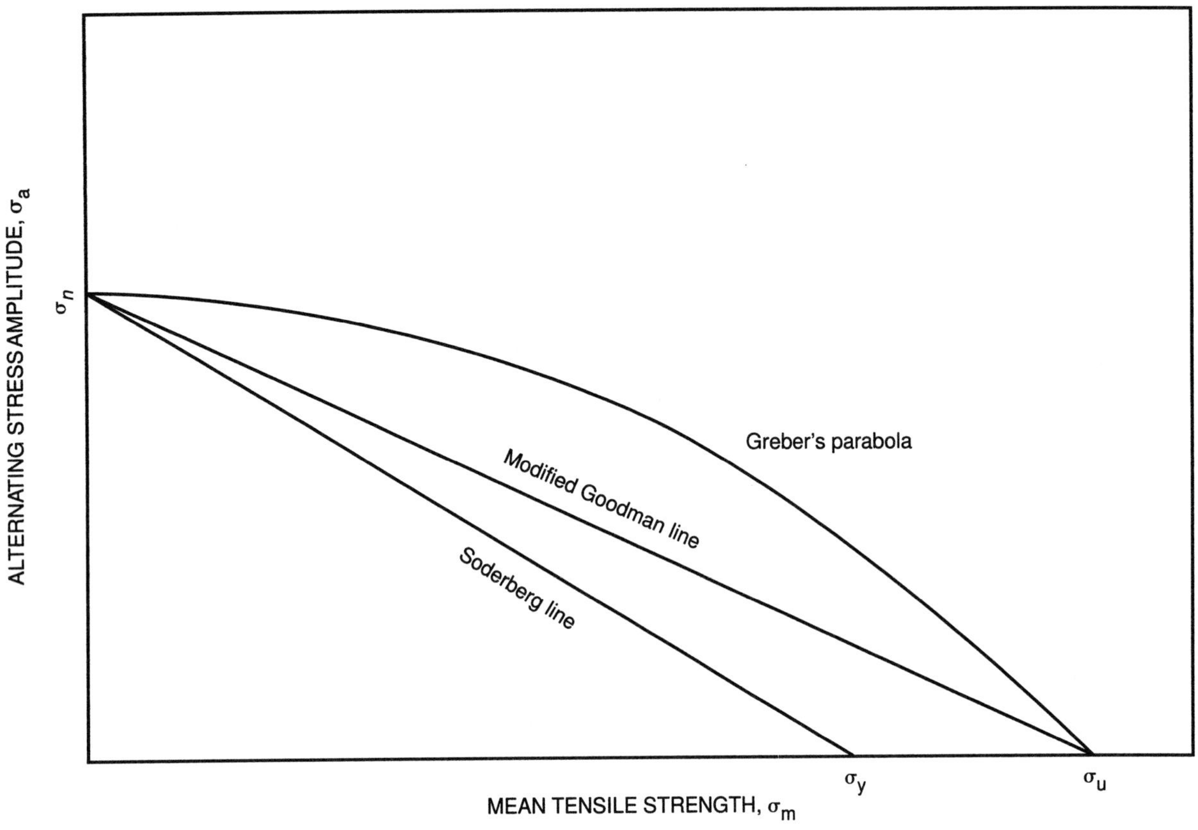

Fig. 1.8.8 Fatigue strength–static strength (or *R–M*) *diagram*
Reprinted from Forrest Fatigue of Metals, copyright 1962, with permission from Elsevier Science Ltd, The Boulevard, Langford Lane, Kidlington OX5 1GB, UK.

Fig. 1.8.8

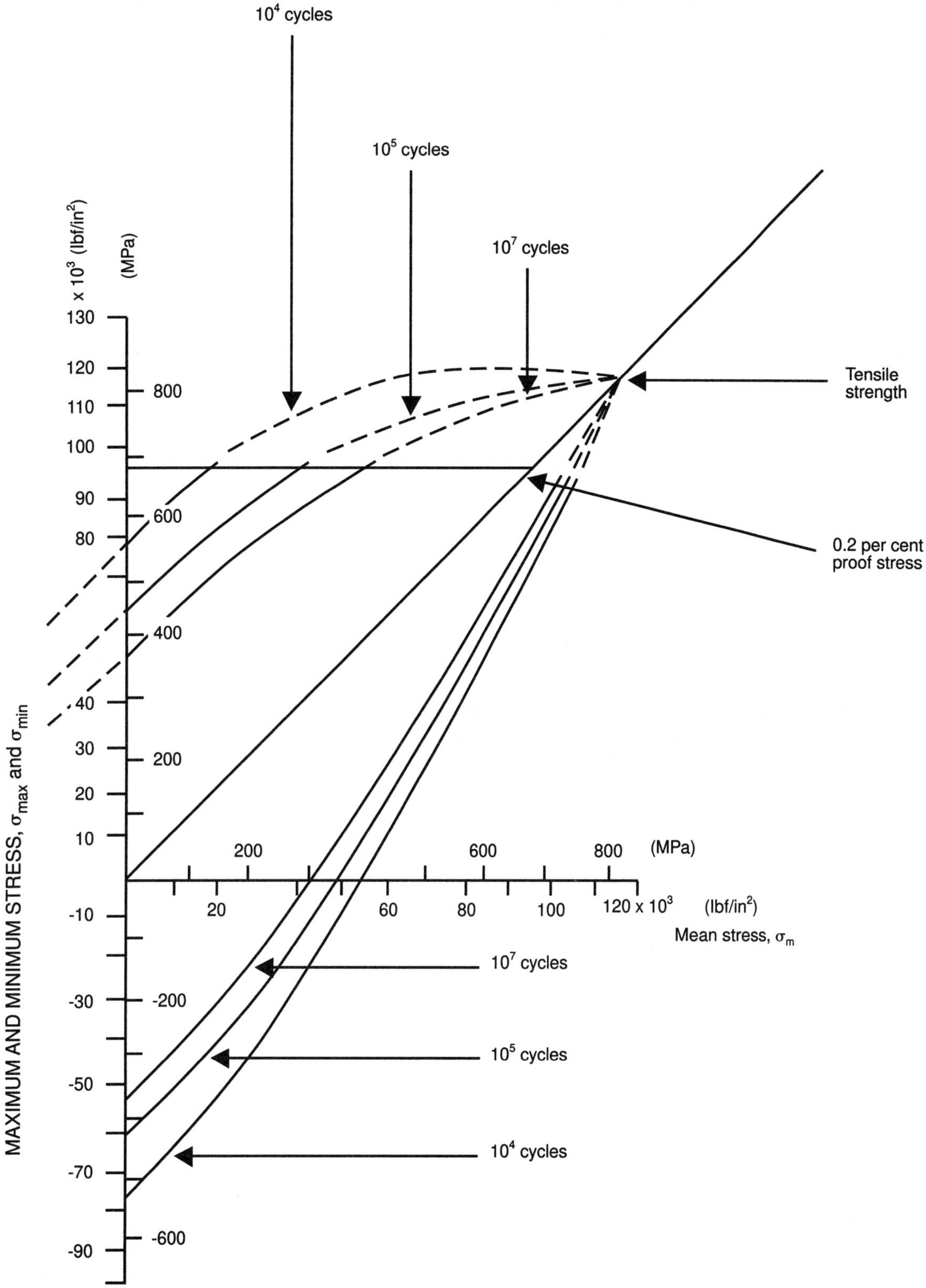

Fig. 1.8.9 Maximum and minimum stresses, σ_{max} and σ_{min} plotted against mean stress σ_m – BS 3518: Part 1: 1962

Fig. 1.8.9

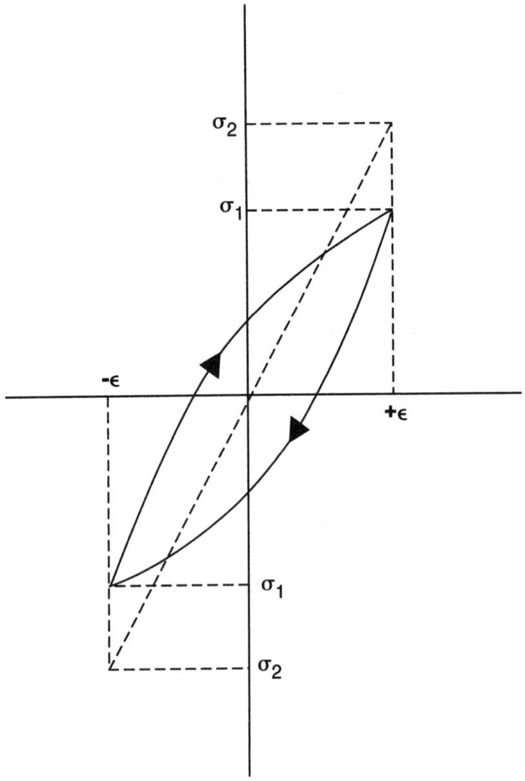

Fig. 1.8.10 Stress–strain hysteresis loop

Fig. 1.8.10

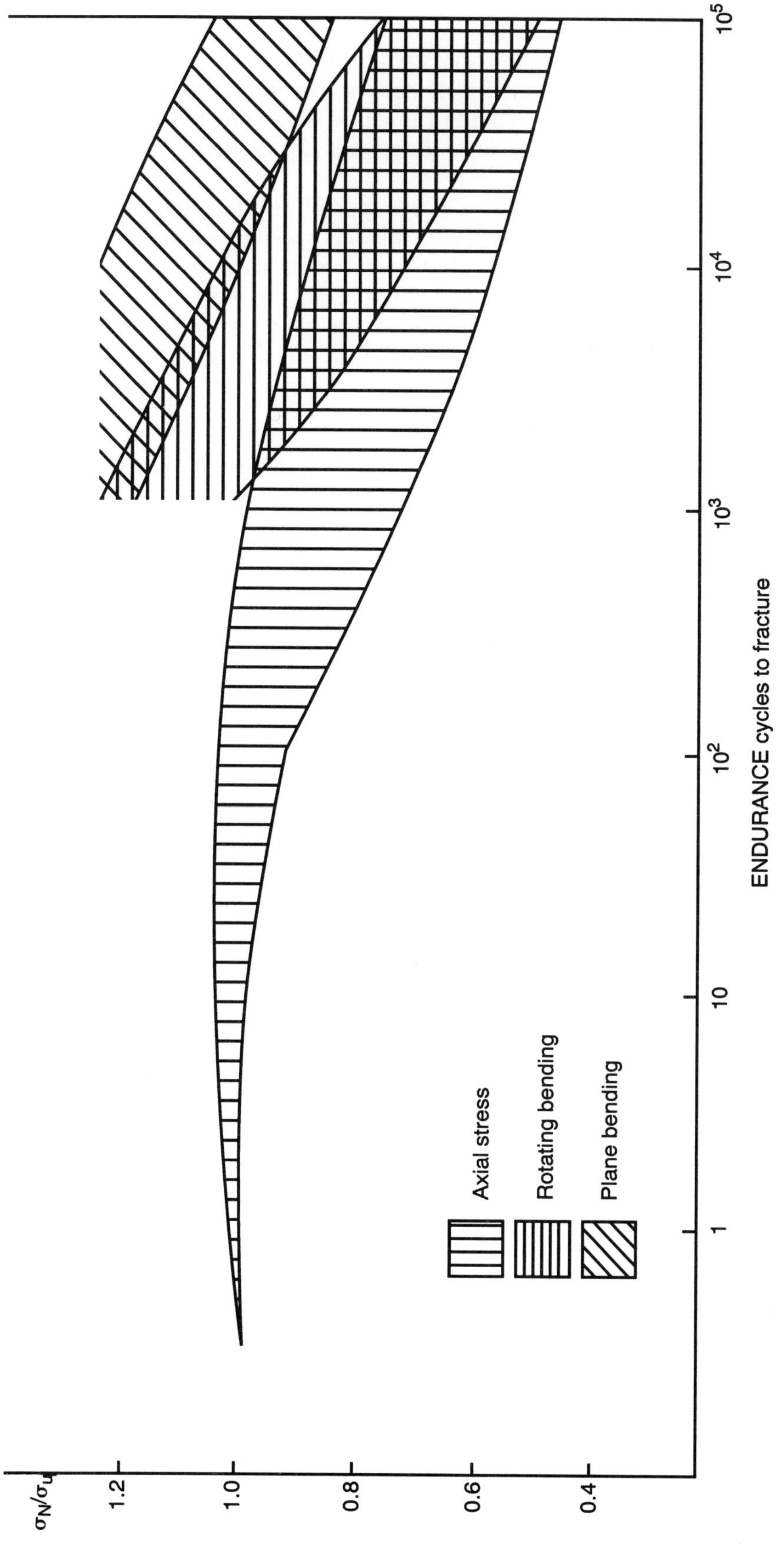

Fig. 1.8.11 Alternating fatigue strength/tensile strength (metals)

Fig. 1.8.11

1.8.5 Impact testing

Impact tests, in which a specimen is struck a blow, are carried out to simulate the response of the material to high rates of loading. The impact energy is not, however, an intrinsic property of the material and cannot be determined by any method which is independent of the test method since it depends, for example, on the notch geometry and specimen size.

Before discussing the significance of this type of test data, the various tests that are used will be described. Impact tests with metallic specimens usually employ a notched specimen, whereas such tests on plastics materials may use either notched or unnotched test-pieces. The usual manner of imparting the blow to the test-piece is either through a striker mounted on a pendulum or by means of a falling weight. Two principal tests employing the former method are the *Izod* and *Charpy* tests.

1.8.5.1 The Izod test

This test consists of measuring the energy absorbed in breaking a notched test-piece by one blow from a striker carried on a pendulum. The test-piece is gripped vertically with the root of the notch in the same plane as the upper face of the grips. The blow is struck on the same face as the notch and at a fixed height above it. Tests are usually performed at the ambient temperature of the test house for metals.

For plastics (see for example BS2782, ASTM D256) testing should be carried out at 20 ±5°C unless otherwise specified in the relevant standard for the material.

When standard metallic test-pieces are used, the following symbols are recommended for the reporting of the results of Izod tests: I for Izod, S for square section, R_s for circular section with straight notch, and R_c for circular section with curved notch. For example 'I160S: W J' indicates that an energy of W J was absorbed by a square section test-piece during an Izod test with a striking energy of 160 J.

For plastics the energy absorbed by the fracture should be less than 85% of the total energy available. The impact strength for plastics is expressed as the energy absorbed in breaking the specimen for unit width of notch.

Specimen sizes and the testing conditions employed for impact testing to the standards for metals and plastics vary and the results are not generally comparable.

1.8.5.2 The Charpy test

In this test, the energy absorbed in breaking a central notched specimen by one blow from a pendulum is measured. The specimen is supported at each end. The notch may be of U, V or keyhole form. Substandard test-pieces are used when the material thickness does not permit full-size specimens. The values of impact energy for fracture obtained from substandard specimens cannot be compared with full-size specimens; nor can the values obtained from different notches be compared.

Rectangular bars of plastics materials are tested in the notched or unnotched state.

The results of tests on standard metallic specimens are expressed in the form 'C320V: W J at T°C', that is an energy of W J was obtained at T°C from a Charpy V-notch specimen struck with an energy of 320 J. The results of tests on plastics are given in the form of impact strength equal to the absorbed energy divided either by the cross-section of the specimen for unnotched specimens, or by the cross-section behind the notch in notched specimens.

Conditions of testing are different for metals and plastics.

1.8.5.3 The falling weight test for plastics

In this test, described in full in BS 2782: Part 3: Methods 306B and C, ASTM D2444, and

D3029, a guided or unguided weight is allowed to fall on to the centre of a disc of the material under test or on to a square of the material resting on an annular support. Moulded plastics and full-thickness samples of sheet material are tested. The impact strength of the material is defined as the energy of a blow that would be expected to fracture half of a large sample of specimens. Methods B and C differ only in the height of the fall, that is 600 mm or 300 mm, respectively. Both methods specify a test temperature of 23±2°C unless otherwise specified in the relevant standard for the material.

Trials are first conducted, in each of which the impact energy is changed by a fixed amount until an energy has been reached which reverses the behaviour of the previous trials, that is an energy which is either no longer sufficient to break the specimen or else is sufficient if the previous trials had not broken it.

On completing whichever trial is appropriate, the remainder of the specimens are tested using for each an energy of blow less by another fixed amount than that of the blow imparted to the previous specimen if it broke, or greater by the same amount if the previous specimen was unbroken.

The impact strength of the material (W_1) is then calculated from the following equation, which assumes a total of 20 test specimens.

$$W_1 = (W_{m+1} + W_{m+2} + \ldots + W_{21})/(21 - m)$$

where

m	=	the number of blows in the trial run;
W_{m+1}	=	the impact energy of the first blow of the testing run;
W_{m+2}	=	the impact energy of the second blow of the testing run, and
W_{21}	=	the impact energy of the twentieth blow decreased or increased by the fixed amount depending on whether the twentieth specimen did or did not break, respectively.

There are also specific tests for plastics fabrications, for example falling-weight impact tests for pipes.

1.8.5.4 Response of materials to impact

The modes of failure observed under conditions of impactive loading can be broadly classified into two: typically *brittle failure* and typically *ductile failure*. If a force–displacement diagram were recorded during the impact test, the two types of failure would be exemplified by the upper and lower traces in Fig. 1.8.12.

Fig. 1.8.12 *Force–displacement curves for typical modes of failure: (a) brittle, (b) intermediate, (c) ductile*

With the brittle type of failure a crack is initiated and propagates during the linear rising part of the force–displacement trace prior to gross plastic yielding. Ductile failures, however, show considerable plastic deformation in the region of the crack and the failure occurs well after the maximum in the force–displacement trace. Since the area under the curve represents the work done on the specimen, it can be seen that brittle fractures are associated with very low absorbed energies compared with the ductile failures. In practice, intermediate-type failures often occur, when a limited amount of yielding takes place in the region of the fracture prior to propagation of the crack. The failure point on the trace will then be close to the maximum value of force and the energy to failure will be variable and lie somewhere between the extremes of fully brittle and fully ductile.

The appearance of the fracture surface of an impact specimen can often be correlated with the type of failure. The fractured surface of plastics specimens giving brittle failures may be smooth and glassy or somewhat splintered and irregular (often described as *conchoid*). In plastics specimens giving ductile failure, deformation and yielding occur to an appreciable depth over the whole fracture face, often accompanied by stress whitening. The texture of the surface may range from fairly smooth to rough, depending on the structure of the material. With metallic materials that fail in a ductile fashion the fracture surface is rough and fibrous in appearance and there is usually a considerable diminution in cross-section at the fracture. The fracture surface of a metallic material with a brittle fracture is faceted and crystalline in appearance, and either follows specific crystal planes across the grains in cleavage fractures or is more irregular when it is intercrystalline and follows the grain boundaries. In specimens of both plastics and metallic materials giving the intermediate type of failure, there may be a transition (ductile–brittle transition) from the first type of fracture surface (ductile) to the second (brittle) as the crack moves through the specimen. It is customary when impact testing metallic materials to report what percentage of the fracture surface is crystalline and what plastic. This appraisal is not required in routine plastics tests, but it can be invaluable for troubleshooting purposes.

In terms of impact resistance, ductile behaviour in itself is usually regarded as a criterion of satisfactory performance, the actual value of the ductile fracture energy probably having only limited significance in the design and selection stages compared with other properties. The change from ductile to brittle behaviour, which occurs at a *transition temperature,* and the factors that cause a significant shift in this temperature are more important criteria in comparing the impact behaviour of materials.

FACTORS INFLUENCING IMPACT STRENGTH

Temperature effect

A decrease in temperature may cause a change in the mode of impact behaviour resulting in an appreciable decrease in energy absorbed in fracturing the specimen; this is not always the case and data for the temperature range appropriate to the anticipated service conditions should be examined.

Speed of impact

An increase in the speed of impact may also cause a change towards a more brittle mode of failure with a corresponding drop in energy. An increase of test speed is analogous to a decrease in test temperature because both lead to an increase in flow stress (through the rate sensitivity of flow stress) and thus favour crack propagation at the expense of plastic deformation. With plastics it should be noted that a logarithmic scale of testing speed is necessary to obtain transitions in impact energy of equivalent sharpness to those observed with a linear fall in temperature. Thus, the changes in temperature that occur in practice are more likely to cause differences in impact energy for failure than are variations in the speed of impact. Changes in speed are only likely to lead to significant differences in impact energy for failures occurring close to the transition temperature. As a rough guide for themoplastics, a 10°C fall in temperature is equivalent to a tenfold increase of speed of impact.

For most purposes, the effect of changing the testing speed is best observed from data obtained over a range of temperatures at a constant testing speed. For applications in which the speed of impact is much higher than that used in the standard test, special tests become desirable.

Stress concentrations

High stress concentrations result in an increased tendency for fracture to occur in the brittle mode.

Anisotropy

Any degree of anisotropy of the material resulting from fabrication procedures (particularly a uniaxial orientation) can lead to an increased tendency towards cracking in a direction parallel to that of the orientation. This is particularly perceptible in plastics produced by injection moulding, where the hot plastics melt is forced at high speeds into a narrow, cold cavity, and in heavily worked steel plate, for example, where the impact resistance to cracking is least along the rolling plane of the sheet (delamination).

Other factors

The thickness of the specimen can also affect the impact energy; thicker sections show an increased tendency to brittleness because of increased triaxial constraint on plastic deformation. With plastics materials the moisture content may also affect the impact energy. With steels and other metallic materials small differences in composition and microstructure can have profound effects on the impact strength.

DUCTILE–BRITTLE TRANSITIONS IN METALS

With designs involving metallic materials there is less emphasis than for plastics on the ability of the component to withstand impactive loading during service. Impact test data is usually presented in the form of a fracture energy–temperature curve determined from standard tests of the type discussed at the beginning of this section, the most common test now uses the Charpy V-notched specimen.

Whilst the transition from ductile to brittle behaviour is not abrupt, and there is no single temperature below which the material is fully brittle, or above which it is fully ductile, the concept of a transition temperature is useful as a guide to temperatures at which brittle behaviour may be expected.

The occurrence of a ductile-to-brittle transition is restricted to certain metallic materials: the body-centred cubic and close-packed hexagonal crystal structures exhibit the phenomenon, but not the face-centred cubic structure. The low strain rate sensitivity of the latter class of metals leads to little increase in flow stress during the impact test, so that yielding is not inhibited and replaced by brittle fracture. Of the common engineering materials: (1) ferritic steels exemplify the body-centred cubic structure, (2) magnesium, titanium, and zinc base alloys are of the close-packed hexagonal structure, and (3) aluminium, copper, and nickel base alloys, together with austenitic steels, represent the face-centred cubic structure. Impact testing of class (3) materials is not carried out frequently since ductile behaviour can usually be expected under most circumstances. However, most steel specifications call for a minimum value of impact energy.

1.8.5.5 Significance of impact testing of plastics

The results of the standard impact tests on plastics are expressed either as the energy absorbed during fracture of a standard notched specimen or as the energy calculated for unit width of specimen. For design purposes, it is not practicable to specify a single specimen size for the whole range of materials. In order to rationalize results from tests on material of different dimensions, the impact energy is often expressed as the energy absorbed by unit area of specimen fractured. Although this is in line with the European practice for metallic materials, the fracture energy for unit area is not truly a material property, since the energy absorbed in fracturing the specimen also depends on the section thickness and the stress concentration associated with the notch of the specimen. However, it has been shown that for a range of plastics materials tested in different laboratories and with a variety of notch tip radii this method of calculation gives the same overall pattern of behaviour in both Izod and Charpy tests. When comparing values obtained from such different sources small differences in level may not be significant, provided that the pattern of behaviour is similar.

During the interpretation of impact data, emphasis should be placed on the occurrence of the ductile–brittle transition and the factors that mainly influence its temperature. In the ductile fracture regime, these considerations assume greater significance than the actual level of the impact energy for failure. It is important that multipoint data should be used to define areas of avoidance for brittle failure under different circumstances. In selecting from the multipoint data the curve most relevant to the particular application, both the design and the degree of anisotropy associated with the processing step should be considered.

As an example of the design considerations, compare the behaviour under impact of a hollow sphere, such as a fishing float, with that of an intricate injection moulding, such as a bottle crate. Because of its shape the material of the sphere is compressed initially in the region of the impact, and since compressive stressing leads to ductile rather than to brittle failure, the practical response of the sphere to impact loading is better than that indicated by the production tests on unnotched specimens. The many points of stress concentration at the corners and changes in cross-section of the crate, however, would lead to a better correlation between the impact response of the crate and the results of pendulum tests on notched specimens. The relationship between the stress concentration factor and the impact strength indicates the improvement in performance that it is possible to achieve by providing generous radii at the points of potential stress concentration.

Some materials that show a marked difference in impact strength between tests with different notch radii are notch-sensitive and, especially at service temperatures close to the ductile–brittle transition, may fail sporadically in a brittle fashion even in designs with no apparent stress concentrations. An internal or surface imperfection can act as a stress raiser during impact and can initiate brittle fracture.

The main use of impact strength data for plastics materials is to provide a guide to their performance in practice. The impact behaviour of a product depends on the design used, the method of manufacture, and the in-service conditions as well as on the basic properties of the material. It is not surprising, therefore, that a satisfactory comparison between materials cannot be made on the basis of a single impact test value. However, these tests do have a value in the routine control of quality, once it is known that a material of a given impact strength is satisfactory for a particular application.

Because, even with the use of multipoint data, it is still not possible to predict quantitatively the impact performance of a plastics article, drop-impact tests are frequently used to simulate in-service conditions more closely. Such tests often show up weak points of a design and they can also serve as control tests for monitoring quality of production.

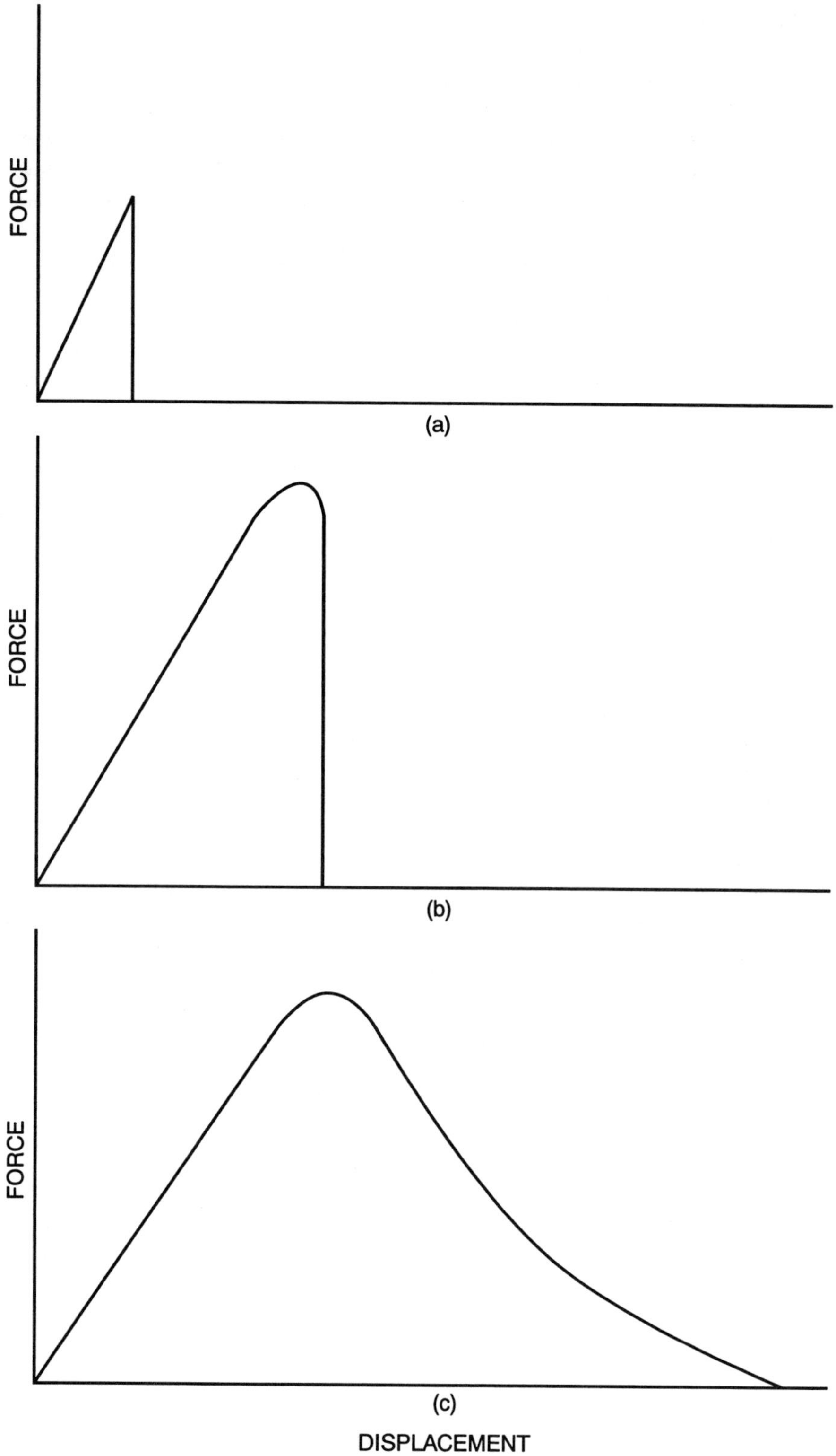

Fig. 1.8.12 Force–displacement curves for typical modes of failure: (a) brittle, (b) intermediate, (c) ductile

1.8.6 Temperature effects

The effect of temperature on the tensile properties of materials and the sensitivity of flow stress to temperature have been discussed in the earlier sections on tensile properties and creep. Reference to the effect of temperature on impact resistance and fatigue properties has also been made. It is self-evident that design stresses should be based on properties determined at the service temperature and relevant to the expected life. Some other indirect effects of temperature are not so obvious.

1.8.6.1 Metallic materials

Many metallic materials are put into service in a state of thermodynamic non-equilibrium, for example cold worked copper sheet, quenched and tempered steels, and aluminium alloys aged to peak hardness. When these materials are used for long periods at ambient temperature, their properties do not vary with time. However, if they are maintained at elevated temperatures over a period, they may revert to an equilibrium state indicated, for example by softening caused by recrystallization of the copper sheet or over-ageing of the aluminium alloys. In such circumstances, although the design may well have been based on short-term test properties determined at the appropriate temperature, the service life could be prematurely terminated by failure. It is, therefore, essential to consider the long-term stability of the microstructure of materials and their properties. Such considerations will apply to all structure sensitive properties. Apart from the structural changes already indicated, long service lives can also result in embrittlement due to the eventual precipitation of an equilibrium phase. Long-term creep testing will reflect these changes in structure by a change in creep rate or creep ductility and strength. Short-term hot tensile testing of materials after prolonged exposure to service conditions will also reveal the effect of any structural changes on the properties.

Welds are a special case where structural changes may occur and cause a degradation of properties. During the welding process the localized heating of the parent material can produce temperatures that may be sufficient to produce a deleterious change in microstructure, thus necessitating a post-weld heat treatment to rectify the damage.

Increasing the temperature can also have the effect of increasing the rate at which a material reacts with its environment. Thus, although most engineering metallic materials oxidize slowly at ambient temperature and may indeed form protective films so that *oxidation* is limited at higher temperatures the rate, or even the mechanism, of oxidation may change so that considerable scaling can take place. In certain instances the scale may exfoliate, baring fresh surfaces of metal, and there may be a considerable reduction in thickness of the material, which in extreme cases can lead to a reduction of load-bearing capacity and failure. Sometimes oxidation may be localized at grain boundaries or a particular constituent of the microstructure may be selectively attacked. This type of attack affords the possibility of points of stress concentration being introduced into the part and could thus be the cause of premature failure, particularly under fatigueloading conditions.

Materials that are subjected to temperature cycles in service may undergo microstructural and volume changes, causing dimensional changes of the component. In the absence of such phase changes, a material will still suffer dimensional changes as a result of thermal expansion and contraction. If the material is constrained in any way, changes caused by thermal expansion will not be reversible and an internal stress system will arise in the part or structure. Such an internal stress system will have the same effect as an externally applied stress and can lead to a fatigue failure if a sufficient number of temperature, and hence stress, cycles occur. This is known as *thermal fatigue.* As an indication of the magnitude of such thermal stresses, consider a low carbon steel with a linear expansion coefficient of 15×10^{-6}/°C. A change of 1°C will cause a strain of 1.5×10^{-5} if the expansion is restrained. Since the elastic modulus of the steel is around 200 GPa this strain corresponds to a stress of 3 MPa. Thus, assuming that the thermal expansion is fully restrained, a temperature change of 70°C would cause a stress of 210 MPa, which is equal to the yield strength of steel at, say, 100°C.

Although a part is unlikely to be fully constrained, a similar situation can arise where parts of a structure or component undergoing a temperature cycle are made from materials having different coefficients of thermal expansion, since the mismatch can cause considerable stresses to be generated from the restraint of one part on the other. Apart from the possibility of fatigue failures, there is also the possibility of distortion and deformation of one of the parts taking place through stress relaxation or creep occurring if the temperature cycles are of long duration. Thus, in designs that in service suffer temperature cycles, consideration should be given to the freedom of parts to expand and to the mismatch of expansion of the various parts if large thermal stresses are to be avoided.

1.8.6.2 Plastics

In the common practice of replacing metals with plastics it is necessary to remember that most of the commonly used thermoplastics have softening points defined as the temperature at which some physical property is reduced (usually) to some arbitrarily defined value, which is low by most standards. Typical values are given in:

Table 1.8.1 *Softening points (°C) by various standard methods*

It is more important, however, to remember that few thermoplastics have a sharp melting point. They soften gradually, and the maximum safe design temperature will be well below the softening point. This is to some extent shown by the two centre columns of Table 1.8.1 where the test methods vary only in that the arbitrarily chosen critical stress level is different. Maximum service temperatures for thermoset materials are usually higher, but the limitation on service temperature still exists.

Plastics materials also exhibit a loss of flexibility, or embrittlement, as their temperatures are reduced, and a variety of tests are used to indicate their characteristics in this respect, some showing (as with softening point) the temperature at which some property (usually torsional modulus) reaches a specified value (typically Method 104B *Cold flex temperature* of BS 2782). Others relate to the temperature at which embrittlement, as defined by the ability to resist impact, occurs (ASTM D746 is typical of these).

As with metals, plastics oxidize, and usually at much lower temperatures than metals. Degradation may also occur: the longchain molecules of many plastics materials gradually 'unzip' at quite moderate temperatures, although this phenomenon can be delayed or stopped by the incorporation in their compounding of various protective chemicals, antioxidants, stabilizers, etc., which react with the chain ends to impede the degradative process.

Continuing the analogy with metals, thermal expansion can be an even larger problem, as the actual coefficient is considerably greater. Typical values per °C are:

aluminium	2.5×10^{-5}
copper	1.7×10^{-5}
steel	1.1×10^{-5}
polystyrene	$6\text{–}8 \times 10^{-5}$
polythene	18×10^{-5}
phenol formaldehyde resin	$5\text{–}15 \times 10^{-5}$.

The forces involved in restraining plastics components subject to fluctuating temperatures are not great, and the resulting fatigue failure is not yet recognized as a problem. Nevertheless, designers must take this relatively large expansion into consideration, especially when it can combine with other special properties of the material to produce phenomena unknown with metals. In recent years plastics have been introduced widely in building and civil engineering applications. For example, plastic gutter not only expands far more than the traditional cast iron it replaced, but because of its low thermal capacity it expands more rapidly, and systems have to designed to accommodate this. The increasing use of room-sized wall sections has to some extent been made possible by plastics, but they cannot be butt-joined and pointed; elaborate gaskets providing adequate expansion relief must be incorporated in such structures. When low thermal capacity skins separated by thermally insulating cores are used on the exterior of buildings, buckling induced by temperature differences across the thickness can result.

Table 1.8.1 Softening points (°C) by various standard methods

Material	*Cantilever BS 2782: M. 102C*	*Deflection temperature under load (DTL) (ASTM D648)*		*Vicat BS 2782: M. 102D*
		1.8 MPa (264 lbf/in^2) fibre stress	*0.45 MPa (66 lbf/in^2) fibre stress*	
Polystyrene	95	90	97	98
Toughened polystyrene	84	72	85	86
ABS	94	84	96	95
Polymethyl methacrylate	95	80	97	90
Cellulose acetate	76	64	77	72
Rigid PVC	78	70	82	82
Polythylene (low density)	(too flexible)	(too flexible)	45	85
(medium density)	90	35	69	105
(high density)	115	45	75	125
Polypropylene	145	60	140	150
Nylon 66	180	75	183	185
Acetal	170	120	165	175

1.8.7 Flammability and fire testing

Properties of materials such as density, tensile strength, etc., can be measured by a variety of methods, any of which will give approximately the same results. The results of fire tests, however, are acutely dependent on the method used, and situations can easily be established in which metals will burn, or (at the other extreme) in which wood will not. As a result, no clear-cut statements can be made concerning the flammability, combustibility, etc., of most materials, and the best that can be achieved is a statement of the way in which a particular material or material combination (which must be closely defined) will behave in a particular fire test. The full significance of the data related to its particular test environment must therefore be appreciated if the risk of fire is to be minimized. The tests from which the data are obtained generally fall into four categories:

(1) small-scale, conventional laboratory tests, whose original purpose was probably to yield information on the fire risk of using a certain material, but in practice can only give assurance that a particular batch of material is no more liable to burn than one previously found acceptable (they are, at best, a crude form of quality control),
(2) small-scale tests, which are rather more scientifically based, generally to provide information in a very limited field (for example, to furnish an estimate of the heat or smoke generated when a standard sample burns, under prescribed conditions),
(3) relatively large-scale tests in which large specimens are subjected to temperatures and heat fluxes that tend to be similar to those occurring in real fire situations (the interpretation of results of such tests still needs considerable experience and intelligence if they are to be used to the best advantage, but they are likely to provide the best data available for many years to come),
(4) full-scale tests to fulfil the need to simulate as nearly as possible real fire conditions, which have in recent years included several major experimental fires using houses due for demolition or occasionally purpose-constructed buildings. For obvious reasons, relatively few data from this source are available, and for commercial and logistical reasons it is unlikely that individual products will generally be evaluated by such techniques. The technique is valuable because it will gradually enable predictions based on the laboratory tests to be confirmed (or otherwise) in practical situations.

1.8.7.1 Conventional laboratory tests

Many tests are included in British Standard, Underwriters Laboratory, ASTM and other authoritative specifications. As previously described they are useful as quality-control techniques and can provide an indication to the knowledgeable user of those materials that may ignite very easily, for example, from a match flame. They give no indication of the behaviour of materials subjected to larger ignition sources, extrapolation from such small tests is not valid and reports now have to state:

'The following test results relate only to the behaviour of the test specimens under the particular conditions of test; they are not intended as a means of assessing the potential fire hazard of the material in use.'

In addition, the reports are now limited to factual statements describing the performance of the material under test. However, the designer should beware of the many items of sales literature produced from earlier tests, where data derived from the tests cited in the first paragraph of this section could justify such sweeping conclusions as 'that the material is self-extinguishing' or 'the material is of very low flammability'.

1.8.7.2 Small-scale tests

The more scientific of the small-scale tests can be exemplified by BS 476: Part 6 (*Fire propagation test for materials*), in which a 228 mm square specimen of the material or composite of materials is exposed to radiation from electrical heating elements and small gas flames whilst enclosed in a small, asbestos box. The hot gases produced by the heat sources and by any combustion of the specimen rise through a chimney where their temperature is measured at intervals and recorded. Prior to this stage of the test a similar run is made using a standard, non-combustible specimen, and the temperatures throughout the range are compared. It is obvious that a comparison of this type can yield positive evidence of the amount of heat the specimen will emit under this sort of fire situation, but in practice the processing of the results is limited to the production of four numbers: i_1, i_2, and i_3, which are the sub-indices at 3, 10, and 20 min, and I the fire propagation index, which is the total of the sub-indices.

The significance of the results in pure number form is not immediately apparent, but it is generally accepted that for most situations values of i_1 less than 6 and I less than 12 are required for plastics materials. Conditions covering the softening point of the material may also be imposed.

This type of test, yielding numerical results related to the combustion of a material under conditions that are, in practice, very carefully controlled, represents a significant advance on the majority of the 'simple' tests. While the form of the results as reported precludes further prediction of the behaviour of the sample in an actual conflagration, acceptance levels are soundly based. None the less it must be borne in mind that the heat emitted (and therefore the result of the test) is a function of both the thermal capacity of the specimen and the rate at which it emits heat of combustion under the standardized test conditions. Thus, the results of the tests can only be related in practice to exactly similar materials or composites; for instance, a 'fire-resistant' decorative laminate applied to a chipboard substrate would yield results which would depend on the following:

1. The thickness of both components. If the laminate were thick, it could protect the chipboard from radiation for long enough for the indices to be acceptable. If the chipboard were thin, its calorific value could be low enough to result in acceptable indices. In extreme circumstances, even the nature and thickness of the glueline could be critical.
2. The density of the chipboard. An increase of density might be expected to increase the quantity of combustible material available. At the same time a denser board might be consumed more slowly.
3. The colour of the laminate. A white or reflective surface would obviously protect the substrate from radiation for a longer period than a matt-black one.

The combustion properties of other materials, however, may be subject to different considerations from those already discussed; and, as a further example, consider an incombustible substrate, say, a steel sheet to which a thin, highly flammable decorative surface has been applied. At an early stage in the test the flammable component will burn, but as it is present in a small quantity only its heat contribution will be small, and in all probability the acceptable performance indices will not be exceeded. A similar combination submitted to the surface spread of flame test for materials, which is detailed under '1.8.7.3 Large-scale tests', is highly unlikely to be acceptable under the criterion applied to that test.

Thus, it can be seen that the fire propagation test for materials performs a useful function in defining the performance of a material in a specific (and rather artificial) fire situation, but that its results must be used with caution and cannot be extensively extrapolated. Many tests of similar value and having similar limitations are in use.

1.8.7.3 Large-scale tests

For most purposes the results obtained from the relatively large-scale tests are the most relevant, and thus these will have rather more attention paid to them than to the results obtained from other tests.

The most commonly quoted result in this category is for the *Surface spread of flame test for materials* (see for example BS 476: Part 7). This test procedure is usually applied to lining materials for ceilings or walls, and measures the rate at which a flame front will progress along an exposed surface in standard conditions. In essence, a 900 × 230 mm panel of the material is mounted, with the long axis horizontal and the short axis vertical, at right angles to one edge of a 900 mm square furnace, which is operated under standard conditions. The tendency of the material to support the spread of flame across its surface is assessed from the rate and distance of flame spread along the test-piece, according to the four classes given in Table 1.8.2. One specimen only in a sample is allowed to exceed the limits shown by the tolerance given.

Table 1.8.2 *Flame-spread classification – BS 476*

The various species of plastics materials behave in different ways in this test and some, those which deform easily and rapidly at elevated temperature, present special difficulties as they fall away from the test position during the first minute of test and thus cannot be classified. It is therefore not surprising that flame-spread data are not offered for some quite common materials. Materials other than plastics are commonly subjected to this test, which originated long before today's widespread use of plastics was even envisaged. Timber and timber products properly treated by various fire-resistant processes commonly achieve good ratings and in general this test is applied to any 'new' surface that might be used in the interior of a building unless it is completely obvious that it cannot burn.

Once again it is usually necessary to test materials in the combination that will be used in practice. There are a number of reasons for this; for example a substrate might be ignited through a surface layer and then initiate surface flaming, and it is apparent that the thermal mass and conductivity of a substrate will influence the temperature achieved by the surface and thus the speed at which flame will be initiated and subsequently spread.

The second most frequently used of the large-scale test procedures is the *External fire exposure roof test* (see for example BS 476: Part 3). This is applied to roofing materials to measure (a) their capacity to resist penetration by fire when the external surface is exposed to radiation, and (b) the distance of the spread of flame on the outer surface of the roof covering under certain conditions. Tests of two types are applied (after a preliminary test to eliminate materials that have no chance of passing them). In the test of the first type, the upper (external) surfaces of three 33 × 33 inch specimens are exposed to thermal radiation of a specified intensity for up to 1 h, and after the first 5 min a flame is also applied for 1 min directly to the surface of the specimen. The prime objective of this test is to ensure that the specimen is not penetrated by fire within the hour, or, if it is, to note the length of time the specimen will survive without penetration. In the test of the second type three similar specimens are exposed to thermal radiation of an intensity which varies over the surface of the specimen. The test flame is applied for a longer period than in the previous test and performance is assessed on the basis of spread of flame over the specimens.

Many roofing systems nowadays are constructed from combinations which include combustible materials: glass-reinforced plastics for light transmission, expanded plastics or wood fibre in various forms for thermal insulation, and timber and timber products for lightness. In addition to ensuring that the test data available apply exactly to the roofing system proposed for use, similar care must be taken to ensure that the jointing system used between the various sections of the roof has been properly evaluated.

Finally, the third large-scale test procedure, which is given in BS 476: Part 8 (*Test methods and criteria for the fire resistance of elements of building construction*), can take several forms depending on the particular structure to be evaluated, but the test for non-load-bearing walls and partitions is the most likely to be applied to a plastics material or to other 'combustible' sheet products. For this a specimen 2.5 × 2.5m or larger, is built into a restraint frame (unless the material or system of materials is such that it could move in service to accommodate thermal expansion) and the assembly so produced is used to close the open face of a furnace, which when operated produced a small, positive pressure in its upper part. The furnace is ignited at the start of the test and the temperature is increased in accordance with the relationship $T - T_o = 345\log_{10}(8t + 1)$, where T is the furnace temperature in °C at time t in min, and T_o is the starting temperature. Thus, after 30 min $T - T_o$ = 821°C, at 1 h 925°C, at 2 h 1029°C, and so on. During the test the specimen is inspected continuously, and assessed for the following properties:

1. Stability. Any deformation, collapse, or other factor affecting the stability of the specimen should be recorded.
2. Integrity. Elements which have a separating function (that is party walls as distinct from external walls) should be able to prevent ignition of a standard ignitable test-piece (cotton wool) at the 'cold' face.
3. Insulation. Separating elements must ensure that the mean temperature on the 'cold' face of the specimen does not increase more than 140°C and that no point increase is greater than 180°C.

Constructions are considered to have failed when any one of these criteria ceases to be satisfied, and the fire resistance of the assembly is reported as the duration of the test to the time of the first failure.

Once again great care is needed to ensure that any test certificate is strictly applicable to the construction proposed for use. It is relatively easy to design a building panel that will meet the requirement of BS 476: Part 8 over all or most of its area. However, the insulation and integrity requirements apply not only to the whole panel but separately to each small part, and to any joining (either between panels or between panel and floor or ceiling) as this is a particular cause of failure. In a simple structure the surfaces of the panel may be nailed or screwed to a timber frame and, unless great care is employed in the use of this type of fastening, hot spots leading to insulation failure or even integrity failure may result. There is an even greater chance of insulation failure with systems of construction that require panels to be slotted into metal sections.

There is also a continuing interest in newer methods of assessing fire hazards. An example is the oxygen index test used to determine the ratio of oxygen to nitrogen that, in a mixture of these gases, will just support combustion at 'ambient' temperature. A recent development of this test determines the temperature at which a 1 : 4 mixture of these gases will just support combustion.

There is a growing awareness of the hazard of smoke in fire situations as its presence may cause suffocation and certainly impedes both escape and the fighting of the fire.

In assessing fire test results for plastics and similar materials it should be remembered that they are all quite specific to the test situation and that quite small changes in formulation, thickness, method of application, etc., may significantly alter performance. Most plastics and similar materials will burn if the combustion conditions are severe enough, yet the greater proportion can be modified by the inclusion of additives or by clever design so that they will satisfy the fairly stringent conditions of the large-scale British Standard tests. The inclusion of additives to reduce fire risk is usually accompanied by a deterioration in some other property, such as strength, or weathering resistance, or the production of more smoke and toxic gases when the material finally does burn.

Table 1.8.2 Flame-spread classification BS 476

Classification	*Flame spread at 90 s*		*Final flame spread*	
	Limit (mm)	*Tolerance for one specimen in sample (mm)*	*Limit (mm)*	*Tolerance for one specimen in sample (mm)*
Class 1	165	25	165	25
Class 2	215	25	455	45
Class 3	265	25	710	75
Class 4	(exceeding Class 3 limits)			

1.8.8 Electrical properties of plastics

1.8.8.1 Dielectric strength – ASTM D149

SIGNIFICANCE

This test is an indication of the electrical strength of a material as an insulator. The dielectric strength of an insulating material is the voltage gradient at which electric failure or breakdown occurs as a continuous arc (the electrical property analogous to tensile strength in mechanical properties). The dielectric strength of materials varies greatly with several conditions, such as humidity and geometry, and it is not possible to directly apply the standard test values to field use unless all conditions, including specimen dimension, are the same. Because of this, the dielectric strength test results are of relative rather than absolute value as a specification guide.

The dielectric strength of polyethylenes is usually around 200×10^2 kV/m. The value will drop sharply if holes, bubbles, or contaminants are present in the specimen being tested.

The dielectric strength varies inversely with the thickness of the specimen.

SPECIMEN

Specimens are thin sheets or plates having parallel plane surfaces and of a size sufficient to prevent flashing-over. Dielectric strength varies with thickness and therefore specimen thickness must be reported.

Since temperature and humidity affect results, it is necessary to condition each type of material as directed in the specification for that material. The test for dielectric strength must be run in the conditioning chamber or immediately after removal of the specimen from the chamber.

PROCEDURE

The specimen is placed between heavy cylindrical brass electrodes which carry electrical current during the test. There are two ways of running this test for dielectric strength:

Short-time test

The voltage is increased from zero to breakdown at a uniform rate – 0.5 to 1.0 kV/s. The precise rate of voltage rise is specified in governing material specifications.

Step-by-step test

The initial voltage applied is 50% of breakdown voltage shown by the short-time test. It is increased at rates specified for each type of material and the breakdown level noted.

Breakdown by these tests means passage of sudden excessive current through the specimen and can be verified by instrument and visible damage to the specimen.

Fig. 1.8.13 *Dielectric strength apparatus*

1.8.8.2 Dielectric constant and dissipation factor – ASTM D150, BS 4618: Parts 2.1 and 2.2

SIGNIFICANCE

Dissipation factor is a ratio of the real power (in-phase power) to the reactive power (power 90° out of phase). It is defined also in other ways:

Dissipation factor is the ratio of conductance of a capacitor of which the material is the dielectric to its susceptance.

Dissipation factor is the ratio of its parallel reactance to its parallel resistance. It is the tangent of the loss angle and the cotangent of the phase angle.

The dissipation factor is a measure of the conversion of the reactive power to real power, showing as heat.

Dielectric constant is the ratio of the capacity of a condenser made with a particular dielectric to the capacity of the same condenser with air as the dielectric. For a material used to support and insulate components of an electrical network from each other and ground, it is generally desirable to have a low level of dielectric constant. For a material to function as the dielectric of a capacitor on the other hand, it is desirable to have a high value of dielectric constant, so the capacitor may be physically as small as possible.

Loss factor is the product of the dielectric constant and the power factor, and is a measure of total losses in the dielectric material.

SPECIMEN

The specimen may be a sheet of any size convenient to test, but should have uniform thickness. The test may be run at standard room temperatures and humidity, or in special sets of conditions as desired. In any case, the specimens should be pre-conditioned to the set of conditions used.

PROCEDURE

Electrodes are applied to opposite faces of the test specimen. The capacitance and dielectric loss are then measured by comparison or substitution methods in an electric bridge circuit. From these measurements and the dimensions of the specimen dielectric constant and loss factor are computed.

Fig. 1.8.14 *Dielectric constant apparatus*

1.8.8.3 High voltage, low current, dry arc resistance of solid electrical insulation – ASTM D495

SIGNIFICANCE

The test is a high voltage-low current test which simulates those existing in AC current circuits at low current. Types of failure for plastics and elastomers include ignition, tracking and carbonization.

SPECIMENS

Test specimens shall be 0.125 ± 0.01 inch (3.17 ± 0.25 mm) in thickness and during the test no part of the arc is closer than 1/4 inch (6.6 mm) to the edge or closer than 1/2 inch (12.7 mm) to a previously tested area. Surfaces should be clean.

PROCEDURE

Electrodes are applied and internal current steps are applied until failure occurs. The failure is defined as the point at which a conducting path is formed across the sample and the arc completely disappears into the material.

Fig. 1.8.15 *Arc resistance apparatus*

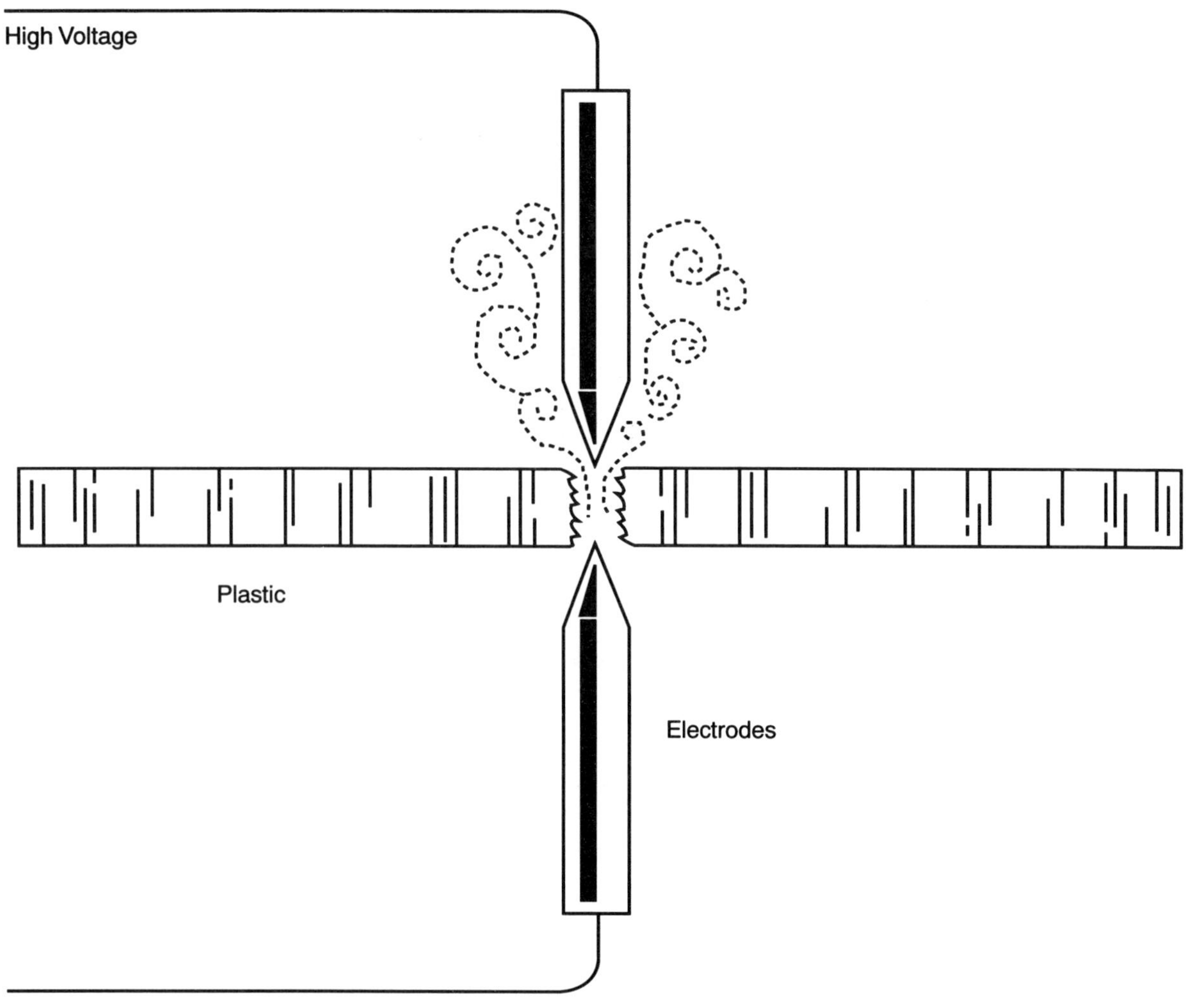

Fig. 1.8.13 Dielectric strength apparatus

Fig. 1.8.13

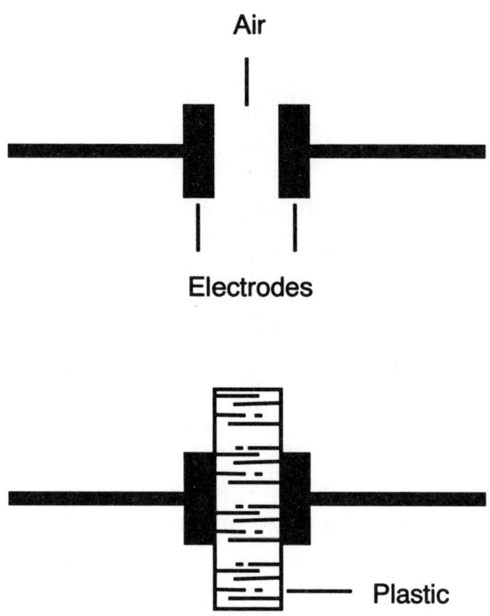

Fig. 1.8.14 Dielectric constant apparatus

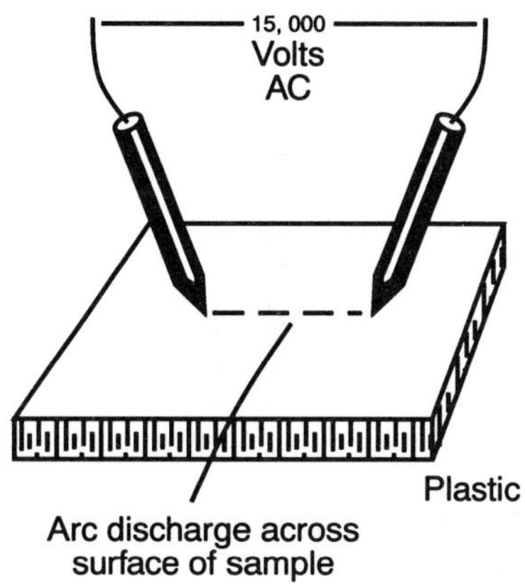

Fig. 1.8.15 Arc resistance apparatus

1.8.9 Processing (plastics)

1.8.9.1 Mould shrinkage – ASTM D955

SIGNIFICANCE

The test is to record initial shrinkage, i.e. not for any shrinkage after the first 48 h. Under any of the standard methods of moulding, the mould shrinkage will vary according to design and operation of the mould. Some further comments below refer to specific processes.

Compression moulding

Shrinkage will be at a minimum where there is a maximum of material being forced solidly into the mould cavity, and vice versa. The plasticity of the material may affect shrinkage insofar as it effects the retention and compression of the charge given during the moulding.

Injection moulding

In addition to type, size and thickness of the piece, mould shrinkage here will vary with the nozzle size of the mould, the operating cycle, temperature and the length of time that follow-up pressure is maintained. As with compression moulding, shrinkages will be much higher where the charge must flow into the mould cavity but does not receive enough pressure to be forced firmly into all of the recesses.

Transfer moulding

The comments for compression and injection moulding also apply; it should be noted that the direction of flow is not as an important a factor as would be expected.

SPECIMEN

The full test describes detailed methods of preparing specimens of various bar and disc shapes in a series of compression moulds, injection moulds, transfer moulds, etc.

PROCEDURE

The materials are moulded under carefully controlled conditions (sizes, rates of heat, etc.), discharged from the mould, cooled for a short period of time, and then measured. The difference in dimension size and mould size is recorded as the mould shrinkage.

1.8.10 Optical properties (plastics)

Plastics materials find extensive application in engineering and related fields for their optical properties.

1.8.10.1 Refractive index

The refractive index n of a material for a given wavelength, in vacuum, of electromagnetic radiation is defined by the ratio of the velocity of the radiation in vacuum and the velocity in the material. The refractive index is a function of wavelength and temperature.

Birefringence is a measure of optical anisotropy. Some materials are isotropic unless stressed elastically; permanent birefringence may be introduced by processing, or by dispersions of one isotropic material within another.

1.8.10.2 Transparency

Transparency, which is a general quality of a plastics material depending largely upon bulk and superficial homogeneity, is measured by loss of clarity and loss of contrast. These two effects are closely interconnected and constitute the two different aspects of one phenomenon.

For most purposes, transparency can be described by the following characteristics. A certain amount ϕ_{sc} (scattered) of the monochromatic luminous flux ϕ (parallel beam) falling perpendicularly on a translucent film or sheet, will be scattered in all directions.

If ϕ_A is the flux absorbed by the material (ignoring re-emission with changed wavelength – fluorescence) the difference

$$\phi - \phi_A - \phi_{sc} = \phi_{\text{undeviated}}$$

The direct transmission factor T is defined by $T = \dfrac{\phi_{\text{undeviated}}}{\phi}$

For weak scattering $T = e^{-(\sigma + k^{\ell})^{\ell}}$

where ℓ is the material thickness and σ is a measure of the amount of light scattered (scattering coefficient or turbidity). k is a measure of the amount of light absorbed (absorption coefficient). $(\sigma + k)$ is known as the alternation or extinction coefficient.

The haze characterizes the loss of contrast which results when objects are seen through a scattering medium. Deterioration of contrast is mainly due to the light scattered forward at high angles to the undeviated transmitted beam. The 'milkiness' of translucent samples when viewed from the side on which light is incident is largely due to the backward scattering.

The clarity is a measure of the capacity of the sample for allowing details in the object to be resolved in the image which it forms. It is strongly dependent upon angular distribution of scattering intensity and distance between object and sample.

1.8.10.3 Gloss

Gloss is a property of the surface of a plastic specimen. Its magnitude depends on the refractive index of the plastic, the smoothness of the surface and the occurrence of subsurface optical features. The relative importance of these factors depends upon the angle of incidence of light falling on the surface.

1.8.10.4 Light transfer

The total transmission factor (transmittance), τ, is defined by the ratio of the total transmitted flux ϕ_t and the incident flux ϕ.

$$\tau = \frac{\phi_t}{\phi}$$

The total reflection factor (reflectance), ρ, is defined (for normal incidence) as the ratio between the backward scattered flux ϕ^b_{sc} and the incident flux ϕ.

$$\rho = \frac{\phi^b_{sc}}{\phi}$$

1–9

Materials for electromagnetic compatibility

Contents

List of tables

1.9.1 Definition

Electromagnetic compatibility (EMC) describes that property of a device, system or machine which ensures that:

1. it does not radiate or transmit signals above defined maximum levels, and
2. it is not susceptible to interference by externally radiated or transmitted signals below defined minimum levels.

1.9.2 General background

Until recently only those designers working in specialized areas such as high-power RF machines and safety-critical control systems would have considered electromagnetic compatibility as a normal part of their work. However from 1 January 1996 the EC introduced mandatory legal EMC requirements for all systems or machines with electrical or electronic components (The EMC Directive 89/336/EEC). The manufacturer or importer is required to show that the product complies, and place upon the product a 'CE' mark to declare this compliance.

These regulations are sufficiently strict that many units have had to be redesigned to comply. This combined with the increasing complexity of control systems (microprocessors are now found in many white goods), the rapidly rising clock frequencies of microprocessors (now up to 200 MHz), and the increasing sensitivity of devices as operating voltages fall, makes careful design and correct materials choice much more important.

Interference can leave a piece of equipment by two separate paths: it can be radiated or conducted. Radiated emissions occur as fields from wires, printed-circuit boards (PCBs) or components which act as aerials. Conducted emissions are those that travel along interconnecting cables between different pieces of equipment. Similarly, a device can be susceptible to external interference picked up by wires, PCBs and components, as well as to that conducted along interconnecting cables.

There are many design techniques to reduce or eliminate generation of, or susceptibility to, signals at source. These are not discussed here. If, however, the problem cannot be eliminated at source, there are a number of methods available to ensure compliance with the requirements. There is no single 'best' method, but rather a range of options depending upon the form of the signal, the frequency and the level of attenuation required. Commercial pressure in this area means that new solutions are constantly being developed.

1.9.3 Enclosures

The best solution is to place the system in a conducting box. Shields work by both reflecting the incident energy and by absorbing it. The absorption loss, A_{dB}, is given by:

$$A_{dB} = 132t\,(f\mu_r\sigma_r)^{0.5}$$

where:
t = shield thickness;
f = frequency;
μ_r = permeability relative to copper;
σ_r = conductivity relative to copper.

The reflection loss, R_{dB}, is caused by the impedance mismatch between the incident wave and the shield. Since the source wave impedance depends on both the type of field (electric or magnetic) and the distance from the source, there are three expressions for R_{dB}:

$$R_{dB} = 168 - 10\log_{10}\left(\frac{f\mu_r}{\sigma_r}\right) \quad \text{for plane waves;}$$

$$R_{dB} = 323 - 10\log_{10}\left(f^3 r^2 \mu_r \sigma_r\right) \quad \text{for an electric field source;}$$

$$R_{dB} = 15 - 10\log_{10}\left(\frac{\mu_r}{r^2 f \sigma_r}\right) \quad \text{for a magnetic field source}$$

where:
r = distance from shield to source.

A list of commonly used shield materials and their shielding properties relative to copper are given in:
Table 1.9.1 *Shield materials and their properties*

From Table 1.9.1 it can be seen that A_{dB} for iron is some 14 times that for copper, while R_{dB} is about 37 dB lower. It would thus appear that iron is to be preferred to copper as a shield material. However, at high frequencies, the permeability of ferromagnetic materials decreases, making them less suitable for shielding purposes. High permeability materials, such as Mu-metal, are also sensitive to distortion and mechanical shock, which can considerably decrease their permeability.

A good metallic box will give an inherent attenuation of >130 dB, but remember:

- A folded metal box will 'leak' where edges are folded together. The best solution to this is a continuously welded joint, but conductive sealants (see Section 1.9.6) will improve performance.
- Boxes with lids or doors will 'leak' at the seams. This problem increases dramatically as frequencies rise. Seals for doors, or gaskets for less frequently used openings will give improvement, but the quality of the matching faces is also important. It is also vital that good electrical connection exists between the door/lid and the main body of the box.
- Wires entering or leaving the box will carry the radiation/transmission with them and require separate attenuation measures.
- Windows for visual inspection will break the screen. Conductive windows (see Section 1.9.5) can be used but will always result in a deterioration in performance. They should be as small as possible.

Where conventional metal boxes cannot be used for design or cost reasons, there are a number of alternatives:

1. Some alloys (such as 78% Zn–22% Al) can be vacuum–formed at 250–300°C, but still retain sufficient conductivity to give an attenuation >100 dB over a wide band of frequencies. This process has the benefit of avoiding joins in complex shapes so that overall performance may exceed folded metal.
2. Plastic formed parts can be coated with a metallic film. Because the coating processes (evaporation, sputtering, etc) produce thin films, very high conductivity is important. Although single layer films are used, particularly aluminium, and can give attenuations up to 70 dB, the best results are obtained by a three layer sandwich – a keying layer to bond to the plastic (normally aluminium, nickel or a steel), a high conductivity layer (normally copper but sometimes silver), and a protective outer layer (normally the same as the bonding layer). These systems can give attenuation >80 dB over a wide range. However, using these techniques, complex shapes can lead to variations in film thickness (or no film at all) with a resulting degradation in attenuation.

   Electroless plating can also be used, giving attenuations in the range 50–80 dB, rather better coverage of complex shapes, and improved wear resistance.

3. Cases formed from conductive plastics (either inherently conductive, or with conductive fillers) can also be used, but attenuation levels are highly variable and when fillers are used they can be strongly affected by filler orientation and by weld-lines. Attenuation in the best cases with metallic fillers can exceed 50 dB, but is more commonly below 40 dB. Other fillers, such as conductive carbon, give attenuations >25 dB.

1.9.4 Seals and gaskets

These can take a wide range of forms with performances matched to specific requirements. In all cases achieving the best performance from the seal/gasket is dependent upon having good, clean surfaces on both faces, having good electrical contact between the part and the seal/gasket at all points along the face, and on avoiding over sharp bends (as a general rule, where the seal/gasket is required to bend, the radius should exceed 1 cm wherever possible).

1.9.4.1 Sprung metal contact strips

These are typically used where a wide band, high attenuation seal is required for a door which is repeatedly opened. A wide variety of designs is available giving attenuation >100 dB, although performance in the microwave band falls to about 90 dB. They are normally made from beryllium–copper and maintain good electrical contact through a wiping action on closure. Marginal improvements can be made by gold-plating the contact areas. Best performance is dependent upon careful use – snapping the individual fingers results in rapid performance degradation and the component must be replaced.

1.9.4.2 Conductive elastomers

These materials are metallic particles in an elastomer matrix. They become conductive when compressed (recommended operating compression for flat parts being 5–15%). The elastomer is normally silicone-based for wear and environmental reasons. The particulate content has a strong influence upon the attenuation characteristics, most materials performing best in a band about a decade wide. Care must also be taken to minimize the risk of electrochemical interaction between the case and gasket materials since the by-product of such a reaction will significantly impair the screening function. A number of typical results for material fabricated from sheets or strips are tabulated in:

Table 1.9.2 *Effect of particle type on electrical properties of conductive elastomers*

The same types of material can be used to extrude special forms, but effective conductivity is reduced by a factor of 2–3, with a commensurate affect upon the attenuation.

Wire grids and oriented arrays of wires, encapsulated in elastomers or foams are also available. For grids, the mesh size controls the upper limit of effective operation. Below this limit attenuations better than 90 dB can be achieved, but continuous contact along the seal can be a problem. Above the frequency limit attenuation can fall below 30 dB. Oriented arrays of wires can give excellent attenuation, >120 dB at 10 MHz, in one plane. Such arrays make excellent flat gaskets.

All the elastomer/foam seals/gaskets only work well for a limited number of compressions. Repeated compression will cause fatigue failure in the matrix resulting in loss of attenuation.

1.9.4.3 Metallic seals/gaskets

It is also possible to produce gaskets from a variety of malleable forms of metal. They are generally only suited to a single use or a very small number of uses. Forms which can be used include:

1. Woven wire, sometimes silver-coated, punched to shape. Attenuation at 10 MHz>120 dB, but falling rapidly above a knee–point determined by mesh size.
2. Foamed metals, punched from sheet. Attenuation at 10 MHz>110 dB, but falling rapidly above a knee–point determined by 'hole' size.
3. Sintered metallic pads. These are usually sintered to shape and give an improved performance at higher frequencies, achieving attenuations as high as 70 dB at 10 GHz.
4. Spring gaskets. These are generally designed to fit in slots in the mating faces. They can give attenuations as high as 120 dB at low frequencies, but are susceptible to damage or movement during assembly. Canted springs can be used to improve high frequency attenuation and reduce the risks of damage or poor bedding.

Galvanic corrosion can be minimized by producing the shield and gasket from metals close in the galvanic series.

1.9.5 Windows and vents

These are the two items which cause greatest problems in maintaining a high level of signal attenuation. Both should be kept as small as possible.

For windows, the surface conductivity should be kept as high as possible. Since most windows use metallized layers this results in a direct trade-off between conductance (screening) and transmittance (visibility through the window). The layer can take the form of an ultra-fine grid laminated into the window, or a thin continuous metallic layer generated by vacuum deposition or sputtering. A laminated grid can give an attenuation at 10 MHz of 120 dB, but typically falling to 40 dB at 10 GHz, whilst continuous films can give an attenuation at 10 MHz of 90 dB falling rapidly to 30 dB at 1 GHz.

For vents, it is possible, if a low airflow is required, to use a fine mesh giving an attenuation at 10 MHz of >120 dB, falling at 10 GHz to 55 dB, but for higher airflows or better high frequency performance, a baffled honeycomb structure generally fabricated in steel, brass or aluminium, is used. These can give an attenuation at 10 MHz of >130 dB, falling only to 110 dB at 10 GHz for small panels (less than 15×15 cm).

1.9.6 Coatings, sealants and adhesives

A wide variety of materials are available to act as single use sealants. Most use a silver, or silver-coated, conducting filler at sufficiently high proportions to ensure conduction when dry. They are therefore inherently expensive. The medium which carries this filler is highly varied and includes polyolefin, silicone and epoxy resins for applications where gluing or filling cracks is required, silicone grease for applications where a rotary or longitudinal drive must enter/leave the case, and acrylic and polyurethane paints for covering larger areas. The surfaces to be treated must be carefully prepared to ensure maximum conductive contact and then these materials can provide a valuable means of curing minor problems.

Whilst the ideal performance of these materials is excellent, it is preferable to avoid their use by careful design and use of the other techniques outlined above because performance can vary with application and drying conditions, and they are subject to mechanical damage.

1.9.7 Connectors and cabling accessories

One of the most common causes of EMC failure is very high frequency noise on data-lines which pass through the case. Whilst formal filtration techniques can be used at the ports, this is an issue of electronics design and will not be dealt with here. A simpler technique is the use of ferrite filters to attenuate signals above a given frequency. These are now commonly seen on video cables as an integral part of the cable.

These ferrite filters are available both as clip-on filters for flat or round cables, or integrated into the panel-mounting plugs where they actually lie in the screening plane (probably the ideal position).

Table 1.9.1 Shield materials and their properties

Metal	μ_r	σ_r	$(\mu_r\sigma_r)^{0.5}$	$10\log_{10}(\mu_r/\sigma_r)$
Copper	1	1	1	0
Aluminium	1	0.6	0.8	2.3
Magnesium	1	0.4	0.6	4
Zinc	1	0.3	0.5	5
Nickel	1	0.2	0.4	7
Lead	1	0.08	0.3	11
Monel	1	0.04	0.2	28
Iron	1000	0.2	14	37
Steel	1000	0.1	10	40
Mu-metal	80 000	0.03	50	64

Table 1.9.2 Effect of particle type on electrical properties of conductive elastomers

Particle type	***Attenuation at 10 Mhz (dB)***	***Attenuation at 1 Ghz (dB)***	***Attenuation at 10 Ghz (dB)***	***Compressive force[a] (kg)***
Silver-coated inert particles	>130	100	90	150
Silver-coated copper	>120	115	110	110
Silver-plated nickel	120	110	100	270
Silver-plated aluminium	>120	110	100	90
Aluminium	85	105	100	350
Carbon	85	75	65	180

[a]Force required per m on a 3 mm diameter strip to achieve recommended compression

1–10

Design, materials and manufacturing process route selection for minimal environmental impact

Contents

List of tables

List of figures

1.10.1 Introduction

Environmental pressures are exerting an increasing influence on design, materials and manufacturing process route selection. The pressures are of three types:

- **Legislative.** National Environmental Protection Acts (EPA) now impose restrictions on the permissible levels of gaseous, liquid or solid waste products. This results in certain processes for certain materials either being prohibited or prohibitively expensive to control (see below for examples).
- **Economic.** The cost of production waste disposal or a legislative requirement to recover and recycle products at the end of their service life means that for many designs, materials and manufacturing processes it has become 'economical to be ecological'.
- **Consumer.** Consumers, often as a result of publicity by environmental pressure groups, may boycott products which are perceived (rightly or wrongly) to be environmentally unfriendly.

For product makers it is no longer permissible to pass environmental problems on to their suppliers or their customers. The product designer is responsible for the impact of the product from the start of the supply chain of the materials to the recycling or reuse of the end-product.

A number of tools and techniques have been developed to assist designers and product makers to assess the environmental impact of a particular design, material or manufacturing route. The most important of these are:

Life cycle assessment[1]
Environmental audits[2]

Details on the application of these techniques are given in the references below. These techniques can be applied during the design reviews which are held at each of the four stages of design.

In most countries, the environmental best practice covering design, materials selection and manufacturing processes is governed by BATNEEC:

Best
Available
Technology
Not
Entailing
Excessive
Cost

1.10.2 Energy content of materials

All materials consume energy during their life cycle. The early stages of the life cycle, namely extraction and refining, are the most energy intensive. Relatively little energy is consumed in converting as-received materials into a component shape and recycling generally costs much less in energy terms than primary extraction and refining.

The energy content of the generic classes of materials is given in:

Table 1.10.1 *Energy contents of materials*

Materials with the highest energy content are likely to have a greater impact on the environment due to the pollution caused by primary energy consumption unless that

[1]SETAC (Society of Environmental Toxicology and Chemistry) 1993 'Guidelines for Life-Cycle Assessment: A Code of Practice', SETAC, Brussels (B) and Pensacola (USA).

[2]Clift R (1995) 'The New Industrial Ecology', pp 129-143 in Advanced Materials in the Market Place, ed J E Castle and M J Kelly, The Institute of Materials, London, UK.

energy is produced in an environmentally friendly way, (for example aluminium smelters sited near to hydroelectric power stations).

It is possible to derive comparative materials performance indices for energy content in the same way as was done for performance against cost, or weight in Sections 1.1.14 and 1.1.6.

For example, the selection of a material for a beam which must meet a stiffness constraint, at minimum energy content. The energy content per kilogram of a material is q (data: Table 1.8.1). Then a beam of square section A and length l contains energy

$$Q = Al\rho q \qquad (1)$$

The stiffness, S, of the beam is specified; it is related to modulus E and second moment of area I by the equation

$$S = \frac{C_1 EI}{l^3} \qquad (2)$$

where C_1 is a constant which depends on the way the loads are applied. For a square section beam $I = A^2/12$. Eliminating A in equation (1) gives

$$Q = \left(\frac{12 S l^5}{C_1}\right)^{1/2} \frac{\rho q}{E^{1/2}}$$

The energy content is minimized by maximizing

$$M_1 = \frac{E^{1/2}}{\rho q} \qquad (3)$$

As this suggests, performance criteria for minimizing energy content are the same as those for minimizing weight, with density ρ replaced by ρq. The stiff tie of minimum energy content is best made of a material of high $E/\rho q$; the stiff plate, of material with high $E^{1/3}/\rho q$. Strength works the same way. The best choice of material for a beam of specified bending strength, and minimum energy content, is that with the highest value of:

$$M_2 = \frac{\sigma_f^{2/3}}{\rho q}$$

and so on.

Figs 1.10.1 and 1.10.2 are a pair of materials selection charts for minimizing energy content per unit of function. The first shows modulus, E, plotted against energy content, $q\rho$; the design guidelines give the slopes for three of the commonest performance indices. The second shows strength σ_f against $q\rho$ again; design guidelines give the slopes. The charts are used in exactly the same way as before, see Section 1.1.6.

Fig. 1.10.1 *Young's modulus vs energy chart*

Fig. 1.10.2 *Strength vs energy chart*

Energy consumption, and the potential for saving, are significant when large quantities of material are used, as they are in civil construction. It can quickly be established that the most energy efficient beam, whether the design is based on stiffness or on strength, is that made of wood. Steel, even with a large shape factor, consumes far more energy. Columns of brick or stone are more energy efficient than concrete, though more labour intensive.

Most polymers are derived from oil. This leads to statements that they are 'energy-intensive', with implications for the future. Charts 1.10.1 and 1.10.2 show that, per unit of function in bending (the commonest mode of loading), most polymers are less energy intensive than aluminium, and that several are competitive with steel. On an energy basis, magnesium is poor, titanium worse.

Energy content, as stated earlier, is only one measure of the environmental impact of material usage. In many circumstances it is not the most important one; the emission of a toxic by-product, the difficulty of recycling, or the resistance to biodegrading can be the real environmental threat.

1.10.3 Recycling and re-use of materials

Legislation and the increasing costs of incineration and landfill disposal are increasing the pressure to recycle or re-use materials. Primary scrap, that is waste material produced during the manufacture of components (for example the sprues of injection moulding and the runners and risers from metal castings), are normally re-processed in controlled amounts within the same production site.

Recycling or re-use of the materials in products and components at the end of their lifetime is more problematical and, in the absence of legislation, often less attractive economically than manufacture with virgin material or virgin material plus a controlled amount of primary scrap.

In some cases worn components can be restored with surface treatments, e.g. electro or spray deposition which build up the parts to their original dimensions. In many cases however, the original designs will have been superseded and the old design is obsolete. In most cases the use of surface coatings makes a component more difficult or expensive to recycle or re-use.

The method of assembly and the materials' construction in that assembly are very important influences on the recyclability or re-use of materials. In many cases the attempts to optimize materials selection by splitting the functions of a component and choosing two materials instead of one will increase the difficulties of recycling.

Mechanically-fastened assemblies are, in principle, the easiest to dismantle and recycle, but this option may be the most expensive design.

Adhesively bonded components may be separated by heat or chemical attack but environmental pollution constraints may make these methods illegal or expensive.

Welded assemblies will be the most difficult to separate but welded assemblies are most frequently of the same material and recycling is therefore much simpler.

If it is expected that legislation will require the recycling of the material content of a product, then that product must be designed for recycling from the outset, i.e. this becomes one of the primary functions of the product, and at each design review of each design stage, the ease or difficulty of recycling will influence the selection of materials and manufacturing route.

1.10.4 Toxicity

The toxicity of certain materials makes their processing, application and disposal or re-use extremely difficult and/or expensive. The toxicological problems of structural materials and surface treatments are covered in the appropriate sections of the Materials Selector. The materials which are causing most concern on toxicological grounds are as follows:

Asbestos	Restrictions regarding the processing of blue and brown asbestos make them practically unusable. White asbestos can still be employed where no effective substitute exists (see Section 3.8.8 for more details).
Cadmium	Cadmium plating, which is still needed for some fasteners in the aerospace industry can in most countries be conducted in environmentally isolated plants. Such plants are a special purpose design which has no release or build-up of cadmium containing effluent. The cadmium is extracted and recycled, liquid losses are topped up.

Chromium	Hexavalent chromium which occurs in chromium plating baths is a known carcinogen when ingested. Skin contact leads to burns and ulceration. Chromium plating when conducted according to best practice (which should involve multiple washing stages and a neutralization stage) converts any remaining hexavalent chromium on the plated components to the relatively harmless trivalent chromium.
Lead	There are pressures to eliminate lead from components and products. Lead is found in bearings and solders. In most cases, substitute lead-free compositions now exist but these are generally more expensive.

Table 1.10.1 Energy content of materials

Class	*Material*	*Energy* *q (MJ/kg)*	*Energy* *ρq(GJ/m³)*
Metals	Titanium and alloys	555–565	2400–2880
	Magnesium and alloys	410–420	717–756
	Cast irons	60–260	468–1500
	Aluminium and alloys	290–305	754–884
	Stainless steels	110–120	825–972
	Copper and alloys	95–115	712–1035
	Zinc and alloys	67–73	348–525
	Carbon steels	50–60	390–468
	Lead and alloys	28–32	300–360
Polymers	Nylon 66	170–180	187–216
	Polypropylene	108–113	95–102
	H.D. polyethylene	103–120	97–116
	L.D. polyethylene	80–104	73–94
	Polystyrene	96–140	96–154
	PVC	67–92	87–147
	Synthetic rubber	120–140	108–126
	Natural rubber	5.5–6.5	5–6
Ceramics Glasses	Glasses	13–23	32–57
	Glass fibres	38–64	95–160
	Bone china	270	540–580
	Bricks	3.4–6.0	6.8–12
	Refractories	1–50	3–100
	Pottery	6–15	12–30
	Cement	4.5–8.0	9–18
	Concrete	3–6	7–15
	Stone	1.8–4.0	4–8.8
Composites (estimates)	GFRP	90–120	160–220
	CFRP	130–300	230–540
Other	Hard and soft woods	1.8–4.0	1.2–3.6
	Reinforced concrete	8–20	20–50
	Gravel/stone	0.1	0.2–0.4
	Oil	44	38–40
	Coal	29	27–30

Table 1.10.1

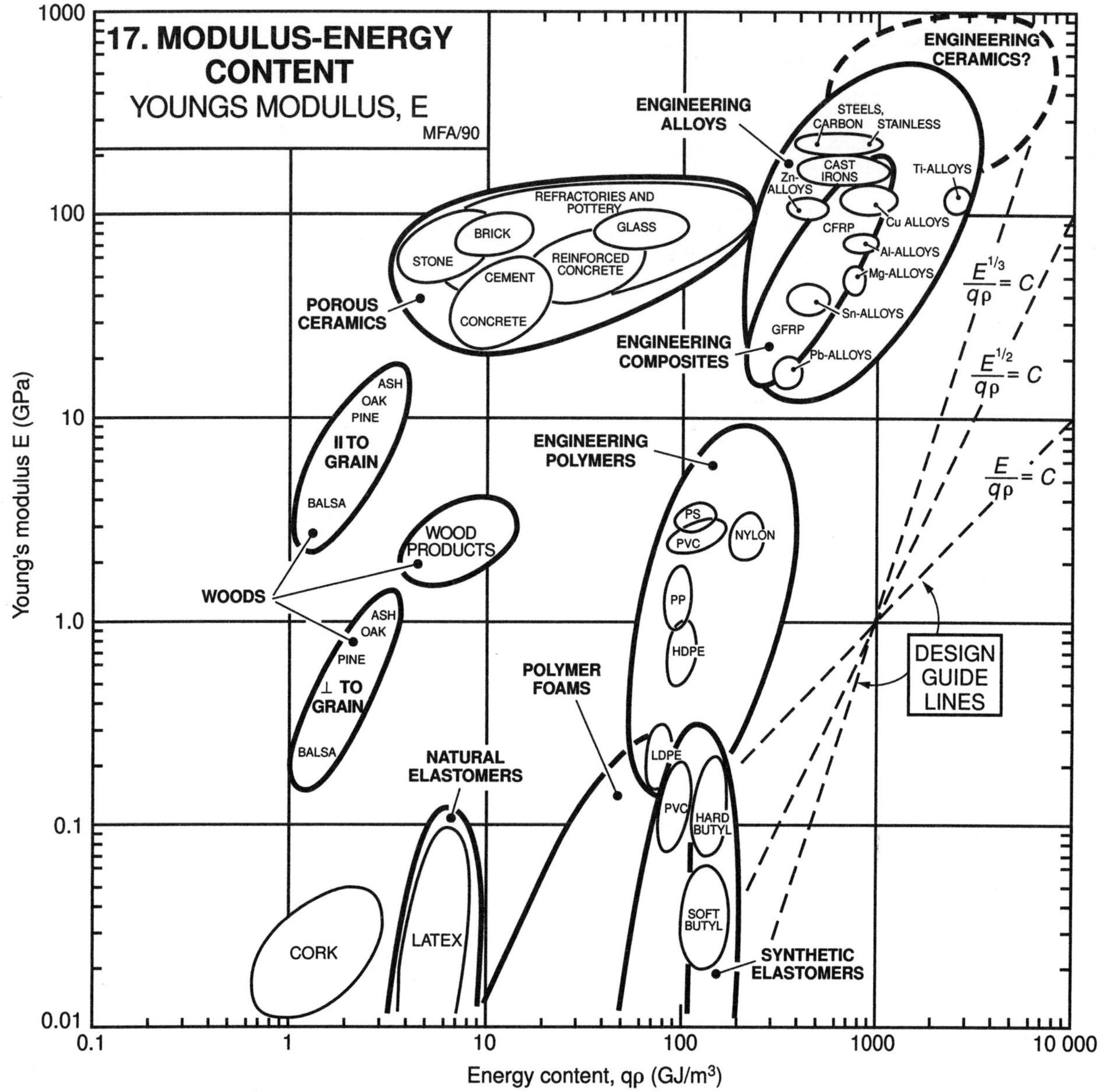

Fig. 1.10.1 Young's modulus vs energy content chart, with guidelines for selecting materials for stiff structures at minimum energy content. It is an example of charts which allow selection to minimize environmental impact.

Fig. 1.10.1

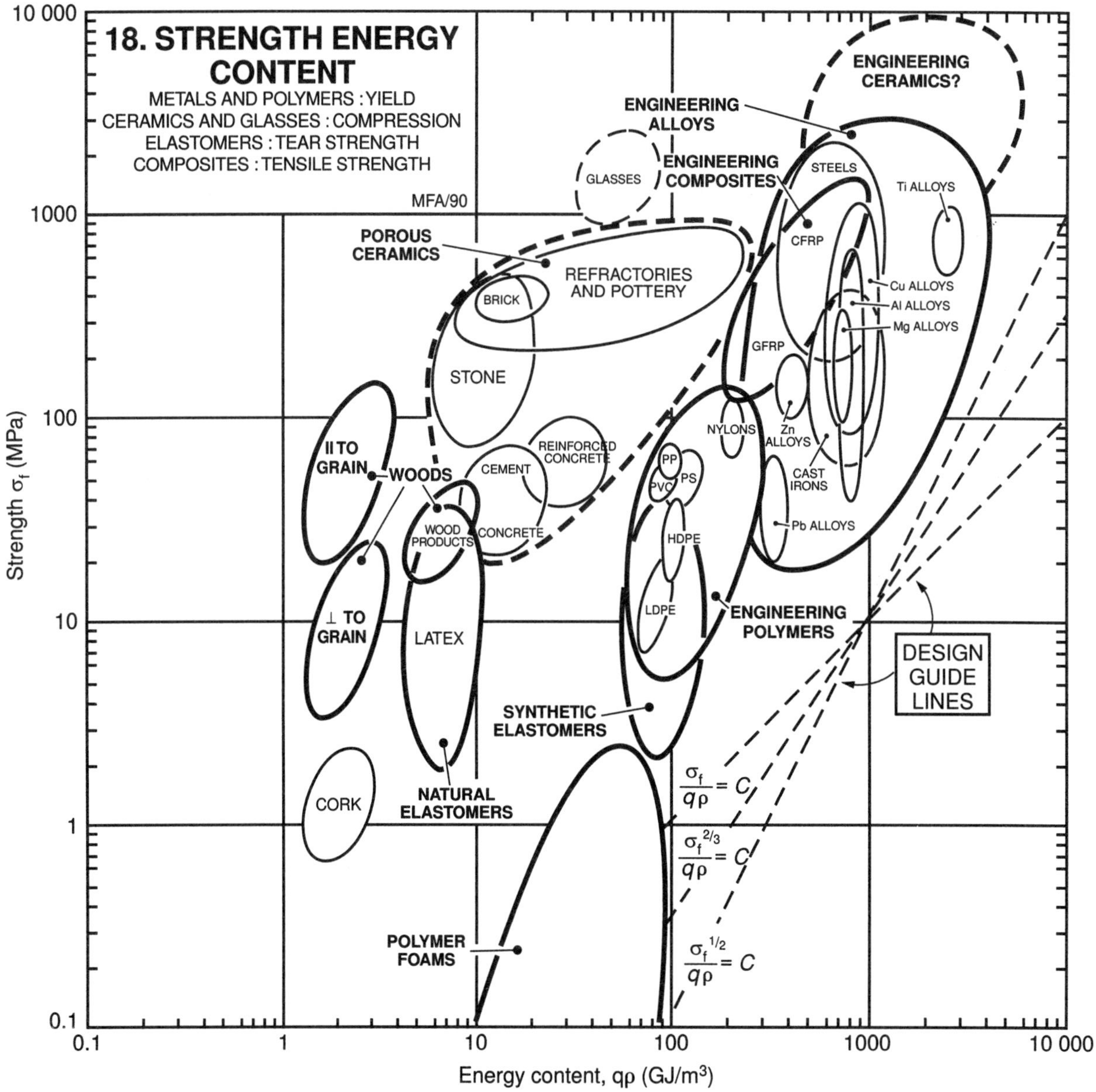

Fig. 1.10.2 Strength vs energy content chart, with guidelines for selecting materials for strong structures at minimum energy content.

Fig. 1.10.2

1–11

Units and conversion factors

Contents

List of tables

List of figures

1.11.1 SI units, conversion factors and constants

1.11.1.1 SI units and conversion factors

Units according to the Systeme Internationale (SI units) are used throughout the Selector. The most common units and their conversion factors are given in:

Table 1.11.1 *Frequently needed factors for conversion to SI units*

A comprehensive list of SI units and conversion factors for a large number of materials properties is given in:

Table 1.11.2 *SI units and conversion factors*

Units are covered by national and international standards, which should be consulted for further details. The relevant British Standards are:

BS 1637:1950	Memorandum on the MKS system of electrical and magnetic units
BS 3763:1964	International System (SI) Units
BS 350, Part 1:1959	Conversion Factors and Tables
BS 5775, Parts 0–13:1979 onwards.	Specification for quantities, units and symbols.

1.11.1.2 Allowed multiples of SI units

The recommended and permissible multiples of SI units are given in:

Table 1.11.3 *Multiples and submultiples of SI units*

1.11.1.3 Mathematical and physical constants

Frequently used mathematical and physical constants are given in:

Table 1.11.4 *Mathematical constants*

Table 1.11.5 *Physical constants*

Table 1.11.1 Frequently needed factors for conversion to SI units

Property	*Dimensional form*	*SI unit*	*Conversion factor*
Length	[L]	m (metre)	1 μm = 10^{-6} m 1 thou = 2.54×10^{-5} m 1 in = 2.54×10^{-2} m 1 ft = 3.048×10^{-1} m
Mass	[M]	kg (kilogramme)	1 oz (av) = 2.83495×10^{-2} kg 1 lb = 4.5359237×10^{-1} kg 1 t = 1.01605×10^{3} kg
Force	$[MLT^{-2}]$	N = kgm/s^2 (newton)	1 dyne = 10^{-5} N 1 lbf = 4.44822 N 1 kgf = 9.80665 N (= kilopond) 1 tonf = 9.96402×10^{3} N
Pressure (stress)	$[ML^{-1}T^{-2}]$	Pa = N/m^2 (pascal)	1 tf/in^2 = 1.54443×10^{7} Pa 1 hectobar = 10^{7} Pa 1 kgf/mm^2 = 9.80665×10^{6} Pa 1 atmosphere = 1.01325×10^{5} Pa 1 lbf/in^2 = 6.89476×10^{3} Pa 1 torr = 1.33322×10^{2} Pa 1 millibar = 10^{2} Pa 1 micron of Hg = 1.33322×10^{-1} Pa 1 $dyne/cm^2$ = 10^{-1} Pa
Energy (work) (heat)	$[ML^2T^{-2}]$	J = N m (joule)	1 Btu = 1.05506×10^{3} J 1 cal_{IT} = 4.1868 J 1 ft lbf = 1.35582 J 1 erg = 10^{-7} J 1 eV = 1.6021×10^{-19} J

Table 1.11.2 SI units and conversion factors

Property	***Dimensional form***	***SI unit***	***Conversion factor***	
Acceleration	$[LT^{-2}]$	m/s^2	1 cm/s^2 1 ft/s^2 1g	$= 10^{-2}$ m/s^2 $= 3.048 \times 10^{-1}$ m/s^2 $= 9.80665$ m/s^2
Action	$[ML^2T^{-1}]$	J s	1 erg s 1 ft poundal s	$= 10^{-7}$ J s $= 4.21401 \times 10^{-2}$ J s
Admittance (as for conductance)				
Angle		rad (radian)	1 sec 1 min 1 grade 1 degree 1 right angle 1 revolution	$= 4.84814 \times 10^{-6}$ rad $= 2.90888 \times 10^{-4}$ rad $= 1.57080 \times 10^{-2}$ rad $= 1.74533 \times 10^{-2}$ rad = 1.57080 rad = 6.2832 rad
Angular momentum	$[ML^2T^{-1}]$	kg/m^2 per s	1 g cm^2/s 1 lb ft^2/s	$= 10^{-7}$ kg m^2/s $= 4.21401 \times 10^{-2}$ kg m^2/s
Angular velocity	$[T^{-1}]$	rad/s	1 deg/s 1 rad/min 1 rev/min 1 rev/s	$= 1.74533 \times 10^{-2}$ rad/s $= 1.66667 \times 10^{-2}$ rad/s $= 1.04720 \times 10^{-1}$ rad/s = 6.28319 rad/s
Area	$[L^2]$	m^2	1 mm^2 1 cm^2 1 in^2 1 ft^2 1 yd^2 1 acre 1 ha 1 sq mile	$= 10^{-6}$ m^2 $= 10^{-4}$ m^2 $= 6.4516 \times 10^{-4}$ m^2 $= 9.29030 \times 10^{-2}$ m^2 $= 8.36127 \times 10^{-1}$ m^2 $= 4.04686 \times 10^{3}$ m^2 $= 10^{4}$ m^2 $= 2.58999 \times 10^{6}$ m^2
Calorific value (as for latent heat)				
Calorific value by volume	$[ML^{-1}T^{-1}]$	J/m^3	1 $kcal/m^3$ 1 Btu/ft^3	$= 4.1868$ J/m^3 $= 3.72589 \times 10$ J/m^3
Capacity	$[M^{-1}L^{-2}T^4I^2]$	F = A s/V (farad)	1 esu 1 emu	$= 1.11265 \times 10^{-12}$ F $= 10^{9}$ F
Charge	[TI]	C = A s (coulomb)	1 esu 1 emu	$= 3.33564 \times 10^{-10}$C = 10 C
Compressibility	$[M^{-1}LT^{-2}]$	1/Pa	1 cm^2/dyne	= 10/Pa
Concentration (see density)				
Conductance	$[M^{-1}L^{-2}T^3I^2]$	$1/\Omega$ = S (siemens)	1 esu 1 emu	$= 1.11265 \times 10^{-12}/\Omega$ $= 10^{9}/\Omega$
Current density	$[L^{-2}I]$	A/m^2	1 esu 1 emu 1 A/cm^2 1 A/in^2 1 A/ft^2	$= 3.33564 \times 10^{-6}$ A/m^2 $= 10^{5}$ A/m^2 $= 10^{4}$ A/m^2 $= 1.5500 \times 10^{3}$ A/m^2 $= 1.0764 \times 10$ A/m^2

Table 1.11.2 SI units and conversion factors – *continued*

Property	*Dimensional form*	*SI Unit*	*Conversion factor*	
Density	[ML^{-3}]	kg/m^3	1 mg/litre 1 g/m^3 1 oz/UK gal 1 oz/US gal 1 lb/ft^3 1 oz/UK pint 1 lb/UK gal 1 lb/US gal 1 g/ml (= 1 kg/litre) 1 g/cm^3 1 lb/in^3	= 9.99972 × 10^{-4} kg/m^3 = 10^{-3} kg/m^3 = 6.23603 kg/m^3 = 7.48915 kg/m^3 = 1.60185 × 10 kg/m^3 = 4.98882 × 10 kg/m^3 = 9.97764 × 10 kg/m^3 = 1.19826 × 10 kg/m^3 = 9.99972 × 10^2 kg/m^3 = 10^3 kg/m^3 = 2.76799 × 10^4 kg/m^3
Diffusion coefficient	[L^2T^{-1}]	m^2/s	1 cm^2/s	= 10^{-4} m^2/s
Dipole moment	[LTI]	C m	1 esu 1 emu	= 3.3564 × 10^{-12} C m = 10^{-1} Cm
Displacement	[L^{-2} TI]	C/m^2	1 esu 1 emu	= 2.6544 × 10^{-7} C/m^2 = 7.9578 × 10^3 C/m^2
EMF (as for electric potential)				
Elastic compliance (as for compressibility)				
Elastic (stiffness) constant (as for pressure)				
Electric current	[I]	A (ampere)	1 esu 1 emu	= 3.33564 × 10^{-10} A = 10 A
Electric field	[MLT^{-3} I^{-1}]	V/m	1 esu 1 emu 1 V/cm	= 2.9979 × 10^2 V/m = 10^{-6} V/m = 10^2 V/m
Electrical potential	[ML^2T^{-3} I^{-1}]	V = W/A (volt)	1 emu 1 esu	= 10^{-8} V = 2.9979 × 10^2 V
Electrical conductivity	[M^{-1} L^{-3} T^3 I^2]	1/Ω per m	1 esu 1 emu 1 mho/cm	= 1.11265 × 10^{-10}/Ω per m = 10^{11}/Ω per m = 10^2/Ω per m
Electro optical coefficient (as for Piezoelectric strain coefficient)				
Energy	[ML^2T^{-2}]	J = N m (joule)	1 therm 1 kW h 1 horsepower hour 1 Btu 1 litre atm. 1 kgf m 1 cal$_{IT}$ 1 cal$_{15}$ 1 cal (thermochem) (US) 1 ft lbf 1 foot poundal	= 1.05506 × 10^8 J = 3.6 × 10^6 J = 2.68452 × 10^6 J = 1.05506 × 10^3 J = 1.01328 × 10^2 J = 9.80665 J = 4.1868 J = 4.1855 J = 4.814 J = 1.35582 J = 4.21401 × 10^{-2} J

Table 1.11.2 – *continued*

Table 1.11.2 SI units and conversion factors – *continued*

Property	***Dimensional form***	***SI unit***	***Conversion factor***	
Energy *(continued)*			1 eV 1 erg 1 inch lbf 1 eV (per atom) 1 quad (= 10^{15} Btu)	= 1.6021×10^{-19} J = 10^{-7} J = 1.1299×10^{-1} J = 9.6487×10^{4} J/mole = 1.05506×10^{18} J
Entropy	$[ML^2 T^{-2} \Theta^{-1}]$	J/K	1 cal/degC 1 Btu/degF 1 cal (therm)/degC	= 4.1868 J/K = 1.89911×10^{3} J/K = 4.184 J/K
Fluidity	$[M^{-1}LT]$	1/Pa per s	1 ft^2/lbf s 1 ft^2/pdl s 1 $poise^{-1}$	= 2.0885×10^{-2}/Pa per s = 6.720×10^{-1}/Pa per s = 10/Pa per s
Force	$[MLT^{-2}]$	N = kgm/s^2 (newton)	1 dyne 1 poundal 1 ozf 1 lbf 1 kgf (= kilopond) 1 tonf	= 10^{-5} N = 1.38255×10^{-1} N = 2.78014×10^{-1} N = 4.44822 N = 9.80665 N = 9.96402×10^{3} N
Fracture toughness	$[ML^{-1/2} T^{-2}]$	$N/m^{3/2}$	1 $(kgf/cm^2)\sqrt{cm}$ 1 ksi $\sqrt{in}$ 1 $(ton/in^2)\sqrt{in}$ 1 hbar $(mm)^{1/2}$ 1 hbar $(cm)^{1/2}$	= 9.80665×10^{2} $N/m^{3/2}$ = 1.09885×10^{6} $N/m^{3/2}$ = 2.4614×10^{6} $N/m^{3/2}$ = 3.1623×10^{5} $N/m^{3/2}$ = 10^{6} $N/m^{3/2}$
Frequency	$[T^{-1}]$	Hz = 1/s (hertz)	1 cycle/s	= 1 Hz
Gas permeability	$[M^{-1} L^3 T]$	m^2/Pa per s	1 cm^2/(s cmHg)	= $7.5006 \times 10^{-8} m^2$/Pa per s
Hardness (as for pressure)	See Tables 1.9.6–1.9.10			
Heat (see energy)				
Heat flow rate (see power)				
Heat flow rate per unit area	$[MT^{-3}]$	W/m^2	1 Btu/ft^2 h 1 cal/cm^2s	= 3.15459 W/m^2 = 4.1868×10^{4} W/m^2
Heat release rate	$[ML^{-1} T^{-3}]$	W/m^3	1 Btu/ft^3h 1 cal/cm^3s	= 1.03497×10 W/m^3 = 4.1868×10^{6} W/m^3
Heat transfer coefficient (as for thermal conductance)				
Illumination		lx = lm/m^2 (lux)		
Impact strength (Izod-plastics)	$[MLT^{-2}]$	J/m (joules per metre of notch)	1 J/cm 1 ft lb/in	= 10^{2} J/m = 5.337864×10 J/m
Impact strength (Izod-metals)	$[ML^2T^{-2}]$	J	See energy	

Table 1.11.2 – *continued*

Table 1.11.2 SI units and conversion factors – *continued*

Property	*Dimensional form*	*SI unit*	*Conversion factor*	
Impact strength (Charpy – metals)	$[ML^2T^{-2}]$	J	See energy	
Impact strength (Charpy – plastics)	$[MT^{-2}]$	J/m^2	1 J/cm^2 1 $(ft\ lb)^2$	= 10^4 J/m^2 = 2.1015 × 10^6 J/m^2
Impedance (as for resistance)				
Impulse (see momentum)				
Inductance	$[ML^2T^{-2}I^{-2}]$	H = V s/A (henry)	1 esu 1 emu	= 8.9876 × 10^{11} H = 10^{-9} H
Joule–Thomson coefficient	$[M^{-1}LT^{-2}\ \Theta]$	m^2/N K	1 cm^2 degC/dyne	= $10m^{-2}$/N K
Kinematic viscosity	$[L^2T^{-1}]$	m^2/s	1 stokes 1 in^2/h 1 ft^2/h 1 m^2/h 1 in^2/s 1 ft^2/s	= 10^{-4} m^2/s = 1.79211 × 10^{-7} m^2/s = 2.58064 × 10^{-5} m^2/s = 2.7778 × 10^{-4} m^2/s = 6.4516 × 10^{-4} m^2/s = 9.29030 × 10^{-2} m^2/s
Latent heat	$[L^2T^{-2}]$	J/kg	1 J/g 1 Btu/lb 1 cal_{IT}/g 1 ft lbf/lb 1 kgf m/kg 1 cal (thermochem)/g	= 10^3 J/kg = 2.326 × 10^3 J/kg = 4.1868 × 10^3 J/kg = 2.98907 J/kg = 9.80665 J/kg = 4.184 × 10^3 J/kg
Leak rate	$[ML^2T^{-3}]$	J/s	1 lusec	= 1.33322 × 10^{-4} J/s
Length	[L]	m (metre)	1 Å 1 μm 1 thou (= mil) 1 inch 1 cm 1 foot 1 yard 1 mile	=10^{-10} m = 10^{-6} m = 2.54 × 10^{-5} m = 2.54 × 10^{-2} m = 10^{-2} m = 3.048 × 10^{-1} m = 9.144 × 10^{-1} m = 1.60934 × 10^3 m
Luminance		cd/m^2		
Luminous flux		lm = cd sr (lumen)		
Lumininous intensity		cd (candela)		
Magnetic field	$[L^{-1}\ I]$	A m^{-1}	1 esu 1 oersted 1 A turn/cm 1 A turn/in	= 2.6544 × 10^{-9} A/m = 7.9578 × 10 A/m = 10^2 A/m = 3.9370 × 10 A/m
Magnetic flux	$[ML^2\ T^2\ I^{-1}]$	Wb = V s (weber)	1 esu 1 Maxwel	= 2.9979 × 10^2 Wb = 10^{-8} Wb

Table 1.11.2 – *continued*

Table 1.11.2 SI units and conversion factors – *continued*

Property	*Dimensional form*	*SI unit*	*Conversion factor*	
Magnetic flux density	$[MT^{-2} I^{-1}]$	T = Wb/m^2 (tesla)	1 esu 1 emu	= 2.9979 × 10^2 T = 10^{-4} T
Magnetic induction	$[MT^{-2} I^{-1}]$	Wb/m^2	1 esu 1 Gauss	= 2.9979 × 10^6 Wb/m^2 = 10^{-4} Wb/m^2
Magnetic moment	$[ML^3 T^{-2} I^{-1}]$	Wb m	1 esu 1 emu 1 Maxwell cm.	= 3.7673 × 10 Wb m = 1.2566 × 10^{-9} Wb m = 10^{-10} Wb m
Magnetic vector potential	$[MLT^{-2} I^{-1}]$	Wb/m	1 esu	= 2.9979 × 10^4 Wb/m
Magnetization	$[MT^{-2} I^{-1}]$	Wb/m^2	1 esu 1 emu 1 Maxwell/cm^2	= 3.7673 × 10^7 Wb/m^2 = 1.2566 × 10^{-3} Wb/m^2 = 10^{-4} Wb/m^2
Magnetostatic potential	[I]	A turn	1 esu 1 emu	= 2.6544 × 10^{-11} A turn = 1 Gilbert = 7.9578 × 10^{-1} A turn
Mass	[M]	kg (kilogramme)	1 grain 1 carat 1 gramme 1 oz (av) 1 oz (tr = apoth) 1 lb 1 stone 1 short cwt 1 cwt 1 short t 1 tonne 1 t 1 t (USA) is normally 1	= 6.479891 × 10^{-5} kg = 2.0 × 10^{-4} kg = 10^{-3} kg = 2.83495 × 10^{-2} kg = 3.11035 × 10^{-2} kg = 4.5359237 × 10^{-1} kg = 6.350 kg = 4.53592 × 10 kg = 5.08023 × 10 kg = 9.07185 × 10^2 kg = 10^3 kg = 1.01605 × 10^3 kg short t
Mass per unit area	$[ML^{-2}]$	kg/m^2	1 lb/acre 1 oz/yd^2 1 lb/ft^2	= 1.2085 × 10^{-4} kg/m^2 = 3.39057 × 10^{-2} kg/m^2 = 4.88243 kg/m^2
Mass per unit length	$[ML^{-1}]$	kg/m	1 lb/ft 1 lb/in	= 1.48816 kg/m = 1.78580 × 10 kg/m
Mass rate of flow	$[MT^{-1}]$	kg/s	1 lb/h 1 kg/h 1 lb/min 1 UK ton/h	= 1.25998 × 10^{-4} kg/s = 2.77778 × 10^{-4} kg/s = 7.55988 × 10^{-3} kg/s = 2.82235 × 10^{-1} kg/s
Moment of inertia	$[ML^2]$	kg m^2	1 g cm^2 1 lb in^2 1 lb ft^2	= 10^{-7} kg m^2 = 2.92640 × 10^{-4} kg m^2 = 4.21401 × 10^{-2} kg m^2
Momentum	$[MLT^{-1}]$	kg m/s	1 g cm/s 1 lb ft/s	= 10^{-5} kg m/s = 1.38255 × 10^{-1} kg m/s
Peltier coefficient (as for electric potential)				
Permeability	$[MLT^{-2} I^{-2}]$	H/m	1 esu 1 emu	= 1.1294 × 10^{-15} H/m = 1.2566 × 10^{-6} H/m

Table 1.11.2 – *continued*

Table 1.11.2 SI units and conversion factors – *continued*

Property	Dimensional form	SI unit	Conversion factor	
Permittivity	$[M^{-1} L^{-3} T^4 I^2]$	F/m	1 esu 1 emu	$= 8.8541 \times 10^{-10}$ F/m $= 7.9578 \times 10^9$ F/m
Piezoelectric strain coefficient	$[M^{-1} L^{-1} T^3 I]$	m/V	1 cm/esu	$= 3.336 \times 10^{-3}$ m/V
Piezoelectric stress coefficient	$[M^{-1} L^{-1} T^{-1} I]$	C/N	1 esu/dyne	$= 3.33564 \times 10^{-5}$ C/N
Piezo optical coefficient (as for compressibility)				
Polarization	$[L^{-2} TI]$	C/m^2	1 esu 1 emu 1 coulomb/cm^2	$= 3.3356 \times 10^{-6}$ C/m^2 $= 10^5$ C/m^2 $= 10^4$ C/m^2
Pole strength	$[ML^2 T^{-2} I^{-1}]$	Wb = V s (weber)	1 esu 1 emu 1 maxwell	$= 3.7673 \times 10^{-3}$ Wb $= 1.2566 \times 10^{-7}$ Wb $= 10^{-8}$ Wb
Power	$[ML^2 T^{-3}]$	W = J/s (watt)	1 erg/s 1 Btu/hour 1 kcal/hour 1 ft.lbf/s 1 cal/s 1 hp	$= 10^{-7}$ W $= 2.93071 \times 10^{-1}$ W = 1.163 W = 1.35582 W = 4.1868 W $= 7.45700 \times 10^2$ W
Pressure	$[ML^{-1} T^{-2}]$	Pa = N/m^2	1 kilobar 1 tonf/in^2 1 hectobar 1 kgf/mm^2 1 tonf/ft^2 1 atmosphere 1 bar 1 kgf/cm^2 (kp) 1 lbf/in^2 1 inch of Hg 1 ft of water 1 in of water 1 torr 1 millibar 1 lbf/ft^2 1 poundal/ft^2 1 micron of Hg 1 dyne/cm^2	$= 10^8$ Pa $= 1.54443 \times 10^7$ Pa $= 10^7$ Pa $= 9.80665 \times 10^6$ Pa $= 1.07252 \times 10^5$ Pa $= 1.01325 \times 10^5$ Pa $= 10^5$ Pa $= 9.80665 \times 10^4$ Pa $= 6.89476 \times 10^3$ Pa $= 3.38639 \times 10^3$ Pa $= 2.98907 \times 10^3$ Pa $= 2.49089 \times 10^2$ Pa $= 1.33322 \times 10^2$ Pa $= 10^2$ Pa $= 4.78803 \times 10$ Pa = 1.48816 Pa $= 1.33322 \times 10^{-1}$ Pa $= 10^{-1}$ Pa
Pyroelectric coefficient	$[L^{-2} TI\Theta^{-1}]$	C/m^2 per K	1 esu °C	$= 3.3356 \times 10^{-6}$ C/m^2 per K
Reactance (see resistance)				
Reluctance	$[M^{-1} L^{-2} T^2 I^2]$	1/H	1 esu 1 emu	$= 1.11265 \times 10^{-12}$ H^{-1} $= 10^9$ H^{-1}
Resistance	$[ML^2 T^{-3} I^{-2}]$	Ω = V/A (ohm)	1 esu 1 emu	$= 8.9876 \times 10^{11}$ Ω $= 10^{-9}$ Ω

Table 1.11.2 – *continued*

Table 1.11.2 SI units and conversion factors – *continued*

Property	*Dimensional form*	*SI unit*	*Conversion factor*	
Resistivity	$[ML^3 T^{-3} I^{-2}]$	Ω m	1 esu 1 emu 1 Ω cm	$= 8.9876 \times 10^9$ Ω m $= 10^{-11}$ Ω m $= 10^{-2}$ Ω m
Second moment of area	$[L^4]$	m^4	1 cm^4 1 in^4 1 ft^4	$= 10^{-8}$ m^4 $= 4.16231 \times 10^{-7}$ m^4 $= 8.63097 \times 10^{-3}$ m^4
Solid angle		sr (steradian)	Solid angle subtended by sphere = 1.25664 × 10 sr	
Specific entropy (see Specific heat by mass)				
Specific heat by mass	$[L^2 T^{-2} \Theta^{-1}]$	J/kg per K	1 J/(g deg C) 1 cal/(g deg C) 1 Btu/(lb deg F)	$= 10^3$ J/kg per K $= 4.1868 \times 10^3$ J/kg per K $= 4.1868 \times 10^3$ J/kg per K
Specific heat by volume	$[ML^{-1} T^{-1} \Theta^{-1}]$	J/m^3 per K	1 Btu/ft^3 deg F 1 cal/cm^3 deg C	$= 6.70661 \times 10^4$ J/m^3 per K $= 4.1868 \times 10^6$ J/m^3 per K
Specific modulus (as for specific strength)				
Specific strength (strength/density)	$[L^2 T^{-2}]$	N m/kg = m^2/s^2	1 lbf in/lb 1 hbar m^3/kg 1 hbar cm^3/g	$= 2.4909 \times 10^{-1}$ N m/kg $= 10^7$ N m/kg $= 10^4$ N m/kg
Specific surface	$[M^{-1} L^2]$	m^2/kg	1 ft^2/lb	= 4.88243 m^2/kg
Specific volume	$[M^{-1} L^3]$	m^3/kg	1 ft^3/lb	$= 6.24280 \times 10^{-2}$ m^3/kg
Stress (as for pressure)				
Stress intensity factor (see fracture toughness)				
Stress optical coefficient	$[M^{-1} LT^2]$	1/N per m^2	1 Brewster 1 in^2/lbf	$= 10^{-12}$/N per m^2 $= 1.45038 \times 10^{-4}$/N per m^2
Surface energy (Surface tension)	$[MT^{-2}]$	N/m = J/m^2	1 dyne/cm 1 lbf/ft 1 lbf/inch	$= 1$ erg/cm^2 $= 10^{-3}$ N/m $= 1.4594 \times 10$ N/m $= 1.7513 \times 10^2$ N/m
Temperature	$[\Theta]$	K (kelvin)	Θ°C t°F r°R $\frac{F - 32}{9} = \frac{C}{5} =$	= (Θ + 273.15) K = (0.5556t + 255.37) K = (0.5556r) K $\frac{K - 73.15}{5}$
Temperature interval	$[\Theta]$	K	1 deg C 1 deg F 1 deg Rankine	= 1 K = 0.55556 K = 0.55556 K
Thermal conductance	$[MT^{-3} \Theta^{-1}]$	W/m^2 per K	1 Btu/ft^2 h deg F 1 cal/cm^2s deg C	= 5.67826 W/m^2 per K $= 4.1868 \times 10^4$ W/m^2 per K

Table 1.11.2 – *continued*

Table 1.11.2 SI units and conversion factors – *continued*

Property	*Dimensional form*	*SI unit*	*Conversion factor*	
Thermal conductivity	$[MLT^{-3}\,\Theta^{-1}]$	W/m per K	1 btu in/(ft^2 hr deg F) 1 kcal/(m hr deg C) 1 Btu/(ft hr deg F) 1 W/(cm deg C) 1 cal/(cm sec deg C)	= 1.44228 × 10^{-1} W/m per K = 1.163 W/m per K = 1.73073 W/m per K = 10^2 W/m per K = 4.1868 × 10^2 W/m per K
Thermal diffusivity (as for kinematic viscosity)				
Thermal expansion coefficient	$[\Theta^{-1}]$	1/K	1 (deg C)$^{-1}$ 1 (deg F)$^{-1}$	= 1/K = 1.8/K
Thermal resistivity	$[M^{-1}\,L^{-1}\,T^3\,\Theta]$	m K/W	1 cm s deg C/cal 1 ft h deg C/Btu 1 ft^2 h deg C/Btu in	= 2.38846 × 10^{-3} m K/W = 5.77789 × 10^{-1} m K/W = 6.93347 m K/W
Thermal stress resistance factor $R = \frac{S}{E\,\alpha}(1-\nu)$ (as for temperature interval)				
Thermal stress resistance factor $R', R' = \frac{ks}{E\,\alpha}(1-\nu)$	$[MLT^{-3}]$	W/m	1 cal/(cm sec) 1 Btu/(ft hr) 1 Btu in/(ft^2 hr)	= 4.1868 × 10^2 W/m = 9.6152 × 10^{-1} W/m = 8.0127 × 10^{-2} W/m
Thermal stress resistance factor $R'', R'' = \frac{ks}{E\,\alpha\,\rho}C(1-\nu)$	$[L^2\,T^{-1}\,\Theta]$	m^2/s K	1 deg F in^2/s 1 deg F ft^2/hr 1 deg F ft^2/s 1 deg C cm^2/s	= 3.5842 × 10^{-4} m^2/s K = 1.4337 × 10^{-5} m^2/s K = 5.1613 × 10^{-2} m^2/s K = 10^{-4} m^2 s K
Thermoelectric power	$[ML^2\,T^{-3}\,I\Theta^{-1}]$	V/K	1 emu/deg C	= 10^{-8} V/K
Thomson coefficient (as for thermoelectric power)				
Time	[T]	s (second)	1 minute 1 hour 1 day 1 week 1 year (calendar) 1 year (astron.)	= 6.0 × 10 s = 3.6 × 10^3 s = 8.64 × 10^4 s = 6.048 × 10^5 s = 3.1558 × 10^7 s = 3.1557 × 10^7 s
Torque (as for energy)				
Velocity	$[LT^{-1}]$	m/s	1 cm/s 1 km/h 1 ft/s 1 mile/h 1 UK knot 1 int. knot 1 ft/min	= 10^{-2} m/s = 2.77778 × 10^{-1} m/s = 3.048 × 10^{-1} m/s = 4.4704 × 10^{-1} m/s = 5.14773 × 10^{-1} m/s = 5.14444 × 10^{-1} m/s = 5.08 × 10^{-3} m/s

Table 1.11.2 – *continued*

Table 1.11.2 SI units and conversion factors – *continued*

Property	***Dimensional form***	***SI unit***	***Conversion factor***	
Viscosity	$[ML^{-1} T^{-1}]$	Pa s	1 poise 1 poundal s/ft^2 1 kg f s/m^2 1 lbf s/ft^2	= 10^{-1} Pa s = 1.48816 Pa s = 9.80665 Pa s = 4.78803 × 10 Pa s
Volume	$[L^3]$	m^3	1 cm 1 ml 1 in^3 1 UK fl oz 1 US fl oz 1 US pint (liq) 1 UK pint 1 litre 1 US gallon 1 UK gallon 1 ft^3 1 yd^3 1 Barrel (= 42 US gal)	= 10^{-6} m^3 = 1.000028 × 10^{-6} m^3 = 1.63871 × 10^{-5} m^3 = 2.84130 × 10^{-5} m^3 = 2.95735 × 10^{-5} m^3 = 4.73176 × 10^{-4} m^3 = 5.68261 × 10^{-4} m^3 = 1.000028 × 10^{-3} m^3 = 3.78541 × 10^{-3} m^3 = 4.54609 × 10^{-3} m^3 = 2.83168 × 10^{-2} m^3 = 7.64555 × 10^{-1} m^3 = 1.58945 × 10^{-1} m^3
Volume rate of flow	$[L^3 T^{-1}]$	m^3/s	1 l/h 1 UK gal/h 1 l/min 1 UK gal/min 1 ft^3/min 1 l/s 1 ft^3/s	= 2.7779 × 10^{-7} m^3/s = 1.26280 × 10^{-6} m^3/s = 1.66671 × 10^{-5} m^3/s = 7.57681 × 10^{-5} m^3/s = 4.71947 × 10^{-4} m^3/s = 1.000028 × 10^{-3} m^3/s = 2.83168 × 10^{-2} m^3/s

Table 1.11.2 – *continued*

Table 1.11.3 Multiples and submultiples of SI units

Value	*Name*	*Symbol*	*Notes*
10^{12}	tera	T	recommended
10^{9}	giga	G	
10^{6}	mega	M	
10^{3}	kilo	k	
10^{-3}	milli	m	
10^{-6}	micro	μ	
10^{-9}	nano	n	
10^{-12}	pico	p	
10^{-15}	femto	f	
10^{-18}	atto	a	
10^{2}	hecto (a)	h	permissible
10	deca	da	
10^{-1}	deci	d	
10^{-2}	centi	c	

Table 1.11.4 Mathematical constants

Constant	*Value*
π	3.141 592 653 6
e	2.718 281 828 5
γ	0.577 215 664 9
$\log_{10} \pi$	0.497 149 872 7
$\log_{10} e$	0.434 294 481 9
$\log_{e} 10$	2.302 585 092 9

Table 1.11.5 Physical constants

Constant	*Value*
R	8.3143 J mole^{-1} K^{-1}
k	1.38054×10^{-23} J K^{-1}
N	6.02252×10^{-23} mole^{-1}
h	6.6256×10^{-34} J s
F	9.64870×10^{4} C mole^{-1}
m electron	9.1091×10^{-31} kg
e	1.60210×10^{-19} C
e/m	1.758796×10^{11} C kg^{-1}
m H atom	1.67343×10^{-27} kg
m proton	1.67252×10^{-27} kg
m neutron	1.67482×10^{-27} kg
molar volume at STP	2.24136×10^{-2} m^{3} mole^{-1}
c	2.997925×10^{8} m s^{-1}
σ	5.6697×10^{-8} m^{-2} K^{-4}
g (Teddington)	9.81181 m s^{-2}
m_{o}	1.25664×10^{-6} H m^{-1}
K_{o}	8.85414×10^{-12} F m^{-1}

Table 1.11.3, Table 1.11.4 and Table 1.11.5

1.11.2 Hardness scales and conversion factors

Hardness is a measure of a material's resistance to deformation, and is generally measured in a scratch or indentation test. There are several standard hardness test methods, which each measure slightly different materials properties and are therefore not strictly comparable. However, approximate conversions from one hardness scale to another are possible. Examples are given in:

Table 1.11.6 *Typical variations in the comparison of hardness scales*

Table 1.11.7 *Approximate equivalent hardness numbers, tensile and compressive strengths for steels*

Table 1.11.8 *Approximate equivalent hardness numbers for aluminium, its alloys and brass*

Table 1.11.9 *Approximate equivalent hardness numbers for sintered carbide cermets*

Table 1.11.10 *Approximate equivalent Vickers indentation and Mohs scratch hardness for selected minerals*

Hardness is related to compressive strength. Compressive strength may be estimated from hardness and Young's modulus using the nomogram in:

Fig. 1.11.1 *Nomogram for approximating compressive strength of metallic materials from Vickers hardness and Young's modulus*

Table 1.11.6 Typical variations in the comparison of hardness scales

Vickers diamond pyramid hardness	*Brinell hardness no. 3000 kg load 100 mm ball*	*Rockwell hardness No.*		
		A-scale 60 kg load diamond cone	*B-scale 100 kg load 1/16 in diam. ball*	*C-scale 150 kg load diamond cone*
20	15–25			
100	80–100		47–61	
200	175–205	58–60	93–95	
240				18–23
300	280–300	65–68		27–33
400	370–395	70–72		38–42
500	445–480	73–76		46–50
600	515–550	75–79		52–56
700	580–620	76–80		57–61
800		77–83		60–64
900		78–84		63–67
1000				65–69
1250		87–90		
1400		90–93		

Data from BS 860.

Table 1.11.7 Approximate equivalent hardness numbers, tensile and compressive strengths for steels

Vickers diamond pyramid hardness 50 kg load (kg/mm)	Brinell Hardness 3000 kg load 10 mm ball		Rockwell Hardness No.					Rockwell superficial Hardness No. Diamond Cone			Knoop hardness No. 500 g load and greater	Shore Schlero scope hardness No.	Approximate steel strengths	
	Standard ball	Tungsten carbide ball	A-scale 60 kg load diamond cone	B-scale 100 kg load 1/16 in diam. ball	C-scale 150 kg load diamond cone	E-scale 100 kg load 1/8 in diam. ball	F-scale 60 kg load 1/16 in diam. ball	15N scale 15 kg load	30N scale 30 kg load	45N scale 45 kg load			UTS (GPa)	Compressive stress at 10% strain (GPa)
940			85.6		68.0	76.9		93.2	84.4	75.4	920	97		
920			85.3		67.5	76.5		93.0	84.0	74.8	908	96		
900			85.0		67.0	76.1		92.9	83.6	74.2	895	95		
880		(767)	84.7		66.4	75.7		92.7	83.1	73.6	882	93		
860		(757)	84.4		65.9	75.3		92.5	82.7	73.1	867	92		
840		(745)	84.1		65.3	74.8		92.3	82.2	72.2	852	91		
820		(733)	83.8		64.7	74.3		92.1	81.7	71.8	837	90		
800		(722)	83.4		64.0	73.8		91.8	81.1	71.0	822	88		
780		(710)	83.0		63.3	73.3		91.5	80.4	70.2	806	87		
760		(698)	82.6		62.5	72.6		91.2	79.7	69.4	788	86		
740		(684)	82.2		61.8	72.1		91.0	79.1	68.6	772	84		
720		(670)	81.8		61.0	71.5		90.7	78.4	67.7	754	83		
700		(656)	81.3		60.1	70.8		90.3	77.6	66.7	735	81		
690		(647)	81.1		59.7	70.5		90.1	77.2	66.2	725			
680		(638)	80.8		59.2	70.1		89.8	76.8	65.7	716	80	2.45	2.80
670		(630)	80.6		58.8	69.8		89.7	76.4	65.3	706		2.40	2.74
660		620	80.3		58.3	69.4		89.5	75.9	64.7	697	79	2.36	2.68
650		611	80.0		57.8	69.0		89.2	75.5	64.1	687	78	2.32	2.62
640		601	79.8		57.3	68.7		89.0	75.1	63.5	677	77	2.26	2.56
630		591	79.5		56.8	68.3		88.8	74.6	63.0	667	76	2.23	2.50
620		582	79.2		56.3	67.9		88.5	74.2	62.4	657	75	2.19	2.45
610		573	78.9		55.7	67.5		88.2	73.6	61.7	646		2.14	2.40
600		564	78.6		55.2	67.0		88.0	73.2	61.2	636	74	2.09	2.35
590		554	78.4		54.7	66.7		87.8	72.7	60.5	625	73	2.05	2.30
580		545	78.0		54.1	66.2		87.5	72.1	59.9	615	72	2.02	2.24
570		535	77.8		53.6	65.8		87.2	71.7	59.3	604		1.99	2.19
560		525	77.4		53.0	65.4		86.9	71.2	58.6	594	71	1.95	2.14
550	(505)	517	77.0		52.3	64.8		86.6	70.5	57.8	583	70	1.90	2.09
540	(496)	507	76.7		51.7	64.4		86.3	70.0	57.0	572	69	1.86	2.04
530	(488)	497	76.4		51.1	63.9		86.0	69.5	56.2	561	68	1.83	1.99
520	(480)	488	76.1		50.5	63.5		85.7	69.0	55.6	550	67	1.79	1.94
510	(473)	479	75.7		49.8	62.9		85.4	68.3	54.7	539		1.75	1.89
500	(465)	471	75.3		49.1	62.2		85.0	67.7	53.9	528	66	1.70	1.85
490	(456)	460	74.9		48.4	61.6		84.7	67.1	53.1	517	65	1.66	1.80
480	(448)	452	74.5		47.7	61.3		84.3	66.4	52.2	505	64	1.62	1.75
470	441	442	74.1		46.9	60.7		83.9	65.7	51.3	494		1.57	1.70
460	433	433	73.6		46.1	60.1		83.6	64.9	50.4	482	62	1.54	1.66
450	425	425	73.3		45.3	59.4		83.2	64.3	49.4	471		1.50	1.61
440	415	415	72.8		44.5	58.8		82.8	63.5	48.4	459	59	1.46	1.57
430	405	405	72.3		43.6	58.2		82.3	62.7	47.4	447	58	1.41	1.53
420	397	397	71.8		42.7	57.5		81.8	61.9	46.4	435	57	1.37	1.48
410	388	388	71.4		41.8	56.8		81.4	61.1	45.3	423	56	1.33	1.44
400	379	379	70.8		40.8	56.0		80.8	60.2	44.1	412	55	1.29	1.40
390	369	369	70.3		39.8	55.2		80.3	59.3	42.9	400		1.25	1.35
380	360	360	69.8	(110.0)	38.8	54.4		79.8	58.4	41.7	389	52	1.21	1.31

Table 1.11.7

Table 1.11.7 Approximate equivalent hardness numbers, tensile and compressive strengths for steels – *continued*

370	350	350	69.2		37.7	53.6		79.2	57.4	40.4	378	51	1.17	1.27
360	341	341	68.7	(109.0)	36.6	52.8		78.6	56.4	39.1	367	50	1.13	1.22
350	331	331	68.1		35.5	51.9		78.0	55.4	37.8	356	48	1.10	1.18
340	322	322	67.6	(108.0)	34.4	51.1		77.4	54.4	36.5	346	47	1.07	1.13
330	313	313	67.0		33.3	50.2		76.8	53.6	35.2	337	46	1.03	1.09
320	303	303	66.4	(107.0)	32.2	49.4		76.2	52.3	33.9	328	45	1.01	1.05
310	294	294	65.8		31.0	48.4		75.6	51.3	32.5	318		0.98	1.01
300	284	284	65.2	(105.5)	29.8	47.5		74.9	50.2	31.1	309	42	0.95	0.96
290	275	275	64.5	(104.5)	28.5	46.5		74.2	49.0	29.5	300	41	0.92	0.92
280	265	265	63.8	(103.5)	27.1	45.3		73.4	47.8	27.9	291	40	0.89	0.88
270	256	256	63.1	(102.0)	25.6	44.3		72.6	46.4	26.2	282	38	0.85	0.84
260	247	247	62.4	(101.0)	24.0	43.1		71.6	45.0	24.3	272	37	0.83	0.81
250	238	238	61.6	99.5	22.2	41.7		70.6	43.4	22.2	262	36	0.79	0.77
240	228	228	60.7	98.1	20.3	40.3		69.6	41.7	19.9	253	34	0.77	0.73
230	219	219		96.7	(18.0)						243	33	0.73	0.69
220	209	209	(60)	95.0	(15.7)						234	32	0.70	0.65
210	200	200		93.4	(13.4)						226	30	0.67	0.62
200	190	190	(59)	91.5	(11.0)		(111)				216	29	0.63	0.58
190	181	181	(57)	89.5	(8.5)		(110)				206	28	0.61	0.55
180	171	171	(56)	87.1	(6.0)		(109)				196	26	0.58	0.51
170	162	162	(54)	85.0	(3.0)		(107.5)				185	25	0.54	0.48
160	152	152	(53)	81.7	(0.0)		(106)				175	23	0.52	0.44
150	143	143	(52)	78.7			(104.5)				164	22	0.49	0.40
140	133	133	(50)	75.0			(103)				154	21	0.46	
130	124	124		71.2			(100.5)				143	20	0.43	
120	114	114	(47)	66.7			98				133	18	0.39	
110	105	105		62.3			95.5				123		0.37	
100	95	95	(43)	56.2			92				112	17	0.34	
95	90	90		52.0			89.5				107		0.32	
90	86	86		48.0			88				102		0.31	
85	81	81		41.0			85.5				97		0.29	
80	75	75		39.0			83							

N.B. The data in this table applies to carbon and low alloy steels in the annealed, normalized and quenched and tempered conditions. It does not apply to non-ferrous metals and surface treated steels and is less accurate for cold-worked and austenitic steels. The data in parenthesis is beyond the normal hardness range. Data from ASTM E140 and after Studman, C. J., Moore, M. A., and Jones, S. E. (1977) *J. Phys. D: Appl. Phys.*, **10**, 949–56.

Table 1.11.8 Approximate equivalent hardness numbers for aluminium, its alloys and brass

Vickers diamond pyramid hardness 10 kg load (kg/mm²)	*Aluminium and its alloys*				*Brass*					
	Brinell hardness No. (see note)	Rockwell Superficial Hardness No. 1/16 in diam. ball			Brinell hardness No. 500 kg load 10 mm ball	Rockwell hardness No.		Rockwell superficial hardness No. 1/16 in diam ball		
		15T-scale 15kg load	30T-scale 30 kg load	45T-scale 45 kg load		B-Scale 100 kg load 1/16 in diam ball	F-Scale 60 kg load 1/16 in diam ball	15T-scale 15 kg load	30T-scale 30 kg load	45T-scale 45 kg load
200	185.0	90.6	79.7	66.2						
190	175.8	90.0	78.4	64.0	164	92.5	(109.0)		76.5	64.5
180	166.6	89.4	77.0	61.7	156	90.0	(107.5)		75.0	62.0
170	157.4	88.7	75.5	59.0	147	87.0				59.5
160	148.2	88.0	73.7	56.0	139	83.5			71.5	56.5
150	139.0	87.1	71.7	52.6	131	80.0		86.5	69.5	53.5
140	129.8	86.2	69.4	48.8	122	76.0	(100.5)	85.5	67.0	50.0
130	120.6	85.0	66.8	44.3	114	72.0	98.0	84.0	64.5	45.5
120	111.4	83.7	63.8	39.0	106	67.0	95.5		61.0	41.0
110	102.2	82.2	60.2		97	62.0	92.6	80.5	58.0	35.5
100	93.0	80.3	55.9		88	56.0	89.0	78.5	53.5	29.5
96	89.4	79.5	54.0		85	53.0	87.2	77.5	51.5	26.5
92	85.7	78.6	51.8		82	49.5	85.4	76.5	49.0	23.0
88	82.0	77.6	49.4		79	46.0	83.5	75.0	47.0	19.0
84	78.3	76.5	46.9		76	42.0	81.2	73.5	44.0	14.5
80	74.6	75.3	44.1		72	37.5	78.6	72.0	41.0	10.0
76	71.0	74.0	41.0		68	32.5	76.0	70.5	38.0	4.5
72	67.3	72.5	37.6		64	27.5	73.2	69.0	34.0	
68	63.6	70.8	33.7		62	21.5	70.0	67.0	30.0	
64	59.9	69.0	29.4		59	(15.5)	66.8	65.0	25.5	
60	56.2	66.9	24.5		55	(10.0)	63.0	62.5	20.5	
56	52.6	64.5	19.1		52		58.8	60.0	15.0	
52	48.9	61.8	12.6		48		53.5	57.0		
48	45.2	58.6	5.2		45		47.0	53.5		
44	41.5	54.8								
40	37.8	50.3								
38	36.0	47.7								
36	34.2	45.0								
34	32.3	41.7								
32	30.5	38.2								
30	28.6	34.0								
28	26.8	29.5								
26	25.0	24.4								
24	23.1	18.1								
22	21.3	11.1								
20	19.4	2.3								

N.B. For aluminium and its alloys the ratio load/(diam. of ball)^2 was 5 for Brinell Hardness from 20 to 60 and 10 for hardnesses above 60. The data in parentheses is beyond the normal hardness range. Data from ASTM E140–65 and *Metals Reference Book*, ed. C.J. Smithells and E.A. Brandes, 5th edn. Butterworth, London, 1976.

Table 1.11.8

Table 1.11.9 Approximate equivalent hardness numbers for sintered carbide cermets

Vickers diamond pyramid hardness 50 kg load (kg/mm^2)	*Rockwell hardness No*	
	A–Scale 60 kg load diamond cone	*C–Scale 150 kg load diamond cone*
1750	92.4	80.5
1700	92.0	79.8
1650	91.7	79.2
1600	91.2	78.4
1550	90.9	77.7
1500	90.5	77.0
1450	90.1	76.2
1400	89.7	75.4
1350	89.3	74.6
1300	88.9	73.8
1250	88.5	73.0
1200	88.1	72.2
1150	87.6	71.3
1100	87.0	70.4
1050	86.4	69.4
1000	85.7	68.2
950	85.0	66.6
900	84.0	64.6
850	82.8	

Data from *Metals Reference Book*, ed. C.J. Smithells and E.A. Brandes, 5th edn. Butterworth, London, 1976.

Table 1.11.10 Approximate equivalent Vickers indentation and Mohs scratch hardness for selected minerals

Materials	*Vickers diamond pyramid Hardness (kg/mm^2)*	*Mohs scratch hardness no.*
Diamond	10 000	10
Corundum	2000–2200	9
Topaz	1200–1650	8
Quartz	1040–1300	7
Orthoclase	710–800	6
Apatite	540–850	5
Fluorite	160–250	4
Calcite	105–260	3
Gypsum	35–75	2
Talc	2–50	1

Reproduced with permission from J.H. Westbrook and H. Conrad, *The Science of Hardness Testing*, published by ASM International, 1973.

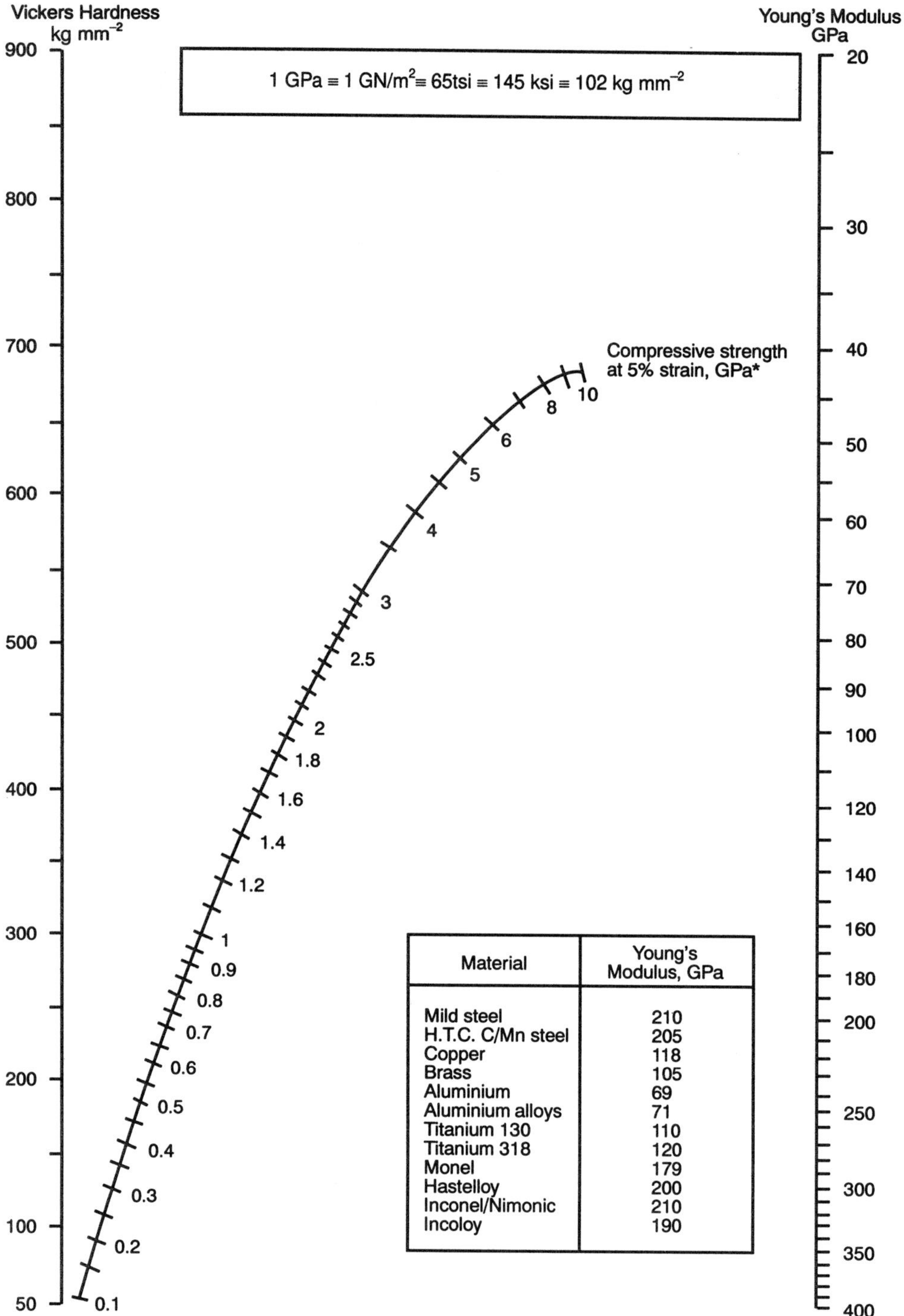

Material	Young's Modulus, GPa
Mild steel	210
H.T.C. C/Mn steel	205
Copper	118
Brass	105
Aluminium	69
Aluminium alloys	71
Titanium 130	110
Titanium 318	120
Monel	179
Hastelloy	200
Inconel/Nimonic	210
Incoloy	190

Join the hardness to the Young's modulus; the point of intersection with the centre curve indicates compression strength
* This strain is additional to strain by cold work prior to hardness measurement

Fig. 1.11.1 Nomogram for approximating compressive strength of metallic materials from Vickers hardness and Young's modulus

Fig. 1.11.1

Index

Volumes 1–3

Page numbers appearing in **bold** refer to tables and page numbers appearing in *italic* refer to figures. Roman numerals (I, II, III) at the beginning of page number sequences refer to volume numbers. For tables and figures longer than one page, the first page of the table/figure is given; please look through for the particular material or property.